ANNALS OF THE NEW YORK ACADEMY OF SCIENCES

Volume 589

EDITORIAL STAFF
Executive Editor
BILL BOLAND
Managing Editor
JUSTINE CULLINAN
Associate Editor
STEFAN MALMOLI

The New York Academy of Sciences
2 East 63rd Street
New York, New York 10021

THE NEW YORK ACADEMY OF SCIENCES
(Founded in 1817)

BOARD OF GOVERNORS, 1990

LEWIS THOMAS, *Chairman of the Board*
CHARLES A. SANDERS, *President*
DENNIS D. KELLY, *President-Elect*

Honorary Life Governors

H. CHRISTINE REILLY   IRVING J. SELIKOFF

Vice-Presidents

DAVID A. HAMBURG   CYRIL M. HARRIS
PETER D. LAX   CHARLES G. NICHOLSON

HENRY A. LICHSTEIN, *Secretary-Treasurer*

Elected Governors-at-Large

JOSEPH L. BIRMAN   FLORENCE L. DENMARK   LAWRENCE R. KLEIN
GERALD D. LAUBACH   LLOYD N. MORRISETT   GERARD PIEL

WILLIAM T. GOLDEN, *Past Chairman*   HELENE L. KAPLAN, *General Counsel*

OAKES AMES, *Executive Director*

BIOCHEMICAL ENGINEERING VI

ANNALS OF THE NEW YORK ACADEMY OF SCIENCES
Volume 589

BIOCHEMICAL ENGINEERING VI

Edited by Walter E. Goldstein, David DiBiasio, and Henrik Pedersen

The New York Academy of Sciences
New York, New York
1990

Copyright © 1990 by the New York Academy of Sciences. All rights reserved. Under the provisions of the United States Copyright Act of 1976, individual readers of the Annals are permitted to make fair use of the material in them for teaching or research. Permission is granted to quote from the Annals provided that the customary acknowledgment is made of the source. Material in the Annals may be republished only by permission of the Academy. Address inquiries to the Executive Editor at the New York Academy of Sciences.

Copying fees: For each copy of an article made beyond the free copying permitted under Section 107 or 108 of the 1976 Copyright Act, a fee should be paid through the Copyright Clearance Center, Inc., 21 Congress St., Salem, MA 01970. For articles of more than 3 pages, the copying fee is $1.75.

♾ The paper used in this publication meets the minimum requirements of American National Standard for Information Sciences–Permanence of Paper for Printed Library Materials, ANSI Z39.48-1984.

Cover: The cover shows a scanning electron micrograph of the cross section of an activated yeast bead (see page 277).

Library of Congress Cataloging-in-Publication Data

Biochemical engineering VI/edited by Walter E. Goldstein, David DiBiasio, and Henrik Pedersen.
    p.   cm.—(Annals of the New York Academy of Sciences, ISSN 0077-8923; v. 589)
    Papers presented at the Sixth Biochemical Engineering Conference, held Oct. 2–7, 1988, in Santa Barbara, Calif., organized by the Engineering Foundation with the support of the National Science Foundation and the New York Academy of Sciences.
    Includes bibliographical references.
    ISBN 0-89766-561-9 (alk. paper)—ISBN 0-89766-562-7 (pbk.: alk. paper)
    1. Biochemical engineering—Congresses. I. Goldstein, Walter E. II. DiBiasio, David. III. Pedersen, Henrik. IV. National Science Foundation (U.S.). V. Engineering Foundation (U.S.). VI. New York Academy of Sciences. VII. Biochemical Engineering Conference (6th: 1988: Santa Barbara, Calif.). VIII. Title: Biochemical engineering 6. IX. Series.
    [DNLM: 1. Biochemistry—congresses. 2. Chemical Engineering—congresses. W1 AN626YL v. 589/TP 248.3 B615 1988]
Q11.N5 vol. 589
[TP248.3]
500 s—dc20
[660'.63]
DNLM/DLC
for Library of Congress
                                                        90-5802
                                                          CIP

SP
Printed in the United States of America
ISBN 0-89766-561-9 (cloth)
ISBN 0-89766-562-7 (paper)
ISSN 0077-8923

ANNALS OF THE NEW YORK ACADEMY OF SCIENCES

Volume 589
May 20, 1990

BIOCHEMICAL ENGINEERING VI[a]

Editors and Conference Chairmen
WALTER E. GOLDSTEIN, DAVID DIBIASIO,
and HENRIK PEDERSEN

CONTENTS

Preface. *By* W. E. GOLDSTEIN, D. DIBIASIO, and H. PEDERSEN xi

Part I. Manipulation and Analysis of Metabolic Pathways

Strategies and Challenges in Metabolic Engineering. *By* J. E. BAILEY, S. BIRNBAUM, J. L. GALAZZO, C. KHOSLA, and J. V. SHANKS 1

Genetic Engineering of Metabolic Pathways Applied to the Production of Phenylalanine. *By* KEITH BACKMAN, MARY JANE O'CONNOR, AIKO MARUYA, EDWIN RUDD, DIANE MCKAY, R. BALAKRISHNAN, M. RADJAI, V. DIPASQUANTONIO, DIANE SHODA, RANDOLPH HATCH, and K. VENKATASUBRAMANIAN 16

Stoichiometry and Kinetics of Xylose Fermentation by *Pichia stipitis*. *By* P. J. SLININGER, L. E. BRANSTRATOR, J. M. LOMONT, B. S. DIEN, M. R. OKOS, M. R. LADISCH, and R. J. BOTHAST 25

Improvement of Plasmid Stability by Immobilization of Recombinant Microorganisms. *By* JEAN-NOËL BARBOTIN, SAMI SAYADI, MONCEF NASRI, FADWA BERRY, and DANIEL THOMAS 41

Regulation and Dynamics of Elicitor Response in Suspension-cultured Plant Cells. *By* S. Y. BYUN, H. PEDERSEN, and C-K. CHIN 54

Methods for Cloning Key Primary Metabolic Enzymes and Ancillary Proteins Associated with the Acetone-Butanol Fermentation of *Clostridium acetobutylicum*. *By* JEFFREY W. CARY, DANIEL J. PETERSEN, GEORGE N. BENNETT, and E. T. PAPOUTSAKIS 67

A Highly Structured Model for Simulation of Batch and Continuous Cultures of *B. subtilis* and Examination of Cellular Differentiation. *By* JINWOOK JEONG and MOHAMMAD M. ATAAI 82

Part II. Processes Involving Genetically Engineered Organisms

Effects of Culture Conditions on Plasmid Stability and Production of a Plasmid-encoded Protein in Batch and Continuous Cultures of *Escherichia coli* JM103[pUC8]. *By* WEN RYAN and SATISH J. PARULEKAR 91

---

[a]The papers in this volume were presented at the Sixth Biochemical Engineering Conference, held on October 2–7, 1988, in Santa Barbara, California. The conference was organized by the Engineering Foundation with the support of the National Science Foundation and the New York Academy of Sciences.

Effects of Promoter Induction and Copy Number Amplification on Cloned Gene Expression and Growth of Recombinant Cell Cultures. *By* MICHAEL J. BETENBAUGH and PRASAD DHURJATI...... 111

Optimal Induction of Protein Synthesis in Recombinant Bacterial Cultures. *By* WILLIAM E. BENTLEY and DHINAKAR S. KOMPALA...... 121

Expression, Purification, and Immobilization of a Protein A–β-Lactamase Hybrid Protein. *By* GEORGE GEORGIOU and FRANÇOIS BANEYX 139

Part III. Protein Separations and Downstream Processing

A Novel Immunoaffinity Chromatography System for the Purification of Therapeutic Proteins. *By* PETER GRANDICS, ZSOLT SZATHMARY, and SUSAN SZATHMARY 148

New Approaches to the More Efficient Purification of Proteins and Enzymes. *By* N. J. TITCHENER-HOOKER, M. HOARE, and P. DUNNILL 157

Gradient Elution in Preparative Liquid Chromatography. *By* FIROZ D. ANTIA and CSABA HORVÁTH 172

Purification of β-Galactosidase by Combined Frontal and Displacement Chromatography. *By* ABRAHAM LIAO and CSABA HORVÁTH 182

Part IV. Membrane-based Reactions and Separations

Novel Membrane-based Immobilization Technique for Bioreactors. *By* W. K. KANG, R. SHUKLA, and K. K. SIRKAR 192

Separation of Amino Acids Using Composite Ion Exchange Membranes. *By* BINAY K. DUTTA and SUBHAS K. SIKDAR 203

Studies of Transport Processes Coupled with Reaction in Membrane-sandwiched Yeast Cell Reactors. *By* YONG S. JEONG, W. R. VIETH, and TAKESHI MATSUURA 214

Chitin-Chitosan Membranes: Separations of Amino Acids and Polypeptides. *By* JOHN J. PELLEGRINO, STUART GEER, KAREN MAEGLEY, RAPHAEL RIVERA, DARLENE STEWARD, and MYONG KO 229

Diffusion of Proteins in Porous Membranes. *By* RUTH E. BALTUS and ZHONG LU 245

A Continuous Enzyme Membrane Reactor Retaining the Native Nicotinamide Cofactor NAD(H). *By* MICHAEL W. HOWALDT, KLAUS D. KULBE, and HORST CHMIEL 253

Covalently Attached GRGD on Polymer Surfaces Promotes Biospecific Adhesion of Mammalian Cells. *By* STEPHEN P. MASSIA and JEFFREY A. HUBBELL 261

Part V. Bioreactors I (Microbial Systems)

Immobilization of Growing Cells and Its Application to the Continuous Ethanol Fermentation Process. *By* JOHN J. JOUNG and G. P. ROYER 271

Mechanisms of Oxygen Transfer Enhancement during Submerged Cultivation in Perfluorochemical-in-Water Dispersions. *By* JAMES D. MCMILLAN and DANIEL I. C. WANG 283

The Hyperthermophilic Archaebacterium, *Pyrococcus furiosus*: Development of Culturing Protocols, Perspectives on Scaleup, and Potential Applications. *By* I. I. BLUMENTALS, S. H. BROWN, R. N. SCHICHO, A. K. SKAJA, H. R. COSTANTINO, and R. M. KELLY 301

Diauxic Metabolism of *Hansenula polymorpha*: Steady- and Unsteady-State Considerations. *By* JAMES D. BRYERS and TIMOTHY YEH 315

Nutrient Transport and Cellular Morphology in Immobilized Cell Aggregates. *By* J. D. FOWLER and C. R. ROBERTSON 333

Fermentation Development of Recombinant *Pichia pastoris* Expressing the Heterologous Gene: Bovine Lysozyme. *By* R. A. BRIERLEY, C. BUSSINEAU, R. KOSSON, A. MELTON, and R. S. SIEGEL 350

Large-Scale Growth of *Bordetella pertussis* for Production of Extracellular Toxin. *By* N. ANDORN, J. B. KAUFMAN, T. R. CLEM, R. FASS, and J. SHILOACH 363

Part VI. Bioreactors II (Higher Eukaryotic Systems)

Growth Kinetics of Free and Immobilized Insect Cell Cultures. *By* S. N. AGATHOS, Y-H. JEONG, and K. VENKAT 372

Bioreactor Development for Production of Viral Pesticides or Heterologous Proteins in Insect Cell Cultures. *By* M. L. SHULER, T. CHO, T. WICKHAM, O. OGONAH, M. KOOL, D. A. HAMMER, R. R. GRANADOS, and H. A. WOOD 399

Production of Baculovirus in a Continuous Insect-Cell Culture: Bioreactor Design, Operation, and Modeling. *By* J. TRAMPER, E. J. VAN DEN END, C. D. DE GOOIJER, R. KOMPIER, F. L. J. VAN LIER, M. USMANY, and J. M. VLAK. 423

Effects of Microcarriers and Serum on Local Hydrodynamics within an Airlift Column. *By* G. T. JONES, L. E. ERICKSON, and L. A. GLASGOW 431

Continuous Cell Cultures in Fluidized-Bed Bioreactors: Cultivation of Hybridomas and Recombinant Chinese Hamster Ovary Cells Immobilized in Collagen Microspheres. *By* N. G. RAY, A. S. TUNG, E. G. HAYMAN, J. N. VOURNAKIS, and P. W. RUNSTADLER, JR. 443

Part VII. Biosensors and Analytical Methods in Bioprocessing

Nuclear Magnetic Resonance Methods for Observing the Intracellular Environment of Mammalian Cells. *By* ERIK J. FERNANDEZ, ANTHONY MANCUSO, MARILEE K. MURPHY, HARVEY W. BLANCH, and DOUGLAS S. CLARK 458

Scanning Tunneling Microscopy and Atomic Force Microscopy of Biological Surfaces. *By* JOSEPH A. N. ZASADZINSKI and PAUL K. HANSMA 476

Uses of Fluorescence Sensors: For the Monitoring of Immobilized Cell Culture Fluorescence and as Optical Biosensors. *By* K-D. ANDERS, W. MÜLLER, T. SCHEPER, and A. F. BÜCKMANN 492

Part VIII. Bioprocess Optimization and Control

Multivariable Control of Continuous and Fed-Batch Bioreactors. *By* DANIEL WEI, SATISH J. PARULEKAR, and WILLIAM A. WEIGAND 508

Intelligent Purification of Monoclonal Antibodies. *By* P. W. THOMPSON, A. C. KENNEY, P. MOULDING, and D. WORMALD 529

L-Aspartic Acid Production Using Immobilized *E. coli* Cells in a Packed-Bed Reactor: Design of Reactor and Its Optimal Operation. *By* HIROYASU SEKO, SHINOBU TAKEUCHI, KAZUYOSHI YAJIMA, MASARU SENUMA, and TETSUYA TOSA 540

Modeling and Control of the Biocatalytic Conversion of Hydantoins to D–Amino Acids. *By* L. TRANCHINO and F. MELLE 553

An Expert System for Cultivating Operations. *By* H. ASAMA, T. NAGAMUNE, M. HIRATA, A. HIRATA, and I. ENDO 569

Frequency Response Analysis of Naphthalene Biotransformation Activity. *By* JAMES W. BLACKBURN 580

Analysis of Performance Limitations in Immobilized Cell Fermentors. *By* C. WEBB, G. A. DERVAKOS, and J. F. DEAN 593

Part IX. Mixing and Scaleup of Biological Reactors

Development and Scaleup of a High-Rate Biogas Process for Treatment of Organically Polluted Effluents. *By* A. AIVASIDIS and C. WANDREY 599

Scaleup and Optimization of Oxygen Transfer in Fermentors: Newtonian and Non-Newtonian Systems. *By* V. SINGH, R. FUCHS, W. HENSLER, and A. CONSTANTINIDES 616

Design and Scaleup of an Anchorage-dependent Mammalian Cell Bioreactor. *By* EDWARD L. PAUL, SR. 642

Fluctuating Environmental Conditions in Scaled-up Bioreactors: Heating and Cooling Effects. *By* GEOFFREY HAMER and ARMIN HEITZER 650

Comparison of Cephalosporin C Production in Stirred-Tank and Airlift Tower Loop Reactors. *By* T. BAYER, T. HEROLD, K. HOLZHAUER, W. ZHOU, and K. SCHÜGERL 665

Phase Holdup and Dispersion in a Three-Phase Fluidized-Bed Bioreactor with Low-Density Gel Beads. *By* BRIAN H. DAVISON 670

Part X. Immobilized Enzyme Bioreactors

Hydration of Cyanopyridine to Nicotinamide by Immobilized Nitrile Hydratase. *By* JACOB EYAL and MARVIN CHARLES 678

Immobilized Pig Brain NAD Glycohydrolase for the Preparation of NAD Analogues. *By* MARIO PACE, PIER GIORGIO PIETTA, DARIO AGNELLINI, PIER LUIGI MAURI, and SILVIA GHEZZI 689

Microbial Decarboxylation of Succinate to Propionate: Kinetic Studies. *By* NISSIM S. SAMUELOV, RATHIN DATTA, MAHENDRA K. JAIN, and J. GREGORY ZEIKUS 697

Bacterial Enzymes in Halogenation Processes. *By* WOLFGANG WIESNER, MANFRED KARL OTTO, and KLAUS DIETER KULBE 705

Index of Contributors 713

The New York Academy of Sciences believes it has a responsibility to provide an open forum for discussion of scientific questions. The positions taken by the participants in the reported conferences are their own and not necessarily those of the Academy. The Academy has no intent to influence legislation by providing such forums.

Preface

This volume contains the papers presented at the Sixth Biochemical Engineering Conference held in Santa Barbara, California, during October 1988. The theme of this sixth conference was Biochemical Engineering for Technology Leading to Processes and Products of Economical Value. Process and product applications covered at the conference included enzymatic, microbial, plant, and mammalian systems, and products covered the range from therapeutic to specialty chemicals to commodity chemicals.

The Sixth Biochemical Engineering Conference was organized by the Engineering Foundation with the financial support of the National Science Foundation and the New York Academy of Sciences. The support of these organizations is gratefully acknowledged. We are also grateful for the financial contributions of several companies, including Ajinomoto, Pharmacia LKB, Bio-Technical Resources, Novo Laboratories, H. J. Heinz, Novo Biochemical, American Cyanamid, Eastman Kodak, AmGen Incorporated, Abbott Labs, Biogen NV, Beecham Pharmaceuticals, and Merck & Company.

The conference planning and programming was carried out by the following individuals: Chairman—Walter E. Goldstein, Escagenetics Incorporated, California; Program Chairmen—David DiBiasio, Worcester Polytechnic Institute, and Henrik Pedersen, Rutgers University; Executive Committee—Shuichi Suzuki, Saitama Institute of Technology; William A. Weigand, University of Maryland; James Bailey, California Institute of Technology; Wolf Vieth, Rutgers University; K. Venkat, H. J. Heinz; Daniel Wang, Massachusetts Institute of Technology; Michael Shuler, Cornell University; Christian Wandrey, Institut für Biotechnologie, Jülich; Larry E. Erickson, Kansas State University; and Harold A. Comerer, Engineering Foundation.

The session chairmen and cochairmen were: G. Georgiou, University of Texas, Austin; D. Ryu, University of California, Davis; A. H. Ramel, Genentech, Incorporated; J. Lopez, Sepracor, Incorporated; J. D. Bryers, Duke University; S. Agathos, Rutgers University; J. Rollings, Worcester Polytechnic Institute; D. Clark, University of California, Berkeley; W. Weigand, University of Maryland; G. Stephanopoulos, Massachusetts Institute of Technology; and C. Wandrey, Jülich.

Larry E. Erickson chaired a large poster session and organized the poster program for the conference. These papers have been included in the regularly scheduled sessions of the meeting wherever possible. An additional session has been included in these proceedings, entitled Immobilized Enzyme Bioreactors, for the poster presentations not easily accommodated elsewhere.

W. E. Goldstein
D. DiBiasio
H. Pedersen

PART I. MANIPULATION AND ANALYSIS OF METABOLIC PATHWAYS

Strategies and Challenges in Metabolic Engineering[a]

J. E. BAILEY, S. BIRNBAUM, J. L. GALAZZO, C. KHOSLA,
AND J. V. SHANKS

*Department of Chemical Engineering
California Institute of Technology
Pasadena, California 91125*

INTRODUCTION

Metabolic activities of cells underlie a tremendous variety of processes ranging from waste treatment to the production of chemicals and antibiotics to synthesis of cloned proteins to food production. The metabolism of the native organism is almost never optimized for its process application. Furthermore, bioprocesses often employ cells after extensive genetic modification, which disrupts the original metabolic balance in the cell, or in unnatural configurations, such as immobilized within a gel matrix. It follows that manipulation of the metabolism of cells is a vital component in obtaining a feasible, and eventually a superior, bioprocess.

Cell metabolism can be manipulated externally or internally. Biochemical engineers have long practiced environmental manipulation of cell performance, choosing operating conditions during batch processes to improve growth and productivity. Internal changes have traditionally been achieved by random mutagenesis and selection, searching for mutations that render the cell more active for the desired metabolic process.

The advent of recombinant DNA technology opens a new era in metabolic programming that has been called metabolic engineering, cell engineering, pathway engineering, and other related names. A central concept here is the power of contemporary genetic engineering methods to achieve highly specific additions or deletions of particular protein activities in cells, or to modify the regulation of gene expression such that the mapping from cell environment to cell metabolic activity is altered in a desirable way.

The purpose of this report is to illustrate some of the challenges and opportunities in metabolic engineering as well as to suggest approaches that should prove valuable in progressing from trial-and-error approaches to a systematic engineering methodology for cell improvement through metabolic engineering. Two approaches that are conceptually quite different, but that are highly complementary, will be illustrated for

---

[a] This research was supported by the Energy Conversion and Utilization Technologies (ECUT) Program of the United States Department of Energy and the National Science Foundation. J. L. Galazzo was supported in part by a fellowship from the National Research Council–CONICET, Argentina, and S. Birnbaum was supported in part by the Sweden-America Foundation, the Swedish Institute, and the Royal Physiophralical Society of Lund.

selecting which genes to clone and which proteins to introduce, augment, or delete in order to improve cell performance. One is based upon detailed biochemical and mathematical analysis of a particular metabolic reaction network, arriving at the ability to compute the sensitivity of overall function of a desired pathway to the activities of each important step in the network. The second approach is based upon using examples from nature to identify genetic manipulations to obtain cells that are more robust and more active under process conditions.

Another primary theme developed in this presentation is the potential complexity of cell response to genetic engineering manipulations and to alteration in the normal amounts and types of protein activities found in the wild-type organism. Both genetic-level and enzyme-level control systems in the cell respond to genetic engineering manipulations in ways that can substantially complicate the process of metabolic engineering design. However, powerful analytical tools for multicomponent analysis of cellular proteins and metabolites can be employed to expand greatly the understanding of the cellular system and its response mechanisms.

FLUX CONTROL IN FERMENTATION BY *SACCHAROMYCES CEREVISIAE*

The pathways of metabolic reactions that transform glucose into glycerol, ethanol, and intracellular reserves of trehalose and glycogen in yeast *S. cerevisiae* are well known and have been extensively studied. Based upon previous investigations, the reactions indicated in FIGURE 1 are important in kinetic control of fermentation conducted by nongrowing yeast cells; other reactions that operate near equilibrium *in vivo* are not considered. Recent studies have elucidated the relative contributions of the steps indicated in FIGURE 1 in controlling the flux to ethanol end product in yeast under several different conditions.[1,2]

To accomplish this analysis, a detailed kinetic model for the pathway, including complete rate expressions for all of the reactions in FIGURE 1, was formulated and tested. Two different types of experiments were conducted in order to establish this model. First, yeast cells under different conditions were suddenly exposed to glucose, and the rates of glucose consumption and ethanol and glycerol production were measured. These rates remained constant for a period in excess of 30 minutes after glucose addition under all conditions studied, indicating that kinetic behavior of the cells is independent of extracellular glucose concentration, which is declining during the experiment, as well as extracellular ethanol and glycerol concentrations, which are increasing with time. Assuming quasi–steady state for metabolic intermediates (this was validated in separate experiments summarized next), the rates of all of the intracellular steps shown in FIGURE 1 may be calculated from pathway structure and stoichiometry in terms of the observed extracellular rates. The general principles employed in such calculations have been illustrated in several previous studies.[3,4]

The second type of measurement necessary to establish intracellular kinetics is the determination of the intracellular concentrations of all substrates and effectors that enter into the rate expressions for the key reactions in the pathways of interest. For these particular pathways, almost all of the needed concentrations can be obtained by phosphorous-31 nuclear magnetic resonance (NMR) spectroscopy. Applying recently developed methods for estimating individual sugar phosphate concentrations from the broad sugar phosphate resonance obtained in an *in vivo* ^{31}P NMR measurement of yeast,[5] the intracellular concentrations of glucose-6-phosphate, fructose-6-phosphate, β-fructose, 1,6-diphosphate, and 3-phosphoglycerate may be estimated. Also, by using glucose labeled by carbon-13, ^{13}C NMR can be used to measure uptake of glucose and

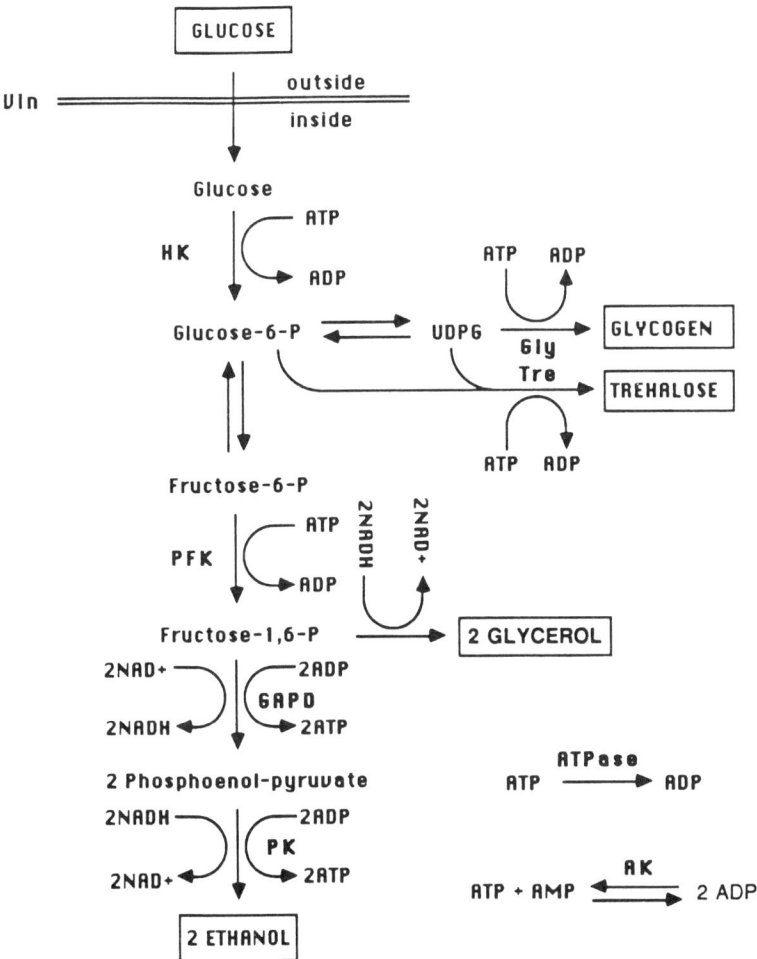

FIGURE 1. Anaerobic fermentation pathways of nongrowing yeast *Saccharomyces cerevisiae* from glucose to ethanol, glycerol, and polysaccharides. The enzymes of the intermediate steps not shown are assumed to catalyze very fast reactions so that equilibrium is maintained at all times.[2]

its conversion into various end products. Using these methods and auxiliary calculations based upon equilibrium in the adenylate kinase reaction, detailed rate expressions were evaluated. For example, the rate of phosphorylation of fructose-6-phosphate catalyzed by phosphofructokinase (PFK) has been reported based upon *in vitro* experiments to be the following:[6]

$$V_{PFK} = V_{PFK}^{max} \cdot v_{PFK}(\lambda_1, \lambda_2, \gamma), \quad (1)$$

$$v_{PFK} = \frac{g_R \lambda_1 \lambda_2 R^{n-1} + qLg_T c_1 c_2 \lambda_2 T^{n-1}}{R^n + LT^n},$$

where

$\lambda_1 = F6P/K_{R,F6P}$

$\lambda_2 = ATP/K_{R,ATP}$

$\gamma = AMP/K_{R,AMP}$

$c_j = K_{R,j}/K_{T,j}$

$q = V_{T,max}/V_{R,max}$

$L = L_0 \cdot \left(\frac{1 + c_\gamma \gamma}{1 + \gamma}\right)^n$

$R = 1 + \lambda_1 + \lambda_2 + g_R \lambda_1 \lambda_2$

$T = 1 + c_1\lambda_1 + c_2\lambda_2 + g_T c_1 c_2 \lambda_1 \lambda_2$

L_0 is pH dependent

with parameter values of

$K_{R,F6P} = 1.0$	$c_{F6P} = 0.0005$	$q = 0.0$
$K_{R,ATP} = 0.06$	$c_{ATP} = 1.0$	$g_R = 10.0$
$K_{R,AMP} = 0.025$	$c_{AMP} = 0.019$	$g_T = 1.0$
		$V_{PFK}^{max} = 31.7.$

The parameter values indicated here were taken from the same source,[6] and the pH dependence of L_0 is given in reference 2. However, the maximum velocity parameter, which depends upon the amount of active PFK in the cell, was calculated from one of the available measurements; that is, under one condition, the observed PFK rate was used on the left-hand side of equation 1, all of the measured effector concentrations under the same conditions were substituted into the rate expression on the right-hand side (which was also evaluated at the pH value corresponding to the cytoplasm), and the single remaining unknown—the parameter V_{PFK}^{max}—was then calculated. This same value was employed in subsequent calculations because all of the yeasts used in these experiments were prepared identically and therefore are expected to have the same cellular machinery at the protein level. FIGURE 2 shows the rates of PFK calculated from this model versus the corresponding measurements of these rates.[2] The agreement is excellent, thus indicating that the kinetics established for this enzyme *in vitro* here performed quite satisfactorily *in vivo*.

By knowing rate expressions for all of the pertinent pathway steps and by having measurements of the substrates and effectors that enter into these rates, it is possible to calculate the fractional change in flux to ethanol that will result for a corresponding fractional increase in the rate of one metabolic step. This corresponds to a different rate of transport for the glucose uptake step and to a change in the amount of enzyme for all of the enzyme-catalyzed steps. These sensitivity coefficients have been called flux control coefficients in the theories elaborated by Kacser and Burns,[7] Heinrich and Rapoport,[8] and subsequent investigators. One useful result from these theoretical developments is a general formalism for calculating these sensitivities. Such methods were employed to determine the flux control coefficients for the production of ethanol in these experiments, which considered identically grown yeast cells in four different environments. The environments considered were suspended cells in medium of pH 4.5 and in medium of pH 5.5, as well as yeast entrapped in calcium alginate exposed to media at these same two pH values. Parenthetically, it should be noted that the absence of external and intraparticle mass transfer effects on rates measured for the immobilized cells was verified. TABLE 1 lists the flux control coefficients so determined for each of these cases.[2]

These results provide important insights into the influence of the different environ-

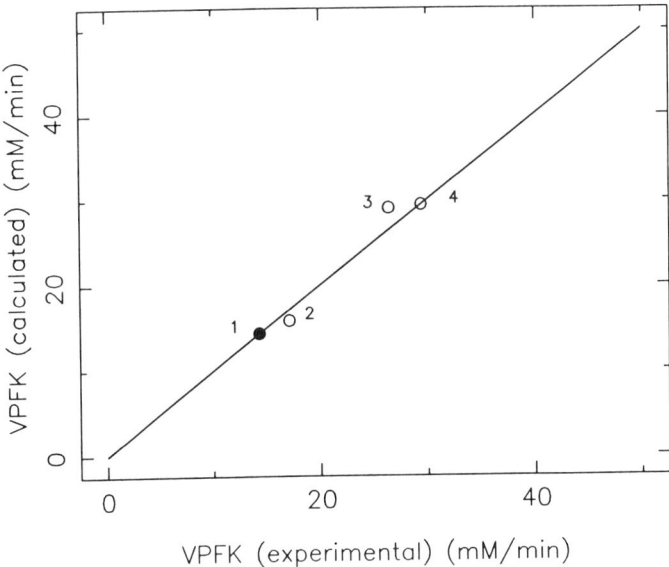

FIGURE 2. Comparison of experimental (V_{PFK}^{exp}) and calculated (V_{PFK}^{calc}) *in vivo* phosphofructokinase reaction rates.[2] Experimental ethanol and glycerol production rates were used to obtain V_{PFK}^{exp}, and the kinetic expression of equation 1 and *in vivo* NMR pH and metabolite concentration data were used to obtain V_{PFK}^{calc}. Point (●) was used to calculate V_{PFK}^{max}, which was assumed to be the same for all of the other experiments. Notation: (1) suspended cells, $pH^{ex} = 5.5$; (2) suspended cells, $pH^{ex} = 4.5$; (3) immobilized cells, $pH^{ex} = 5.5$; (4) immobilized cells, $pH^{ex} = 4.5$. Immobilized cells were entrapped in 2% Ca-alginate.

ments considered in these experiments and they also provide guidance for future metabolic engineering activities to genetically modify specific steps in order to improve ethanol productivity. The glucose uptake step exerts large flux control for suspended cells at both pH values, whereas the flux control of this step is substantially reduced for the immobilized cells. This finding is physically reasonable because the only difference between the suspended and immobilized cells, given the way these experiments were conducted, is the processing of the cells to place them in the alginate and their alginate location thereafter. Cells were not grown at all in the immobilized state in these experiments, so no adaptation or response of the cells in terms of altered gene expression is possible. Apparently, immobilization has altered the outer envelope of the cell, permitting glucose to enter more rapidly and increasing the rate of glucose uptake and the rate of ethanol production by a factor of two compared to the suspended cells at both pH values considered.

The changes in flux control coefficients with different pH values are also interesting. At high pH, the ATPase-catalyzed step exerts significant flux control, indicating that more rapid hydrolysis of ATP would increase the flux. Lowering extracellular pH increases the pH difference across the cytoplasmic membrane and increases the energy that the cell must expend in order to maintain intracellular pH at values near 7. This greater expenditure of ATP at extracellular pH 4.5 results in a greater ethanol production rate and a greater glucose uptake rate than observed at pH 5.5. In other words, decreasing the pH increases the ATP maintenance demand of the cell, which is supplied under these anaerobic conditions by increasing the rate of substrate-level phosphorylation.

TABLE 1. Flux Control Coefficients for the Production of Ethanol by Anaerobic Nongrowing *S. cerevisiae*[a]

	Suspended Cells pH^{ex} 5.5	Suspended Cells pH^{ex} 4.5	Immobilized Cells pH^{ex} 5.5	Immobilized Cells pH^{ex} 4.5
In	0.646	0.831	0.297	0.316
HK	0.008	0.011	0.007	0.007
PFK	0.098	0.115	0.336	0.497
GAPD	0.002	0.002	0.009	0.016
PK	0.001	0.001	0.003	0.003
Pol	0.006	0.000	0.109	0.157
Gol	0.001	0.001	0.001	0.002
ATPase	0.238	0.039	0.237	0.002

[a]Abbreviations in the left-hand column refer to pathway steps in FIGURE 1. Cells were immobilized in 2% Ca-alginate prior to kinetic analysis at the indicated pH values of the reaction medium.

The implications of these results for genetic manipulation of particular enzyme activities are also quite interesting. The exact steps that should be accelerated through metabolic engineering vary somewhat depending upon the particular conditions in which the cell is to be used. In addition, under some conditions, such as the immobilized cells at pH 5.5, three different steps, namely, glucose uptake, phosphorylation of fructose-6-phosphate, and ATPase, all exert about the same level of flux control. Calculations using the kinetic model discussed earlier show that increasing the rate of only one of these steps results in a rapid decline in its flux control and an increase in the flux control of the other two steps, and that the total extent to which the flux can be increased by enhancing only one step of these three is quite small. The model indicates in such a situation that it is necessary to increase the activities of several steps simultaneously in order to obtain a significant improvement in the overall rate of ethanol production. This example illustrates the importance of such a systematic modeling and sensitivity analysis approach in some situations. Trial-and-error efforts at improvements under some of the circumstances studied here would almost certainly fail.

All of the experiments discussed so far involved nongrowing yeasts that were then exposed to glucose in order to conduct fermentation. Of course, in a growing culture, the metabolic balance could be significantly different due to utilization of cofactors and intermediates from the fermentation pathways in biosynthetic processes. Accordingly, the flux control analysis, which is also based on glucose fermentation by nongrowing cells, could give significantly different results for the growing cell case. Efforts are currently in progress to establish a suitable physical configuration, as well as improved data processing methods, to enable NMR analysis of metabolite levels in growing cultures.[9,10]

USING MODELS FROM NATURE TO ENGINEER BETTER CELLS: THE *VITREOSCILLA* HEMOGLOBIN EXAMPLE

In a search for enzymes active at high temperature and under alkaline conditions for use in the detergent industry, microbiologists looked for environments in nature that had similar characteristics and then isolated microorganisms from those environments in order to find enzymes with the desired properties. A similar strategy exists for genetic improvement of cells for use in bioprocessing. Special metabolic attributes of cells that have adapted for growth under conditions similar to those encountered in a bioprocess serve as candidates for transfer to production strains. By cloning genes from the naturally adapted cells or by otherwise genetically modifying the production strain to mimic the metabolic properties of the natural model, new strains that perform better under process conditions may be created.

A project now in progress at Caltech illustrates the potential utility of this approach. Oxygen limitation is an extremely common situation in bioprocessing employing aerobic cellular metabolism. Attainment of the maximum possible cell density is an almost universal goal because of the volumetric productivity benefits that high cell densities provide. Using controlled nutrient feeding and sometimes employing contemporary technology for cell retention, it is possible to grow cells to very high densities such that oxygen supply to the dense cell population becomes difficult, if not impossible. All previous efforts to address this question have concentrated on improving in some sense the transport of oxygen from an oxygen-containing gas to the outer surface of the cell.

An example from nature suggests a different approach that is based upon changing the cell itself. A bacterium of the genus *Vitreoscilla*, an obligate aerobe, synthesizes large quantities of a heme-containing oxygen-binding protein that has been shown to have striking amino acid sequence homology to eukaryotic hemoglobins.[11] Although the physiological role of this hemoglobin protein has not been proven in *Vitreoscilla*, the biochemical properties of the molecule combined with the conditions under which its cofactor is highly expressed strongly suggest that the presence of this hemoglobin enhances the aerobic metabolism of this bacterium, which grows in natural environments that are frequently oxygen-poor.[11]

Motivated by this hypothesis, the gene for *Vitreoscilla* hemoglobin has been cloned and functionally expressed in *Escherichia coli*.[12,13] The nucleotide sequence of the open reading frame in a 2.2-kb *Hin*dIII fragment of genomic DNA from *Vitreoscilla* has a nucleotide sequence corresponding exactly to the published amino acid sequence of *Vitreoscilla* hemoglobin (VHb).[13] VHb is expressed in *E. coli* from initiation signals also contained in the 2.2-kb fragment cloned from *Vitreoscilla*.

Initial evaluation of the physiological effect of the presence of VHb in *E. coli* was conducted in a series of shake-flask experiments in which cells were grown to stationary phase in an initial batch cultivation and subsequently fed by addition of concentrated nutrient to the culture. The results of this experiment employing the VHb gene and its own promoter carried in a pUC plasmid in *E. coli* JM101 are summarized in TABLE 2. These data clearly indicate a significant effect of the presence of VHb. Cells containing the gene grow to higher cell densities than either plasmid-carrying or plasmid-free controls, and they also exhibit more rapid growth rates during the second growth phase following feed addition to the flasks. This augmentation of growth in the second phase of batch cultivation is consistent with the hypothesis that hemoglobin should be beneficial under oxygen-limited conditions especially.

Many additional experiments have now been conducted to investigate further the influence of VHb on *E. coli* metabolism and growth. One of the experiments involved a fed-batch cultivation in which, after reaching an initial stationary phase, concentrated medium was added continuously to an agitated, sparged laboratory fermentor. This experiment, which utilized the same strains as those listed in TABLE 2, again showed significant growth rate and maximum cell density benefits in the hemoglobin-containing strain. Samples of cells taken from this cultivation were examined for respiratory activity in a respirometer, with results shown in FIGURE 3.[12] The hemoglobin-containing strain (labeled JM101:pRED2) respires more rapidly than the plasmid-containing control (JM101:pUC9) over a broad range of dissolved oxygen (DO) values. Furthermore, the hemoglobin-containing strain has a constant respiration rate from high DO values down to DO values approaching zero. On the other hand, the respiration rate of the plasmid-carrying control declines as the dissolved oxygen falls and ceases respiration at a higher dissolved oxygen value than does the hemoglobin-containing strain.

TABLE 2. Results of Shake-Flask Cultivations of Plasmid-free (JM101), Plasmid-containing/Hemoglobin-free (JM101:pUC9), and Plasmid-containing/Hemoglobin-containing (JM101:pRED2) *E. coli*[a]

	JM101:pRED2	JM101:pUC9	JM101
OD_{600} before nutrient replenishment	0.937	0.737	0.945
OD_{600} final	1.230	0.880	0.985
Maximum attained dry weight	1.5 g/L	0.85 g/L	1 g/L
Relative heme content	5.5	1	—
Relative hemoglobin activity	5	1	—
Specific growth rate after feeding	0.04/h	0.01/h	0.009/h

[a]Cultures were grown in complex medium until stationary phase (first row) and then replenished with 1 volume % of concentrated nutrient broth. Rows two through four pertain to final culture properties when growth ceased after this single nutrient feeding.

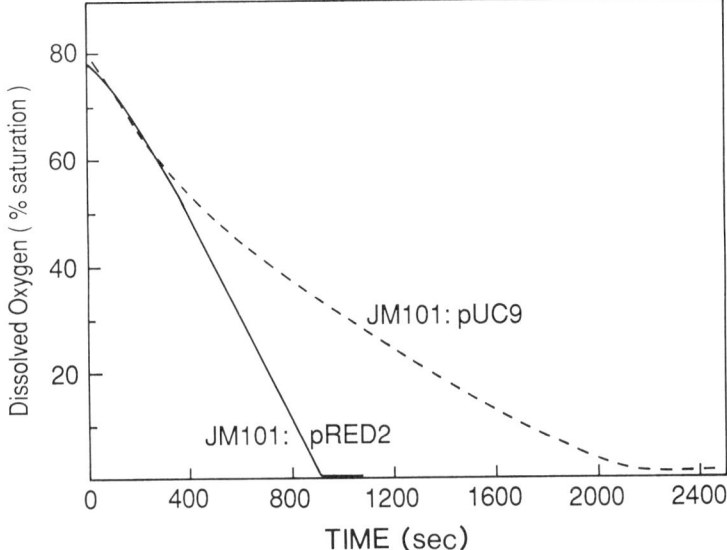

FIGURE 3. Respiration activities of hemoglobin-containing *E. coli* (JM101:pRED2) and a plasmid-containing control (JM101:pUC9). Cell samples from an exponential-phase aerobic culture in a laboratory fermentor were placed in the respirometer; dissolved oxygen concentration was monitored thereafter and recorded continuously as a function of time.[12]

Several different configurations and contexts for expression of VHb have been prepared and studied to some degree. The VHb protein has been expressed in active, functional form using cloned *E. coli* promoters (e.g., *trp* and *tac*) in place of the native VHb promoter. Functional VHb has been expressed from plasmids of both the pUC and pBR series, and several different *E. coli* hosts including JM101, HB101, and a wild-type strain have all successfully expressed active hemoglobin. Improved growth under oxygen-limited conditions due to the presence of VHb has been observed in many of these constructs and remains to be evaluated in others.

AN OXYGEN-REGULATED PROMOTER FOR EXTERNAL METABOLIC CONTROL

In most bioreactor processes, it is desirable to alter the metabolism of the cells as the process proceeds, for example, reconfiguring the cells from a metabolism optimized for growth early in the cultivation to a metabolism directed towards product synthesis at a later stage in a batch process. Thus, the design problem to be addressed by the metabolic engineer is not a single, static objective, but one of configuring a complex chemical plant that can be effectively switched from one kind of product to another while doing both the growth and product formation tasks very efficiently. Essential to any successful strategy for solving this problem is a variety of genetic control elements that the metabolic engineer can place ahead of appropriate genes to enable the process engineer to turn expression of different genes up and down at particular points in the process through manipulation of extracellular conditions.

The promoter system cloned with the VHb gene is an interesting and important addition to the genetic switches available to the metabolic engineer as well as to those interested in high-level expression of a particular protein. The *Vitreoscilla* hemoglobin promoter system is regulated in *E. coli* by dissolved oxygen.[12,13] When cells are grown under oxygen-limited conditions, the gene downstream of that promoter is expressed to high levels, whereas relatively little expression occurs under well-aerated conditions. Recent experiments have shown that this promoter sequence provides oxygen-dependent regulation of transcription and that the promoter sequence can be fused to a gene different from that for VHb to provide expression control dependent upon dissolved oxygen content of the culture. Although conditions for maximal expression have not been determined, expression of *Vitreoscilla* hemoglobin using this promoter in a multicopy plasmid routinely reaches levels of 10–20% of total cell protein with copious inclusion body formation. Smaller quantities of cloned protein, appropriate for metabolic engineering applications, can be obtained by varying the number of cloned gene copies or the dissolved oxygen level.

IMPACT OF METABOLIC REGULATION ON GENETIC ENHANCEMENT OF A KEY METABOLIC STEP: CLONED HEXOKINASES IN *S. CEREVISIAE*

Whenever metabolic engineering is undertaken, results may differ from those expected due to responses of the cell at the metabolite or protein level. Experiments in which hexokinase activities have been manipulated by genetic engineering of the yeast *S. cerevisiae* provide an example of unexpected results. Wild-type yeasts possess three different enzymes that catalyze phosphorylation of glucose: glucokinase (GLK), hexokinase PI (HXK1), and hexokinase PII (HXK2). In order to study the effect of genetic alterations in these activities, three different genetically engineered *S. cerevisiae* strains were investigated.[14] The host cell involved in all three cases is a mutant that lacks all three glucose phosphorylation activities. Each strain investigated carried one of the kinases expressed from the corresponding cloned gene on a multicopy plasmid. All of the genetic constructions were performed in the laboratory of D. G. Fraenkel and all of the strains studied were obtained from that laboratory.[15]

The strains considered were cultivated aerobically, and exponential-phase cells were sampled and assayed *in vitro* for glucose phosphorylation activities, with results as listed in TABLE 3. Also measured were the specific growth rates of each strain in exponential-batch growth as well as growth and product yields (TABLE 3).[14]

It is interesting to note that the strains that grow most rapidly are not the strains with the highest *in vitro* glucose phosphorylation activity. Therefore, enzyme-level regulation by metabolites of the glucose phosphorylation step *in vivo* must be playing a significant role in the *in vivo* operation of these enzymes. In order to seek further information on metabolite levels in these cells, ^{31}P NMR measurements were conducted of dense suspensions of each of the three strains, with results as shown in FIGURE 4. Differences in intracellular pH are manifested by the chemical shift positions of the intracellular inorganic phosphate peak P_i, and differences in intracellular sugar phosphates and ATP + ADP levels are also evident in these spectra. Further analysis of these measurements suggests that intracellular magnesium-free ATP levels may be sufficiently high to inhibit hexokinase PII, thereby contributing to the disparity between *in vitro* glucose phosphorylation and *in vivo* glucose uptake activities in the engineered strain carrying HXK2.[14] This information helps somewhat in understanding the *in vivo* behavior of these systems, but a complete explanation remains to be determined.

TABLE 3. Growth Properties of Recombinant *S. cerevisiae* Strains Expressing Different Glucose Phosphorylation Enzymes from Cloned Genes[a]

	DFY437: pHXK1	DFY437: pHXK2	DFY437: pGLK
Batch Fermentation Properties			
specific growth rate (h^{-1})	0.41	0.29	0.33
ethanol yield from glucose (mol/mol)	1.47	1.33	1.30
biomass yield from glucose (g cell dry weight/mol)	32	22	39
In Vitro Enzyme Activities			
glucose phosphorylation (U/mg of protein)	0.5	5.1	0.5
fructose phosphorylation (U/mg of protein)	2.2	6.3	0.1

[a] Aerobic cultures were grown in YPD medium at 30 °C. Extracts from exponential-phase cultures were assayed for glucose phosphorylation activities (bottom).[14]

HOST-CELL GENETIC-LEVEL RESPONSES TO THE PRESENCE OF PLASMIDS AND CLONED GENE EXPRESSION

Introduction of additional protein activity into a cell or deletion of an existing activity can cause reallocation of cellular resources for macromolecular synthesis and also can introduce perturbations in intracellular metabolite levels. An indirect consequence of both types of perturbations is changes in the levels of expression of host-cell proteins. Thus, a genetically manipulated cell may respond with its own internal controls to reconfigure its own metabolism somewhat, giving rise to a much different cell than was anticipated based upon the initial genetic manipulation. In order to investigate this possibility, analytical two-dimensional gel electrophoresis has been employed to analyze the relative levels of numerous proteins in *E. coli* HB101 and in *E. coli* HB101 harboring plasmids propagated at different copy numbers.[16] Autoradiograms of two-dimensional gels for three different *E. coli* strains, two of which carry plasmids at different copy numbers, are shown in FIGURE 5. Data from these gels have been digitized and analyzed using the PDQUEST computer analysis system (Protein Databases). This software system provides not only quantitative analysis of individual gels, but is specifically designed to enable comparisons of the relative amounts of the same protein on different gels.

Based upon this analysis, differences in levels of numerous enzymes and other cellular proteins have been found between the plasmid-free host and the plasmid-carrying recombinant cells. Some of these differences are listed in TABLE 4. These data are based on replicate analyses of each strain. Considering the variation in replicate experiments, values here that deviate by more than 10–20% from the HB101 values are significant.

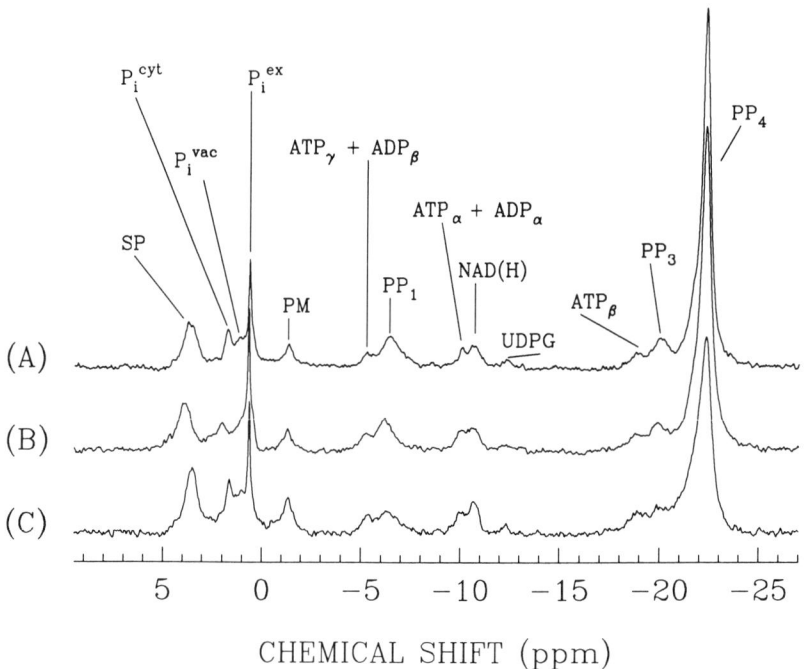

FIGURE 4. ^{31}P NMR spectra at 202.46 MHz of anaerobic *S. cerevisiae* suspensions at 20 °C during quasi–steady state glycolysis: (A) *S. cerevisiae* DFY437:pHXK1, (B) *S. cerevisiae* DFY437:pHXK2, and (C) *S. cerevisiae* DFY437:pGLK. Cells were grown to mid-log phase on YPD and then harvested and resuspended in MES buffer for the NMR experiment in which 80 mM glucose was added. Spectra represent one-minute time accumulations and were recorded using 70° pulses (0.5-s acquisition time). Spectra are scaled so that peak areas can be compared for a given intracellular resonance between spectra. Comparisons within a spectrum cannot be made between any two resonances except for polyphosphates. Abbreviations: SP, sugar phosphate; P_i^{cyt}, cytoplasmic inorganic phosphate; P_i^{vac}, vacuolar inorganic phosphate; P_i^{ex}, extracellular inorganic phosphate; PM, phosphomannan; ADP, adenosine diphosphate; ATP, adenosine triphosphate; NAD(H), nicotinamide adenine dinucleotide; PP_1, pyrophosphate and terminal phosphates of polyphosphate; UDPG, uridine diphosphoglucose; PP_3, penultimate phosphates of polyphosphate; and PP_4, middle phosphates of polyphosphate.[14]

Further experiments of this kind are important in order to characterize more completely how cells respond to efforts to engineer their metabolism so that a comprehensive systems design procedure, incorporating the genetic- and enzyme-level regulatory responses of the host cell, can be realized.

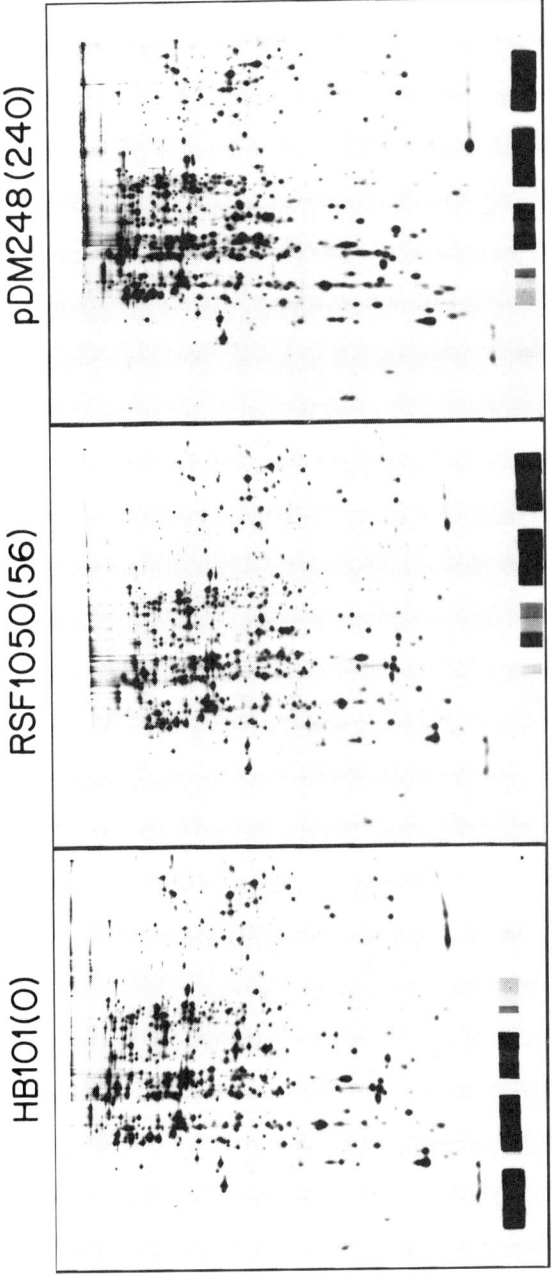

FIGURE 5. Autoradiograms of two-dimensional gel analyses of the cellular proteins contained in plasmid-free *E. coli*[16] (left), *E. coli* HB101 carrying RSF1050 (middle), and *E. coli* pDM248 (right).

DISCUSSION

Metabolism is a complex system of transport and catalytic reaction processes. Relative to networks of homogeneous chemical reactions, extraordinary complications develop due to couplings and feedbacks arising from modulation of protein activities by various effectors as well as due to genetic-level responses that modulate the quantities of different proteins present and that can substantially alter even the topological structure of the metabolic reaction and transport network.

Fortunately, some of the same molecular biology methods that facilitate and enable specific genetic intervention in cells also serve as tools for multicomponent analysis of numerous metabolites and arrays of proteins and genetic elements. The full power of these experimental techniques must be mobilized for the experimental definition of the metabolic system in a fashion sufficiently detailed and quantitative to be used in a

TABLE 4. Amounts of *E. coli* Proteins relative to Their Amounts in the Plasmid-free HB101 Host and in HB101 Carrying Plasmids RSF1050 and pDM248 as Determined by Analytical Two-dimensional Gel Electrophoresis[a]

Protein	RSF1050	pDM248
succinate thiokinase (β)	2.3	1.1
PEP carboxylase	0.99	1.6
pyruvate kinase I	0.97	0.65
aspartate transcarbamoylase	2.0	0.57
RNA polymerase (α)	0.89	0.97
ribosomal proteins		
S2	1.4	1.4
S6b	1.1	0.31
L5	0.29	0.09
heat shock proteins		
HtpG	1.5	2.1
GrpE	1.0	2.3

[a]Under the culture conditions used to prepare samples for analysis, the copy numbers (number of plasmids per chromosome) of RSF1050 and pDM248 are 56 and 240, respectively.[16]

systems engineering framework. Of course, even more powerful experimental methods in terms of sensitivity, resolution, and access to numerous simultaneous components would also be extremely useful in this endeavor.

Methods for the mathematical description of well-defined reaction and transport networks and for the systems analysis of the parametric properties of these networks are well in hand. However, it is unlikely in the foreseeable future that sufficient experimental information will be available for full quantitative definition of a large and complex metabolic structure. Further original thought and basic theoretical development are likely required to provide new conceptual approaches to mathematical description and analysis of metabolic systems.

These formidable technical challenges invite the best efforts of biochemical engineers, molecular biologists, and mathematical theoreticians in the years ahead. Besides the intellectual and aesthetic rewards of penetrating these problems, the emergence of a systematic paradigm for metabolic engineering will transform the present pharmaceutical, food, and chemical industries.

ACKNOWLEDGMENTS

D. G. Fraenkel kindly provided the *S. cerevisiae* strains and plasmids used in this research. Computer analysis of digitized gel patterns was conducted at the facilities of the Center for the Development of an Integrated Protein and Nucleic Acid Biotechnology supported by the National Science Foundation.

REFERENCES

1. GALAZZO, J. L. & J. E. BAILEY. 1989. *In vivo* nuclear magnetic resonance analysis of immobilization effects on glucose metabolism of yeast *Saccharomyces cerevisiae*. Biotechnol. Bioeng. **33**: 1283–1289.
2. GALAZZO, J. L. & J. E. BAILEY. 1990. Fermentation pathway kinetics and metabolic flux control in suspended and immobilized *Saccharomyces cerevisiae*. Enzyme Microb. Technol. In press.
3. PAPOUTSAKIS, E. T. 1984. Equations and calculations for fermentations of butyric acid bacteria. Biotechnol. Bioeng. **26**: 174–187.
4. REARDON, K. F., T-H. SCHEPER & J. E. BAILEY. 1987. Metabolic pathway rates and culture fluorescence in batch fermentations of *Clostridium acetobutylicum*. Biotechnol. Prog. **3**: 153–167.
5. SHANKS, J. V. & J. E. BAILEY. 1988. Estimation of intracellular sugar phosphate concentrations in *Saccharomyces cerevisiae* using ^{31}P nuclear magnetic resonance spectroscopy. Biotechnol. Bioeng. **32**: 1138–1152.
6. HESS, B. & T. PLESSER. 1978. Temporal and spatial order in biochemical systems. Ann. N.Y. Acad. Sci. **316**: 203–213.
7. KACSER, H. & J. A. BURNS. 1973. The control of flux. Symp. Soc. Exp. Biol. **27**: 65–104.
8. HEINRICH, B. & T. A. RAPOPORT. 1974. A linear steady-state treatment of enzymatic chains: general properties, control, and effector strength. Eur. J. Biochem. **42**: 89–95.
9. GALAZZO, J. L. & J. E. BAILEY. 1989. Application of linear prediction singular value decomposition for processing *in vivo* NMR data with low signal-to-noise ratio. Biotechnol. Tech. **3**: 13–18.
10. GALT, S. & J. E. BAILEY. 1989. Studies of flowing suspensions of *Escherichia coli* using phosphorous-31 nuclear magnetic resonance spectroscopy. Biotechnol. Tech. In press.
11. WAKABAYASHI, S., H. MATSUBARA & D. A. WEBSTER. 1986. Primary sequence of a dimeric bacterial hemoglobin from *Vitreoscilla*. Nature **322**: 481–483.
12. KHOSLA, C. & J. E. BAILEY. 1988. Heterologous expression of a bacterial haemoglobin improves the growth properties of recombinant *E. coli*. Nature **331**: 633–635.
13. KHOSLA, C. & J. E. BAILEY. 1988. The *Vitreoscilla* hemoglobin gene: molecular cloning, genetic expression, and its effect on *in vivo* heme metabolism in *Escherichia coli*. Mol. Gen. Genet. **214**: 158–161.
14. SHANKS, J. V. & J. E. BAILEY. 1989. ^{31}P NMR and ^{13}C NMR studies of recombinant *S. cerevisiae* with altered glucose phosphorylation activities. Submitted.
15. WALSH, R. B., G. KAWASAKI & D. G. FRAENKEL. 1983. Cloning of genes that complement yeast hexokinase and glucokinase mutants. J. Bacteriol. **154**: 1002–1004.
16. BIRNBAUM, S. & J. E. BAILEY. 1990. Plasmid presence changes the relative levels of many host cell proteins and ribosome components in recombinant *Escherichia coli*. Submitted.

Genetic Engineering of Metabolic Pathways Applied to the Production of Phenylalanine

KEITH BACKMAN,[a] MARY JANE O'CONNOR,[a]
AIKO MARUYA,[a] EDWIN RUDD,[a] DIANE McKAY,[a]
R. BALAKRISHNAN,[a] M. RADJAI,[a]
V. DiPASQUANTONIO,[a] DIANE SHODA,[a]
RANDOLPH HATCH,[a] AND
K. VENKATASUBRAMANIAN[b]

[a] BioTechnica International
Cambridge, Massachusetts 02140

[b] H. J. Heinz Company
Pittsburgh, Pennsylvania 15230
and
Department of Chemical and Biochemical Engineering
Rutgers University
Piscataway, New Jersey 08854

INTRODUCTION

We desired the development of a commercializable process for the production of the amino acid L-phenylalanine, which has uses in the manufacture of aspartame and in parenteral nutrition. In order to approach the problem, we set performance targets for our envisioned process based upon the technical and scientific constraints on phenylalanine biosynthesis and upon the commercial constraints on the value of the product. One mole of phenylalanine is synthesized from two moles of glucose and one mole of ammonia in common bacterial systems. This establishes minimum raw material costs. We wished to develop an organism that could produce high concentrations of phenylalanine with relatively few by-products (including cell mass) by efficiently using raw materials and capital equipment (fermentors and downstream processing equipment). In order to do this, we pursued a systematic and thorough strategy covering organism selection, optimization of biosynthetic capacity, and development of fermentation and recovery processes.

ORGANISM SELECTION

It was our belief that many organisms might be made to produce high titers of phenylalanine through suitable genetic manipulation. We therefore sought to choose an organism that might give a commercial advantage in terms of the rate at which it could produce the product. This would indicate an organism capable of very rapid growth on simple defined media, for which technology existed that could be used to alter specifically those genetic aspects related to phenylalanine formation without negatively influencing the rapid growth properties. Additionally, it would be of value in the development effort to employ an organism for which the genetics, physiology, and enzymology of phenylalanine biosynthesis were well researched and reported in the literature; this would allow us to avoid having to make many basic developments and discoveries. Based upon these criteria, we selected *Escherichia coli* as our production organism.

TABLE 1. Steps in Phenylalanine Biosynthesis

Intermediate(s)	Enzyme	Gene(s)
Erythrose-4-phosphate + Phospho-enol-pyruvate	DAHP synthetase	*aro*F,G,H
Deoxyarabinoheptulosonate-7-phosphate	Dehydroquinate synthetase	*aro*B
Dehydroquinate	Dehydroquinate dehydratase	*aro*D
Dehydroshikimate	Dehydroshikimate reductase	*aro*E
Shikimate	Shikimate kinase	*aro*L
Shikimate phosphate + Phospho-enol-pyruvate	EPSP synthetase	*aro*A
Enolpyruvoylshikimate phosphate	Chorismate synthetase	*aro*C
Chorismate	Chorismate mutase	*phe*A
Prephenate	Prephenate dehydratase	*phe*A
Phenylpyruvate	Transaminase(s)	*tyr*B, *ilv*E, *asp*C
Phenylalanine		

OPTIMIZATION OF BIOSYNTHETIC CAPACITY

Phenylalanine is formed from the glycolytic intermediates erythrose-4-phosphate and phospho-enol-pyruvate in a series of steps, many of which are common to the formation of the aromatic amino acids tyrosine and tryptophan (TABLE 1; reviewed in reference 1). The formation of phenylalanine (and the other aromatic amino acids) is regulated at several points and in diverse ways. The first committed step in aromatic biosynthesis is catalyzed by three isozymes, each regulated by a different aromatic amino acid. The regulation of these isozymes occurs both by control of enzyme formation (via regulation of transcription of the genes for the isozymes) and by regulation of enzyme activity (by feedback inhibition of already formed enzyme). The isozymes are determined by the genes *aro*F, *aro*G, and *aro*H; regulation of these genes and their enzyme products is effected by tyrosine, phenylalanine, and tryptophan, respectively. Within the common steps of aromatic amino acid synthesis, the formation of shikimate kinase, determined by *aro*L, is regulated at the level of enzyme formation. Finally, the conversion of the common aromatic intermediate chorismate to the specific amino acid phenylalanine is also tightly regulated. The enzyme chorismate mutase–prephenate dehydratase (CMPD) is determined by the gene *phe*A and is regulated both at the level of enzyme formation and at the level of enzyme activity.

In order to systematically improve the biosynthetic capacity of cells making phenylalanine, we cloned from *E. coli* all of the genes involved in phenylalanine biosynthesis except for the transaminases. We independently determined that the *E. coli* strains we were working with were capable of rapidly converting high levels of phenylpyruvate to phenylalanine and concluded that enhancement of this step was not necessary. Our cloning efforts were helped by the fact that several of the genes involved in aromatic amino acid biosynthesis had already been cloned and their sequences reported.[2-6] Each of the genes was engineered for increased expression and the effect of such increased expression on phenylalanine formation was assessed. As expected, not all genes improved phenylalanine formation because their enzyme products were not limiting in the biosynthetic pathway. In some cases (TABLE 2), particular genes would even prove detrimental to phenylalanine formation. This illustrates the improbity of attempting to produce more product merely by overproducing all of the enzymes involved in its biosynthesis. Based on the results of these overexpression experiments, we have incorporated engineered copies of appropriate genes into our production organism.

TABLE 2. Effect of Cloned *aro* Gene on Phenylalanine Formation

Construction	Phenylalanine Titer (g/L)
optimal plasmid	45
optimal plasmid + *aro*"X"	34

Having identified those genes that were important in the rate of phenylalanine biosynthesis, we next wished to remove from the regulated genes the sensitivity to regulation by accumulated phenylalanine. We focused our attention first on the regulation of gene expression, leaving regulation of enzyme activity for subsequent investigation. Because several of the genes relevant to phenylalanine biosynthesis are normally regulated by the repressor protein encoded by the *tyr*R gene,[7] we generated *tyr*R mutations and introduced them into our production strain. This results in the derepression and consequent constitutive expression of several genes, including *aro*F, *aro*G, *aro*L, and genes involved in transport and transamination of phenylalanine. By attacking this regulation issue at the regulatory protein, we avoided the more complex task of altering the regulatory sites associated with each of those regulated genes.

The *phe*A gene is also regulated at the level of gene expression. In this instance, however, the gene is subject to multiple forms of regulation.[2,8] The promoter has an associated operator at which a phenylalanine-specific repressor protein can bind to regulate the initiation of transcription. In addition, there is a leader-attenuator sequence that controls the elongation of the RNA transcript into *phe*A in response to the levels of phenylalanine available (FIGURE 1). In view of this more complex regulation, we decided to alter the expression of *phe*A by providing an entirely new promoter rather than by attempting to address the points of transcriptional regulation

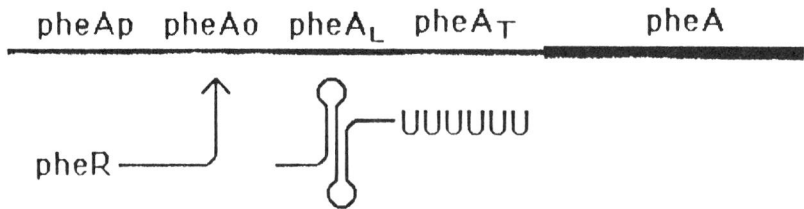

FIGURE 1. Regulation of *phe*A expression. A line represents (not to scale) the *phe*A gene and its associated regulatory apparatus. The structural gene is indicated by a thickened line. Transcription initiation is regulated from the promoter (*phe*Ap) by a repressor protein determined by the *phe*R gene that binds at an operator (*phe*Ao). When excess phenylalanine is present, it alters the repressor protein so that it binds the operator and prevents transcription. After transcription initiates, the RNA can adopt one of two "stem and loop" configurations, depending on the efficiency of translation of a peptide (determined by *phe*A_L). The peptide contains an extraordinarily high proportion of phenylalanine residues. If phenylalanine is present in abundance, the peptide is translated efficiently and the passage of the ribosome along the RNA forces the RNA to adopt a stem and loop configuration leading to termination of transcription at a stretch of uridine (U) residues within the terminator (*phe*A_T). In the absence of phenylalanine, the ribosome is halted within the portion of the RNA requiring phenylalanine for translation, thereby allowing the alternative stem and loop structure to form and consequently preventing termination of transcription at *phe*A_T.

of *phe*A one at a time. We used various nucleolytic procedures to remove the naturally associated regulatory apparatus from *phe*A and joined the remaining structural gene to new promoters. As summarized in TABLE 3, we found that the level of expression of *phe*A in such constructions was strongly influenced by the nature of the promoter to which the gene was joined. It is worth noting that some of the entries in TABLE 3 represent the same promoter joined to the *phe*A gene at slightly different positions. We tested these various promoter-*phe*A fusions for their influence on phenylalanine formation and incorporated the optimal arrangement into our production strain.

We next turned our attention to the issue of feedback inhibition of enzymes in the biosynthetic pathway. CMPD is subject to feedback inhibition by phenylalanine. We discovered that a particular tryptophan residue in the protein is crucially involved in the manifestation of feedback inhibition. Genetic alterations that result in substitutions at that residue or deletions including that residue result in the formation of a CMPD enzyme that is no longer sensitive to feedback inhibition.[9] We call such mutations *phe*A'. We have included a *phe*A' gene (fused to an optimal promoter) in our production strain.

TABLE 3. Effect of Various Promoters on *phe*A Expression

Promoter	Relative Expression
none	0
promoter 1	1×
promoter 2	3×
promoter 3	3×
promoter 4	6×
promoter 5	10×

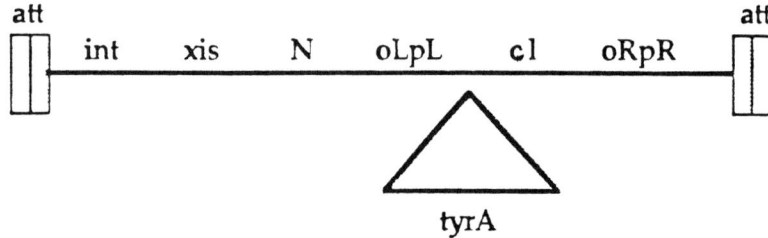

FIGURE 2. Excision vector technology. An excision vector is represented as a line between two boxes. Above the line and boxes are indicated those genes from bacteriophage lambda that are carried on the vector. oRpR is required for the expression of a repressor determined by cI. This repressor binds to oRpR and oLpL to prevent the expression of other lambda genes. The other genes collectively act to form a recombination activity that allows the vector to integrate into or excise from the bacterial chromosome at the sites (indicated by the boxes) named *att*. Other genes, such as *tyr*A (indicated below the excision vector), can be cloned into the excision vector prior to its introduction into a target cell and can thereby be present or absent in the cell in coordination with the vector. Upon entering a cell, the genes for recombination (*N*, *xis*, and *int*) are expressed because there is not yet any repressor. As repressor accumulates, it shuts off the expression of those genes. In a fraction of the recipient cells, the recombination enzymes cause the vector to integrate into the cell chromosome before those enzymes decay away. That cell and all of its progeny inherit the vector and any gene it might carry (such as *tyr*A). If the repressor is inactivated, such as by high temperature, new recombination enzymes are formed that excise the vector from the chromosome. In such a cell and all of its progeny, the vector and the gene(s) it might carry are lost.

We initially expected to circumvent the feedback inhibition effected by phenylalanine on DAHP synthase by employing the tyrosine-specific isozyme encoded by *aro*F. This strategy worked up to a point, but several complications ensued. Because the enzyme was sensitive to tyrosine, we had to limit the availability of tyrosine to the cell in order to stimulate the formation of phenylalanine. This could be accomplished in a variety of ways. We could make the strain a tyrosine auxotroph and feed exogenous tyrosine at a limiting rate. Alternatively, we could control the internal synthesis of tyrosine. One way of doing this was by use of excision vector technology (FIGURE 2), invented expressly to address this problem.[10,11] An excision vector is a genetic element that can carry a cloned gene (in this case, *tyr*A for tyrosine biosynthesis) and can both integrate into and be excised from the bacterial chromosome. The excision event can be made to occur in response to an outside signal or condition, such as temperature. As shown in TABLE 4, strains carrying our excision vector stably maintain the ability to synthesize their own tyrosine at low temperatures, but they lose that ability completely

TABLE 4. Temperature-dependent Loss of Excision Vector Yields Tyrosine Auxotrophy

Growth Temperature of Culture	Total Cell Count (with tyrosine)	Prototrophic Cell Count (without tyrosine)
30 °C	2.9×10^9	2.5×10^9
42 °C	1.0×10^9	$<10^5$

TABLE 5. Phenylalanine Inhibits the Tyrosine-specific DAHP Synthetase

Addition	Relative DAHP Synthetase (tyr) Activity
none	100%
10 mM tyrosine	<5%
60 mM phenylalanine	50%

at slightly higher temperatures. This provides us with a convenient way to cause starvation for tyrosine at a time in the growth cycle of our choice.

All of these approaches to simply utilize *aro*F as the source of enzyme were confounded by a simple fact: contrary to conventional wisdom, the tyrosine-specific DAHP synthase determined by *aro*F is in fact inhibited by phenylalanine. The effect has not been noted previously because the levels of phenylalanine required to manifest the inhibition are vastly greater than physiological levels in ordinary bacteria (TABLE 5). However, we were dealing with bacteria that already could produce in excess of 10 grams of phenylalanine per liter of growth medium. In the face of this phenomenon, we adopted an historically proven approach and isolated feedback-inhibition-insensitive mutations in the gene(s) for DAHP synthase by means of resistance to toxic amino acid analogues. Appropriate analogue resistance mutations have been incorporated into our production strain.

By the means discussed earlier, we systematically identified those genes that could be engineered to contribute to phenylalanine formation and, within that subset of genes that was subject to regulation by accumulated phenylalanine, we eliminated regulation both at the level of enzyme formation and at the level of enzyme activity. Having accomplished this, several important issues remained to be resolved. It was important

TABLE 6. By-products Found in Broth[a]

Compound	Concentration (g/L)
Organic acids	
acetate	<2
formate	<1
pyruvate	<0.1
lactate	<0.1
valerate	NQ
Aromatic intermediates	
shikimate	<0.2
dihydroxybenzoate	0.03
trihydroxybenzoate	0.11
p-hydroxybenzoate	0.05
prephenate	BDL
phenylpyruvate	BDL
Aromatic amino acids	
tryptophan	BDL
tyrosine	BDL

[a]NQ = not quantified; BDL = below detection limit (0.005 g/L).

to determine whether the phenylalanine produced by our engineered strains would be retained within the cells or excreted. To our satisfaction, the product appeared primarily in the culture broth without any need for intervention on our part. It was also important to establish that biosynthetic intermediates or by-products were not released in any significant quantities. As summarized in TABLE 6, our strains did not produce

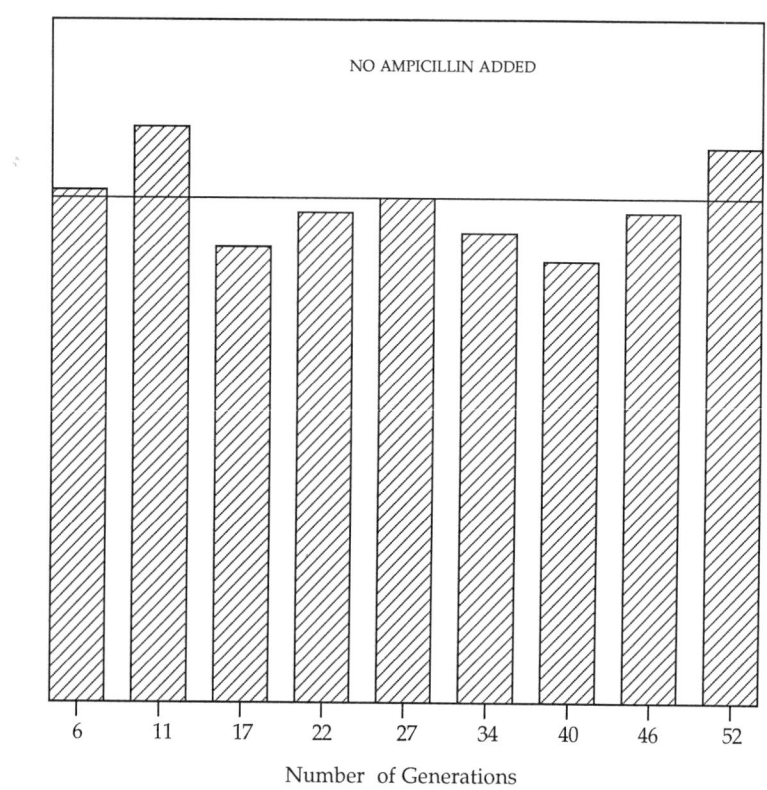

FIGURE 3. Indication of genetic stability by phenylalanine added. The plasmid-carrying production strain was grown nonselectively for 52 generations by serial transfer. Transfers to fresh medium were performed when cultures reached early stationary phase, and phenylalanine accumulation in the culture supernatant was determined at the time of transfer. The constancy of yield over time is within experimental fluctuation; the yield shown here is somewhat lower than reported in TABLE 7 because these experiments were conducted at the shake-flask level.

significant by-products. We attribute this to the limited and precise nature of the genetic changes we introduced into the strain without causing the kind of concomitant damage that can result from heavy chemical mutagenesis. Finally, and especially in view of the fact that some of the genetically engineered changes in our strains are carried on plasmids, we were concerned that our strains might be genetically unstable.

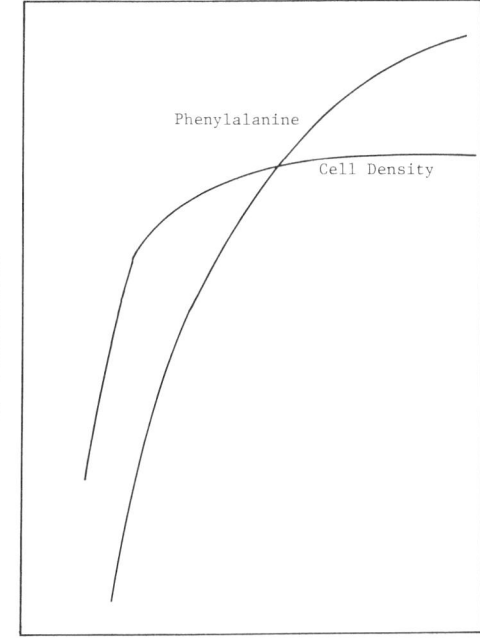

FIGURE 4. A typical growth curve for our production organism is shown, reporting the accumulation of cell mass and phenylalanine over time.

Stability tests were carried out by serial subculturing and it was determined that, both in terms of plasmid retention and in terms of ability to produce phenylalanine, our strains were completely stable. The same result was obtained whether or not antibiotic was included in the growth medium to select for plasmid maintenance (FIGURE 3).

FERMENTATION AND PROCESS ISSUES

Having developed a strain that was capable of unregulated biosynthesis of phenylalanine, it was necessary to develop a process to harness the capabilities of the strain. Although we discuss this in a separate section, the process development and the strain development took place in parallel, with results from each effort providing guidance and direction to the other. It was found that a fed-batch process yielded the best

TABLE 7. Summary of Typical Process Parameters

Final phenylalanine titer	50 g/L
Phenylalanine yield on glucose	0.23 g/g
Fermentation time	36 hours
Final cell density (dry weight)	23 g/L

performance of the strain. The feed stream was used to control the concentration of glucose during growth. We were able to develop an inexpensive, simple, defined medium that not only gave excellent performance of the strain, but also minimized the medium components that might have to be removed during downstream processing of the phenylalanine. Oxygen transfer requirements were determined, and aeration and stirring were controlled to keep phenylalanine production from falling due to anoxia. Automatic control of pH and mechanical foam breakers were also employed in the fermentors. A typical growth curve is shown in FIGURE 4. Note that the phenylalanine production occurs both during and after active growth, and the great majority of product accumulates during stationary phase.

STRAIN AND PROCESS PERFORMANCE

We selected *E. coli* because we expected it would be suitable for manipulation without loss of its ability to grow rapidly on simple media. As evidenced by our results (TABLE 7), we have realized our expectations for this organism. The organism can achieve titers of 50 grams of phenylalanine per liter in about 36 hours of vessel time. In doing this, the strain produces over 2 grams of product per gram of cell mass formed and about 0.25 grams of product per gram of glucose consumed. Moreover, the fermentation broth is remarkably free from by-products at the end of the production cycle. Thus, our process meets the criteria we established at the outset of our research.

REFERENCES

1. METZLER, D. E. 1977. Biochemistry, p. 849. Academic Press. New York.
2. ZURAWSKI, G., K. BROWN & C. YANOFSKY. 1978. Proc. Natl. Acad. Sci. U.S.A. **75:** 4271.
3. HUDSON, G. S. & B. E. DAVIDSON. 1984. J. Mol. Biol. **180:** 1023.
4. DUNCAN, K., A. LEWENDON & J. R. COGGINS. 1984. FEBS Lett. **170:** 59.
5. MILLAR, G. & J. R. COGGINS. 1986. FEBS Lett. **200:** 11.
6. DEFEYTER, R. C., B. E. DAVIDSON & J. PITTARD. 1986. J. Bacteriol. **165:** 233.
7. CORNISH, E. C., V. P. ARGYROPOULOS, J. PITTARD & B. E. DAVIDSON. 1986. J. Biol. Chem. **261:** 403.
8. GOWRISHANKAR, J. & J. PITTARD. 1982. J. Bacteriol. **152:** 1.
9. BACKMAN, K. & R. BALAKRISHNAN. 1988. United States patent no. 4,753,883.
10. BALAKRISHNAN, R. & K. BACKMAN. 1988. Gene **67:** 97.
11. BACKMAN, K. & R. BALAKRISHNAN. 1988. United States patent no. 4,743,546.

Stoichiometry and Kinetics of Xylose Fermentation by *Pichia stipitis*[a]

P. J. SLININGER,[b] L. E. BRANSTRATOR,[b]
J. M. LOMONT,[b,c]
B. S. DIEN,[b,c] M. R. OKOS,[c,d] M. R. LADISCH,[c,d]
AND R. J. BOTHAST[b]

[b]*Northern Regional Research Center*
Agricultural Research Service
United States Department of Agriculture
Peoria, Illinois 61604

[c]*Departments of Chemical and Agricultural Engineering*
[d]*Laboratory of Renewable Resources Engineering*
Purdue University
West Lafayette, Indiana 47907

INTRODUCTION AND THEORETICAL CONSIDERATIONS

Conclusions of previous investigations have led us to focus on *Pichia stipitis* as a yeast with high potential for producing ethanol from xylose-rich, wood-processing wastes.[1–5] Given 150 g/L xylose in complex medium, strain Y-7124 functions optimally at 25–26 °C and pH 4–7 to accumulate 56 g/L ethanol with negligible xylitol by-production.[6] In a past report, we cited the need for an optimal bioreactor system; toward this end, we put oxygen uptake, growth, and death kinetics into mathematical form.[7] The present report builds on our previous work as the pathways and stoichiometry of xylose metabolism are examined and models of xylose uptake and ethanol production are identified.

Yeasts begin xylose metabolism with the sequence, xylose → xylitol → xylulose → xylulose 5-phosphate.[8] In balanced form, this sequence is:

$$6\ C_5H_{10}O_5 + 6f\ NADPH,H^+ + 6(1-f)\ NADH,H^+$$
$$\rightarrow 6\ C_5H_{12}O_5 + 6f\ NADP^+ + 6(1-f)\ NAD^+ \quad (1)$$

$$6\ C_5H_{12}O_5 + 6\ NAD^+ \rightarrow 6\ C_5H_{10}O_5' + 6\ NADH,H^+ \quad (2)$$

$$6\ C_5H_{10}O_5' + 6\ ATP \rightarrow 6\ C_5H_{11}O_5PO_3 + 6\ ADP. \quad (3)$$

Here, $C_5H_{10}O_5'$ designates xylulose, an isomer of xylose. Verduyn *et al.*[9] have shown that *Pichia stipitis* has a xylose reductase with dual cofactor specificity. The parameter f designates the fraction of reductase activity supported by NADPH, and the difference $(1-f)$ designates the fraction supported by NADH. Our stoichiometry model

---

[a]The mention of firm names or trade products does not imply that they are endorsed or recommended by the United States Department of Agriculture over other firms or similar products not mentioned.

presumes that metabolism of xylulose 5-phosphate can continue via any of four processes: assimilation, pentose phosphate oxidation, respiration, and ethanolic fermentation.

The following equation describes xylose assimilation for cell synthesis and maintenance:

$$883\ C_5H_{10}O_5 + 280\ NH_2CONH_2 + (912 + 883f)\ NADPH,H^+$$
$$+ (1322 + 883f)\ NAD^+ + 3219.2\ [ATP + H_2O]$$
$$\to 1000\ C_4H_{7.05}N_{0.56}O_{2.20} + 695.5\ CO_2 + (912 + 883f)\ NADP^+$$
$$+ (1322 + 883f)\ NADH,H^+ + 1105\ H_2O + 3219.2\ [ADP + P_i]. \quad (4)$$

In building this equation, we assumed that the empirical formula for Y-7124 biomass was $C_4H_{7.05}N_{0.56}O_{2.20}$ (98 g/mole) based on elemental analysis of lyophilized cells (Galbraith Laboratories); that 0.736 glucose + 0.56 NH_3 + 3.07 ATP + 0.9124 · $NADPH,H^+$ + 1.322 NAD^+ produces 0.4155 CO_2 + 1 mole yeast plus the complementary cofactors;[10,11] that urea hydrolysis obeys the stoichiometry, NH_2CONH_2 + $H_2O \to 2\ NH_3 + CO_2$; and that xylose and glucose are related by

$$6\ C_5H_{10}O_5 + 6f\ NADPH,H^+ + 6f\ NAD^+ + [ATP + H_2O]$$
$$\to 5\ C_6H_{12}O_6 + 6f\ NADH,H^+ + 6f\ NADP^+ + [ADP + P_i]. \quad (5)$$

Equation 6 accounts for the conversion of xylulose 5-phosphate to glucose 6-phosphate via the pentose phosphate pathway:

$$6\ C_5H_{11}O_5PO_3 + H_2O \to 5\ C_6H_{13}O_6PO_3 + P_i. \quad (6)$$

Equation 7 compensates for the difference in ATP required to phosphorylate glucose and the carbon-equivalent of xylose:

$$5\ C_6H_{13}O_6PO_3 + 5\ ADP \to 5\ C_6H_{12}O_6 + 5\ ATP. \quad (7)$$

Equation 5 is the sum of equations 1–3, 6, and 7. Thus, the stoichiometry for xylose assimilation was derived from current knowledge of pentose and glucose metabolism in yeasts. Note that there is a small (<1%) hydrogen deficit on the left side of equation 4 that stems from the H:O ratio fixed by our empirical cell formula.

Pentose phosphate oxidation, respiration, and ethanolic fermentation can be classified as dissimilatory processes, which supply the cofactors and energy needed for assimilation. Three major enzyme systems participate, including the pentose phosphate (PP), Embden-Meyerhof-Parnas (EMP), and tricarboxylic acid (TCA) pathways.[8,12] Anaerobic cycling of the PP pathway oxidizes xylose to CO_2 as ATP is consumed and reduced cofactors are produced:

$$6\ C_5H_{10}O_5 + 6f\ NAD^+ + (60 - 6f)\ NADP^+ + 6\ [ATP + H_2O]$$
$$+ 30\ H_2O \to 30\ CO_2 + (60 - 6f)\ NADPH,H^+$$
$$+ 6f\ NADH,H + 6\ [ADP + P_i]. \quad (8)$$

As a primary function, the PP pathway supplies $NADPH,H^+$ to assimilation. However, it also produces the intermediates, glyceraldehyde 3-phosphate and fructose 6-phosphate, which can enter the EMP pathway to form pyruvate. The pyruvate formed either enters the TCA cycle or is fermented to ethanol. Under aerobic conditions, respiration of xylose occurs via the path, PP → EMP → TCA, which adds up to ATP production with no net change in reducing equivalents:

$$6\ C_5H_{10}O_5 + 6f\ NADPH,H^+ + 6f\ NAD^+ + 30\ O_2$$
$$+ 180\ [ADP + P_i] \rightarrow 30\ CO_2 + 6f\ NADH,H^+$$
$$+ 6f\ NADP^+ + 180\ [ATP + H_2O] + 30\ H_2O. \quad (9)$$

During oxygen limitation, fermentation via the sequence, PP → EMP → ethanol, is expected to predominate, even though it yields less energy than respiration:

$$6\ C_5H_{10}O_5 + 6f\ NADPH,H^+ + 6f\ NAD^+$$
$$+ 10\ [ADP + P_i] \rightarrow 10\ C_2H_5OH + 10\ CO_2$$
$$+ 6f\ NADH,H^+ + 6f\ NADP^+ + 10\ [ATP + H_2O]. \quad (10)$$

Optimistically, the fraction of xylose sent through each of the four processes is regulated such that production and consumption of ATP, $NADH_2^+$, and $NADPH_2^+$ are balanced. If this does not occur, metabolism may stall as intermediates accumulate. For example, xylitol accumulation is believed to occur because of imperfect recycling of $NADH_2^+$ and NAD^+ between xylose reductase and xylitol dehydrogenase activities.[1] If xylose reductase activity uses $NADPH,H^+$ instead of $NADH,H^+$ (i.e., $1 \geq f > 0$), an NAD^+ shortage is indicated by the two-step reaction (equations 1 and 2) in which xylose is isomerized to xylulose. Cofunctioning of other pathways can lessen the likelihood of this imbalance. For instance, assimilation consumes $NADPH,H^+$ and produces $NADH,H^+$, thereby increasing the relative availability of $NADH,H^+$ for use by the reductase.

If we consider batch culture kinetics, xylose uptake and ethanol production are stoichiometrically related as follows:

$$dX/dt = (1/Y_{E/X})dE/dt, \quad (11)$$

where X and E are xylose and ethanol concentrations, respectively, and $Y_{E/X}$ is the yield of ethanol per xylose consumed. Previously, we verified that growth is an obligately aerobic process for strain Y-7124[7] and, although growth is expected to stimulate ethanol production, the necessary presence of oxygen brings on respiration as well. Considering that respiration provides more ATP than fermentation, we might expect $Y_{E/X}$ to drop with increasing μ.

Batch culture ethanol concentration increases at a rate proportional to the viable biomass concentration (b) by the specific productivity (p_E):

$$dE/dt = p_E b. \quad (12)$$

Because fermentation is coupled to assimilation through ATP production, we propose that p_E is at least partly growth-associated. Although originally applied to lactate production, Luedeking and Piret's model[13] expresses this behavior in the following form:

$$p_E = Y_{E/b}\mu + m_E. \quad (13)$$

In the context of ethanol production, $Y_{E/b}$ represents the ethanol yielded per biomass formed, μ is the specific growth rate based on viable cells, and m_E is the specific ethanol productivity that provides maintenance energy to resting cells. The rate p_E and the parameters $Y_{E/b}$ and m_E are expected to vary with ethanol and xylose concentration.[14,15]

Experiments were designed to test the application of these concepts to modeling of Y-7124 during oxygen-limited growth and to evaluate the parameters of useful kinetic expressions. Continuous culture yields indicated how well our stoichiometry model (equations 4 and 8–10) accounted for production and consumption of carbon and cofactors. Overall stoichiometry and p_E were studied as functions of μ in cultures operated at various dilution rates, and dependences of p_E and $Y_{E/X}$ on E and X were obtained from initial batch performances of concentrated cell populations.

MATERIALS AND METHODS

Organism and Media

Lyophilized *Pichia stipitis* NRRL Y-7124 (CBS 5773) was acquired from the ARS Culture Collection (Northern Regional Research Center, Peoria, Illinois). Stock cultures maintained on agar slants were used to prepare fermentor inocula adapted to xylose broth.[6] The complex media for slants (YM) and liquid cultures (CCY) have been described in detail.[16] CCY medium contained yeast extract, urea, potassium phosphate buffer, mineral salts, and xylose, and it was supplemented in fermentors with 1 g/L Hodag FD-62 antifoam.

Continuous Culture Evaluation of Ethanol Productivity and Stoichiometry as Functions of Growth

Fermentation rates and stoichiometry were studied in oxygen-limited continuous cultures operated at 25 °C, pH 4.5, and various dilution rates (D). Aeration and stirring rates were set to provide specific oxygen transfer coefficients ($K_l a$), which were evaluated prior to inoculation from semilog plots of oxygen saturation time courses. The configuration of B. Braun Biostat 2ER (2-L) fermentors and our method for evaluating $K_l a$ have been described.[7] The differential ethanol balance for this system was as follows:

$$dE/dt = -DE + p_E b. \tag{14}$$

Given that the subscript "s" designates concentrations measured at steady state (i.e., $dE/dt = 0$), the specific ethanol productivity (p_E) was calculated from the equation

$$p_E = DE_s/b_s. \tag{15}$$

In general, specific growth rate (μ) is related to D according to

$$\mu = D[1 + (b_{d,s}/b_s)], \tag{16}$$

which was derived previously by applying the steady-state criterion to differential balances on viable (b) and dead (b_d) biomass concentrations.[7] Note that $\mu = D$ only if $b_{d,s} = 0$ or $b_{d,s} \ll b_s$. We did not find this to be true for our system.

The oxygen to xylose uptake ratio ($Y_{OX/X}$) and the yields of ethanol ($Y_{E/X}$), biomass ($Y_{b/X}$), and carbon dioxide ($Y_{CO_2/X}$) were calculated as follows:

$$Y_{OX/X} = K_1 a(C_{ox}^* - C_{ox,s})/[D(X_f - X_s)], \quad (17)$$

$$Y_{E/X} = E_s/(X_f - X_s), \quad (18)$$

$$Y_{b/X} = b_{T,s}/(X_f - X_s), \quad (19)$$

$$Y_{CO_2/X} = Q_g F_{CO_2} d_{CO_2}/[V_1 D(X_f - X_s)]. \quad (20)$$

The parameters $C_{ox,s}$ and C_{ox}^* denote steady-state dissolved oxygen concentration and oxygen solubility, respectively; X_f represents the feed xylose concentration of 40 g/L; Q_g is the volumetric rate of gas flow from the fermentor; F_{CO_2} is the mole fraction of CO_2 in the exit gas; d_{CO_2} is the density of CO_2; and V_1 is the culture liquid volume. The amount of CO_2 arising from the PP cycle was the total CO_2 minus fermentative, assimilative, and respirative CO_2. The yield of PP-cycle-produced CO_2 per xylose consumed ($Y_{CO_2(P)/X}$) was thus calculated from the available yield data, the stoichiometry of equation 4, and the formula weights (g/mole) of CO_2 (44), ethanol (46), biomass (98), and O_2 (32):

$$Y_{CO_2(P)/X} = Y_{CO_2/X} - Y_{E/X}(44/46)$$
$$- Y_{b/X}(695.5/1000)(44/98) - Y_{OX/X}(44/32). \quad (21)$$

Fractions of xylose used in assimilation (f_a), PP cycling (f_p), respiration (f_r), and ethanolic fermentation (f_e) were calculated as ratios of the observed yield to that theoretically possible if all xylose was metabolized by a given process:

$$f_a = (Y_{b/X})/(0.738 \text{ g biomass per g xylose}), \quad (22)$$

$$f_p = (Y_{CO_2(P)/X})/(1.47 \text{ g } CO_2 \text{ per g xylose}), \quad (23)$$

$$f_r = (Y_{OX/X})/(1.066 \text{ g oxygen per g xylose}), \quad (24)$$

$$f_e = (Y_{E/X})/(0.51 \text{ g ethanol per g xylose}). \quad (25)$$

Theoretical yields used in the denominators of equations 22–25 were based on the stoichiometry model (equations 4 and 8–10).

If our stoichiometry model is correct in accounting for carbon metabolism by assimilation (equation 4), pentose phosphate oxidation (equation 8), respiration (equation 9), and ethanol fermentation (equation 10), then our yield data should allow us to calculate that

$$f_a + f_p + f_r + f_e = 1. \quad (26)$$

Given this is true, the overall equation for a particular fermentation can be constructed by summing the four model equations, which have been scaled to reflect the fraction of xylose used by each path:

overall equation = $[f_a \times (\text{equation 4})/883] + [f_p \times (\text{equation 8})/6]$
$$+ [f_r \times (\text{equation 9})/6] + [f_e \times (\text{equation 10})/6]. \quad (27)$$

Division of stoichiometric coefficients by 883 or 6 normalized the equations for reaction of one mole of xylose.

Maintenance Ethanol Production by Concentrated Cells

Cell Cultivation and Harvest

Cells were grown in a New Brunswick Fermacel CF50 batch fermentor and centrifuged after 48 h, as described earlier.[6] Portions of the harvested cell paste were transferred to fermentors or flasks for resuspension in fresh CCY such that cell concentrations of 5–7 g/L (dry weight) were achieved. These concentrated batch cultures were used to study effects of X and E on maintenance ethanol yield and productivity (m_E), that is, p_E when $\mu = 0$.

Effect of X on Productivity and Yield

New Brunswick Microferm fermentors with 2-L working volumes were equipped with antifoam control, operated at pH 4.5 and 25 °C, and aerated at $K_l a = 0.175$ min^{-1}. Initial specific ethanol productivities were measured as a function of the xylose concentration provided ($X_o = 10$–150 g/L). Volumetric productivity (P_E) was obtained by linear regression of the early ethanol time course (0 to ~10 h) and p_E was calculated as P_E/b_{av}. Because of slow oxygen transfer relative to the large population of viable cells present, $\mu \simeq 0$ over this interval and an average value of the biomass concentration (b_{av}) was applied in the calculations. $Y_{E/X}$ was calculated as $E/(X_o - X)$.

Effect of E on Productivity and Yield

Cell suspensions in stirred flask cultures were sparged with CO_2 to remove oxygen and to allow equilibration of CO_2 and liquid phases at barometric pressure (P_o). Once 20 g/L xylose and 3–80 g/L ethanol were added to start the fermentation, CO_2 sparging was stopped and flasks were sealed except for connection to an open u-tube Hg manometer. We assumed that CO_2 evolution was solely responsible for the pressure increase and calculated the moles of CO_2 formed at any time as the difference between the current (n) and the initial moles (n_o) in gas and liquid phases:

$$n - n_o = [V_{g,o}(P - P_o)/RT] \\ + [(V_g - V_{g,o})(P - P_w)/RT] + V_l k_H (P - P_o), \quad (28)$$

where V_g and P represent the volume and absolute pressure of gas-phase CO_2 currently in the system, the subscript "o" designates initial conditions, P_w is water vapor pressure (0.031 atm), T is incubator temperature (298 K), R is the ideal gas constant (0.0821 L-atm/K-mole), V_l is the volume of liquid culture, and k_H is the solubility of CO_2 in CCY medium (0.0274 moles/L-atm) [Sigma Chemical Company Diagnostic Kit no. 130-A]. Note that V_g depended on manometer mercury position. Assuming equimolar production of ethanol and CO_2, we calculated volumetric ethanol productivity as $P_E =$

(the slope of the initial two-hour $n - n_o$ time course)/V_1 and the specific productivity as $p_E = P_E/b_{av}$. The yield $[Y_{E/X} = (E - E_o)/(X_o - X)]$ was calculated from chromatography analyses of initial and final (6-h) broth samples.

Analyses

Total and Viable Biomass Concentrations

These were evaluated by light absorbance[6] and staining techniques.[7]

Dissolved Oxygen Concentration

This was monitored throughout the course of each fermentation as the product of Ingold electrode response (in terms of the fraction of oxygen saturation) and oxygen solubility (C_{ox}^*). Given air (21% O_2) at 25 °C and 760 mmHg, C_{ox}^* (mg/L) was determined as the following function of X (g/L):[7] $C_{ox}^* = (0.21)(1.08)(34.6 - 0.0644X + 0.000156X^2)$.

Ethanol, Xylitol, and Xylose Measurements

Filtered samples were analyzed by gas and liquid chromatography methods.[6] However, samples with E < 1 g/L were analyzed on a Hewlett Packard 5890 GC with a 30-m megabore column of 1-micron DB 1701 stationary phase (J&W Scientific). The carrier gas to the column was helium flowing at 6.79 mL/min, split to 1/13. Oven, injector, and flame ionization detector temperatures were 150, 175, and 250 °C, respectively.

Exit Gas CO_2

The mole fraction of CO_2 in the gas exiting our continuous fermentors was measured by gas chromatography.[17]

RESULTS AND DISCUSSION

Stoichiometry as a Function of Growth

Pathway Usage

Oxygen-limited continuous cultures were run at various dilution rates to test the effect of growth rate on pathway usage during xylose metabolism. TABLE 1 summarizes these results in terms of the fractions of xylose used by each pathway. The sum of pathway usage fractions was always near 1, indicating that the proposed stoichiometry model (equations 7–10) adequately accounted for xylose utilization. The metabolic state of the yeast varied with specific growth rate. As μ increased, metabolism shifted from fermentation and the pentose phosphate cycle to assimilation and respiration, that is, from anaerobic to aerobic processes. In conjunction with this observation, earlier data have shown that specific oxygen uptake (or respiration) rate increases linearly with growth rate.[7]

It is notable that xylitol did not accumulate in any of our continuous cultures although small amounts were observed in batch cultures.[6] This finding was consistent with studies of labeled xylose uptake[18] suggesting that xylose reductase is predominately NADH-dependent and that $f \simeq 0$ in our model. A question arises, however, because our f_p values (TABLE 1) were 3–10 times higher than anticipated from our model, which assumed carbon flow through the PP cycle to be regulated solely by the $NADPH,H^+$ demanded for assimilation.

Overall Stoichiometry

As we pursued this problem, an analysis of the overall stoichiometry for each growth condition (TABLE 2) allowed us to make another interesting observation. This analysis suggested that a net production of 1.42–1.63 moles of ATP and 0.63–1.03 moles of reducing equivalents ($NADH,H^+$ + $NADPH,H^+$) occurred regardless of the value of f. Inaccurate estimations of cofactors required by assimilation (equation 4), which applied to only 10–20% of xylose metabolism, could not account for the large excesses observed. For example, consumptions of ATP and $NADPH,H^+$ would have to be raised by factors of 4 and 2–10, respectively, whereas NAD^+ would have to be halved (at $f = 0$). Such large adjustments in $NADPH,H^+$ and NAD^+ would bring our assimilation equation into serious violation of mass and charge conservation principles. The ability of mitochondrial electron transport to relieve an H^+ imbalance in the cytoplasm has been mentioned as a possible explanation for the obligate aerobic growth of xylose-fermenting yeasts.[8] Although such a process might consume the reducing equivalents, it would require twice the oxygen demand observed and further augment the ATP surplus.

TABLE 1. Pathway Usage as a Function of Specific Growth Rate ($K_1a = 0.049$ min^{-1})

Specific Growth Rate (h^{-1})	Fraction of Xylose Used by Each Pathway[a]				
	f_a	f_p[b]	f_r	f_e	sum
0.055	0.093	0.101 (0.009)	0.020	0.83	1.04
0.089	0.142	0.101 (0.013)	0.027	0.77	1.04
0.118	0.144	0.059 (0.013)	0.029	0.75	0.98
0.161	0.160	0.059 (0.015)	0.031	0.70	0.95
0.195	0.161	0.053 (0.015)	0.034	0.71	0.96
0.250	0.220	0.068 (0.021)	0.044	0.66	0.99
				average sum =	0.99

[a] The fraction of xylose consumed for xylitol production, $f_{XOH} = Y_{XOH/X}/1.01$, was 0 for all D. The parameters $f_a, f_p, f_r,$ and f_e represent the fractions of xylose consumed by assimilation, pentose phosphate oxidation, respiration, and ethanolic fermentation, respectively.

[b] (...) = value of f_p calculated from f_a assuming that $f = 0$ and that the pentose phosphate cycle supplies all $NADPH,H^+$ for assimilation.

TABLE 2. Overall Stoichiometry of Xylose Metabolism as a Function of Specific Growth Rate

	Moles of Reactants or Products					
	Specific Growth Rate (h^{-1})					
	0.055	0.089	0.118	0.161	0.195	0.250
Reactants						
xylose	1.00	1.00	1.00	1.00	1.00	1.00
urea	0.03	0.04	0.05	0.05	0.05	0.07
O_2	0.10	0.13	0.15	0.16	0.18	0.22
NAD^+	$0.13 + f$	$0.20 + f$	$0.22 + f$	$0.25 + f$	$0.25 + f$	$0.33 + f$
$NADP^+$	$0.88 - f$	$0.83 - f$	$0.45 - f$	$0.45 - f$	$0.38 - f$	$0.46 - f$
$ATP + P_i$	1.48	1.42	1.57	1.53	1.63	1.56
H_2O	0.28	0.18	0	0	0	0
Products						
biomass	0.10	0.15	0.17	0.19	0.19	0.25
ethanol	1.32	1.23	1.27	1.22	1.24	1.11
CO_2	1.97	1.96	1.84	1.83	1.82	1.85
$NADH,H^+$	$0.13 + f$	$0.20 + f$	$0.22 + f$	$0.25 + f$	$0.25 + f$	$0.33 + f$
$NADPH,H^+$	$0.88 - f$	$0.83 - f$	$0.45 - f$	$0.45 - f$	$0.38 - f$	$0.46 - f$
$ATP + H_2O$	1.48	1.42	1.57	1.53	1.63	1.56
H_2O	0	0	0.03	0.06	0.11	0.16

However, this apparent cofactor imbalance was consistent with recent evidence that xylose is transported via a proton symport system. Kilian and Van Uden have shown that xylose transport by *P. stipitis* is accompanied by proton uptake (pH rise) and is likely to be energy-dependent.[19] In this context, averaged data from TABLE 2 suggest that *P. stipitis* generates 1.53 moles of ATP to maintain the chemiosmotic gradient required for cotransport of 0.8 moles of H^+ and 1 mole of xylose. The fact that f_p values were higher than needed for assimilation alone suggests possible involvement of the pentose phosphate cycle in proton uptake (or acceptance) during transport.

Ethanol Yield and Its Impact on Xylose Uptake Kinetics

Concentration Effects

Batch cultures provided with high concentrations of nongrowing cells were used to examine the effects of xylose and ethanol concentration on yield. As calculated from initial time courses, the ethanol yield on xylose ($Y_{E/X}$) showed no trend on 10–150 g/L xylose or 3–80 g/L ethanol, but averaged 0.41 (0.054 standard deviation). A similar value was predicted by extrapolating our continuous culture results to $\mu = 0$ as shown later.

Dependence of Ethanol Yield on Growth

In agreement with observed shifts in the overall stoichiometry, $Y_{E/X}$ declined gradually with increasing growth rate (FIGURE 1). The empirical relationship that fit this behavior was the following:

$$Y_{E/X} = 0.421 - 0.343\mu. \qquad (29)$$

Hence, in view of equations 11–13, xylose uptake rate was weakly dependent on μ via $Y_{E/X}$, if not through p_E as well. This expression predicts that $Y_{E/X}$ will reach zero at $\mu = 1.23$ h^{-1}; however, the highest specific growth rates observed in our experiments were ca. 0.55 h^{-1},[7] which corresponds to $Y_{E/X} = 0.23$ g/g.

Ethanol Production Kinetics

Dependence of Productivity on Growth

Specific ethanol productivity and growth rate, measured in continuous cultures at steady state, were plotted against one another as shown in FIGURE 2. Regardless of $K_l a$ setting, the data were compatible with Luedeking and Piret's model at $Y_{E/b} = 1.73$ g/g and $m_E = 0.254$ g/g/h (equation 13). The fact that this expression was independent of $K_l a$ is related to our earlier finding that the specific oxygen uptake rate (q_{ox}) is controlled by μ. At any particular D, q_{ox} remains unchanged with shifts in $K_l a$ as b_s shifts proportionately.[7]

The observation that ethanol production is partly growth-associated impacts on our interpretation of the literature in various ways. It provides a kinetic basis for early references to xylose fermentations as "oxygen stimulated". Growth is obligately aerobic and usually oxygen-limited,[7] so p_E, being a function of μ, is subject to oxygen

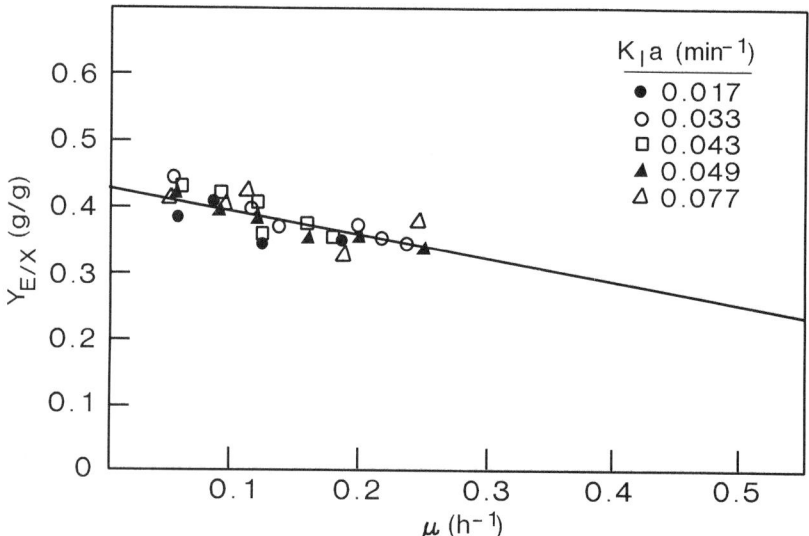

FIGURE 1. Linear dependence of ethanol yield ($Y_{E/X}$) on specific growth rate (μ). Regression provided a slope (-0.35 h g/g) and intercept (0.42 g/g) with a correlation coefficient of 0.80, given variable oxygen transfer ($K_l a$).

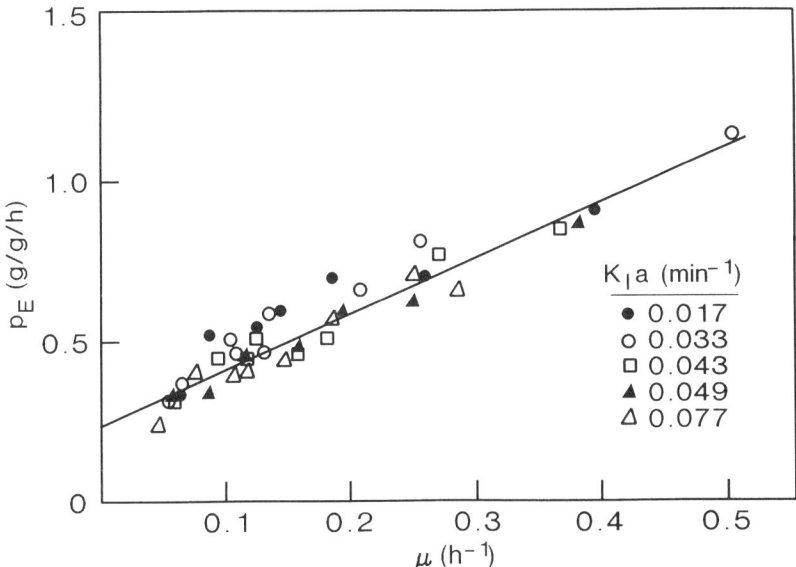

FIGURE 2. Linear dependence of specific ethanol productivity (p_E) on growth (μ). The yield provided a slope ($Y_{E/b}$) of 1.73 g/g and an intercept (m_E) of 0.25 g/g/h with a correlation of 0.95 despite variable oxygen transfer ($K_l a$).

limitation kinetics. Our combined expressions for p_E and μ can also account for the curious observation that the slope of E versus t (equaling $p_E b$) is constant in early batch cultures.[4] Rising biomass concentration tends to increase dE/dt, but this effect is offset by declining oxygen concentration, which lowers μ and p_E. As indicated by the large value of m_E, ethanol production continues even in the absence of oxygen and growth. In batch cultures, the average value of p_E is skewed toward m_E because depletion of oxygen concentration lowers μ early on. Therefore, average specific ethanol productivities measured in batch cultures tend to be ≤0.25 g/g/h, depending on concentrations of xylose and ethanol.

Dependence of Maintenance Productivity on X

As a standard practice, *P. stipitis* was grown batchwise 48 h on 150 g/L xylose and then resuspended at high concentration to measure maintenance ethanol production. Time courses in FIGURE 3 indicate that our practice of growing cells at the extreme xylose concentration was key to eliminating adaptation effects from subsequent measurements of m_E. When resuspended in 150 g/L xylose, cells grown on 20 g/L xylose produced ethanol at one-tenth the rate of those grown on 150 g/L. This finding was consistent with adaptation effects reported by other workers.[20]

Raising xylose to 35–80 g/L stimulated m_E, but higher concentrations inhibited the fermentation (FIGURE 4). Curves through the data are discussed in detail by Slininger[25] and represent best fits of models reviewed by Edwards.[15] Performing nearly as well as four-parameter equations, the three-parameter empirical models of Aiba *et al.* and Edwards adequately simulated our data (TABLE 3). However, Edwards' model converged on the most realistic parameter values; to cast it in terms of xylose-dependent maintenance productivity, we gave it the following form:

$$m_E = m_{E,max}[\exp(-X/K_i') - \exp(-X/K_x')]. \tag{30}$$

Dependence of Maintenance Productivity on E

The effects of xylose on m_E were compounded by ethanol inhibition. End-product inhibition has received much notoriety and it has been described by numerous mathematical models.[14] As a noncompetitive inhibitor, ethanol affects the maximum

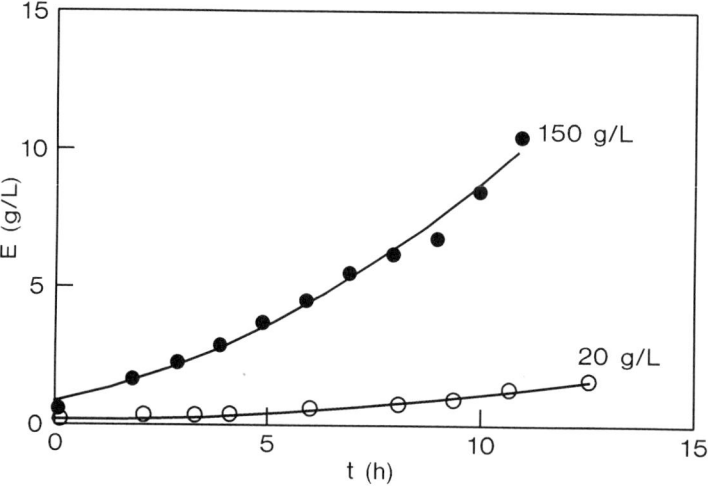

FIGURE 3. Ethanol production on 150 g/L xylose by cells pregrown on 20 versus 150 g/L xylose.

specific rate and not saturation kinetics. Consequently, Luong's empirical model has been generally useful for modeling inhibition as follows:[14]

$$m_{E,i}/m_{E,o} = 1 - (E/E_m')^B. \tag{31}$$

In the context of this study, $m_{E,i}$ and $m_{E,o}$ are rates in the presence and absence of ethanol as inhibitor, E_m' is the ethanol concentration that prohibits further production, and B is an empirical constant. In a previous study, we showed that Luong's model was applicable to the growth of *P. stipitis*. Furthermore, linearization of the data (FIGURE 5) described m_E as a function of E when E_m' and B were 189 g/L and 0.935, respectively. These parameters were analogous, but not equal to those determined for growth ($E_m = 64.3$ g/L, $A = 1.324$).[7] The higher ethanol sensitivity of μ (compared with p_E) was also characteristic of *Pachysolen tannophilus*.[8]

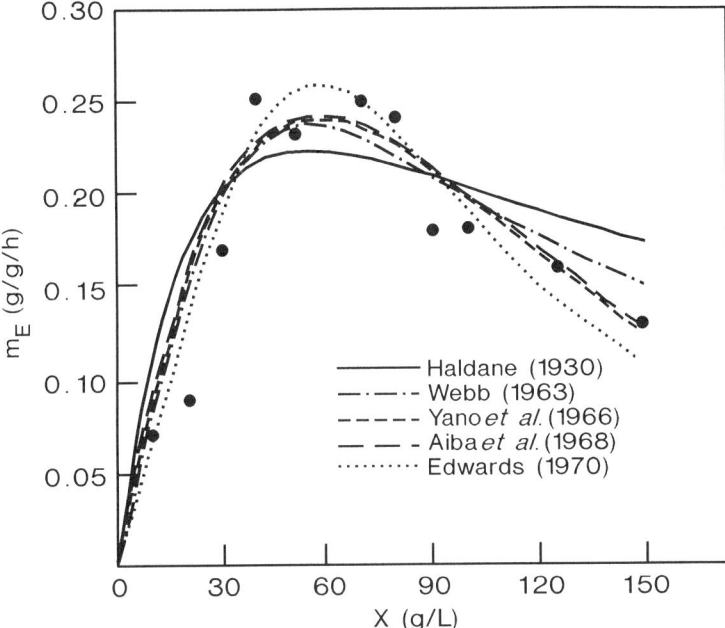

FIGURE 4. Dependence of maintenance productivity (m_E) on xylose concentration (X). Curves through the data represent best fits of various substrate inhibition models. Parameter values are given in TABLE 3.

CONCLUSIONS

In summary, observed carbon balances were consistent with modeling xylose metabolism as a combination of four process equations: assimilation (equation 4), pentose phosphate oxidation (equation 8), respiration (equation 9), and ethanolic fermentation (equation 10). Cofactor balances indicated excesses of 0.8 moles H^+ (as NADPH,H^+ and NADH,H^+) and 1.53 moles ATP per mole xylose consumed. This condition was consistent with xylose transport by energy-dependent proton symport,

TABLE 3. Optimized Parameters and Prediction Errors Associated with Substrate Inhibition Models

Equation	$m_{E,max}$ (g/g/h)	K'_x (g/L)	K'_i (g/L)	K (g/L)	Average Squared Error ($\pm$g/g/h)
Haldane[21]	0.867	53.4	56.10	—	0.038
Webb[22]	12.300	1350.0	2.09	2.83×10^8	0.027
Yano et al.[23]	3.220	434.0	185.50	4.92	0.021
Aiba et al.[24]	5.880	514.0	63.70	—	0.028
Edwards[15]	1.430	45.9	72.70	—	0.028

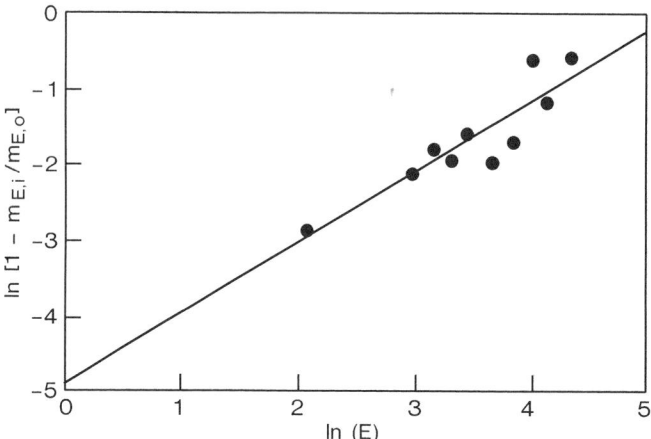

FIGURE 5. Linearization of Luong's model for the dependence of maintenance ethanol productivity (m_E) on ethanol concentration (E). Regression indicated 0.90 correlation and provided model parameters (equation 31) via the slope ($B = 0.935$) and intercept $[-B \ln(E'_m) = -4.903]$.

and high values of f_p (relative to f_a) suggested involvement of the pentose phosphate cycle in proton uptake. Although these cofactor excesses remained relatively constant, the overall stoichiometry gradually shifted from fermentation to respiration as specific growth rate increased. Because xylose uptake was proportional to ethanol production (equation 11), these shifts could be accounted for in our kinetic model by a growth-dependent yield coefficient, $Y_{E/X}$ (equation 29). Specific ethanol productivity was also growth-associated as described by Luedeking and Piret's model (equation 13) at $Y_{E/b} = 1.73$ g/g. Maintenance productivity (m_E) was adequately represented by the combination of Luong's two-parameter model of ethanol inhibition (equation 31) and any of the three-parameter functions of xylose concentration (equation 30).

ACKNOWLEDGMENT

We are grateful to Roy Butterfield for computer optimizations of parameters in substrate inhibition models.

REFERENCES

1. BRUINENBERG, P. M., P. H. M. DE BOT, J. P. VAN DIJKEN & W. A. SCHEFFERS. 1984. Appl. Microbiol. Biotechnol. **19**: 256–269.
2. TOIVOLA, A., D. YARROW, E. VAN DEN BOSCH, J. P. VAN DIJKEN & W. A. SCHEFFERS. 1984. Appl. Environ. Microbiol. **47**: 1221–1223.
3. DU PREEZ, J. C. & B. A. PRIOR. 1985. Biotechnol. Lett. **7**: 241–246.
4. SLININGER, P. J., R. J. BOTHAST, M. R. OKOS & M. R. LADISCH. 1985. Biotechnol. Lett. **7**: 431–436.
5. PAREKH, S. R., R. S. PAREKH & M. WAYMAN. 1987. Process Biochem. **22**: 85–91.
6. SLININGER, P. J., R. J. BOTHAST, M. R. LADISCH & M. R. OKOS. 1989. Biotechnol. Bioeng. In press.

7. SLININGER, P. J., L. E. BRANSTRATOR, B. S. DIEN, R. J. BOTHAST, M. R. OKOS & M. R. LADISCH. 1990 (January). Biotechnol. Bioeng. Submitted.
8. SLININGER, P. J., P. L. BOLEN & C. P. KURTZMAN. 1987. Enzyme Microb. Technol. **9:** 5–15.
9. VERDUYN, C., R. VAN KLEEF, J. FRANK, H. SCHREUDER, J. P. VAN DIJKEN & W. A. SCHEFFERS. 1985. Biochem. J. **226:** 669–677.
10. STOUTHAMER, A. H. 1973. Antonie van Leeuwenhoek; J. Microbiol. Serol. **39:** 545–565.
11. BRUINENBERG, P. M., J. P. VAN DIJKEN & W. A. SCHEFFERS. 1983. J. Gen. Microbiol. **129:** 953–964.
12. LEHNINGER, A. L. 1975. Biochemistry. Worth. New York.
13. LUEDEKING, R. & E. L. PIRET. 1959. J. Biochem. Microbiol. Technol. Eng. **1:** 393–412.
14. LUONG, J. H. T. 1985. Biotechnol. Bioeng. **27:** 280–285.
15. EDWARDS, V. H. 1970. Biotechnol. Bioeng. **12:** 679–712.
16. SLININGER, P. J., R. J. BOTHAST, J. E. VAN CAUWENBERGE & C. P. KURTZMAN. 1982. Biotechnol. Bioeng. **24:** 371–384.
17. RAMSTACK, J. M., E. B. LANCASTER & R. J. BOTHAST. 1979. Process Biochem. **14:** 2–4.
18. LIGTHELM, M. E., B. A. PRIOR, J. C. DU PREEZ & V. BRANDT. 1988. Appl. Microbiol. Biotechnol. **28:** 293–296.
19. KILIAN, S. G. & N. VAN UDEN. 1988. Appl. Microbiol. Biotechnol. **27:** 545–548.
20. PAREKH, S. R., S. YU & M. WAYMAN. 1986. Appl. Microbiol. Biotechnol. **25:** 300–304.
21. HALDANE, J. B. S. 1930. Enzymes. Longmans, Green. New York. (Also: 1965. MIT Press. Cambridge, Massachusetts.)
22. WEBB, J. L. 1963. Enzyme and Metabolic Inhibitors. Academic Press. New York.
23. YANO, T., T. NAKAHARA, S. KAMIYAMA & K. YAMADA. 1966. Agric. Biol. Chem. **30:** 42–48.
24. AIBA, S., M. SHODA & M. NAGATANI. 1968. Biotechnol. Bioeng. **10:** 845–864.
25. SLININGER, P. J. 1988 (December). Xylose fermentation: analysis, modeling, and design. Ph.D. thesis, Purdue University, West Lafayette, Indiana.

APPENDIX

Nomenclature

Symbols

B = exponent governing ethanol inhibition of fermentation (dimensionless)
b = viable biomass concentration (g/L)
b_{av} = average viable biomass concentration (g/L)
b_d = dead biomass concentration (g/L)
$b_{d,s}$ = steady-state dead biomass concentration (g/L)
b_s = steady-state viable biomass concentration (g/L)
b_T = total biomass concentration (g/L)
$b_{T,s}$ = steady-state total biomass concentration (g/L)
C_{ox} = dissolved oxygen concentration (mg/L)
$C_{ox,s}$ = steady-state dissolved oxygen concentration (mg/L)
C_{ox}^* = oxygen solubility (mg/L)
D = dilution rate (h^{-1})
d_{CO_2} = density of CO_2 (g/L)
E = ethanol concentration (g/L)
E'_m = maximum ethanol concentration allowing fermentation (g/L)
E_o = initial ethanol concentration in a batch culture (g/L)
E_s = steady-state ethanol concentration (g/L)
F_{CO_2} = mole fraction of CO_2 in the fermentor exit gas (dimensionless)

f = fraction of xylose reductase activity supported by NADPH, as opposed to NADH (dimensionless)
f_a = fraction of xylose used in assimilation (dimensionless)
f_e = fraction of xylose used in ethanol fermentation (dimensionless)
f_p = fraction of xylose used in pentose phosphate cycle oxidation (dimensionless)
f_r = fraction of xylose used in respiration (dimensionless)
$K_l a$ = lumped oxygen mass transfer coefficient (h^{-1})
K_i' = parameter governing substrate inhibition of fermentation (g/L)
K_x' = saturation constant governing xylose-limited fermentation (g/L)
k_H = Henry's law constant (moles/L/atm)
m_E = specific ethanol productivity for maintenance (g/g/h)
$m_{E,i}$ = maintenance productivity in the presence of inhibitor (g/g/h)
$m_{E,max}$ = maximum specific maintenance productivity (g/g/h)
$m_{E,o}$ = maintenance productivity in the absence of inhibitor (g/g/h)
n = total moles of CO_2 in gas and liquid phases
n_o = initial total moles of CO_2 in gas and liquid phases
P = absolute pressure (atm)
P_E = volumetric ethanol productivity (g/L/h)
P_o = barometric pressure (atm)
P_w = water vapor pressure (atm)
p_E = specific ethanol productivity (g/g/h)
Q_g = volumetric gas flow rate from the fermentor (L/h)
q_{ox} = specific oxygen uptake rate (mg/g/h)
R = ideal gas constant (0.0821 L-atm/K-mole)
T = temperature (K)
t = time (h)
V_g = culture headspace gas volume (L)
$V_{g,o}$ = initial culture headspace gas volume (L)
V_l = culture liquid volume (L)
X = xylose concentration (g/L)
X_f = concentration of xylose in continuous culture feed (g/L)
X_o = concentration of xylose initially present in a batch culture (g/L)
X_s = steady-state xylose concentration (g/L)
$Y_{b/X}$ = yield of biomass per xylose consumed (g/g)
$Y_{CO_2/X}$ = yield of CO_2 per xylose consumed (g/g)
$Y_{CO_2(P)/X}$ = yield of pentose phosphate cycle–produced CO_2 per xylose consumed (g/g)
$Y_{E/b}$ = yield of ethanol per biomass produced (g/g)
$Y_{E/X}$ = yield of ethanol per xylose consumed (g/g)
$Y_{OX/X}$ = oxygen to xylose uptake ratio (g/g)
μ = specific growth rate (h^{-1})

Abbreviations

EMP = Embden-Meyerhof-Parnas
PP = pentose phosphate
TCA = tricarboxylic acid

Improvement of Plasmid Stability by Immobilization of Recombinant Microorganisms

JEAN-NOËL BARBOTIN, SAMI SAYADI, MONCEF NASRI,
FADWA BERRY, AND DANIEL THOMAS

Laboratoire de Technologie Enzymatique
Université de Technologie de Compiègne
60206 Compiègne, France

INTRODUCTION

The development of recombinant DNA technology during the past decade has allowed the production of useful peptides and proteins at the industrial level. However, the importance of recombinant DNA instability must be recognized for the scaleup of fermentation processes.[1-3] A variety of experimental approaches to overcoming plasmid instability have been proposed. One of the most widely used methods is the addition of an antibiotic to the culture medium; however, in a large-scale preparation, this would be expensive. A recently developed method involves the use of a plasmid that gives a host the ability to produce bacteriocins that are lethal to plasmid-free bacteria of the same host type.[4] Other techniques involving the use of temperature-sensitive systems or two-stage continuous culture reactors,[2,5,6] or the insertion of the Sop^+ gene of the F plasmid and the par B^+ of the R_1 plasmid[7] and the use of substrate inhibition effects on the growth of plasmid-free bacteria,[8] have also been exploited.

On the other hand, the application of immobilized-cell reactors has been emphasized in order to increase cell concentration, prevent cell washout, and increase the productivity of the plasmid-coded gene product. Inloes *et al.*,[9] Mosbach *et al.*,[10] Dhulster *et al.*,[11] Georgiou *et al.*,[12] and Kanayama *et al.*[13] have studied recombinant protein production by immobilized bacteria. Some authors have shown that the immobilization of recombinant microorganisms may increase plasmid stability.[14-22]

The main objective of this work was to examine the maintenance conditions of a stable population of plasmid-carrying cells during continuous culture without selection pressure by using the immobilization of recombinant *E. coli* within carrageenan gel beads.

MATERIALS AND METHODS

Microorganisms and Plasmids

Bacterial strains used in this study were *E. coli* W3101 (*rec A13*,' *E. coli* B. The nonconjugative plasmid pTG201 (a generous gift from France) is an $Ap^R Tc^S$ *E. coli* cloning vector, derivative of pBR 322 gene Xyl E, from *Pseudomonas putida*, which codes for catecho'

(Cat-O_2-ase), under transcriptional control of the λ PR and CI857 repressor. Plasmid pBR 328 was purchased from Boehringer Mannheim.

Media and Culture Conditions

The medium used throughout the experiments was LB medium (10 g/L tryptone, 5 g/L yeast extract, 5 g/L NaCl, pH 7.3). For immobilized cells, LB medium was supplemented with 0.1 M KCl. Continuous cultures were carried out, in the absence of antibiotic selection, in 50-mL volume in 100-mL glass vessels maintained at 37 °C with an aeration rate of about 170 mL/min. In classical experiments, the dilution rates were

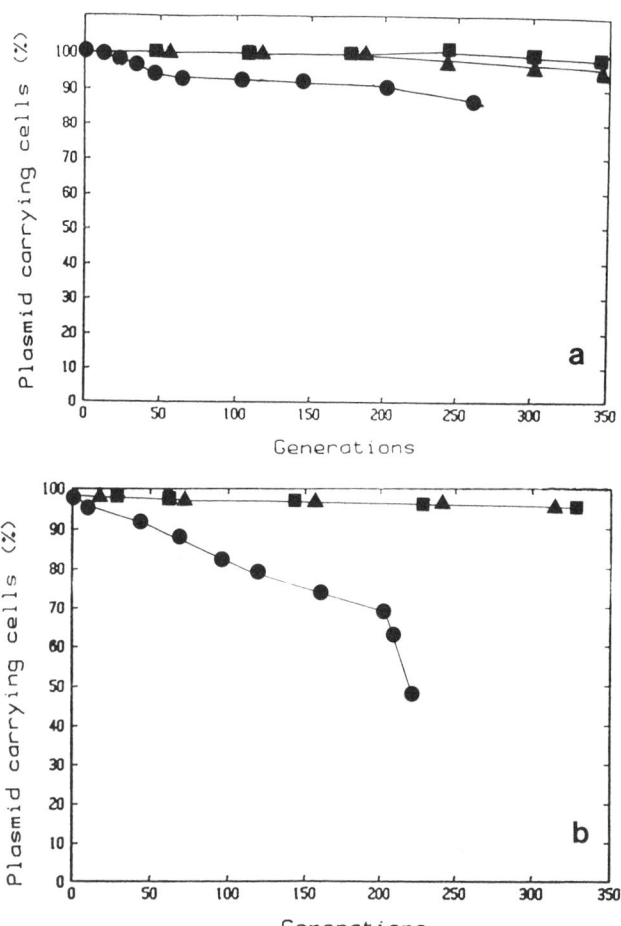

FIGURE 1. Stability of plasmid pTG201 in continuous cultures of free and immobilized *E. coli* 3101 (part a) and *E. coli* B (part b). Symbols: ● = free cells, ■ = immobilized cells, and ▲ = sed cells.

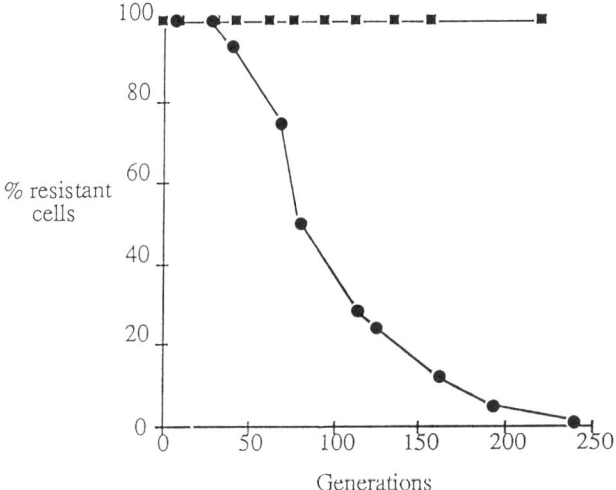

FIGURE 2. Stability of plasmid pBR 328 in continuous cultures of free and immobilized *E. coli* W3101. Symbols: ● = free cells, ■ = immobilized cells.

about 1.35 h^{-1} for free continuous cultures and 3.9 h^{-1} for immobilized-cell systems. Experiments dealing with dilution-rate effects were performed in M9 minimal medium.

Immobilization Procedure

Cells were immobilized as follows: 0.5 mL of precultured bacteria at a given concentration mixed with 9.5 mL of 2.1% (w/v) K-carrageenan solution (E 407, CECA, France) at 42 °C. Gel beads were formed by adding the mixture dropwise to 0.3 M KCl solution.

Plasmid Stability Test

The presence of Cat-O_2-ase activity was detected by spraying bacterial colonies (free cells, released cells, or immobilized cells after gel disruption in sodium citrate) with 0.5 M catechol. A yellow color indicated a Cat-O_2-ase–producing colony.[23] Quantitative measurement of Cat-O_2-ase activity was performed as previously described.[14]

RESULTS AND DISCUSSION

Influence of Immobilization on the Stability of pTG201 Plasmid in E. coli *W3101 and B*

When *E. coli* W3101/pTG201 was cultivated in free-cell continuous culture in the absence of antibiotic selection, loss of pTG201 was detected after 25–30 generations

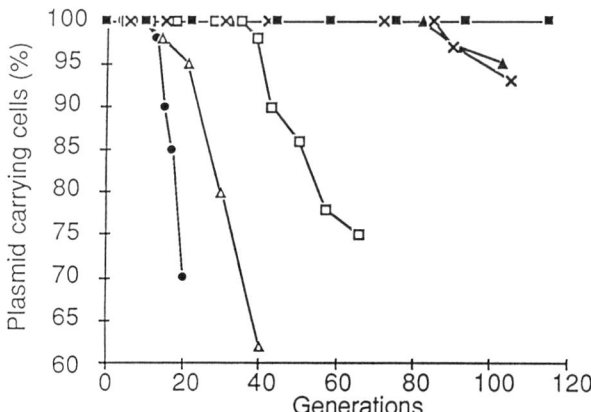

FIGURE 3. Stability of plasmid pTG201 in continuous cultures of free and immobilized *E. coli* W3101 in M9 minimal medium. Free cells: dilution rates (h^{-1}) were $D = 0.66$ (▲), $D = 0.47$ (×), $D = 0.25$ (□), $D = 0.13$ (△), and $D = 0.06$ (●). Immobilized and released cells: dilution rate was $1.3\ h^{-1}$ (■).

(FIGURE 1a). After prolonged incubation, the proportion of plasmid-containing cells (P^+ cells) gradually decreased. In contrast, when the strain was cultivated in immobilized-cell culture, pTG201 was completely stable. No loss of pTG201 was observed after more than 300 generations for either released or immobilized cells (FIGURE 1a).

Similar experiments were performed with *E. coli* B/pTG201 (FIGURE 1b). The plasmid pTG201 was found to be extremely unstable in free-cell continuous culture. However, in the immobilized-cell system, high plasmid stability was observed and no loss of pTG201 occurred even after 300 generations (FIGURE 1b).

On the other hand, we also have shown[24] that plasmid-free cells (P^- cells), when coimmobilized and grown in competition with P^+ cells, could not overrun the culture.

In these experiments, a high dilution rate ($D = 3.9\ h^{-1}$) was used for the immobilized-cell systems. This value is significantly greater than the maximum specific growth rates measured in batch cultures ($1.85\ h^{-1}$ and $2.5\ h^{-1}$ for *E. coli* W3101 and B, respectively). Under these conditions, the growth rate of cells immobilized within the outer layer of the gel beads was at least equal to the maximum growth rate of free cells. Furthermore, due to the rapid cell growth near the gel surface, cells were continuously released from the gel beads, but could not divide in the chemostat because the residence time (15 min) was much shorter than the minimum doubling time.

The number of cell generations was calculated from $D_{rate} \times t/\ln 2$ [where D_{rate} is the dilution rate (h^{-1}) and t is the time (h)] in the case of free-cell cultures and from $\mu\max(P^+) \times t/\ln 2$ for immobilized-cell cultures.[14] For immobilized cells, this value is not an accurate representation of the actual number of generations produced in the gel bead. However, taking into account the high flow rate used in this case, the production of released cells was higher than that observed with free cells (data not shown). Thus, as has been found for other immobilized-cell cultures, cells were generated at a higher rate than in a free-cell system.

Influence of Immobilization on the Stability of pBR 328 Plasmid in E. coli W3101

Immobilization of P^+ cells was found to increase the stability of other vectors such as the pBR 322–related plasmids.[25] FIGURE 2 shows that the stability of pBR 328 in *E. coli* W3101 was strongly affected during free-cell continuous culture and P^- cells appeared after a lag period of 28 generations. In contrast, P^- cells were not detected in the immobilized-cell culture even after 250 generations.

Effect of Dilution Rate on Stability, Plasmid Copy Number, and Catechol 2,3-Dioxygenase in Free and Immobilized E. coli W3101/pTG201 Cells

In order to better understand the high plasmid stability in immobilized recombinant *E. coli* cells, the effect of dilution rate (from 0.06 to 0.66 h^{-1}) on pTG201 plasmid stability, copy number (number of plasmids per chromosome), and enzyme production was studied in free *E. coli* W3101 continuous cultures in minimal media.[26] The kinetics of plasmid loss (FIGURE 3) showed that the plasmid was considerably more stable at the higher dilution rates (0.47 and 0.66 h^{-1}) than at the lower dilution rates. This type of phenomenon has previously been reported by other authors.[5,27] At high dilution

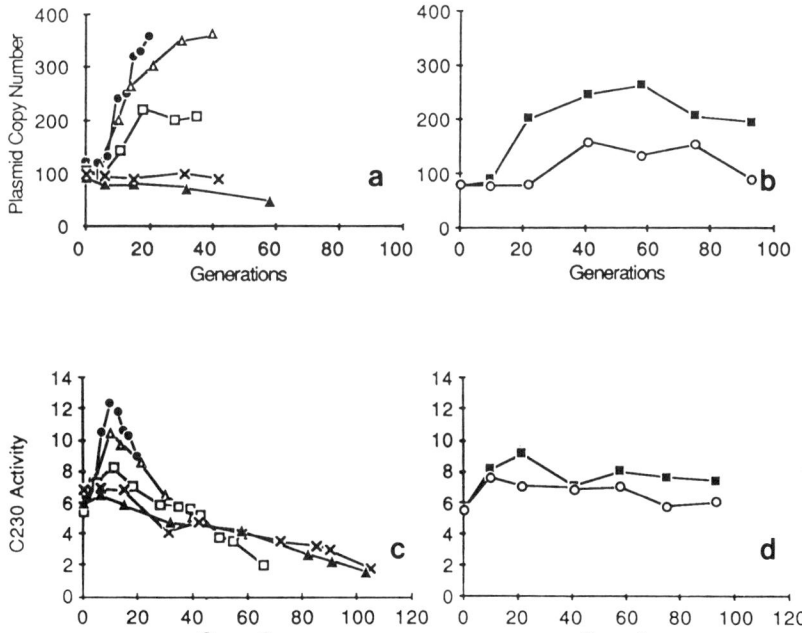

FIGURE 4. Plasmid copy number per recombinant cell and Cat-O_2-ase activity (specific activity expressed as $\Delta A_{375}/\min$) in free (a, c) (see caption to FIGURE 3) and immobilized (b, d) cultures (■ = beads, ○ = released cells).

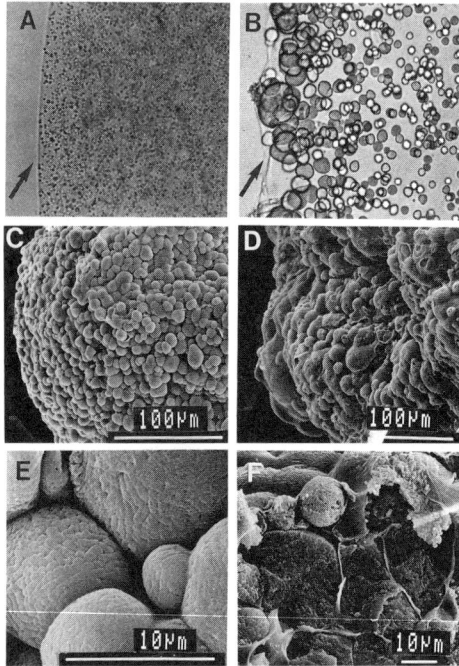

FIGURE 5. Optical micrographs of sliced K-carrageenan gel beads (glutaraldehyde fixation 0.5% during 4 h) containing *E. coli* microcolonies after 1–5 h (A) and 10 h (B) of incubation. Arrows indicate the edge of the beads. Scanning electron micrographs showing the surface of the gel bead after 6 h (C and E) and 24 h (D) of incubation. In this last case, the fracture section allows the visualization of cavities full of cells (F).

rates, the plasmid copy number (FIGURE 4a) and the Cat-O_2-ase activity (FIGURE 4c) exhibited a slow decrease. The plasmid content increased from 100 to 350 copies during 40 generations, but the dilution rate decreased from 0.47 to 0.06 h^{-1} (FIGURE 4a). Seo and Bailey[28] have also found that plasmid content exhibits a maximum at low dilution rates. At low dilution rates, the Cat-O_2-ase activity increased for the first 10–13 generations and then decreased (FIGURE 4c). This phenomenon can be explained by the decrease in the percentage of P^+ cells. Increasing the number of plasmid gene copies in the cell may also lead to a decrease in translation efficiency.

In contrast, both plasmid stability and enzyme production in immobilized-cell continuous cultures were maintained constant at high levels for more than 100 generations (FIGURES 3 and 4d). The dilution rate did not influence plasmid stability in the gel beads. The evolution of plasmid copy number (FIGURE 4b) showed an increase from the initial value until it reached a maximum value of 260 copies for gel bead cells and 160 copies for released cells. Such a difference can be explained by the distribution of growth inside the gel beads (FIGURES 5 and 6). The growth rate of immobilized cells depended on their location in the beads. The internal cell population, which grew at a slower rate, likely exhibited a high plasmid copy number. The external cell population, located close to the gel bead surface, had nearly the same specific growth rate as free cells and thus probably exhibited a lower copy number. The plasmid copy number determined for entrapped cells should therefore be a mean value and the gel beads can be considered as a reservoir of plasmid-carrying cells.

Effect of Inoculum Density on Biomass Production and Plasmid Stability of Immobilized E. coli W3101/pTG201

As indicated earlier, studies have shown that immobilized cells grew near the surface of the gel beads, where they formed microcolonies.[29] In particular, it was noted that cell growth was limited to the outer 50–150 μm of the gel beads (FIGURES 5 and 6). With a small inoculum density (4.7×10^3 cells/mL), the number of cells in the gel beads increased remarkably and 24–26 generations of cells were required to form large microcolonies before an eventual gel disruption (FIGURE 5, TABLE 1). In contrast, the use of a very high inoculum density (2.1×10^{10} cells/mL) resulted in growth limited to the outer layer of the gel beads and only 3–5 generations were required before cavity disruption (FIGURE 6, TABLE 1). In both cases, the final biomass in the gel beads was of the same order.[30] As shown in FIGURE 7, the length of the period corresponding to the decrease of P^+ cells appeared to be a function of inoculum density. Overall, pTG201 was found to be more stable with a high inoculum size, that is, when the average number of cell divisions within the volume occupied by the active bacteria was lower.

These results are in accord with the previously reported hypothesis[17] suggesting that the increased plasmid stability in immobilized cells is essentially due to the mechanical properties of the gel bead system, which may only allow a limited number of cell divisions to occur in each cavity before the clones escape from the gel beads. In the immobilized system, pTG201-free cells may appear during growth inside cavities,

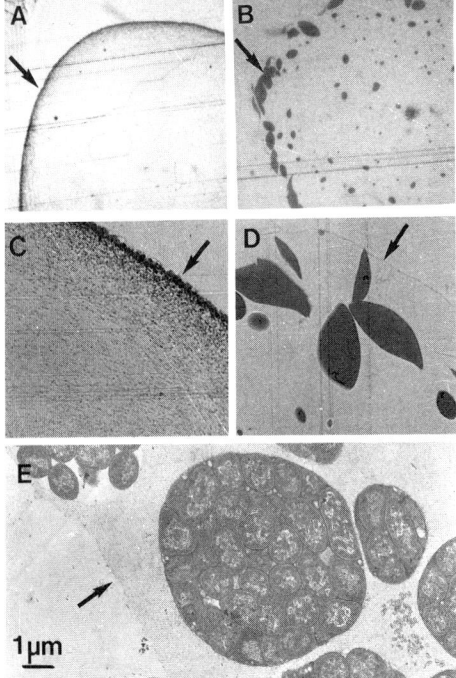

FIGURE 6. Optical micrographs of sections of K-carrageenan gel beads containing *E. coli* B/pTG201 microcolonies after 48 h of incubation. The inoculum concentration was 2.1×10^{10} cells/mL (A, C) and 4.7×10^3 cells/mL (B, D). A transmission electron micrograph showing the cell density inside a cavity is shown in part E.

TABLE 1. Effect of Inoculum Density on Biomass Production

Inoculum size[a]	2.10×10^{10}	4.70×10^{8}	6.00×10^{6}	4.7×10^{3}
Cell divisions	$4-5^b$	$9-10^b$	$11-15^c$	26^c
Immobilized cells[d]	3.80×10^{10}	2.89×10^{10}	2.32×10^{10}	2.00×10^{10}
Released cells[d]	3.70×10^{8}	2.05×10^{8}	2.11×10^{8}	nd[e]

[a] Number of viable immobilized cells per mL of gel bead before the onset of incubation.
[b] Average number of cell divisions within the volume actually occupied by active bacteria.
[c] Average number of cell divisions within the gel beads assuming a uniform growth of cells throughout the gel beads.
[d] Viable cell number of immobilized cells is expressed per mL of gel bead and that of released cells is expressed per mL of liquid medium.
[e] nd = not determined.

but competition between P^+ and P^- cells exists only approximately 16 generations, after which cell growth stops. This limitation of competition improves net plasmid stability, as can be concluded from the effect of decreasing inoculum density. (The maximum number of generations was measured from the maximum cavity size, which was determined using microscopy.)

Development of a Two-Stage Chemostat with Free and Immobilized E. coli *W3101/pTG201 for Production of Catechol 2,3-Dioxygenase*

The use of a temperature-sensitive system[5,6,31] such as pTG201 offers the possibility of switching on the synthesis of recombinant protein at will. However, pTG201 plasmid

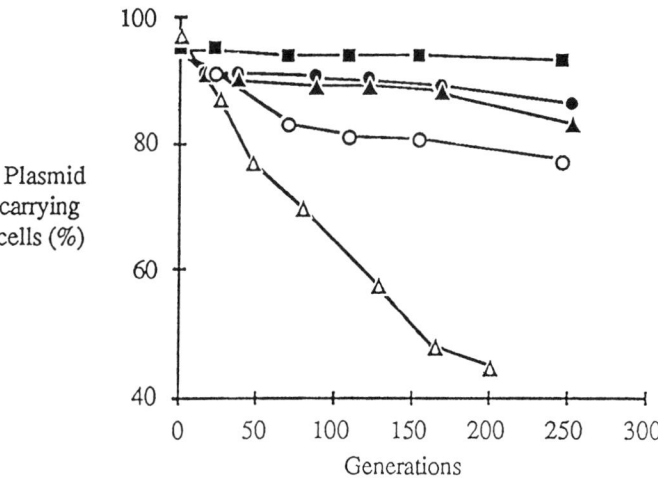

FIGURE 7. Effect of inoculum density on the stability of pTG201 plasmid in continuous cultures with immobilized *E. coli* B cells: (■) 2.10×10^{10}; (●) 4.70×10^{8}; (▲) 6.00×10^{6}; (○) 4.70×10^{3}; and (△) free cells.

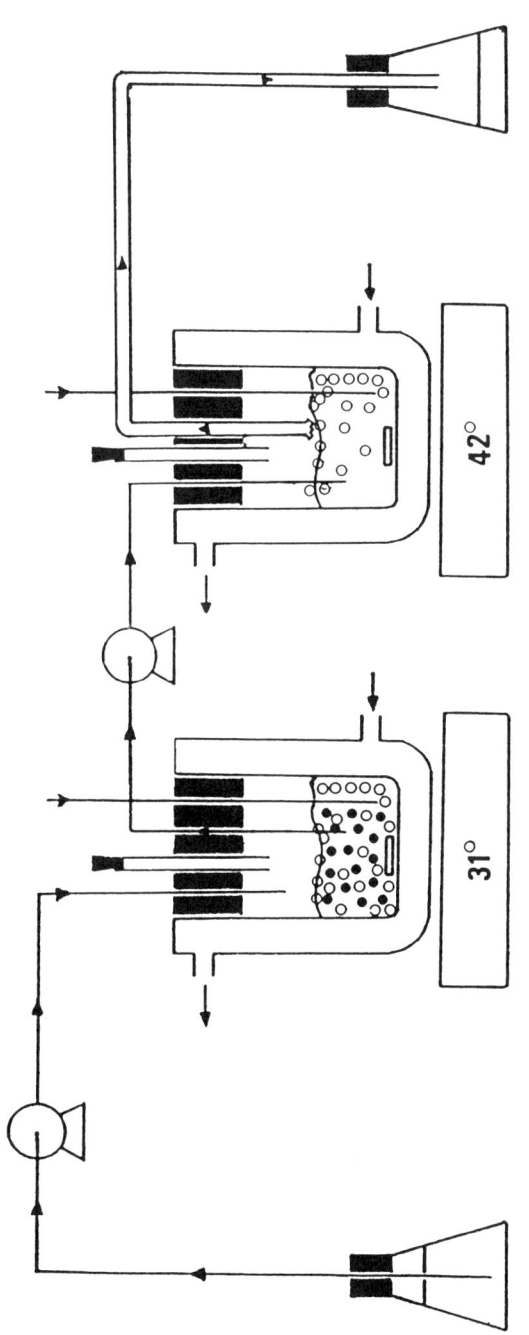

FIGURE 8. Schematic representation of the configuration of a two-stage continuous culture.

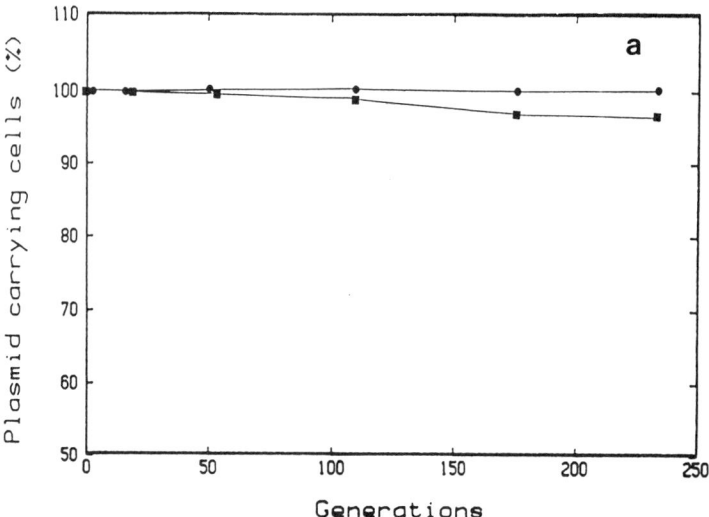

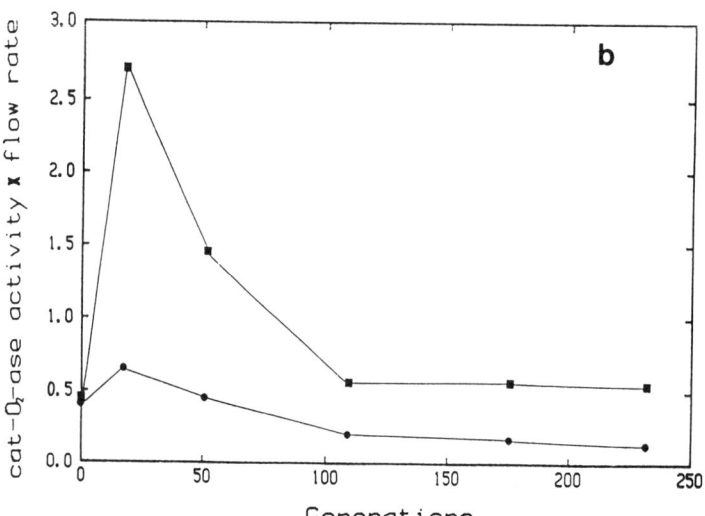

FIGURE 9. Stability of pTG201 (a) and Cat-O_2-ase productivity (b) in a two-stage continuous culture of *E. coli* W3101 (pTG201). Symbols: ● = first stage, ■ = second stage.

stability was strongly affected by high expression of the cloned Xyl E gene.[31] As immobilization considerably increased both the stability of recombinant plasmids and the productivity of the cloned gene (due to the high cell concentration obtained within the gel bead) and minimized the deleterious effects of high expression, we separated the phases of growth and expression of the Xyl E gene (FIGURE 8).

In the first stage, immobilized cells were grown in the repressed state (31 °C) and the released cells were then continuously transferred to the second stage where derepression was induced (42 °C). Such a two-stage process coupled with a temperature-sensitive system resulted in an increase in the productivity of catechol 2,3-dioxygenase (FIGURE 9).

CONCLUSIONS

The data presented here outline the role played by immobilization in improving the stability of vectors used to overproduce gene products. Furthermore, phenomena such as periodic genetic changes that occurred in the plasmid or host strain after long-term growth of bacteria in free-cell continuous culture could be avoided through cell immobilization.[32] We found that plasmid-harboring strains could be stably maintained over two weeks. Due to the mechanical properties of gel beads that allow a limited number of cell divisions, P^+ cells may be partially saved from competition with P^- cells.

With the eventual goal of resolving the limitation of oxygen diffusion in immobilized cells, the effect of dissolved oxygen on recombinant bacterial properties has recently been studied[33,34] and this approach could be expanded by the use of anaerobic bacteria.

REFERENCES

1. ENSLEY, B. D. 1986. Stability of recombinant plasmids in industrial microorganisms. CRC Crit. Rev. Biotechnol. **4:** 263–277.
2. RYU, D. D. Y. & R. SIEGEL. 1986. Scale-up of fermentation processes using recombinant microorganisms. Ann. N.Y. Acad. Sci. **469:** 73–82.
3. ZABRISKIE, D. W. & E. J. ARCURI. 1986. Factors influencing productivity of fermentations employing recombinant microorganisms. Enzyme Microb. Technol. **8:** 706–717.
4. LAUFFENBURGER, D. A. 1987. Bacteriocin production as a method of maintaining plasmid-bearing cells in continuous culture. Trends Biotechnol. **5:** 87–89.
5. CAULLOTT, C. A. & M. RHODES. 1986. Temperature-induced synthesis of recombinant proteins. Trends Biotechnol. **4:** 142–146.
6. SIEGEL, R. & D. D. Y. RYU. 1985. Kinetic study of instability of recombinant plasmid pPLc23 trp Al in *E. coli* using two-stage continuous culture system. Biotechnol. Bioeng. **27:** 28–33.
7. BOE, L., K. GERDES & S. MOLIN. 1987. Effects of genes exerting growth inhibition and plasmid stability on plasmid maintenance. J. Bacteriol. **169:** 4646–4650.
8. SERESSIOTIS, A. & J. E. BAILEY. 1987. Optimal gene expression and amplification strategies for batch and continuous culture. Biotechnol. Bioeng. **29:** 392–398.
9. INLOES, D. S., W. J. SMITH, D. P. TAYLOR, N. COHENS, A. S. MICHAELS & C. R. ROBERTSON. 1983. Hollow-fiber membrane bioreactors using immobilized *E. coli* for protein synthesis. Biotechnol. Bioeng. **25:** 2653–2681.
10. MOSBACH, K., S. BIRNBAUM, K. HARDY, J. DAVIES & L. BULOW. 1983. Formation of proinsulin by immobilized *Bacillus subtilis*. Nature **302:** 543–545.
11. DHULSTER, P., J. N. BARBOTIN & D. THOMAS. 1984. Culture and bioconversion use of plasmid-harboring strain of immobilized *E. coli*. Appl. Microbiol. Biotechnol. **20:** 87–93.
12. GEORGIOU, G., J. J. CHALMERS, M. L. SHULER & D. B. WILSON. 1985. Continuous immobilized recombinant protein production from *E. coli* capable of selective protein excretion: a feasibility study. Biotechnol. Prog. **1:** 75–79.

13. KANAYAMA, H., K. SODE & I. KARUBE. 1988. Continuous hydrogen evolution by immobilized recombinant *Escherichia coli* using a bioreactor. Biotechnol. Bioeng. **32:** 396–399.
14. DE TAXIS DU POET, P., P. DHULSTER, J. N. BARBOTIN & D. THOMAS. 1986. Plasmid inheritability and biomass production: comparison between free and immobilized cell cultures of *Escherichia coli* BZ18 (pTG201) without selection pressure. J. Bacteriol. **165:** 871–877.
15. JAOUA, S., A. M. BRETON, G. YOUNES & J. F. GUESPIN-MICHEL. 1986. Structural instability and stabilization of Inc P1 plasmids integrated into the chromosome of *Myxococcus xanthus*. J. Biotechnol. **4:** 313–323.
16. DE TAXIS DU POET, P., Y. ARCAND, R. BERNIER, JR., J. N. BARBOTIN & D. THOMAS. 1987. Plasmid stability in immobilized and free recombinant *Escherichia coli* J.M. 105 (pKK223-200): importance of oxygen diffusion, growth rate, and plasmid copy number. Appl. Environ. Microbiol. **53:** 1548–1555.
17. NASRI, M., S. SAYADI, J. N. BARBOTIN & D. THOMAS. 1987. The use of the immobilization of whole living cells to increase stability of recombinant plasmids in *Escherichia coli*. J. Biotechnol. **6:** 147–157.
18. JOSHI, S. & H. YAMAZAKI. 1987. Stability of plasmid pBR 322 in *Escherichia coli* immobilized on cotton cloth during use as resident inoculum. Biotechnol. Lett. **9:** 825–830.
19. ORIEL, P. 1988. Amylase production by *Escherichia coli* immobilized in silicone foam. Biotechnol. Lett. **10:** 113–116.
20. ORIEL, P. 1988. Immobilization of recombinant *Escherichia coli* in silicone polymer beads. Enzyme Microb. Technol. **10:** 518–523.
21. SODE, K., T. MORITA, A. PETERMANS, F. MEUSSDOERFFER, K. MOSBACH & I. KARUBE. 1988. Continuous production of α-peptide using immobilized recombinant yeast cells. A model for continuous production of foreign peptide by recombinant yeast. J. Biotechnol. **8:** 113–122.
22. WALLS, E. L. & J. L. GAINER. 1988. Production of enzymes by recombinant yeast. Presented at this conference.
23. ZUKOWSKI, M. N., D. F. GAFFNEY, D. SPECK, M. KAUFFMAN, A. FINDELLI, A. WISECUP & J. P. LECOCQ. 1983. Chromogenic identification of genetic regulator signals in *Bacillus subtilis* based on expression of a cloned *Pseudomonas* gene. Proc. Natl. Acad. Sci. U.S.A. **80:** 1101–1105.
24. NASRI, M., S. SAYADI, J. N. BARBOTIN, P. DHULSTER & D. THOMAS. 1987. Influence of immobilization on the stability of pTG201 recombinant plasmid in some strains of *Escherichia coli*. Appl. Environ. Microbiol. **53:** 740–744.
25. SAYADI, S., F. BERRY, M. NASRI, J. N. BARBOTIN & D. THOMAS. 1988. Increased stability of pBR 322-related plasmids in *Escherichia coli* W3101 grown in carrageenan gel beads. FEMS Microbiol. Lett. **56:** 307–312.
26. SAYADI, S., M. NASRI, J. N. BARBOTIN & D. THOMAS. 1989. Effect of environmental growth conditions on plasmid stability, plasmid copy number, and catechol 2,3-dioxygenase activity in free and immobilized *Escherichia coli* cells. Biotechnol. Bioeng. **33:** 801–808.
27. CAULCOTT, C. A., A. DUNN, H. A. ROBERTSON, N. COOPER, M. F. BROWN & P. M. RHODES. 1987. Investigation of the effect of growth environment on the stability of low-copy number plasmids in *Escherichia coli*. J. Gen. Microbiol. **133:** 1881–1889.
28. SEO, J. H. & J. E. BAILEY. 1986. Continuous cultivation of recombinant *Escherichia coli*: existence of an optimum dilution rate for maximum plasmid and gene product concentration. Biotechnol. Bioeng. **28:** 1590–1594.
29. MARIN INIESTA, F., P. DE TAXIS DU POET, P. DHULSTER, D. THOMAS & J. N. BARBOTIN. 1987. Immobilized bacteria and plasmid stability. Ann. N.Y. Acad. Sci. **501:** 317–329.
30. BERRY, F., S. SAYADI, M. NASRI, J. N. BARBOTIN & D. THOMAS. 1988. Effect of growing conditions of recombinant *E. coli* in carrageenan gel beads upon biomass production and plasmid stability. Biotechnol. Lett. **10:** 619–624.
31. SAYADI, S., M. NASRI, F. BERRY, J. N. BARBOTIN & D. THOMAS. 1987. Effect of temperature on the stability of plasmid pTG201 and productivity of Xyl E gene product in

recombinant *Escherichia coli*: development of a two-stage chemostat with free and immobilized cells. J. Gen. Microbiol. **133:** 1901–1908.
32. NASRI, M., F. BERRY, S. SAYADI, D. THOMAS & J. N. BARBOTIN. 1988. Stability fluctuations of plasmid-bearing cells: immobilization effects. J. Gen. Microbiol. **134:** 2325–2331.
33. MARIN INIESTA, F., S. SAYADI, P. DHULSTER, J. N. BARBOTIN & D. THOMAS. 1988. Influence of oxygen supply on the stability of recombinant plasmid pTG201 in immobilized *E. coli* cells. Appl. Microbiol. Biotechnol. **28:** 455–462.
34. HUANG, J., P. DHULSTER, J. N. BARBOTIN & D. THOMAS. 1989. Effects of oxygen diffusion on recombinant *E. coli* B (pTG201) plasmid stability, growth rate, biomass production, and enzyme activity in immobilized and free bacteria during continuous culture. J. Chem. Technol. Biotechnol. **45:** 259–269.

Regulation and Dynamics of Elicitor Response in Suspension-cultured Plant Cells

S. Y. BYUN,[a] H. PEDERSEN,[a,b] AND C-K. CHIN[c]

[a]Department of Chemical and Biochemical Engineering
[c]Department of Horticulture and Forestry
Rutgers, The State University of New Jersey
New Brunswick, New Jersey 08903

INTRODUCTION

Elicitor-induced accumulation of secondary metabolites in plant tissue culture has lately received increasing attention.[1] This is due, in part, to the fact that elicitation can improve the efficiency of secondary metabolite accumulation in systems where product formation appears near or after the late growth phase. In "elicited" cells, it may therefore be possible to reduce the time required to obtain a product. The technique also serves to eliminate media exchanges as a means of promoting secondary product formation. Metabolites typically accumulate not only within the cell, but, in several instances, large amounts are found excreted into the medium. Furthermore, elicited cells may be used for enhancing biotransformation reactions due to the induction of key enzymes.[2,3]

The induction of plant biosynthetic pathways by specific compounds added extracellularly to the media has been demonstrated in only a few tissue culture systems so far. Nevertheless, some general mechanisms have emerged from these studies. The rapid transient expression of transcriptional activity after elicitor treatment has been demonstrated and the corresponding translation of enzymes has been reported.[4-6] These results point out that elicitor-induced gene expression is specific, but complex. The translation of an enzyme or enzymes along a biochemical pathway can lead to product accumulation.

Suspension cultures of *Eschscholtzia californica* accumulate the benzophenanthridine alkaloids, sanguinarine, chelirubine, chelerythrine, and macarpine, all of which are known to be constituents of the *Eschscholtzia* plant.[7] The benzophenanthridine alkaloids have recently been the subject of increasing interest because of their dental and medical uses.[3,8] The biosynthesis of benzophenanthridine alkaloids has been well studied and some of the complex precursor relationships have been deduced. For example, Battersby and co-workers[9] have made a thorough study of benzophenanthridine alkaloid biosynthesis and have identified the key intermediate after dopamine as (+)reticuline, which is cyclized oxidatively to (−)scoulerine. The next step is the addition of two methylenedioxy groups to produce stylopine. Although no intermediates have been isolated for the subsequent stages, they are thought to involve 6-hydroxylation of the stylopine metho salt. The subsequent steps, though, have been

---

[b]To whom all correspondence should be addressed.

investigated by Takao et al.[10] in feeding experiments to define the biosynthetic pathway from stylopine metho salt to macarpine as shown in FIGURE 1.

This work examines elicitor-induced accumulation of alkaloids in *Eschscholtzia californica* and develops a mathematical model of *de novo* enzyme synthesis for key regulatory enzymes.

THEORETICAL DEVELOPMENT

The conceptual framework on which the mathematical model is based is illustrated schematically in FIGURE 2. A second messenger may be generated by elicitor contact

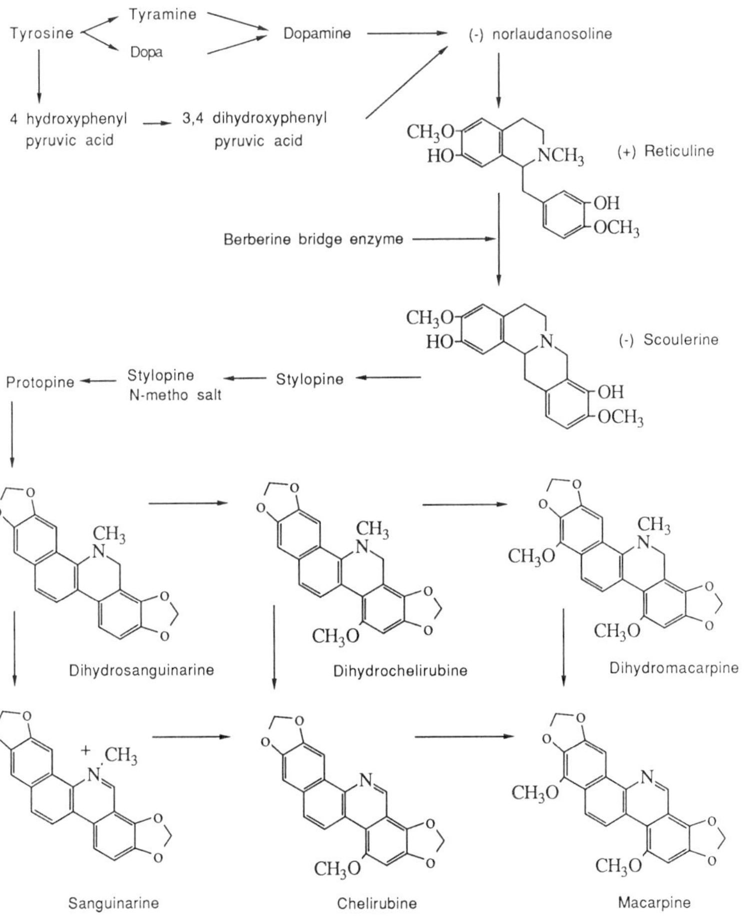

FIGURE 1. Biosynthetic sequence for the benzophenanthridine alkaloids in *Eschscholtzia californica*.

with cell membrane receptors and is transmitted to the nucleus to trigger transcription of mRNA. The corresponding translation of enzymes establishes biochemical pathways involved in product accumulation. The mathematical model developed here is based on some assumptions. First, the second messenger evoked by elicitation induces transcription and the transcription rate is dependent on the strength of the messenger signal.[11] Second, the gene concentration is constant during elicitation. Third, the product formation rate is mainly dependent on a limiting enzyme and the product accumulation further regulates the transcription of mRNA.

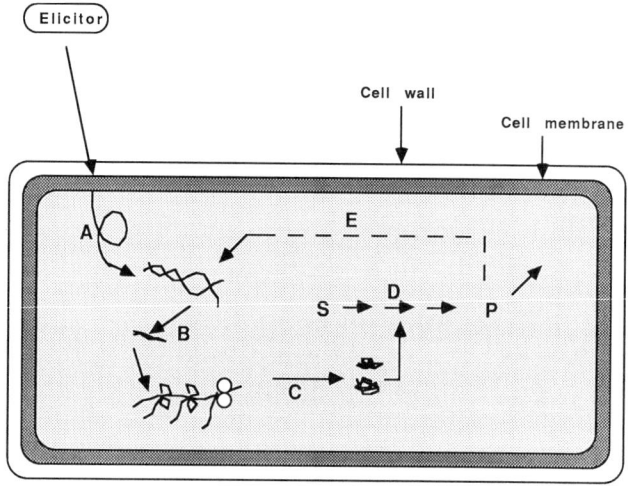

FIGURE 2. Schematic representation of secondary metabolic production in response to elicitor— A: second messenger; B: transcription; C: translation; D: product formation; E: product downregulation; S: substrate; P: product.

Transmission of Signal

Let α and M denote the elicitor concentration and the concentration of second messenger signal, respectively. Then, the following equation is written for the transmission of messenger by elicitation:

$$\frac{d[M]}{dt} = f(\alpha) - k_1[M] - \mu[M]. \tag{1}$$

In this linear relation, $f(\alpha)$ denotes the generation of second messenger, which is a function of elicitor concentration. The parameters k_1 and μ are the rate constant for loss of M and the specific growth rate of the cell, respectively. The third term in equation 1 represents the dilution of M by the cell mass increase.

Transcription and Translation

The formation of a specific messenger RNA (mRNA) is assumed to depend on the specific M concentration and on the DNA concentration. However, mRNA induction often does not continue even though the elicitor remains in the culture medium.[12] The production of mRNA is down-regulated by secondary metabolite formation. A mass balance on mRNA concentration may be written as

$$\frac{d[mRNA]}{dt} = \eta \frac{[M][G]}{g(P)} - k_2[mRNA] - \mu[mRNA], \qquad (2)$$

where $[G]$ and k_2 denote the DNA concentration and the decay rate constant of $[mRNA]$, respectively, and η represents a transcription efficiency. The function $g(P)$ in equation 2 represents the regulation of transcription by product accumulation.

In a similar fashion, a material balance can be written for the intracellular concentration of enzyme due to translation. In this case, the synthesis term contains the concentration of the mRNA for a specific enzyme E. The required mass balance is given by

$$\frac{d[E]}{dt} = k_3[mRNA] - k_4[E] - \mu[E], \qquad (3)$$

where k_3 denotes the translation rate constant, which includes a translation efficiency, and k_4 represents the decay rate constant of E.

Product Formation

Presuming constant primary substrate concentrations in the product synthesis pathway, the rate of product formation can be written as a first-order reaction of the form,

$$\frac{d[P]}{dt} = k_5[E] - \mu[P], \qquad (4)$$

where k_5 denotes the rate constant of $[P]$ formation.

Cell Growth and Substrate Consumption

The cell growth rate is evaluated in terms of the simple expression

$$\frac{dX}{dt} = \mu X. \qquad (5)$$

The specific growth rate μ is

$$\mu = \mu_{max} \frac{S}{K_s + S}, \qquad (6)$$

where μ_{max} and K_s are the maximum specific growth rate and the Monod constant, respectively.

For constant yield factors, the substrate consumption rate can be obtained from a mass balance equation of the form,

$$\frac{dS}{dt} = -k_6 \frac{dX}{dt} - k_7 \frac{dP}{dt}, \qquad (7)$$

where k_6 is the sum of the yield factors due to cell growth and k_7 is the yield factor for product formation. The second term in equation 7 is typically considered to be negligible. Whereas X and S are extracellular (volumetric) concentrations, note that all other components are referenced to the cell mass and represent intracellular concentrations.

MATERIALS AND METHODS

Cell Cultures

Cultures of *Eschscholtzia californica* were kindly provided by Peter Brodelius (Institute of Biotechnology, ETH, Zurich, Switzerland) and were maintained on B5 medium[13] supplemented with 5 μM 2,4-D and 0.5 μM kinetin on a gyrotary shaker (180 rpm) at 26 °C. Subculturing was carried out every seven days by transferring 16 g of cells (fresh cell weight) into 200 mL of medium in 500-mL Erlenmeyer flasks. We used 125-mL Erlenmeyer flasks containing 50 mL of growth medium for experimental batch cultures where 4.0 g of cells (fresh cell weight) was inoculated into each flask.

Chemicals

B5 media was obtained from Gibco Laboratories. Sanguinarine nitrate was from Research Plus (Bayonne, New Jersey) and chelerythrine was supplied by Atomergic Chemicals (Farmingdale, New York). All other chemicals involved in this study were reagent grade.

Preparation of Elicitors

The yeast elicitor was prepared from yeast extract (DIFCO Laboratories, Detroit, Michigan) according to the literature.[14] The strains used to prepare fungal elicitor were *Phytophthora megasperma* f. sp. *glycinea* (ATCC 28001), *Colletotrichum lindemuthianum* (ATCC 11225), and *Verticillium dahliae* (ATCC 7611). The fungal elicitor was mycelial wall release, which was isolated[15] after growth on suitable media.[16-18] The carbohydrate concentration of applied elicitor solution was determined by the orcinol–sulfuric acid procedure.[19]

Preparation of Alkaloid Standards

Macarpine was extracted from cultured cell mass because a commercial supply was not available. Approximately 500 g of filtered cell mass of *Eschscholtzia califor-*

nica was dispersed in 1.5 liters of methanol. The mixture was stirred overnight at room temperature and filtered. The filtered cells were then washed with 1 liter of methanol. The combined methanol extracts were evaporated under reduced pressure. A 200-mL solution of acetic acid and water (50:50) was added to the residue and the mixture was again filtered. The filtrate was extracted with petroleum ether (200 mL) to remove colored materials and the aqueous layer was then made alkaline with 15% ammonium hydroxide and extracted with chloroform (400 mL). Evaporation of the organic solvent gave a crude mixture of *Eschscholtzia californica* alkaloids.

The crude extract was dissolved in methanol and separated by FPLC (Pharmacia, Piscataway, New Jersey). Reverse phase silica C-18 (PepRPC™, Pharmacia) was used as a separation column. A programmed mixture of acetonitrile and water was used for the mobile phase where the water phase included 1 mM tetrabutyl ammonium phosphate with the pH adjusted to 2.0 using phosphoric acid. Macarpine peaks were further isolated by solvent extraction with methylene chloride and were positively identified with a mass spectrometer.

Alkaloid Analysis

For analysis, a slightly different procedure was used. The cells were harvested by vacuum filtration and 1.0 g of cells was extracted with 10 mL of pure methanol. All extracts were filtered and 10 µL of the solution was injected for the measurement of alkaloid concentration. An HPLC system (Spectra-Physics, San Jose, California) equipped with a UV detector (Spectroflow 773, Kratos Corporation, Ramsey, New Jersey) was used to determine benzophenanthridine alkaloids. The sample was chromatographed on a Supelco C-18 column (15 cm × 4.6 mm) equipped with a precolumn. The mobile phase was a mixture of water (65%) and acetonitrile (35%). The aqueous phase contained 1 mM of tetrabutyl ammonium phosphate and was adjusted to pH 2.0 with phosphoric acid. The absorbance of the alkaloid was measured at 280 nm.

Sugar and Growth Measurement

Sucrose in the culture media was rapidly degraded by a cellular invertase to glucose and fructose. To determine these sugars, an HPLC system with an NH_2-silica column (25 cm × 4.6 mm) and a refractive index detector (Perkin Elmer, Wilton, Connecticut) was used. The mobile phase was a mixture of acetonitrile (75%) and water (25%).

Cell growth was determined as dry cell weight (DCW). After determination of the fresh weight, filtered cells were washed with distilled water and dried overnight at 60 °C to a constant weight for DCW determination.

Calculation of Model Equations

Various parameters in the model equations were obtained by a nonlinear regression analysis.[20] However, the maximum specific growth rate, μ_{max}, and the Monod constant, K_s, were derived directly from experimental data.

RESULTS AND DISCUSSION

Response to Different Elicitors

Elicitation of secondary metabolite production in cell cultures of *Eschscholtzia californica* is a function of the source material used for elicitor preparation. FIGURE 3 shows the response to four different biotic elicitors. Elicitors prepared from *Colletotrichum lindemuthianum*, *Verticillium dahliae*, and yeast extract caused induction and accumulation of macarpine, whereas the elicitor from *Phytophthora megasperma* resulted in no response. The differences in the total amounts in FIGURE 3 may represent differences in the optimum elicitor concentration for each elicitor type. In some cases, different elicitors may also result in unique responses within the same cell culture. For instance, a varied response to different elicitors was noted in cultured cells of *Petroselinum hortense*.[21] Irradiation with UV light induced formation of flavonoids, whereas addition of fungal elicitors, in contrast, caused formation and accumulation of furanocoumarins. The elicitors are obviously representative of different types of stress and result in production of active compounds against UV light and pathogens, respectively. However, this distinction is not expected to play a role in our system.

Dependence of Macarpine Accumulation on Elicitor Concentration

The elicitor concentration is a factor that strongly affects the intensity of the response. FIGURE 4 shows the quantitative accumulation of macarpine in response to different elicitor concentrations. The accumulation pattern of macarpine versus elicitor concentration demonstrates a saturated phenomenon. The accumulation rate was highly affected by elicitor concentration at low elicitor concentration, but was virtually unaffected at high elicitor concentration. The maximum accumulation of macarpine was obtained at 60 μg of yeast elicitor per gram of fresh cell weight (μg ye/g fcw). The

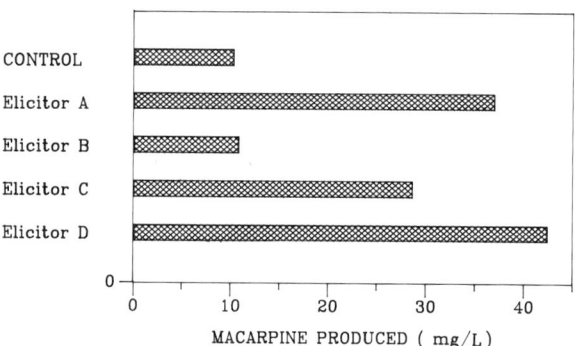

FIGURE 3. Macarpine accumulation in *Eschscholtzia californica* in response to four different biotic elicitors. Seventy μg of elicitor per gram of fresh cell weight was dosed at exponential growth phase. Cultures were harvested after 48 hours—elicitor A: yeast extract; elicitor B: *Phytophthora megasperma* f. sp. *glycinea*; elicitor C: *Colletotrichum lindemuthianum*; elicitor D: *Verticillium dahliae*.

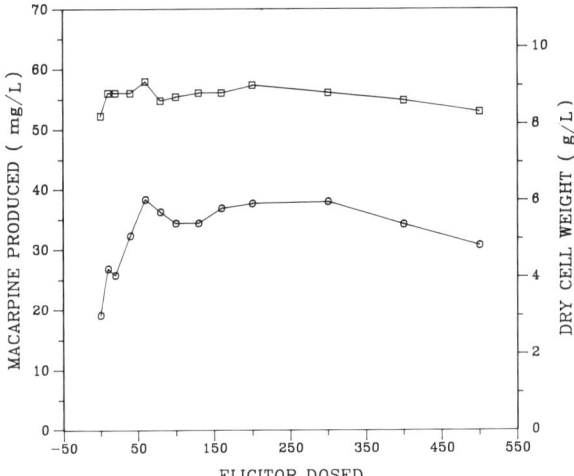

FIGURE 4. Effects of elicitor concentration on macarpine accumulation of *Eschscholtzia californica*. Yeast elicitor was dosed at exponential growth phase. Cultures were harvested after 16 hours. The unit of elicitor dosed is μg of yeast elicitor per gram of fresh cell weight. Symbols: ○ = macarpine produced; □ = dry cell weight.

accumulation patterns of other secondary metabolites, namely, sanguinarine and chelerythrine, were different. The maximum value of sanguinarine accumulation was obtained at 20 μg ye/g fcw, whereas it was 40 μg ye/g fcw for chelerythrine. FIGURE 4 also demonstrates that "overloading" of elicitor has adverse effects. Inhibition by overdosed elicitor reduced the accumulation of macarpine as well as cell growth. Similar results with respect to cell growth have been reported for *Petroselinum hortense*.[21]

Kinetics of Secondary Metabolite Accumulation

The kinetics of second messenger, mRNA, enzyme, and product in response to treatment with elicitor can be calculated by equations 1–7. However, the experimental data to validate these kinetic equations have been reported in a few tissue culture systems only. For *Petroselinum hortense* (parsley) as well as *Glycine max* (soybean), the kinetics of transcription, translation, and product formation were available in response to biotic and abiotic elicitors.[6,21–23] Irradiation of *Petroselinum hortense* with UV light induced flavonoid production, which is known to absorb UV light.[4,24] A key enzyme in the metabolic pathway leading to flavonoids is phenylalanine ammonialyase (PAL). The corresponding mRNA, PAL activity, and flavonoid accumulation data were used to verify the kinetic model equations developed. Although no data related to a second messenger were available, equation 1 leads to a lag period that was experimentally observed in the initial part of the mRNA activity data. A similar lag period in mRNA activity was also reported for *Glycine max*[6,23] as well as *Phaseolus vulgaris* (bean).[5,25] The lag period commonly appears in these cell lines and could represent the events associated with mobilization of intracellular synthesis machinery, all of which are considered to be part of the signal transmission.

The rapid changes in amount and activity of mRNA in response to elicitor are due to a correspondingly rapid transient increase in transcription rate of the respective genes.[26] The transcription rate depends on elicitor contact; however, mRNA synthesis does not continue even though the elicitor remains in the culture medium. Once the mRNA has been induced by elicitation and subsequently disappears, the same pattern of induction and disappearance could not be observed again over time scales on the order of the growth rate. These observations lead to an assumption that the mRNA synthesis could be somehow regulated by the product. A regulation term is used in equation 2, $g(P)$, to acknowledge this and is expressed as a function of product concentration in a simple way:

$$g(P) = 1 + P/K_i^1 + P^2/K_i^2, \qquad (8)$$

where K_i^1 and K_i^2 are inhibition constants. When we applied equation 8 as a regulation term in equation 2, we could accurately calculate the changes of mRNA activity, enzyme activity, and product accumulation versus time for *Petroselinum hortense* (FIGURE 5). The dynamics of mRNA activity, that is, a rapid decline in mRNA, requires the "regulating" term; without it in equation 2, the mRNA activity calculated would be quite different from experimental data (data not shown).

The enzyme synthesis depends on mRNA activity as expressed in equation 3. The lag period in enzyme activity data was similarly observed. The corresponding product formation is known to depend on both enzyme concentration as well as a precursor substrate, often times referring to various amino acids. However, the flavonoid

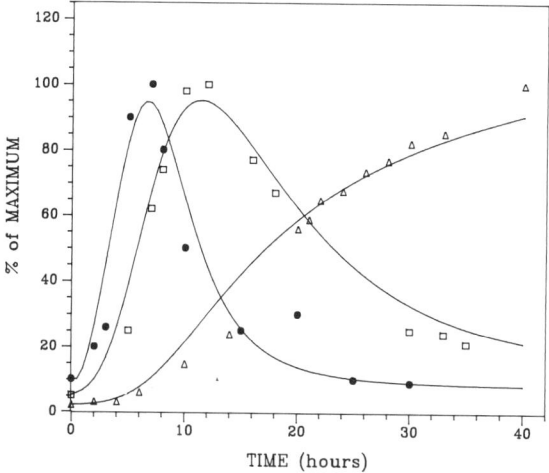

FIGURE 5. UV light–induced changes in mRNA and enzyme activity of PAL and in flavonoid accumulation of *Petroselinum hortense*: ● = PAL mRNA activity; □ = PAL activity; △ = flavonoid accumulation; — = calculation.

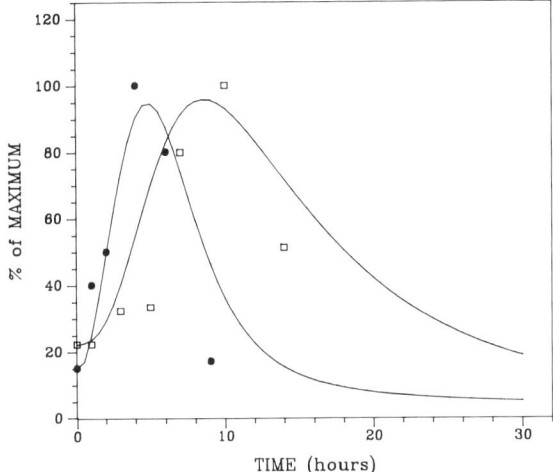

FIGURE 6. Time course of mRNA and enzyme activity of chalcone synthase induced in cultured *Glycine max* cells after treatment with fungal elicitor: ● = chalcone synthase mRNA activity; ○ = chalcone synthase activity; —— = calculation.

formation in *Petroselinum hortense* mainly depends on the PAL activity and the amino acid levels are not greatly perturbed. The calculated results for flavonoid formation agree with experimental data as shown in FIGURE 5.

FIGURE 6 shows the results from another plant cell culture, *Glycine max*, to validate the kinetic model equations. The calculated mRNA and enzyme activity levels were fairly consistent with experimental data. Chalcone synthase was used as the key enzyme in the phytoalexin accumulation pathway of *Glycine max*.[6,23]

Macarpine is one of the benzophenanthridine alkaloids accumulated in *Eschscholtzia californica*. It is the final product in the metabolic pathway. To produce macarpine, many enzymes are involved in the metabolic pathway; however, tyrosine decarboxylase, dihydrobenzophenanthridine oxidase,[27] and the berberine bridge enzyme[28] are reported as key factors. Experimental data for the berberine bridge enzyme activity were collected[28] and were used for comparison with the theoretical results. Our experimental data on macarpine accumulation, cell growth, and substrate consumption were obtained under identical experimental conditions as those of the berberine bridge enzyme activity data. Theoretical results calculated with equations 1–8 agree well with experimental data as seen in FIGURE 7. The kinetic model equations accurately explain the changes in both the berberine bridge enzyme activity and the macarpine accumulation, even though experimental data for second messenger and mRNA were not available. Other alkaloids, which are intermediates of the metabolic pathway, show different accumulation patterns. For instance, sanguinarine remained almost constant during the culture period, whereas chelerythrine showed a maximum accumulation around 10 hours from elicitation and then decreased to a constant low level.

CONCLUSIONS

When we compare theoretical results with experimental data, we find that our kinetic (structured) model equations are useful to explain the changes of mRNA, enzyme activity, and product formation in *Eschscholtzia californica*. This is also true for *Petroselinum hortense* and *Glycine max*. Supplementary verifications are under way to make the model equations more general. A different formulation for regulating effects may be required for different cell culture systems as well. More insight into the underlying molecular basis of our assumptions and more cell culture systems with experimental data are clearly needed. However, our framework should prove useful for guidance of quantitative analysis.

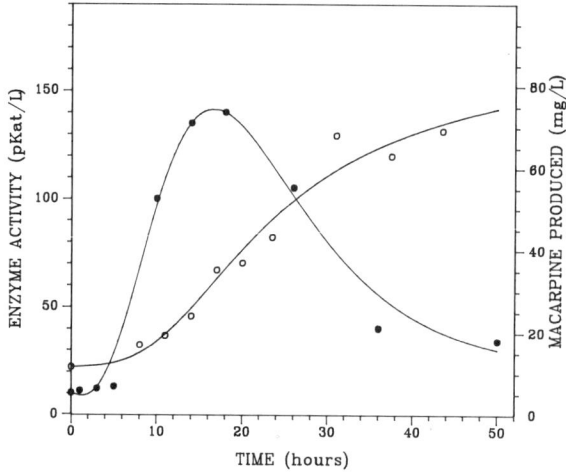

FIGURE 7. Elicitor-induced changes in berberine bridge enzyme activity and macarpine accumulation in *Eschscholtzia californica*. Sixty µg of yeast elicitor was dosed per gram of fresh cell weight. Symbols: ● = berberine bridge enzyme activity; ○ = macarpine accumulation; —— = calculation.

REFERENCES

1. DiCosmo, F. & M. Misawa. 1985. Eliciting secondary metabolism in plant cell cultures. Trends Biotechnol. **3:** 318–322.
2. Eilert, U., W. Kurtz & F. Constabel. 1987. Alkaloid accumulation in plant cell cultures upon treatment with elicitors. *In* Plant Tissue and Cell Culture, p. 213–219. Alan R. Liss. New York.
3. Cordell, G. A. 1981. Introduction to Alkaloids—A Biogenetic Approach, p. 509–517. Wiley. New York.
4. Kuhn, D. N., J. Chappell, A. Boudet & K. Hahlbrock. 1984. Induction of phenylalanine ammonia-lyase and 4-coumarate:CoA ligase mRNA in cultured plant cells by UV light or fungal elicitor. Proc. Natl. Acad. Sci. U.S.A. **81:** 1102–1106.
5. Ryder, T. B., C. L. Cramer, J. N. Bell, M. P. Robbins & R. A. Dixon. 1984. Elicitor

rapidly induces chalcone synthase mRNA in *Phaseolus vulgaris* cells at the onset of the phytoalexin defense response. Proc. Natl. Acad. Sci. U.S.A. **81:** 5724–5728.
6. EBEL, J., W. E. SCHMIDT & R. LOYAL. 1984. Phytoalexin synthesis in soybean cells: elicitor induction of phenylalanine ammonia-lyase and chalcone synthase mRNAs and correlation with phytoalexin accumulation. Arch. Biochem. Biophys. **232:** 240–248.
7. BERLIN, J., E. FORCHE, V. WRAY, J. HAMMER & W. HÖSEL. 1983. Formation of benzophenanthridine alkaloids by suspension cultures of *Eschscholtzia californica*. Z. Naturforsch. **38c:** 346–352.
8. SOUTHARD, G. L., R. T. BOULWARE, D. R. WALBORN, W. J. GROZNIK, E. E. THORNE & S. L. YANKELL. 1984. Sanguinarine, a new antiplaque agent: retention and plaque specificity. J. Am. Dent. Assoc. **108:** 338–341.
9. BATTERSBY, A. R., R. J. FRANCIS, M. HIRST, E. A. RUVEDA & J. STAUNTON. 1975. Biosynthesis. Part XXI. Investigations on the biosynthesis of stylopine in *Chelidonium majus*. J. Chem. Soc. Perkin Trans. 1 **12:** 1140–1146.
10. TAKAO, N., M. KAMIGAUCHI & M. OKADA. 1983. Biosynthesis of benzo[*c*]phenanthridine alkaloids, sanguinarine, chelirubine, and macarpine. Helv. Chim. Acta **66:** 473–484.
11. BERRIDGE, M. J. 1985. The molecular basis of communication within the cell. Sci. Am. **252:** 142–152.
12. HAHLBROCK, K. 1976. Regulation of phenylalanine ammonia-lyase activity in cell-suspension cultures of *Petroselinum hortense*. Eur. J. Biochem. **63:** 137–145.
13. GAMBORG, O. L., R. A. MILLER & K. OJIMA. 1968. Nutrient requirements of suspension cultures of soybean root cells. Exp. Cell Res. **50:** 151–158.
14. GOLDSTEIN, W. E. 1983. Large-scale processing of plant cell culture. Ann. N.Y. Acad. Sci. **413:** 394–408.
15. AYERS, A. R., J. EBEL, B. VALENT & P. ALBERSHEIM. 1976. Host-pathogen interactions—X. Fractionation and biological activity of an elicitor isolated from the mycelial walls of *Phytophthora megasperma* var. *sojae*. Plant Physiol. **57:** 760–765.
16. ANDERSON-POUTRY, A. J. & P. ALBERSHEIM. 1975. Host-pathogen interactions—VIII. Isolation of a pathogen-synthesized fraction rich in glucan that elicits a defense response in the pathogen's host. Plant Physiol. **56:** 286–291.
17. AYERS, A. R., E. JURGEN, F. FINELLI, N. BERGER & P. ALBERSHEIM. 1976. Host-pathogen interactions—IX. Quantitative assays of elicitor activity and characterization of the elicitor present in the extracellular medium of cultures of *Phytophthora megasperma* var. *sojae*. Plant Physiol. **57:** 751–759.
18. IKUTA, A. & H. ITOKAWA. 1982. Berberine and other protoberberine alkaloids in callus tissue of *Thalictrum minus*. Phytochemistry **21:** 1419–1421.
19. FRANÇOIS, C., R. D. MARSHALL & A. NEUBERGER. 1962. Carbohydrates in protein—the determination of mannose in hen's-egg albumin by radioisotope dilution. Biochem. J. **83:** 335–341.
20. METZLER, C. M., G. L. ELFRING & A. J. MCEWEN. 1974. A package of computer programs for pharmaco-kinetic modeling. Biometrics **30:** 562–563.
21. HAHLBROCK, K., C. J. LAMB, C. PURWIN, J. EBEL, E. FAUTZ & E. SCHÄFER. 1981. Rapid response of suspension-cultured parsley cells to the elicitor from *Phytophthora megasperma* var. *sojae*. Plant Physiol. **67:** 768–773.
22. HAHLBROCK, K., F. KREUZALER, H. RAGG, E. FAUTZ & D. N. KUHN. 1982. Phytoalexin accumulation through mRNA and enzyme induction in cultured plant cells. *In* Biochemistry of Differentiation and Morphogenesis, p. 33–43. Springer-Verlag. Berlin/New York.
23. SCHMELZER, E., H. BÖRNER, H. GRISEBACH, J. EBEL & K. HAHLBROCK. 1984. Phytoalexin synthesis in soybean. FEBS Lett. **172:** 59–63.
24. SCHRÖDER, J., F. KREUZALER, E. SCHÄFER & K. HAHLBROCK. 1979. Concomitant induction of phenylalanine ammonia-lyase and flavanone synthase mRNAs in irradiated plant cells. J. Biol. Chem. **254:** 57–65.
25. CRAMER, C. L., J. N. BELL, T. B. RYDER, J. A. BAILEY, W. SCHUCH, G. P. BOLWELL, M. P.

ROBBINS, R. A. DIXON & C. J. LAMB. 1985. Coordinated synthesis of phytoalexin biosynthetic enzymes in biologically stressed cells of bean. EMBO J. **4:** 285–289.
26. CHAPPELL, J. & K. HAHLBROCK. 1984. Transcription of plant defense genes in response to UV light or fungal elicitor. Nature **311:** 76–78.
27. SCHUMACHER, H. M. & M. H. ZENK. 1988. Partial purification and characterization of dihydrobenzophenanthridine oxidase from *Eschscholtzia californica* cell suspension cultures. Plant Cell Rep. **7:** 43–46.
28. SCHUMACHER, H. M., H. GUNDLACH & M. H. ZENK. 1987. Elicitation of benzophenanthridine alkaloid synthesis from *Eschscholtzia californica* cell cultures. Plant Cell Rep. **6:** 410–413.

Methods for Cloning Key Primary Metabolic Enzymes and Ancillary Proteins Associated with the Acetone-Butanol Fermentation of *Clostridium acetobutylicum*[a]

JEFFREY W. CARY,[b] DANIEL J. PETERSEN,[b,c]
GEORGE N. BENNETT,[b] AND E. T. PAPOUTSAKIS[d,e]

[b]*Department of Biochemistry*
Rice University
Houston, Texas 77251-1892

[d]*Department of Chemical Engineering*
Northwestern University
Evanston, Illinois 60208

INTRODUCTION

The main interest and applications of the first-generation recombinant DNA technologies have been in the expression of enzymes or polypeptides (of mammalian-cell origin primarily) directly coded by DNA sequences. For future applications, there is great interest in indirectly coded products, such as primary and secondary metabolites, whereby the organisms are literally used as biocatalysts for the transformation of a variety of feedstocks (such as various carbohydrates) into other valuable bulk or specialty chemicals. The approach may be termed metabolic engineering. The objective is to enhance (by protein amplification) existing metabolic pathways, delete undesirable pathways, and create new biosynthetic pathways. Due to the direct effect of the new, amplified, or deleted proteins on cell metabolism (because the proteins participate in pathways of primary or secondary metabolism) and due to the strict regulation mechanisms employed by the cells, metabolic engineering presents unusual challenges and opportunities.

Clostridium acetobutylicum fermentations, aside from their potential industrial importance, are a good model system for understanding the regulation and molecular biology of complex primary metabolism, as well as for applications and studies on metabolic engineering.

Among the first problems that need to be addressed towards achieving beneficial

---

[a]This research was supported by a grant from the National Science Foundation (No. ECE-8613077).
[c]D. J. Petersen was supported by a National Science Foundation graduate fellowship.
[e]To whom all correspondence should be addressed.

genetic manipulations of *C. acetobutylicum* and related species are (i) the ability to efficiently and stably transform these cells with foreign DNA and (ii) the cloning, characterization, and regulation of the important genes necessary for the formation of acid and solvent products. Our laboratories are engaged in solving both of these critical problems and we report here on our efforts to clone and characterize important genes for acid and solvent product formation.

Over the years, three basic strategies for identifying cloned genes have been developed based on functional protein expression, immunological screening, or nucleic acid hybridization. Traditionally, gene cloning has been accomplished through complementation of genetically defined mutants because this method is an inherently simple process, allowing identification by either positive or negative selection for the gene of interest based on expression of the foreign protein. Screening by direct genetic selection, by the observation of a detectable phenotype by replica plating or on indicator plates, or by direct enzyme activity assays of cell extracts allows a multiplicity of detection methods to be fashioned.

Immunological screening methods are often employed when genetically defined mutants are not available, when inadequate protein expression limits complementation ability, or when the protein confers no detectable phenotype upon the recipient strain. Although existing high copy number vectors or expression vectors can aid expression, there is often the need to extensively manipulate the cloned DNA fragment in order to achieve increased expression of the foreign protein. Problems may also be encountered when the inserted gene lacks necessary regulatory control genes. Additionally, the absence of specific cofactors required for enzyme activity in the host cell may hinder complementation or enzyme detection. Immunological detection using polyclonal or more defined monoclonal antibodies raised to the particular protein allows testing for the presence of even minute quantities of the protein regardless of its activity. Lysed colonies, phage plaques, and fractionated cell extracts can all be used in filter-based screening protocols. The size and number of antigenic components can be determined using Western blots of SDS-polyacrylamide gel–separated proteins. However, antibody screening can give a positive signal if only an antigenic portion of the protein is present. This is useful, but it provides only a first step towards successful cloning of the desired coding region. Other problems include degradation of certain foreign proteins by the host and possible cross-reactivity of the antibody with other host proteins. The technique may also fail in certain instances where the protein is toxic to the cell or if the cloned DNA segment codes for only one subunit of the protein that is not very antigenic.

Identification of genes by hybridization with radioactivity-labeled oligonucleotide probes circumvents many problems inherent in the other two methods because it relies solely on the presence of the proper DNA segment. It is thus independent of protein expression and function. Provided with an oligonucleotide, or mixture of oligonucleotides, of sufficient specificity, a gene can be recognized by its nucleotide base sequence alone. Oligonucleotides for use as probes are synthesized based either on a known region of the protein of interest (e.g., an amino-terminal sequence) or on the known sequence of a similar protein from another organism with expected high homology to the protein of interest.

APPLICATION IN CLONING CLOSTRIDIAL GENES

Several clostridial genes for solvent and acid production have already been cloned using variations of the traditional complementation approach. Contag and Rogers[1] have reported the cloning of the *C. acetobutylicum* B643 butyraldehyde dehydrogenase gene by positive selection of a known *E. coli* mutant. An *hfl* A150Tn*10* derivative of *E. coli* PRC436, an aldehyde dehydrogenase–deficient mutant (*adh*E) with normal alcohol dehydrogenase activity, was infected with a *C. acetobutylicum* genomic library constructed in lambda *gt*11. Screening was conducted by plating host cells on an indicator media composed of M9-minimal media containing butanol and tetrazolium (TTC). Colonies that became red on this media, but not on media containing ethanol in place of butanol were thus shown to contain clostridial DNA expressing a butanol-specific aldehyde dehydrogenase.

Youngleson *et al.*[2] reported the cloning of an alcohol dehydrogenase gene from *C. acetobutylicum* P262 utilizing a "suicide substrate" selection technique in *E. coli*. Alcohol dehydrogenase–defective mutants of *E. coli* HB101 were generated by selection for resistance to allyl alcohol. One mutant was shown to lack NAD-dependent butanol dehydrogenase activity. This mutant, HB101-*adh*1, was transformed using a pEcoR251 plasmid library of *C. acetobutylicum* genomic DNA, and clones harboring *adh* genes were identified by enhanced sensitivity to allyl alcohol.

In our laboratory, the butyrate kinase (BK) and phosphotransbutyrylase (PTB) genes have also been cloned by complementation of an *E. coli* mutant.[3] Whereas wild-type *E. coli* is normally unable to grow on butyrate, regulatory mutations that alter the expression of fatty acid degradation enzymes allow the short-chain fatty acids to be used as the sole carbon source.[4] The inability of the *ato* D mutant LJ32 to utilize butyrate was shown to be due to a defect in the enzyme acetyl-CoA:acetoacetyl-CoA transferase (AA-CoA transferase).[5] We constructed a *C. acetobutylicum* plasmid library by inserting fragments generated by partial *Sau* 3*A* 1 digestion of genomic DNA into the *Bam* H1 site of pBR322. This plasmid library was used to transform competent LJ32 cells, which were subsequently plated on M9-minimal agar containing butyrate as the sole carbon source. An ampicillin-resistant (Ap^R) transformant able to grow on butyrate alone (But^+) was selected and sonicated cell extracts from cultures of this transformant were used to test for AA-CoA transferase activity. Assays for this enzyme activity were negative, whereas assays for phosphotransbutyrylase and butyrate kinase were positive, indicating that complementation of the enzymatic deficiency of the mutant was due to a reversal of the BK-PTB pathway normally operating in *C. acetobutylicum* (FIGURE 1). Restriction enzyme analysis of plasmid DNA isolated from this But^+ transformant showed the presence of a plasmid of ~8.7 kb. A Southern blot experiment demonstrated that the recombinant plasmid, designated pJC7, contained ~4.4 kb of *C. acetobutylicum* DNA inserted into the *Bam* H1 site of pBR322 (FIGURE 2). Tn*5* mutagenesis further localized the BK-PTB genes to a small 2.3-kb region of the insert and suggested that the genes were being transcribed as an operon. SDS-polyacrylamide gel electrophoresis confirmed that the two proteins, phosphotransbutyrylase and butyrate kinase, were encoded by the plasmid and were presumably transcribed from a promoter of clostridial origin (FIGURE 3).

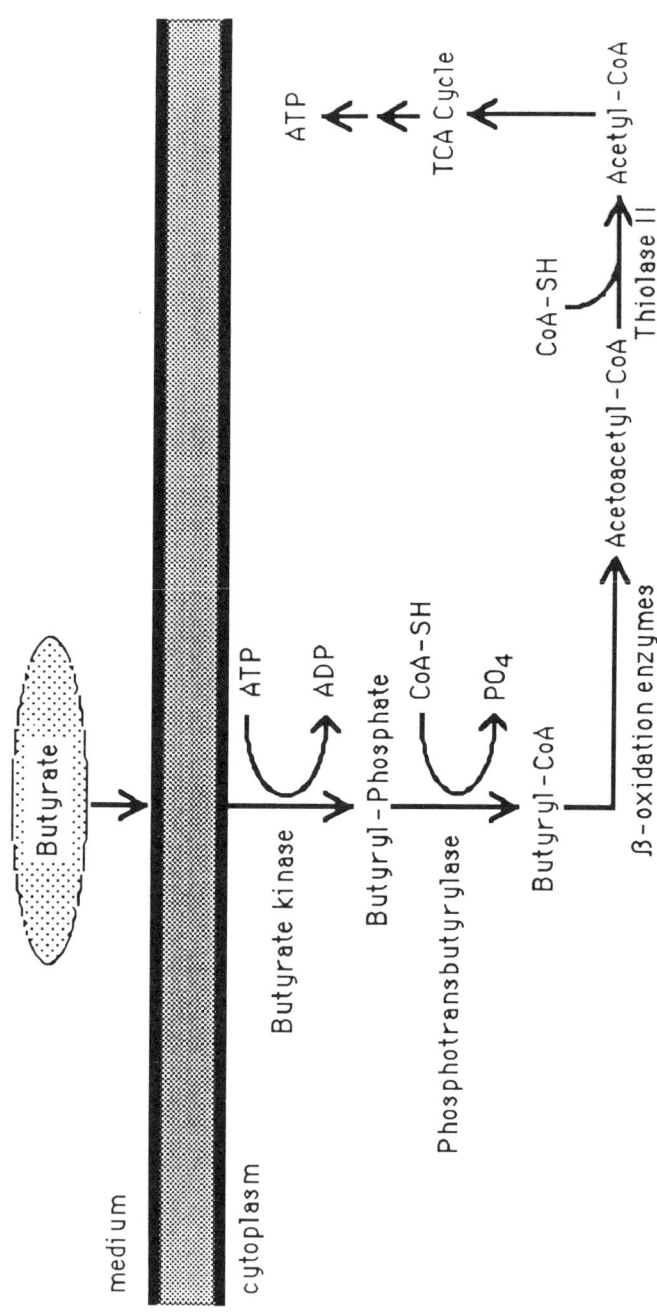

FIGURE 1. Schematic representation of butyrate metabolism by *E. coli* carrying cloned PTB and BK genes of *C. acetobutylicum*.

Although Youngleson et al.[2] were able to generate an *E. coli* mutant for use in complementation and we were able to indirectly select for genes coding for proteins involved in acid production via complementation, such methods are not always possible or experimentally feasible. Indeed, the uniqueness of some clostridial metabolic pathways, such as the conversion of acetoacetyl-CoA to acetone, does not allow for the use of *E. coli* mutants in complementation studies, thus making oligonucleotide probe screening the logical method of choice.

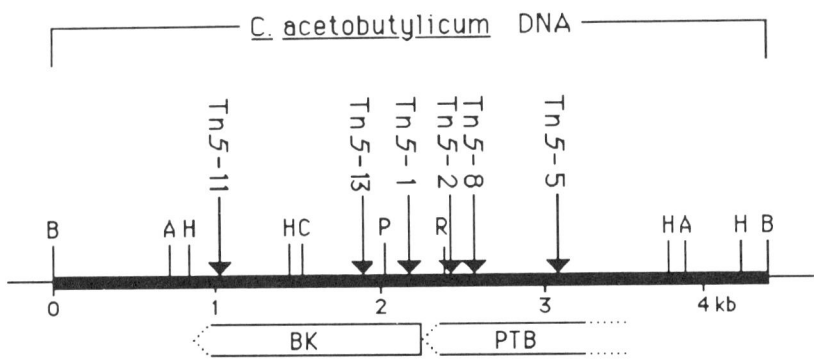

FIGURE 2. Physical and Tn5 insertion map of pJC7. The locations of restriction endonuclease sites within the 4.4-kb *Bam* H1 insert of *C. acetobutylicum* DNA are shown. pJC7:Tn5 mutants were isolated by infecting cells harboring the plasmid pJC7 with λb221 (*rex*::Tn5)[3] and testing for ampicillin and kanamycin resistance and the loss of PTB or BK activity. The locations of Tn5 insertion sites were determined by restriction mapping to within ±50 bp. The approximate locations of the PTB and BK structural genes as well as their presumptive direction of transcription are indicated. Abbreviations: B, *Bam* H1; A, *Ava* I; H, *Hin*d III; C, *Cla* I; P, *Pvu* II; and R, *Eco* R V.

OLIGONUCLEOTIDE HYBRIDIZATION METHODS

Probe Design

To begin construction of an oligonucleotide, we must first have an amino acid sequence for the protein of interest (FIGURE 4). This is most easily accomplished by obtaining an amino-terminal sequence if the protein has been purified. However, the sequence used can be virtually any portion of the peptide without regard to location. In fact, active-site sequences reported in earlier enzymological experiments are quite sufficient and may provide a higher degree of specificity. If none of these are available, the amino acid sequence from a conserved region of the protein from a related species might be considered.

Having obtained the sequence from the protein of interest, several parameters must be evaluated before selecting an oligonucleotide. The "reverse translation" of the amino acid sequence is the critical step in the hybridization screening procedure. Details of the effects of length, G + C content, and self-complementarity have already

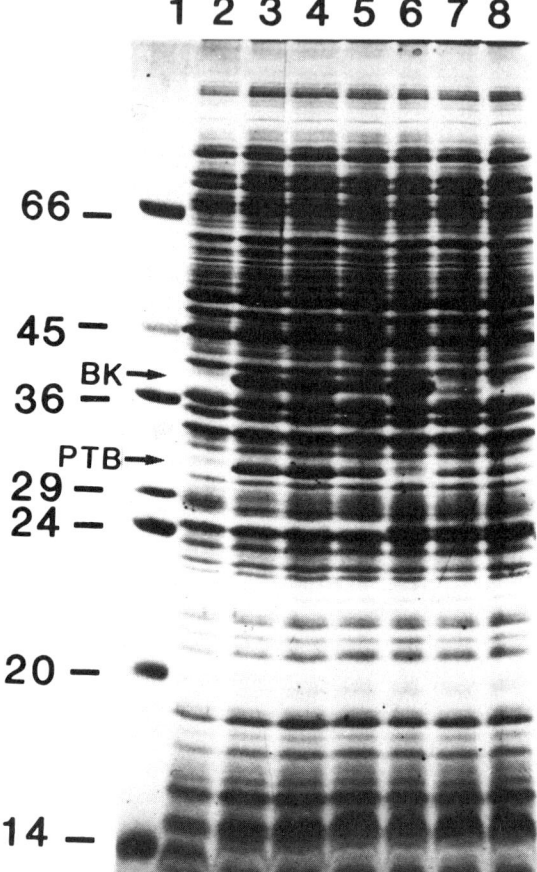

FIGURE 3. Analysis of proteins in whole cell extracts. *E. coli* K2006 harboring pJC7 and its derivatives were grown overnight in Luria broth. Aliquots of cells were lysed in 1× sample buffer and electrophoresed in a 12.5% SDS-polyacrylamide gel. Protein bands were visualized by Coomassie blue staining. Lane 1 contains the molecular weight markers (bovine albumin, 66 kDa; ovalbumin, 45 kDa; glyceraldehyde-3-phosphate dehydrogenase, 36 kDa; carbonic anhydrase, 29 kDa; trypsinogen, 24 kDa; trypsin inhibitor, 20.1 kDa; and lactalbumin, 14.2 kDa). Lanes 2–8 are extracts of *E. coli* K2006 harboring pBR322 (lane 2); pJC6, a plasmid containing 13.6 kb of *C. acetobutylicum* DNA from which the 4.4-kb insert of pJC7 was subcloned (lane 3); pJC7*, a plasmid in which the 4.4-kb insert is in the opposite orientation as in pJC7 (lane 4); pJC7 (lane 5); pJC7::Tn5-8 (lane 6); pJC7::Tn5-11 (lane 7); and pJC7::Tn5-13 (lane 8). Arrows point to bands representing phosphotransbutyrylase (PTB) and butyrate kinase (BK). These bands are not found in extracts of *E. coli* K2006 (pBR322) alone.

been examined;[6] however, because *C. acetobutylicum* has a low G+C of ~28%,[7] these parameters must be reassessed.

Obviously, the amino acid sequence that has the least number of possible codons is the easiest to "reverse translate". Protein regions containing high concentrations of methionine or tryptophan are optimal, whereas regions containing leucine, arginine, or serine generally make for poor selection as coding segments for generating oligonucleotide probes because each has six possible codons. From evaluation of the available amino acid sequence, several possible sequences can be derived for further investigation.

A + T bias is considered next to reduce the complexity of the mixture. Due to the high A + T concentration of *C. acetobutylicum* DNA, the majority of bases in the third position of the codon (the "wobble" base) will be adenine or thymidine. Janssen *et al.*[8] reported that the AGA codon was used in 100% of the codons for arginine in the *C. acetobutylicum gln* A gene, and the published sequence demonstrates the codon usage preferences for other amino acids. Accordingly, codon usage charts for bacillus or related clostridial species should be consulted, considering a codon bias to help reduce the degeneracy of the oligonucleotide mixture. However, it must be cautioned that codon usage varies not only from species to species, but between different proteins of the same species as well, depending on the protein's concentration, cellular location, and function.[8-11]

The importance of oligonucleotide probe length has been studied with regard to its effect on hybridization.[12] In general, the longer the probe, the more stringent the hybridization conditions can be, resulting in higher specificity of the oligonucleotide-DNA interaction. The formula for stability of duplexes formed by oligonucleotide

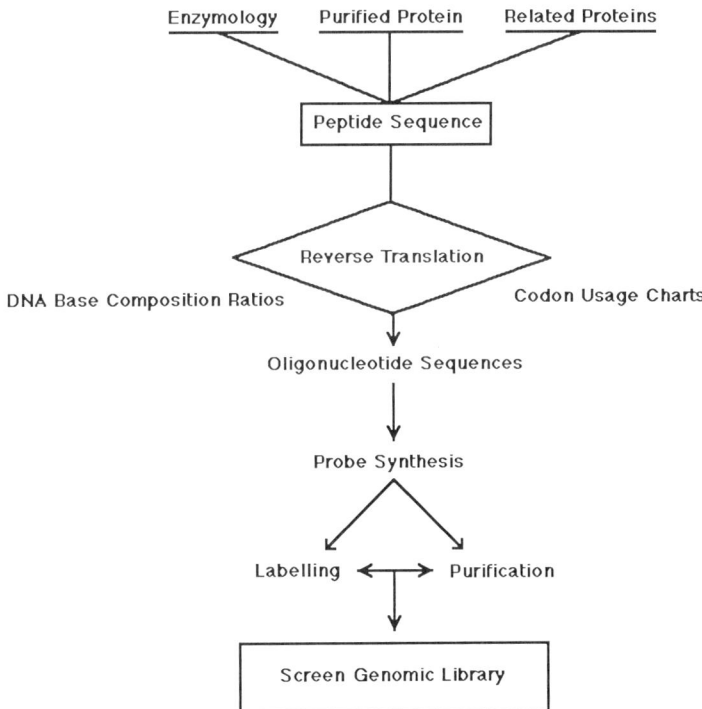

FIGURE 4. Outline of the strategy used to generate oligonucleotide probes for hybridization screening.

hybridization under the buffer conditions generally employed is[13]

$$2°(A + T) + 4°(G + C) = T_M + T_H + 5,$$

where T_M is the melting temperature at which only 50% of the oligonucleotides exist as duplex structures and T_H is the recommended hybridization temperature. A probe with a T_M less than 25 °C would not form sufficient duplexes even at room temperature. With the high A+T content of *C. acetobutylicum*, we have found that probes less than 17 bases in length do not work well. Protocols have been developed for increasing the hybrid stability of A+T probes during washes through the use of TMA (tetramethyl ammonium chloride)[14] to allow the hybridization to be a function of probe length rather than sequence. Sequences enriched in amino acids that have G or C as the first two bases of the codon are preferred in choosing a protein sequence. Still, probe specificity is the primary controlling element and should not be sacrificed in favor of length.

Self-complementarity cannot be avoided with high A+T content, but it does not appear to be a problem in our studies because the hybridization temperature utilized appears to easily "melt" these short, partially mismatched duplexes. Finally, because relatively low hybridization and wash temperatures are used, the oligonucleotide mixture should be as unambiguous as possible. As the complexity increases, the hybridization specificity decreases due to the possibility of other sequences in the mixed probe annealing to areas of the chromosome that have a nearly complementary sequence. In addition, the abundance of the single correct probe in the mixture is lessened considerably, leading to the need for the addition of higher amounts of total label to the hybridization solution to obtain the desired positive signal and introducing a concomitant increase in background.

Oligonucleotides were synthesized using a Cyclone 8600 DNA synthesizer (Biosearch) and were purified by electrophoresis on denaturing 10% polyacrylamide gels, followed by UV shadowing to visualize the bands, excision of the bands, and elution of the oligonucleotides overnight by shaking at room temperature.[15] To remove all traces of polyacrylamide, the eluted oligonucleotides were column purified using DE-51 (Whatman) to bind the probe and using 2 M NH_4OH to elute the DNA from the column. The purified mixture was radiolabeled with [γ-^{32}P]-ATP (sp. act. ~ 3000 Cu/mmol, ICN) using T4-polynucleotide kinase (Promega Biotech) by the method of Maxam and Gilbert.[16] Probe purity was reconfirmed by electrophoresis on denaturing 20% polyacrylamide gels followed by gel autoradiography.

Hybridization Conditions

As stated earlier, present research being conducted in our laboratory is concerned with the cloning and characterization of genes required for the production of acids and solvents in *C. acetobutylicum*, with the long-term goal of gaining a better understanding of the regulation of these enzyme pathways.

Initial experiments were designed to clone the gene encoding CoA-transferase of this organism. Very little is known about the regulatory mechanisms that control the switch from the acidogenic to solventogenic phase of growth in *C. acetobutylicum*. Various groups have suggested that expression of CoA-transferase plays an important role in this metabolic event.[17,18] In butanol-forming clostridia, the CoA-transferase serves as a survival mechanism by detoxifying the medium through the removal of acetate and butyrate excreted during the acidogenic phase of growth.[18] This provides acetoacetate for decarboxylation to acetone and butyryl-CoA for subsequent conversion to butanol during solventogenesis.

Previously, we attempted to clone the CoA-transferase gene of *C. acetobutylicum* 824 via complementation of *E. coli* mutants defective in acetyl-CoA:acetoacetyl-CoA transferase (*ato*⁻). The *ato* mutants were transformed with a *C. acetobutylicum* genomic library constructed in the plasmid pBR322. Numerous screenings were conducted on media selective for transformants expressing CoA-transferase activity, but these failed to produce the desired clone. An alternative procedure was then utilized in which synthetic deoxyoligonucleotides containing sequences deduced from the amino terminus of each subunit of the CoA-transferase were used to screen a *C. acetobutylicum* genomic library constructed in the lambda phage EMBL3. The purified protein is a heterotetramer with subunit molecular weights of approximately 23 and 26 kDa.[19] Approximately 5 μg of each subunit was required to obtain the amino-terminal amino acid sequence utilizing the methods of Matsudaira.[20] Owing to the degeneracy of the genetic code, but often times biasing the "wobble" position of the codon towards A or T, it was determined that the 23-kDa subunit oligonucleotide probe consisted of a 4-fold degenerate 18-mer sequence with a 78% AT content. The 26-kDa subunit oligonucleotide probe was a 12-fold degenerate 20-mer with an 87% AT content. The oligonucleotide probes were synthesized and purified as described in the previous section.

Probe specificity was tested using genomic DNA dot blots of positive and negative controls bound to nitrocellulose filters (0.45 μm, Schleichter & Schuell, Incorporated). In our case, *C. acetobutylicum* 824 DNA was the positive control, whereas phage vector and recipient host cell DNAs were the negative controls. It is extremely important that conditions are established such that the oligonucleotide does not hybridize to the host or vector used to screen the genomic library. The DNA was heat-denatured and then dotted onto the filters. After air-drying, the filters were soaked in 2× SSPE (0.36 M NaCl; 20 mM NaPO$_4$, pH 7.7; 2 mM EDTA) and placed in heat-sealable pouches. Prehybridization solution consisting of 6× SSPE containing 0.5% SDS and 5× Denhardt's solution[21] was added to the pouch, which was then sealed and incubated with gentle agitation for 30 min at 34 °C. The prehybridization solution was then removed and replaced with fresh solution, along with 100 μg/mL denatured sheared salmon sperm DNA and ~10⁶ cpm radiolabeled oligonucleotide probe, followed by overnight incubation at the desired temperature. Hybridizations were performed at several different temperatures including the T_H to determine the optimal hybridization temperature (34 °C for the 23- and 26-kDa CoA-transferase probes) for the probe of interest. Because noncomplementary oligonucleotide:DNA duplexes are more easily destabilized by high temperatures, the hybridization temperature used should be as high as possible. Following hybridization, the nitrocellulose filters were washed in a solution of 2× SSPE containing 0.1% SDS for 20 min at the hybridization temperature. If further washing was needed, the temperature was raised in small increments or the SSPE concentration was lowered (or both). Following washing, the filters were air-dried, marked with ink for alignment purposes, and then autoradiographed. When the hybridization temperature and wash stringency have been optimized, it should be possible to obtain relatively strong signals from the positive control with little background.

Having tested the specificity of the oligonucleotide probe and established stringency conditions, hybridizations to blots of *C. acetobutylicum* DNA were performed by the method of Southern.[22] Genomic DNA was completely digested with various restriction enzymes and separated by electrophoresis on 0.8%–1.0% agarose gels. After transfer to nitrocellulose filters, the filters were hybridized using the parameters already optimized with the DNA dot blots. Both the hybridization temperature and washing conditions may need to be adjusted somewhat from the conditions used for the genomic DNA dot blots in order to achieve the proper stringency (FIGURE 5). Upon washing under optimal conditions, a single defined band in each lane is obtained. If more than one band consistently appears, it is either because some labeled component in the probe hybridizes equally well with another gene or because the sequence of interest is present in multicopy form.

SCREENING PROCEDURES

A phage library was prepared by partially digesting *C. acetobutylicum* DNA with *Sau* 3A 1 to generate fragments in the 9–30 kb region. The digested chromosomal DNA was ligated to the *Bam* H1-cleaved, phosphatased arms of lambda EMBL3 (Promega Biotech) and the recombinant phage was assembled using Packagene packaging extracts (Promega Biotech). The library was titered by infecting *E. coli* NM 534 and then small aliquots were used to infect *E. coli* NM 519 cells to amplify the library. The phage library was prepared for screening by infecting NM 519 cells with enough phage to produce nearly confluent plaques. Phage DNA was bound to nitrocellulose filters using the plaque lift method of Benton and Davis.[23] Filters were baked and prepared for hybridization as described previously. Washes were performed as for Southern hybridization experiments with some modifications to account for the differences in DNA concentrations and the background A+T content differences that affected binding stringency (FIGURE 5).

Individual plaques corresponding to strong hybridization signals from autoradiographs were taken from their plates as agar plugs and transferred into one mL of lambda diluent (per liter: Tris base, 1.21 g; $MgSO_4$, 1.04 g; gelatin, 0.1 g; pH 7.4). These individual phage stocks were again used to infect *E. coli* at a concentration that would give approximately 100–200 plaques per plate. Phage DNA was again transferred to filters and then hybridized, washed, and autoradiographed as before. This plaque purification procedure was repeated (usually one time) until all plaques gave a positive hybridization signal (FIGURE 6). DNA was isolated from the phage of interest using the plate lysate method described by Maniatis *et al.*[24] The phage DNA was digested with *Hin*d III and electrophoresed followed by a Southern blot hybridized with the 23-kDa oligonucleotide probe. As expected, the autoradiograph in most cases demonstrated strong hybridization signals from the phage DNA that corresponded to the signal obtained from complete *Hin*d III digestion of *C. acetobutylicum* genomic DNA (FIGURE 7).

A restriction fragment derived from the insert of one of these positive phages was shown to hybridize with both the 23-kDa and 26-kDa probes. Subcloning this fragment into the plasmid vector pUC19 resulted in significant expression of CoA-transferase activity in an *E. coli* AA-CoA transferase mutant. Additional studies are being undertaken to confirm the origin of this DNA fragment.

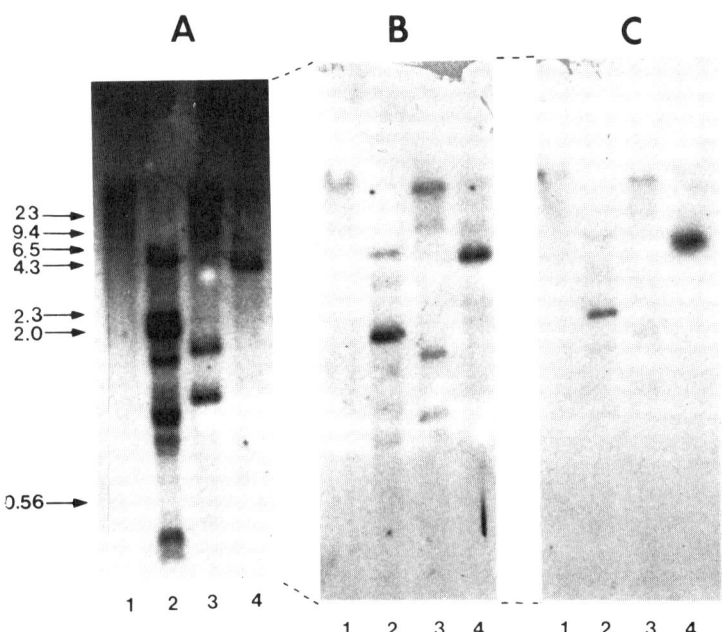

FIGURE 5. Thermal instability of noncomplementary oligonucleotide–chromosomal DNA duplexes. *C. acetobutylicum* genomic DNA was digested with the restriction enzymes, *Kpn* I, *Hin*d III, *Eco* RI, and *Eco* RV, and the resulting fragments were electrophoresed on a 1.0% agarose gel (lanes 1–4, respectively). Following transfer to nitrocellulose filters as described in the text, hybridization was performed with a 16-fold degenerate mixture of 44-mer oligonucleotides whose synthesis was based on the amino-terminal amino acid sequence of a purified *C. acetobutylicum* surface protein and end-labeled using [γ-^{32}P]-ATP. After hybridization overnight at 34 °C using a buffer of 6× SSPE containing 0.5% SDS and 5× Denhardt's solution, filters were washed to remove nonspecific hybrids. Washes were conducted in 2× SSPE containing 0.1% SDS with gentle agitation. Wash temperatures and times were varied: panel A, 25 °C for 25 min; panel B, as in A with an additional wash at 37 °C for 15 min; and panel C, as in B with an additional wash at 45 °C for 15 min. The sizes of *Hin*d III–digested lambda DNA markers in kilobases (kb) are shown at left.

SUMMARY

The unavailability of genetically defined mutants for complementation has intensified the problems inherent in cloning genes from *C. acetobutylicum*. The uniqueness of some of the pathways of this organism coupled with the relative inefficiency of

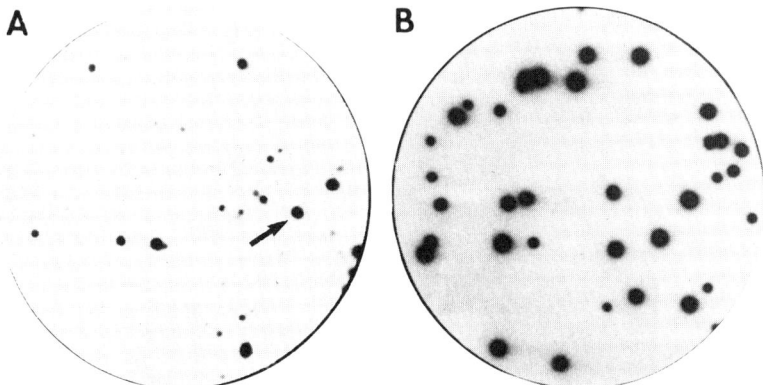

FIGURE 6. Plaque purification of recombinant phage. A *C. acetobutylicum* DNA library was generated in the lambda phage EMBL3 and was then used to infect *E. coli* NM 519 cells for plaque hybridization screening as described in the text. Phage DNA bound to nitrocellulose filters was hybridized using a 4-fold degenerate mixture of [γ-^{32}P]-ATP–labeled 18-mer oligonucleotides, whose synthesis was based on the amino-terminal amino acid sequence of the purified 23-kDa subunit of *C. acetobutylicum* CoA-transferase. Hybridization overnight at 34 °C was followed by three washes at 46 °C for 30 min in 2× SSPE containing 0.1% SDS followed by a single wash at 51 °C for 20 min using the same strength wash buffer. Phages from plaques that gave a positive hybridization signal (A) were subjected to further plaque purification as described in the text. The recombinant phage identified by the arrow in part A was successfully purified, as evidenced in part B where all plaques of the original isolate gave a positive hybridization signal.

transformation of clostridia and few characterized mutants in these pathways have made cloning these genes by traditional complementation methods impractical.

Oligonucleotide hybridization techniques have been shown to circumvent many problems involved in detecting protein expression. The ease of hybridization screening of plaques allows phage libraries to be examined more readily than is generally the case with colony screening techniques. Recombinant lambda phages also contain more DNA per insert than most plasmid vectors can maintain, thus further decreasing the amount of screening necessary. Cosmid libraries, offering even greater length of individual inserts, can be screened in a similar manner, although such screening incorporates the limitations of colony screening techniques.

It is true that the technique hinges on the ability to obtain an amino acid sequence from which an oligonucleotide can be designed. In the past, the ability to obtain sequences was limited because the quantity and number of purified proteins were limited or the proteins were amino-terminally blocked. However, recent technological advances in this area, such as high-resolution gel separation techniques coupled with microsequencing, have opened the door to proteins previously inaccessible. Deformylation methods have been developed to deblock amino-terminally formylated proteins,[20,25] and successful internal amino acid sequence analysis by *in situ* protease digestion[26] has also been reported using only picomolar quantities of proteins separated by one- or two-dimensional gel electrophoresis. Protein and DNA sequence data banks have been significantly upgraded in the past few years. A proposed oligonucleotide sequence can be evaluated to determine what other possible sequences have similar homology; moreover, protein similarity comparisons between related species might possibly supplant the need for protein isolation if regions of highly conserved amino acid sequences are found.

To our knowledge, this represents the first reported use of oligonucleotide probe hybridization screening technology as a strategy for cloning solvent pathway genes of *C. acetobutylicum*. Despite the deleterious effects on hybridization inherent in the high

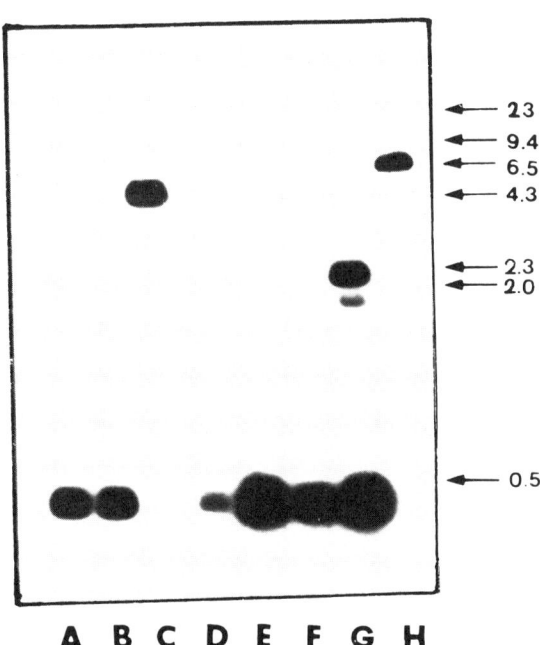

FIGURE 7. Southern hybridization to recombinant phage DNA. Phage DNA was isolated from plate lysates using the method of Maniatis et al.[24] DNA from six different positive phages was digested with *Hin*d III and electrophoresed on a 0.8% agarose gel. Following Southern transfer to nitrocellulose filters, hybridization was performed using a [γ-^{32}P]-ATP–radiolabeled mixture of the 23-kDa 18-mer oligonucleotides of the CoA-transferase. Hybridization was performed at 34 °C overnight followed by three washes in 2× SSPE containing 0.1% SDS for 30 min each at 37 °C and then one wash in 1× SSPE containing 0.1% SDS for 20 min at 42 °C. Comparison of the hybridization bands from the six recombinant phages (lanes A, B, C, E, F, and G) to that of the *C. acetobutylicum* genomic DNA digest (lane D) shows that four of the six phage isolates contain a restriction fragment of ~0.4 kb corresponding to the same size restriction fragment that yields a positive hybridization signal in the lane containing the genomic digest. Lane G shows two upper bands of lower signal intensity, which may correspond to incompletely digested phage DNA. Lane C shows an upper band of ~5.2 kb, which may correspond to the proper 0.4-kb fragment fused with one of the phage lambda arms; this suggests that the hybridizing sequence in this phage is located near the end of the insert. *Hin*d III–digested lambda DNA standards were run in lane H and their sizes in kb are shown at left. The oligonucleotide probe appears to have some homology with the 6.5-kb restriction fragment of phage lambda, but this region is not present in the arms of phage EMBL3.

A+T content of *C. acetobutylicum* gene specific–directed oligonucleotides, the technique has been shown to function with few modifications to previously recorded systems.

ACKNOWLEDGMENTS

We thank D. Wiesenborn for a sample of purified CoA-transferase from *C. acetobutylicum* and R. Cook for the protein sequence analysis.

REFERENCES

1. CONTAG, P. R. & P. ROGERS. 1988. The cloning and expression of the *Clostridium acetobutylicum* B643 butyraldehyde dehydrogenase by complementation of an *Escherichia coli* aldehyde dehydrogenase deficient mutant. Abstract 88th Annual Meeting, Amer. Soc. Microbiol. H-146, p. 169.
2. YOUNGLESON, J. S., J. D. SANTANGELO, D. T. JONES & D. R. WOODS. 1988. Cloning and expression of a *Clostridium acetobutylicum* alcohol dehydrogenase gene in *Escherichia coli*. Appl. Environ. Microbiol. **54:** 676–682.
3. CARY, J. W., D. J. PETERSEN, E. T. PAPOUTSAKIS & G. N. BENNETT. 1988. Cloning and expression of *Clostridium acetobutylicum* phosphotransbutyrylase and butyrate kinase genes in *Escherichia coli*. J. Bacteriol. **170:** 4613–4618.
4. PAULI, G. & P. OVERATH. 1972. ato operon: a highly inducible system for acetoacetate and butyrate degradation in *Escherichia coli*. Eur. J. Biochem. **29:** 553–562.
5. JENKINS, L. S. & W. D. NUNN. 1987. Genetic and molecular characterization of the genes involved in short-chain fatty acid degradation in *Escherichia coli*: the ato system. J. Bacteriol. **169:** 42–52.
6. WALLACE, R. B. & C. G. MIYADA. 1987. Oligonucleotide probes for the screening of recombinant DNA libraries. Methods Enzymol. **152:** 432–442.
7. CUMMINS, C. S. & J. L. JOHNSON. 1971. Taxonomy of the clostridia: wall composition and DNA homologies in *Clostridium butyricum* and other butyric acid–producing clostridia. J. Gen. Microbiol. **67:** 33–46.
8. JANSSEN, P. J., W. A. JONES, D. T. JONES & D. R. WOODS. 1988. Molecular analysis and regulation of the *gln*A gene of the gram-positive anaerobe *Clostridium acetobutylicum*. J. Bacteriol. **170:** 400–408.
9. GRAVES, M. C., G. T. MULLENBACH & J. C. RABINOWITZ. 1985. Cloning and nucleotide sequence determination of the *Clostridium pasteurianum* ferredoxin gene. Proc. Natl. Acad. Sci. U.S.A. **82:** 1653–1657.
10. GREPINET, O. & P. BEGUIN. 1986. Sequence of the cellulase gene of *Clostridium thermocellum* coding for endoglucanase B. Nucleic Acids Res. **14:** 1791–1799.
11. JOLIFF, G., P. BEGUIN & J-P. AUBERT. 1986. Nucleotide sequence of the cellulase gene *cel* D encoding endoglucanase D of *Clostridium thermocellum*. Nucleic Acids Res. **14:** 8605–8613.
12. WALLACE, R. B., J. SHAFFER, R. F. MURPHY, J. BONNER, T. HIROSE & K. ITAKURA. 1981. Hybridization of synthetic oligodeoxyribonucleotides to ØX174 DNA: the effect of single base pair mismatch. Nucleic Acids Res. **11:** 3543–3557.
13. SUGGS, S. V., T. HIROSE, T. MIYAKE, E. H. KAWASHIMA, M. J. JOHNSON, K. ITAKURA & R. B. WALLACE. 1981. Using purified genes. ICN-UCLA Symp. Dev. Biol. Volume 23. D. D. Brown, Ed.: 683–693. Academic Press. New York.
14. DILELLA, A. G. & S. L. C. WOO. 1987. Hybridization of genomic DNA to oligonucleotide probes in the presence of tetramethyl ammonium chloride. Methods Enzymol. **152:** 447–451.
15. VERMERSCH, P. V. & G. N. BENNETT. 1988. Synthesis and expression of a gene for a mini type II dihydrofolate reductase. DNA **7:** 243–251.

16. MAXAM, A. M. & W. GILBERT. 1980. Sequencing end-labeled DNA with base-specific chemical cleavages. Methods Enzymol. **65:** 499–560.
17. YAN, R-T., C-X. ZHU, C. GOLEMBOSKI & J-S. CHEN. 1988. Expression of solvent-forming enzymes and onset of solvent production in batch cultures of *Clostridium beijerinkii* (*"Clostridium butylicum"*). Appl. Environ. Microbiol. **54:** 642–648.
18. HARTMANIS, M. G. N., T. KLASON & S. GATENBECK. 1984. Uptake and activation of acetate and butyrate in *Clostridium acetobutylicum*. Appl. Microbiol. Biotechnol. **20:** 66–71.
19. WIESENBORN, D. P., F. B. RUDOLPH & E. T. PAPOUTSAKIS. 1989. Coenzyme A transferase from *Clostridium acetobutylicum* ATCC 824 and its role in the uptake of acids. Appl. Environ. Microbiol. **55:** 323–329.
20. MATSUDAIRA, P. 1987. Sequence from picomole quantities of proteins electroblotted onto polyvinylidene difluoride membranes. J. Biol. Chem. **262:** 10035–10038.
21. DENHARDT, D. T. 1966. A membrane-filter technique for the detection of complementary DNA. Biochem. Biophys. Res. Commun. **23:** 641–646.
22. SOUTHERN, E. 1975. Detection of specific sequences among DNA fragments separated by gel electrophoresis. J. Mol. Biol. **98:** 503–517.
23. BENTON, W. D. & R. W. DAVIS. 1977. Screening λgt recombinant clones by hybridization to single plaques *in situ*. Science **196:** 180–182.
24. MANIATIS, T., E. F. FRITSCH & J. SAMBROOK. 1982. Molecular cloning: a laboratory manual. Cold Spring Harbor Laboratory, Cold Spring Harbor, New York.
25. ELSON, N. A., H. B. BREWER & W. F. ANDERSON. 1974. Hemoglobin switching in sheep and goats. III. Cell-free initiation of sheep globin synthesis. J. Biol. Chem. **249:** 5227–5235.
26. AEBERSOLD, R. H., J. LEAVITT, R. A. SAAVEDRA, L. E. HOOD & S. B. H. KENT. 1987. Internal amino acid sequence of proteins separated by one- or two-dimensional gel electrophoresis after *in situ* protease digestion on nitrocellulose. Proc. Natl. Acad. Sci. U.S.A. **84:** 6970–6974.

A Highly Structured Model for Simulation of Batch and Continuous Cultures of *B. subtilis* and Examination of Cellular Differentiation[a]

JINWOOK JEONG AND MOHAMMAD M. ATAAI

Department of Chemical Engineering
University of Pittsburgh
Pittsburgh, Pennsylvania 15261

INTRODUCTION

A comprehensive model for the growth process of the bacterium *Bacillus subtilis* has been developed.[1,2] *Bacillus subtilis* cells undergo an interesting differentiation (sporulation) process in response to some nutrient limitations or adverse environmental conditions. In this context, *B. subtilis* represents a primordial prototype to understand questions of developmental biology. Particularly important is the analogy that can be drawn between sporulation in *B. subtilis* and the mechanisms of differentiation in higher cellular systems. The *B. subtilis* model that we have developed is a highly structured model for the growth process.[1] The model accounts explicitly for 35 cellular components including metabolic intermediates of glycolysis, gluconeogenesis, the TCA cycle, and a detailed representation of purine nucleotide metabolism. The driving force for inclusion of purine metabolism into the model was the hypothesis relating some of the components of nucleotide metabolism to sporulation initiation. One of the benefits of structured models at the whole cell level is the ease of incorporation of details of any subcellular process of interest. The model cell is represented by 35 nonlinear and coupled differential equations containing nearly 200 parameters. A great deal of effort was required to evaluate or extract the model parameters from independent experiments reported in the literature on various aspects of *B. subtilis* growth. In addition to the differential equations representing the net rate of production of model components, the model includes the criteria for cell division, DNA replication, and septum synthesis. These auxiliary functions were the same as in the Cornell model of *E. coli*.[3] The model was used to simulate batch and continuous growth processes. Simulation of the batch growth process agrees with experimental observations.[1] The model is, to our knowledge, the most detailed representation of bacterial cell growth and metabolism that has been attempted. The degree of model complexity depends on the intended application of the model. Models aimed at understanding cellular regulation require far greater detail than those designed for process development and control. Examples of complex models of the growth process include the pioneering work of Shuler's group[3]

---

[a]This research was supported in part by a grant from the Pittsburgh Supercomputing Center through the NIH Division of Research Resources Cooperative Agreement (No. U41 RR04154).

on the development of the Cornell model of *E. coli* and the extended version of the model by Peretti and Bailey.[4] The larger complexity included in our model for *B. subtilis* regarding cellular metabolism is driven by our eventual goal that such a model will be used for understanding the coordination between growth and metabolism that may be characteristic of the differentiation process.

Due to the very complex interrelation among various cellular components, interpretation of experimental results is a very difficult task. A reliable mechanistic model for the cell growth process should be indispensable for better understanding of cellular regulation and eventually the sporulation process. The need for model development, the model structure and parameter evaluation, and simulation results for the postexpo-

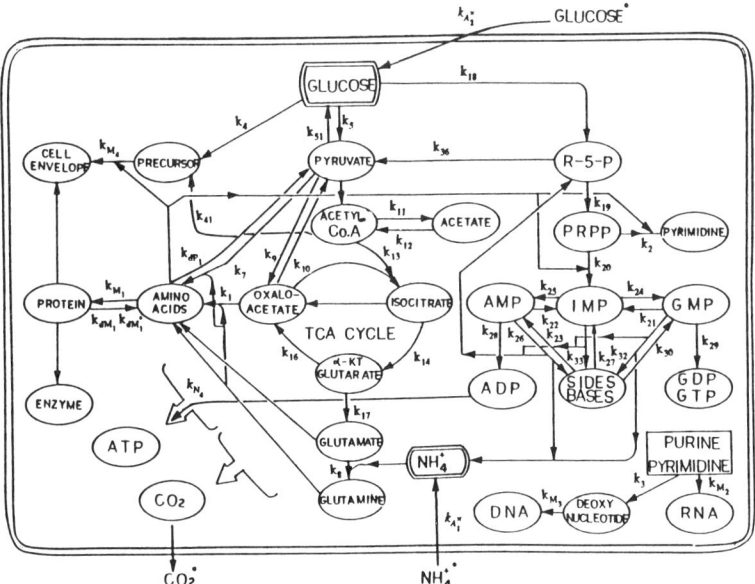

FIGURE 1. The model structure.

nential phase of a batch growth process have been described in reference 1. The values for the model parameters were obtained from an extensive literature search and the model predicted the transition of growth from the exponential to the stationary phase of growth in a synchronized batch culture. In this report, we present the model predictions for *B. subtilis* growth in chemostat and batch cultures.

MODEL STRUCTURE

FIGURE 1 depicts our computer model for *B. subtilis* in glucose minimal media. As examples, the pseudochemical reactions around oxaloacetate and amino acids are described. Oxaloacetate is synthesized from pyruvate by the pyruvate carboxylase reaction and also from α-ketoglutarate as part of the TCA cycle reactions. A portion of oxaloacetate is incorporated into the aspartic family of amino acids and the rest is condensed with acetyl-CoA into citrate (or isocitrate). Following glucose depletion, oxaloacetate can be synthesized from isocitrate via the glyoxylate path. Under this condition, oxaloacetate becomes a gluconeogenic metabolite via PEPCK (phosphoenolpyruvate carboxykinase). Amino acids other than glutamine and glutamate are synthesized from pyruvate and oxaloacetate. Utilization of amino acids takes place through protein, cell envelope, and nucleotide synthesis. Similarly, other interactions among model components can be deduced from FIGURE 1. Because some components of nucleotide metabolism have been speculated to be involved in sporulation initiation, perhaps through activation or derepression (or both) of sporulation-specific sigma factors of RNA polymerase, we have included the description of nucleotide metabolism in our model structure. The structure included in FIGURE 1 for nucleotide metabolism was derived from the literature information on purine nucleotide metabolism with a particular emphasis on the regulation of these enzymes in *B. subtilis*. A detailed discussion regarding the structure of the model can be found in references 1 and 2.

MODEL EQUATIONS AND PARAMETER EVALUATION

The list of the model components and the equations is given in references 1 and 2. Each equation for a model component contains a synthesis term, a degradation term, and terms representing the incorporation to other cellular components. Equations are written as multiple saturation–type kinetics. Multiple saturation–type kinetics have been found satisfactory for several enzymes and were applied successfully to the *E. coli* model.[3] Some of the saturation and inhibition constants are calculated based on the values of the normal intracellular concentrations. The values for normal intracellular concentrations are also summarized in references 1 and 2. In this section, one representative model equation is briefly described.

The example is for the assimilation of inorganic nitrogen into an organic form by glutamine synthetase. The overall balance for glutamine can be written as

$$\frac{dP_{12}}{dt} = R_{P_{12}P_{14}} - R_{M_1P_{12}},$$

where

$R_{P_{12}P_{14}}$: rate of glutamine (P_{12}) formation from glutamate (P_{14}) and
$R_{M_1P_{12}}$: rate of glutamine (P_{12}) incorporation into protein (M_1).

The reaction involved in $R_{P_{12}P_{14}}$ can be written as

$$NH_4^+ + glutamate + ATP \rightarrow glutamine + ADP + P_i$$

and can be represented by the following equation:

$$R_{P_{12}P_{14}} = k_8 \left(\frac{P_{14}/V}{K_{P_{12}P_{14}} + P_{14}/V}\right)\left(\frac{A_1/V}{K_{P_{12}A_1} + A_1/V}\right)\left(\frac{N_4/V}{K_{P_{12}N_4} + N_4/V}\right)\left(\frac{K_{iP_{12}}}{K_{iP_{12}} + P_{12}/V}\right)V,$$

where

k_8: maximum rate of reaction,
A_1: ammonium ion (mmol/cell),
N_4: ATP (mmol/cell),
$K_{P_{12}P_{14}}$, $K_{P_{12}A_1}$, $K_{P_{12}N_4}$: saturation constants, and
$K_{iP_{12}}$: feedback inhibition constant.

The numerical value of k_8 is estimated as:

k_8 = requirement for protein synthesis + glutamine expansion rate

= 0.16(Wt/V × 0.6 × 0.69/140 × 1000.0) + 5.6/1000 × 0.69

= 0.15 mmol/h-cm^3.

The numerical values are: 140.0 for the average molecular weight of amino acids, 0.16 for the bound glutamine composition in protein,[5] 0.69 for the maximum growth rate (h^{-1}), 5.6 for the normal intracellular concentration (mM), and 0.6 for the protein composition (g protein per g cell).[1,2] The symbols, Wt and V, represent cell dry weight and volume, respectively.

Deuel and Stadtman[6] reported saturation constants for each of the substrates in the above reaction. The values of $K_{P_{12}P_{14}}$, $K_{P_{12}A_1}$, and $K_{P_{12}N_4}$ were given as 8.4×10^{-4}, 4.0×10^{-4}, and 2.0×10^{-5} mmol/cm^3, respectively. Deuel and Prusiner[7] showed that glutamine exerted strong inhibitory effects. The reported value for the intracellular concentration of glutamine[8] is 5.6 mM. We selected a value of 9 mM for $K_{iP_{12}}$ to have the proper inhibitory role. Values equal to intracellular concentrations or higher were chosen for unmeasured inhibition constants. Significant inhibition will result only if the level of the metabolite rises much above its normal intracellular concentration.

RESULTS AND DISCUSSION

Most of our computations have been performed at the Pittsburgh Supercomputing Center on a Cray. We have vectorized the entire program to be consistent with supercomputer codes for efficient utilization of CPU power of supercomputers. Our program is executed efficiently by the Cray.

For the simulation of synchronized batch growth, we define the external glucose concentration and begin the simulation with a low inoculum concentration. The glucose concentration decreases as the result of the glucose uptake rate and the cell mass increases. The simulation proceeds very efficiently during the exponential phase of growth; however, as the glucose level drops to exceedingly low values, the cell growth rate is reduced drastically (stationary phase of growth begins) and the required computation time increases significantly. It should be noted that we have not developed a population model to simulate the batch growth process, but have assumed that all the cells in the population are identical.

An initial extracellular glucose concentration of 1 g/L was used for batch simulations. Three hours after the beginning of simulation, the growth process was stabilized and the culture entered the exponential phase of growth. During the exponential phase, the time-dependent variations of the intracellular concentrations of the model compo-

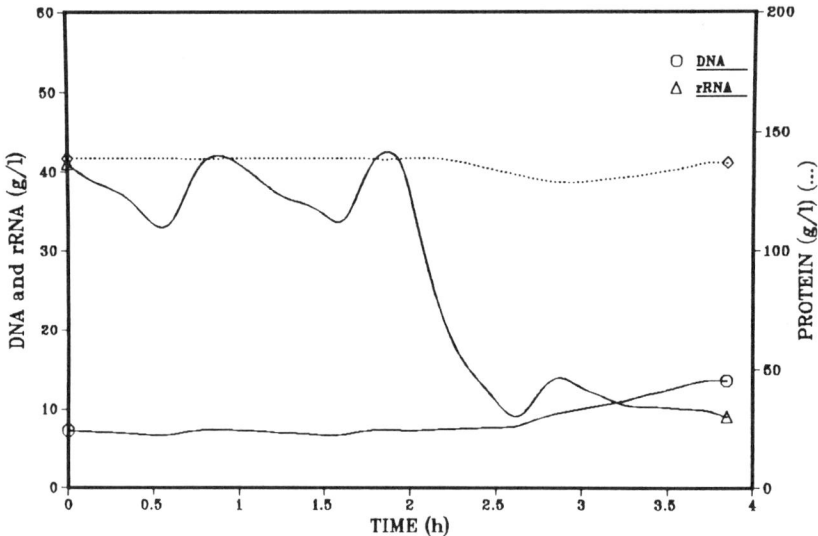

FIGURE 2. Batch culture simulations: variations in intracellular concentrations for protein, RNA, and DNA.

nents were similar from one generation to another. For simulation of a continuous culture, a value for the extracellular glucose concentration was fixed. Steady state is assumed when the concentrations of all the model components as well as cell length, width, volume, the timing of initiation of the chromosome, and cell doubling time are essentially identical at any two corresponding points during the two subsequent cell cycles. The growth rate is calculated using the value of the cell doubling time at steady state. By changing the values for the external glucose concentration, the values of the growth rate can be varied.

FIGURES 2 and 3 show the predicted profiles for the intracellular concentrations of a number of model components for the last four hours of the batch growth process. Time 0.0 in these figures indicates the beginning of the last cell cycle of the exponential phase of growth. The predicted cell cycle time during the exponential phase of growth is one hour. The variation in all the intracellular components during the exponential growth prior to time 0.0 follows the same trend as the one shown for the time interval of 0 to 1. After two hours, the extracellular glucose was almost completely utilized and the glucose concentration reached a very low value, leading to slower growth rates.

The compositions of the macromolecules are presented in FIGURE 2. The overall trends of these curves are in good agreement with the experimental data. Chow and Takahashi[9] observed that the DNA content for *B. subtilis* remained almost constant during the transition period from the exponential to the stationary phase in a batch culture, whereas the RNA content decreased over the same period. A similar observation has also been reported by Nelson and Kornberg[10] for *B. megaterium*. Furthermore, it has been reported that a carbon step-down reduces the stable RNA accumulation due to both the restriction of synthesis and the degradation.[11] The simulation curve for RNA concentration clearly shows the observed trend. FIGURE 3 shows the simulation results for the concentrations of GTP (guanosine 5′-triphosphate), GMP (guanosine 5′-monophosphate), pyruvate, and extracellular glucose. The amplitude of the GTP fluctuation during the exponential phase is higher than that of any other nucleotides. After glucose depletion, cells cannot maintain such a high fluctuation. The GTP level decreases immediately upon glucose depletion (at about 2 h, which corresponds to the time that glucose has been depleted) and shows less severe fluctuations at the lower levels and finally goes up at 3 h. The decrease in GTP level is mainly due to the gluconeogenesis. Glucose-synthesizing reactions require not only ATP as a high-energy molecule, but also GTP. The explanation is more clear if we recognize the coincidence of the timing of pyruvate depletion, which leads to a much lower gluconeogenesis activity and the sudden GTP increase at 3 h. GTP may be consumed in the synthesis of ppGpp (guanosine 3′-diphosphate 5′-diphosphate).[12] However, because the ppGpp pool size is much smaller than GTP,[12] it is unlikely that ppGpp synthesis contributed to such a large decrease in the magnitude of GTP. The predicted trend for the changes in the intracellular concentrations of the majority of the model components (not shown here) is in agreement with experimental observation. Moreover, the predicted values for the intracellular concentrations of a large number of the model components during the exponential phase of growth are reasonably close to the normal intracellular concentrations extracted from the literature.

FIGURES 4 and 5 show the predicted growth rate dependency for a few of the model components in continuous culture. Although there are no experimental observations to compare with the model predictions, the results seem reasonable. For example, the

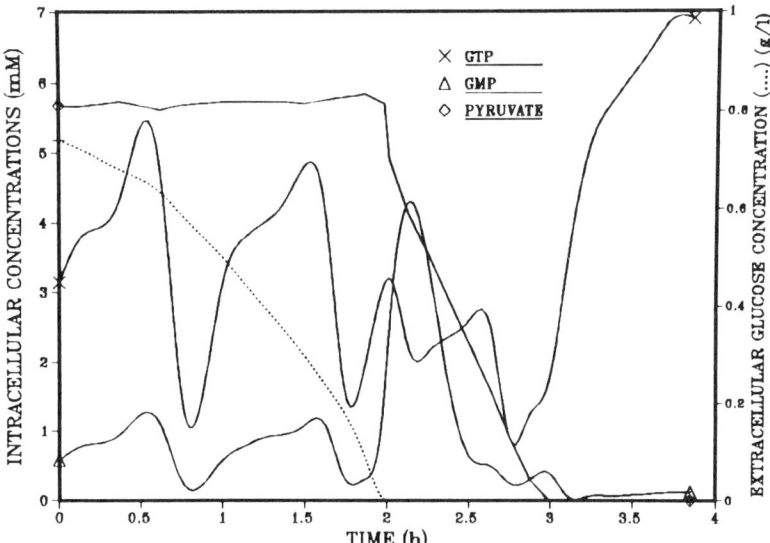

FIGURE 3. Batch culture simulations: variations in intracellular concentrations for GTP, GMP, pyruvate, and extracellular glucose.

drop in GTP level as the growth rate decreases is in general agreement with the observation that a lower level of GTP may be required for the sporulation initiation.

UTILITY OF THE MODEL

We simulated a synchronized batch culture in glucose minimal media.[1] The model simulates the growth process for the postexponential phase of growth. At the end of the exponential phase, the external glucose levels drop to significantly low values, which leads to a dramatic increase in computation time. Simulation, however, continues with no numerical instability. The point here is that the model structure allows for the simulation of the transition of growth from the exponential to the stationary phase and predicts the quantitative changes that occur in values of each model component with time. Thus, the model has the capability to examine complex metabolic issues.

An important application of the model is for testing the sequences of metabolic changes leading to the initiation of sporulation. Particularly important is the existence of various metabolic mutants. Some of the mutations do not affect the sporulation process and some prevent the initiation of sporulation. If a component is identified as a possible molecule for initiation of sporulation, then simulation of various mutants with the capability to sporulate must not affect the predicted trend for that component. In contrast, simulation results for asporogenous mutants must predict a different trend. For example, if lower levels of GTP signal the initiation process, then lower levels of GTP must result from simulation of all the sporogenous mutants, and higher or relatively constant values for GTP concentration should be characteristic of simulation

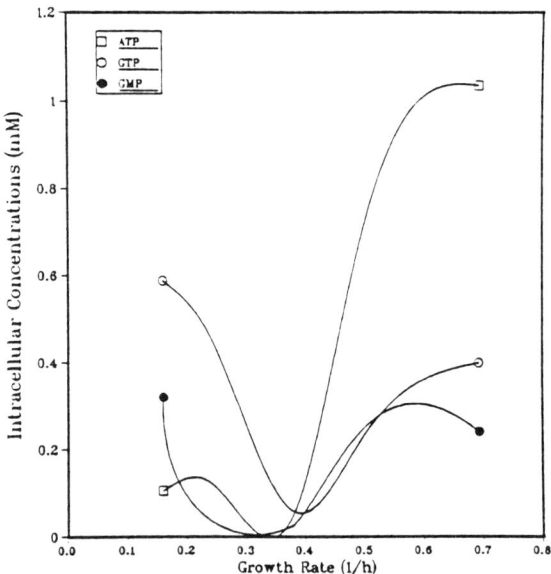

FIGURE 4. Continuous culture simulations: variations in intracellular concentrations for ATP, GTP, and GMP as a function of growth rate.

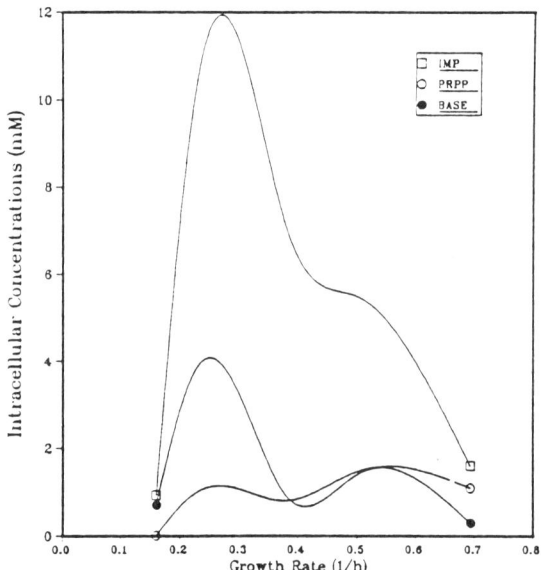

FIGURE 5. Continuous culture simulations: variations in intracellular concentrations for IMP, PRPP, and purine bases as a function of growth rate.

of asporogenous mutants. A large number of mutants in the TCA cycle and of mutants in glycolysis or in part of the purine metabolism pathway can be incorporated into the model. Simulating the behavior of such mutants should reveal metabolic patterns of the mutants that are distinct from the wild type.

This model can also serve as an important tool for metabolic engineering applications. In biochemical processes for an excreted product, higher values for the flux of such products are desirable; on the other hand, for the intracellular components, achieving high levels of intracellular concentration is the target of cellular manipulations. Owing to the significant amount of metabolic information embedded in the model structure, useful predictions can be made for developing strategies to alter either the flux through a reaction sequence or the intracellular concentration of a metabolite. We have applied the model for metabolic engineering applications, and an example of metabolic alterations that could be used for decreasing the rate of formation of organic by-products such as acetate and lactate is discussed elsewhere.[2] We are currently conducting experiments using batch and continuous cultures and are in the process of evaluating the intracellular concentrations of model components under well-defined conditions. Experimental measurements and simulations of various mutants will be the subject of a future manuscript.

REFERENCES

1. JEONG, J. W., J. SNAY & M. M. ATAAI. 1990. A mathematical model for examining growth and sporulation processes of a bacterium *Bacillus subtilis*. Biotechnol. Bioeng. In press.
2. JEONG, J. W. 1989. Development of a highly structured mathematical model for the

metabolism and sporulation of *Bacillus subtilis*. Ph.D. thesis. Department of Chemical and Petroleum Engineering, University of Pittsburgh.
3. DOMACH, M. M., S. K. LEUNG, R. E. CHAN, G. G. COCKS & M. L. SHULER. 1984. Biotechnol. Bioeng. **26:** 203.
4. PERETTI, S. W. & J. E. BAILEY. 1986. Biotechnol. Bioeng. **28:** 1672.
5. NELSON, D. L. & A. KORNBERG. 1970. J. Biol. Chem. **245:** 1128.
6. DEUEL, T. F. & E. R. STADTMAN. 1970. J. Biol. Chem. **245:** 5206.
7. DEUEL, T. F. & S. PRUSINER. 1974. J. Biol. Chem. **249:** 257.
8. BERNLOHR, D. A. & R. L. SWITZER. 1983. J. Bacteriol. **153:** 937.
9. CHOW, C. T. & I. TAKAHASHI. 1972. J. Bacteriol. **109:** 1175.
10. NELSON, D. L. & A. KORNBERG. 1970. J. Biol. Chem. **245:** 1137.
11. GALLANT, J. & R. A. LAZZARINI. 1975. *In* Protein Synthesis—A Series of Advances. E. H. McKonkey, Ed. Dekker. New York.
12. FORTNAGEL, P. & R. BERGMAN. 1974. Biochem. Biophys. Res. Commun. **56:** 264.

PART II. PROCESSES INVOLVING GENETICALLY ENGINEERED ORGANISMS

Effects of Culture Conditions on Plasmid Stability and Production of a Plasmid-encoded Protein in Batch and Continuous Cultures of *Escherichia coli* JM103[pUC8][a]

WEN RYAN AND SATISH J. PARULEKAR
Department of Chemical Engineering
Illinois Institute of Technology
Chicago, Illinois 60616

INTRODUCTION

The realization of the potential offered by recombinant microorganisms for production of large amounts of certain desired proteins requires, among other things, optimization of cultivation conditions. A microorganism containing a recombinant plasmid is compelled to allocate a fraction of its limited resources for maintenance and replication of the plasmid and for synthesis of plasmid-encoded products.[1-5] This metabolic burden leads to a reduction in growth rate of the recombinant cells as the plasmid DNA content or synthesis of plasmid-encoded protein(s) (or both) is increased.[2,4] This resource allocation strongly influences plasmid DNA amplification, expression of plasmid-encoded genes, and stability of the plasmid. In view of this, cultivation conditions supporting better host growth may not bring about high-level expression of plasmid-carried genes.[6,7] As this study shows and a few previous studies[7-10] have indicated, high plasmid copy number levels need not necessarily lead to excessive product formation and increased plasmid stability.

For maximizing expression of cloned genes, it is essential to develop and understand the interrelationships among cell growth, plasmid content, and expression of plasmid-carried genes (formation of plasmid-encoded proteins). These relationships are influenced at a fundamental level by cultivation conditions in a complex fashion. With this in mind, this study was undertaken to investigate the effects of environmental conditions on plasmid stability, plasmid DNA amplification, and production and excretion of a plasmid-encoded protein in batch and continuous cultures of a recombinant microorganism harboring a plasmid at high copy number levels. The environmental parameters studied were culture pH, inorganic phosphate concentration in the medium, and dissolved oxygen level (altered via variation in agitation speed).

---

[a]Partial support of this research was provided by the Engineering Foundation (Grant No. RI-A-86-11), the Amoco Foundation, and the Public Health Service (BRSG Grant No. 2-S07-RR07027-22).

MATERIALS AND METHODS

Bacterial Host and Plasmid

The bacterial host used in this study was *E. coli* JM103.[11] It is sensitive to ampicillin. The plasmid employed, pUC8 (a relaxed plasmid), is harbored in JM103 at high copy numbers and encodes a nonfunctional β-galactosidase besides the β-lactamase, the latter conferring the host resistance to ampicillin. The calcium treatment method due to Maniatis *et al.*[12] was used for host transformation. Stock cultures of plasmid-bearing cells were maintained on LB-Ap plates at 4 °C and LB-Ap-glycerol medium at -80 °C. The host is capable of synthesizing functional β-galactosidase through intragenic complementation in the presence of plasmid pUC8. Plasmid-bearing cells can therefore grow on lactose, whereas plasmid-free cells cannot.

Bioreactor Operation

A Bellco spinner flask with 0.8-L working volume was used for batch experiments, whereas a BioFlo I bioreactor with 0.5-L working volume was used for all continuous culture experiments. A minimal medium M9P (M9 supplemented with 12 g/L Na_2HPO_4 and 6 g/L KH_2PO_4)[6] was used in all experiments except those dealing with the study of phosphate effects. Unless mentioned otherwise, the experiments were conducted in the absence of any selection pressure at 37 °C, pH 7.4, agitation speed of 800 rpm, aeration rates of 5.12 L/min for batch (Bellco) and 2.25 L/min for continuous culture (BioFlo I) experiments, and dilution rate of 0.375 h^{-1} for continuous culture experiments. The total cell mass concentration and the concentrations (number densities) of recombinant and host cells were measured in terms of the optical density of the culture at 550 nm and in terms of colony counts of culture samples on LB and LB-Ap plates.

Enzyme Assay

The activity of the plasmid-encoded protein, β-lactamase, released from the cells by osmotic shock,[6,13,14] was determined by measuring the rate of hydrolysis of cephalothin, a penicillin analogue.[15] The colorimetric method due to King was used for measuring the phosphate concentration in the culture samples.[16,17] Glucose concentration was measured using a YSI (Yellow Springs Instrument) glucose analyzer.

Copy Number Measurement

Agarose gel electrophoresis and densitometry were employed for plasmid DNA measurements.[12] Samples for these measurements, taken in duplicate, were subject to cleavage by *Eco*RI restriction enzyme (Sigma Chemical) for two hours at 37 °C under conditions suggested by the enzyme supplier.

Quantification of plasmid DNA content (plasmid monomer equivalents) was carried out by the method of Maniatis *et al.*[12] Cleaved DNA samples were loaded to a

1.0% agarose gel prepared in TPE buffer (100 mM Tris-base, 2 mM EDTA, pH 8.0).[12] Five standards, containing 20, 37.5, 75, 150, and 300 ng of pUC8, respectively, cleaved in the same manner as the culture samples, were also loaded to each gel. The gel was run horizontally at 3 volts/cm, followed by its staining with ethidium bromide for 30 min and destaining in distilled deionized water for 30 min. The DNA was visualized under UV light and photographs were taken using a Polaroid Type 55 film with an orange filter. The negatives were then scanned on a densitometer (E-C Apparatus Corporation, St. Petersburg, Florida, Model EC910) at 540 nm and the pUC8 DNA content of each sample was determined via the linear correlation obtained from the five internal standards of pUC8 DNA on each gel.

RESULTS AND DISCUSSION

In the experiments discussed next, only one environmental parameter was changed in a particular set of experiments. In batch experiments, the parameter was kept uniform in a particular experiment, but varied from experiment to experiment. For the sake of fair comparison among different continuous culture experiments, the batch portions of all of these were conducted under identical conditions. In continuous cultures, the parameter of interest was fixed at the desired value immediately after switching from batch operation to continuous operation.

The five indicators of performance of the recombinant system in a particular batch experiment are the maximum specific cell growth rate (MSG), the maximum cell mass concentration (X_{max}) or overall cell mass-to-substrate yield ($Y_{X/S}$), the maximum bulk enzyme activity (MBEA), the maximum specific enzyme activity (MSEA), and the plasmid copy number (plasmid monomer equivalents per plasmid-bearing cell). The bulk enzyme (β-lactamase) activity (BEA) is determined using the enzyme assay. The specific enzyme activity (SEA), which is a good indicator of synthesis of the plasmid-encoded enzyme, is obtained by dividing the bulk enzyme activity by the concentration of plasmid-bearing cells. In a batch culture, maxima in BEA and SEA need not occur at the same time. As the results to be presented next indicate, the plasmid pUC8 is maintained at high copy number and is therefore segregationally stable in batch cultures. Whereas no significant variation in copy number was observed in the exponential growth phase, the plasmid copy number increased in the entire growth phase of each batch experiment as the specific growth rate of plasmid-harboring cells decreased.

This plasmid is, however, subject to severe segregational instability in continuous cultures. The five indicators of performance of the recombinant system in continuous culture experiments are the cell mass-to-substrate yield ($Y_{X/S}$), SEA, BEA, the fraction of the total cell population that is plasmid-bearing [C_+/C, where C_+ and C represent the concentrations (number/L) of plasmid-bearing cell population and total cell population, respectively], and the plasmid copy number based on the total cell population. With continual dilution of the plasmid-bearing cell population because of segregational instability of pUC8, the precise determination of plasmid copies (monomer equivalents) per plasmid-bearing cell becomes increasingly susceptible to errors primarily due to sensitivity of colony counts of plasmid-bearing cells on LB-Ap plates.

Hence, the copy number for continuous cultures is expressed as plasmid monomer equivalents per cell in the total (plasmid-bearing plus plasmid-free) population.

Transition from batch to continuous operation in many experiments resulted in establishment of quasi-stationary states. The total cell mass concentration and the dissolved oxygen level remained fairly constant in each quasi-stationary state, with the culture being dominated by plasmid-bearing cells despite the slow appearance of segregants. Changes in some of the other parameters such as β-lactamase activity and plasmid copy number with time were not significant during each quasi-stationary state. Each such state was followed by a period of fast transients where the fraction of the total cell population that was plasmid-bearing declined more rapidly due to increased formation of plasmid-free cells. Significant transients in other culture parameters were also observed during this period. The plasmid copy number, being defined on the basis of the total cell population, decreased with time during this transient period (and eventually vanished) as the plasmid-bearing cells were continually "washed out" due to plasmid instability (see FIGURES 3, 6, and 10 later). Each continuous bioreactor operation appeared to be headed eventually to a stationary state for plasmid-free cells.

Before discussing the results for individual experiments, it is pertinent to consider the multiple effects that the principal plasmid-related functions, namely, plasmid replication and synthesis of plasmid-encoded protein(s), can have on the physiology of the plasmid-bearing host. Whereas increased plasmid content may be desirable for increasing expression of plasmid-carried genes and decreasing the probability of formation of plasmid-free cells, such an increase or an increase in the synthesis of plasmid-encoded proteins can lead to increased diversion of cellular resources from chromosome-related (growth-related) activities to plasmid-related activities. This leads to an increased growth disadvantage of plasmid-bearing cells with respect to plasmid-free cells (i.e., increased μ_-/μ_+, where μ_- and μ_+ represent specific growth rates of plasmid-free cells and plasmid-bearing cells, respectively). The probability of formation of plasmid-free cells from plasmid-bearing cells decreases with increasing plasmid copy number. For plasmids existing in the host in monomeric form, plasmid DNA content is proportional to plasmid copy number. In a separate and more recent investigation,[18] significant formation of multimers of pUC8 has been observed in both batch and continuous cultures. Such multimer formation results in a reduction in the number of distinct plasmid units in a host cell, increasing the probability of formation of plasmid-free cells in the process. This multimer formation may thus be partially responsible for the severe segregational instability observed (despite large plasmid DNA content) in continuous cultures in the absence of a selection pressure. As the results discussed next repeatedly indicate, these seem to be the prime reasons for faster accumulation of plasmid-free cells in continuous cultures once a plasmid-free cell appears in the culture. On the unit mass basis, the energy requirement for synthesis of a protein is severalfold higher than that for synthesis of DNA. As a result, enhanced production of plasmid-encoded proteins places much more burden on the host resources as compared to enhanced plasmid replication (higher plasmid DNA content). In agreement with this, the results presented next demonstrate that enhanced β-lactamase production causes much faster accumulation of plasmid-free cells (faster "washout" of plasmid-bearing cells) in continuous cultures.

Effects of Culture pH

Culture pH strongly affects many enzymatic processes and the transport of several species across the cell membrane. It has been reported that performance of many cellular processes can be optimized with respect to pH.[19-21] Competition among subpopulations in a mixed culture (such as that of recombinant cells and host cells) may be affected by the different pH preferences of the subpopulations.[22] Bailey et al.[23] have reported that the presence of plasmids alters the transmembrane pH difference in E. coli. In a recent study,[6] it has been shown that production of β-lactamase by a recombinant strain of E. coli JM103 grown in a complex medium can be maximized with respect to pH.

Profiles of plasmid copy number and specific β-lactamase activity in batch experiments conducted at three different pH's are shown in FIGURE 1. The maximum specific cell growth rate as well as the overall cell mass-to-substrate yield (proportional to the maximum cell mass concentration for fixed initial substrate concentration) decrease with increasing pH (see TABLE 1). The high copy numbers at pH 8.0 are partially responsible for the low specific growth rate at this pH. The profiles of specific β-lactamase activity in FIGURE 1 and the data in TABLE 1 on bulk β-lactamase activity indicate that, among the three pH's studied, production of the plasmid-encoded protein is promoted the most at the lowest pH. It is interesting to note that the copy numbers for this pH are much smaller than those for pH 8.0. Very high copy numbers (such as those observed for pH 8.0) therefore do not necessarily lead to increased production of β-lactamase. The results from continuous culture experiments conducted at these three pH's are presented next. Profiles of the fraction of the total cell population that is plasmid-bearing and bulk β-lactamase activity are presented in FIGURE 2, whereas profiles of plasmid copy number and total cell mass concentration are presented in FIGURE 3. Time zero in these figures (as well as in FIGURES 5, 6, 9, and 10 later) corresponds to initiation of continuous bioreactor operation.

The results in FIGURES 2 and 3 can be interpreted conveniently by close examination of these over a period of 20 hours after the start of each continuous operation. The trends exhibited by variables such as total cell mass concentration (X), plasmid copy number, and bulk β-lactamase activity during this portion of continuous operation are the same as those exhibited by the respective variables in batch cultures. In this period, with the culture comprising primarily plasmid-bearing cells, the total cell concentration and the plasmid copy number based on total cell population are close to the concentration of plasmid-bearing cells and plasmid monomer equivalents per plasmid-bearing cell, respectively. The decline in X with a switch from pH 7.4 (batch) to pH 8.0 (continuous) and the increase in X with a switch from pH 7.4 to pH 6.8 vis-à-vis X for pH 7.4 (continuous) during this period are indicative of a decline in growth of plasmid-bearing cells with an increase in pH in the range 6.8–8.0, which is a trend also observed in batch cultures.

Consistent with observations in batch experiments, the bulk and specific β-lactamase activities during the initial portions of the continuous culture experiments increase with decreasing pH. For a particular pH, it must be observed that the average specific enzyme activity during such an initial portion is higher than the maximum specific enzyme activity in the corresponding batch experiment. The specific growth rate of plasmid-bearing host decreases with increasing plasmid replication or with

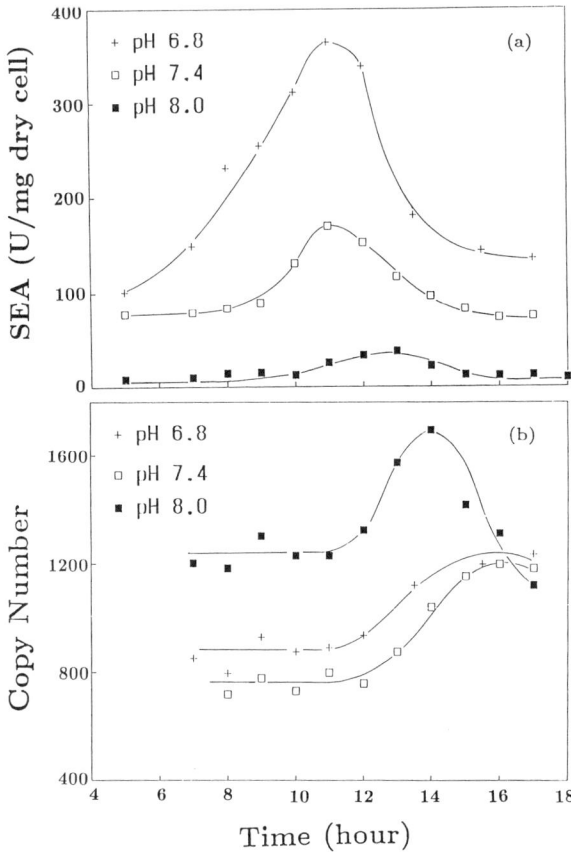

FIGURE 1. Effect of pH on (a) specific β-lactamase activity (SEA) and (b) plasmid copy number in batch cultures.

increasing synthesis of plasmid-encoded protein(s) (or both). Once a plasmid-free cell is formed, the rate of "washout" of plasmid-bearing cells (due to segregational instability of the plasmid) is dictated by the growth rate differential between plasmid-bearing and plasmid-free cells. The larger the differential, the faster the "washout". Both specific β-lactamase activity and plasmid copy number (plasmid monomer equivalents per plasmid-bearing cell) are higher for pH 6.8 than those for pH 7.4 (during the first 20 hours after the start of continuous operation). It is thus expected that plasmid-bearing cells would be at increased disadvantage compared to plasmid-free cells at pH 6.8 than at pH 7.4. As a result, noticeable accumulation of plasmid-free cells and eventual takeover of the culture by these occur much earlier at pH 6.8. Production of plasmid-encoded protein(s) is energetically much more taxing on the host than is replication of plasmid DNA. Although the plasmid DNA content of plasmid-bearing cells in the continuous culture at pH 8.0 is much larger than that in the continuous culture at pH 7.4, the specific β-lactamase activity for pH 8.0 is much lower than that for pH 7.4. Plasmid-bearing cells are therefore at increased growth disadvantage vis-à-vis plasmid-free cells at pH 7.4 than at pH 8.0. The accumulation of plasmid-free cells in the culture is faster as a result for pH 7.4. In consistency with observations of Imanaka et al.,[24,25] the results in FIGURES 2 and 3 demonstrate that the higher the expression of plasmid-encoded gene(s), the faster the appearance of segregants (plasmid-free cells) and the shorter the retention of plasmid-bearing cells in the culture.

Quasi-stationary states in which the total cell mass concentration remained nearly time-invariant were established in continuous cultures at pH 6.8 and 7.4 few hours after the start of continuous operation. Changes in β-lactamase activity and plasmid copy number were not significant in each quasi-stationary state. Such quasi-stationary states have also been observed by other investigators.[26] The significant differences among the total cell mass concentrations in a quasi-stationary state and the corresponding eventual stationary state for plasmid-free cells are indicative of higher mass yields for plasmid-free cells vis-à-vis plasmid-bearing cells. The results of batch and continuous culture experiments reveal that, among the three pH's, growth of plasmid-bearing cells is promoted the most at pH 6.8, whereas the growth of plasmid-free cells is promoted the most at pH 7.4. These observations are clear indicators of the different pH preferences by plasmid-bearing cells and plasmid-free cells as far as their growth is concerned.

Effects of Inorganic Phosphate

The effects of inorganic phosphate in the medium on cell growth and β-lactamase production were studied by supplementing M9 medium with various amounts of inorganic phosphate. Phosphate plays a vital role as an effector of a large number of enzymatic reactions of primary metabolism.[27] Excess inorganic phosphate, at levels comparable to those employed here, has been reported to stimulate glucose utilization, cell growth, and enzyme production in wild-type and recombinant microorganisms.[27–29] Additionally, Godwin and Slater[3] and Melling et al.[30] have shown that plasmid loss can result from phosphate limitation.

TABLE 1. Effect of pH on Performance of Batch Cultures of E. coli JM103[pUC8][a]

Performance Indicators	pH		
	6.8	7.4	8.0
μ_{max} (h^{-1})	0.410	0.402	0.328
$Y_{X/S}$ (g cell/g glucose)	0.364	0.326	0.246
MSEA (U/mg cell)	365	171	38
MBEA (U/mL)	219	93	14
copy number	892	760	1232

[a]The copy number (plasmid monomer equivalents per plasmid-bearing cell) represents the average value during the exponential growth phase of each experiment.

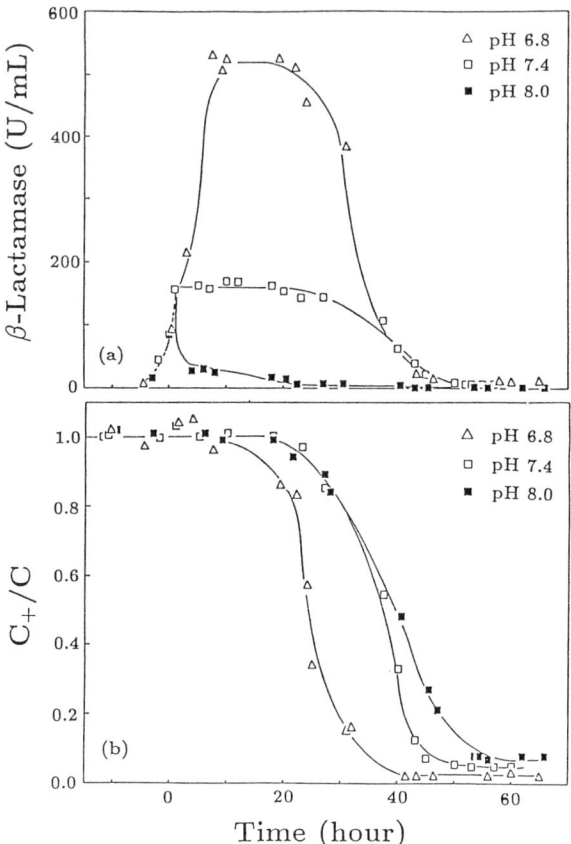

FIGURE 2. Profiles of (a) bulk β-lactamase activity and (b) plasmid-bearing fraction of the total cell population (C_+: number of plasmid-bearing cells/mL; C: total number of cells/mL) in continuous cultures under different pH. Time zero corresponds to initiation of continuous operation.

A summary of batch experiments conducted with three different excess inorganic phosphate levels in the growth medium is provided in TABLE 2. Both specific and bulk β-lactamase activities are highest (among the three phosphate concentrations considered) at the intermediate excess phosphate concentration (64 mM). Whereas the copy numbers in the exponential growth phase are considerably lower for the experiment with 64 mM excess phosphate than those for the experiment with 128 mM excess phosphate, the profiles in FIGURE 4 reveal that the copy numbers reached at the termination of growth phase in these experiments are comparable. The higher the plasmid DNA content in the host in a batch experiment, the greater the burden on the host resources and the lower the overall cell growth. Increasing plasmid copy numbers in the order 64, 128, and 32 mM excess phosphate are therefore partially responsible for decreasing X_{max} in the same order.

The results for continuous cultures supplied with comparable excess levels of inorganic phosphate are presented in FIGURES 5 and 6. The batch portions of these experiments were conducted using 130 mM excess inorganic phosphate in the initial medium. Continuing to the continuous operation with 64 mM excess phosphate in the feed led to substantially higher β-lactamase production (both in terms of bulk and specific activities), in comparison to continuous cultures fed with 39 mM and 130 mM excess phosphate, during the first 25 hours. The specific and bulk β-lactamase activities during the first 25 hours were comparable for experiments involving 39 mM and 130 mM excess phosphate. The copy numbers (plasmid monomer equivalents per plasmid-bearing cell) for continuous cultures fed with 64 mM and 130 mM excess phosphate during this period were comparable and much lower than those for the continuous culture fed with 39 mM excess phosphate (see FIGURES 5 and 6).

The growth disadvantage of plasmid-bearing cells with respect to plasmid-free cells increases with increased plasmid DNA content of the former or with increased

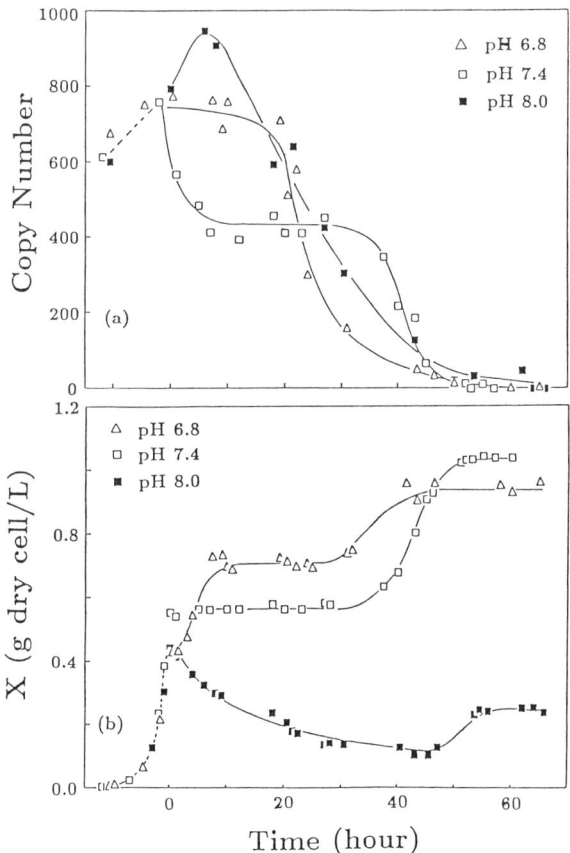

FIGURE 3. Effect of pH on (a) plasmid copy number and (b) total cell mass concentration (X) in continuous cultures.

TABLE 2. Effect of Initial Excess Inorganic Phosphate Concentration on Performance of Batch Cultures of *E. coli* JM103[pUC8][a]

Performance Indicators	Excess Inorganic Phosphate Concentration		
	32 mM	64 mM	128 mM
μ_{max} (h^{-1})	0.317	0.399	0.402
X_{max} (g cell/L)	0.551	0.631	0.609
$Y_{X/S}$ (g cell/g glucose)	0.304	0.336	0.326
MSEA (U/mg cell)	182	199	171
MBEA (U/mL)	69	117	93
copy number	1076	498	760

[a]The copy number (plasmid monomer equivalents per plasmid-bearing cell) represents the average value during the exponential growth phase of each experiment.

expression of plasmid-encoded gene(s) (or both). This leads to a faster appearance of plasmid-free cells (increased instability of plasmid-bearing cells) once a plasmid-free cell is generated in the culture. Increased production of β-lactamase in the case of culture fed with 64 mM excess phosphate and increased plasmid DNA content of recombinant cells for culture fed with 39 mM excess phosphate are responsible for increased instability of plasmid-bearing cells in those experiments. Production of plasmid-encoded protein(s) places more burden on the host resources than does maintenance and replication of plasmid DNA. Increased β-lactamase production (64 mM) is therefore more damaging to the plasmid-bearing host than increased plasmid replication (39 mM). Plasmid-bearing cells are thus retained for a shorter time in the culture fed with 64 mM excess phosphate.

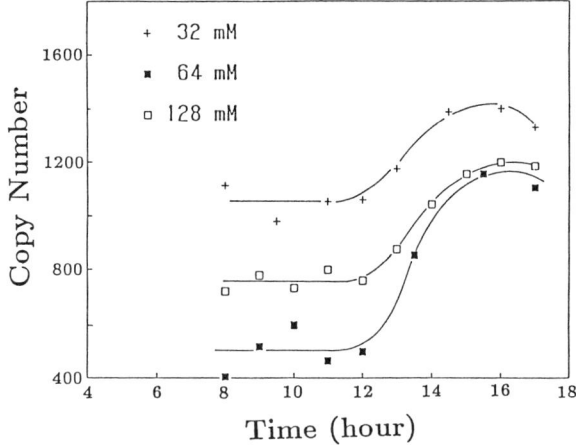

FIGURE 4. Profiles of plasmid copy number in batch cultures at different initial excess inorganic phosphate concentrations.

The cell yields in the initial portions of the continuous experiments (cultures being dominated by plasmid-bearing cells) exhibit the same trend as that observed in the corresponding batch cultures. The eventual cell yield in a continuous operation increases with increasing phosphate concentration. These observations indicate the differences in stimulation of growth of plasmid-free cells and plasmid-bearing cells by

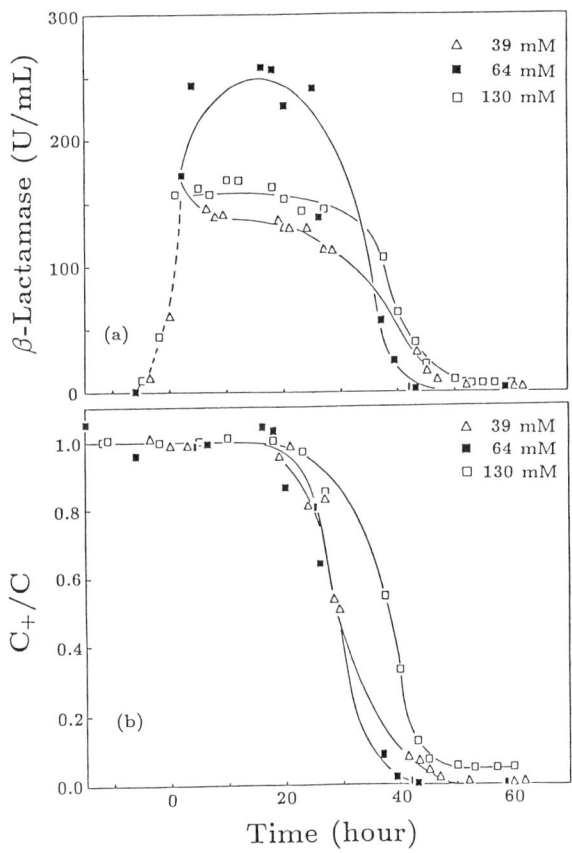

FIGURE 5. Profiles of (a) bulk β-lactamase activity and (b) plasmid-bearing fraction of the total cell population in continuous cultures under different excess inorganic phosphate concentrations in the feed.

inorganic phosphate. The eventual increase in total cell concentration in each of the three experiments is indicative of the higher yield of plasmid-free cells vis-à-vis plasmid-bearing cells. The results in FIGURES 5 and 6 clearly indicate that quasi-stationary states were established in continuous cultures fed with 64 and 130 mM excess phosphate. The total cell mass concentration remained unchanged and the

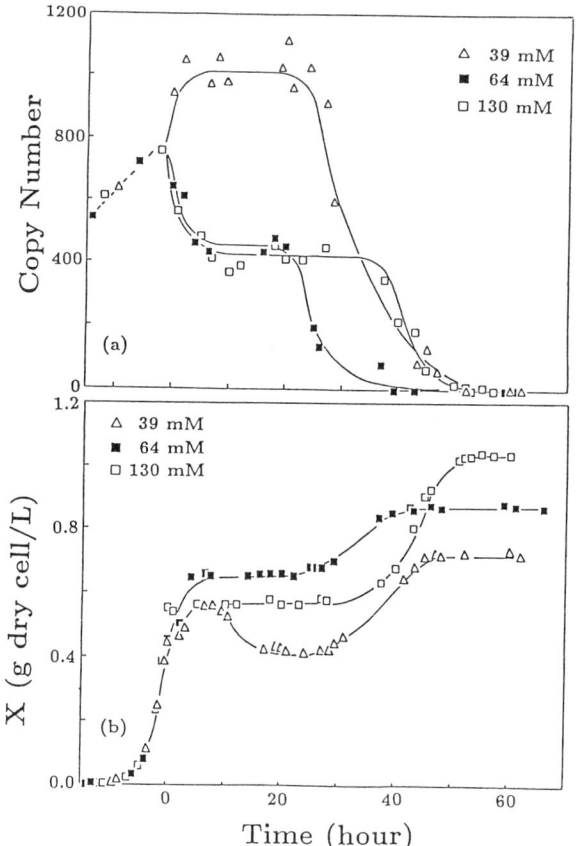

FIGURE 6. Effect of excess inorganic phosphate concentration on (a) plasmid copy number and (b) total cell mass concentration (X) in continuous cultures.

plasmid copy number and bulk β-lactamase activity did not change significantly in each such state, with the cell population being dominated by plasmid-bearing cells.

It must be pointed out that phosphate consumption was insignificant compared to the amount of phosphate added in all batch and continuous culture experiments. Phosphate was therefore neither a limiting nutrient nor consumed substantially for β-lactamase production.

Effects of Dissolved Oxygen

E. coli can grow under both aerobic and anaerobic conditions. A transition in the metabolic pathways may occur when dissolved oxygen (DO) levels are shifted beyond "a critical DO level".[31] A dissolved oxygen shock, introduced in the culture by eliminating the airflow to the bioreactor for a limited period, has been observed to lead to severe plasmid instability for a recombinant *E. coli* strain.[32] The presence of foreign genes is known to increase the oxygen requirement of the host.[33] A moderate reduction in aeration rate has been shown by us to lead to enhanced β-lactamase production in recombinant *E. coli* grown on LB.[6] The dissolved oxygen level was varied in the experiments discussed next by changing the agitation speed.

The results from two batch experiments conducted using different agitation speeds are presented in TABLE 3. A substantial reduction in agitation speed such as the one considered here leads to higher plasmid copy numbers, increased bulk and specific β-lactamase activities, and decreases in the maximum specific cell growth rate and the overall cell mass yield with the growth phase being considerably prolonged. Maintenance of plasmid DNA at higher levels at the lower agitation speed requires an increased allocation of cellular resources for plasmid-related activities, thereby leading to a reduction in specific cell growth rate. The differences in the copy numbers in the two experiments are the most significant during exponential growth phase and diminish as the stationary phase is reached (see FIGURE 7 and TABLE 3).

The results from three continuous culture experiments conducted at different agitation rates are discussed next. The agitation rate in batch operation of each experiment was kept at 800 rpm. Upon switching to continuous operation, the agitation rate in a particular experiment was fixed at the desired value. The agitation rates of 300, 500, and 800 rpm produced dissolved oxygen levels of (a) between 0 and 25%, (b) between 50% and 65%, and (c) close to 100%, respectively (see FIGURE 8). Profiles of the plasmid-bearing fraction of the total cell population and bulk β-lactamase activity are presented in FIGURE 9, whereas profiles of plasmid copy number (based on total cell population) and total cell mass concentration are presented in FIGURE 10.

As the profiles of total cell mass concentration indicate, a decrease in agitation speed from 800 to 500 rpm causes a slight decrease in the total cell growth rate. On the other hand, a decrease in agitation speed from 800 to 300 rpm causes a significant reduction in total cell growth rate and, as a result, the total cell mass concentration decreases substantially in the first 10 hours of continuous operation. The plasmid copy

TABLE 3. Effect of Agitation Speed on Performance of Batch Cultures of *E. coli* JM103[pUC8][a]

Performance Indicators	Agitation Speed (rpm)	
	200	800
μ_{max} (h^{-1})	0.355	0.402
X_{max} (g cell/L)	0.571	0.609
$Y_{X/S}$ (g cell/g glucose)	0.301	0.326
MSEA (U/mg cell)	222	171
MBEA (U/mL)	113	92
copy number	1270	760
minimum DO (% saturation)	32	60

[a]The copy number (plasmid monomer equivalents per plasmid-bearing cell) represents the average value during the exponential growth phase of each experiment.

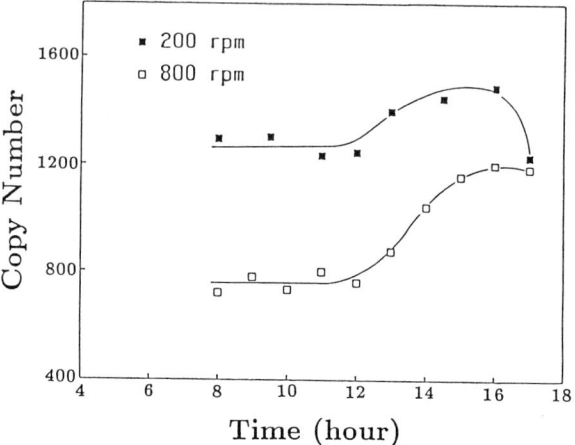

FIGURE 7. Profiles of plasmid copy numbers in batch cultures with different agitation speeds.

numbers (on the per plasmid-bearing cell basis) in the first 25 hours of continuous operation are much higher for the 300 and 500 rpm experiments than those for the 800 rpm experiment. The significant reduction in specific cell growth rate in the 300 rpm experiment is due to severe oxygen limitation and increased allocation of cellular resources for plasmid-related activities. Plasmid instability effects are initiated in the order: 500 rpm, 300 rpm, and 800 rpm.

The continuous culture behavior depicted in FIGURES 9 and 10 can be interpreted by a close examination of the data obtained for the first 20 hours of each continuous

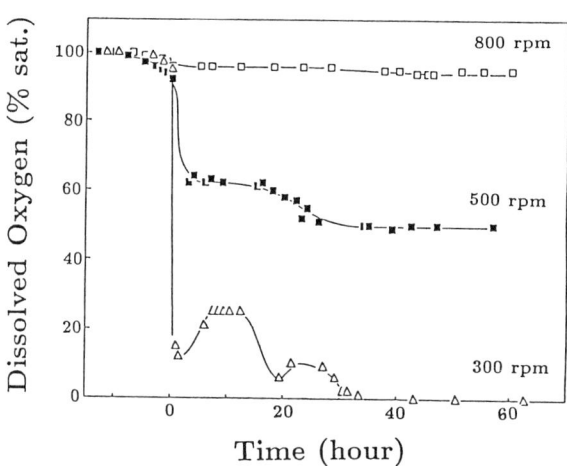

FIGURE 8. Profiles of dissolved oxygen level in continuous culture experiments with different agitation speeds.

operation. The plasmid DNA contents on the per plasmid-bearing cell basis during this period are comparable in the 300 and 500 rpm experiments and are substantially higher than those in the 800 rpm experiment. Furthermore, the specific β-lactamase activity during this period is higher in the 500 rpm experiment than in the 800 rpm experiment. Increased plasmid DNA content and increased β-lactamase synthesis in

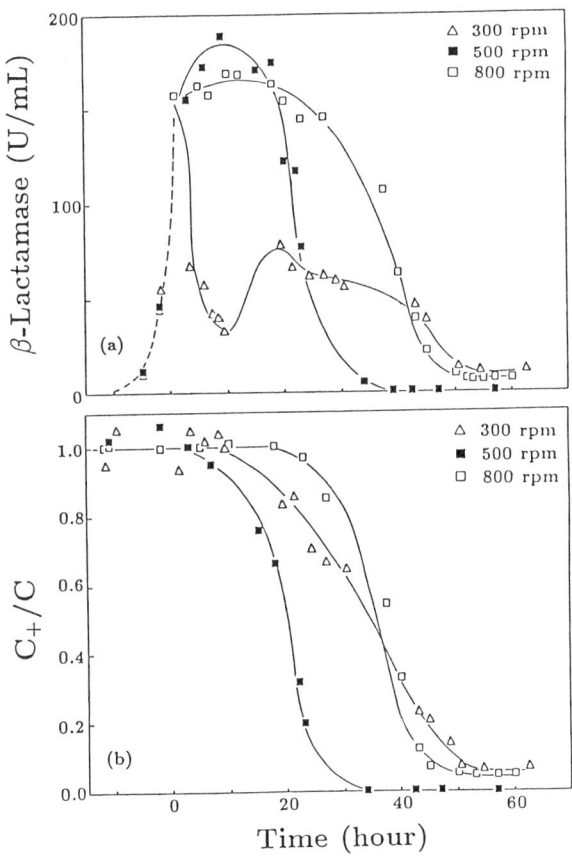

FIGURE 9. Profiles of (a) bulk β-lactamase activity and (b) plasmid-bearing fraction of the total cell population in continuous cultures with different agitation speeds.

the 500 rpm experiment are responsible for the faster appearance of segregants (faster "washout" of plasmid-bearing cells) in this experiment as compared to the 800 rpm experiment (see FIGURE 9).

Plasmid-bearing cells are under increased stress at 300 rpm (than at 800 rpm) because of increased burden on the cellular resources due to maintenance of higher

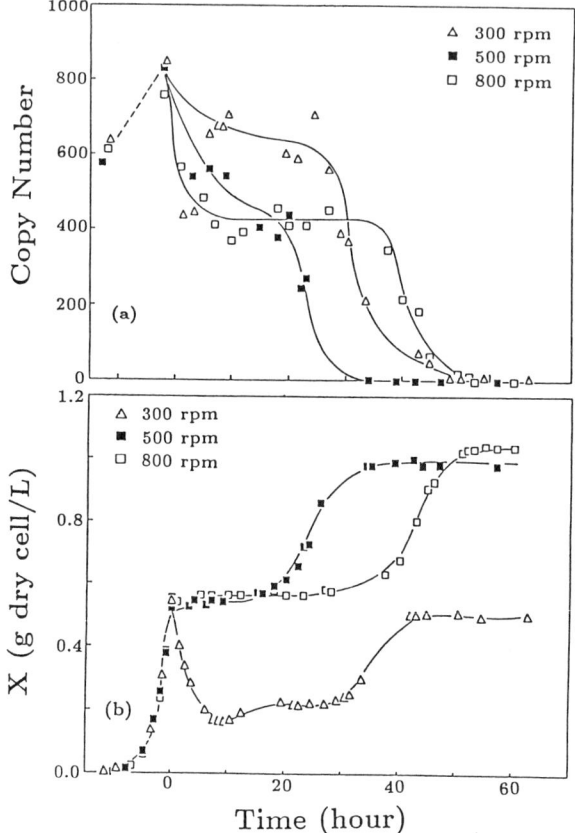

FIGURE 10. Effect of agitation speed (dissolved oxygen level) on (a) plasmid copy number and (b) total cell mass concentration (X) in continuous cultures.

plasmid DNA content (compared to that observed in the 800 rpm experiment) and severe oxygen limitation. Moreover, the specific enzyme (β-lactamase) activities, which are a good indicator of enzyme synthesis rates, are not substantially different in these experiments. Considering this, the slower "washout" of plasmid-bearing cells in the 300 rpm experiment (when compared to the 800 rpm experiment)—despite the earlier appearance of plasmid-free cells—cannot be explained in a simple manner because the dissolved oxygen (DO) level in this experiment exhibited an oscillatory behavior rather than the gradual slow changes observed in DO levels in the other two experiments. The specific growth rates of plasmid-bearing and plasmid-free cells are very sensitive to variations in the DO level in this range (0–25% saturation). As expected, variations in DO level and total cell concentration are out of phase (i.e., the DO level decreases with an increase in total cell concentration due to an increase in oxygen demand and vice versa) in this experiment. Further interpretation of the culture behavior in the 300 rpm experiment is not possible due to a lack of information on how the specific growth rates of plasmid-bearing and plasmid-free cells and the growth advantage of plasmid-free cells over plasmid-bearing cells changed with time.

In summary, we observe that there are no simple correlations between the time after the batch-to-continuous transition at which plasmid-free cells begin to appear in significant numbers, the rate of reversion of plasmid-bearing cells to plasmid-free cells, and DO level. This is presumably a result of two competing tendencies: (1) for plasmid-bearing cells growing at a lower DO level to have a higher copy number (connected with the reduced specific growth rate at a lower DO level) and (2) for plasmid-bearing cells to compete less well with plasmid-free cells at a lower DO level, where greater energy needs of plasmid-bearing cells and lower yield of ATP from a given amount of substrate are increasingly at odds with each other.

Quasi-stationary states were reached in the continuous cultures operated at 500 and 800 rpm as indicated by the near constancy of DO level and total cell mass concentration over a finite time interval soon after the start of continuous operation (see FIGURES 8 and 10). Comparison of initial behavior (dominance by plasmid-bearing cells) and final behavior (dominance by plasmid-free cells) of each of these cultures as far as the total cell mass concentration is concerned reveals the following: The concentrations of both plasmid-bearing and plasmid-free cells increase with increasing DO level. Under comparable operating conditions, the yield of plasmid-free cells is greater than that of plasmid-bearing cells.

A phase plane of the average plasmid copy number and the specific cell growth rate during exponential growth phase in the batch experiments discussed earlier shows a nice relation between the two when the specific growth rate of the plasmid-harboring cells is varied by altering pH, phosphate content of the medium, or agitation rate (see FIGURE 11).

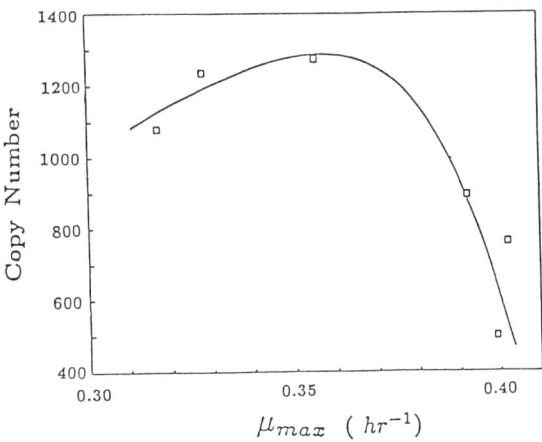

FIGURE 11. Correlation between the average plasmid copy number and specific cell growth rate during the exponential growth phase in batch cultures represented by FIGURES 1, 4, and 7 and TABLES 1–3.

β-Lactamase Excretion

Excretion of recombinant proteins into the extracellular medium is highly desirable from several points of view. First, the recovery and purification of the proteins are simplified because excretion eliminates the need to rupture the cells to harvest the product. The cell mass is removed by centrifugation or filtration and the product is purified from a relatively dilute suspension of contaminating material. Excretion can also stabilize the product from proteolytic degradation,[34] provided the host organism does not release proteolytic enzymes into the culture fluid. Finally, excessive accumulation within the cells of certain proteins/enzymes may be toxic to the host. Excretion of enzymes/proteins will help in partially overcoming any such toxic effects.

Varying degrees of excretion of β-lactamase were observed in the experiments discussed in this paper. Under certain environmental conditions, extracellular β-lactamase activity was as much as 70% (in batch cultures) and 52% (in the initial portions of continuous cultures) of the total (intracellular plus extracellular) β-lactamase activity. Although the reasons for this protein excretion are as yet unclear, the extent of the enzyme excretion may be related to the physiological state of the cells, as suggested by the dramatic increase in the percentage of extracellular β-lactamase following the cessation of exponential growth phase. A close examination of the data in TABLES 4 and 5 reveals that the percentage of the total β-lactamase that is present in the abiotic phase increases with a reduction in the growth of plasmid-bearing cells or with an increase in the enzyme synthesis (or both), with reduction in cell growth being more effective for increased enzyme excretion. When the cell growth is reduced, it is possible that the plasmid-bearing cells excrete an increased percentage of β-lactamase produced as a protection mechanism. Increased enzyme excretion in situations resulting in enhanced enzyme synthesis may be due to an effort on the part of plasmid-bearing cells to reduce the burden on their resources due to excessive accumulation of the enzyme (β-lactamase). It will be necessary to further investigate the mechanism of protein transport in *E. coli* JM103[pUC8] in order to take advantage of this feature.

CONCLUSIONS

The effects of culture pH, inorganic phosphate concentration, and dissolved oxygen level (agitation speed) on host growth and plasmid-related activities have been

TABLE 4. Overall β-Lactamase Excretion in Batch Cultures of *E. coli* JM103[pUC8] under Different Cultivation Conditions[a]

Conditions	EE/TE × 100 (%)
reference[b]	33
pH 8.0	52
pH 6.8	50
32 mM P_i	70
64 mM P_i	53
200 rpm	42

[a] EE = β-lactamase activity in the abiotic phase (U/mL); TE = total β-lactamase activity [same as bulk β-lactamase activity (U/mL)]; P_i = excess inorganic phosphate.
[b] Reference conditions: pH 7.4, 128 mM P_i, and 800 rpm.

TABLE 5. Excretion of β-Lactamase in Continuous Cultures of *E. coli* JM103[pUC8] under Different Cultivation Conditions (Dilution Rate = 0.375 h^{-1})^a

Conditions	β-Lactamase (U/mL)			EE/TE × 100 (%)
	IE	EE	TE	
reference^b	151	9	160	6
pH 8.0	11	2	13	16
pH 6.8	477	53	530	10
39 mM P$_i$	89	42	131	32
64 mM P$_i$	230	26	256	10
500 rpm	152	21	173	12
300 rpm	21	22	43	52

^{a}The data tabulated represent average values during the initial portions of the continuous cultures. IE, EE, and TE denote intracellular, extracellular, and total (or bulk) β-lactamase activities, respectively; P$_i$ denotes excess inorganic phosphate.
^{b}Reference conditions: pH 7.4, 130 mM P$_i$, and 800 rpm.

examined in this work using the plasmid pUC8, harbored at high DNA content in *E. coli* JM103. Whereas no plasmid-free cells were detected in batch cultures, the plasmid is subject to severe segregational instability in continuous cultures. Washout of plasmid-bearing cells was accelerated with an increased growth rate differential between plasmid-free and plasmid-bearing cells resulting from increased synthesis of the plasmid-encoded protein (β-lactamase) or plasmid DNA (or both). Excessive synthesis of the plasmid-encoded protein had more damaging effect in this regard than maintenance of high plasmid DNA content. The results presented here demonstrate that judicious selection of environmental conditions is essential for longer retention of plasmid-harboring cells in continuous cultures. Whereas an increase in plasmid DNA content always led to a reduction in specific growth rate of the host, plasmid maintenance at high DNA levels did not always result in increased production of β-lactamase. In all experiments discussed here, plasmid-bearing cells were retained in continuous cultures for residence times in excess of 12 after the start of each continuous operation. Much faster washout rates of plasmid-bearing cells have been observed for other recombinant systems. For example, for a recombinant *E. coli* system with an average copy number near 50, Ryu and co-workers[26,35] have reported that, at a dilution rate of 0.358 h^{-1}, plasmid-bearing cells could be retained for only 2.4 residence times.

Significant excretion of β-lactamase was observed under certain environmental conditions. Increased excretion of the plasmid-encoded protein was observed under conditions resulting in reduced cell growth or increased accumulation of the protein (or both). In view of the detrimental effect of the intracellular accumulation of the plasmid-encoded protein on plasmid stability, it would be of considerable interest to further investigate the mechanism of protein transport across the cell membrane especially when the protein is being overproduced.

ACKNOWLEDGMENT

We are grateful to Benjamin C. Stark of the Department of Biology of the Illinois Institute of Technology for technical discussions.

REFERENCES

1. BAILEY, J. E., N. A. DASILVA, S. W. PERETTI, J-H. SEO & F. SRIENC. 1986. Ann. N.Y. Acad. Sci. **469**: 194–211.
2. ZUND, P. & G. LEBEK. 1980. Plasmid **3**: 65–69.
3. GODWIN, D. & J. H. SLATER. 1979. J. Gen. Microbiol. **111**: 201–210.
4. HELLING, R. B., T. KINNEY & J. ADAMS. 1981. J. Gen. Microbiol. **123**: 129–140.
5. LEE, S. B. & J. E. BAILEY. 1984. Biotechnol. Bioeng. **26**: 66–73.
6. RYAN, W., S. J. PARULEKAR & B. C. STARK. 1989. Biotechnol. Bioeng. **34**: 309–319.
7. SEO, J-H. & J. E. BAILEY. 1985. Biotechnol. Bioeng. **27**: 156–165.
8. UHLIN, B. E., S. MOLIN, P. GUSTAFSSON & K. NORDSTROM. 1979. Gene **6**: 91–106.
9. LEE, S. B. & J. E. BAILEY. 1984. Biotechnol. Bioeng. **26**: 1372–1382.
10. CHEAH, U. E., W. A. WEIGAND & B. C. STARK. 1987. Plasmid **18**: 127–134.
11. MESSING, J. 1983. Methods Enzymol. **101**: 20–78.
12. MANIATIS, T., E. F. FRITSCH & J. SAMBROOK. 1982. Molecular cloning. Cold Spring Harbor Laboratory, Cold Spring Harbor, New York.
13. ANRAKU, Y. & L. A. HEPPEL. 1967. J. Biol. Chem. **242**: 2561–2569.
14. NEU, H. C. & J. CHOU. 1967. J. Bacteriol. **94**: 1934–1945.
15. O'CALLAGHAN, C. H., P. W. MUGGLETON & G. W. ROSS. 1968. Antimicrob. Agents Chemother., p. 57.
16. BREWER, J. M., A. J. PESCE & R. B. ASHWORTH. 1974. Experimental Techniques in Biochemistry. Prentice–Hall. Englewood Cliffs, New Jersey.
17. KING, E. J. 1932. Biochem. J. **26**: 292–297.
18. RYAN, W. & S. J. PARULEKAR. Biotechnol. Bioeng. Submitted.
19. ROUKAS, T. & L. HARVEY. 1988. Biotechnol. Lett. **10**: 289–294.
20. LALLAI, A., G. MURA, R. MILIDDI & C. MASTINU. 1988. Biotechnol. Bioeng. **31**: 130–134.
21. EROSHIN, V. K., I. S. UTKIN, S. V. LADYNICHEV, V. V. SAMOYLOV, V. D. KURSHINNIKOV & G. K. SHRYABIN. 1976. Biotechnol. Bioeng. **18**: 289–295.
22. MEERS, J. L. 1973. Crit. Rev. Microbiol. **2**: 139–184.
23. BAILEY, J. E., D. D. AXE, P. M. DORAN, J. L. GALAZZO, K. F. REARDON, A. SERESSIOTIS & J. V. SHANKS. 1987. Ann. N.Y. Acad. Sci. **506**: 1–23.
24. IMANAKA, T., H. TSUNEKEWA & S. AIBA. 1980. J. Gen. Microbiol. **118**: 253.
25. IMANAKA, T. 1986. Adv. Biochem. Eng./Biotechnol. **33**: 1–27.
26. LEE, S. B., D. D. Y. RYU, R. SIEGEL & S. H. PARK. 1988. Biotechnol. Bioeng. **31**: 805–820.
27. MARTIN, J. F. 1977. Adv. Biochem. Eng. **6**: 105–127.
28. PERLMAN, D. & G. H. WAGMAN. 1952. J. Bacteriol. **63**: 253.
29. LIU, C. M., L. E. MCDANIEL & C. P. SCHAFFNER. 1975. Antimicrob. Agents Chemother. **7**: 196–202.
30. MELLING, J., D. ELLWOOD & A. ROBINSON. 1977. FEMS Microbiol. Lett. **2**: 87–89.
31. HARRISON, D. E. F. 1973. Crit. Rev. Microbiol. **2**: 185–228.
32. HOPKINS, D. J., M. J. BETENBAUGH & P. DHURJATI. 1987. Biotechnol. Bioeng. **29**: 85–91.
33. KHOSLA, C. & J. E. BAILEY. 1988. Nature **331**: 633–635.
34. GEORGIOU, G. 1988. AIChE J. **34**: 1233–1245.
35. SIEGEL, R. & D. D. Y. RYU. 1985. Biotechnol. Bioeng. **27**: 28–33.

Effects of Promoter Induction and Copy Number Amplification on Cloned Gene Expression and Growth of Recombinant Cell Cultures[a]

MICHAEL J. BETENBAUGH[b] AND PRASAD DHURJATI

Department of Chemical Engineering
University of Delaware
Newark, Delaware 19716

INTRODUCTION

Two methods for enhancing recombinant gene expression at the transcriptional level are promoter induction and plasmid copy number amplification. Promoter induction enhances recombinant product yields by increasing the efficiency of transcriptional initiation. Copy number amplification improves recombinant yields by increasing the number of plasmid copies that are available for transcription. One goal of this project was to compare these two tools as separate techniques for enhancing cloned gene expression. However, cloned gene expression and plasmid amplification may also have negative effects on recombinant cell behavior, so another goal of this study was to identify and quantify the effects of these two parameters on recombinant cell growth kinetics.

To accomplish these goals, a recombinant system was chosen in which plasmid copy number and promoter induction were controlled using independent mechanisms. *E. coli* was the host and the plasmid utilized was pVH106/172. This temperature-sensitive plasmid was utilized because the copy number of the plasmid could be changed by adjusting culture temperature, and the *lac* promoter, which controlled expression of the recombinant genes, could be activated independently by adding chemicals to the medium. A series of batch fermentations were performed to identify the effects of promoter induction and amplified plasmid levels on recombinant product yields and cell growth kinetics.

A final goal of this study was to compare the experimental results for product yields and cell growth under conditions of promoter induction and plasmid amplification with the predictions of mathematical models in the literature. A number of mathematical models have been constructed to describe the growth and gene expression for recombinant systems. However, very few of these models have been compared with experimental measurements. In this study, the predictions of a combination of two literature mathematical models were compared with experimental measurements of cell growth and recombinant protein synthesis for CSH22 with pVH106/172. Such a comparison is worthwhile in identifying the range of applicability of such mathematical models.

---

[a]This work was partially supported by the National Science Foundation.
[b]Current address: Department of Chemical Engineering, The Johns Hopkins University, Baltimore, Maryland 21218.

RECOMBINANT SYSTEM

Bacterial Host

The bacterial strain used in these experiments was *E. coli* strain CSH22[*trpR* Δ(*lacpro*) *thi*]. This strain contains the sex factor F'*lacZ proA+,B+*.[1]

Plasmid

The plasmid, pVH106/172, was provided by P. Valentin-Hansen of Odense University.[2] A map of plasmid pVH106/172 (11.8 kb) is shown in FIGURE 1. This plasmid, a derivative of R1, was chosen because it contains regions for independent control of plasmid copy number and promoter induction. Recombinant genes were contained on the *lacZ'*, *lacY*, and *lacA* genes, which code for the synthesis of β-galactosidase, lactose permease, and β-galactoside acetyltransferase, respectively. Several nonessential codons were deleted from the *lacZ* gene to give the *lacZ'* gene.[3] Expression of these genes was under the control of the *lac* promoter located upstream of the gene. The *lac* promoter was induced by adding the chemicals, IPTG (isopropyl-β-D-thiogalactopyranoside) and cyclic AMP (adenosine 3',5'-cyclic monophosphate), to the medium.

The plasmid copy number of pVH106/172 was changed by altering culture temperature because the plasmid contains a temperature-sensitive replicon derived from pOU61.[4] Below 38 °C, the λ_{P_R} promoter is relatively inactive as a result of binding by the cI857 repressor protein. However, raising the culture temperature to 42–43 °C denatures the repressor and allows the λ_{P_R} promoter to be activated. This action increases the synthesis of *repA* protein, which is a critical positive-acting element in plasmid replication. As a result, at higher temperatures, the plasmid content is considerably amplified.

PROCEDURES

All fermentations were performed using 1 liter of M9 minimal media supplemented with glucose in a 2-liter LH Systems Fermentor (Model 501). Agitation was maintained at 600 rpm with an LH Systems Magnetic Agitator (Model 502). Temperature was controlled using a Watlow 808 temperature controller, and the airflow was maintained at 0.5 liters/minute using a Matheson rotameter. Optical density was measured continuously using a modified Bausch and Lomb Spectrophotometer (Spec 20).[5] A correlation between dry cell mass and optical density was obtained and the on-line optical density measurements were then converted to dry cell mass levels. All on-line variables were monitored using an ISAAC Cyborg analog to digital converter and an IBM XT personal computer.[6]

Total DNA content was measured on an HPLC with a Nucleogen DEAE 4000-10 column[7] using the method outlined by Coppella *et al.*[8] The fraction of DNA present as plasmid DNA was then determined from a densitometer scan of chromosomal and plasmid DNA fractions on an agarose gel. Plasmid DNA content in μg/g cell mass was determined from this plasmid DNA measurement and the dry cell mass measurement.

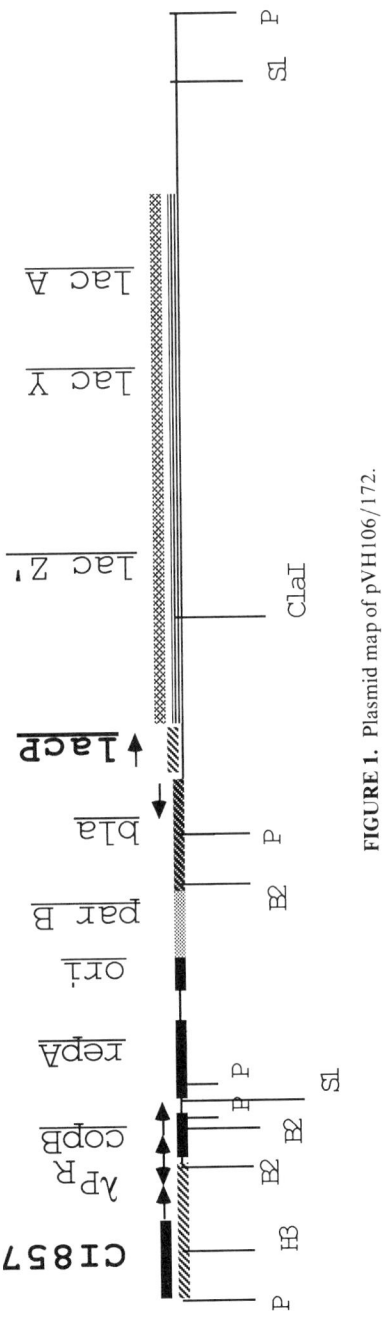

FIGURE 1. Plasmid map of pVH106/172.

Plasmid DNA content in copies per cell was calculated from the plasmid DNA determination and the recombinant cell counts.

Recombinant β-galactosidase activity was measured in activity units per milliliter using the standard assay of Pardee et al.[9]

PROMOTER INDUCTION

To identify the effects of promoter induction on cloned gene expression and specific growth rate, batch fermentations were performed with *E. coli* CSH22-bearing plasmid pVH106/172. In these experiments, expression of the *lac* genes on pVH106/172 was induced by adding chemicals to the fermentation medium. Also, a control experiment (CONTROL) was performed using CSH22-bearing pVH106/172 in which no chemicals were added to the fermentation medium. Three different types of chemical induction experiments were performed and the results from these experiments were compared to one another and to the CONTROL experiment. In one experiment, cyclic AMP (C-AMP) at 5×10^{-3} M was added to the medium. Another experiment was performed in which the inducer IPTG at 1×10^{-3} M was added to the medium. A final promoter induction experiment, C-AMP/IPTG, was performed in which cyclic AMP was added to the medium first, followed by IPTG after 10 minutes.[10] A number of fermentation variables were followed for these experiments including dry cell mass, recombinant β-galactosidase activity, plasmid levels, and recombinant cell counts. These measurements were then used to calculate specific growth rate, recombinant β-galactosidase product yield, and plasmid content.

The growth rate and recombinant β-galactosidase product yield results from all the promoter induction experiments are summarized in TABLE 1.[11] Plasmid content was constant for this series of fermentations. Adding the different chemical inducers caused the product yield to change over various magnitudes. Adding cyclic AMP alone caused only a slight improvement in the product yield over the CONTROL case. The chemical inducer IPTG was a much more effective inducer of the *lac* promoter. The product yield for the IPTG induction experiment was 700 times greater than the CONTROL level. Combining both chemical induction techniques in experiment C-AMP/IPTG resulted in the highest product yield of any chemical induction experiment at $630 \times 10^{+3}$ units/gram. This yield was 1850 times greater than the CONTROL yield.

Because the chemical inducers, cyclic AMP and IPTG, induced the *lac* promoter to express recombinant β-galactosidase to different levels, it was possible to identify the

TABLE 1. Recombinant Product Yield and Growth Rate for Promoter Induction Experiments[11]

Experiment[a]	Product Yield ($\times 10^{-3}$ units/g cell mass)	Factor	Growth Rate (h^{-1})
CONTROL	0.34	1.0	0.68
C-AMP	0.70	2.0	0.68
IPTG	240.0	700.0	0.69
C-AMP/IPTG	630.0	1850.0	0.42

[a]All experiments were performed at 38 °C.

TABLE 2. Product Yield, Plasmid Content, and Growth Rate for T-SHIFT, CONTROL, and IPTG Experiments[11]

Experiment	Product Yield ($\times 10^{-3}$ units/g cell mass)	Factor	Plasmid Content (μg/g cell mass)	Growth Rate (h^{-1})
T-SHIFT	33.0	100	1900.0	0.52
CONTROL	0.34	1	60.0	0.68
IPTG	240.0	700	65.0	0.69

effects of various expression levels of the *lac* operon genes on cell growth. Effects on cell growth were identified by measuring the specific growth rate in all of these chemical induction experiments. The specific growth rates are also listed in TABLE 1. The specific growth rates for the CONTROL, C-AMP, and IPTG experiments were equal within the range of experimental error at 0.68–0.69 h^{-1}. However, the specific growth rate for the experiment in which both cyclic AMP and IPTG were added to the medium, C-AMP/IPTG, was nearly 40% below the other fermentations at 0.41 h^{-1}. This lowered growth rate was caused by the expression of high amounts of recombinant *lac* proteins in the C-AMP/IPTG experiment either as a result of the "metabolic burden" caused by synthesizing such large quantities of protein or as a result of product toxicity. The toxicity effect is supported by recombinant cell count data[11] in which the number of viable recombinant cells declined at later times in the fermentation. Zabeau and Stanley[12] also observed a decrease in recombinant cell viability following expression of large amounts of the recombinant *lacY* protein.

COPY NUMBER AMPLIFICATION

Another batch fermentation was performed with CSH22-bearing pVH106/172 to identify the effects of plasmid amplification on recombinant product yields and cell growth. In this fermentation (T-SHIFT), the culture temperature was shifted to 43 °C following inoculation at 38 °C. Because the replicon of plasmid pVH106/172 is temperature-sensitive, an increase in culture temperature will promote an amplification of cell copy number. The same fermentation variables were measured for this T-SHIFT experiment as in the chemical induction experiments and the results from this experiment were compared with the CONTROL experiment. The CONTROL experiment was run with the same recombinant system at 38 °C, which is a low copy number culture temperature.

The same fermentor variables were followed in the temperature-shift experiment as were followed in the chemical induction experiments. The results for the T-SHIFT experiment are summarized in TABLE 2,[11] where the product yield, plasmid content, and specific growth rate are listed along with results for the IPTG and CONTROL experiments. Because the product yield for the T-SHIFT experiment was 100 times greater than the product yield for the CONTROL experiment, plasmid amplification alone must have been sufficient to induce expression of the *lac* operon genes. This amplification in the product yield was caused by the inability of the *lac* repressors to bind all available copies of the *lac* promoter at high copy numbers. Caulcott *et al.*[13]

observed a similar "leaking" effect using the *trp* promoter on the multicopy plasmid pCT70 in *E. coli*.

That the plasmid content increased for this high temperature experiment can be seen by comparing the plasmid content for the T-SHIFT experiment with that of the CONTROL and IPTG experiments. The plasmid content reached 1900 μg/g cell mass for the T-SHIFT experiment, which was 30 times greater than the plasmid content for experiments at 38 °C. For the T-SHIFT experiment, the plasmid content in copies per cell was determined to be approximately 200.

The specific growth rate for the T-SHIFT experiment was 0.52 h^{-1}, which was 20% below the CONTROL growth rate. This lowered growth rate was not caused by the expression of the recombinant *lac* proteins in the T-SHIFT experiment because the product yield for the IPTG experiment was much larger and its growth rate was unaffected. Similarly, the lowered growth rate was not caused by a difference in culture temperature because host *E. coli* growth rates were equal at the two temperatures. The reduction in growth rate must have been due to the effects of amplified plasmid levels or the expression of other genes (besides the *lac* genes) on plasmid pVH106/172 at high copy numbers.

COMBINED PROMOTER INDUCTION AND COPY NUMBER AMPLIFICATION

A final batch fermentation was performed in which inducing chemicals were added to the medium and the culture temperature was shifted concurrently in order to identify the effects of combined promoter induction and plasmid amplification on cloned gene expression and cell growth. Cyclic AMP and IPTG were added to the medium in the same concentrations as in the chemical induction experiments, and the culture temperature was raised to 43 °C as in the T-SHIFT experiment. Cell mass concentrations, β-galactosidase activity, and plasmid content were followed.

The results for this fermentation, C-AMP/IPTG/T-SHIFT, are summarized in TABLE 3.[11] The average recombinant product yield was $1500 \times 10^{+3}$ units/gram, which was more than 4400 times greater than the CONTROL experiment. Combined chemical and temperature induction was the most efficient method for increasing recombinant product yields because yields for this experiment were higher than for either the promoter induction (C-AMP/IPTG) or the plasmid amplification

TABLE 3. Product Yield and Growth Rate for the Simultaneous Promoter Induction and Plasmid Amplification Experiment[11] (C-AMP/IPTG/T-SHIFT)^a

Experiment	Product Yield ($\times 10^{-3}$ units/gram)	Factor	Growth Rate (h^{-1})
CONTROL	0.34	1.0	0.68
T-SHIFT	33.0	100.0	0.52
C-AMP/IPTG	630.0	1850.0	0.42
C-AMP/IPTG/T-SHIFT	1500.0	4400.0	0.21

^{a}Other experimental results have been included for comparison.

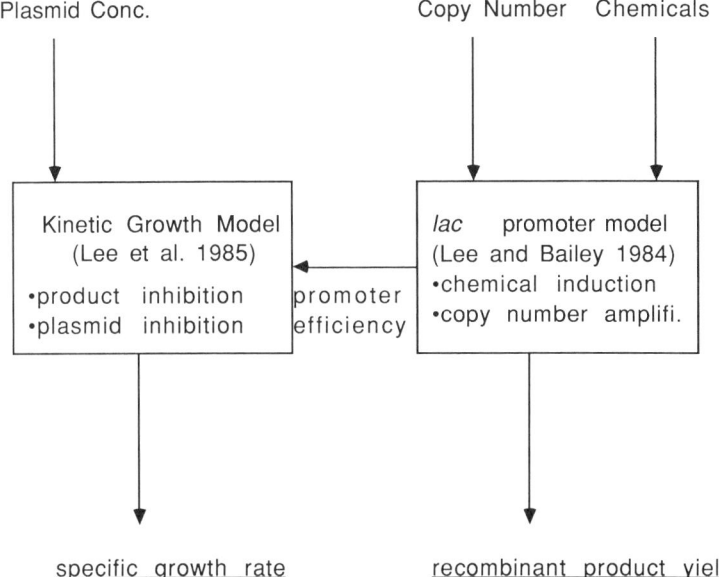

FIGURE 2. Schematic diagram of the combined mathematical model.

(T-SHIFT) experiment. However, the reduction in the specific growth rate was also more significant in this experiment. The specific growth rate was $0.21\ h^{-1}$, which was 70% below the CONTROL rate and also more than 50% below the rate for the chemical and temperature induction experiments. The reduction in specific growth rate was most likely caused by the combined effects of plasmid amplification and recombinant gene expression. From the previous experiments, we have noted that recombinant gene expression and plasmid amplification can separately affect cell growth. In the C-AMP/IPTG/T-SHIFT experiment, the plasmid content was amplified to 800 $\mu g/g$ cell mass and recombinant product yield increased by a factor of 4400, so both recombinant gene expression and plasmid amplification contributed to a lower growth rate in this experiment.

MATHEMATICAL MODELING

A final objective of this study was to examine the ability of literature mathematical models to predict recombinant product formation and cell growth under conditions of copy number amplification and promoter induction. If these models can predict actual experimental cell behavior, the models may then be used for the optimization and control of recombinant bioreactors. In order to adequately describe the cell growth and product formation conditions that prevailed for the experimental recombinant system of CSH22 with plasmid pVH106/172, two literature mathematical models were incorporated into a single combined model as shown in FIGURE 2. The Kinetic Model for Product Formation of Unstable Recombinants by Lee, Seressiotis, and Bailey[14] was included because this model considers the kinetics of recombinant cell growth as well as the inhibition of cell growth due to plasmid amplification and cloned gene expression. The equation, which incorporates the effects of plasmid amplification and product inhibition on cell growth, is shown below:

$$\mu = \mu_0 * g(G) * h(P), \qquad (1)$$

where μ_0 is the maximum specific growth rate and $g(G)$ and $h(P)$ are growth inhibition functions of the intracellular plasmid content, G, and intracellular product concentration, P, respectively. The form of these inhibition functions is shown below in equations 2 and 3:

$$g(G) = (1 - G/G_{max}) \qquad (2)$$

and

$$h(P) = (1 - P/P_{max}), \qquad (3)$$

where G_{max} and P_{max} are the maximum levels of plasmid content and product concentration that the cell can tolerate. As the intracellular product and plasmid levels increase, the cell growth rate is reduced below its maximum value.

The Structured Model of the *lac* Promoter-Operon on Multicopy Plasmids by Lee and Bailey[15,16] was also included because this model contains a complex mathematical description of the *lac* promoter on a multicopy plasmid. This model can be used to predict product amplification following induction with the chemical inducers, cyclic AMP and IPTG, and following plasmid amplification.

The predictions of the combined model were compared with experimental measurements for cell growth and product yield observed for the previous chemical induction and plasmid amplification experiments. Product yield results from the CONTROL experiment were used to normalize the transcription efficiency in the combined model so that model predictions would be comparable to experimental measurements for all other induction experiments. One parameter, the initial dry cell mass, was adjusted between the simulations because this parameter varies from one experiment to the next. The value was chosen to minimize the error between dry cell mass measurement and model dry cell mass prediction.

The model predictions were compared to experimental measurements for all the chemical induction and plasmid amplification experiments. For the chemical induction cases, the concentration of the inducer chemical in the model, cyclic AMP and IPTG, was changed from its CONTROL concentration to the induction concentration at the same time as the chemical was added to the fermentation broth experimentally. For the plasmid amplification simulations, the plasmid content, G, and the copy number, N, were altered to reflect experimentally measured levels of the plasmid copy number at high temperatures. The model was then used to predict recombinant β-galactosidase activity and yields. In FIGURE 3, a comparison between model prediction and experimental measurement of β-galactosidase activity for the chemical induction experiments is shown. Agreement between the model predictions and experimental observations is significant due to the range of magnitudes over which the β-galactosidase activity varied.

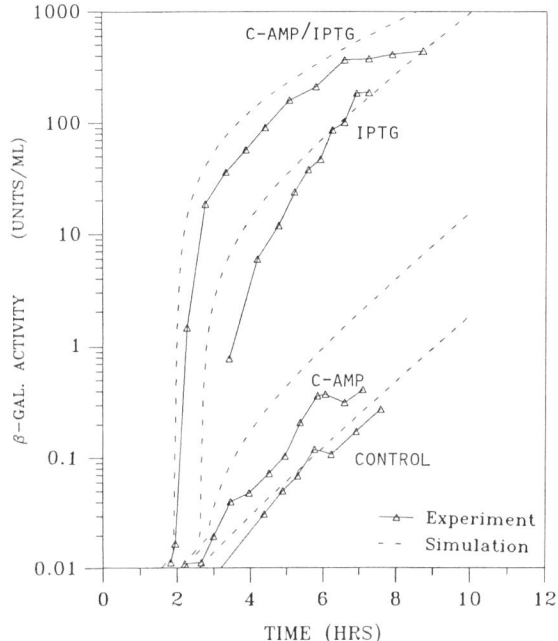

FIGURE 3. Model prediction and experimental measurement of recombinant β-galactosidase activity for promoter induction experiments.[17]

A comparison between model simulation and experimental measurement for product yields for all experiments is shown in TABLE 4.[17] The model successfully predicted experimental product yields following promoter induction and plasmid amplification even though these values covered nearly four orders of magnitude. The combined model failed only in its prediction of product yield for the copy number amplification (T-SHIFT) case, in which the model overestimated the product yield increase following plasmid amplification.

TABLE 4. Model Prediction and Experimental Measurement of Recombinant β-Galactosidase Product Yield[17]

Simulation	Product Yield ($\times 10^{-3}$ units/gram)	
	Model Prediction	Experiment
CONTROL	0.34	0.34
C-AMP	2.0	0.7
IPTG	170	240
C-AMP/IPTG	900	660
T-SHIFT	210	33
C-AMP/IPTG/T-SHIFT	1300	1500

CONCLUSIONS

In this study, a model recombinant system was used to compare and contrast many of the important features of promoter induction and plasmid amplification. The important observations were:

(1) Promoter induction using the *lac* promoter was a more effective tool than plasmid amplification for increasing recombinant product yields for CSH22 with plasmid pVH106/172.
(2) Both recombinant gene expression of the *lac* operon genes and plasmid amplification caused detrimental effects on cell growth rate.
(3) Simultaneous induction of the *lac* promoter and plasmid amplification produced a combination of the individual induction effects on recombinant product yields and cell growth rate.
(4) Literature mathematical models made reasonable predictions for experimental product amplification following promoter induction and plasmid amplification even though the yields covered more than four orders of magnitude in activity.

REFERENCES

1. MILLER, J. H. 1972. Experiments in Molecular Genetics. Cold Spring Harbor Laboratory. Cold Spring Harbor, New York.
2. VALENTIN-HANSEN, P. 1986. Personal communication.
3. VALENTIN-HANSEN, P., B. ALBRECHTSEN & J. E. L. LARSEN. 1986. EMBO J. **5:** 2015–2021.
4. LARSEN, J. E. L., K. GERDES, J. LIGHT & S. MOLIN. 1984. Gene **28:** 45–54.
5. LEE, G. & H. LIM. 1980. Biotechnol. Bioeng. **22:** 639–642.
6. COPPELLA, S. J. & P. DHURJATI. 1987. Biotechnol. Bioeng. **29:** 679–689.
7. COLPAN, M. & D. RIESNER. 1984. J. Chromatogr. **296:** 339–353.
8. COPPELLA, S. J., C. M. ACHESON & P. DHURJATI. 1987. J. Chromatogr. **402:** 189–199.
9. PARDEE, A. B., F. JACOB & J. MONOD. 1958. J. Mol. Biol. **1:** 165.
10. BECKWITH, J. R. & D. ZIPSER, Eds. 1970. The Lactose Operon. Cold Spring Harbor Laboratory. Cold Spring Harbor, New York.
11. BETENBAUGH, M. J., C. BEATY & P. DHURJATI. 1989. Biotechnol. Bioeng. **33:** 1425–1436.
12. ZABEAU, M. & K. K. STANLEY. 1982. EMBO J. **1:** 1217–1224.
13. CAULCOTT, C. A., *et al.* 1985. J. Gen. Microbiol. **131:** 3355–3365.
14. LEE, S. B., A. SERESSIOTIS & J. E. BAILEY. 1985. Biotechnol. Bioeng. **27:** 1699–1709.
15. LEE, S. B. & J. E. BAILEY. 1984. Biotechnol. Bioeng. **26:** 1372–1382.
16. LEE, S. B. & J. E. BAILEY. 1984. Biotechnol. Bioeng. **26:** 1383–1389.
17. BETENBAUGH, M. J. & P. DHURJATI. 1990. Biotechnol. Bioeng. In press.

Optimal Induction of Protein Synthesis in Recombinant Bacterial Cultures[a]

WILLIAM E. BENTLEY[b] AND DHINAKAR S. KOMPALA

Department of Chemical Engineering
University of Colorado
Boulder, Colorado 80309-0424

INTRODUCTION

The yield of recombinant biologicals depends not only on the functional capability of the host cell, but also on the biosynthetic capacity of the cell. Many researchers have alluded to the importance of this capacity.[1-3] Indeed, the elimination of biological bottlenecks and the directed metabolic flow through any given pathway is being pursued by many researchers for many products, recombinant proteins notwithstanding.[1,4,5]

In recent years, significant progress has been made towards characterizing, circumventing, and eliminating plasmid instability in recombinant bacterial systems.[6] For example, by inserting a gene into the expression plasmid that encodes the synthesis of an essential enzyme, whose activity has been deleted in the host cell, we can prevent growth of plasmid-free cells born by uneven partitioning. Many researchers have proposed methods for estimating and potentially ensuring plasmid or culture stability;[7-10] however, few have addressed the capacity of a cell to produce a recombinant product.

Our previously reported model for recombinant *E. coli*[10] contains sufficient metabolic detail to determine the influence of rDNA and its concomitant activity on the metabolism of the host so that the biosynthetic capacity of the host can be evaluated. This work uniquely and explicitly calculates the specific growth rate of a bacterial cell mass as a dynamic function of the metabolic burdens of foreign protein translation and plasmid replication. We illustrate its power here by predicting the influence that induced protein synthesis and induced "runaway" replication have on the growth rate and, in turn, the influence that the change in growth rate has on the product expression and eventual yield in batch cultures.

MODEL OVERVIEW

The mathematical model was developed to provide sufficient flexibility and structural complexity for calculating the separable effects of plasmid replication and cloned gene product (cgp) translation on cell growth rate. It is not numerically overwhelming

---

[a]This work was supported by the National Science Foundation (Grant Nos. EET-8611305 and BCS-8857719).
[b]Present address: Center for Agricultural Biotechnology and Chemical Engineering Program, University of Maryland, College Park, Maryland 20742.

and is intended for general use by design engineers. Furthermore, the general framework can be used for analysis of any biological cell culture and it has been used subsequently in simulations of hybridoma culture dynamics.[11]

Modeling Framework

The approach for directly calculating the instantaneous specific growth rate of a microorganism is summarized here. This general approach and our specific model for *E. coli* are discussed in detail in a separate paper (see reference 10).

The entire cell mass is divided into an appropriate number, n, of constituent pools, C_i, so that the sum of their concentrations, given in mass fraction, is unity at all times:

$$\sum_{i=1}^{n} C_i = 1.0. \quad (1)$$

Furthermore, the time derivative of this sum is zero at all times:

$$\frac{d}{dt}\sum_{i=1}^{n} C_i = 0. \quad (2)$$

A differential mass balance equation is written for each pooled constituent in which the rate of change of a constituent concentration is equal to the sum of the rates of synthesis and depletion, r_{ij}. In general, this equation may be written as

$$\frac{dC_i}{dt} = \sum_{j=1}^{n} r_{ij} - \mu C_i. \quad (3)$$

The synthesis rate expressions may be of any mathematically consistent form, such as saturation kinetic terms relating intrinsic cellular components.[12] The summation of all n differential equations yields

$$\sum_{i=1}^{n} \frac{dC_i}{dt} = \sum_{i=1}^{n}\sum_{j=1}^{n} r_{ij} - \mu \left[\sum_{i=1}^{n} C_i\right], \quad (4)$$

where μ is the instantaneous specific growth rate. This mass balance equation was first derived in a slightly different form by Fredrickson.[13] We have reformulated this equation by expressing the intrinsic concentrations as quantities of mass fraction; thus, by substituting equations 1 and 2 into equation 4, it can be readily shown that

$$\mu = \sum_{i=1}^{n}\sum_{j=1}^{n} r_{ij}. \quad (5)$$

This equation enables the development of multicompartmental growth models, which have previously been limited to two or three components.[14] Furthermore, by employing mass fraction, the numerical complexity is substantially reduced without sacrificing model performance. As previously stated, this simple and powerful equation for calculating the instantaneous specific growth rate can potentially be used for any microorganism. Full utilization of this expression requires judicious determination of the intrinsic quantities that are metabolically important to explain the phenomena of interest.

Model for E. coli *Metabolism*

In order to predict the dynamic reduction in growth rate due to the previously mentioned, additional metabolic burdens, our lumped metabolic model should include significant cellular metabolic detail. Palsson and Joshi[15] performed a linear algebraic analysis of the highly detailed *E. coli* single-cell model[16,17] and found that the doubling time of plasmid-free *E. coli* cells could be predicted using a much simpler three-compartment model. However, in order to simulate the metabolism of recombinant *E. coli*, we must differentiate between indigenous and foreign proteins, indigenous and foreign DNA, and the precursors for each. Necessarily, we must have a more developed biochemical structure and, thus, have included eight major intracellular constituent pools. These are: protein, P; foreign protein, P_f; chromosomal DNA, G; plasmid DNA, G_f; ribosomal RNA, R; lipids and other cell membrane material, L; nucleotides, N; and amino acids and TCA-cycle intermediates, A.

A differential mass balance equation is written for each *E. coli* constituent pool. These equations stoichiometrically interrelate the reaction pathways. In models utilizing this structure, the level of metabolic detail and genetic control is incorporated into the synthesis expressions for any given constituent and must utilize our complete knowledge of the system biochemistry. The resultant synthesis expressions for *E. coli* are given in table 2 of reference 10. The instantaneous specific growth rate for recombinant *E. coli* cells, calculated using the general form shown earlier, is given by

$$\mu = \left[\frac{dA}{dt}\right]_s + (1 - \epsilon_1)\left[\frac{dN}{dt}\right]_s + (1 - \epsilon_2)\left[\frac{dL}{dt}\right]_s$$
$$+ (1 - \gamma_1)\left(\left[\frac{dP}{dt}\right]_s + \left[\frac{dP_f}{dt}\right]_s\right) + (1 - \gamma_2)\left(\left[\frac{dG}{dt}\right]_s + \left[\frac{dG_f}{dt}\right]_s + \left[\frac{dR}{dt}\right]_s\right), \quad (6)$$

where $\epsilon_1, \epsilon_2, \gamma_1$, and γ_2 denote stoichiometric constants and the subscript s refers to the synthesis rates of the pooled components.

By eliminating two constituent pools, G_f and P_f, the model characterizes the plasmid-free host. This is accomplished by removing their mass balance expressions from the previous set of eight (equations 4 and 6 in tables 1 and 2 of reference 10) and solving the remaining six simultaneously. The resultant expression for the specific growth rate of the plasmid-free host then becomes

$$\mu = \left[\frac{dA}{dt}\right]_s + (1 - \epsilon_1)\left[\frac{dN}{dt}\right]_s + (1 - \epsilon_2)\left[\frac{dL}{dt}\right]_s$$
$$+ (1 - \gamma_1)\left[\frac{dP}{dt}\right]_s + (1 - \gamma_2)\left(\left[\frac{dG}{dt}\right]_s + \left[\frac{dR}{dt}\right]_s\right). \quad (7)$$

RESULTS AND DISCUSSION

In order to maximize the product yield in recombinant bacterial cultures, it is necessary to minimize culture or plasmid instabilities at the reactor level, while maximizing the product synthesis rate at the cellular level. These are necessarily coupled because the difference in growth rate between the plasmid-bearing and plasmid-free cells increases as the foreign protein synthesis rate increases. This difference, in several host/vector systems, was identified as the dominant factor in determining the culture stability.[18,19]

Induction of Protein Synthesis

In order to maximize the total product expression in batch cultures, we must first identify which factors influence product expression and determine the relative importance of each. An appropriately structured model can serve as a powerful tool in this respect. A comprehensive model, specific for a unique host/vector system, can further be used to uniquely define conditions that optimize product expression.

In systems where the growth rate differential between plasmid-bearing and plasmid-free cells is the dominant factor in determining the population dynamics, it may be advantageous to confine the translational metabolic burden to that associated with the synthesis of the specific product protein. Thus, by controlling translation so that only the specified protein is overproduced and by leaving the remaining plasmid-encoded proteins constitutively expressed at low levels, we minimize the overall growth rate differential. This can be accomplished by inserting controllable promoters, such as λP_L and *tac*, upstream as well as transcriptional and translational stop codons downstream of the product gene so that promoter activity is confined to the product gene.

The differential mass balance equation for the foreign protein pool, P_f, contains a combined transcription and translation rate, μ_4, that characterizes the strength of the promoter and expression level of the plasmid-encoded protein. All constants, except μ_4, and the structure of the foreign protein mass balance equation are identical to those of the indigenous protein pool, P (equation 3 in tables 1 and 2 of reference 10):

$$\frac{dP_f}{dt} = \mu_4 \left[\frac{A}{K_{P_fA} + A} \right] RG_f - K_{TP}P_f - \mu P_f. \tag{8}$$

We have determined this constant for our model experimental system, *E. coli* RR1[pBR329], and have evaluated this and the corresponding plasmid replication constants for several published studies (including those of Bron and Luxen[20] and Siegel and Ryu[21]). The resulting comprehensive host/vector models perform well in describing the continuous culture population dynamics.[19]

This constant, μ_4, may be a function of an external or internal stimulus, such as temperature (using a λP_L promoter) or inducer concentration (with *tac*, *trc*, or *lac* promoters), that can be manipulated in experiments to derepress the product synthesis. In a subsequent work, we develop the functional dependence of μ_4 on the control variable (i.e., inducer concentration, temperature, cell mass, etc.):

$$\mu_4 = f([inducer], T, X). \tag{9}$$

We have assumed that the plasmid-bearing and plasmid-free subpopulations can each be approximated by average cells. Also, the two cell-type models were identical except for the additional recombinant cell equations concerning foreign protein and plasmid DNA. Hence, the eight equations for the recombinant cell mass were coupled with the six equations for the plasmid-free host. These equations, which determine the material fluxes within the cell mass, provide for explicit calculation of μ^+ and μ^-, which, in turn, are incorporated into the reactor material balances:

$$\frac{dX_T}{dt} = \mu^+ X^+ + \mu^- X^- \tag{10}$$

and

$$\frac{dS}{dt} = -\frac{1}{Y_s^+}\mu^+ X^+ - \frac{1}{Y_s^-}\mu^- X^-, \tag{11}$$

where the superscripts plus and minus denote the presence and absence of plasmids, respectively. The total cell mass, the growth-limiting substrate concentration, and the yield coefficients are denoted by X_T, S, and Y_s, respectively. In these simulations, the yield coefficients for the plasmid-bearing and plasmid-free cells were assumed identical.

Simulation Results

The intent of these simulations is to investigate the effects of cgp induction on the host cell growth rate and subsequent product expression. Consequently, we have limited our simulations to batch cultures where population instabilities are dominated by the growth rate differential between the plasmid-bearing and plasmid-free subpopulations. In such host/vector systems, the plasmid copy number is sufficiently high to ensure that the influence of plasmid segregation during the batch culture is negligible on population dynamics.[18] However, the effects that the growth rate dynamics of the product-synthesizing subpopulation have on the overall product expression in a mixed culture are illustrated by assuming 1% plasmid-free cells in the inoculum.

The pulse addition of a strong inducer (e.g., IPTG for the *tac* promoter) to a batch culture is simulated in FIGURE 1. The maximum expression rate of cgp, μ_4, is increased instantaneously from a basal level of 1500 to 6000 h^{-1}, 5.0 hours after inoculation. The growth rate of the recombinant cells, calculated explicitly by equation 6, decreases after inducer addition (the shift in μ_4) and then recovers to a steady, but depressed level. The plasmid-free population, set at 1% in the inoculum, increases, but does not overtake the recombinant population by the end of the batch. The product level steadily increases over the duration and reaches 1.6 g/L when the batch is complete after 7.0 hours due to substrate exhaustion.

In a similar simulation (FIGURE 2), μ_4 is increased from 1500 to 8000 h^{-1}, again after 5.0 hours of uninduced growth. In this case, the recombinant cells essentially stop growing ($\mu \approx 0.08$ h^{-1}), but their intracellular cgp content increases slightly. The batch is overrun by the plasmid-free population and the substrate is exhausted after 9.5 hours. The final cgp concentration is significantly lower (0.2 g/L), which surprisingly

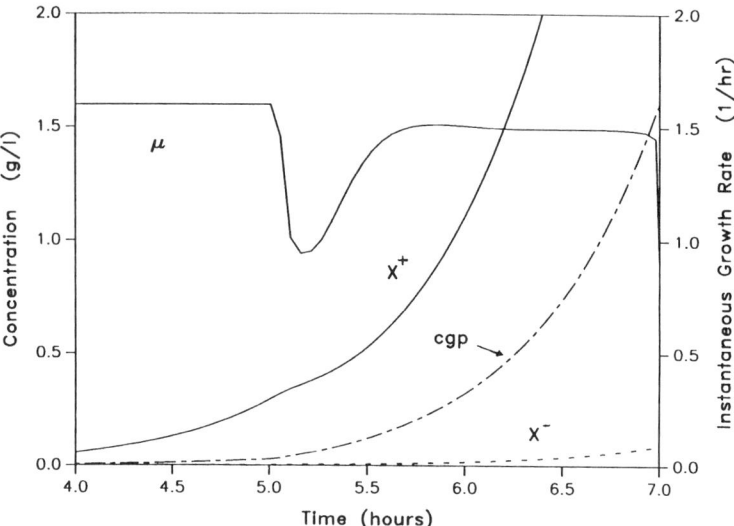

FIGURE 1. Simulation of induced batch culture. Cloned gene product (cgp) induction is simulated by a step jump in the maximum translation rate, μ_4, from 1500 to 6000 h^{-1}. The effects on recombinant cell mass, X^+, wild-type cell mass, X^-, cloned gene product, cgp, and instantaneous specific growth rate, μ, are depicted. A significant perturbation in μ is predicted. The cell mass yield coefficient, Y_s, for both plasmid-free and plasmid-bearing subpopulations was 0.51 g/g · s for all simulations. Initial substrate and cell mass concentrations for all simulations were 10 and 1×10^{-4} g/L, respectively.

suggests that lower levels of induction, as shown in FIGURE 1, enhance overall product synthesis.

In FIGURE 3, the growth rate of plasmid-bearing cells is illustrated as a function of time for several postinduction values of μ_4. In all cases, the growth rate is immediately depressed upon induction; however, in many cases, it rebounds to a new, but somewhat lower level than before induction. When the value of μ_4 is shifted to levels between 3000 and 8000 h^{-1}, growth rate oscillations are observed. Growth rate oscillations, albeit for different circumstances, have been observed experimentally.[22] More importantly, when the induction is quite strong, $\mu_4 \geq 7800$ h^{-1}, the initial drop in growth rate is so great that the cells do not recover by the time the substrate has been exhausted. The sudden increase in product synthesis can be accommodated by the preinduction level of ribosomes and the induced level of mRNA; however, the depletion of the amino acid/precursor pool is so severe that the rate of amino acid synthesis becomes self-limiting. This hypothesis is a result of examining the model synthesis rates and mass fractions in the constituent pools and is not based on experimental evidence. It is, though, consistent with the views of many researchers who have suggested that the production of recombinant proteins and the burden placed on the cells are directly linked to the depletion of precursor molecules.

From these simulations, it becomes apparent that many factors must be considered when inducing protein synthesis. The optimal induction time may depend on the stability of the plasmid and subsequent population dynamics, on the absolute strength of the promoter, and on the extent to which the promoter can be controlled; induction activity may be strictly "on/off" or moderated at intermediate levels.

FIGURE 4 summarizes a series of simulations similar to those in FIGURES 1 and 2. This is a diagram of the total cgp produced by the time the substrate has been depleted as a function of both promoter constant, μ_4, and time of induction. The dashed line is the cgp obtained at basal translation rates ($\mu_4 = 1500$ h^{-1}). When the addition of an inducer quantity yields an immediate and severe growth rate reduction (e.g., $\mu_4 = 8000$ h^{-1}, FIGURE 3), the best induction time is toward the end of the batch, when the recombinant cell mass is high. The cessation of recombinant cell growth due to the product synthesis is tolerated here because the cell mass has reached a sufficiently high value.

Conversely, an increase in μ_4 from 1500 h^{-1} to $\leq$7000 h^{-1}, when made at any time before the substrate becomes significantly reduced, results in high cgp expression. This is indicative of a weak promoter/operator region or a promoter for which intermediate levels of activity can be obtained. Furthermore, as the induction is increased in this regime, the yield is also increased.

It is significant to note that product expression may be in a very sensitive or nearly unstable regime for intermediate induction levels. For example, by inducing foreign protein synthesis to a level just greater than that simulated by $\mu_4 = 7000$ h^{-1}, cgp expression could drop from nearly 2 to 0 g/L. This assumes a linear response in promoter activity with inducer. Simulations indicate that there is a product synthesis rate above which the cells cannot recover. Operation near this regime may be optimal, but not stable.

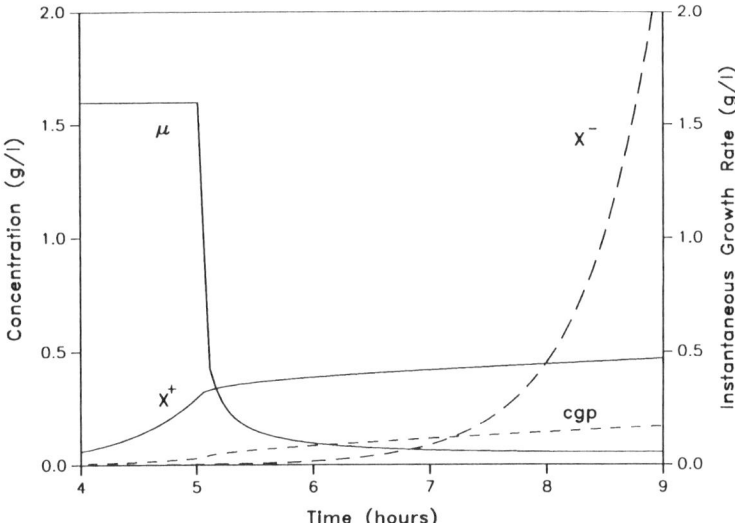

FIGURE 2. Simulation of induced batch culture. The maximum translation rate, μ_4, was changed from 1500 to 8000 h^{-1} after 5 hours of growth. Results indicate near cessation of cell growth brought about by the initiation of a high rate of cgp synthesis. The wild-type cell mass overtakes the batch culture and eventually consumes all remaining substrate. The resultant product yield is significantly lower than that in FIGURE 1.

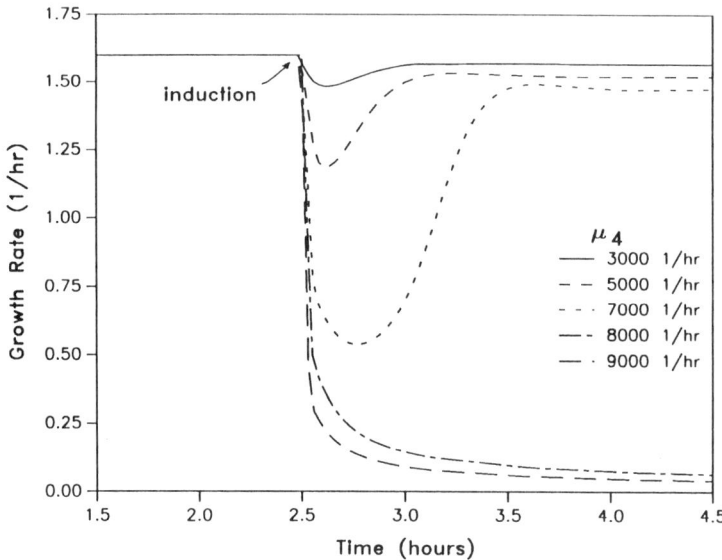

FIGURE 3. Dynamic influence of cgp induction on the instantaneous specific growth rate. Predictions indicate that growth rate is highly sensitive to induction level. An initial drop in growth rate is predicted at all levels of induction (μ_4). The magnitude of this reduction and the time required for recovery increase with the strength of induction. Ultimately, cell growth is arrested for $\mu_4 \geq 8000$ h^{-1}. Experimental evidence from Takagi et al. supports these simulations and substantiates that the range of μ_4 used here is experimentally valid. The addition of 2 mM IPTG in the Takagi et al. experiments resulted in cell death, which is analogous to the simulation with $\mu_4 = 8000$ h^{-1}.

In summary, if the promoter is particularly strong and not easily controlled, then the time of induction may become very important. If protein synthesis is only increased slightly by induction, then the induction time may be less important. In either case, it is apparent that promoter activity is of major importance in determining optimal induction conditions. Moreover, substantial gains in product expression may be realized by controlling and optimizing promoter activity. Consequently, efforts taken to develop a suitable promoter region are warranted when designing inducible expression systems.

Comparison to Experimental Results

The above simulation results are substantiated qualitatively by comparison to a recent study by Takagi et al.,[23] who examined the IPTG-induced expression of subtilisin E in *E. coli*. Their experimental procedure called for the addition of IPTG at various concentrations after a Klett reading of 100 was obtained. Ampicillin was added (50 µg/mL) to maintain a homogeneous plasmid-bearing population. Upon addition of 2 mM IPTG, the cells actually stopped growing and lysed due to the overexpression of subtilisin. This is qualitatively the same response as our simulation result for shifting μ_4

from 1500 to 8000 h^{-1} (FIGURE 2). Furthermore, the addition of IPTG at low concentrations (0.002 mM) brought about a slight reduction in growth rate. This observation is also in excellent agreement with our simulation results (FIGURE 1).

In order to more closely compare model predictions with the Takagi *et al.* experimental results, further simulations were performed for a homogeneous plasmid-bearing population, which Takagi *et al.* maintained by antibiotic selective pressure. Therefore, only the eight equations representing the recombinant cell mass were incorporated into the reactor material balances. FIGURE 5 illustrates the increase in predicted cgp content at several times after induction was initiated (in this case, the step in μ_4 was at 4 hours). The Takagi experiments were sampled once, at 4 hours after IPTG addition. It is apparent that the cgp increases with time in the postinduction period. Furthermore, an increase in promoter activity (μ_4) yields an increase in product synthesis. When μ_4 is increased to $7000 \leq \mu_4 \leq 8000$ h^{-1}, the drop in growth rate is dramatic, as illustrated in FIGURE 3, and the model predictions indicate a substantially delayed recovery in growth rate. In this regime, the product yield continues to increase with μ_4. However, the time required for substrate exhaustion is extended dramatically. Simulations with postinduction $\mu_4 \geq 8000$ h^{-1} do not predict a recovery within 10 hours and are most likely representative of cell death. It is noteworthy that an increase in product synthesis is accompanied by an increase in the overall batch time because the growth rate has been reduced. This factor was not seen in the previous simulations because of the presence of plasmid-free cells that rapidly consume substrate.

As shown in the previous simulations, when the promoter constant, μ_4, is increased, a maximum in product yield is predicted when the batch is sampled at a fixed

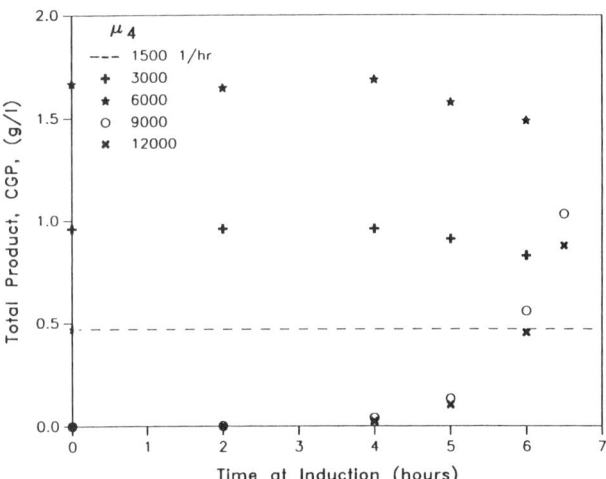

FIGURE 4. Predicted influence of induction time and strength on cgp yield. The reference dashed line represents a simulation of the constitutively expressed system (no change in μ_4). The simulated product yield for the inducible system increases with promoter constant, μ_4, until a maximum is predicted, followed by a dramatic reduction. Simulation of strong "on/off" promoters ($\mu_4 \geq 8000$ h^{-1}) indicates that induction time is important and should occur near the end of the batch.

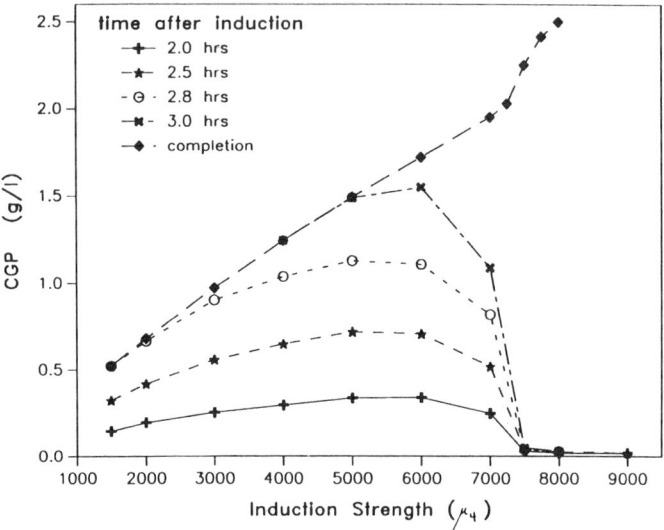

FIGURE 5. Simulated cgp concentration during postinduction period. Product concentration is plotted as a function of promoter constant, μ_4, and time following induction. A significant result is that the length of time for substrate consumption increases with μ_4 because the calculated growth rate is most significantly reduced when promoter activity is highest (high μ_4).

postinduction time. This maximum is followed by a substantial drop as μ_4 is increased further. As previously stated, this drop in cgp yield is due to the drop in growth rate, which, in turn, occurs when the amino acid/TCA-cycle intermediate pool is severely depleted. Takagi *et al.* increased the IPTG concentration (from 0.002 to 0.005 mM) and found an increase and maximum in subtilisin expression. A further increase in IPTG concentration (ultimately to 2.0 mM) yielded a five-times reduction in subtilisin. The same trends were observed in cultures at two different temperatures (37 °C and 23 °C). TABLE 1 includes the simulated cgp content resulting from one specific induction time (e.g., $t = 6.0$ hours, FIGURE 3) as a function of inducer level (μ_4). We obtain the same functional variation. The yield of subtilisin, measured by Takagi *et al.*

TABLE 1. Comparison of Experimental Results and Model Predictions

Induction by IPTG (mM)	Takagi et al. cgp (rel.)		Induction: Increase μ_4 at 6 h[a]	Model Final cgp (g/L)
	37 °C	23 °C		
2	1.0	9.0	12,000	0.45
0.2	1.0	12	9000	0.58
0.02	1.0	14	7500	1.08
0.01	2.5	14	6000	1.58
0.005	5.0	16	3000	0.83
0.002	4.8	9.3		

[a]Terms in this column are in units of h^{-1}.

and normalized to 2 mM IPTG at 37 °C, is included for comparison. The product yield is low for high IPTG concentrations because the cells were killed by subtilisin overproduction. As the IPTG concentration was further reduced, the subtilisin expression increased until a maximum occurred (at 0.005 mM IPTG). A subsequent reduction in IPTG was followed by a drop in product yield.

The agreement between the model predictions and the experimental results is quite good. An induction level at which cells essentially stop growing was observed both experimentally and by simulation. Furthermore, slight levels of induction brought about a reduction in growth rate. Most importantly, however, both experimental results and model predictions indicate that a maximum product yield is obtained for intermediate levels of induction.

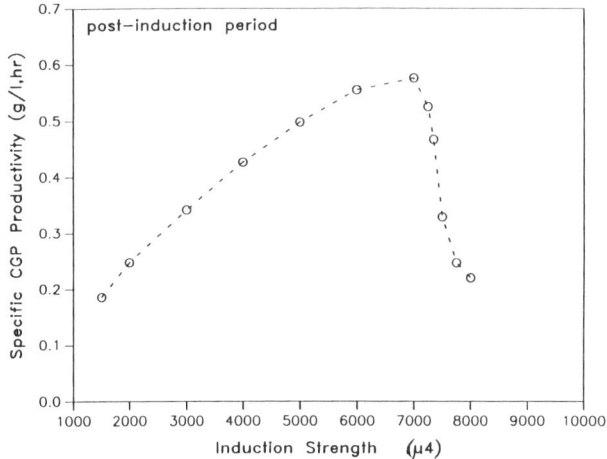

FIGURE 6. Average postinduction productivity versus induction strength, μ_4. The average productivity is the product concentration (g/L) at the completion of the batch divided by the length of time after induction for substrate depletion. A maximum in productivity is predicted. Operation of a batch culture may be desired at the maximum yield (FIGURE 5) or, alternatively, at the maximum productivity (above).

The overall product yield in batch cultures, with selective media, can be maximized by operating an IPTG concentration at or just above the level where a significant growth rate reduction occurs. However, this may not be the optimal IPTG concentration because the completion of this batch culture will be delayed due to a significantly reduced growth rate. By dividing the final product concentration by the time required for batch completion, we obtain an average postinduction productivity. This average productivity was plotted as a function of the postinduction promoter constant, μ_4, in FIGURE 6. The highest productivity corresponds to a μ_4 lower than where the maximum yield was obtained. In this simulation, the maximum productivity was obtained when $\mu_4 = 7000$ h^{-1} and the maximum yield was obtained when $\mu_4 = 8000$ h^{-1}. From these simulations, if reactor time is economically important, it would be best to operate at one specific set of operating conditions. Conversely, if overall product yield determines the product viability, a different operating strategy should be employed. We would like to emphasize here that the exact operating conditions for subtilisin E production have not been established. Instead, we demonstrate the utility of the structured kinetic model that can sift through the factors influencing product expression and ultimately define an optimal reaction strategy.

Runaway Replication

The stimulation of cgp synthesis can also be accomplished by providing multiple copies of the plasmid with the anticipation that an increase in copy number will yield a proportional increase in cgp. For example, by utilizing ColE1-type plasmids, which are present at copy numbers ranging from 10 to 100 per cell, we can achieve much higher expression levels than by employing R1-type plasmids, which are present at only a few copies per cell.

Remaut and co-workers[24] and Larsen and co-workers[25] have constructed plasmid replication systems in which replication is inducible and becomes rampant. Plasmid DNA levels equivalent to over 1000 monomer units were obtained per cell.[24] Larsen *et al.* inserted the λP_R thermoregulated promoter and the cI gene with its p_M promoter just upstream of the *repA* gene in the R1 replication region. Overexpression of the *repA* gene product, brought about by a simple increase in temperature, resulted in runaway replication. The dramatic increase in plasmid content, though, was not accompanied by a proportional increase in the concomitant plasmid-encoded product level. Moreover, in both studies, the cells eventually died within a few hours. These observations were rationalized by the severe limitation of precursors and the limitations of the protein-synthesizing system when the copy number was extremely high.

We have modeled this phenomenon with the structured kinetic model of recombinant *E. coli*. Previous model simulations[10] suggest that plasmid replication, in itself, has no significant deleterious effect on the cell growth rate, so increasing plasmid content does not significantly burden the host cells. These simulations have been substantiated, in part, by DaSilva and Bailey.[3] Increasing plasmid copy number may, however, increase the expression of constitutively produced proteins, which will depress the growth rate because the translation of foreign mRNA into proteins represents a significant metabolic burden. The runaway-replication system itself requires the overexpression of a polypeptide gene product; therefore, we would expect a reduction in growth rate even in the absence of other significant product polypeptides.

Simulation Results and Comparison to Experimental Data

Our plasmid synthesis expression contains a maximum replication rate, μ_5, and a saturation constant describing the dependence of plasmid replication on nucleotide content, $K_{G_f N}$. The overall mass balance expression for the plasmid DNA mass fraction, G_f, is written as

$$\frac{dG_f}{dt} = \mu_5 \left[\frac{N}{K_{G_f N} + N} \right] - \mu G_f, \tag{12}$$

where the synthesis expression is the first term on the right-hand side of this equation. We have estimated the numerical values for these constants, $\mu_s = 5.0 \times 10^{-4}$ h^{-1} and $K_{G_fN} = 1 \times 10^{-9}$, for the ColE1-type plasmid, pBR329, in *E. coli* RR1.[18] The result of a near zero value for K_{G_fN} is zeroth-order reaction kinetics for plasmid replication. This is consistent with the ColE1 replication model by Wittrup and Bailey.[26] However, these constants may be functions of external controllable factors, such as temperature (using a λP_R promoter near the replication origin) or inducer concentration (with *deo*, *trp*, or *lac* promoters), which are varied in experiments to depress the replication.

We have simulated runaway replication using an arbitrarily high value for μ_5 in an analogous fashion to the protein induction simulations with varying μ_4. FIGURE 7 illustrates the response in plasmid copy number and cgp content (g/L) for a batch

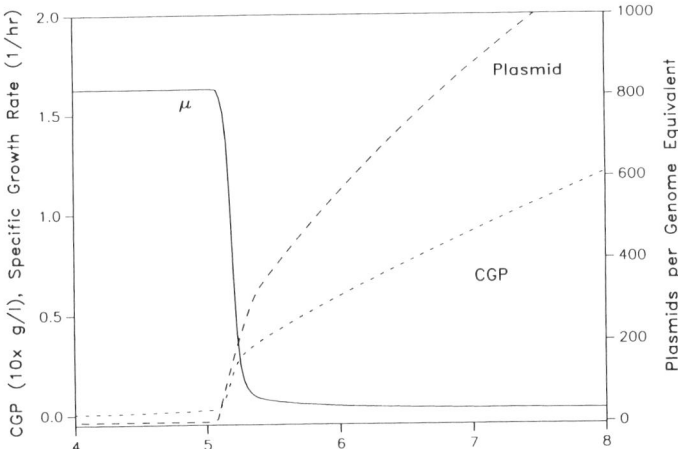

FIGURE 7. Simulation of the runaway-replication expression system. The onset of runaway replication (mathematically brought about by a step jump in μ_5 from 0.0002 to 0.075 h^{-1}) is accompanied by the cessation of cell growth. Simulated plasmid content is amplified several hundred times, whereas the cgp concentration is amplified by less than 50 times. This is in close agreement with observations by Larsen *et al.* and Remaut *et al.*

culture upon the addition of an inducer that initiates runaway replication. At 5 hours, the maximum replication constant is changed from $\mu_5 = 0.0002$ to $\mu_5 = 0.075$ h^{-1}. The corresponding plasmid copy number increases dramatically from 2, which is typical of R1-type plasmids, to upwards of 1000 after 2 hours. This is quantitatively the same as observed by both Remaut *et al.* and Larsen *et al.* The simulated plasmid content was amplified several hundred times. The simulated foreign protein content 7 hours into the batch was 0.93 g/L in the runaway-replication system, whereas the same system allowed to grow in the uninduced state (no step change in μ_5) resulted in 0.03 g/L cgp. Thus, the foreign protein content was effectively amplified less than 50 times. This much lower amplification in protein content was also observed by both Remaut *et al.* and Larsen *et al.* The calculated growth rate, also depicted in this figure, illustrates a

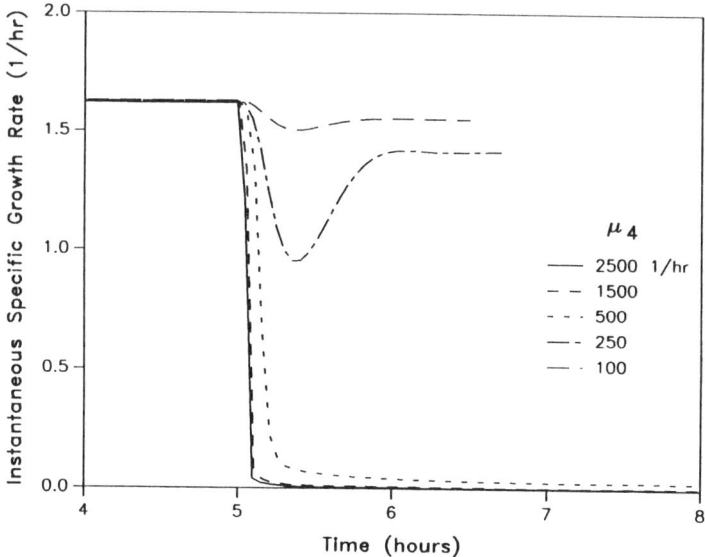

FIGURE 8. Growth rate dynamics accompanying runaway replication. The effect of different levels of background, constitutively expressed protein is illustrated by simulating runaway replication with different values of μ_4, the maximum translation rate of foreign protein. Predictions indicate that runaway replication is possible without cell death for low levels of background protein expression ($\mu_4 \leq 250 \ h^{-1}$).

severe drop within the first 30 minutes after runaway replication was initiated. The onset of cell death indicated in the experimental studies is not addressed by the present model equations. Therefore, we are unable to predict how long the cells will remain viable while at such low predicted growth rates.

However, the potential for runaway-replication vectors may be studied using the present model structure because it successfully predicts the response in copy number and product synthesis. Furthermore, growth rate predictions agree quite well with those experimentally observed. An important criterion in determining this potential is the synthesis of other constitutively expressed plasmid-encoded proteins. As mentioned previously, an increase in copy number should be accompanied by an increase in plasmid-encoded protein. FIGURE 8 illustrates the response in growth rate due to the initiation of runaway replication with varying levels of background, constitutively expressed proteins. In this simulation, the maximum replication rate constant, μ_5, was shifted from 0.0002 to 0.075 h^{-1} (initiating runaway replication), whereas the foreign protein synthesis constant, μ_4, was varied for the different curves. The growth rate responses are not unlike those observed for the protein induction simulations (FIGURES 1–3).

A significant difference, though, is that the response curves that do not result in cell death ($\mu_4 = 100$ and $250 \ h^{-1}$) are presently unrealistic. The preamplification foreign protein levels for these two cases were 0.2% and 0.6% of the total protein, respectively, which are lower than experimentally measured, even for R1 plasmids. Such expression vectors presently encode at least one antibiotic resistance in addition to the constitutively expressed product and the derepressed *repA* gene product, which can account for approximately 15% of the total protein. One obvious conclusion from this simulation is that it will potentially be advantageous to construct an expression vector that does not yield high levels of extraneous proteins.

FIGURE 9 depicts the plasmid content for a series of simulations similar to those illustrated in FIGURE 8. As the plasmid-encoded gene product level is decreased, the rate of plasmid synthesis is increased until a maximum rate of plasmid synthesis is reached ($\mu_4 = 250$ h^{-1}). A further reduction in protein content yielded less plasmid amplification ($\mu_4 = 100$ h^{-1}). In cases where the product desired is plasmid DNA, it would appear that there is an intermediate level of plasmid-encoded protein that would result in a maximum replication rate. Therefore, if a specific gene or segment of DNA is desired in bulk quantity, these simulations suggest that consideration should be given to the construction of the remaining sections of the plasmid so that maximum replication rates can be obtained.

The synthesis of the concomitant plasmid-encoded gene products is depicted in FIGURE 10. The foreign protein is shown in dimension % of total protein because we are not interested in the degree of amplification, but in the final content in the cells after the fermentation has been stopped. Simulation results indicate that there is a maximum amplified foreign protein content corresponding to a preamplified level of constitutively expressed protein. In the results shown, this occurs when $\mu_4 = 250$ h^{-1} or when the unamplified level of protein is around 1% of the total. Moreover, local maxima may exist as illustrated by the difference between the simulations with $\mu_4 =$

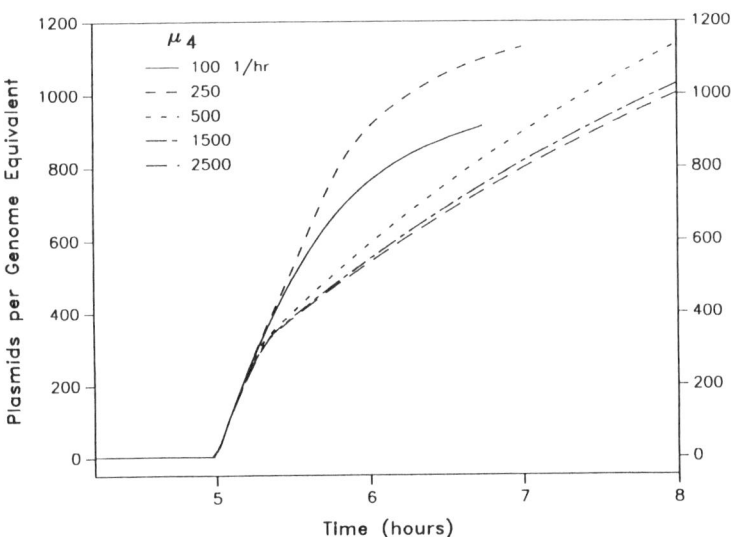

FIGURE 9. Amplification of plasmid copy number. Predictions indicate that maximum amplification is achieved when an intermediate level of background plasmid-encoded protein is expressed (μ_4 near 250 h^{-1}). The corresponding preinduction levels of foreign protein were 5.5%, 3.4%, 1.1%, 0.6%, and 0.2% of total protein for $\mu_4 = 2500, 1500, 500, 250,$ and 100 h^{-1}, respectively.

1500 and 2500 h^{-1}, respectively. For all these simulations, the onset of runaway replication was brought about by a switch in μ_5 from 0.0002 to 0.075 h^{-1}. Thus, the differences are due solely to the influence of foreign protein synthesis.

From this analysis, it is apparent that the level of constitutively expressed proteins is significant when optimizing the induced expression of recombinant proteins. The phenomena that were observed experimentally for runaway-replication vectors have been predicted with the present model structure. The detailed analysis indicates that significant improvements in product expression rates may be possible, with slight genetic modification to the plasmid vector.

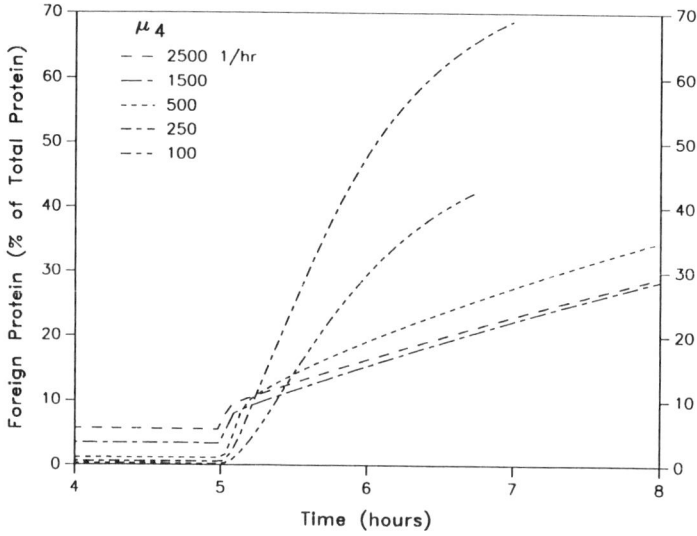

FIGURE 10. Cloned gene product content during runaway replication. This figure illustrates the cgp content for the simulations depicted in FIGURE 9. A maximum product expression rate is also predicted for a low level of background protein expression. Present experimental systems, however, are characterized by higher extraneous protein production followed by cell death upon runaway replication. These simulations suggest that plasmid modifications that reduce constitutive nonproduct foreign proteins may yield substantial improvements in product yield.

SIGNIFICANCE

Many factors were shown to be significant when developing optimal strategies for maximizing product synthesis using inducible expression vectors. These include plasmid stability, promoter activity, the extent of promoter regulation, the timing of induction, and the background level of constitutively expressed proteins. The experimentally observed phenomena agree quite well with the predicted phenomena. It is important to note again that these simulations are based on a model structure that was previously unavailable; hence, the influence of these factors may not have previously been considered. Our intention here is to bring attention to—but not definitively state—which factors are important and where.

The detailed quantitative description and validation or refutation of these predictions is not possible at this time because more comprehensive experimental data are required. The simulations do suggest, however, that significant improvements in product expression can result with carefully controlled reactor conditions as well as with minor genetic modifications to a given host/vector system. In this light, simulations such as these serve as indicators of potential research direction, along with being tools for evaluating the relative influence of system characteristics.

Quantitative agreement with the experimental observations may be thoroughly tested after the required plasmid and cgp constants—μ_4, μ_5, and $K_{G/N}$—are independently determined for the specific host/vector system. In this way, the general host/vector model can be made comprehensive for a unique host/vector system. As previously mentioned, these model constants were evaluated for several published studies in which the cgp was constitutively expressed, and the resultant comprehensive models performed quite well in predicting the experimental phenomena. Experimental elucidation of these constants is presently under way for a specific model host/vector system in which the expression of chloramphenicol-acetyl-transferase is under *trc* control.

The intent of this work, though, is not to develop the optimal conditions for any given system, but to illustrate the various factors that should be considered for maximizing the product yield. We would like to emphasize the utility of the structured kinetic model in identifying and differentiating between the important operating parameters whose optimization will bring about a maximal product yield. The model provides a powerful design tool in that it elucidates the dynamic influence of recombinant DNA on the metabolism of host microorganisms.

REFERENCES

1. ATAAI, M. M. & M. L. SHULER. 1987. A mathematical model for the prediction of plasmid copy number and genetic stability in *Escherichia coli*. Biotechnol. Bioeng. **30:** 389–397.
2. SERESSIOTIS, A. & J. E. BAILEY. 1987. Optimal gene expression and amplification strategies for batch and continuous recombinant cultures. Biotechnol. Bioeng. **29:** 392–398.
3. DASILVA, N. A. & J. E. BAILEY. 1986. Theoretical growth yield estimates for recombinant cells. Biotechnol. Bioeng. **28:** 741–746.
4. BAILEY, J. E., D. D. AXE, P. M. DORAN, J. L. GALAZZO, K. F. REARDON, A. SERESSIOTIS & J. V. SHANKS. 1987. Redirection of cellular metabolism: analysis and synthesis. *In* Biochemical Engineering V. Volume 506. M. L. Shuler & W. A. Weigand, Eds.: 1–23. N.Y. Acad. Sci. New York.
5. RAO, G., P. J. WARD & R. MUTHARASAN. 1987. Manipulation of end-product distribution in strict anaerobes. *In* Biochemical Engineering V. Volume 506. M. L. Shuler & W. A. Weigand, Eds.: 76–83. N.Y. Acad. Sci. New York.
6. IMANAKA, T. 1987. Some factors affecting the copy number of specific plasmids in *Bacillus* species. *In* Biochemical Engineering V. Volume 506. M. L. Shuler & W. A. Weigand, Eds.: 371–383. N.Y. Acad. Sci. New York.
7. IMANAKA, T. & S. AIBA. 1981. A perspective on the application of genetic engineering: stability of recombinant plasmids. Ann. N.Y. Acad. Sci. **369:** 1–14.
8. COUTINHO, T. M. & M. A. HJORTSO. 1988. A new method for quantifying the probability of segregative plasmid loss in *E. coli* B/r. Biotechnol. Tech. **2(2):** 141–146.
9. CAULCOTT, C. A., A. DUNN, H. A. ROBERTSON, N. S. COOPER, M. E. BROWN & P. M.

      RHODES. 1987. Investigation of the effect of growth environment on the stability of low-copy-number plasmids in *Escherichia coli*. J. Gen. Microbiol. **133**: 1881–1889.
10. BENTLEY, W. E. & D. S. KOMPALA. 1989. A novel structured kinetic modeling approach for the analysis of plasmid instability in recombinant bacterial cultures. Biotechnol. Bioeng. **33**: 49–61.
11. BATT, B. C. & D. S. KOMPALA. 1989. A structured kinetic modeling framework for the dynamics of hybridoma growth and monoclonal antibody production in continuous suspension cultures. Biotechnol. Bioeng. **34**: 515–531.
12. SHULER, M. L. 1985. The use of chemically structured models for bioreactors. Chem. Eng. Commun. **36**: 161–189.
13. FREDRICKSON, A. G. 1976. Formulation of structured growth models. Biotechnol. Bioeng. **18**: 1481–1486.
14. HARDER, A. & J. A. ROELS. 1982. Application of simple structured models in bioengineering. *In* Advances in Biochemical Engineering. Volume 21. A. Fiechter, Ed. Springer-Verlag. New York/Berlin.
15. PALSSON, B. O. & A. JOSHI. 1987. On the dynamic order of structured *Escherichia coli* growth models. Biotechnol. Bioeng. **29**: 789–792.
16. SHULER, M. L., S. LEUNG & C. C. DICK. 1979. A mathematical model for the growth of a single bacterial cell. Ann. N.Y. Acad. Sci. **326**: 35–55.
17. DOMACH, M. M., S. K. LEUNG, R. E. CAHN, G. G. COCKS & M. L. SHULER. 1984. Computer model for glucose-limited growth of a single cell of *Escherichia coli* B/r-A. Biotechnol. Bioeng. **27**: 203–216.
18. BENTLEY, W. E. & D. S. KOMPALA. 1989. Using structured kinetic models for analyzing instability in recombinant bacterial cultures. *In* Frontiers in Bioprocessing. S. Sikdar, P. Todd & M. Bier, Eds. CRC Press. Boca Raton, Florida.
19. BENTLEY, W. E. & D. S. KOMPALA. 1989. Stability in continuous cultures of recombinant bacteria: a metabolic approach. Biotechnol. Lett. Submitted.
20. BRON, S. & E. LUXEN. Segregational instability of pUB110-derived recombinant plasmids in *Bacillus subtilis*. Plasmid **14**: 235.
21. SIEGEL, R. & D. D. Y. RYU. 1985. Kinetic study of instability of recombinant plasmid pPlc23*trpAl* in *E. coli* using two-stage continuous culture systems. Biotechnol. Bioeng. **27**: 28–33.
22. MATELES, R. I., D. Y. RYU & T. YASUDA. 1965. Measurement of unsteady state growth rates of micro-organisms. Nature **208**: 263–268.
23. TAKAGI, H., Y. MORINAAGA, M. TSUCHIYA, H. IKEMURA & M. INOUYE. 1988. Control of folding of proteins secreted by a high expression secretion vector, pIN-III-*ompA*: 16-fold increase in production of active subtilisin E in *Escherichia coli*. Bio/Technology **6**: 948–950.
24. REMAUT, E., H. TSAO & W. FIERS. 1983. Improved plasmid vectors with a thermoinducible expression and temperature-regulated runaway replication. Gene **22**: 103–113.
25. LARSEN, J. E. L., K. GERDES, J. LIGHT & S. MOLIN. 1984. Low-copy-number plasmid-cloning vectors amplifiable by derepression of an inserted foreign promoter. Gene **28**: 45–54.
26. WITTRUP, K. D. & J. E. BAILEY. 1988. Mathematical model of recombinational amplification of the 2μ plasmid in the yeast *Saccharomyces cerevisiae*. Presented at the 196th ACS National Meeting, Los Angeles, California.

Expression, Purification, and Immobilization of a Protein A–β-Lactamase Hybrid Protein[a]

GEORGE GEORGIOU AND FRANÇOIS BANEYX

Department of Chemical Engineering
University of Texas at Austin
Austin, Texas 78712

INTRODUCTION

With the progress of recombinant DNA techniques, it is now possible to join the genes coding for two (or more) different proteins together. Expression of these gene fusions in microorganisms such as *Escherichia coli* produces hybrid proteins that frequently retain both functional activities. Hybrid proteins have been used in the investigation of basic biological processes such as gene regulation, cellular differentiation, protein transport, and detection of cloned genes. Other applications of hybrid proteins include the production of bifunctional enzymes[1,2] and protein purification.[3-5] In this report, we demonstrate that bifunctional proteins produced from a gene fusion are useful for the selective immobilization of enzymes. When a hybrid protein is constructed between the desired enzyme and a protein presenting a high affinity for a ligand, the resulting polypeptide can be immobilized on a matrix to which the ligand has been covalently attached. The affinity immobilization of hybrid proteins could offer several advantages. First, if the binding of the hybrid protein to the ligand is very specific, immobilization and purification can be accomplished in a single step by mixing crude cell extract with the ligand-coated insoluble matrix. Second, binding occurs through a part of the hybrid protein that is not involved in the catalytic process. Hence, immobilization should not have a significant effect on the active site of the bifunctional polypeptide. Third, steric hindrances with the polymeric matrix may be reduced because the binding domain of the hybrid protein acts as a spacer between the insoluble carrier and the active site. Overall, the affinity immobilization of hybrid proteins could provide the same benefits that have been demonstrated in the immobilization of proteins through monoclonal antibodies.[6,7] However, the high cost of monoclonal antibodies may be a serious disadvantage for the latter immobilization technique.

As a model system, we constructed a hybrid protein between β-lactamase and protein A. β-Lactamase is a 29,000-dalton monomeric enzyme that hydrolyzes the β-lactam ring of certain antibiotics. The β-lactamase gene was fused to the protein A gene from *Staphylococcus aureus* (SpA). SpA binds with high affinity to the Fc part of immunoglobin G (IgG) from several mammalian species. Its gene has been cloned and

---

[a]This work was supported in part from NSF by a Presidential Young Investigator award to George Georgiou (No. CBT-8657471).

expressed in *E. coli*, and a series of vectors suitable for gene fusions to protein A have been developed.[4,5,8,9] Fusions to protein A have been previously used for affinity purification using immobilized IgG. We have shown that the protein A–β-lactamase hybrid protein retains its enzymatic activity and its ability to bind IgG. The very strong binding between IgG and the hybrid protein (the dissociation constant is equal to 1.4×10^{-7} M in solution[10]) was exploited to bind the protein on IgG Sepharose. The immobilized hybrid protein exhibited higher enzymatic activity compared to covalently immobilized β-lactamase.

RESULTS AND DISCUSSION

Construction of the Protein A–β-Lactamase Gene Fusion

The plasmid pFB3 carrying the gene coding for the protein A–β-lactamase hybrid protein was constructed as described in FIGURE 1. The ampicillin resistance gene of plasmid pRIT12[11] was inactivated by insertion of an *Hin*c II–digested kanamycin resistance cartridge (Pharmacia) at its unique Sca I. The resulting plasmid was designated pRIT12K and conferred a kan^r and amp^s phenotype. A 2.0-kb EcoR I fragment containing the complete β-lactamase gene was isolated from plasmid pJG105[12] and ligated into the unique EcoR I site of the plasmid pRIT12K. The resulting plasmid pFB3 was isolated from ampicillin-, kanamycin-, and chloramphenicol-resistant transformants. The structure of plasmid pFB3 was confirmed by restriction analysis. Cell extracts from HB101(pFB3) exhibited β-lactamase activity and hydrolyzed penicillin G. Incubation of the cell extracts with IgG Sepharose 4B caused the complete removal of the enzymatic activity from the soluble fraction, indicating that the hybrid β-lactamase could bind IgG.

Localization of Protein A–β-Lactamase

Because both protein A and β-lactamase are efficiently secreted into the periplasmic space of *E. coli*, it was expected that the hybrid protein A–β-lactamase would also be secreted. The cellular location of the hybrid protein was investigated in cells harboring the plasmid pFB3. Approximately 10% of the total β-lactamase activity was excreted in the medium. However, the osmotic shock treatment of Neu and Heppel,[13] which is routinely used to recover proteins secreted in the periplasmic space, reproducibly released only about one-third of the total β-lactamase activity (data not shown). The low level of release by osmotic shock treatment was surprising because hybrid proteins between protein A and alkaline phosphatase or epidermal growth factor are efficiently secreted in the periplasmic space.[9] The localization of protein A–β-lactamase was investigated in detail.[10] It was shown that more than 70% of the protein is secreted through the inner membrane. However, about 25% of the secreted protein A–β-lactamase remained associated with the periplasmic face of the inner membrane and could not be released by osmotic shock.

Production and Purification

Expression of the gene fusion in HB101 resulted in low β-lactamase specific activities. Growth in minimum media or at lower temperature (30 °C) somewhat enhanced the total specific activities in this strain (TABLE 1), indicating that the hybrid

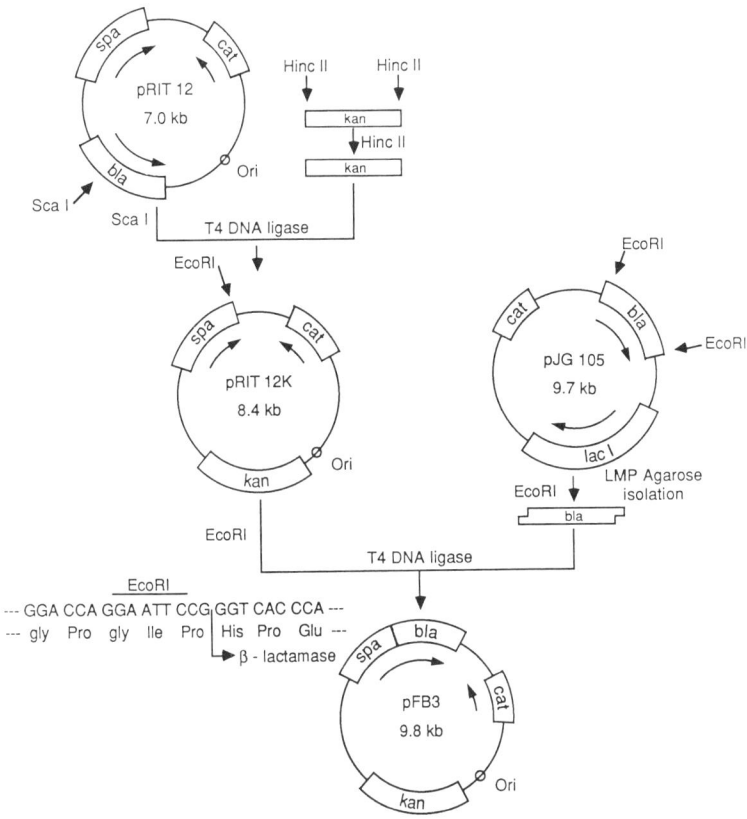

FIGURE 1. Construction of pFB3. Restriction sites and the location of the origins of replication are indicated. Boxes show the relative position of the genes coding for β-lactamase (bla), protein A (SpA), chloramphenicol acetyl transferase (cat), and lac I. The nucleotide sequence at the junction between protein A and the beginning of the mature β-lactamase is also shown. (Reprinted from reference 10.)

protein may be susceptible to proteolytic degradation. Analysis of the proteins in the osmotic shock fraction by immunoblotting with β-lactamase-specific antisera showed the presence of several proteolytic products and confirmed that the hybrid protein was susceptible to degradation.[10] Strauch and Beckwith[14] recently isolated an *E. coli* mutation (*deg*P), which was shown to confer increased stability to periplasmic proteins

TABLE 1. Effect of Host Strain and Growth Conditions on the Production of the Protein A–β-Lactamase Hybrid Protein[a]

Plasmid/Strain	Medium	Total Specific Activity (U/mg)	
		30 °C	37 °C
pFB3/HB101	M9	8.5 (ND)	2.7 (14.3)
pFB3/KS474	M9	7.5 (ND)	9.8 (14.0)
pFB3/HB101	LB	6.8 (5.0)	1.1 (16.0)
pFB3/KS474	LB	32.7 (0.5)	41.1 (10.0)

[a]Cells were grown in shake flasks in either LB media supplemented with 50 μg/mL ampicillin and 0.2% glucose or M9 media supplemented with the above concentrations of ampicillin and glucose and 0.2% casein amino acid hydrolysate. The total specific activity is the ratio of the total β-lactamase activity to the sum of protein in the supernatant, periplasmic fraction, and extract. The values in parentheses represent the percent of total activity excreted in the supernatant. ND stands for nondetected. Reprinted from reference 10.

sensitive to proteolysis. When the *deg*P strain KS474 (KS272, *deg*P41, lpp$^+$) was transformed with pFB3, a significant increase in the β-lactamase specific activity was observed (TABLE 1). Under optimal conditions for *E. coli* (LB medium and 37 °C), a 40-fold increase in total specific activity was obtained compared to HB101(pFB3). Nevertheless, proteolysis was not totally eliminated in KS474 as confirmed by immunoblotting (data not shown).

The effect of fermentation conditions on the production of the hybrid protein was investigated. As discussed earlier, shake-flask experiments indicated that growth in LB media resulted in higher production relative to M9 media. Therefore, KS474(pFB3) was grown in LB broth supplemented with 0.2% glucose and 50 μg/mL ampicillin. Sugimura and Higashi[15] recently detected a new outer membrane–associated protease in *E. coli* that was responsible for the degradation of recombinant human interferon-γ. This protease was completely inhibited by addition of 0.5 mM $ZnCl_2$. Because preliminary shake-flask experiments indicated a positive effect of zinc chloride, 0.5 mM $ZnCl_2$ was added to the growth medium. Fermentations were conducted in a Bioflo III fermentor (New Brunswick) with a working volume of two liters.

Production of the hybrid protein was lower when the pH was maintained at a constant value relative to fermentations run without pH control. At a constant pH of 6.0 or 7.0, the specific activity of β-lactamase reached 30 U/mg, corresponding to about 1% of the total soluble protein. In variable pH fermentations, the maximum specific activity was more than four times higher. Under these conditions, the hybrid protein accounted for about 4% of the total soluble protein of the cell. From these results, it is evident that the effect of the culture pH on the production of the hybrid protein is complex.

Hybrid polypeptides containing a functional protein A domain have been purified previously by affinity chromatography based on their affinity for IgG antibodies. However, due to the very strong binding between protein A and IgG, harsh conditions are required for elution. With the protein A–β-lactamase, elution with 0.5 M glycine-HCl, pH 3.0, or 100 mM lithium diiodosalicylate[9,16] caused nearly complete deactivation. Furthermore, the presence of proteolytic fragments containing a functional protein A moiety—and therefore able to bind IgG—complicated the isolation of the intact hybrid protein. For these reasons, the hybrid protein was purified to near homogeneity by the following procedure. The osmotic shock fraction of KS474(pFB3) was used for purification in order to take advantage of its higher specific activity relative to the total cell extract. The protein A–β-lactamase was precipitated with ammonium sulfate and further purified by two consecutive ion-exchange HPLC steps. Details of the purification process have been provided elsewhere.[10] The purified fraction collected after the second HPLC step was about 95% pure as judged by SDS-PAGE and had a specific activity of 2600 U/mg.

Functional Characterization of Protein A–β-Lactamase in Solution

The kinetics of hydrolysis of penicillin G of different preparations were compared with the native β-lactamase (TABLE 2). The k_{cat} and K_m values for the hydrolysis of penicillin G were essentially the same for the soluble β-lactamase, hybrid protein and for the hybrid protein:rabbit IgG complex (1:2 mol/mol ratio) in solution. It can be concluded that fusion of a 35-kDa polypeptide at the amino terminal of β-lactamase had essentially no effect on its catalytic function. The other function of the hybrid protein, that is, the ability of the protein A domain to form a complex with IgG, was also investigated.[10] The formation of the complex can be written as

protein A–β-lactamase + rabbit IgG ⇔ protein A–β-lactamase:rabbit IgG.

Briefly, different amounts of rabbit IgG were incubated with a known concentration of hybrid protein. The formation of protein A–β-lactamase:rabbit IgG complex

TABLE 2. Kinetic Parameters for the Hydrolysis of Penicillin G by Different Preparations of β-Lactamase and Protein A–β-Lactamase[a]

Enzyme	Buffer	K_m (μM)	k_{cat} (s^{-1})	$10^{-6} \times k_{cat}/K_m$ (M^{-1}s^{-1})
Soluble β-lactamase	50 mM potassium phosphate	53	1700	32
Soluble hybrid protein	50 mM potassium phosphate	79	2300	29.1
Soluble IgG:hybrid protein complex	phosphate buffer saline (PBS)	56	1990	35.5
CNBr immobilized β-lactamase	50 mM potassium phosphate	1190	920	0.8
IgG immobilized hybrid 2 mg IgG/mL of swollen gel	phosphate buffer saline (PBS)	478	1465	3.1
	PBS + 500 mM NaCl	463	1575	3.4
IgG immobilized crude extract 2 mg IgG/mL of swollen gel	phosphate buffer saline (PBS)	545	—	—

[a]The values of k_{cat} and K_m were obtained from Eadie-Hofstee plots using three progress curves at five different substrate concentrations. The rate constant is apparent for the immobilized enzyme preparations and is referred to as k'_{cat} in the text. Reprinted from reference 20.

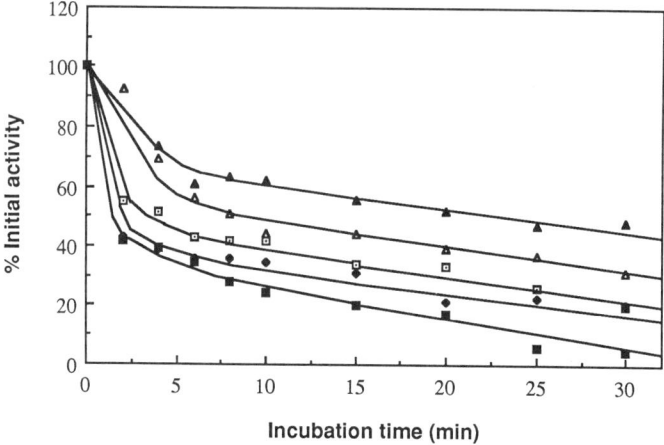

FIGURE 2. Thermal stability against irreversible deactivation at 60 °C of soluble β-lactamase (□), soluble protein A–β-lactamase (■), soluble rabbit IgG:protein A–β-lactamase complex (1:2 mol/mol) (♦), CNBr immobilized β-lactamase (△), and immobilized protein A–β-lactamase (2 mg IgG matrix) (▲). The different enzyme preparations were incubated in a constant temperature water bath (60 °C). Samples were withdrawn at different times, transferred on ice, and assayed. (Reprinted from reference 20.)

was allowed to reach equilibrium. Subsequently, an excess of goat antirabbit IgG covalently linked to agarose beads was added to remove all the unbound rabbit IgG from the solution. After one hour of incubation, the beads were precipitated by centrifugation. Protein A, and therefore the hybrid protein, has a weak affinity for IgG antibodies raised in goat[17] and binds poorly to the immunobeads. Less than 2% of the hybrid protein is bound to the beads in the absence of rabbit antibodies. However, the protein A–β-lactamase that has formed a complex with rabbit IgG could be precipitated specifically. Consequently, the β-lactamase activity remaining after removal of the beads is a direct measure of the free hybrid protein in solution. Using this simple procedure, the dissociation constant for the formation of the complex was estimated as 1.4×10^{-7} M. This value differs from the $K_d = 2 \times 10^{-8}$ M reported by Jonsson and Kronvall,[18] but is in good agreement with Lancet et al.,[19] who obtained a $K_d = 3.3 \times 10^{-7}$ M by fluorescence quenching measurements.

Although the hybrid protein was fully bifunctional, its stability was considerably lower compared to authentic β-lactamase. For example, it was inactivated faster at a temperature of 60 °C (FIGURE 2). Interestingly, a slightly higher stability to thermal deactivation was observed when the hybrid was incubated in the presence of an excess rabbit IgG (the binding of rabbit IgG and protein A is stable at this temperature[16]). The stability of the soluble enzymes against proteolytic degradation by trypsin is presented in FIGURE 3. The hybrid protein was totally degraded in 60 minutes, whereas the soluble β-lactamase retained about 80% of its initial activity under the same conditions. Consequently, the hybrid protein is much more sensitive to proteases *in vitro* as well as *in vivo*, as discussed above.

Immobilization on IgG Sepharose

For the immobilization of the hybrid protein, 2 mg of rabbit IgG/mL of swollen gel was covalently coupled to CNBr-activated Sepharose 4B. The hybrid protein could be immobilized quantitatively simply by mixing the crude cell lysate with IgG Sepharose. However, at least one other protein was also bound tightly to the support (data not shown). This protein was probably a degradation product containing a nearly intact protein A domain and was therefore still able to bind the Fc fragment of IgG. As a control, the native β-lactamase was covalently immobilized on CNBr-activated Sepharose 4B and the effect of immobilization on the kinetics of penicillin G hydrolysis as well as the stability profiles were compared.

The intrinsic k'_{cat} and K_m for the two immobilized enzyme preparations were determined in a batch microreactor. Care was taken to eliminate external and internal diffusional limitations by optimizing the size of the support and the load of enzyme.[20] The results of these studies are shown in TABLE 2. Covalent immobilization of β-lactamase on CNBr-activated Sepharose caused a 20-fold increase in K_m and a 54% decrease in k_{cat} compared to the soluble enzyme. On the other hand, immobilization of 0.058 mg of purified hybrid protein per mL of swollen gel on IgG Sepharose resulted in a less dramatic increase in K_m. Furthermore, the apparent k'_{cat} was 37% higher as compared to the covalently immobilized enzyme, but it was also lower than the value obtained for the soluble hybrid protein. Overall, affinity immobilization of the hybrid protein on IgG Sepharose enhanced the catalytic efficiency, k'_{cat}/K_m, by 3.9-fold compared to the covalently immobilized β-lactamase.

The kinetic parameters for the affinity immobilized hybrid protein were more remote from those of the soluble hybrid protein than expected. Because neither K_m nor

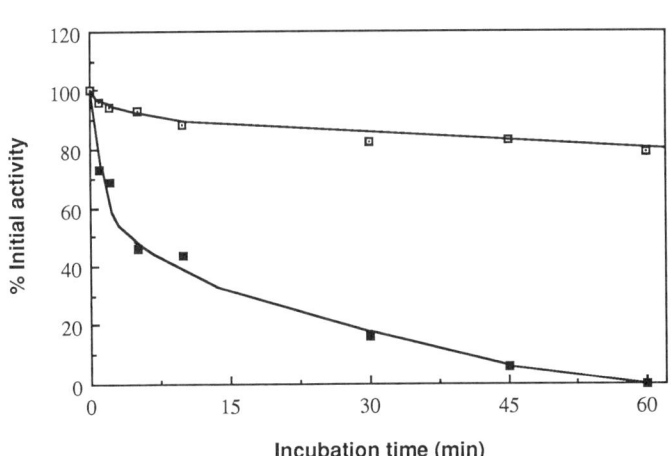

FIGURE 3. Stability of soluble β-lactamase (□) and soluble protein A–β-lactamase (■) against proteolytic degradation. Ten μL of enzyme solution was incubated with 10 μL of a 0.4 mg/mL trypsin solution for increasing time periods. The reaction was stopped by addition of 10 μL of 0.8 mg/mL trypsin inhibitor. The tubes were then transferred on ice and the residual β-lactamase activity was measured.

k_{cat} was drastically modified when the hybrid protein was incubated with rabbit IgG in solution, the change in the kinetic parameters of the immobilized hybrid protein could not be attributed to conformational modification resulting from the binding to IgG (TABLE 2). However, an increase in the amount of covalently immobilized IgG to 10 mg per mL of swollen gel increased the catalytic efficiency $(k'_{cat}/K_m)_{\text{affinity imm.}}$ 5.6 times relative to $(k'_{cat}/K_m)_{\text{covalent imm.}}$. It is likely that the preparation of the affinity matrix can be optimized so that the behavior of the immobilized hybrid enzyme resembles that of the native β-lactamase in solution.

The thermal stability profiles at 60 °C are shown in FIGURE 2. Immobilization of β-lactamase on CNBr Sepharose led to stabilization of the enzyme against thermal deactivation. As has been shown with other enzymes, covalent attachment can inhibit deactivating processes such as unfolding and intramolecular associations.[21] Surprisingly, immobilization of the hybrid protein to IgG Sepharose resulted in even higher stability against thermal deactivation despite the absence of covalent attachment. The basis of this phenomenon is not yet understood.

CONCLUSIONS

The genes for protein A and β-lactamase were fused in-frame using recombinant DNA techniques. The gene fusion was expressed in *E. coli* to give a 63-kDa hybrid polypeptide with both protein A and β-lactamase activities. The hybrid protein is secreted through the inner membrane. However, a fraction of the protein remains associated with the inner membrane and is not readily released from the cell. It was shown that the production of protein A–β-lactamase is limited by proteolytic degradation within the cell. The yield can be increased about 100-fold by using a suitable protease-deficient *E. coli* strain and by optimizing the fermentation conditions. The purified protein A–β-lactamase hybrid protein exhibits functional properties that are nearly identical to the unfused protein A and β-lactamase. Whereas the fusion of the two polypeptides had no apparent effect on either function, the stability of the hybrid protein was markedly decreased.

The hybrid protein can be simultaneously purified and immobilized in a single step by exploiting the affinity of protein A for IgG bound on Sepharose. We have observed that the binding is very strong and that no dissociation of the protein occurs even after several weeks of storage at 4 °C and physiological pH. Under these conditions, the immobilized hybrid protein retained about 60% of its initial activity after 24 days of storage. Furthermore, the immobilized hybrid protein has a higher specific activity and a lower K_m compared to covalently immobilized β-lactamase. These results demonstrate that affinity immobilization of engineered proteins is a promising alternative to conventional immobilization techniques. The additional genetic engineering work necessary for the construction of the gene fusion is largely compensated by the higher catalytic efficiency compared to covalent coupling and by the fact that purification and immobilization of the hybrid protein can be accomplished in one step.

REFERENCES

1. BÜLOW, L., P. LJUNGCRANTZ & K. MOSBACH. 1985. Bio/Technology **3**: 821–823.
2. WARREN, R. A. J., B. GERHARD, N. R. GILKES, J. B. OWOLABI, D. G. KILBURN & R. C. MILLER. 1987. Gene **67**: 421–427.

3. ULLMAN, A. 1984. Gene **29**: 27–31.
4. MOKS, T., L. ABRAHMSÉN, B. OSTERLOF, S. JOSEPHSON, M. OSTLING, S. O. ENFORS, I. PERSSON, B. NILSSON & M. UHLÉN. 1987. Bio/Technology **5**: 379–382.
5. MOKS, T., L. ABRAHMSÉN, E. HOLMGREN, M. BILICH, A. OLSSON, M. UHLÉN, G. POHL, C. STERKY, H. HULTBERG, S. JOSEPHSON, A. HOLMGREN, H. JORNVALL & B. NILSSON. 1987. Biochemistry **26**: 5239–5244.
6. SOLOMON, B., R. KOPPEL, G. PINES & E. KATCHALSKI-KATZIR. 1986. Biotechnol. Bioeng. **28**: 1213–1221.
7. FUSEK, M., J. TURKOVÁ, J. STOVICKOVÁ & F. FRANEK. 1988. Biotechnol. Lett. **10**: 85–90.
8. UHLÉN, M., B. GUSS, B. NILSSON, S. GATENBECK, L. PHILIPSON & M. LINDBERG. 1984. J. Biol. Chem. **259**: 1695–1702.
9. NILSSON, B., L. ABRAHMSÉN & M. UHLÉN. 1985. EMBO J. **4**: 775–780.
10. BANEYX, F. & G. GEORGIOU. 1989. Enzyme Microb. Technol. **11**: 559–567.
11. LOWENADLER, B., B. NILSSON, L. ABRAHMSÉN, T. MOKS, L. LJUNGQVIST, E. HOLMGREN, S. PALEUS, S. JOSEPHSON, L. PHILIPSON & M. UHLÉN. 1986. EMBO J. **5**: 2393–2398.
12. GHRAYEB, J., H. KIMURA, M. TAKAHARA, H. HSIUNG, Y. MASUI & M. INOUYE. 1984. EMBO J. **3**: 2437–2442.
13. NEU, H. C. & L. A. HEPPEL. 1965. J. Biol. Chem. **240**: 3685–3691.
14. STRAUCH, K. L. & J. BECKWITH. 1988. Proc. Natl. Acad. Sci. U.S.A. **85**: 1576–1580.
15. SUGIMURA, K. & N. HIGASHI. 1988. J. Bacteriol. **170**: 3650–3654.
16. MACSWEEN, J. M. & S. L. EASTWOOD. 1981. Methods Enzymol. **73**: 459–471.
17. LANGONE, J. J. 1982. Immunology **32**: 157–252.
18. JONSSON, S. & G. KRONVALL. 1974. Eur. J. Immunol. **4**: 29–33.
19. LANCET, D., D. ISENMAN, J. SJÖDAHL, J. SJÖQUIST & I. PECHT. 1978. Biochem. Biophys. Res. Commun. **85**: 608–614.
20. BANEYX, F., C. SCHMIDT & G. GEORGIOU. 1990. Enzyme Microb. Technol. In press.
21. KLIBANOV, A. M. 1983. Science **219**: 722–727.

PART III. PROTEIN SEPARATIONS AND DOWNSTREAM PROCESSING

A Novel Immunoaffinity Chromatography System for the Purification of Therapeutic Proteins

PETER GRANDICS, ZSOLT SZATHMARY, AND
SUSAN SZATHMARY

Sterogene Biochemicals
Arcadia, California 91006

Immunoaffinity chromatography is a powerful method for protein separation and is used in bioprocessing to isolate protein products in a highly purified form from a dilute mixture containing numerous contaminants. Some of these contaminants may be present in much higher concentration than the concentration of the desired product. However, immunoaffinity chromatography can introduce contaminants into the product stream that need to be monitored and eliminated in order to satisfy regulatory requirements.

The potential sources of product contamination due to immunoaffinity chromatography itself are as follows:

(a) chromatography support (bead, membrane),
(b) chemical linkages between the support and ligand,
(c) low-molecular-weight "leaving groups",
(d) proteolytic fragmentation of the ligand, and
(e) dissociation of subunits of immobilized protein.

First, contamination from the chromatography support: The appropriate chemical and mechanical stability of the chromatography support is a paramount issue for any bioprocess. We have studied a variety of chromatography supports for performance in affinity chromatography. Besides polysaccharide-based supports like agarose or cellulose, silica matrices and polymeric supports, such as hydroxymethyl-metacrylate (HEMA) or polystyrene resins, have been evaluated. Silica and HEMA matrices exhibit a pH-dependent release of ligand due to matrix disintegration. Support materials (beads or membranes) containing ester linkages are the source of matrix leaching at low or high pH.

Polymeric matrices may exhibit another type of matrix leaching. Due to incomplete polymerization, monomers or low-molecular-weight polymers are released from the matrices into the product stream. This has been recognized early on with polymer-based ion exchange media. Some of the monomers are mutagenic. The USP VI test for extractables provides information on the toxicity or irritation caused by the extracted materials, but fails to provide any information on the effects of monomers on the protein product. This needs to be examined and validated on an individual basis.

Chemical derivatization of membranes may also cause matrix leaching by changing the composition of the polymeric chains.

Second, contamination from unstable chemical bonds: The stability of the chemical linkage between the support and ligand is of great significance for an immunoaffinity chromatography separation. The currently used methods are known to be unstable and release the immobilized ligand over a period of time. Ligands released by affinity resins contaminate process streams and further purification steps would be required to eliminate the leached ligand from the product, thus increasing the cost of purification. Leached ligands may associate with purified proteins, thereby altering protein conformation and its function.

One of the reasons for the instability of current immobilization chemistries is that linkages of different stability are established between the support and different groups on ligands (proteins) or contaminants, such as endotoxin, DNA, or RNA, by reaction of the reactive groups on the support with amino, thiol, hydroxyl, or carboxyl groups on the respective ligands. For instance, the N-hydroxysuccinimide ester–activated groups react in a side reaction with hydroxyls to form unstable esters. The epoxide chemistry, which has been thought to exclusively establish stable ether or alkylamine linkages, also forms unstable esters by reaction of epoxide and carboxyl groups, particularly around neutral pH. These observations are suggestive of why immobilized proteins leach from affinity matrices, that is, due to the multiplicity or poorly defined nature of the chemical linkages (or both).

Third, contamination from leaving groups: It is also important to exclude those coupling chemistries having leaving groups that could covalently attach to proteins or that could interact hydrophobically with proteins. An example of covalent modification is the CNBr chemistry. CNBr-activated resins shed reactive, toxic isocyanates and, for that reason, are excluded from use in immobilized enzyme systems for food processing. The isocyanate modification may introduce new antigenic sites or alter protein function. Proteins may also associate hydrophobically with cyclic, unsaturated molecules like imidazole from the CDI coupling or methyl-pyridone from FMP coupling. In addition, the toxicity of these leaving groups has not been investigated. Therefore, validation protocols demonstrating that these leaving groups are not associated with the protein need to be developed.

Fourth, contamination from proteolytic degradation of the ligand: Proteolytic degradation of immobilized protein ligand is also an important factor that has received little attention. The presentation of Kato and Sada recently highlighted this issue[1] and demonstrated that antibodies immobilized to a CNBr-activated resin can be extensively proteolyzed. They suggested that the way that antibodies are immobilized may determine protease sensitivity. It is known that lengthy alkaline immobilization methods, characteristic of conventional protein immobilization techniques, may lead to inactivation (denaturation) of immobilized monoclonal antibodies.[2] It is well known that denatured proteins exhibit an increased sensitivity to proteolysis. This may be the reason why antibodies immobilized by the lengthy alkaline CNBr coupling exhibit a noticeable degree of protease sensitivity.

Lastly, product contamination may also derive from subunit loss or exchange of immobilized oligomeric proteins, like Con A. This can be prevented if the subunits are appropriately cross-linked following immobilization of the protein.

ActiSep IMMUNOAFFINITY CHROMATOGRAPHY SYSTEM

Current immunoaffinity chromatography techniques suffer from several shortcomings including leaching of the ligand from the support, significant loss of immunoreactivity upon immobilization, and denaturing of elution conditions which lead to the loss of immunosorbent capacity and denaturation of eluted proteins to some degree. We have addressed the issues underlying the development of a stable chemistry for an activated resin along with the conditions necessary to dissociate high-affinity antigen-antibody interactions. Our objective was to develop chromatography media and purification systems for protein biotherapeutics that would eliminate the hazards of existing technologies and simplify scaleup and process validation while increasing the economy of operations.

Consideration for the Support

The interaction between the protein and the chromatography support is very important. Until now, there has not been much concern about what happens to the structure of the protein as it passes through the column or the membrane cartridge. This is an important issue because, if a protein structure is changed during passing through the column or membrane, it is also possible that its function will change. These changes could unknowingly alter the outcome of the research and clinical studies. Experimental data have shown that hydrophobic interactions between the protein and the chromatography support can induce changes in the protein folding that could lead to alterations in either immunological characteristics or the function of the product (or both). Hydrophobic supports may also cause high nonspecific binding, leading to significant loss of protein during chromatography. Hydrophilic supports are therefore preferred to minimize changes in protein conformation as well as to minimize product loss. Polysaccharide chromatography media are highly hydrophilic and exhibit excellent chemical stability. We have solved the problem of low mechanical stability of polysaccharide matrices by developing a hardening method that allows the preparation of beaded agarose chromatography supports capable of withstanding a linear flow rate of 3000 cm/h.

Consideration for the Chemistry

It is of great importance to develop an immobilization chemistry that establishes uniform, stable linkages, preferably with a single functionality on proteins, such as an amino group. As the aldehyde chemistry allows coupling in a wide pH range and has no leaving group other than water, we therefore decided to optimize the aldehyde chemistry. We have found that the leaching is a function of the aldehyde group reactivity. In the case of glutaraldehyde, extensive polymerization occurs, forming a diene structure that significantly enhances the reactivity of the aldehyde groups. The reactivity of the aldehyde group is strongly affected by the chemistry of neighboring groups. The highly reactive aldehydes, such as glutaraldehyde, form strong Schiff bases that, however, are not as stable as secondary amines and that cannot be completely reduced to a stable secondary amine with $NaCNBH_3$. Therefore, these

Schiff bases slowly release immobilized ligands. Ligands attached to partially polymerized aldehydes also leach for the same reason. We have developed an aldehyde group of very low reactivity so that it does not polymerize or retain proteins in the absence of NaCNBH$_3$. The Schiff bases formed can be readily reduced with NaCNBH$_3$ to a stable secondary amine.

The coupling reaction is very rapid and the effective pH range for coupling is between pH 3 and 10. This allows protein immobilization at physiological pH, which is necessary for optimizing retained bioactivity. This is a very important feature for the immobilization of protein. For instance, monoclonal antibodies have been shown to be sensitive to extremes of pH.[2] When kept at pH 10 for 30 min, the monoclonals lost up to 60–70% of their immunoreactivity. Alkaline coupling conditions can clearly diminish the immunoreactivity of monoclonal antibodies. By selecting the most appropriate

TABLE 1. Leakage Study on Actigel A[a]

Treatments	Released Radioactivity (cpm)
0.01 M NaOH	55
0.1 M NaOH	92
1 M CH$_3$COOH, pH 2.0	48
0.2 M Tris, pH 8.5	63
1 M NH$_4$SCN, pH 7.0	69
3 M NH$_4$SCN, pH 7.0	60
2 M NaCl	65
ActiSep Elution Medium, pH 7.0	68

[a]Actigel A, reacted with 0.1 M [^{14}C]ethanolamine, was aliquoted into test tubes and a 20-fold excess of the above reagents was added. Samples were taken from the supernatant and were counted for radioactivity. Total radioactivity bound in each sample: 64,605 cpm. Background radioactivity: 40–60 cpm (to be subtracted from the values above).

coupling pH, the biological activity of immobilized protein can be maximized. The mild reducing agent NaCNBH$_3$ does not affect protein structure. An important advantage to the aldehyde coupling chemistry is that the leaving group is water.

Actigel A–activated resin is so stable that it can be sterilized by autoclaving, which is the most stringent, but widely used method of sterilization. Endotoxin removal by NaOH treatment is also feasible. None of the other activated resins exhibit similar characteristics.

The stability of the linkage between the ligand and Actigel A has been demonstrated by treating [^{14}C]ethanolamine-coupled Actigel A with various agents that may either induce the release of immobilized ligand or destabilize the support particles (TABLE 1). Actigel A, reacted with 0.1 M [^{14}C]ethanolamine, pH 8.5, was aliquoted into test tubes and a 20-fold excess of the above reagents was added. Samples were taken from the supernatants and counted for radioactivity. The radioactivity in the supernatant was found to be in the range of the background. No leakage of immobilized ligand was observed over a period of one year.

Considerations for the Elution Conditions

The high affinity of antibodies for antigens generally precludes elution in the solvent of application. Desorption of antigen from immunoadsorbents is usually achieved by reagents, such as chaotropic salts or denaturing agents like acids, urea, guanidine, aliphatic acids, and alcohols (or a combination of these compounds). These kinds of reagents, though, adversely affect the immunoreactivity of both the antigen and the immunoadsorbent, leading to the deterioration of immunosorbent capacity and the denaturation of eluted proteins to some degree. In spite of being strong denaturants, most of these reagents are unable to fully displace antigens from immunoadsorbents. The deterioration of expensive immunoadsorbents as well as the denaturation of purified proteins have limited the application range of immunoaffinity chromatography.

Protein denaturation occurring upon purification is an important issue and we have addressed protein denaturation on immunoaffinity chromatography by developing a neutral, nondenaturing elution medium (ActiSep) that allows the recovery of antibodies from immunoadsorbents with practically no loss of immunoreactivity. In the future, a strong emphasis will be placed on protein denaturation during purification because altered protein structure may cause experimental artifacts and severe side effects in clinical trials.

Several biotherapeutics are currently being manufactured by recombinant DNA methods and monoclonal antibody production is also rapidly growing. Immunoaffinity purification would enhance the overall economy of these manufacturing processes by permitting the recovery of the protein products rapidly in a high yield and purity.

IMMUNOAFFINITY CHROMATOGRAPHY BY USING ActiSep

Mouse IgG was immobilized to Actigel A and antimouse IgG antibodies were purified. In parallel experiments, the recovery of bound immunoglobulin was carried out by using the mild, neutral elution medium we have developed (ActiSep), acidic (0.1 M or 1 M acetic acid) elution media, and chaotropic (3 M NH_4SCN, pH 7.0) elution medium. Among all the elution media, the maximum recovery of bound antibody (96–98%) was obtained (TABLE 2). Among the acidic elution media, 0.1 M acetic acid (pH 2.8) failed to displace antibody efficiently from the column and was replaced by 1 M acetic acid (pH 2.2), thus permitting 82–86% recovery of bound anti-IgG. Chaotropic elution medium (3 M NH_4SCN, pH 7.0) was less efficient in this regard, recovering only 73–78% of bound immunoglobulin.

Immunoreactivity of eluted anti-IgG was determined[3] and found to be the highest with ActiSep elution medium (TABLE 2): 96–98% of the initial (reference sample). Antibody eluted with 1 M acetic acid displayed 27–32% of the initial immunoreactivity. Elution with 3 M NH_4SCN affected antibody activity less severely with the

TABLE 2. Comparative Elution of Antimouse IgG from Immunoadsorbent[a]

Elution Medium	1 M Acetic Acid	3 M NH$_4$SCN	ActiSep Elution Medium
Antibody bound (mg)	0.91	0.91	0.91
Antibody eluted (mg)	0.76	0.68	0.88
Yield (%)	84	75	97
Immunoreactivity of eluted IgG (% of initial value)	29	43	97

[a] Mouse IgG was immobilized to Actigel A and 2-mL columns were packed. The columns were saturated with antimouse IgG. After washing, the columns were eluted by the respective elution media and fractions were collected. The eluate was analyzed for protein and immunoreactivity determined by ELISA.

retention of 41–46% of the initial immunoreactivity. Note that acidic elution not only leads to significant losses of antibody activity, but also to subtle changes in the conformation of IgG, detectable by its enhanced susceptibility to peptic digestion.[4] This allows complete digestion of isolated IgG by pepsin to F(ab')$_2$ in 2 hours as opposed to 16 hours for the control antibody.[4] The deleterious effect of extreme pH on the immunoreactivity of immunoglobulin has been noticed and leads to the deterioration of immunosorbent capacity upon continued recycling.[5] ActiSep elution medium

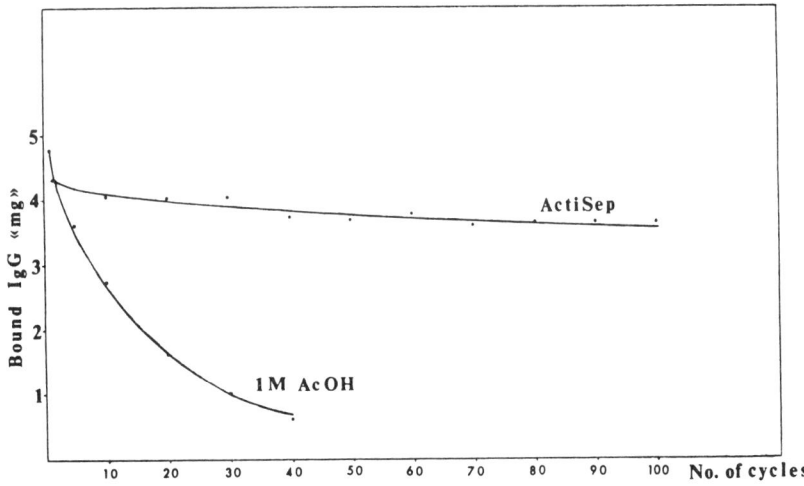

FIGURE 1. Comparative cycling of immunoadsorbents. Affinity-purified antirabbit IgG was immobilized to Actigel A and 5-mL columns were packed. Rabbit IgG was applied to the columns at saturating concentration and unbound protein was measured. After washing with 25 mM phosphate/1 M NaCl, pH 7.2, bound antibody was eluted either with ActiSep or 1 M CH$_3$COOH, pH 2.2, and protein was measured.

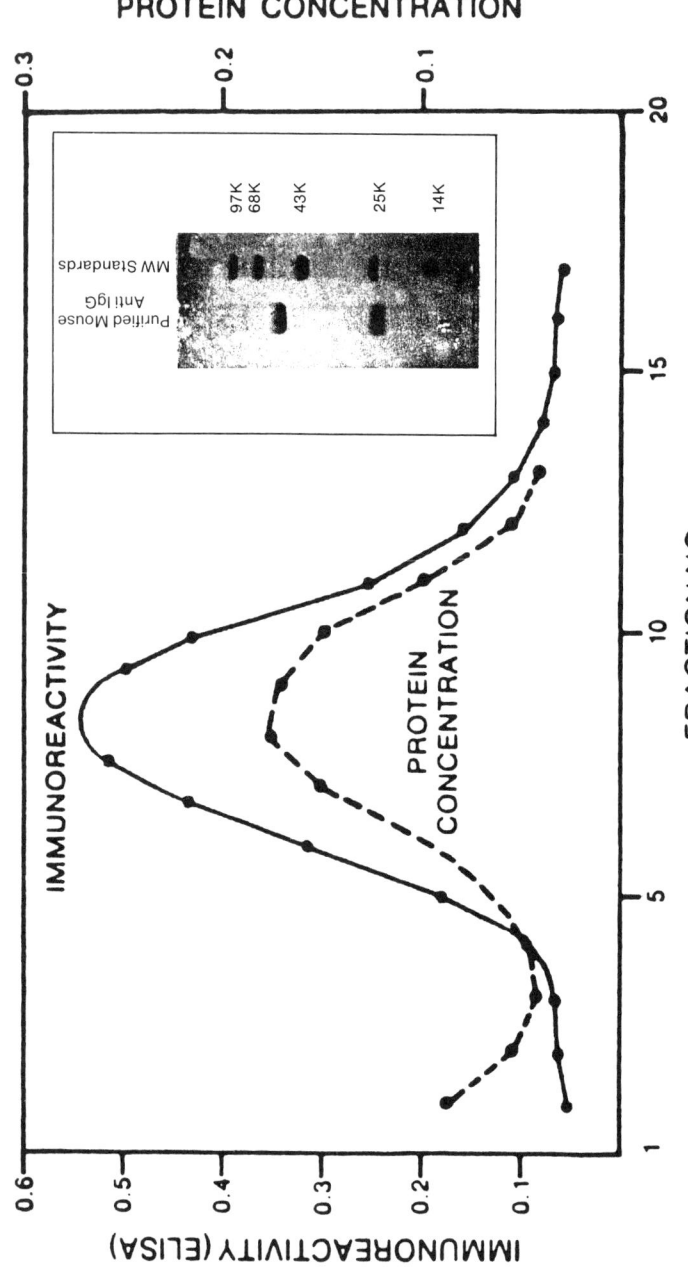

FIGURE 2. Immunoaffinity chromatography of antimouse IgG. Antimouse IgG was eluted from immunoadsorbent, and protein concentration and immunoreactivity were determined in peak fractions. Combined peak fractions were analyzed by SDS–polyacrylamide gel electrophoresis.

has been found to elute anti-IgG from the immunoadsorbent with the highest yield and immunoreactivity. The results suggest that ActiSep elution medium has the potential of becoming the nondenaturing eluant that could significantly expand the application range of immunoaffinity chromatography.

The effect of ActiSep elution medium on the long-term performance of an immunoadsorbent has also been studied. Affinity-purified antirabbit IgG was immobilized to Actigel A. The immunoadsorbents were saturated with rabbit IgG and both bound and eluted IgG concentrations were measured. The immunoadsorbents were cycled in a continuous operation mode. Acidic elution medium [1 M CH_3COOH (AcOH), pH 2.2] was used as a control (FIGURE 1). A very low, approximately 5–10% loss of the binding capacity of the immunoadsorbent was observed on 100 cycles when using ActiSep elution medium. The control immunoadsorbent, eluted by the acidic elution medium, lost over 60% of the initial binding capacity upon 20 cycles.

TABLE 3. Economy of Immunoaffinity Chromatography Purification of Monoclonal Antibodies

Cost of fast-flow support	$2.00/mL resin
Cost of polyclonal antibody immobilized (2 mg/mL)	$10.00/mL resin
10-liter column cost	$120,000
Total cycle life	+100
Column cost per cycle	$1200
Cost of ActiSep elution medium per cycle	$600
Purification cost per cycle	$1800
Average yield per cycle	14 g (IgG)
Ab purified on 100 cycles	1.4 kg
Purification cost per gram of product	$128

The purity of antibodies from immunoaffinity chromatography is very high (FIGURE 2). Purity of higher than 99% can be obtained in this step when purifying antibody from serum. From serum containing culture medium (10% FBS), even higher purity can be achieved (unpublished data).

The ActiSep Immunoaffinity Chromatography System (patent pending) can purify therapeutic proteins to pharmacopoeia specifications. Because monoclonal antibodies or recombinant DNA proteins derive from organisms having oncogenic potential, product purity and safety are the primary issues. Besides the removal of unwanted biologically active substances, the final product should be devoid of antibody leaching from the immunoadsorbent and of column matrix constituents. This must be properly documented. Actigel A is proven to be a stable and safe affinity support. ActiSep elution medium contains no toxic ingredients and components of the medium can be readily separated from the purified protein. The high cycle-life of the ActiSep Immunoaffinity Chromatography System significantly reduces the immunoadsorbent cost per cycle (TABLE 3) and, along with its safety features, it has successfully addressed those problems preventing the widespread application of immunoaffinity chromatography in bioprocesses.

REFERENCES

1. KATO, S & E. SADA. 1988. The Engineering Foundation Conferences on Recovery of Bioproducts, Kailua-Kona, Hawaii.
2. UNDERWOOD, P. A. & P. A. BEAN. 1985. J. Immunol. Methods **80:** 1892.
3. LEW, A. M. 1984. J. Immunol. Methods **72:** 171.
4. ROUSSEAUX, J., R. ROUSSEAUX-PREVOST & H. BAZIN. 1983. J. Immunol. Methods **141:** 141.
5. EVELEIGH, J. W. & D. E. LEVY. 1977. J. Solid-Phase Biochem. **2:** 45.

New Approaches to the More Efficient Purification of Proteins and Enzymes

N. J. TITCHENER-HOOKER, M. HOARE, AND P. DUNNILL

SERC Center for Biochemical Engineering
Department of Chemical and Biochemical Engineering
University College London
London WC1E 7JE, England

INTRODUCTION

The recovery and purification of proteins and enzymes from fermentation broths requires the use of sequences of unit operations, each of which must be designed to cope with inherent variabilities in the fermentation products. There is a need to operate such multistage processes efficiently and this requires that a rational engineering design approach to the processing problems be adopted.

This report will concentrate on the recovery and purification of soluble intracellular products derived principally from yeast. Although each particular material presents a number of individual processing problems, for instance, in terms of the nature of any contaminant species, it is possible to describe the general processing of such products by a common flow sheet.

In a typical flow sheet (FIGURE 1), a fermentation fed from a continuous sterilizer and operating as a chemostat is used to provide product for the downstream recovery train. The process flow scheme is engineered so as to minimize the loss of product due to degradation, such as that caused by protease attack, by employing continuous residence time processing. Rapid in-line broth cooling is followed by cell harvesting, cell disruption with cooling of the homogenate in a recycle loop, and then cell debris removal by centrifuge employing both frame and chamber cooling. Subsequently, the supernatant material is heat-treated to promote autolytic nuclease attack before centrifugal removal of the remaining debris and the heat-precipitated protein. Finally, the initial stages of protein purification are implemented in the form of fractional protein precipitation using in-line static mixing to contact both rapidly and efficiently the precipitating agent and protein streams. The precipitate is then recovered as product. Alternatively, a two-cut process is operated whereby the precipitate is discarded and the supernatant is further purified by precipitation and finally chromatographic separation by a sequence of ion exchange, affinity chromatography, and gel filtration (not shown). The basic process described has been operated with β-galactosidase from *E. coli* as the material.[1]

In the production of soluble intracellular protein products, the early stages of downstream processing are critical. Over the past years at University College London, a number of novel techniques have been developed that enhance the type of purification sequence described above by ensuring continuous operation throughout and by providing an integrated engineering approach to the problems of these early stages of processing. This report will address three of these techniques and examine how they

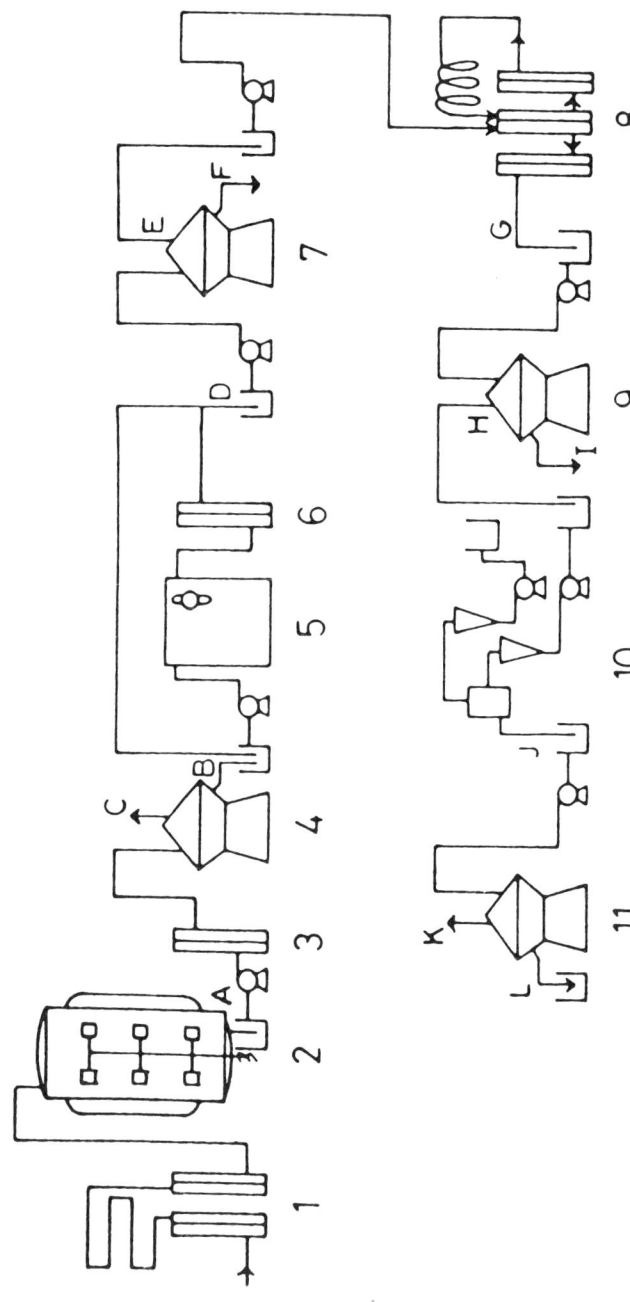

FIGURE 1. Flow diagram for the continuous isolation of β-galactosidase from *E. coli*: (1) continuous-flow plate media sterilizer; (2) 1000-L fermentor; (3,6) heat exchangers; (4,7,9) disk-bowl centrifuges; (5) homogenizer; (8) heat exchanger and holding coil; (10) continuous-flow mixer; and (11) disk-bowl centrifuge or tubular centrifuge.

may be combined in order to optimize such a process. The three areas are highly selective flocculation of cell debris away from soluble protein, the engineering description of fractional precipitation, and the use of low-frequency acoustic conditioning for enhancing the centrifugal recovery of fine material.

PROCESSING OF BIOLOGICAL PARTICLES

One major feature of the downstream processing of fermentation broths is the need to recover or remove a wide variety of biological particles. The particles may represent a product in the form of cells or protein precipitates or they may be a major contaminant such as cell debris. In addition, the concentration of particles will vary from trace amounts to concentrated suspensions. This will determine the physical characteristics of the process stream, for example, in terms of fluid viscosity and the extent of dewatering achievable, as well as influencing the most appropriate choice of separation mechanism, be it perhaps filtration or centrifugation. Furthermore, the size of the particles to be processed may range from submicron microsomal fractions to micron-sized precipitate aggregates.

The properties of biological particles and the ease with which they may be recovered by centrifugation may be shown (FIGURE 2) in a generalized manner.[2,3] By plotting a group incorporating a shape factor, K, the density difference, $\Delta \rho$, and the suspension viscosity, μ, against the particle diameter, it is possible to compare the centrifugal recovery of a wide range of biological products. Clearly, the density and viscosity values are determined by the concentration of the suspension and the properties of the suspending medium, and representative values have been assumed in this figure, for example, cell homogenates of 45 kg dry wt/m^3 and protein precipitate suspensions of 25 kg protein/m^3. This figure does not account for the sensitivity of particles to shear breakup, the number of particles at any given size, or their surface properties, all of which may be expected to have a bearing on their centrifugal recovery.

Protein precipitates as formed in a range of reactors exhibit a wide size distribution, with the random close packing arrangement leading to decreased density with increased size. As may be expected, protein precipitates prepared by salting out with high concentrations of $(NH_4)_2SO_4$ result in low density differences. However, the major problem arises from the shear sensitivity of these particles to breakup in pumps and feed zones to continuous centrifuges.

Protein inclusion bodies as prepared in *E. coli* are found to have a tight size distribution, whereas the *E. coli* debris, which must be separated, is only of slightly smaller size, leading to a difficult particle characterization procedure. In the case of yeast, a wide size distribution is observed ranging from ghosts of whole cells to finely divided debris and microsomal material. It is the recovery of yeast debris from soluble products that will be discussed further.

SELECTIVE FLOCCULATION

The problem of removal of cell debris from soluble protein may be investigated by examining the performance of a wide range of centrifuge types (disk stack, tubular

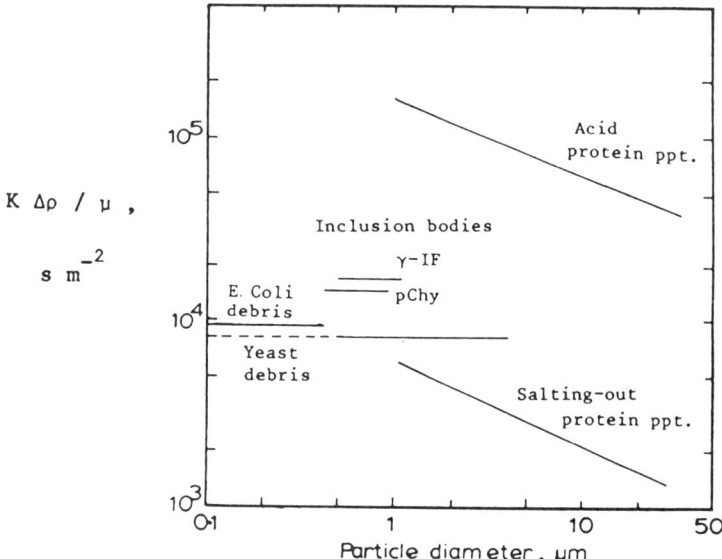

FIGURE 2. Settling properties of biological particles:[2,3] $\Delta\rho$ = density difference, K = shape factor, and μ = suspension viscosity.

bowl, and multichamber bowl) (FIGURE 3).[1,4] The percentage of solids not recovered in the sediment is plotted on a probability scale against the centrifuge throughput, Q, divided by the centrifuge sigma factor. The data for whole *E. coli* serve as a reference.

The material that proves particularly problematic when processing homogenized baker's yeast is represented by the solids (overflow ~10% of total), even when low centrifuge throughputs are employed. The inability to recover this fine cell debris will impose a serious limitation on the subsequent process stages. Analysis of the yeast cell debris chemistry indicates that borate ions may act as a selective flocculating agent for the carbohydrate cell in the presence of debris components of a number of soluble protein products.[5]

The effects of using such a flocculation technique are shown in FIGURE 4.[5] The 100% supernatant turbidity value in this figure corresponds to the turbidity obtained after the removal of the large particles at 2000g for 30 seconds. Any decrease in turbidity represents a removal of the residual fine debris particles. Using 10,000g for 20 minutes to define the limit of debris that may be removed, ~95% of the fine debris particles are removed by borax flocculation at pH 8.2 followed by low-speed (2000g) centrifugation. This represents an overall >99% removal of debris. Higher pH values are avoided in order to prevent enzyme inactivation and the flocculation process is complete at a modest 10 mM borax concentration.

A major problem with other flocculating agents is that the protein is brought down with the debris. However, in the case of borax, the protein and enzyme products remain in the supernatant due to the highly specific nature of the borate anion interactions. For a 10 mM borax flocculation, 100% recovery of the soluble fumarase, ADH, and also a glycoprotein invertase is obtained. However, as may be expected, invertase recovery is lower at higher borax concentrations (FIGURE 5)[5] because it has a carbohydrate component.

By using this form of highly selective flocculation, it is also possible to use high capacity industrial scroll decanter centrifuges for separating the flocculated debris away from the soluble protein product while still achieving a high degree of dewatering of the sediment. Such dewatering is needed to minimize losses of soluble product entrapped within the interstices of the centrifuge paste.[6]

We are presently examining other selective reagents and preliminary indications are that we shall be able to remove remaining, fine colloidal material and other nonprotein components selectively.

ANALYSIS AND CONTROL OF FRACTIONAL PRECIPITATION

By the use of flocculation, it is possible to ensure the maximum yield of soluble protein product for further downstream purification. In the process flow under examination (FIGURE 1), fractional precipitation is used to effect primary purification of the soluble protein product.[1] In the past, the forms of data presentation used to describe fractional precipitation processes have not been particularly useful for process engineering purposes and, hence, the control of fractionation processes has been poor, resulting in substantial time and effort being expended in attempting to develop processes based on fractional precipitation.

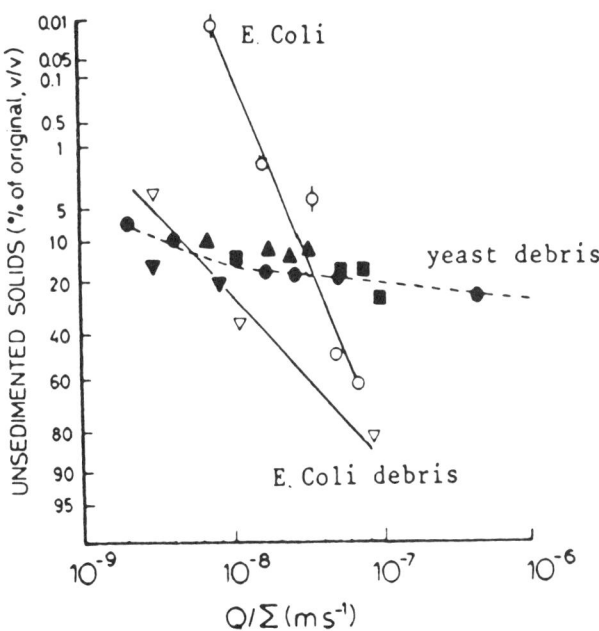

FIGURE 3. Influence of flow rate (Q) and centrifuge equivalent settling area (Σ) on the unsedimented solids in the centrifuge outflow.[14]

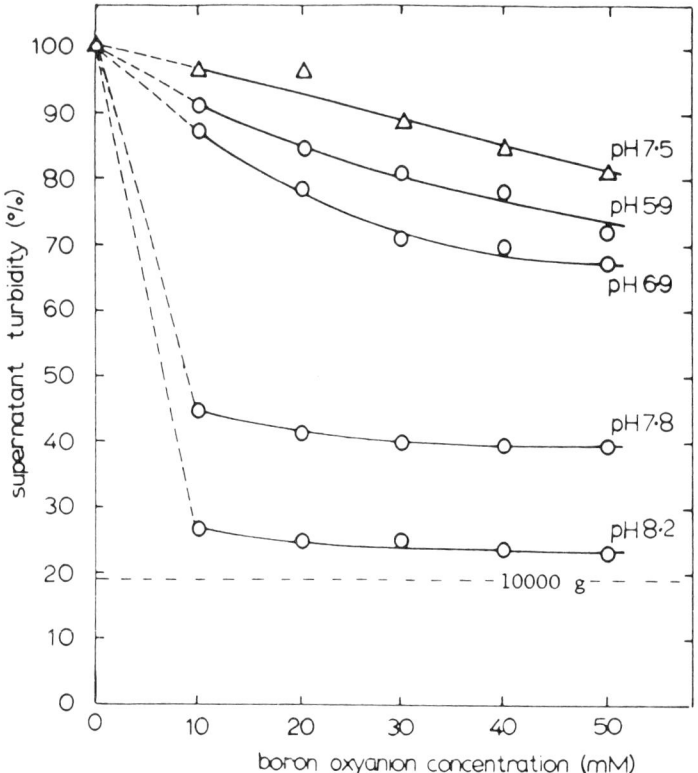

FIGURE 4. The effect of borate anion concentration and pH on the clarification of a yeast homogenate[5]—△: sodium borate; ○: borax (sodium tetraborate).

The ability to operate reproducible precipitation steps at both high yield and purification is essential for a production process. Furthermore, there needs to be the ability to alter the process parameters in order to cope with variations in the feed composition to the precipitation stage.

Using the traditional Cohn-type form of data presentation, a salting-out profile for the purification of ADH from yeast proteins displays typically steep fractionation profiles (FIGURE 6). The position of the fractionation profiles is a function of numerous factors such as feed stream composition and the operating procedures used. Typically, the fractionation process would be run so as to initially remove the low-solubility proteins, for example, a first cut at 40% ammonium sulfate, followed by a second cut for high-solubility proteins at, say, 60% ammonium sulfate.

It is difficult to quantify the performance of such a fractionation process from this form of diagram. However, in order to overcome this problem, the fractionation profiles may be translated into a fractionation diagram (FIGURE 7) where the fraction of ADH in solution is plotted against the fraction of protein remaining in solution for a whole range of ammonium sulfate concentrations.[7] In this form of data presentation, the position at the top right-hand corner of the diagram corresponds to all protein and ADH remaining in solution. At a 40% $(NH_4)_2SO_4$ concentration, protein starts to precipitate from solution followed later by ADH. Thus, by the time that 90% of the ADH is out of solution, a residual 50% of the protein still remains in solution.

The virtue of this diagram lies in its ability to describe the performance of a fractionation process. By indicating the positions of the two cuts taken within the process (FIGURE 7), E_1 and E_2, it is possible to estimate the overall yield of the process and also the purification factor (given by the final purity divided by the initial purity), which is equal to the slope of the line between these two cut points:[7]

$$\text{overall ADH yield} = E_1 - E_2 = 0.8, \qquad (1)$$

$$\text{purification factor} = \frac{(E_1 - E_2)}{(P_1 - P_2)} = 3.5. \qquad (2)$$

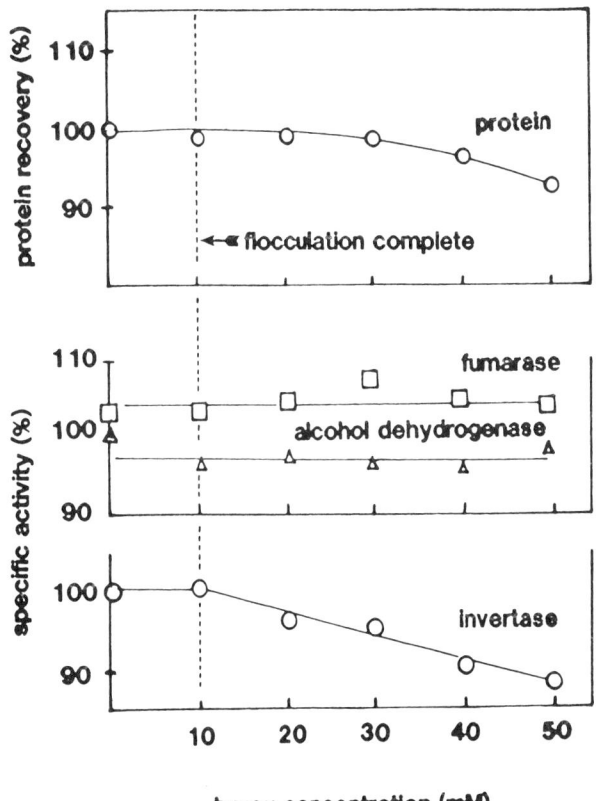

FIGURE 5. Enzyme specific activity in the supernatant following clarification of a yeast homogenate (pH 7.5) with different concentrations of borax[5]—△: ADH; □: fumarase; and ○: invertase.

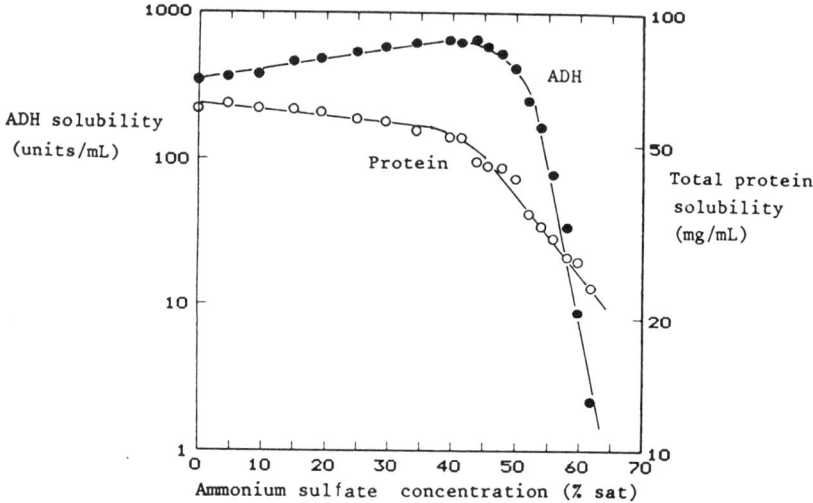

FIGURE 6. Cohn-type salting-out curve for ADH away from yeast proteins.

It is also possible to investigate the influence of moving the position of the two cuts relative to one another in order to maintain the yield and maximize the purification factor, that is, the slope of the line connecting (E_1, P_1) and (E_2, P_2). By so doing, it is possible to obtain a performance curve in terms of the maximum purification factor against yield.

This form of fractionation diagram may be used to examine the effect of altering the process conditions on the fractionation obtained. The effect of many variables may be considered. These would include concentration, the method of precipitant contact (be it batch or continuous), and on what scale.[8]

For example, an examination of earlier data for the purification of salicylate hydroxylase from a pseudomonad fermentation using PEG precipitation utilizing the new method of representation shows quite clearly how the presence of cell debris markedly alters the fractionation obtained (FIGURE 8).[9] The control case is for a well-washed cell harvest that has been disrupted and efficiently washed in order to remove cell debris. This gives a steep fractionation profile and hence a high purification factor, whereas the effect of debris or broth constituents is to significantly alter the performance of the process.

This form of fractionation data presentation has also been used as the basis for on-line continuous analysis, control, and optimization of a fractionation process.[10] The analysis system consists of precipitate preparation with the flows being determined and controlled by variable speed gear pumps and magnetic flow meters linked to a supervisory microcomputer (FIGURE 9). The flow passes through a ratio controller in order to achieve a fixed point on the precipitation profile. An element of the flow is then taken into a miniature high-speed centrifuge in which the solids and supernatant are separated. A sample of supernatant is automatically removed for subsequent flow injection analysis and the solids within the centrifuge head are rinsed out, again automatically; in addition, the process is repeated at different precipitating conditions. The system can generate a complete fractionation profile in about 30 minutes or it may

LOW-FREQUENCY ACOUSTIC CONDITIONING

As noted earlier (FIGURE 1) in a pilot or large-scale process, a continuous centrifugal separation stage follows each fractional precipitation operation. In the first precipitation cut, a fine protein precipitate material is separated from the suspending fluid. A high efficiency of fine particle removal and dewatering is required in order to maximize purification achieved and also to reduce the loading of fine insoluble particles that can alter the performance of subsequent fractionation stages. In the second precipitation cut, it is important to ensure complete recovery of the precipitated protein material in the sediment phase.

Membrane filtration has been proposed as an alternative method for the separation of such particles. However, research at University College London has shown that the transmission of soluble proteins through such membranes is time-dependent, falling as the membrane fouls.[11] This is clearly unacceptable because loss of soluble protein with the retentate will reduce the process yields.

Protein precipitates are typically small ($<30\ \mu$m) and of a low density difference compared with the liquor in which they are suspended ($\Delta\rho < 100\ \text{kg/m}^3$) (FIGURE 2).

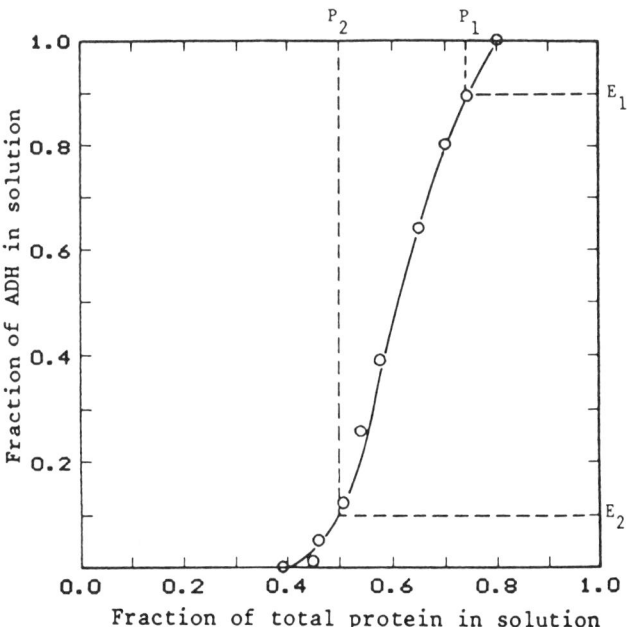

FIGURE 7. Fractionation diagram showing two-cut fractional precipitation of ADH away from yeast proteins.[7]

Moreover, the precipitate suspension on entry to the feed zone of a high-speed centrifuge is subject to a rapid acceleration up to the speed of rotation of the centrifuge and this, in turn, causes the particles to be exposed to a high level of shear with the net result being that particles are broken down to yield fine entities that the centrifuge has problems in recovering.[12,13]

In order to address these problems of size, density, strength, and the proportion of fine particles, it has been traditional to include an extended batch aging step after precipitation during which the necessary improvements in the particle properties are effected.[13] This form of batch aging is incompatible with continuous processing and is furthermore both difficult to control and prone to produce losses due to time-dependent degradation or to contamination. An alternative in the form of a low-residence-time,

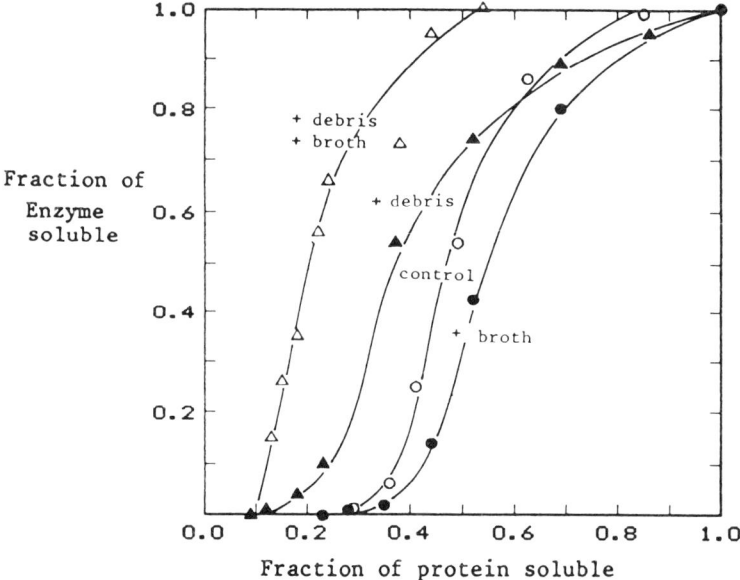

FIGURE 8. The effect of cell debris and broth constituent contaminant in the fractionation of salicylate hydroxylase using PEG as precipitant.[9]

continuous process based upon the use of low-frequency sound waves has been researched at University College London.[14,15]

The continuous flow cell of the conditioning unit has one wall mounted on a flexible membrane and linked to a vibrating drive. The residence time of nascent precipitate entering the cell is less than one minute. During passage through the cell, the suspension is exposed to low-frequency oscillations caused by the movement of the flexibly mounted wall. It is the frequency and amplitude of these oscillations that are critical to the conditioning process.[15]

As the chamber wall moves outwards because there is a positive feed to the device, so too is fluid drawn back into the chamber via the outlet pipe. Either the contents of the chamber and that of the outlet pipe all move down together en masse—and, as a

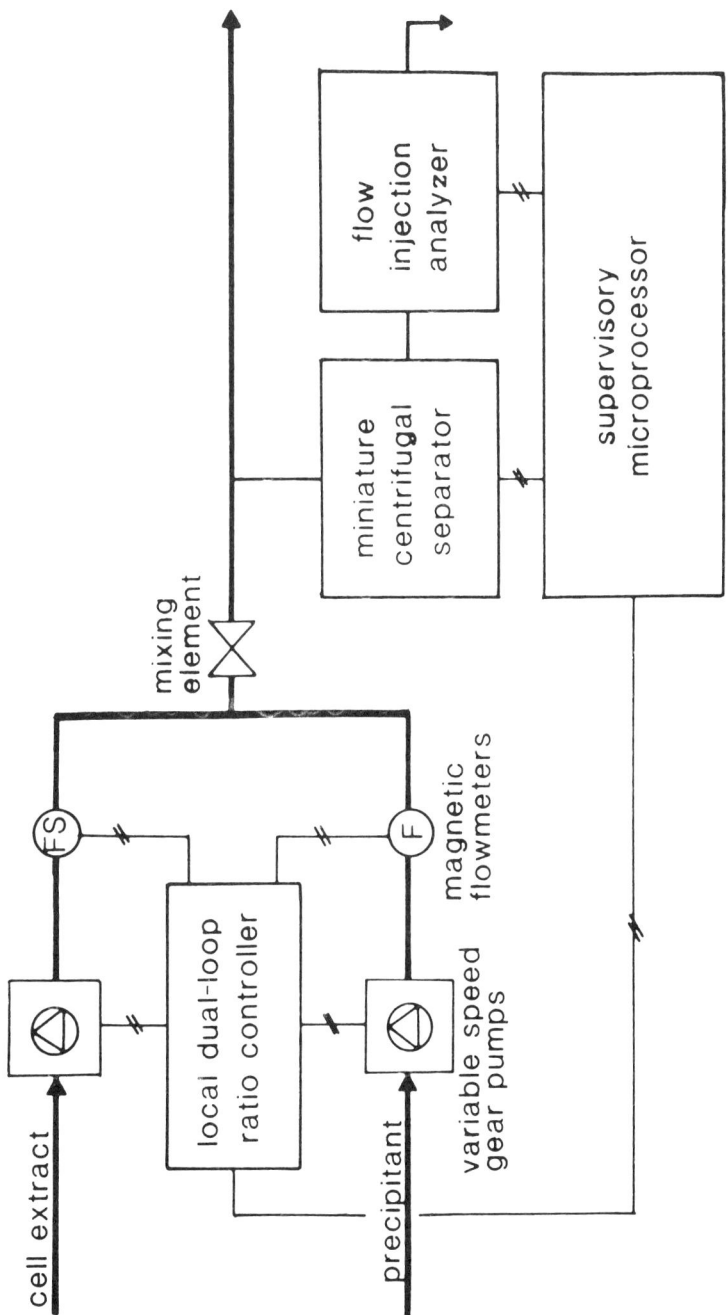

FIGURE 9. Flow sheet for continuous fractional precipitation and on-line analysis.[10]

result, no back-mixing within the chamber is produced—or conditions are such that a high-velocity fluid jet is generated that flows countercurrent to the chamber flow, entraining mass as it dissipates its energy into the stagnant chamber volume and thereby generating good overall mixing within the volume. This mixing is characterized by being turbulent in nature, but taking place in a slow moving stream and hence not exposing sensitive aggregates to extremes of shear breakup.

The formation of a countercurrent mixing jet is, for a given fluid and chamber geometry, dependent upon the frequency of operation.[15] At a low frequency of operation (FIGURE 10), the velocity and pressure of flow back into the chamber are synchronous and, as a result, a high-velocity jet of fluid that penetrates back into the stagnant fluid volume is generated. However, at higher frequencies (FIGURE 10) of operation, due to the momentum of inertia dissipation from the walls of the outlet tube, the velocity and pressure waves are out of phase and, as a consequence, no jet is formed.[15]

An analysis of the fluid flow within the chamber indicates that the effects of low-frequency conditioning may be correlated with the fluid jet acceleration into the chamber as given by

$$\text{fluid jet acceleration} \propto f^2 d\sigma, \quad (3)$$

that is, the frequency of operation, f, squared, the amplitude of displacement, d, and the σ factor that accounts for the effectiveness of the system in forming a mixing jet and where $\sigma = 1$ at low frequency and approaches zero at high frequency.[15]

The effects of conditioning are best examined in terms of the fine particles present in a size distribution (FIGURE 11). The size of these fine particles may be defined by the D_{95} (volume basis) statistic, which corresponds to the equivalent particle diameter for which 95% by volume of the particles are greater in size. The upper curve demonstrates how the conditioning process increases the size of the fine particles from ~2.1 µm up to ~3.0 µm. This improvement, in theory, would yield almost a doubling in the centrifuge throughput for 95% recovery of the particles. As the value of $f^2 d\sigma$ is increased, an optimum in the fine particle size is realized. Increasing the conditioning input still further results in disruption and breakage of the particles and the benefits of conditioning are lost. The lower curve refers to material that has been conditioned and then subsequently exposed to conditions of capillary shear ($\dot{\gamma}_{AV} = 1.7 \times 10^4 \, s^{-1}, \tau = 0.065$ s) in order to simulate the shear breakage effects encountered in the feed zone of a recovering centrifuge.[13] Whereas the benefits of conditioning are reduced, yielding an approximate 50% increase in centrifuge throughput at the optimum, the resultant aggregates are still significantly larger than the corresponding unconditioned sample. This implies that the strength of the aggregates has been increased and this is complemented by separate studies showing that the aggregate has been formed into a more compact and hence denser structure.

CONCLUSIONS

This report has examined three novel techniques that may be combined within a process train for the recovery of a soluble intracellular protein or enzyme product. Using selective flocculation, it has been shown that it is possible to remove the cell

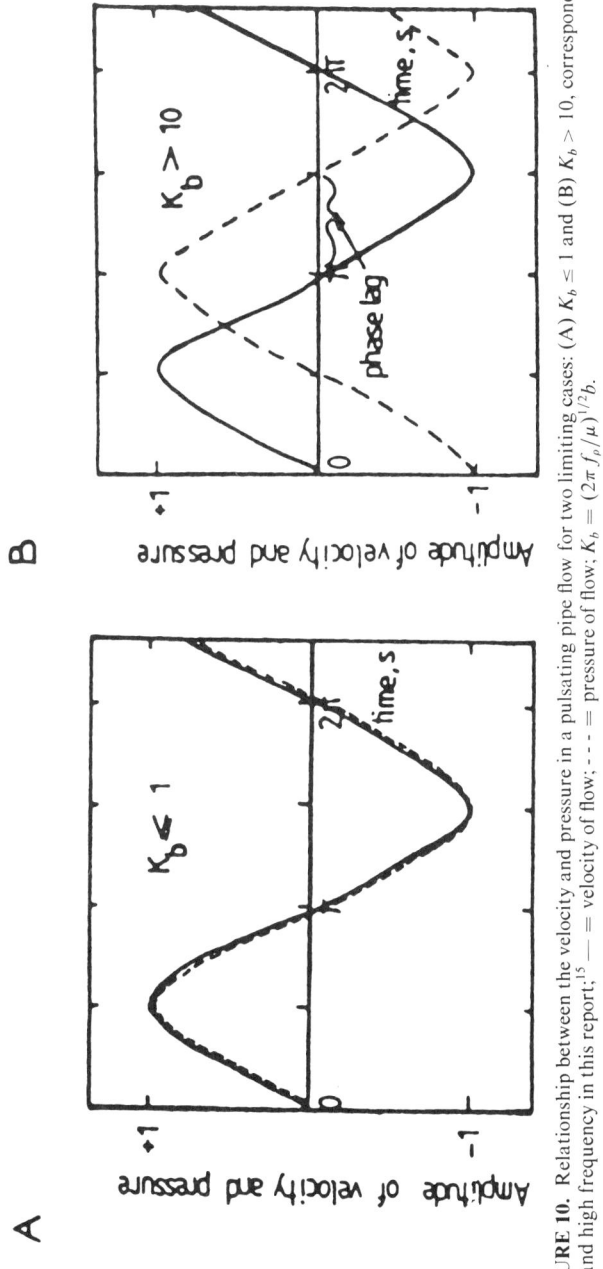

FIGURE 10. Relationship between the velocity and pressure in a pulsating pipe flow for two limiting cases: (A) $K_b \leq 1$ and (B) $K_b > 10$, corresponding to low and high frequency in this report;[15] —— = velocity of flow; - - - = pressure of flow; $K_b = (2\pi f_p/\mu)^{1/2} b$.

debris from soluble protein product in a continuous rather than a batchwise centrifugation. This low-residence-time operation will help minimize product loss due to time-dependent degradative reactions and will also prepare a soluble protein stream more suited to subsequent purification. The first such purification step is often protein precipitation[16] and the use of a fractionation diagram to optimize the performance of such a step has been examined. Such a representation of fractionation performance is a necessary component for the on-line monitoring of the precipitation process. A further aspect of protein precipitation is the need to improve the properties of the precipitate particles with a view to their subsequent centrifugal recovery. Low-frequency continu-

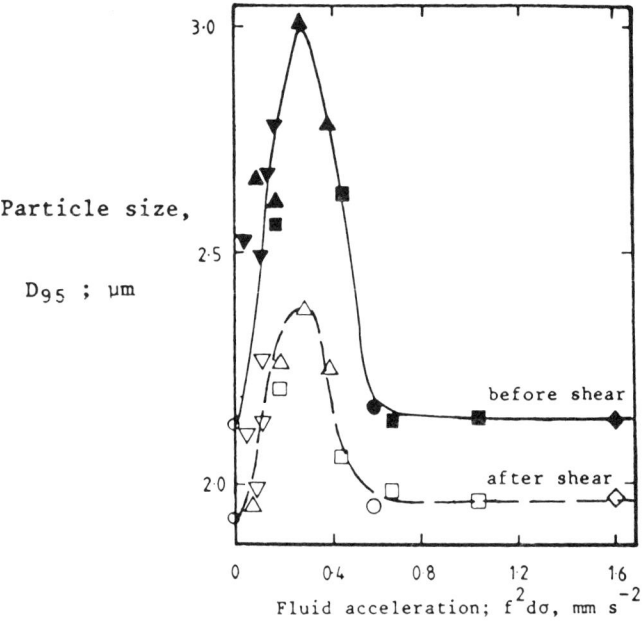

FIGURE 11. The effect of conditioning in increasing the size of fine precipitate particles together with the effect of subsequent exposure to capillary shear:[15] — = improvements before shear; - - - = improvements after shear.

ous conditioning has been shown to achieve such an improvement and, as with the fractionation procedure, the suitable biochemical engineering principles have been developed to allow process optimization.

REFERENCES

1. HIGGINS, J. J., D. J. LEWIS, W. H. DALY, F. G. MOSQUIERIA, P. DUNNILL & M. D. LILLY. 1978. Biotechnol. Bioeng. **XX:** 159–182.
2. BELL, D. J., D. M. HEYWOOD-WADDINGTON, M. HOARE & P. DUNNILL. 1982. Biotechnol. Bioeng. **XXIV:** 127–141.

3. TAYLOR, G., M. HOARE, D. R. GRAY & F. A. O. MARSTON. 1986. Bio/Technology **4:** 553–557.
4. MOSQUIERA, F. G., J. J. HIGGINS, P. DUNNILL & M. D. LILLY. 1981. Biotechnol. Bioeng. **XXIII:** 335–343.
5. BONNERJEA, J., J. JACKSON, M. HOARE & P. DUNNILL. 1988. Enzyme Microb. Technol. **10:** 357–360.
6. BENTHAM, A. C., J. BONNERJEA, M. HOARE & P. DUNNILL. Biotechnol. Bioeng. In preparation.
7. RICHARDSON, P., M. HOARE & P. DUNNILL. In preparation.
8. RICHARDSON, P. 1987. Ph.D. thesis. London University.
9. RUSSEL, A. J. 1980. Ph.D. thesis. London University.
10. RICHARDSON, P., R. RAVENHALL, M. FLANNAGAN, J. MALLOY, M. HOARE & P. DUNNILL. In preparation.
11. DEVEREUX, M. & M. HOARE. 1986. Biotechnol. Bioeng. **XXVIII:** 422–431.
12. BELL, D. J., M. HOARE & P. DUNNILL. 1983. Adv. Biochem. Eng./Biotechnol. Volume 26. A. Fiechter, Ed.: 1–72. Springer-Verlag. Berlin/New York.
13. BELL, D. J. & P. DUNNILL. 1982. Biotechnol. Bioeng. **XXIV:** 2319–2336.
14. HOARE, M., N. J. TITCHENER-HOOKER & P. R. FOSTER. 1987. Biotechnol. Bioeng. **XXIV**(no. 1): 24–32.
15. TITCHENER-HOOKER, N. J., M. HOARE & P. R. FOSTER. Biotechnol. Bioeng. In preparation.
16. BONNERJEA, J., S. OH, M. HOARE & P. DUNNILL. 1984. Bio/Technology **2**(7): 623–627.

APPENDIX

Nomenclature

b = diameter of outlet tube to acoustic chamber (m)
d = peak amplitude of oscillation (m)
E = fraction of enzyme remaining soluble (–)
f = frequency of oscillation (s^{-1})
g = acceleration due to gravity (m s^{-2})
K = shape factor (–)
K_b = low-frequency system factor[15] (–)
P = fraction of protein remaining soluble (–)
Q = centrifuge throughput (m^3 s^{-1})

Greek Variables

ρ = fluid density (kg m^{-3})
$\Delta\rho$ = particle-suspension density difference (kg m^{-3})
σ = effectiveness factor describing the response of a low-frequency system in forming a mixing jet (–)
μ = suspension viscosity (N s m^{-2})
$\dot{\gamma}_{AV}$ = average shear rate at wall of capillary tube[12] (s^{-1})
τ = mean residence time (s)

Gradient Elution in Preparative Liquid Chromatography[a]

FIROZ D. ANTIA AND CSABA HORVÁTH

Department of Chemical Engineering
Yale University
New Haven, Connecticut 06520

INTRODUCTION

Chromatography is a widely used technique for the purification of drugs, therapeutic proteins, and other compounds of chemical and biochemical interest. This communication presents part of an effort to study, evaluate, and compare the different means by which column liquid chromatography may be most effectively utilized in preparative/process scale separations.

Simply put, the goals of a purification method are to obtain from a mixture the largest amount of pure material in the shortest possible time with due regard to cost. To achieve this in chromatography, it is often advantageous to operate with large amounts of feed under conditions where the adsorption isotherms are strongly nonlinear, that is, under nonlinear or overloaded conditions. This is in sharp contrast with analytical or linear chromatography, where components are present at low concentrations and obey Henry's law. Effluent concentration profiles in nonlinear chromatography are considerably different from the quasi-Gaussian peaks generally observed in linear chromatography.

Preparative chromatography may be carried out in one or a combination of three fundamental modes:[1] frontal, displacement, and elution. In the frontal mode, the feed is continuously introduced into the column. A pure fraction of the least-retained component breaks through and is collected before other mixed fractions appear. In the displacement mode, the feed is followed by a continuous stream of the "displacer", which has higher affinity for the stationary phase than any of the feed components. The displacer causes desorption of the feed species, which travel through the column ahead of the displacer front and separate into adjacent bands. In the elution mode, the flow of the mobile phase, which has a lower affinity for the sorbent than any of the feed species, causes the components to migrate at different rates through the column and thus separate. It has been put forward that displacement chromatography is most suitable for the preparative separation of closely related compounds.[1,2] However, due to its greater familiarity, the elution mode is most widely used.

Elution is isocratic when the eluant strength is kept constant during the separation; on the other hand, in gradient elution, the eluant strength is gradually increased by changing the concentration of an appropriate mobile-phase modifier. The gradient shape depends on how the modifier concentration is changed with time and may be

---

[a]This study was supported by Grant Nos. GM 20993 and CA 21948 from the National Institutes of Health and Human Resources.

linear or nonlinear. For the sake of clarity, though, we will use the term linear (or nonlinear) gradient elution to refer only to the shape of the governing adsorption isotherm and not to that of the gradient.

In analytical chromatography, gradient elution is used to increase the peak capacity[3] and to analyze samples containing a number of components with widely disparate retention behavior. Because there are usually only a few rather closely related components in preparative applications, particularly in the final stages of a purification process, it is of interest to determine whether or not gradient elution can be employed profitably. Here, with the help of computer simulations, we compare isocratic and gradient elution separations of a binary mixture in terms of production rate, product concentration, and yield. The effects of sorption kinetics are also considered in some calculations in order to illustrate the consequences of slow sorption rates.

MATHEMATICAL MODEL AND CALCULATIONS

Nonlinear isocratic elution chromatography of a single component has been studied extensively.[4-7] Binary separations in the nonlinear isocratic elution mode have also been examined both theoretically[8] and by computer simulations.[9] Gradient elution has only been treated theoretically in the case of linear chromatography.[10,11] Our model for gradient elution in nonlinear chromatography is based on these past treatments of isocratic elution in nonlinear chromatography and gradient elution in linear chromatography.

The mathematical description of chromatography is based on a mass balance that, at time t for a species i around a differential segment at a position x along a column of length L, is

$$\frac{\partial c_i}{\partial t} + \phi \frac{\partial q_i}{\partial t} + u \frac{\partial c_i}{\partial x} = \frac{\partial}{\partial x}\left(D_{\text{eff}} \frac{\partial c_i}{\partial x}\right), \quad (1)$$

where q_i and c_i denote the concentrations of i in the stationary and mobile phases, respectively, ϕ is the volume ratio of the two phases in the column, u is the chromatographic velocity, and D_{eff} is an effective band-broadening coefficient. The subscript i implies a set of N equations, one for each eluate. The eluant is considered to be unsorbed and not subject to dispersion.[9] For elution chromatography, the initial and boundary conditions are

$$c_i(0, x) = 0, \quad 0 < x < L$$
$$c_i(t, 0) = \begin{cases} c_{0,i}, & 0 < t < t_{\text{inj}} \\ 0, & t > t_{\text{inj}} \end{cases}$$
$$\partial c_i/\partial x = 0 \quad \text{at} \quad x = L \quad (2)$$

where $c_{0,i}$ is the feed concentration of species i and t_{inj} is the duration of feed introduction.

The coefficient D_{eff} accounts for band-broadening due to axial dispersion and mass-transfer resistances. Under most conditions in linear chromatography, it has been found that D_{eff} can be related to the plate height, H, by[12]

$$D_{\text{eff}} = Hu/2. \quad (3)$$

It is assumed that the D_{eff} is the same for each species i and can be used under nonlinear conditions as well. Because H can be determined from known theoretical relationships,[13] equation 3 may be used to provide reasonable estimates for D_{eff}. Such band-broadening coefficients have been used to model nonlinear chromatography in the past.[7,9] Under nonlinear conditions, band-spreading is caused largely by thermodynamic effects, and inaccuracies stemming from the use of such a lumped band-broadening coefficient are expected to be insignificant.

Rapid Sorption Kinetics

When the kinetics of sorption are sufficiently rapid, the eluting species are considered to be in equilibrium between the mobile and the stationary phase at all times. Then, in order to complete the model, it is necessary only to provide the expression for the multicomponent adsorption isotherm, which governs the equilibrium distribution of all the eluates between the two phases at the prevailing temperature. Although information about multicomponent isotherms in chromatographic systems is limited, the available data[14] suggest the use of the multicomponent Langmuir formalism,

$$q_i = a_i c_i \bigg/ \left(1 + \sum_{j=1}^{N} b_j c_j\right), \qquad (4)$$

where a_i and b_i are determined individually for each species. The parameter a_i is proportional to the retention factor[3] for the species i, k'_i, measured at low concentrations, that is, in linear chromatography,

$$k'_i = \phi a_i, \qquad (5)$$

and the ratio a_i/b_i measures the density of sorption sites on the stationary phase.

In the treatment of gradient elution, we use the assumption[10] that the eluant modifier gradient traverses the column without changing its shape. Furthermore, if the modifier concentration, φ, as a function of time at the gradient-forming device is $f(t)$, then the modifier concentration in the column at a position x is given by $f(t - x/u - \tau)$, where τ is a gradient delay time that is set to t_{inj} in this study. Thus, for a linear gradient shape, we obtain

$$\varphi = \alpha + \beta(t - x/u - \tau), \qquad (6)$$

where α is the modifier content at the start of the gradient and β is its linear rate of change. The applicability of equation 6 has been verified[10] in cases where the modifier is very weakly sorbed.

If the logarithmic retention factor is a linear function of the modifier concentration, then

$$a_i = a_{0,i} \exp(-S_i \varphi), \qquad (7)$$

where $a_{0,i}$ is the value of a_i in the absence of the modifier and S_i is a constant whose physical meaning is discussed in reference 10. This relationship has been found to hold for reversed-phase chromatography with organic modifiers[10] and, with appropriate changes of sign, for hydrophobic interaction chromatography using a salt modifier.[15]

If the modifier is weakly sorbed, its presence is not expected to alter the number of sorption sites available to a given eluate. Thus, we may write that

$$b_i = b_{0,i} \exp(-S_i \varphi), \tag{8}$$

where $b_{0,i}$ is the value of b_i in the absence of the modifier. The effect of organic modifiers on nonlinear adsorption isotherms for hydrophobic sorbents in chromatographic systems[14] supports equation 8.

Equations 1, 2, 4, and 6–8 provide a description of nonlinear gradient elution when sorption kinetics are rapid.

Slow Sorption Kinetics

When the sorption kinetics are slow, the Langmuirian isotherm (equation 4) should be replaced by the corresponding rate expression,

$$\partial q_i / \partial t = k_{a,i} \left(\Lambda - \sum_{j=1}^{N} q_j \right) c_i - k_{d,i} q_i, \tag{9}$$

where Λ is the density of binding sites on the stationary phase and $k_{a,i}$ and $k_{d,i}$ are adsorption and desorption rate constants, respectively. The isotherm parameters are related to the rate constants and the density of binding sites via

$$a_i = k_{a,i} \Lambda / k_{d,i} \quad \text{and} \quad b_i = k_{a,i} / k_{d,i}. \tag{10}$$

It has been observed experimentally in affinity chromatography[16] that the presence of a modifier affects the magnitude of the rate constant for desorption, but not for adsorption, and such behavior is believed to occur in other forms of chromatography as well. With this consideration, it follows from equations 7 and 10 that

$$k_{d,i} = k_{0,d,i} \exp(S_i \varphi), \tag{11}$$

where $k_{0,d,i}$ is the value of $k_{d,i}$ when no modifier is present.

Equations 6, 9, and 11 along with equations 1 and 2 describe the process of gradient elution chromatography with slow sorption kinetics.

Calculations

The partial differential equations were reduced to a set of first-order ordinary differential equations by orthogonal collocation on finite elements[17] and were then integrated numerically[18] as was done in the solution of similar problems in the past.[19] All calculations were performed on a MicroVAX computer in the Yale Chemical Engineering Department. The parameters given in FIGURE 1 reflect conditions typical of biopolymer chromatography. Both S_A and S_B were taken as 20 for the sake of convenience. Although S_i is dependent on the eluate and on the chromatographic system, S values for closely related compounds in a given chromatographic system are often quite similar.[10] Under such circumstances, the gradient can be characterized by a steepness parameter, G, defined as[10]

$$G = t_0 S \beta / 2.3, \tag{12}$$

where $t_0 (= L/u)$ is the mobile-phase holdup time. Different systems having identical G values and identical values for the term $k_{0,d,i} \cdot \exp(-S\alpha)$ have the same behavior.

RESULTS AND DISCUSSION

Three performance measures are useful in analyzing preparative chromatography: (i) production rate or throughput, defined as mass of product of given purity obtained per cycle divided by the cycle time, (ii) product concentration, and (iii) yield, given by

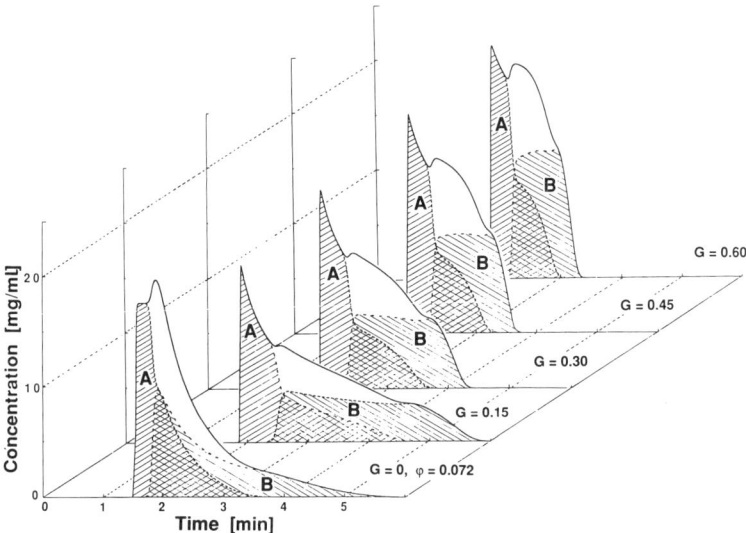

FIGURE 1. Comparison of isocratic and gradient elution profiles for a binary mixture at a load of 10 mg each of components A and B. Isocratic elution is with $G = 0$ and $\varphi = 0.072$ (v/v). Gradients, with different G values, all start at $\varphi = 0$. The solid line is the total concentration envelope. Isotherm parameters: component A—$a_{0,A} = 30$, $b_{0,A} = 0.3$ mL/mg, $S_A = 20$; component B—$a_{0,B} = 50$, $b_{0,B} = 0.5$ mL/mg, $S_B = 20$. Column: 5 × 0.46 cm, $\phi = 0.5$. Mobile-phase holdup time: 0.6 min; $D_{eff} = 0.0833$ cm^2/min [calculated (cf. reference 23) assuming 5-μm totally porous particles and an eluate with a molecular diffusivity of 6 × 10^{-7} cm^2/s in the mobile phase].

the recovered fraction of the feed component of interest. The total cycle time includes that for the feed step, the elution, and the column regeneration or washing, if required. The relative importance of the above performance indicators in a given separation depends on factors such as product cost, the cost of any further concentrating steps, and whether or not impure fractions may be recycled.

In order to compare the effectiveness of isocratic and gradient elution chromatography in preparative separations of a binary mixture, parameters for the calculations were chosen such that, under linear isocratic conditions, the two components were just baseline resolved (resolution,[3] R_s, of unity). This choice of parameters reflects a limiting condition in which gradient elution is anticipated to have minimum benefits over isocratic separation.

Simulations were performed with different values of feed load and gradient steepness, and rapid equilibration was assumed. The column length and diameter were 5 and 0.46 cm, respectively, and ϕ was taken as 0.5. The mobile-phase holdup time was 0.6 min. The feed was a 1:1 mixture of the components at concentrations of 10 mg/mL and the column load was varied by changing the time for which feed was introduced (t_{inj}). All gradients were started at $\varphi = 0$, whereas isocratic elution calculations were run for different values of φ. For the pair of compounds and in the range of load chosen, the highest throughput in isocratic elution was at $\varphi = 0.072$ (v/v). Isocratic results with this eluant composition were compared to those obtained with gradient elution.

In FIGURE 1, we show elution profiles for an isocratic run and several gradient runs at different gradient steepnesses for the same load. The solid line represents the total concentration envelope and the shaded bands reflect the concentration profiles of the individual components. The blank area within each envelope is equal to the area of overlap between the two components. All the profiles show the consequences of the so-called displacement effect[9] that predominates at high column load: the more strongly retained component (B) drives the less-retained band (A) ahead, concentrating it in the process. However, in both isocratic and gradient elution, a small fraction of A that remains unseparated tails into the most concentrated part of the B band, hindering its recovery. It has been shown that such tailing cannot be avoided in nonlinear elution, but it is absent in displacement chromatography.[20]

The results in FIGURE 1 show that larger quantities of the two components can be purified in a shorter time by gradient as compared to isocratic elution. The use of gradient elution has at least two advantages. First, unlike in isocratic elution, the feed can be introduced under conditions where the components have a high affinity for the stationary phase. Thus, the feed components are concentrated into a narrow region at the top of the column before elution begins. Second, during migration down the column, the band is compressed by the gradient.[10] The long tail of the second component is therefore shortened and elution time is significantly reduced.

Chromatograms obtained at different loads for the same gradient steepness illustrate in FIGURE 2 that band overlap increases with load. However, the loss of separation is partly compensated for by the increased amounts of pure product that may be recovered per run. Thus, throughput reaches a maximum although the yield drops monotonically with increasing load. FIGURE 2 also shows that appreciable concentrations of component B are required before the displacement effect described above is manifest.

In FIGURES 3–5, the results of the simulations for different feed loads and gradient steepnesses as well as for isocratic elutions at $\varphi = 0.072$ are presented in terms of throughput, product concentration, and yield. Throughput is calculated for 98% pure products and the cycle time is taken as the elution time for the last trace of the second component. Certain striking trends are evident from the data. Throughputs, yields, and product concentrations are uniformly higher for component A as compared with those for component B. This is due to the displacement effect and the tailing of A discussed earlier. The throughput versus load curves for component A go through a maximum, but the curves for component B appear to have their maxima at loads lower than those shown. Hence, optimal loading for the recovery of the two products differs widely.

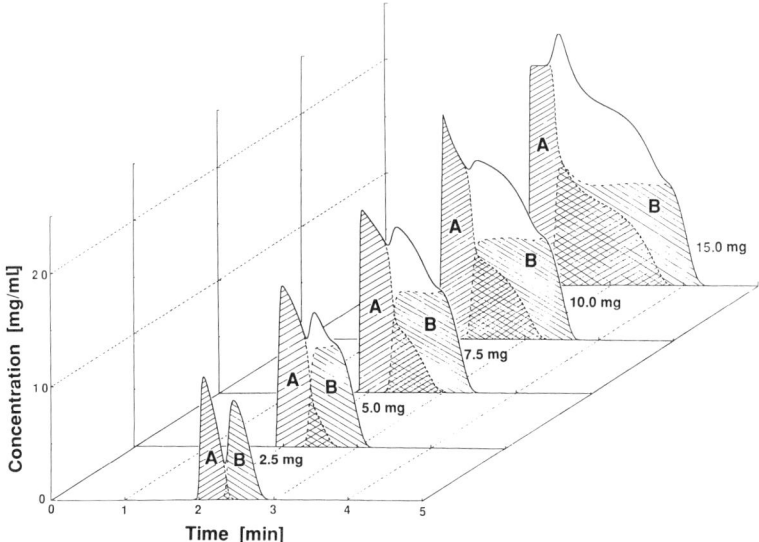

FIGURE 2. Gradient elution of two components having the same mass ratio at $G = 0.45$ and at different column load. Loads shown are for each component. Conditions as in FIGURE 1.

However, if the recovery of both components is important, it may be economically advantageous to operate at higher load.

The concentration versus load curve in FIGURE 4A for isocratic elution ($G = 0$) cuts across the curves for higher G values. This is because the mobile-phase concentration for the isocratic run ($\varphi = 0.072$) is different from that at the start of the gradient runs ($\varphi = 0$). A similar concentration versus load curve drawn for isocratic elution at $\varphi = 0$ would stay beneath the curves for gradient elution.

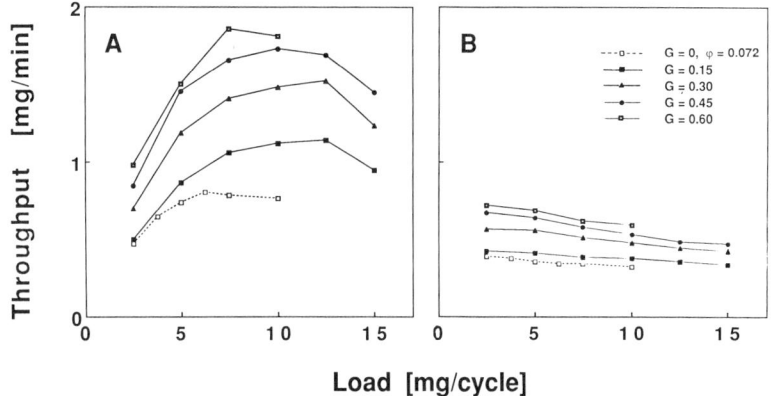

FIGURE 3. Throughput calculated for 98% pure product versus feed load per cycle for the separation of components A and B by isocratic and gradient elution. Conditions as in FIGURE 1.

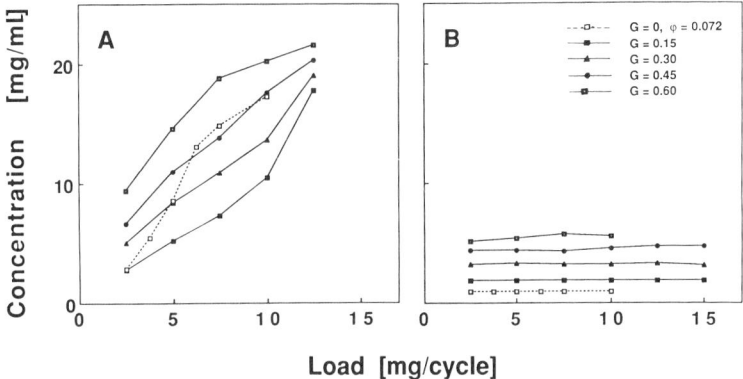

FIGURE 4. Concentration of recovered product versus feed load per cycle for the separation of components A and B by isocratic and gradient elution. Conditions as in FIGURE 1.

The results shown in FIGURES 3–5 demonstrate that production rates, product concentrations, and yields are significantly higher with gradient elution than with isocratic elution. Thus, the band concentration and compression caused by the gradient more than compensate for the loss of separation resulting from the bands being pushed closer together. The throughput as a function of gradient steepness also goes through a maximum, but this effect is not shown. In the examples studied, performance as measured by the throughput and yield deteriorates when the gradient steepness parameter, G, exceeds a value of 0.6.

After a gradient run, it is necessary to return to the initial conditions before starting another. This may be accomplished by washing the column with the starting eluant or by a series of different solvents.[21] In contradistinction, such regeneration steps may not be necessary in the case of isocratic elution because the same eluant is in use at all

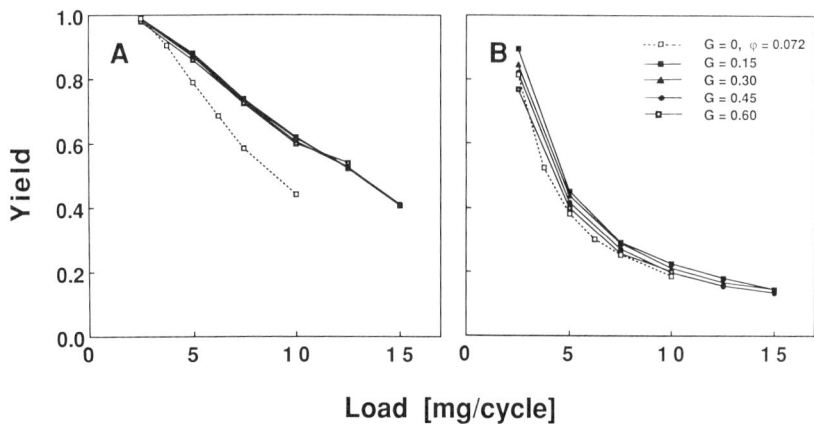

FIGURE 5. Plots of yield versus feed load per cycle for the separation of components A and B by isocratic and gradient elution. Conditions as in FIGURE 1.

times. In the earlier calculations, column regeneration has not been included. According to the results shown in FIGURE 3A, with gradient elution at $G = 0.45$, a wash step of greater than six column volumes, which is usually more than sufficient to reequilibrate the column, is allowed before the production rate falls below that obtained in the corresponding isocratic run. In many applications of preparative chromatography, it is advisable to include a column wash step after an isocratic elution, in which case a gradient elution scheme has even greater advantages.

Effect of Slow Sorption Kinetics

Slow sorption induces severe band-broadening so that chromatographic separations are difficult to perform in the elution mode. If the characteristic time for sorption

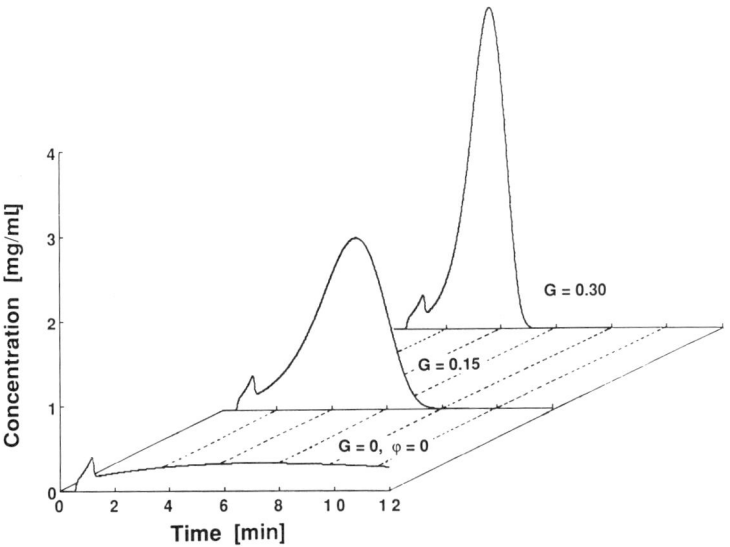

FIGURE 6. Concentration profiles obtained by isocratic and gradient elution under conditions of slow sorption kinetics. Parameters: $a_0 = 50$, $b_0 = 0.5$ mL/mg, and $k_{0,d} = 0.3$ min^{-1}. Other conditions as in FIGURE 1.

is of the same order as the mobile-phase holdup time, some fraction of the eluate remains unretained and forms a "split peak". Gradient elution does not alleviate this because, although the presence of the modifier hastens the desorption process, the gradient has no effect on the early eluting peak. This situation is illustrated in FIGURE 6 for a single component. Under such conditions, separations by elution can only be performed at high modifier concentrations where resolution—and therefore throughput—is severely reduced. In some applications of affinity chromatography, slow kinetics is a limiting factor.[22] However, for most chromatographic separations, the conditions can be chosen to avoid slow sorption. Amongst other possibilities, the temperature can be raised.[23]

CONCLUSIONS

The process of gradient elution chromatography under conditions of high column load where adsorption isotherms are nonlinear has been examined and, with the help of computer simulations, binary separations under gradient and isocratic elution conditions have been compared. Despite a choice of conditions less favorable for gradient elution, this method has been found to be superior to isocratic elution on the basis of production rate, product concentration, and yield. As a function of column load and gradient steepness, production rates go through a maximum that occurs at significantly higher column loads for the less-retained than for the more-retained substance.

The influence of slow sorption on the elution profile of a single component in both isocratic and gradient elution has also been studied. When the characteristic time for sorption is of the same order as the mobile-phase holdup time, a split peak is observed that cannot be avoided by the use of gradient elution.

REFERENCES

1. LEE, A. L., A. W. LIAO & C. HORVÁTH. 1988. J. Chromatogr. **443**: 31–43.
2. HORVÁTH, C. 1985. Displacement chromatography: yesterday, today, and tomorrow. *In* The Science of Chromatography. F. Bruner, Ed.: 179–203. Elsevier. Amsterdam/New York.
3. HORVÁTH, C. & W. R. MELANDER. 1983. Theory of chromatography. *In* Chromatography. E. Heftman, Ed.: A27–A136. Elsevier. Amsterdam/New York.
4. DEVAULT, D. 1943. J. Am. Chem. Soc. **65**: 532–540.
5. THOMAS, H. C. 1948. Ann. N.Y. Acad. Sci. **49**: 161–182.
6. AMUNDSON, N. R. 1950. J. Phys. Colloid Chem. **54**: 812–820.
7. HOUGHTON, G. 1963. J. Phys. Chem. **67**: 84–87.
8. GLUECKAUF, E. 1949. Discuss. Faraday Soc. **7**: 12–25.
9. GHODBANE, S. & G. GUIOCHON. 1988. J. Chromatogr. **440**: 9–22.
10. SNYDER, L. R. 1980. Gradient elution. *In* High Performance Liquid Chromatography: Advances and Perspectives. Volume 1. C. Horváth, Ed.: 207–316. Academic Press. New York; SNYDER, L. R. & M. A. STADALIUS. 1986. HPLC separations of large molecules: a general model. *In* High Performance Liquid Chromatography: Advances and Perspectives. Volume 4. C. Horváth, Ed.: 195–312. Academic Press. New York.
11. GIBBS, S. J. & E. N. LIGHTFOOT. 1986. Ind. Eng. Chem. Fundam. **25**: 490–498.
12. LENHOFF, A. M. 1987. J. Chromatogr. **384**: 285–299.
13. HORVÁTH, C. & H-J. LIN. 1978. J. Chromatogr. **149**: 43–70.
14. JACOBSON, J., J. FRENZ & C. HORVÁTH. 1984. J. Chromatogr. **316**: 53–68.
15. EL RASSI, Z. & C. HORVÁTH. 1986. J. Liq. Chromatogr. **9**(15): 3245–3268.
16. MULLER, A. J. & P. W. CARR. 1986. J. Chromatogr. **357**: 11–32.
17. SCHIESSER, W. E. DSS/2 (Differential System Solver). Lehigh University, Bethlehem, Pennsylvania.
18. HINDEMARSH, A. LSODE (ODE Solver). Lawrence Livermore Laboratories, Berkeley, California.
19. RAGHAVAN, N. S. & D. M. RUTHVEN. 1983. AIChE J. **29**(6): 922–925.
20. HELFFERICH, F. & G. KLEIN. 1970. Multicomponent Chromatography. Dekker. New York.
21. FRENZ, J. & C. HORVÁTH. 1983. J. Chromatogr. **282**: 249–262.
22. WADE, J. L. & P. W. CARR. 1988. J. Chromatogr. **449**: 53–61.
23. ANTIA, F. D. & C. HORVÁTH. 1988. J. Chromatogr. **435**: 1–15.

Purification of β-Galactosidase by Combined Frontal and Displacement Chromatography[a]

ABRAHAM LIAO AND CSABA HORVÁTH

Department of Chemical Engineering
Yale University
New Haven, Connecticut 06520

INTRODUCTION

High performance liquid chromatography (HPLC) in the displacement mode with an analytical anion-exchanger column was successfully used in our laboratory for the separation of β-lactoglobulins A and B from bovine milk:[1] 100 milligrams of the protein mixture was separated on a 3.3-mL column by using chondroitin sulfate as the displacer under conditions of classical displacement development. We have also shown[2] that the separation of β-lactoglobulins A and B can be carried out efficiently by the combined use of frontal chromatography and stepwise elution or displacement. These results prompted a recent examination[3] of the potential of preparative HPLC under nonlinear conditions by displacement chromatography, frontal chromatography, and stepwise elution.

In this study, we apply such a multimodal chromatographic sequence to the purification of the industrial enzyme β-galactosidase from *Aspergillus oryzae*. The enzymes denoted by the generic term β-galactosidase (β-gal) hydrolyze the β-1,4-glycosidic link. They are widely distributed in nature and have been found in various microorganisms, plants, and animals (for reviews, see references 4 and 5). The best-characterized β-gal is the one from *E. coli*, which is used in many bacterial plasmid vectors and as an indicator enzyme in clinical tests. The corresponding enzymes from *Aspergillus niger* and *Aspergillus oryzae* are used in the dairy industry to hydrolyze lactose and are also called lactase.[6] Crude β-gal from *A. oryzae* contains other carbohydrases, such as amylase, N-acetyl-β-D-glucosaminidase, endo-β-N-acetylglucosaminidase, 1,2-α-mannosidase, β-xylosidase, (1,4)-β-xylanase, (1,3)-β-glucanase, β-glucosidase, fucosidase, and cellulase.[7-11] A typical purification sequence of β-gal consists of calcium acetate precipitation of the crude enzyme followed by acetone fractionation, ion exchange, gel filtration, and hydroxyapatite chromatography. The enzyme from *A. oryzae* could not be purified to homogeneity by a single run in ion-exchange chromatography;[8] the molecular weight of the β-gal thus obtained was 112 kDa and this glycoprotein contained 2.6% N-acetylglucosamine and 10% neutral sugars, mainly mannose.[8] The goal of this study is to demonstrate that ion-exchange chromatography in the frontal and displacement mode is a highly promising alterna-

---

[a]This study was supported by Grant Nos. GM 20993 and CA 21948 from the National Institutes of Health, United States Department of Health and Human Resources.

tive to the linear elution mode in large-scale purification of β-gal and possibly other complex protein mixtures.

EXPERIMENTAL WORK

Instruments

The liquid chromatograph was similar to those reported previously.[2] Two Altex model 100A pumps (Beckman Instrument, Fullerton, California) were used for displacement chromatography and a single model 100A pump was used for frontal chromatography. A model 770 variable wavelength spectrophotometer (Kratos, Ramsey, New Jersey) was used to monitor the effluent and a model 7000 Ultrarac fractional collector (LKB, Gaithersburg, Maryland) was used to collect the fractions. In analytical work, an HP 1090 liquid chromatograph (Hewlett-Packard, Palo Alto, California) was used.

Columns

Frontal chromatography was carried out with a 75 × 7.5 mm TSK DEAE-5PW column (Tosoh Company, Tokyo, Japan) and, for displacement chromatography, two such columns were used in series. In analytical chromatography, a 30 × 4.6 mm column packed with 3-μm micropellicular DEAE sorbent was used. It consisted of nonporous polystyrene-divinylbenzene beads with a hydrophilic layer and diethylaminoethyl (DEAE) functions.

Reagents

Crude β-galactosidase from *Aspergillus oryzae*, chondroitin sulfate from shark cartilage, β-lactoglobulins A and B from bovine milk, and o-nitrophenyl-β-D-galactopyranoside (ONPG) were purchased from Sigma (St. Louis, Missouri). Toluidine blue O was from Aldrich (Milwaukee, Wisconsin). Bio-Rad Protein Assay kit was obtained from Bio-Rad (Richmond, California). Other chemicals were of reagent grade from J. T. Baker (Philipsburg, New Jersey). Distilled water was made by a Barnstead Nanopure unit.

Procedures

Analysis for Proteins and Enzymatic Activity

The total protein concentration was measured by the Bradford method[12] using the Bio-Rad Protein Assay with bovine plasma γ-globulin as the standard. The β-gal activity was determined at 22 °C by Lederberg's method[13] with ONPG as the substrate and by measuring the change of absorbance at 420 nm spectrophotometrically. In a typical assay, 2 μL of feed or of each fraction in frontal or displacement chromatography was mixed with 1 mL of buffer containing 100 mM sodium phosphate (pH 7.0), 10 mM potassium chloride, 1 mM magnesium sulfate, and 2.2 mM ONPG. The reaction was allowed to proceed for 2 min with constant mixing and was quenched by adding 1 mL of 1 M sodium carbonate. One unit is defined as the enzyme activity that hydrolyzes 1 nmole of ONPG in 1 min at 22 °C.

Displacement Chromatography

Displacement chromatography was used to separate the crude β-gal proper and the pooled β-gal prepurified by frontal chromatography (see below). In the loading step, crude β-gal (2.1 mg/mL) in 25 mM sodium phosphate, pH 7.0, or β-gal prepurified by frontal chromatography (3.4 mg/mL) was introduced into the columns through a 4-mL sample loop at a flow rate of 1 mL/min of the carrier. After loading the sample, the unretained species were removed by rinsing the tandem column with five empty column volumes of 25 mM sodium phosphate, pH 7.0. Subsequently, chondroitin sulfate (5 mg/mL) in 25 mM sodium phosphate, pH 7.0, was introduced into the column at a flow rate of 0.2 mL/min to displace the bound proteins. The breakthrough of chondroitin sulfate was detected by the metachromasia reaction with toluidine blue O.[14] The column was cleaned by perfusing it with water at a flow rate of 1 mL/min and by injecting seven slugs of 1 M NaOH and then of 0.1 N HCl, with each slug being 4 mL. Lastly, the column was reequilibrated by passing through 20 mL of 25 mM sodium phosphate, pH 7.0, at a flow rate of 1.0 mL/min.

Frontal Chromatography

A solution containing 4.1 g/L crude β-gal in 25 mM sodium phosphate, pH 7.0, was pumped into the column at a flow rate of 0.2 mL/min and at 22 °C. The effluent was monitored at 315 nm and fractions were collected at 1-min intervals and analyzed by HPLC for the presence of proteins eluting after β-gal. The fractions free of such proteins were pooled for subsequent use as feed in the displacement.

HPLC Analysis

The purity of the crude β-gal and of the fractions from frontal or displacement chromatography was analyzed by HPLC on the micropellicular DEAE column with gradient elution. The solution of the crude enzyme and the fraction was diluted 1 to 7 with water and 25 μL of the solution thus obtained was injected into the column. The 4-min linear gradient was from 0 to 0.25 M NaCl in 50 mM Tris/HCl buffer, pH 8.0, at a flow rate of 2 mL/min and at 22 °C. The column effluent was monitored at 230 nm and the peak areas were integrated. The amount of each component was calculated from the peak area relative to both the total area and the total amount of proteins injected, assuming complete protein recovery and identical molar extinction coefficients at 230 nm for all components. The stability of the micropellicular column was periodically tested by measuring the retention times and peak areas of β-lactoglobulins A and B and no changes were observed in the course of this study. In the analysis of the crude β-gal, those species eluting before β-gal were referred to as light end contaminants and those after β-gal were referred to as heavy end contaminants.

RESULTS AND DISCUSSION

The analytical chromatogram of crude β-gal was used as the starting point for designing the separation process. It is seen in FIGURE 1A that the crude enzyme contains two light end contaminants (components I and II) and three heavy end contaminants (components IV–VI). The results of displacement with a load of 92 mg of crude β-gal are shown in FIGURE 2. It is seen that much of the enzymatic activity is located in the leading part of the displacement train that contained several proteins, but no component I because it was removed in the rinsing step before the introduction

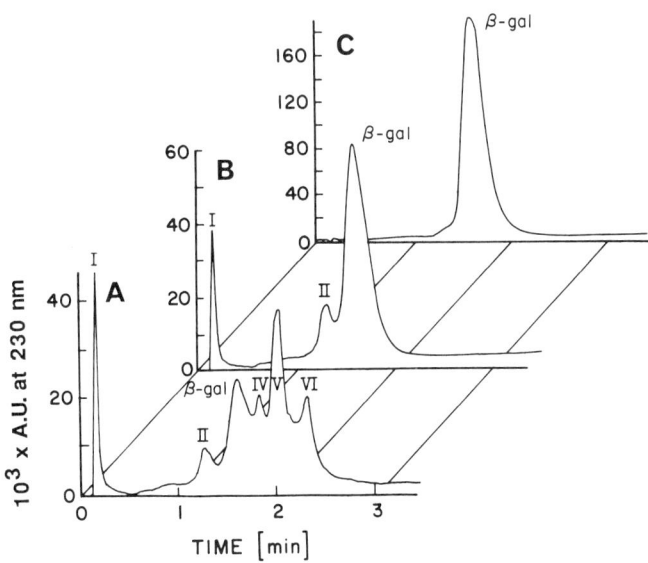

FIGURE 1. Analytical HPLC of β-galactosidase: column, 30 × 4.6 mm micropellicular DEAE on 3-μm polystyrene-divinylbenzene; starting eluant, 50 mM Tris/HCl, pH 8.0; gradient former, 1 M NaCl in starting eluant; linear gradient to 25% (v/v) of gradient former in 4 min; flow rate, 2.0 mL/min; temperature, 22 °C; sample, 25 μL; (A) crude β-gal; (B) β-gal purified by frontal chromatography; (C) β-gal purified by combined frontal and displacement chromatography.

of the displacer. In order to enhance the specific activity of the product, only enzymatically active fractions void of components II or IV were pooled. Because a significant amount of enzyme was discarded, the recovery of enzymatic activity was 47% in the pooled fractions. Displacement using the same chromatographic system and the same amount of feed, but at flow rates of 0.1 and 0.5 mL/min (instead of 0.2 mL/min as above), gave essentially the same results as those depicted in FIGURE 2. The results were the same also upon changing the pH of the carrier from 7.0 to 6.6 and 7.7. Neither reducing the amount of feed from 92 to 31 mg nor increasing the concentration of the displacer from 5 mg/mL to 10 mg/mL improved the separation. As illustrated in

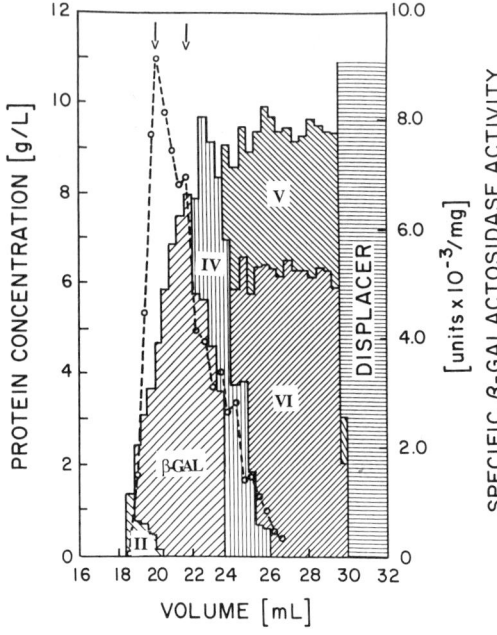

FIGURE 2. Displacement chromatography of crude β-galactosidase: column, two 75 × 7.5 mm TSK DEAE-5PW columns in series; carrier, 25 mM sodium phosphate, pH 7.0; displacer, 5 mg/mL of chondroitin sulfate; flow rate, 0.2 mL/min; fraction volume, 0.4 mL; temperature, 22 °C; feed, 92 mg. Specific β-galactosidase activity is shown by the broken line. Fractions between the arrows were pooled and referred to as the product. The numbering of the components is the same as in FIGURE 1.

FIGURE 2, more than half of the displacement train was occupied by the heavy end; thus, a large part of the separating capacity of the column was wasted and the heavy end protein contaminants could not even be separated from each other by displacement under the conditions discussed above.

In view of earlier work from our laboratory,[2] removal of the heavy end protein contaminants (components IV–VI) by frontal chromatography followed by displacement with a rinsing step to remove the very weakly retained component I offered a tandem separation technique that better utilized the separating capacity of the chromatographic system than direct displacement of the crude enzyme under conditions shown in FIGURE 2.

The results of frontal chromatography of the crude β-gal mixture on a single 75 × 7.5 mm anion-exchanger column are illustrated in FIGURE 3. Component I broke through at the column void volume and was present in every fraction (not shown in FIGURE 3). At the breakthrough of component IV, which is the least retained of the heavy end protein contaminants, frontal chromatography was terminated. The fractions were pooled and used as the feed in subsequent displacement. The results depicted in FIGURE 3 demonstrate that frontal chromatography alone could be used for the removal of the heavy end contaminants (components IV–VI) from the solution of crude β-gal.

After loading the column for the ensuing displacement, first component I was removed by rinsing the column with 25 mM sodium phosphate, pH 7.0, and then the mixture of component II and β-gal was separated by displacement with chondroitin sulfate (FIGURE 4). The purity of the product from the combined frontal and displacement chromatography was compared to that of the pooled fractions from frontal chromatography and to that of the crude β-gal in FIGURE 1. The analytical results are summarized in TABLE 1, which allows a comparison of the various approaches in terms of recoveries, enrichment, and specific enzymatic activities. By direct displacement chromatography of crude β-gal, the specific enzymatic activity more than tripled, but only 15% of the protein mass was recovered in the pooled fractions. Upon removal of the heavy end by frontal chromatography, the specific activity of β-gal almost doubled and 29% of the proteins were found in the pooled fractions. The remaining 71% of the proteins were retained by the column and were desorbed later by a gradient from 25 mM to 500 mM of sodium phosphate, pH 7.0. It is expected that the desorbed proteins, upon dilution to reduce the salt concentration of the solution, can be used as the feed in the next cycle of purification by frontal chromatography.

Large-scale purification of β-gal by elution chromatography has been described by Yamamoto, Nomura, and Sano.[15] They found that ion-exchange chromatography with the same type of stationary phase as was used in the present study was more efficient in the purification of β-gal than hydrophobic interaction and gel filtration chromatography. The detailed literature data permit a comparison of our results to those obtained by traditional chromatography, as shown in TABLE 2. The process was scaled up to a 400 × 600 cm i.d. column that allowed the separation of 400 g of crude β-gal in a linear gradient elution run. The feed occupied one-third of the column volume and, according to the authors, the gradient conditions were optimized. HPLC analysis with a TSK DEAE-5PW column of the pooled product exhibited multiple peaks (figure 4 of

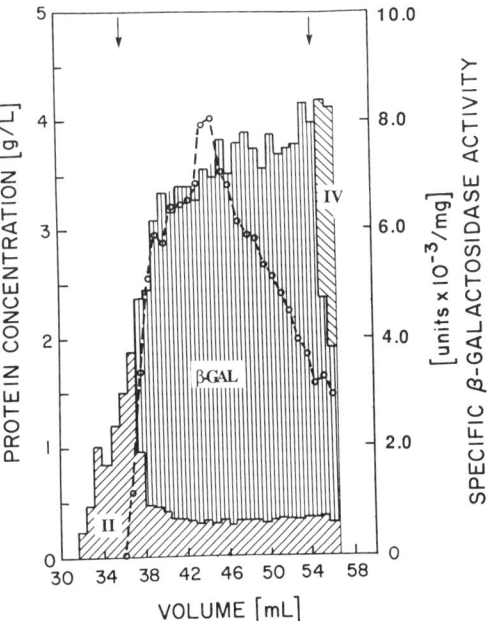

FIGURE 3. Frontal chromatography of crude β-galactosidase: column, 75 × 7.5 mm TSK DEAE-5PW; carrier, 25 mM sodium phosphate, pH 7.0; flow rate, 0.2 mL/min; temperature, 22 °C; fraction volume, 0.2 mL; protein concentration in the feed, 4.1 mg/mL.

reference 15) and only about 80% (w/w) was β-gal. In contradistinction, frontal chromatography was used in the present study to remove the heavy end contaminants and β-gal was further purified by displacement chromatography. Frontal chromatography on a TSK DEAE-5PW column of 3.3 mL allowed 224 mg of crude β-gal to be processed, and 64 mg of ternary mixture (components I and II and β-gal) was pooled. After retaining 13 mg for analysis, 51 mg was separated subsequently by displacement and 28 mg of β-gal was obtained. More importantly, β-gal purified by displacement showed only a single peak both on the micropellicular weak anion-exchanger column (FIGURE 1C) and on the TSK DEAE-5PW column (data not shown). Thus, the product obtained by the present method was significantly purer than that obtained by

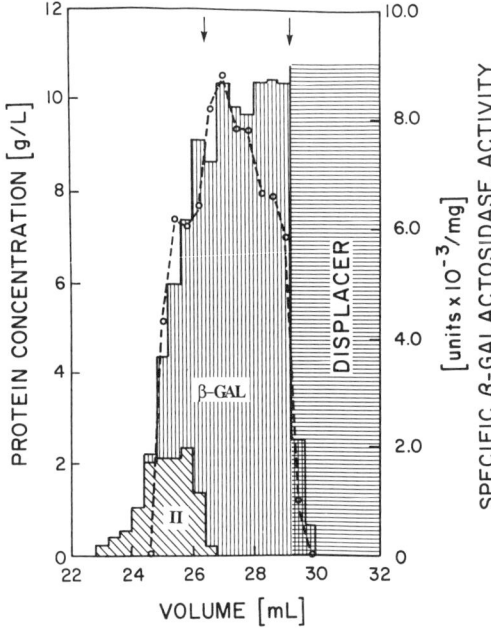

FIGURE 4. Displacement chromatography of β-galactosidase prepurified by frontal chromatography. Conditions as in FIGURE 2 except that the feed was 63 mg.

elution chromatography on the production scale. The efficiency of the combined frontal and displacement chromatography further manifests itself in a 28- and 16-fold enhancement of the throughput, as calculated by using the feed and the product as the respective reference, in comparison to the values with linear gradient elution. The above comparison suggests that a more pure and concentrated product as well as a better utilization of the chromatographic system could be obtained by employing the combined frontal and displacement modes without much optimization than by linear elution chromatography under apparently optimized conditions.[15] Indeed, further improvement in recovery and throughput could be obtained with our system by carrying out frontal chromatography three times and by using the pooled product thus

TABLE 1. Purification of Crude β-Galactosidase

Mode of Chromatography	Volume [mL]		Protein Concentration [mg/mL]		Specific Activity [units $\times 10^{-3}$/mg]		Amount of Protein [mg]		Enzymatic Activity [units $\times 10^{-3}$]		Recovery[a] [%]
	Feed	Product	Feed	Product	Feed	Product	Feed	Product	Feed	Product	
Displacement	44	2.0	2.1	6.7	2.3	7.5	92	13.4	211	100	47
Frontal chromatography	54.6	18.8[b]	4.1	3.4	2.8	5.2	224	64	627	332	53
Displacement after frontal	15[b]	2.8	3.4	10	5.2	7.4	51	28	265	206	78
Combined frontal and displacement[c]	54.6	2.8	4.1	10	2.8	7.4	224	28	627	206	33

[a] Recovery is defined by β-galactosidase activity in the product as percentage of that in the feed.
[b] The difference, 3.8 mL, was consumed for analysis and was retained as a sample.
[c] Overall results of the two steps above.

obtained as the feed in subsequent displacement chromatography. In this way, the capacity of the column in displacement is utilized much better than indicated in TABLES 1 and 2 with a concomitant doubling of the throughput rate.

CONCLUSIONS

Chromatography is poised to become a high-resolution separation technique of great significance in biotechnology for process-scale purification of proteins. However, the linear elution mode that is widely practiced and admirably suitable for laboratory

TABLE 2. Purification of β-Galactosidase on Weak Anion-Exchanger Columns by Combined Frontal and Displacement Chromatography and by Linear Gradient Elution[15]

Operating Conditions	Combined Frontal and Displacement[a]	Linear Gradient Elution[b]
Cycle time[c]	12.1 h	35.3 h
Feed	0.224 g/cycle	400 g/cycle
Mobile phase[c]	0.296 L/cycle	3400 L/cycle
Product	0.028 g/cycle	86.4 g/cycle[d]
Product purity by HPLC	single peak	≈80%[e]
Throughput[f]		
based on feed	2.8 g L^{-1} h^{-1}	0.10 g L^{-1} h^{-1}
based on product	0.35 g L^{-1} h^{-1}	0.022 g L^{-1} h^{-1}

[a]Two 3.3-mL columns packed with 10-μm TSK DEAE-5PW.

[b]A 113-L column packed with 65-μm TSK DEAE 650M.

[c]Volume and time required in the various stages of frontal and displacement chromatography as well as the volume calculated from reference 15 by assuming that column regeneration required the same volume as gradient elution.

[d]The mass of the pooled product was estimated from the peak area relative to both the total area in figure 7 of reference 15 and the total amount of protein injected, assuming complete protein recovery and identical molar extinction coefficients at 254 nm for all components.

[e]The purity of the product from linear gradient elution chromatography was estimated from the peak area relative to the total area in figure 4a of reference 15.

[f]Throughput is the amount of feed processed or product obtained per unit of empty column volume during a cycle.

separations is beset with relatively poor utilization of the separation capacity of the column and, *a fortiori*, is less appropriate for use in manufacturing. Therefore, in the design of separations based on economical considerations, other modes such as nonlinear frontal and displacement chromatography, which offer improved efficiency in process applications, are likely to replace linear elution chromatography when the appurtenant technology will be available. The results of this study illustrate some of the advantages of this approach: better utilization of the chromatographic stationary phase and mobile phase, higher throughput without sacrificing resolution, and economical use of secondary separating agents. In particular, a significantly higher utilization of the separating capacity of the anion-exchanger column can be expected in the purification of crude β-gal when frontal chromatography is employed to remove heavy end protein contaminants and when displacement chromatography is used subsequently to remove light end contaminants. Whereas frontal chromatography alone offered a powerful enrichment method to remove the heavy end contaminants, combination with subsequent displacement chromatography yielded superior results.

ACKNOWLEDGMENTS

We thank Yoshio Kato of Tosoh Company (Tokyo, Japan) for the gift of the TSK DEAE-5PW columns and Yih-Fen Maa for the column packed with micropellicular DEAE.

REFERENCES

1. LIAO, A. W., Z. EL RASSI, D. M. LEMASTER & C. HORVÁTH. 1987. High performance displacement chromatography of proteins: separation of β-lactoglobulins A and B. Chromatographia **24:** 881–885.
2. LEE, A. L., A. W. LIAO & C. HORVÁTH. 1988. Tandem separation schemes for preparative high-performance liquid chromatography of proteins. J. Chromatogr. **443:** 31–43.
3. LEE, A., A. VELAYUDHAN & C. HORVÁTH. 1988. Preparative HPLC. In Proceedings of the 8th Biotechnology Conference, Paris (May 1988), p. 593–610.
4. WALLENFELS, K. & O. P. MALHORTRA. 1961. Galactosidases. Adv. Carbohydr. Chem. **16:** 239–298.
5. WALLENFELS, K. & R. WEIL. 1972. β-Galactosidase. The Enzymes (third edition). **VII:** 617–663.
6. BLAIN, J. A. 1975. Industrial enzyme production. In The Filamentous Fungi. Volume I. J. Smith & D. Berry, Eds.: 193–211. Arnold. London.
7. TANAKA, Y., A. KAGAMISHI, A. KIUCHI & T. HORIUCHI. 1975. Purification and properties of β-galactosidase from Aspergillus oryzae. J. Biochem. **77:** 241–247.
8. MEGA, T. & Y. MATSUSHIMA. 1979. Comparative studies of three exo-β-glycosidases of Aspergillus oryzae. J. Biochem. **85:** 335–341.
9. KATO, Y., J. MATSUSHITA, T. KUBODERA & K. MATSUDA. 1985. A novel enzyme producing isoprimeverose from oligoxyloglucans of Aspergillus oryzae. J. Biochem. **97:** 801–810.
10. YAMAMOTO, K., J. HITOMI, K. KOBATAKE & H. YAMAGUCHI. 1982. Purification and characterization of 1,2-α-mannosidase of Aspergillus oryzae. J. Biochem. **91:** 1971–1979.
11. MEGA, T., T. IKENAKA & Y. MATSUSHIMA. 1970. Studies of N-acetyl-β-D-glucosaminidase of Aspergillus oryzae. J. Biochem. **68:** 109–117.
12. BRADFORD, M. 1976. A rapid and sensitive method for the quantitation of microgram quantities of protein utilizing the principle of protein-dye binding. Anal. Biochem. **72:** 248–254.
13. LEDERBERG, J. 1950. The beta-D-galactosidase of Escherichia coli strain K-12. J. Bacteriol. **60:** 381–392.
14. WALTON, K. W. & C. R. RICKETTS. 1954. Investigation of the histochemical basis of metachromasia. Brit. J. Exp. Pathol. **35:** 227–240.
15. YAMAMOTO, S., M. NOMURA & Y. SANO. 1987. Purification of β-galactosidase by large-scale gradient elution ion-exchange chromatography. J. Chromatogr. **396:** 355–362.

PART IV. MEMBRANE-BASED REACTIONS AND SEPARATIONS

Novel Membrane-based Immobilization Technique for Bioreactors

W. K. KANG, R. SHUKLA, AND K. K. SIRKAR[a]

Department of Chemistry and Chemical Engineering
Stevens Institute of Technology
Hoboken, New Jersey 07030

INTRODUCTION

Whole cells can be immobilized for extended fermentation in a number of ways, for example, surface attachment, entrapment within porous matrices, self-aggregation, and containment behind a barrier.[1] Each technique has its own advantages and limitations. For example, entrapment of cells in hollow fiber devices, either in the fiber lumen or in the extracapillary space, has the advantages of high cell densities, extended cell viability, continuous removal of products and wastes, and cell isolation from substrate. However, hollow fiber wall rupture due to uncontrolled cell growth,[2] diffusional limitations, membrane leakage, gas supply, and removal problems[2] are all significant limitations of hollow fiber devices.

We recently have introduced[3] a novel microporous hollow fiber–based cell immobilization technique that resolves most of the problems of earlier hollow fiber device-based techniques. Chopped microporous hydrophobic or hydrophilic hollow fibers (or bundles of hollow fibers) are used to grow cells in the fiber lumen and the fiber outside surface. Such chopped hollow fibers with cells immobilized in the fiber lumen and outside surface were later utilized in a tubular fermentor to carry on ethanol fermentation by *Saccharomyces cerevisiae* for an extended period. The cell culture was able to produce a very high cell density in the fiber lumen volume.

The chopped hollow fibers with immobilized cells are neutrally buoyant when polypropylene fibers are used.[3] They do not suffer from CO_2-induced bloatings nor need medium pH control as in hydrocolloidal gel–based processes.[4] Furthermore, the chopped hollow fiber support volume is not very high, unlike conventional porous matrix entrapment processes, because hollow fiber wall thicknesses are minimized for membrane separations. Such hollow fiber supports are easily degradable and do not create waste disposal problems as diatomaceous earth particles do. In fact, they can be reused after steam sterilization.[3]

Although a conventional tubular fermentor was used in the first study of this novel technique,[3] a major application of chopped hollow fiber cell immobilization will be in bioreactors having continuous lengths of hollow fibers for gas supply and removal[5] as well as for *in situ* product extraction by dispersion-free solvent extraction.[6,7] Hence, the advantages of matrix entrapment could be grafted independently to the strengths of hollow fiber–based bioreactors. In such applications, it is important to know the immobilized cell density as a function of the microporous chopped fiber dimensions and

---

[a]To whom all correspondence should be sent.

properties. This will enable optimized chopped hollow fiber use in tubular or hollow fiber bioreactors to achieve high cell density and high volumetric productivity.

In this report, we provide detailed experimental evidences of the dependency of immobilized cell density on the chopped hollow fiber diameter, length, and type. Cells of *Saccharomyces cerevisiae* were grown with a view to optimizing the immobilization support for ethanol production. The advantage of this technique is further illustrated by using chopped hydrophilic hollow fibers as support for yeast cells in a conventional tubular bioreactor for ethanol production. The high productivity and cell density achieved are used to demonstrate the usefulness of this technique. Simultaneous fermentation-extraction experiments using this technique for immobilization are also in progress; results will be presented elsewhere.

MATERIALS AND METHODS

Microorganism

The yeast used was *Saccharomyces cerevisiae* (NRRL Y-132) supplied by Northern Regional Research Laboratories (Peoria, Illinois). The composition of the medium is given in the literature.[8]

Materials

Hydrophobic microporous hollow fibers (Celgard X-20 and X-10) were obtained from Questar (Charlotte, North Carolina). Cuprophan microporous hollow fibers that were hydrophilic were procured from Enka (Wuppertal, Federal Republic of Germany). The physical characteristics of these fibers are given in TABLE 1. In all experiments with X-20 fibers, unless otherwise mentioned, X-20 fibers with 240-μm I.D. were used. For those experiments with diameter variations, X-20 fibers with 400-μm I.D. and X-10 fibers with 100-μm I.D. were used. (Note that I.D. stands for inside diameter and O.D. stands for outside diameter.)

Wetting Procedure

The procedure adopted for wetting the hydrophobic Celgard X-20 hollow fibers is given in the literature.[9] This technique is based upon wetting the membranes first with a 60% ethanol solution and then replacing the ethanol slowly by sterile water in an exchange process.

Analytical Methods

Ethanol concentration was determined in a Hewlett Packard gas chromatograph (model 5890A) using Porapak Q (80/100) or Tenax GC (80/100) column and a flame ionization detector. The glucose concentration was analyzed by a YSI model 27 glucose analyzer. Cell concentration in the broth was estimated by measurement of the optical density at 540 nm using a Bausch and Lomb spectrophotometer (model 1001).

TABLE 1. Physical Properties of the Hollow Fiber Membranes Used[a]

Membrane	Material	Pore Size (μm)	O.D. (μm)	I.D. (μm)	Porosity
Celgard X-10	polypropylene hydrophobic	0.03	150	100	0.2
Celgard X-20	polypropylene hydrophobic	0.03	290	240	0.4
Celgard X-20	polypropylene hydrophobic	0.03	450	400	0.4
Cuprophan[b]	regenerated cellulose hydrophilic	—	200^c	140^c	0.55^c

[a] From manufacturer's catalogue.
[b] Pore size not available in manufacturer's catalogue.
[c] Measured in our laboratory.

Cell growth in hollow fibers was measured by subtracting the weight of hollow fibers from the weight of hollow fibers containing cells (these were dried earlier in an incubator overnight at 60 °C). The growth in hollow fibers was observed by SEM (model Jeol JSM-80) using samples appropriately prepared.

Experimental Procedure

Celgard X-20 hollow fibers (obtained from many module residues) were chopped to the required sizes and then wetted. Cuprophan fibers were also chopped to the required sizes. These fibers were transferred to flasks containing medium. After cell growth, fibers were cut at different locations along their length and photographed by SEM. Some fibers were also treated ultrasonically to remove all cells from inside the fibers. An ultrasonicator (model B 2200R-1, Fisher Scientific, Springfield, New Jersey) was used for this purpose.

Tubular Fermentor

A conventional tubular fermentor (1.5-foot-long packed length and 0.5" in diameter) was used to carry out fermentation using chopped hollow fibers as the immobilization support. The tubular fermentor had five sampling ports. The fermentor was packed with chopped hydrophilic hollow fibers (Cuprophan, 0.125" in length, total weight of 10.17 g). Yeast cells were first grown in these fibers outside in shaker flasks. Then, the whole spent broth containing the chopped fibers was poured in the tubular fermentor. A wire mesh (~40 mesh) was used at both ends of the fermentor to retain the fibers while the spent broth was drained. Fresh medium was then passed through the fermentor to carry out fermentation; spent medium was collected in another vessel. A constant pressure nitrogen cylinder was used to pump the medium at a controlled rate. Samples were collected at different time intervals for analysis. Experiments were carried out for a total time of 240 hours. Variations in flow and inlet glucose concentrations were introduced at steady state.

RESULTS AND DISCUSSION

A new immobilization support technique must provide high cell density as well as extended biocatalyst life. These are needed to ensure high conversion and productivity in comparison to other techniques. To achieve these objectives, it is necessary to know the parameters that affect cell growth in, for example, hydrophobic chopped hollow fibers because it has already been demonstrated[3] that cells grow in wetted hydrophobic fibers.

A set of batch experiments were carried out where hydrophobic hollow fibers (Celgard X-20) were cut to 1″ length, wetted, and transferred to a shaker flask containing the medium. This was inoculated with previously grown cells and cells were allowed to grow in hollow fibers. After two days of cell growth, fibers were withdrawn from the flask; these fibers were then cut at $1/32''$ and $1/16''$ along the length from the end of the chopped fiber and photographed. The ends of the withdrawn fibers were also photographed by SEM before cutting and the results are shown in FIGURES 1–3. It is clear from these figures that cells grow from the end to the inside of the fibers, but growth decreases as the fiber middle section is approached. It can be observed from FIGURE 4 of the chopped hollow fiber surface that cells occupy the surface of the fibers as well. It is probable that the substrate and nutrients in the medium cannot diffuse into the fiber lumen from the outside surface because pores may be blocked by the cells on the surface or they are consumed by the surface cells.

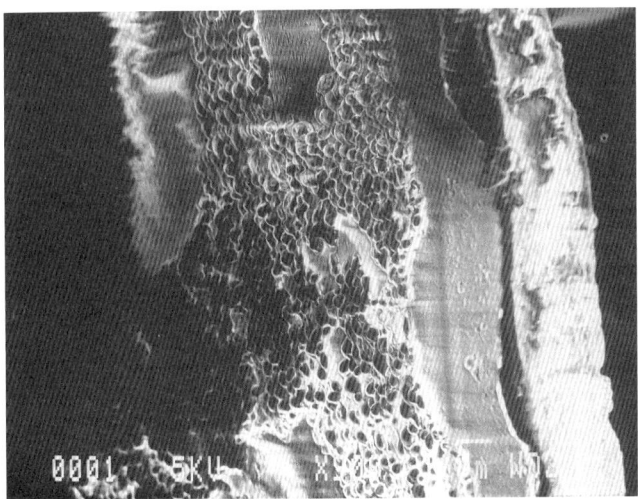

FIGURE 1. SEM photograph of wetted Celgard X-20 fiber after cell growth at the end of the fiber lumen.

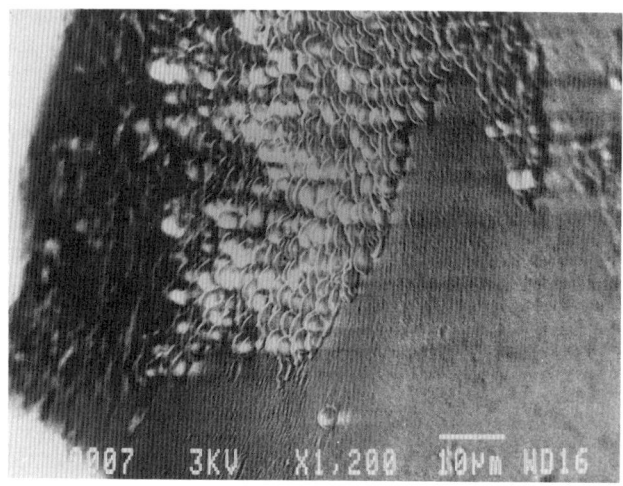

FIGURE 2. SEM photograph of wetted Celgard X-20 fiber after cell growth at $1/32''$ from the end of the fiber lumen.

The length dependency of cell growth was further confirmed by growing the cells in $1/16''$-long and $1/4''$-long chopped hollow fibers in batch culture. These were hydrophobic fibers and the same experimental procedure was followed including wetting. After cell growth, the fibers were taken out and were treated ultrasonically to remove the cells. We were able to remove all the cells; this is confirmed by the SEM photograph (FIGURE

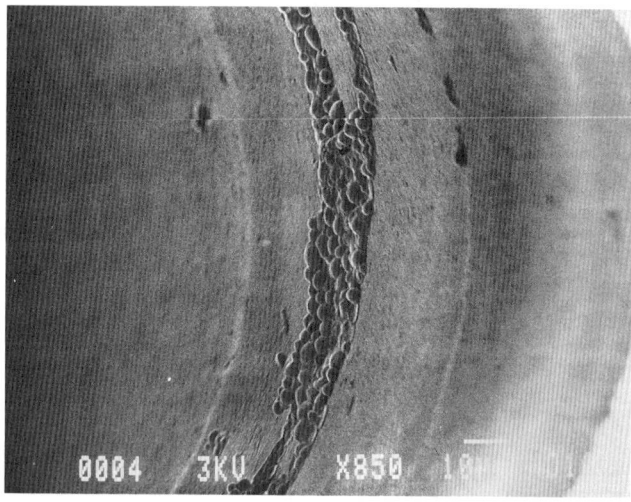

FIGURE 3. SEM photograph of wetted Celgard X-20 fiber after cell growth at $1/16''$ from the end of the fiber lumen.

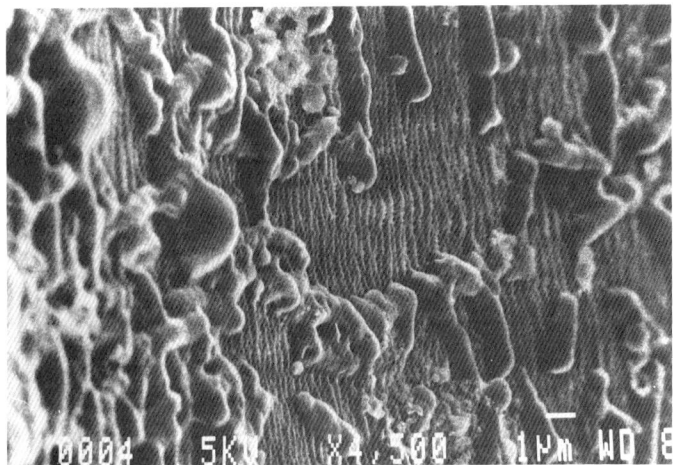

FIGURE 4. SEM photograph of the outer surface of wetted Celgard X-20 fiber after cell growth.

5) of the fiber from which the cells are removed. The data on the length dependency of X-20 (400 μm I.D.) fibers are shown in FIGURE 6. Here, cell concentration per unit of total fiber volume is plotted against time for two different chopped fiber lengths. Cell growth variation with length is quite clear. It shows that diffusional resistance along the fiber length to the supply of substrate and nutrients from the medium at the two ends of the chopped fiber is crucial to cell growth in the fiber lumen.

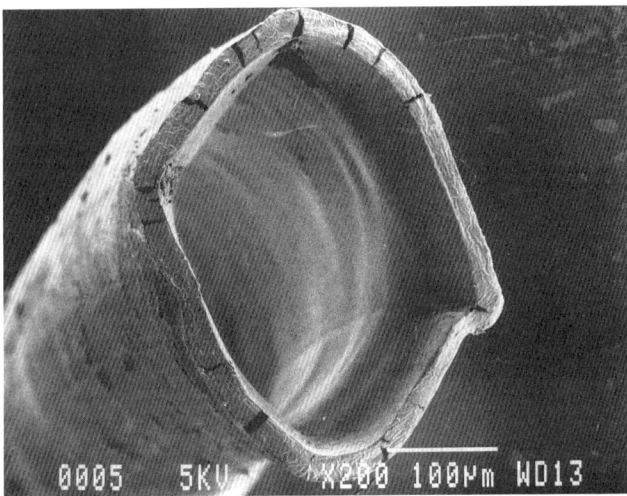

FIGURE 5. SEM photograph of Celgard X-20 fiber at the end of the lumen after all cells are taken out ultrasonically.

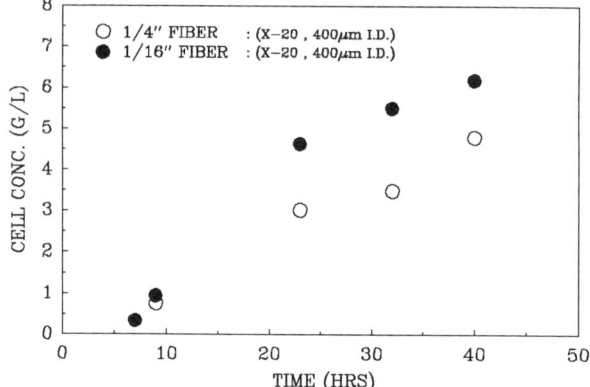

FIGURE 6. Length dependency of yeast cell growth on chopped hollow fibers with time.

The effect of the inside diameter dimension of the fibers on cell growth was similarly studied. Celgard fibers having different diameters were used. The results are shown in FIGURE 7. The cell concentration is based on unit total fiber volume. In fibers with smaller internal diameter, cell growth is considerably higher due most likely to the availability of a higher surface area of attachment per unit volume.

For conventional tubular bioreactor studies, chopped hydrophilic hollow fibers (Cuprophan) of 1/8" length were filled in the tubular fermentor and fermentation was carried out. Because we used hydrophobic chopped fibers in the first study,[3] it was decided to use hydrophilic fibers here. These fibers do not need wetting, unlike the hydrophobic fibers. Furthermore, the problem of a higher specific gravity and therefore settling to the bottom in batch fermentors[3] without vigorous agitation is not important in a tubular fermentor.

The results are shown in FIGURE 8. The substrate flow rate as well as the initial

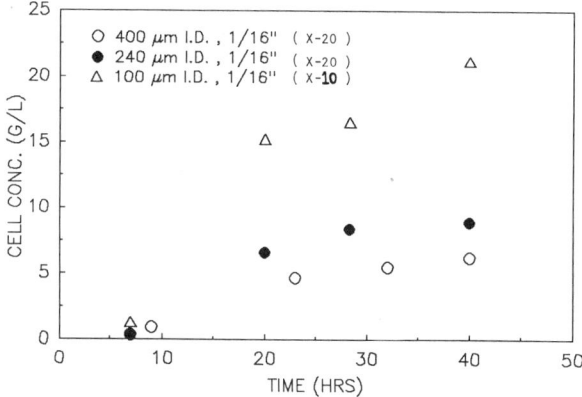

FIGURE 7. Effect of diameter of chopped hollow fibers on cell growth with time.

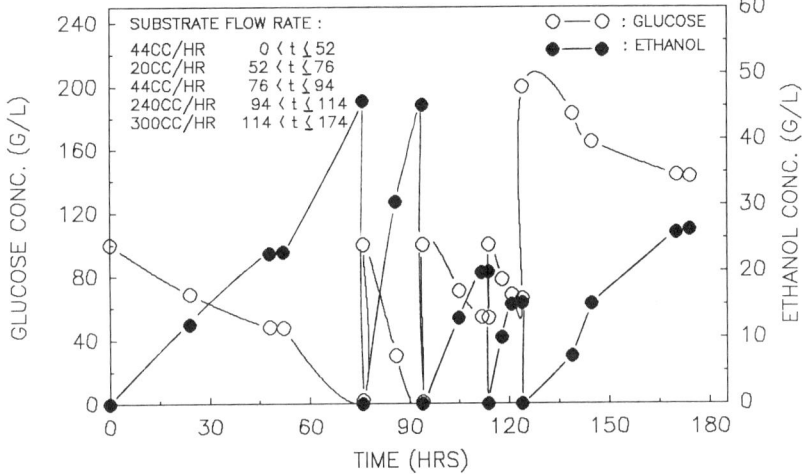

FIGURE 8. Glucose and ethanol concentration profiles in the bioreactor.

glucose concentration were varied. It can be observed that glucose and ethanol concentrations reach steady values for a feed glucose concentration of 200 g/L; however, all glucose is consumed for a feed glucose concentration of 100 g/L. Concentration profiles of glucose and ethanol along the fermentor length at different times are shown in FIGURE 9 for a feed glucose concentration of 100 g/L. It is obvious that fermentation is slow near the end of the fermentor. The productivity of the system

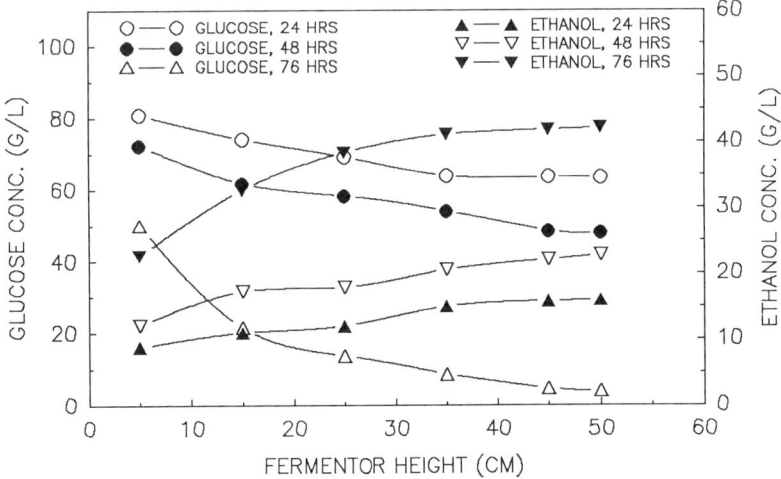

FIGURE 9. Glucose and ethanol concentration profiles along the fermentor length (feed glucose concentration: 100 g/L).

TABLE 2. Ethanol Productivity[a]

Feed Glucose Concentration (g/L)	Flow Rate (mL/h)	Ethanol Productivity (g/L-h)
100	44	28.9
100	240	67.3
100	300	65.0
200	300	107.8

[a] A tubular bioreactor with 0.125-inch-long chopped Cuprophan hollow fibers was used.

is reported in TABLE 2; a very high ethanol productivity of 107.8 g/L-h is obtained. This compares well with some of the best ethanol productivities of 82 and 133 g/L-h reported in the literature.[10,11]

The cell concentration along the fermentor length in the tubular bioreactor is shown in FIGURE 10. The cell density after 24 h of fermentation is almost independent of the fermentor length because cells were grown in these fibers outside in batch mode. With an increase in time, the cell growth near the fermentor inlet is much higher than that at the exit. This is due to the availability of less glucose at the fermentor exit. However, the extent of cell growth at the exit is still considerable.

A maximum cell concentration of 9.2×10^9 cells per unit of fiber lumen volume (4.5×10^9 cells per unit of fiber volume) is obtained near the fermentor entrance at 80 h. The maximum cell concentration obtained in the earlier study[3] with 1/4″ hydrophobic chopped fibers was 9.36×10^9 cells per unit of fiber lumen volume. Thus, the cell densities achieved in these two studies are similar. Moreover, the value is quite high.

The results of the measurement of the rate of cell leakage from the chopped hollow fibers at the fermentor exit are provided in TABLE 3. The effect of the medium flow rate on the amount of cell leakage per unit of immobilized cells can be clearly seen. Cell leakage increased with increased medium flow rate. These cell leakage rates are comparable to the leakage rates of *Zymomonas mobilis* immobilized in k-carrageenan beads for ethanol fermentation.[4]

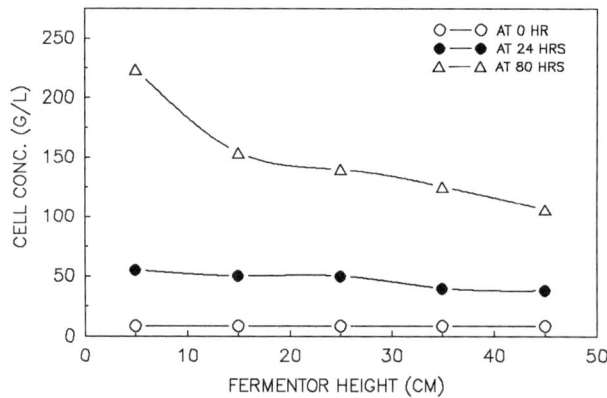

FIGURE 10. Cell concentration profiles in the bioreactor along the fermentor length.

We have seen earlier (FIGURE 6) that, at any given time, immobilized cell density increases as the chopped fiber length decreases. Because cell leakage occurs from the ends of the chopped fibers, it is apparent that cell leakage will increase with a decrease in the chopped fiber length. It is our observation that the leakage rate in our earlier study[3] with $1/4''$ fibers was significantly lower than that in the present study.

CONCLUDING REMARKS

We have illustrated in this report some of the important cell immobilization characteristics of the new chopped hollow fiber–based immobilization technique. The immobilized cell density increases with decreasing hollow fiber diameter; a decrease in chopped fiber length leads to an increased cell density. The utility of this immobilization technique has been convincingly demonstrated with a tubular bioreactor for ethanol fermentation; a high cell density of 9.2×10^9 cells per unit of fiber lumen volume and a high ethanol productivity of 107.8 g/L-h have been obtained with a low cell leak rate. In view of the earlier study of bioreactor performance with chopped hydrophobic hollow fibers, we can conclude that a tubular bioreactor could easily use

TABLE 3. Cell Leakage from Chopped Hollow Fibers in a Tubular Bioreactor

Medium Flow Rate (mL/h)	Leakage Rate (cells leaked/immobilized cell-h)
44	0.007
240	0.029
300	0.038

either wetted chopped hydrophobic fibers or chopped hydrophilic fibers. The question of a dense growth leading to a dense biomass outside the fiber surfaces and a higher pressure drop has not been considered here. Similarly, the relation between chopped fiber length and leakage rate needs to be considered.

ACKNOWLEDGMENTS

We would like to acknowledge the generous supply of Celgard hollow fibers to our research by Robert W. Callahan of Separations Products Division of Hoechst Celanese Corporation, Charlotte, North Carolina. We also thank Enka for supplying us with the Cuprophan hollow fibers. The end residues from many modules provided the source for chopped hollow fibers used in this study.

REFERENCES

1. KAREL, S. F., S. B. LIBICKI & C. R. ROBERTSON. 1985. Chem. Eng. Sci. **40**: 1321–1354.
2. INLOES, D. S., D. P. TAYLOR, S. N. COHEN, A. S. MICHAELS & C. R. ROBERTSON. 1983. Appl. Environ. Microbiol. **46**: 264–278.

3. SHUKLA, R., W. K. KANG & K. K. SIRKAR. 1989. Appl. Biochem. Biotechnol. **20/21:** 571–586.
4. SCOTT, C. D. 1987. Ann. N.Y. Acad. Sci. **501:** 487–493.
5. KANG, W. K., R. SHUKLA, G. T. FRANK & K. K. SIRKAR. 1988. Appl. Biochem. Biotechnol. **18:** 35–51.
6. FRANK, G. T. & K. K. SIRKAR. 1985. Biotechnol. Bioeng. Symp. **15:** 621–631.
7. FRANK, G. T. & K. K. SIRKAR. 1986. Biotechnol. Bioeng. Symp. **17:** 303–316.
8. GENCER, M. A. & R. MUTHARASAN. 1983. Biotechnol. Bioeng. **25:** 2243–2262.
9. BHAVE, R. R. & K. K. SIRKAR. 1987. ACS Symp. Ser. **347:** 138–151.
10. CYSEWSKI, C. R. 1976. Fermentation kinetics and production of ethanol. Ph.D. dissertation. University of California, Berkeley.
11. INLOES, D. S., S. N. COHEN, A. MATIN, A. S. MICHAELS & C. R. ROBERTSON. 1982. Immobilization of bacterial and yeast cells in hollow fiber membrane bioreactors. Paper presented at the AIChE annual meeting in Los Angeles, November 14–19, 1982.

Separation of Amino Acids Using Composite Ion Exchange Membranes

BINAY K. DUTTA[a] AND SUBHAS K. SIKDAR

Center for Chemical Engineering
National Institute of Standards and Technology
Boulder, Colorado 80303

INTRODUCTION

The commercial success of biotechnology in producing specialty chemicals depends to a large extent on developing efficient separation and purification methods. Membrane-based separation processes offer an attractive low-energy approach to bioseparation. In some cases, a membrane separation module can be conveniently coupled to a bioreactor for continuous product recovery without affecting the microbial cells.

In this report, we discuss experimental data representing the recovering and separating of amino acids from aqueous solutions using perfluorosulfonic acid (PFSA) ion exchange membranes. Previously, amino acids were manufactured by hydrolyzing proteins.[1] Lately, their many uses, from food additives to pharmaceuticals, have prompted the development of highly successful microbial manufacturing routes.[2,3] Bioreactors, which are usually run batchwise, can be run continuously for amino acid production (FIGURE 1) provided a satisfactory method of product recovery is available. Previous work in our laboratory[4-6] established the potentiality of commercially available PFSA ion exchange membranes in isolating amino acids from aqueous solutions. However, amino acid permeabilities through these membranes need to be greatly improved before they can be commercially used. We have successfully cast a thin PFSA film on a porous support and have obtained greatly increased fluxes of amino acids. In addition, the composite membrane exhibited good mechanical strength because of the support film. The objective of this report is to assess the utility of this composite membrane and compare its performance with that of the commercial one for recovering amino acids from solutions.

MEMBRANE PREPARATION

The composite membrane used in this study consisted of a thin layer of a perfluorosulfonic acid polymer on a porous support. Films of nylon, porous polypropylene, porous cellulose acetate, and porous polytetrafluoroethylene (PTFE) were used as support materials. The highly porous thin PTFE film was found to be the best backing for the composite membrane.

A commercially available solution of the PFSA polymer (5% w/v) in a mixture of water and lower alcohols was used to prepare the composite membrane. A piece of

---

[a]Binay K. Dutta is a guest scientist from the University of Calcutta, Calcutta, India.

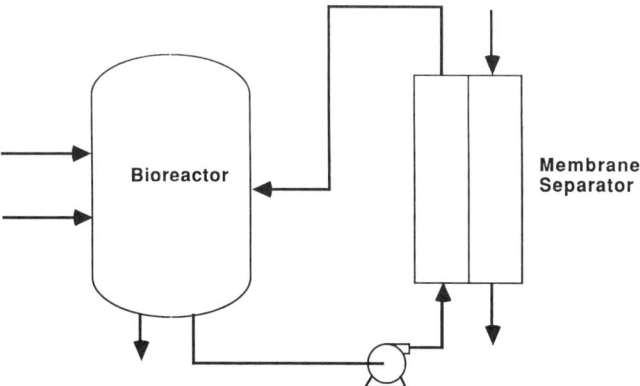

FIGURE 1. Schematic of a bioreactor-separator system.

porous PTFE film was soaked in methanol and then put in a large amount of cold water in which the pores of the hydrophobic support material exchanged their methanol for water. This enabled wetting of the hydrophobic PTFE film with water. The wet backing was then stretched on a leveled glass plate and the solution of the PFSA polymer was spread uniformly over the backing. The thickness of the polymer solution applied was controlled by a glass rod that was rolled over the film to maintain a uniform gap. The liquid film was allowed to dry slowly at room temperature. The composite was then heated to 105 °C for one hour. During this period, phase separation occurred between the hydrophobic fluorocarbon backbone of the polymer and the pendent sulfonic acid groups. This heat treatment yielded a stable and insoluble film of the ion exchange polymer on the PTFE backing. The preparation and characterization of the composite membrane have been discussed in more detail elsewhere.[7]

EXPERIMENTAL SETUP AND PROCEDURE

Permeation rates of amino acids through PFSA membranes were measured using a Plexiglas permeation cell, which is schematically shown in FIGURE 2. The cell consisted of two flanged stirred compartments, 25 mm in radius and 104 mm in length, each provided with suitable openings for inserting stirrer shafts, collecting samples, and inserting temperature probes. The contents of both compartments were kept agitated using stirrers equipped with four flat blades. Stirring eliminated diffusional resistances in the continuous phase. The stirrers were coupled to variable speed motors through flexible cables. After assembling, the permeation cell was kept immersed in a constant temperature bath so as to maintain the temperatures of the liquids in the compartments to within ±0.1 °C.

Before an experiment, the membrane was placed between the flanges of the permeation cell. When the commercial membrane was used, the membrane piece was first treated with 0.5 M sulfuric acid for two hours[6] to insure conversion to the H^+ form. It was then washed repeatedly with water before placing in the permeation cell. While the composite membrane was also treated with dilute acid, it was further soaked in methanol and then put in water for one hour before placing it in the permeation cell. The composite membrane, when wet, was transparent like the commercial membrane.

After placing the assembled permeation cell in the water bath, one compartment (source) was filled with an aqueous amino acid solution of desired concentration and the membrane was allowed to equilibrate with the source solution for about 15 minutes. The other compartment (sink) was quickly filled with water and stirring in both compartments was started. Small volumes of samples (approximately 0.3 mL) were drawn from the sink compartment from time to time to determine the sink concentrations and the permeation rates.

ANALYTICAL PROCEDURE

Samples of solutions containing only one amino acid were analyzed spectrophotometrically after ninhydrin derivatization of the samples.[8] The samples were suitably diluted before derivatization. Absorbance of the ninhydrin derivatives was measured at 570 nm.

Samples of amino acid mixtures were analyzed by HPLC after precolumn derivatization using dansyl chloride.[9] A mixture of acetonitrile and 0.045 M acetate buffer (pH = 4.5) was used as the mobile phase flowing through a C_{18} reverse phase column. The solvent gradient was 20% to 60% acetonitrile over 35 minutes.

RESULTS AND DISCUSSION

Both of the membranes we used were in the H^+ form. Compared to the 170-micron thickness of the commercial membrane (equivalent weight: 1100), the PFSA layer in the composite membrane was only 8–9 microns thick, as determined from its scanning

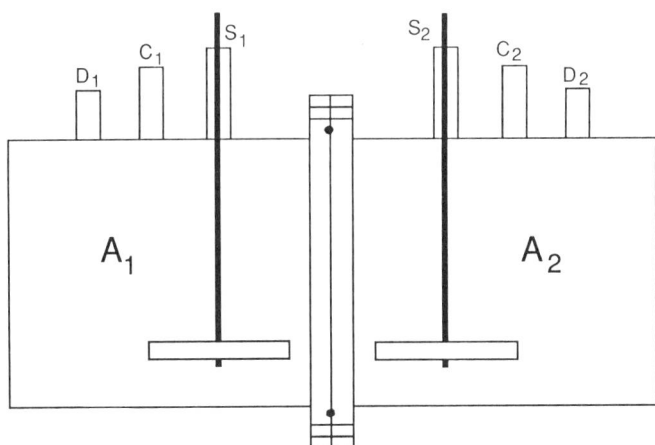

FIGURE 2. Membrane permeation cell (A_1, A_2: source and sink compartments; S_1, S_2: stirrers; C_1, C_2: sampling ports; D_1, D_2: thermocouple ports).

electron micrograph (FIGURE 3). The equivalent weight of the composite membrane material was 1150.

The stirrer speed during permeation was 400 revolutions per minute. At this agitation level, the diffusional resistance at the interface between the membrane and the solution was negligible compared to that offered by the membrane itself. The permeation rate was unaltered even when the stirrer speed was 550 revolutions per minute. However, the membrane support (i.e., the PTFE layer) for the composite membrane offered a diffusional resistance almost equivalent to that of the PFSA layer. This was ascertained by measuring the permeation rate of glycine through the wet PTFE backing itself. In separate control experiments, we noticed that the flow of either water or an aqueous solution through the wet backing under the gradient of a few

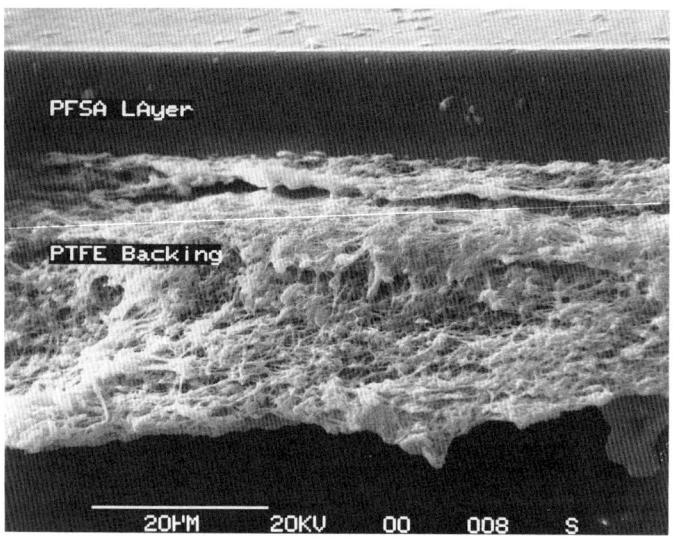

FIGURE 3. SEM photograph of the composite membrane.

inches of water was negligible. FIGURE 4 shows the concentration history of glycine transport through the composite membrane and through the membrane support at 25 °C. The corresponding fluxes were 2.109×10^{-7} and 3.54×10^{-7} mol/cm^2-s for source concentrations of 0.41 and 0.5 mol/L, respectively. The fluxes were calculated from the slopes of the concentration-time curves at times greater than 10 minutes, when the concentrations changed almost linearly with time. The initial curvature is attributed to the fact that the membrane was saturated with the source solution just as an experiment was started.

Permeation results for L-valine, L-threonine, and L-lysine are shown in FIGURES 5–7 for both the commercial and the composite membranes for different initial source concentrations of the permeant species. For all these amino acids, the fluxes for the

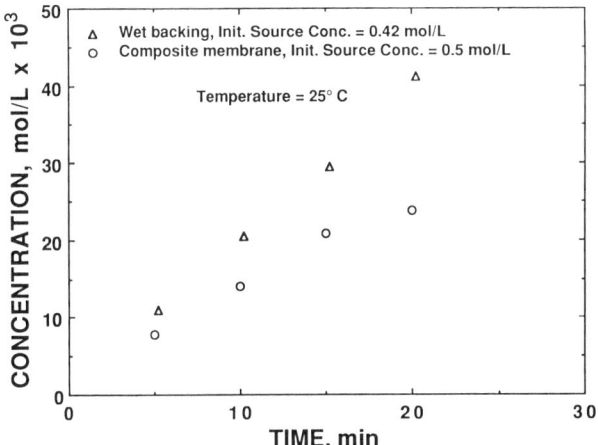

FIGURE 4. Sink concentration history resulting from glycine transport through the composite membrane (O) and the wet PTFE backing (△).

thick commercial membrane leveled off as the source concentration increased. This saturation flux behavior of amino acids through the PFSA membrane was earlier reported by Sikdar.[5] However, the fluxes for the composite membrane increased continuously with the source concentration, without exhibiting the saturation behavior. In addition, compared to the commercial membrane, the composite membrane allowed permeation fluxes larger than an order of magnitude.

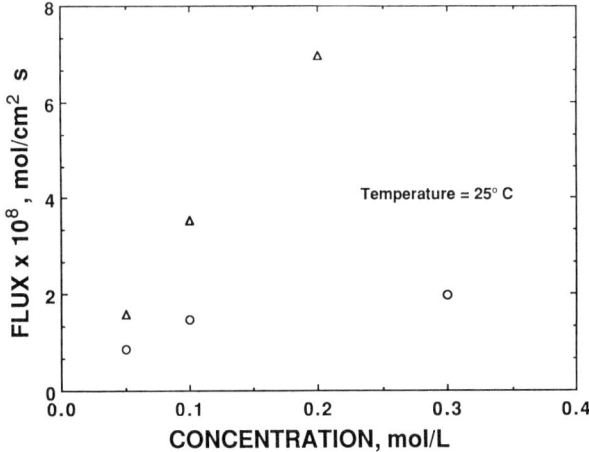

FIGURE 5. Permeation of L-valine through the commercial (O) and the composite (△) membranes.

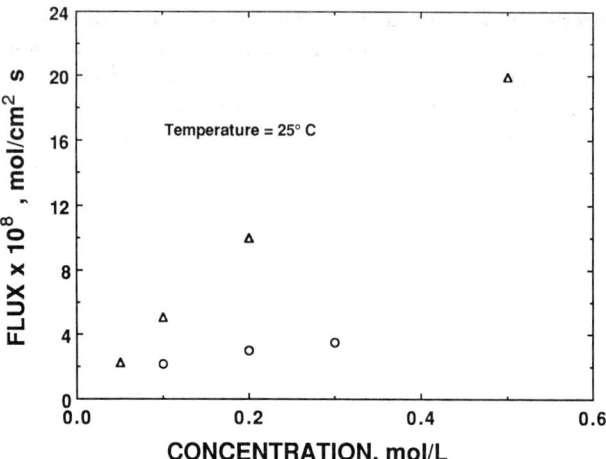

FIGURE 6. Permeation of L-threonine through the commercial (O) and the composite (△) membranes.

Let us compare the performance of our composite membrane with that of the other membranes reported in the literature. Tsukube[10] used a liquid membrane containing ammonium cation–crown ether complex as the carrier for the active transport of L-phenylalanine. Bryzak et al.[11] used a liquid membrane of decalin containing crown ether immobilized on a porous polyethylene film and studied the permeation of L-phenylalanine and L-aspartic acid. The permeation of glycine, L-leucine, and L-phenylalanine through poly(1-alkyl-4-vinylpyridinium iodide-coacrylonitrile) mem-

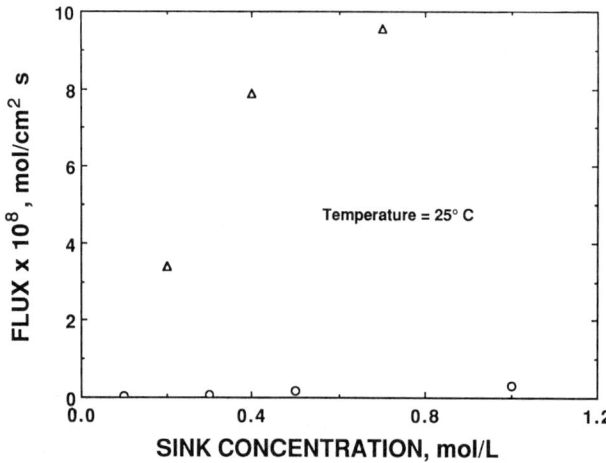

FIGURE 7. Permeation of L-lysine through the commercial (O) and the composite (△) membranes.

brane was reported by Yoshikawa et al.[12] The transport of L-phenylalanine through a chitosan-PVA membrane was studied by Uragami et al.[13] In all these studies, the amino acid fluxes were in the range of 10^{-14} to 10^{-10} mol/cm²-s for source concentrations in the range of 0.025–0.05 mol/L; in contrast, with our composite PFSA membrane, we could get fluxes in the range of 10^{-8}–10^{-7} mol/cm²-s.

The temperature dependence of the amino acid fluxes is shown in FIGURE 8. The

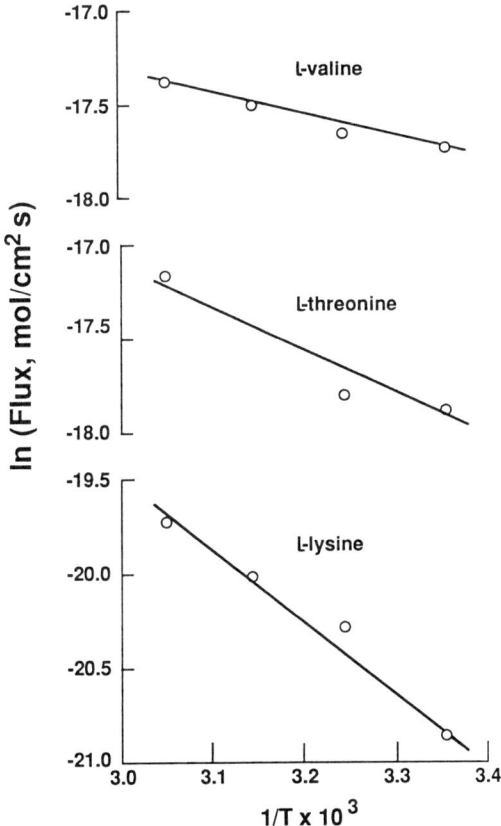

FIGURE 8. Arrhenius plot showing temperature dependence of amino acid fluxes.

activation energies for transport calculated from these Arrhenius plots were 2.2, 4.09, and 7.51 kcal/mol, respectively, for L-valine, L-threonine, and L-lysine. These values are characteristic of diffusive permeation of the amino acids through the membrane.

We also studied the permeation of several mixtures of amino acids through the commercial and the composite membranes. Concentration-time plots for three mixtures—(glycine, L-alanine), (glycine, L-phenylalanine), and (L-phenylalanine,

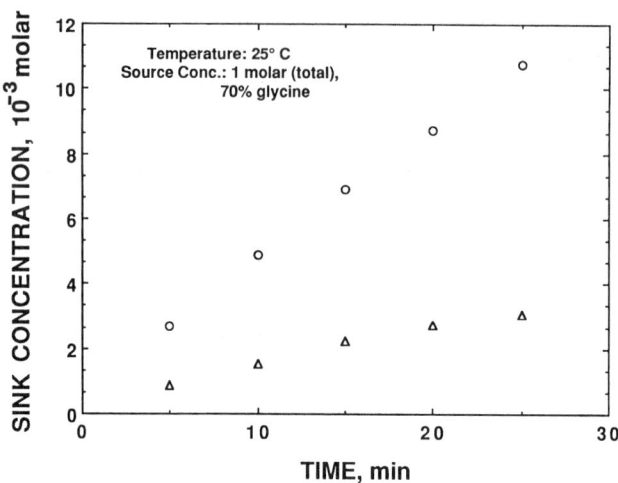

FIGURE 9. Permeation of a mixture of glycine (○) and L-alanine (△) through the composite membrane.

L-aspartic acid)—are given in FIGURES 9–12. These experiments were conducted at the autogenous pH's of the mixtures. Although the membranes exhibited preferential permeation for some of the amino acids, the selectivities calculated from these data were rather low (1.5–4).

A probable mechanism of the amino acid transport from the PFSA membranes is schematically represented in FIGURE 13. Such a mechanism was first suggested by Jin et al.[6] Amino acids in aqueous solutions are present as cations, anions, and bipolar

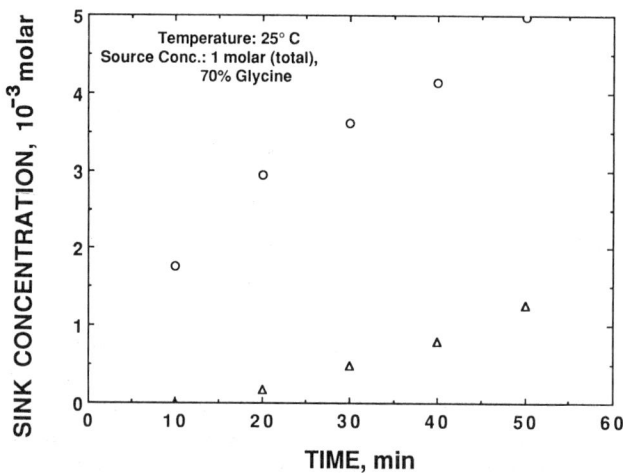

FIGURE 10. Permeation of a mixture of glycine (○) and L-phenylalanine (△) through the commercial membrane.

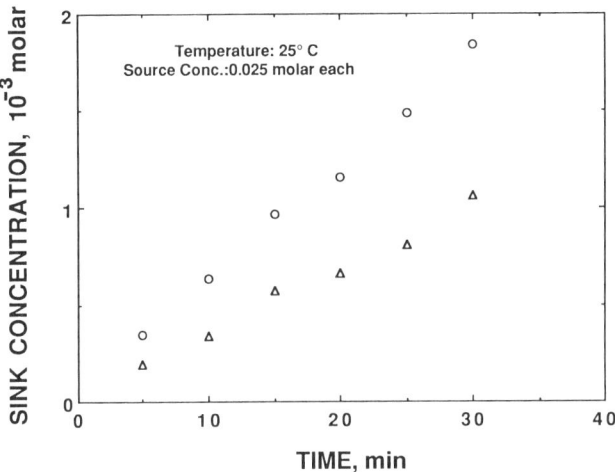

FIGURE 11. Permeation of a mixture of L-phenylalanine (O) and aspartic acid (△) through the composite membrane.

zwitterions depending on pH:

$$HAm + H^+ \leftrightarrow HAmH^+$$

$$HAm \leftrightarrow H^+ + Am^-$$

where HAm is the zwitterionic form (without any net charge) and $HAmH^+$ and Am^- are the positively and negatively charged species, respectively. The acid form of the PFSA membrane used in this study sorbed about 20% of its weight of solution when

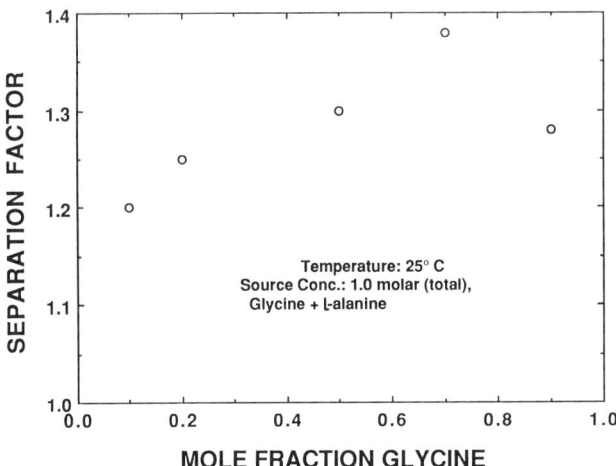

FIGURE 12. Separation factors of mixtures of glycine and L-alanine for the composite membrane.

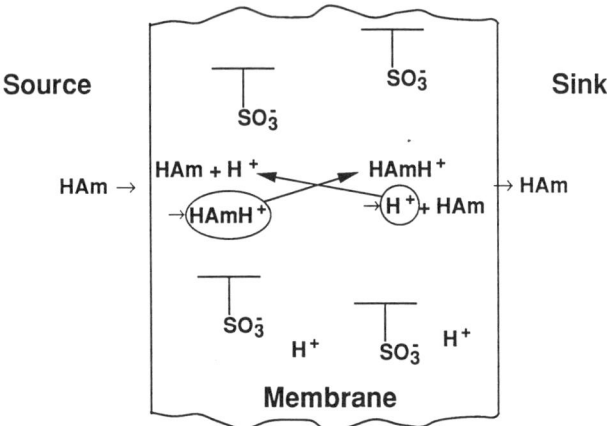

FIGURE 13. Proposed mechanism for amino acid transport through the PFSA membrane.

equilibrated with an aqueous amino acid solution. Sorption studies[14] showed that the amount of an amino acid sorbed by the membrane exceeded the amount that could be expected to be present in the interstitial solution; the excess amount was probably present in the protonated form. The membrane, being of the cation exchange type, promoted the permeation of the protonated species through the membrane matrix. The protonated amino acids were formed at the source solution side of the membrane and were diffusively transported to the sink side of the membrane where deprotonation took place. However, the diffusion of the zwitterionic species also contributed to the total flux. In the case of a mixture, the fraction of the protonated form of a component amino acid depended on its pK, thereby governing its transport.

REFERENCES

1. YAMAMOTO, A. 1980. Amino acids. *In* Encyclopedia of Chemical Technology. Volume 2 (third edition). Kirk & Othmer, Eds. Wiley–Interscience. New York.
2. AIDA, K., I. CHIBATA, K. NAKAYAMA, K. TAKINAMI & H. YAMADA, Eds. 1986. Biotechnology of Amino Acid Production. Elsevier. Amsterdam/New York.
3. KINOSHITA, S. 1987. Thirty years of amino acid fermentation. Proc. 4th European Congress on Biotechnology (Amsterdam) **4:** 679.
4. SIKDAR, S. K. 1985. Amino acid transport from aqueous solutions by a perfluorosulfonic acid membrane. J. Membr. Sci. **24:** 59.
5. SIKDAR, S. K. 1987. Permeation characteristics of amino acids through perfluorosulfonated polymeric membranes. Ind. Eng. Chem. Res. **26:** 170.
6. JIN, M., S. K. SIKDAR & S. D. BISCHKE. 1988. Glycine permeation through Na^+, Ag^+, and Cs^+ forms of perfluorosulfonated ion exchange membranes. Sep. Sci. Technol. **23:** 2293.
7. DUTTA, B. K., D. RANDOLPH & S. K. SIKDAR. 1989. Casting thin and composite membranes of perfluorosulfonated ion exchange polymer to obtain high permeation fluxes. J. Membr. Sci. Submitted.
8. MOORE, S. & W. H. STEIN. 1948. Photometric ninhydrin method for use in the chromatography of amino acids. J. Biol. Chem. **176:** 367.
9. CHANG, J. Y., R. KNECHT & D. G. BROWN. 1983. Amino acid analysis in the picomole

range by precolumn derivatization and high performance liquid chromatography. Methods Enzymol. **91:** 41.
10. TSUKUBE, H. 1983. A proton-driven amino acid pump: lipophilic primary ammonium cation–crown ether complex as a new type of anion-transport-carrier. J. Membr. Sci. **14:** 155.
11. BRYZAK, M., P. WEICZOREK, P. KAFARASKI & B. LEJCZAN. 1988. Crown-ether mediated transport of amino acids through an immobilized liquid membrane. J. Membr. Sci. **37:** 287.
12. YOSHIKAWA, M., M. SUZUKI, K. SANUI & N. OGATA. 1987. Transport of amino acids through synthetic polymer membranes containing pyridinium cation charge sites. J. Membr. Sci. **32:** 235.
13. URAGAMI, T., F. YOSHIDA & M. SUGIHARA. 1988. Studies on synthesis and permeabilities of special polymer membranes. Active transport of organic ions through cross-linked chitosan membrane. Sep. Sci. Technol. **23:** 1067.
14. DUTTA, B. K. & S. K. SIKDAR. Unpublished.

Studies of Transport Processes Coupled with Reaction in Membrane-sandwiched Yeast Cell Reactors[a]

YONG S. JEONG,[b] W. R. VIETH,[b]
AND TAKESHI MATSUURA[c]

[b]Department of Chemical and Biochemical Engineering
Rutgers University
Piscataway, New Jersey 08854

[c]Division of Chemistry
National Research Council of Canada
Ottawa, Canada K1A 0R6

INTRODUCTION

The need for immobilized whole cell reactor systems is widely recognized. Many reactor configurations have been tried using one or more semipermeable membranes to perform bioreaction and separation simultaneously.[1-3] The major problems faced by such systems are high diffusional resistances, substrate depletion, and product inhibition. Over the past few years, many membrane reactors have been proposed that address these problems individually.[4,5]

In our earlier reports, we have proposed a bioreactor in which living yeast cells are sandwiched between an ultrafiltration (UF) and a reverse osmosis (RO) membrane.[6,7] A solution containing glucose substrate and nutrients is in contact with the ultrafiltration membrane and pressure is applied on the feed substrate solution. The substrate permeates through the UF membrane freely together with solvent (water) and arrives at the cell layer, where the bioreaction starts to occur, leading to the product ethanol. When the solution is forced by pressure out of the cell layer from the side that is in contact with the RO membrane, permselection between the substrate glucose and the product ethanol occurs. Whereas ethanol permeates almost freely through the RO membrane, the latter is practically impermeable to glucose substrate.[8] Therefore, it is possible to obtain the product ethanol solution without much contamination from glucose and nutrients. The advantages of such a bioreactor over conventional membrane bioreactors are:

(a) high cell concentration within a limited reactor space,
(b) forced convective flow of substrate to the cell layer, which is much faster than diffusive transport, and
(c) removal of the product ethanol and CO_2 from the cell layer and prevention of the product inhibition.

---

[a]This manuscript was issued as NRCC No. 31036.

A similar device for membrane immobilization has been proposed by Michaels[9] as membrane-moderated immobilized cell bioreactors.

Although the above bioreactor system has been successfully demonstrated[6] and the effect of some operational variables such as operating pressure, feed glucose concentration, and porosities of UF and RO membranes on the reactor performance has been studied,[7] the entire bioreactor system has not been fully understood because the transport characteristics of cell layers were not known for both substrate and product. Hence, it is most urgently needed to establish the transport mechanism and to produce associated parameters, particularly with respect to cell layers.

There are several reports found in the literature on the diffusion coefficients of the substrate glucose and the product ethanol in the cell-immobilized calcium alginate membranes[10] and that of glucose in the cell-immobilized beads.[11] Although it might be interesting to compare transport parameters of the sandwiched membrane system with those in the more conventional form of the cell immobilization, it is, however, out of the scope of this work.

The objective of this work is therefore to establish a transport theory that describes the permeation of the solvent (water) and the substrate glucose and to generate numerical parameters that are associated with transport equations. Although there are two solutes—the substrate glucose and the product ethanol—involved in the bioreaction under study, the transport of glucose alone was chosen because the RO membrane shows semipermeable characteristics of the glucose solution. On the other hand, RO membranes allow free passage of ethanol solute. In order to test the transport theory established for the above system experimentally, a bioreactor system in which dead cells are sandwiched between UF and RO membranes was used so that the transport alone could be isolated. Although the clogging effects of dead cells are different from those of living cells, the transport through a dead cell layer is considered to be the best approximation of that through a living cell layer.

Another objective of this work is to investigate the growth of the yeast cells during the biocatalytic reaction. For this purpose, a continuous reactor system in which a series of four bioreactors were connected was constructed. This system allowed a simultaneous operation of four different reactors and thus facilitated the experiment. The reactors were opened from time to time and the number of yeast cells in the sandwiched cell layer was counted. Another advantage of the continuous reactor system is the improved flow regime in the bioreactor.

MATERIALS AND METHODS

The materials used in the experiment and the experimental methods were described in detail in our previous paper.[6,7] The cellulose acetate membranes, both for RO and UF experiments, were laboratory prepared. The details of the membrane preparation are shown in TABLE 1. It should be noted that the UF-1 membrane was gelled in a gelation medium that is higher in ethanol content than the UF-2 membrane. *Saccharomyces cerevisiae* ATCC 4126 was employed as the biocatalyst in this study because extensive immobilization studies have been done in the past on this cell. Inoculum was prepared using YM broth. A measured volume (20–60 mL) of known concentration (in the exponential phase, about 2×10^7 cells per mL) was centrifuged. The cells were then

TABLE 1. Details of the Cellulose Acetate Membrane Preparation

	UF-1 Membrane	UF-2 Membrane	RO Membrane
Casting solution composition, wt %			
(1) cellulose acetate (E-398-3)		17.0	
(2) acetone		69.2	
(3) magnesium perchlorate		1.45	
(4) water		12.35	
Temperature of casting solution, °C		4	
Temperature of casting atmosphere		room	
Humidity of casting atmosphere		room	
Solvent evaporation period, s		60	
Ethanol/water volume ratio in gelation medium	50/50	40/60	0/100
Temperature of gelation medium, °C		0	
Shrinkage temperature, °C	—	—	72
Shrinkage period, min	—	—	10

brought into 20–60 mL of 50 vol-% ethanol solution and stirred with Vortex-Genie for 30 minutes to homogenize and to kill the yeast. The solution was subsequently centrifuged to remove liquid and the cells were washed with pure water. Staining with methylene blue proved cell death without the lysis of cells.[10] The dead yeast cells were filtered through a 0.2-μm microfilter and then sandwiched between UF and RO membranes. The total number of sandwiched cells was controlled to 10^{10} and the volume of the sandwiched cell layer is about 1 cm^3.

Two kinds of bioreactors were used in this study. A batch-type reactor used for the reverse osmosis experiment was the same as that illustrated in the previous report.[6,7] The membrane was mounted at the bottom of the reactor, the feed solution was loaded, and the pressure was applied by nitrogen gas on the feed solution. A magnetic stirrer was driven in close proximity of the membrane surface so that the development of the concentrated boundary layer was minimized. A continuous-type reactor used for the bioreaction experiment is illustrated in FIGURE 1. The feed solution was pumped into the reactor at a speed sufficient to produce fluid turbulence in the proximity of the membrane surface so that the development of the boundary layer was prevented. After passing a series of four bioreactors, the feed solution was recycled to the feed solution vessel where air in the solution was replaced by nitrogen gas. Each component of the reactor system was sterilized carefully before the start of the bioreaction. All RO experiments were carried out at the laboratory temperature (23–26 °C), at the operating pressure of 2758 kPag (400 psig), and at feed glucose concentrations of 0.27 to 0.76 molal. In each experiment, the fractional solute separation, f, was determined as

$$f = \frac{\text{feed glucose concentration} - \text{permeate glucose concentration}}{\text{feed glucose concentration}}$$

and the product permeation rate (PR), which represents the flux in the presence of glucose in the feed, and the pure water permeation rate (PWP) in kg/h for a given area of the membrane surface (19.64 cm^2 in this work) were determined under the specified experimental conditions. The data on PR and PWP were corrected to 25 °C using the relative viscosity data for pure water. The bioreactor experiment was also carried out at room temperature and at 2758 kPag. The permeate was collected over time intervals of 24 h and the glucose and ethanol concentrations in the permeate were determined. The effective membrane area of the bioreactor was 13.2 cm^2. The glucose concentration was determined using a glucose analyzer (YSI Model 27 Industrial Analyzer) and the ethanol concentration was measured by the gas chromatographic method. The pH value in the permeate solution was also determined.

THEORETICAL WORK

In order to facilitate the understanding of the transport equations below, the symbols shown in FIGURE 2 will be used throughout this report as subscripts. The

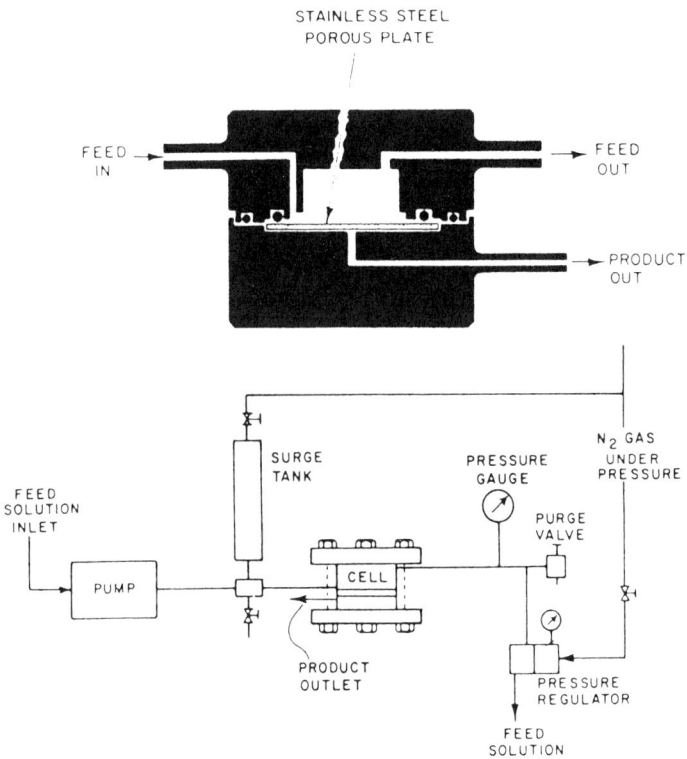

FIGURE 1. Schematic illustration of a flow reactor system.

letters a, b, c, and d indicate the four barrier layers involved and the numbers 1 and 5 indicate the feed and the permeate, respectively. The numbers 2, 3, and 4 indicate the barrier boundaries. Capital letters A and B are also used throughout this report as subscripts to indicate solute (glucose) and solvent (water), respectively. Furthermore, the transport equations given below have been developed on the basis of the following assumptions: both convective and diffusive transports are combined in barrier layers a, b, and c, whereas the solvent convective flow induced by the effective pressure difference $\Delta P - \Delta \pi$ and the solute diffusive flow induced by the solute concentration difference Δc are separate at the barrier d.

Using the symbols defined in FIGURE 2 and on the basis of the assumptions stated above, the transport equations have been developed. The equations that describe the transport through the reverse osmosis membrane alone are given first. FIGURE 3 describes the system, which consists of barriers a and d alone. The feed (1), the surface of contact between the concentrated boundary layer and the RO membrane (2 or 4), and the permeate (5) are considered. The transport equations describing such a system are[12]

$$N_B = A_d(P_2 - P_5 - \pi_2 + \pi_5), \tag{1}$$

$$N_A = (D_{AM}/K\delta)_d(c_{A2} - c_{A5}), \tag{2}$$

$$\frac{c_{A2} - c_{A5}}{c_{A1} - c_{A5}} = \exp\left(\frac{v}{k_a}\right). \tag{3}$$

Equation 1 shows that the solvent flux is proportional to the effective pressure drop at the barrier d (reverse osmosis membrane). Equation 2 shows that the solute flux

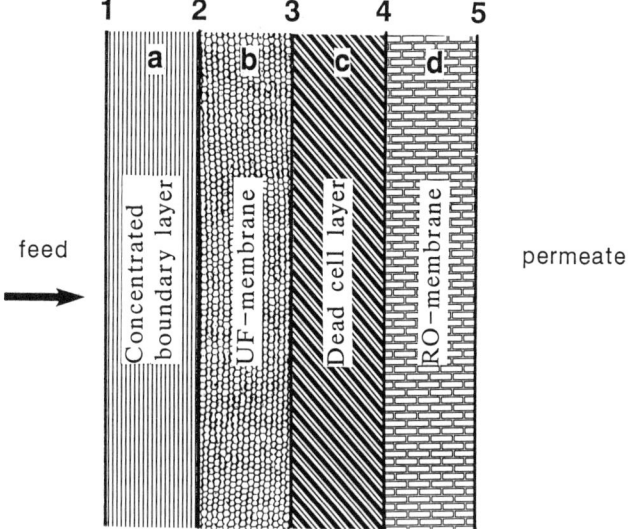

FIGURE 2. Schematic illustration of barrier layers involved in a sandwiched UF membrane/dead cell layer/RO membrane system.[13]

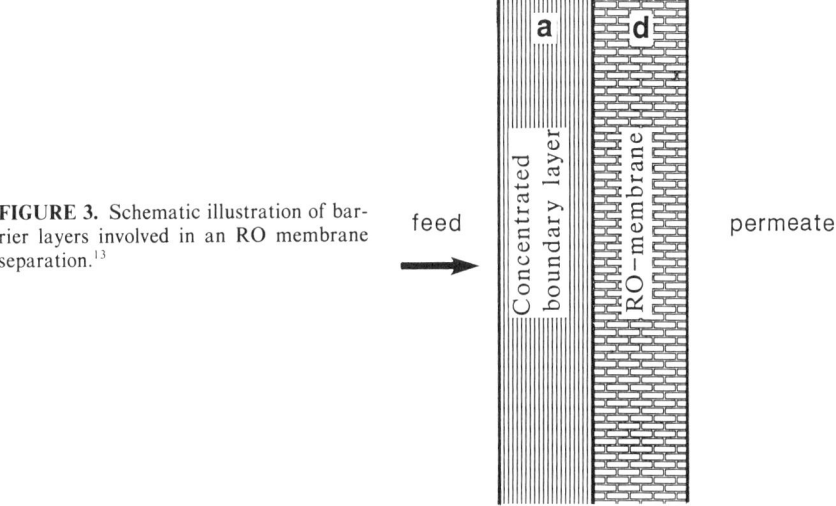

FIGURE 3. Schematic illustration of barrier layers involved in an RO membrane separation.[13]

through barrier d is proportional to the concentration difference on both sides of barrier d, and equation 3 is the concentration polarization equation by which the boundary layer concentration c_{A2} can be calculated. The quantity k_a is the mass-transfer coefficient in the barrier a (the concentrated boundary layer) and is equal to D_{AB}/δ_a according to the film theory. The quantity v is the permeation velocity through barrier d and can be written as

$$v = \frac{N_A + N_B}{c} \div \frac{N_B}{c}, \tag{4}$$

where c is the total molar concentration including solute and solvent; note that c is almost constant in the entire solute concentration range under study. Of course,

$$P_1 = P_2. \tag{5}$$

The system in which a dead cell layer is sandwiched between reverse osmosis and ultrafiltration membranes is schematically represented by FIGURE 2 and consists of barriers a, b, c, and d. The feed (1); the surface of contact between barriers a and b (2), between barriers b and c (3), and between barriers c and d (4); and the permeate (5) are all included in this system. The transport equations describing this system can be written as

$$N_B = A_b(P_2 - P_3) \tag{6}$$

$$= A_c(P_3 - P_4) \tag{7}$$

$$= A_d(P_4 - P_5 - \pi_4 + \pi_5), \tag{8}$$

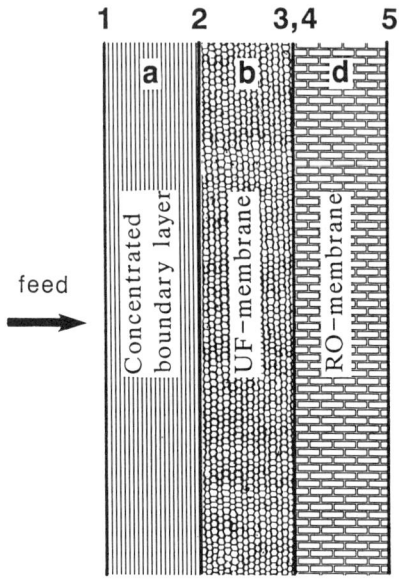

FIGURE 4. Schematic illustration of barrier layers involved in a combined UF membrane/RO membrane system.[13]

$$N_A = (D_{AM}/K\delta)_d(c_{A4} - c_{A5}), \qquad (9)$$

$$\frac{c_{A2} - c_{A5}}{c_{A1} - c_{A5}} = \exp\left(\frac{v}{k_a}\right), \qquad (10)$$

$$\frac{c_{A3} - c_{A5}}{c_{A2} - c_{A5}} = \exp\left(\frac{v}{k_b}\right), \qquad (11)$$

$$\frac{c_{A4} - c_{A5}}{c_{A3} - c_{A5}} = \exp\left(\frac{v}{k_c}\right). \qquad (12)$$

Equations 4 and 5 are valid in this system as well.

Equations 6–12 were developed in analogy to equations 1–3 and they show that there are seven parameters—A_b, A_c, A_d, $(D_{AM}/K\delta)_d$, k_a, k_b, and k_c—that characterize the transport through a system in which a dead cell layer is sandwiched between reverse osmosis and ultrafiltration membranes. It has been shown that the first three parameters can be obtained numerically by conducting the pure water permeation experiment with a RO membrane alone, with a UF membrane alone, and with a sandwiched UF/dead cell layer/RO membrane system (step no. 1). The rest of the parameters can be determined by conducting reverse osmosis separation of the glucose solution with a reverse osmosis membrane (step no. 2), with a combined UF/RO membrane (FIGURE 4, step no. 3), and with a sandwiched UF/dead cell layer/RO membrane system (step no. 4). The details of the method have been given elsewhere.[13]

RESULTS AND DISCUSSION

Typical performance data for glucose separation by the reverse osmosis membrane alone, the ultrafiltration membrane alone, a combined UF/RO membrane, and a sandwiched UF/dead cell layer/RO membrane system are shown in TABLE 2. All flux data are considered as steady-state data. The experiments were conducted under the operating pressure of 2758 kPag (=400 psig) and from 0.4 to 0.5 molal glucose concentration. The data show that the pure water permeation rate decreases significantly from the UF to the RO membrane, but the decrease is less remarkable from the RO to a combined UF/RO membrane, thus indicating that RO film resistance is dominant in the latter system. The pure water permeation rate decreases further from a combined UF/RO membrane to a sandwiched UF/dead cell layer/RO membrane system. As for the product permeation rate, which is the permeation rate in the presence of glucose in the feed, there is a significant decrease from the UF to the RO membrane. The decrease from the RO to a combined UF/RO membrane is also remarkable, reflecting a strong concentration polarization caused by the UF membrane. The decrease in the product permeation rate from a combined UF/RO membrane to a UF/dead cell layer/RO membrane system is not so significant because the increase in concentration polarization by the presence of the cell layer is not quite significant, as explained later. As for glucose separation data, the latter value decreases from the RO to a combined UF/RO membrane and a sandwiched UF/dead cell layer/RO membrane system progressively, whereas the UF membrane alone shows practically no separation to the glucose solute.

All numerical parameters obtained from the above experimental data are listed in TABLE 3 with respect to the combination of UF-1/RO and UF-2/RO membranes, all of which have been laboratory prepared under conditions given in TABLE 1, together with steps involved in the calculation of numerical values. Although there are some discrepancies between values obtained from different steps, TABLE 3 is informative about the contribution of the individual barrier component to the overall mass transport.

TABLE 2. Some Reverse Osmosis and Ultrafiltration Performance Data[a]

Membranes	Feed Concentration (molal)	PWP[b] (g/h)	PR[b] (g/h)	Solute Separation (%)
RO membrane alone	0.484	61.6	35.7	99.4
UF-1 membrane alone	0.497	1137.4	730.0	1.5
UF-1/RO combination	0.425	30.3	5.10	95.7
UF-1/dead cell layer/RO combination	0.407	15.2	3.18	94.5
UF-2 membrane alone	0.470	370.7	258.9	4.2
UF-2/RO combination	0.409	23.8	3.84	95.3
UF-2/dead cell layer/RO combination	0.394	9.5	2.94	87.6

[a]Operating pressure = 2758 kPag (=400 psig).
[b]PWP = pure water permeation rate, PR = product permeation rate in the presence of glucose in the feed, effective membrane area = 19.64 cm^2, and 1 g/h = 1.418 × 10^{-7} m^3/m^2·s.

TABLE 3. Some Transport Parameters Obtained from RO and UF Experiments

	Dimension	Step 1	Step 2	Step 3	Step 4
		Combination of UF-1 and RO Membranes			
A_b	k-mol/m²·s·kPa	3.24×10^{-6}			
A_c	k-mol/m²·s·kPa	5.54×10^{-8}			
A_d	k-mol/m²·s·kPa	1.75×10^{-7}			
k_a	m/s		1.30×10^{-6}		
k_b	m/s			1.16×10^{-6}	
k_c	m/s				3.04×10^{-4}
$(D_{AM}/K\delta)_d$	m/s		0.6×10^{-8}	1.33×10^{-8}	1.28×10^{-8}
		Combination of UF-2 and RO Membranes			
A_b	k-mol/m²·s·kPa	10.55×10^{-7}			
A_c	k-mol/m²·s·kPa	3.30×10^{-8}			
A_d	k-mol/m²·s·kPa	1.75×10^{-7}			
k_a	m/s		1.30×10^{-6}		
k_b	m/s			5.01×10^{-7}	1.05×10^{-6}
k_c	m/s				3.05×10^{-4}
$(D_{AM}/K\delta)_d$	m/s		0.6×10^{-8}	1.03×10^{-8}	2.91×10^{-8}

Comparing A values of each barrier component gives

$$A_b > A_d > A_c,$$

indicating a more intensive resistance against solvent flow from the cell layer than that from either the UF membrane or the RO membrane. As for A values of two different UF membranes,

$$A_{\text{UF-1}} > A_{\text{UF-2}},$$

indicating that the UF-1 membrane has pore sizes larger than those of the UF-2 membrane. This result is in agreement with those reported earlier.[12] Comparing the mass-transfer coefficients k_a, k_b, and k_c listed in TABLE 3 gives

$$k_c \gg k_a > k_b$$

for both UF-1/RO and UF-2/RO combinations. It is understandable that the mass-transfer coefficient of the high concentration boundary layer, k_a, is greater than that of the UF membrane, k_b, because the solute diffusion in the UF membrane is more restricted than in the boundary layer solution. The significantly greater mass-transfer coefficient of the dead cell layer, k_c, in comparison to both k_a and k_b values seems, on the other hand, very puzzling. Probably, the agglomeration of cell particles and the generation of local turbulence in the convective flow occurring between cell particle agglomerates cause a higher mass-transfer coefficient in the cell layer. Furthermore, the order in the pure water permeation constant is $A_b \gg A_c$, whereas the order in the mass-transfer coefficient is $k_c \gg k_b$, which seems contradictory at a first glance. Besides, A_c is calculated to be about three orders of magnitude higher than the experimental value on the basis of 1-mm thickness of closely packed yeast cells. This may, however, be understood by considering that the high resistance ($1/A_c$) against the solvent flow is not due to the cell layer itself, but due to the blocking of the UF membrane pores by cell particles. Remember that the porous sublayer of the UF membrane is in contact with the cell layer. The sizes of the pores on the porous sublayer are sufficiently large to accommodate cell particles and, hence, the blocking of the UF membrane pores from underneath is possible. The resistance against the solvent flow expressed as that from the cell layer $(1/A_c)$ is therefore not necessarily contributed from the cell layer itself, but is, in fact, the resistance contributed from the boundary between the UF membrane and the dead cell layer.

FIGURE 5 shows the pressure profile and the concentration profile across barrier layers calculated using the parameters shown in TABLE 3. Profiles for the combined UF-1/RO membrane and the sandwiched UF-1/dead cell layer/RO membrane system are shown to demonstrate the effect of the inserted cell layer. The pressure and concentration profiles illustrated in FIGURE 5a for the combined UF-1/RO membrane show that the pressure drop takes place primarily at the RO membrane and the

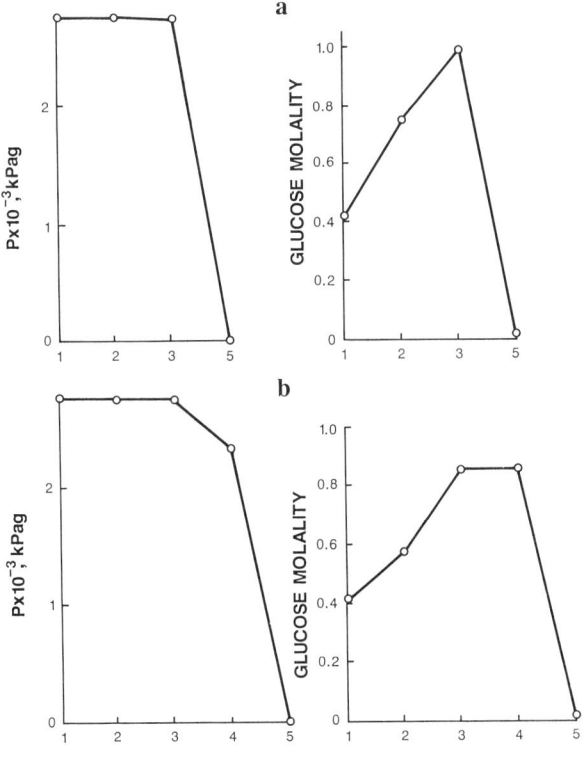

FIGURE 5. Pressure and glucose concentration change at different barrier boundaries—a: combined UF-1/RO membrane, $v = 7.23 \times 10^{-7}$ m/s; b: sandwiched UF-1/dead cell layer/RO membrane, $v = 5.45 \times 10^{-7}$ m/s.[13]

concentration polarizations at both the concentrated boundary layer and the UF membrane are very severe. The profiles illustrated in FIGURE 5b for the sandwiched UF-1/dead cell layer/RO membrane system indicate that the pressure drop takes place also in the cell layer (or the UF membrane/cell layer boundary) and the concentration polarization is less severe than that for a combined UF-1/RO membrane. The less severe concentration polarization is primarily due to a lower permeation velocity of solvent because of a high resistance against solvent flow resulting from the presence of the cell layer and the high mass-transfer coefficient in the cell layer.

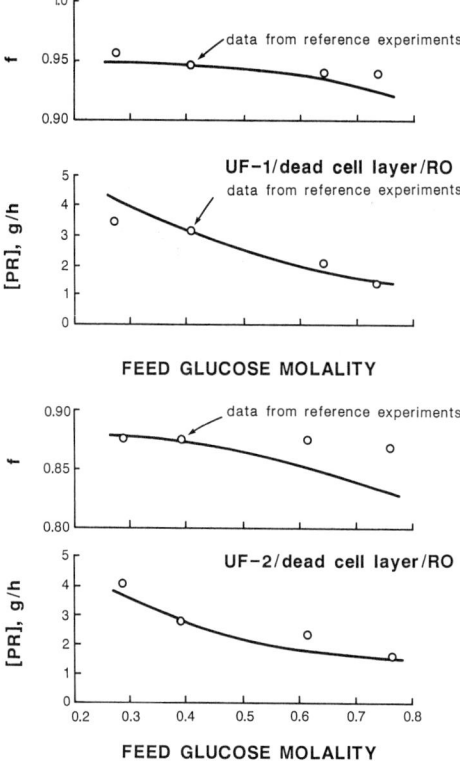

FIGURE 6. Comparison of calculated (line) and experimental (symbol) values for the RO performance of sandwiched UF/dead cell layer/RO membranes: f = solute separation, [PR] = product permeation rate, feed molality of the reference experiment = 0.4, operating pressure = 2758 kPag (=400 psig), effective film area = 19.64 cm^2, and 1 g/h = 1.418 × 10^{-7} m^3/m^2·s.[13]

FIGURE 6 shows some performance data of sandwiched UF-1/dead cell layer/RO and UF-2/dead cell layer/RO membrane systems. The solid line is the calculated data on the basis of the numerical values listed in TABLE 3, whereas the open circles are the experimental data points. The differences between calculated and experimental data are within the experimental error range (solute separation, ±2%; product rate, ±0.3 g/h) except for the separation data of the UF-2/dead cell layer/RO system at 0.76 molal. It should be noted that all transport parameters were generated on the basis of

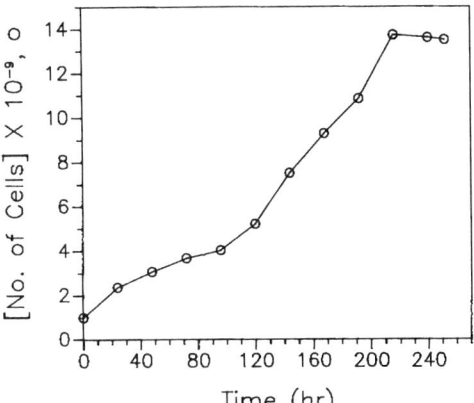

FIGURE 7. Cell number versus reaction time: membranes used, UF-1 and RO films; feed glucose concentration, 1.4 molal; and operating pressure, 2758 kPag (400 psig).

reference experiments, with the data from these also being shown in FIGURE 6. Thus, the calculated values for experimental conditions other than reference experiments have to be considered as purely predicted values. The agreement between the predicted and experimental values shown in FIGURE 6 therefore testifies to the validity of the transport equations developed in this work and to the associated parameters.

Some results from bioreaction experiments are illustrated in FIGURES 7–10. FIGURE 7 shows that the number of cells has grown from 1.0×10^9 to 1.4×10^{10} during

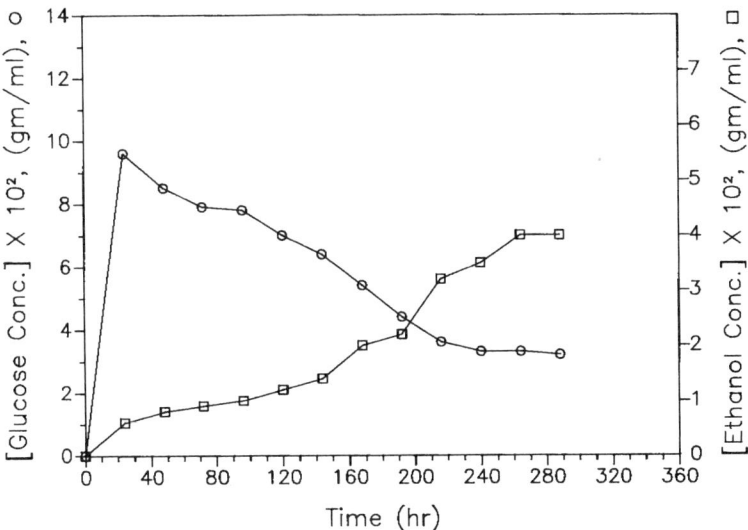

FIGURE 8. Ethanol and glucose concentrations in the permeate versus reaction time. Operating conditions are the same as in FIGURE 7.

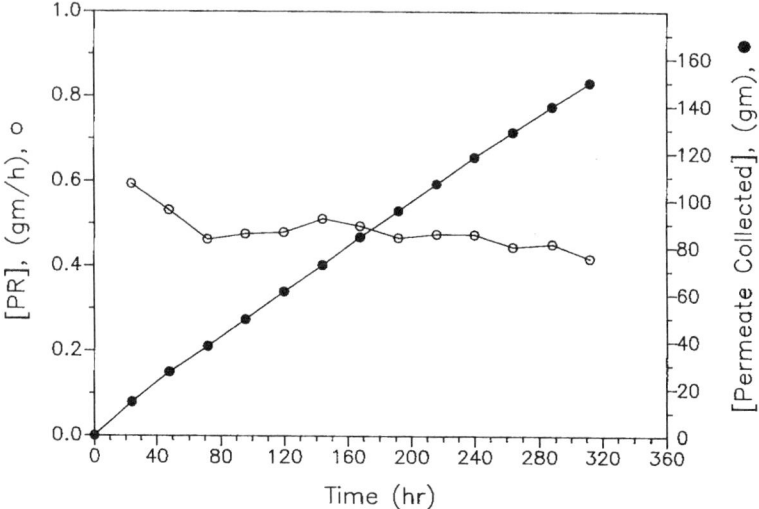

FIGURE 9. Amount of permeate collected and permeation rate versus reaction time. Operating conditions are the same as in FIGURE 7.

the reaction period. FIGURE 8 shows that the ethanol concentration in the permeate increased and the glucose concentration in the permeate decreased with an increase in the cell number of the bioreactor. FIGURE 9 shows that the permeation rate slightly decreased during the reaction period and FIGURE 10 shows that the pH value in the permeate was almost constant. The CO_2 gas was also collected on the permeate side and the volume was measured. It has been found that the mole of CO_2 gas so collected was almost equivalent to the total amount of ethanol collected in the permeate. The analysis of the bioreaction data using the transport parameters obtained in this work is currently under way.

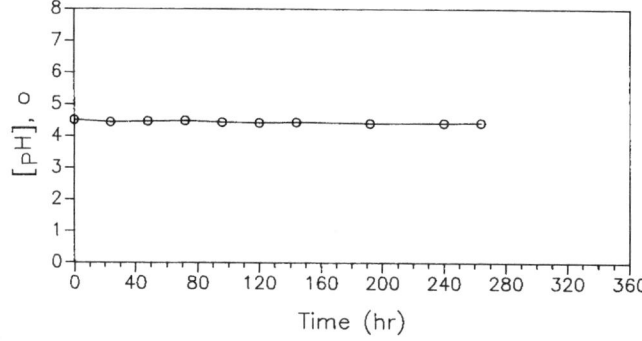

FIGURE 10. pH versus reaction time. Operating conditions are the same as in FIGURE 7.

CONCLUSIONS

From the experimental observations and the subsequent analysis of the experimental data, we have drawn the following conclusions with respect to glucose transport through a sandwiched UF membrane/dead cell layer/RO membrane system:

(a) The transport of solvent (water) and solute (glucose) through the above system can be described by a series of four barrier layers.
(b) Convective and diffusive transports are combined in the concentrated boundary layer, the UF membrane, and the cell layer.
(c) The solvent convective flow and the solute diffusive flow are separate in the RO membrane.
(d) The presence of the dead cell layer decreases the water permeability significantly, probably due to partial closing of the pores of the UF membrane that is in contact with the cell particles.
(e) The mass-transfer coefficient in the cell layer is very high, which results in a small concentration gradient in the dead cell layer.
(f) A useful extension of the work would be to use a glucose/ethanol system to demonstrate the separation effects of the bioreactor.

REFERENCES

1. CHO, T. & M. L. SHULER. 1986. Multimembrane bioreactor for extractive fermentation. Biotechnol. Prog. **2**(no. 1): 53.
2. THARAKAN, J. P. & P. C. CHAU. 1986. A radial flow hollow-fiber bioreactor for the large-scale culture of mammalian cells. Biotechnol. Bioeng. **2**: 329.
3. VIETH, W. R., S. S. WANG & S. G. GILBERT. 1972. Biotechnol. Bioeng. Symp. Ser. **3**: 285.
4. KAREL, S. F., B. L. SHARI & C. R. ROBERTSON. 1985. The immobilization of whole cells: engineering principles. Chem. Eng. Sci. **40**: 1321.
5. VENKATASUBRAMANIAN, K., S. B. KARKARE & W. R. VIETH. 1983. Appl. Biochem. Bioeng. **4**: 321.
6. VASUDEVAN, M., T. MATSUURA, W. R. VIETH & G. K. CHOTANI. 1987. Simultaneous bioreaction and separation using an immobilized yeast membrane bioreactor. Sep. Sci. Technol. **22**: 1651.
7. VASUDEVAN, M., T. MATSUURA, G. K. CHOTANI & W. R. VIETH. 1987. Membrane transport and biocatalytic reaction in an immobilized yeast membrane reactor. Ann. N.Y. Acad. Sci. **506**: 345.
8. DE PINHO, M., T. D. NGUYEN, T. MATSUURA & S. SOURIRAJAN. 1988. Reverse osmosis separation of glucose-ethanol-water system by cellulose acetate membranes. Chem. Eng. Commun. **64**: 113.
9. MICHAELS, A. S. 1980. Membrane technology and biotechnology. Desalination **35**: 329.
10. HANNOUN, B. J. M. & G. STEPHANOPOULOS. 1986. Biotechnol. Bioeng. **28**: 829.
11. CHOTANI, G. K. 1984. Design and analysis of immobilized cell reactors. Ph.D. dissertation. Rutgers University, Piscataway, New Jersey.
12. SOURIRAJAN, S. & T. MATSUURA. 1985. Reverse Osmosis/Ultrafiltration Process Principles. National Research Council of Canada. Ottawa, Canada.
13. JEONG, Y. S., W. R. VIETH & T. MATSUURA. 1988. Study of transport phenomena in an immobilized yeast membrane bioreactor. Ind. Eng. Chem. Res. **28**: 231.

APPENDIX

Nomenclature

- A = pure water permeability constant, k-mol/m²·s·kPa
- c = molar concentration of solution including solute and solvent, mol/m³
- c_A = molar concentration of solute A, mol/m³
- D_{AB} = diffusivity of glucose in water
- $D_{AM}/K\delta$ = solute permeability constant, m/s
- f = fractional solute separation
- K = partition coefficient of solute between solution and membrane phases
- k = mass-transfer coefficient, m/s
- N_A = solute flux, mol/m²·s
- N_B = flux of solvent (water), mol/m²·s
- P = pressure, kPa
- PR = product permeation rate, g/h
- PWP = pure water permeation rate, g/h
- v = permeation velocity through membrane, m/s
- δ = effective membrane thickness, m
- π = osmotic pressure, kPa

Subscripts

- 1, 2, 3, 4, 5 = as illustrated in FIGURE 2
- a, b, c, d = as illustrated in FIGURE 2
- A, B = substrate (glucose) and solvent (water), respectively

Chitin-Chitosan Membranes

Separations of Amino Acids and Polypeptides

JOHN J. PELLEGRINO, STUART GEER,
KAREN MAEGLEY,[a] RAPHAEL RIVERA,
DARLENE STEWARD,[b] AND MYONG KO

National Institute of Standards and Technology
Chemical Engineering Science Division
Boulder, Colorado 80303

INTRODUCTION

Poly(N-acetyl-D-glucosamine), or chitin, is the second most abundant natural polymer after cellulose. The deacetylated form, poly-2-amino-2-deoxy-D-glucose, or chitosan, is made by treating chitin with hot basic solutions. FIGURE 1 presents the structural formulas for chitin and chitosan. Chitin is processed from shellfish waste and recently a few commercial suppliers have begun to provide refined chitin and chitosan. Because the starting material for chitosan is chitin, there are some acetylated amine groups (usually <15%) in the chitosan material used in this and other studies.

Chitosan is a polysaccharide with a molecular mass of approximately 120,000 (it can also be made available with masses from a dimer to >2 million daltons). This will vary with the conditions of the deacetylation and subsequent treatments. It is essentially a primary aliphatic polyamine, with a polyelectrolyte character derived from the high density of regularly spaced amino groups. It has excellent film-forming abilities and will form gels in the presence of cross-linking agents. Due to the free amino groups, it will selectively chelate transition metals. Whereas chitin is insoluble in ordinary solvents, chitosan is soluble in organic acidic solutions and slightly soluble in low concentrations of inorganic acids.

There are many ways of derivatizing chitosan through either the amine or the hydroxyl group. Derivatizations that have been reported have been done for applications in chromatography, mineral recovery, animal feeds, wound healing, and fiber finishing.[1]

Many results[2] have been reported on the use of chitosan as a support medium for cell and enzyme immobilization. Most of this work has involved variations on microencapsulation. Some techniques utilizing absorption have also been successful. Cross-linking of the chitosan produces a stable, insoluble physical structure.

Chitosan films are readily made. Most recent work focuses on the characterization of film tensile properties resulting from variations in formulations, processing conditions, environmental variables, and derivatization.[3-6] Some studies of permeation through chitosan have been made. Muzzarelli[7] reports permeabilities of H_2O, O_2, N_2,

---

[a] Present address: Department of Chemistry, University of California–Santa Barbara, Santa Barbara, California.
[b] Present address: Hauser Chemical Research, Boulder, Colorado.

and CO_2 through 20-μm chitosan films compared to a similar thickness of cellophane. Chitosan films have the same water vapor permeation, but much lower gas permeability. The ideal separation factor of CO_2 transport over N_2 and O_2 was much higher for cellophane than for chitosan. This indicates no significant chemical (solubility) interactions between those gases and the chitosan polymer, and the transport selectivity is probably controlled by diffusion.

The ability of chitosan membranes to function in reverse osmosis (RO) has also been reported.[7] These membranes were made from a modified casting solution that contained organic solvents. Rejections of 72% for NaCl and 80–93% for Ca^{++} have been reported. The absolute water permeances were found to be higher than for

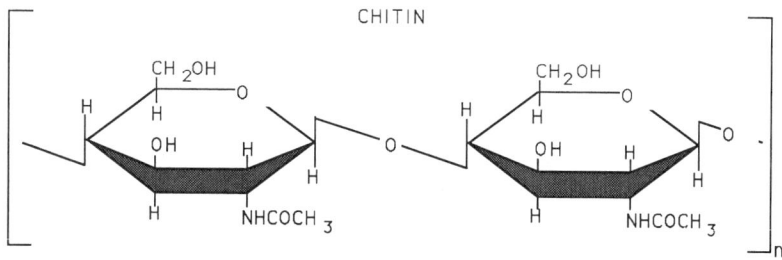

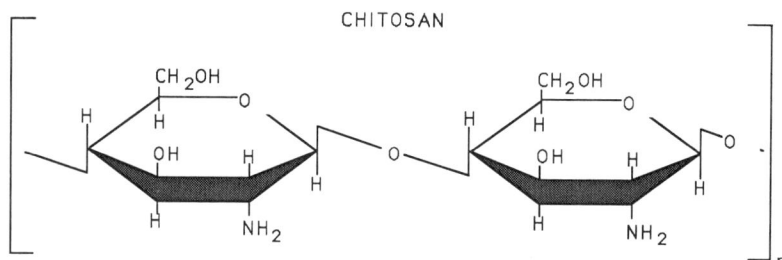

FIGURE 1. Structural formula of chitin and chitosan repeat units.

cellulose acetate RO membranes even though the chitosan was not an asymmetric membrane. It is speculated that the high water permeability is a function of the chitosan molecules being swollen by water and maintaining a stereospecific structure. No discernible pores were visible in these membranes under either light or electron microscope. Additionally, there seems to be greater rejection of bivalent ions than monovalent ones, which leads to the possibility of water-softening applications.

More recently, Yang and Zall[8] studied the effect of the chitosan, casting-solution concentration, using polyethylene glycol as a swelling agent, on the NaCl rejection and membrane permeance. They found that the salt rejection asymptotically approached

79% as the chitosan concentration increased, which is a value very similar to previous work.

Films of predominantly chitin (14–20% deacetylation) have been made by Cho and Lee.[9] These films were tested for molecular weight cutoff with polyethylene glycol and were found to exclude molecules above 10,000 daltons. Rutherford and Dunson[10] tested chitin films derived from a variety of sources for water permeability and the passage of Na, Br, urea, and ethanol. Water permeability was quite high; it was followed by urea, Br, Na, and ethanol in that order. The permeabilities of the latter four solutes were two orders of magnitude lower than the water flux.

Uragami et al.[11] reported results for organic ion transport through a chitosan–polyvinyl alcohol membrane that was slightly cross-linked with glutaraldehyde. Their experiments included benzoate, benzenesulfonate, acetate, phenylalanine, and glycine permeation. They show evidence for active transport driven by a pH gradient across the film. The pH gradient was not stabilized and it diminished with time. There was a major effect of initial pH difference between the two sides of the membrane. They found that the organic anions were actively transported from the acidic to the alkaline side, whereas the amino acids went in the opposite direction.

In this work, we are looking at amino acids as model compounds to study the chemical interactions between chitosan films and permeating solutes. The effects of molecular size, polar nature, and acidity can be conveniently varied with amino acids. The effects of pH, coions, and changes in the chitosan film can be characterized by the permeation and sorption of a consistent family of model compounds.

Selective separation of amino acids is also an industrially important application itself. (World production of amino acids exceeds one million tons/year.) Additionally, by learning more about the interaction of amino acids with chitin-chitosan membranes, we are in a better position to develop membranes to selectively interact with proteins and other polypeptides. Our first steps in that direction have been the development of a supported-gel membrane from chitosan, which we discuss in the latter part of this report.

METHODS

Membrane Preparation: Materials

The starting materials are commercially available chitosan flakes (supplied by Protan[c]). These are classified according to the manufacturer's specifications based on the viscosity of a 1% solution in 1% acetic acid. The primary polymer used in this study was Seacure (2790 cps), a material of high molecular mass (MW). At this point, we have not further characterized the starting material beyond the manufacturer's information.

---

[c]The National Institute of Standards and Technology does not endorse any particular brand of product or company. Commercial names are used only for precise description of experimental materials and procedures.

Membrane Preparation: Film Formation

Following the procedure outlined by Averbach,[12] a 1% solution of chitosan was made by dissolution in a 1% acetic acid solution at 323 K under agitation. The resulting solution was very viscous (high MW) and required removal of insolubles by filtration. We used vacuum filtration with standard cellulosic filter paper. This filtered solution was the starting material for use in casting films, both supported and unsupported.

The basic film was made by pouring chitosan solution onto a clean, flat glass plate. A well was created around the plate by stacked microscope slides in order to control the thickness of the casting solution and a glass rod was drawn along the top to skim off the excess. The casting solution was typically between 1000 to 2000 μm thick and resulted in a final membrane between 25 and 50 μm thick.

The coated plate was allowed to air-dry with a cover to keep dirt off. Drying under these conditions typically takes 24–48 h. The dried film and plate were immersed in 0.2 M NaOH until the film floats off the plate. The film was washed several times with deionized (DI) water and stored wet for later use. The thickness was measured on the wet film.

Membrane Preparation: Gel Film Preparation

The support used in these experiments was a nonwoven made from polyester fiber with hydraulic-needling (no-binders). The support was cut to size and dried in a vacuum oven. The basic chitosan solution was then coated onto the support in a fashion similar to that used when making the basic film. The coated support was allowed to dry for 24 h. After the film was dry, 25 mL of a 0.5% by weight of glutaraldehyde (GA) in DI water solution was added. The covered container with the supported film and the GA solution was put on an orbital shaker for 24 h.

After 24 h, the excess GA solution was poured off and 25 mL of 1 N glycine solution was added. This rinse lasts a maximum of 5 min. The gel was washed with a pH 6 phosphate buffer until the rinse solution was pH 6. The gel should be stored wet to avoid cracking.

Amino Acid Permeation

All the amino acid permeation tests were run in a diffusion apparatus as shown schematically in FIGURE 2 (Jin et al.[13]).

The amino acid solution was placed in the left (feed) compartment and the right (sweep) compartment was filled with unbuffered DI water. The two reservoirs each contain 700 mL to allow frequent sampling of the sweep side without a significant change in concentration. The vessel is made of stainless steel and is temperature controlled by a water recirculation jacket. Internal agitation is provided by a recirculation pumping loop that directs a steady stream of the chamber's solution directly at the membrane. The membrane was mounted vertically between the two chambers. The membrane's permeation area was 19.6 cm^2.

The permeation was allowed to run from 2.5 to 3 h with 2-mL samples of the sweep side taken every 15 min. The samples are analyzed using ninhydrin derivatization,[14] as

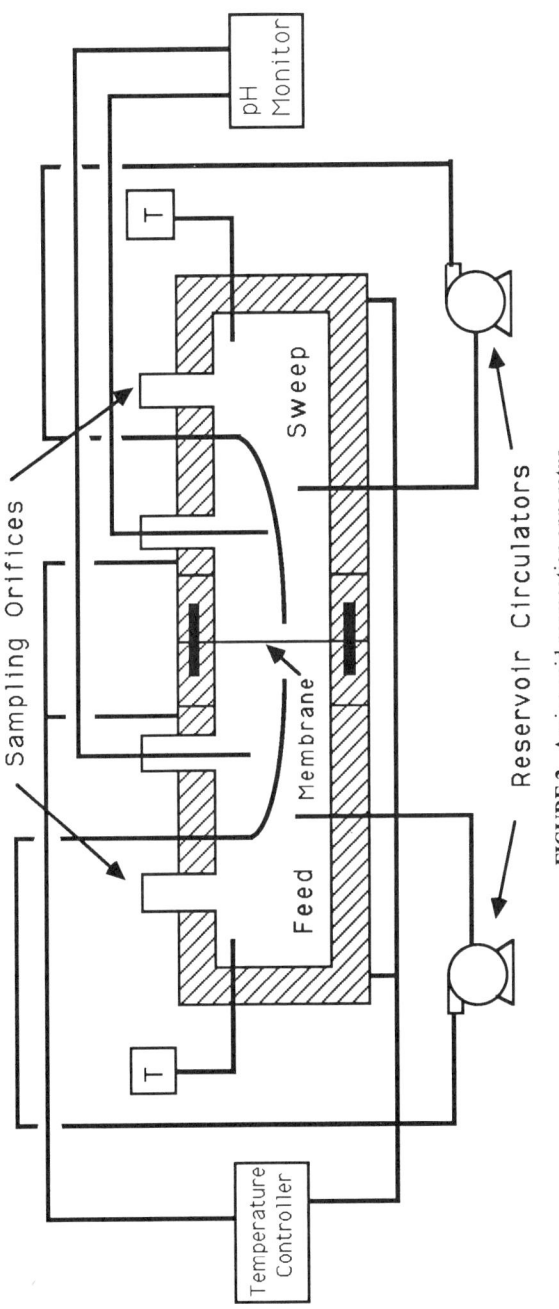

FIGURE 2. Amino acid permeation apparatus.

modified by Pierce Chemical Company's reagent system, and using spectrophotometric absorbance at 570 nm. Samples from the sweep side are diluted so that absorbance measurements can be made on solutions between 0 and 2×10^{-4} M, which was the range of linear response found in our calibration studies.

For multicomponent permeation experiments, the analysis is done using an HPLC ion exchange column and pulsed amperometric detection. The eluant is 0.15 M NaOH. This analysis technique follows the procedure of Polta and Johnson.[15]

Amino Acid Sorption

The membrane was cut into small pieces of approximately 1 cm^2. The pieces to be used in the experiment were then vacuum dried for 12 h and weighed. The amino acid solution (30 mL) was placed in a test tube with the membrane pieces. The test tube was put in a constant temperature bath at 298 K and gently agitated for a period between 10 and 24 h. In preliminary screening tests, we found this incubation time (10 h) to be sufficient to reach equilibrium.

We desorbed by washing the membrane pieces with exactly measured weights of DI water (approximately 10 mL) using the constant temperature bath and an incubation time of 24 h. Prior to desorption, the membrane pieces were quickly patted dry using lab wipes to remove excess solution. Desorption was done at 318 K. Two washes were done for desorption. The second wash only contained trace amounts of amino acid in all the experiments. The concentration of amino acid in the desorption-wash solution was analyzed using the ninhydrin technique, as described earlier.

Protein Permeation

The protein permeation studies were also done using a diffusion cell. The membrane was mounted between two 400-mL chambers. Stirring bars were put in both chambers and the apparatus was suspended above two stirring plates so that the membrane was vertical. The temperature was measured, but not controlled in these experiments. The membrane permeation area was 11.3 cm^2.

Yeast alcohol dehydrogenase (YADH) is the protein used in these experiments. It has a molecular mass of 141,000 and is composed of four subunits of 35,000 each. The feed solution contains 0.2 mg/mL of YADH (0.02% by weight). The YADH is in buffer (pH = 7.1) containing biological buffer (MOPS & MES)^d and acetic acid. This buffer is also on the sweep side. The diffusion experiment is allowed to run for 10 h and a 1-mL sample is taken from both the feed and the sweep side at intervals on the order of 1 h.

The YADH concentration is determined kinetically using the assay of Vallee and Hoch[16] in which the rate of absorbancy at 340 nm resulting from NAD reduction is measured. For the purposes of this study, only relative activity units are reported. The activities of all samples are reported relative to the initial activity of the feed solution.

---

^{d} MOPS = 3-[N-morpholino]propanesulfonic acid; MES = 2-[N-morpholino]ethanesulfonic acid.

RESULTS

The chitosan membranes appear to be homogeneous films when viewed under an electron microscope. There is no evidence of any pore structure down to a resolution of 500 Å. Additionally, no bulk water permeation occurred under a pressure differential up to 262 kPa (38 psig) using supported films. These pressures are sufficient to observe the bulk flow of water through conventional ultrafiltration membranes. We can therefore conclude that a solution-diffusion mechanism will be a major contributor to the amino acid permeation through the membranes. To get a measure of the relative solubility of the amino acids in the membrane phase, we ran sorption studies.

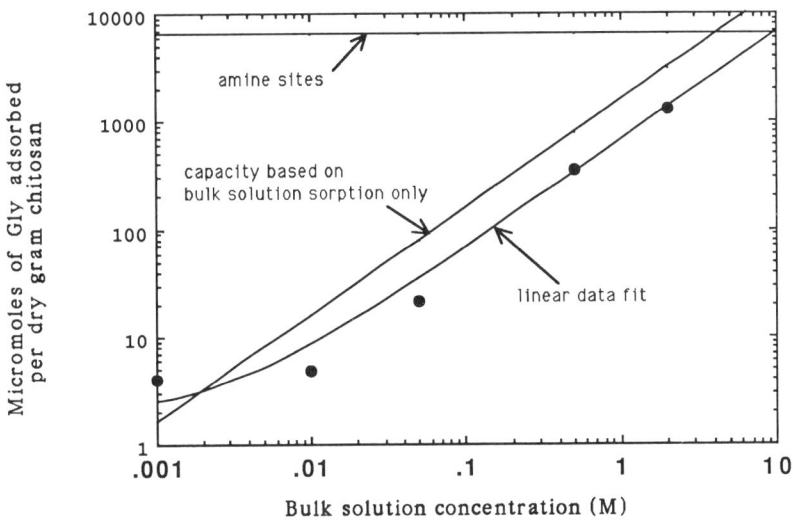

FIGURE 3. Sorption isotherm for glycine (Gly) into chitosan at 298 K.

Amino Acid Sorption

FIGURE 3 presents a sorption isotherm for glycine (Gly) at 298 K from unbuffered solution. The data are also listed in TABLE 1. The upper horizontal line indicates the expected loading if each amine site interacted with a Gly. The line labeled "capacity based on bulk solution sorption only" is based on the water uptake of chitosan and assumes that the water contains the bulk solution's amino acid concentration. The actual data are lower than either of these limits except at the lowest solution concentration.

We did not reach a saturation loading at the highest amino acid concentrations, but there is a possibility that sorptive sites may exist that saturate at very low concentrations. This is indicated by the leveling off of the loading level, for very dilute solutions, at around 4 μmol/g chitosan. In general, the data are well fit by a linear relationship between bulk concentration and membrane loading.

TABLE 1. Glycine Sorption in Chitosan Film[a]

Bulk Solution Concentration (M)	Glycine Sorbed (μmol/g)
2.0	1322.0
0.5	354.0
0.05	21.0
0.01	4.7
0.002	3.5
0.001	4.0

[a] From unbuffered solution at 298 K.

Sorption of other amino acids was also measured at both high (relative to the amino acid's maximum water solubility at 298 K) and low solution concentrations. The data for these measurements are presented in TABLE 2. The values for the sorption from a bulk solution concentration of 2 mM are presented in FIGURE 4. The aromatic amino acids, tryptophan (Trp) and tyrosine (Tyr), and, to a lesser extent, phenylalanine (Phe) seem to be much more strongly adsorbed in the chitosan matrix. The difference between the Tyr and the Phe is especially interesting in that their side groups differ only by the hydroxyl radical.

At this time, we do not have enough data to explain how one amino acid adsorbs to a greater degree than another. However, the difference in the loading between the aromatic residue acids and the other aliphatic ones is suggestive of an interaction with the glucoside units of the chitosan. We think that this is a fruitful area for further study and may provide insights into a basis for more selective separations.

Amino Acid Membrane Permeation

All of our experiments were performed with membranes made from high molecular weight chitosan. We have found that there is a degree of variability between mem-

TABLE 2. Amino Acid Sorption in Chitosan Film[a]

Amino Acid	Bulk Solution Concentration (M)	Amino Acid Sorbed (μmol/g)
Gly (glycine)	2.0	1322.0
	0.002	3.5
L-Ala (alanine)	1.5	839.0
	0.002	1.5
L-Phe (phenylalanine)	0.144	142.0
	0.002	1.8
L-Trp (tryptophan)	0.045	159.3
	0.002	11.9
L-Tyr (tyrosine)	0.002	18.9
	0.0002	4.5
L-Val (valine)	0.002	1.8

[a] From unbuffered solution at 298 K.

branes as we currently make them and we have therefore kept experiments grouped by the specific membrane. The variability between membranes can be due to many factors that will affect both the internal morphology and the chemical nature. These include varying degrees of deacetylation, molecular weight distribution, impurities, and casting conditions.

The data are presented as permeability, which is defined as

$$\text{permeability} = \frac{\text{flux} \cdot \text{thickness}}{\Delta(\text{initial concentration})}$$

and which normalizes the permeation results for variations in thickness and driving force.

FIGURE 5 presents the permeability versus molecular weight for a group of amino acids. These experiments were run with a feed-side amino acid concentration of 0.002 M. Some error bars are quite large in these experiments and they reflect the fact that

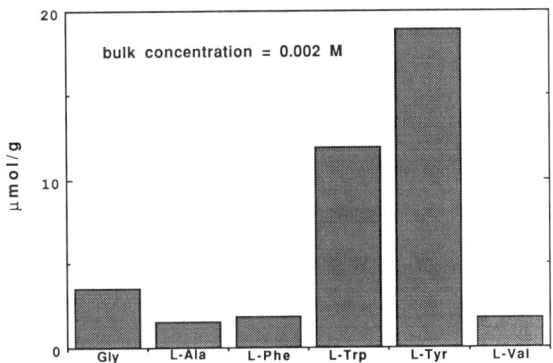

FIGURE 4. Sorption for glycine (Gly), alanine (Ala), valine (Val), phenylalanine (Phe), tyrosine (Tyr), and tryptophan (Trp) into chitosan from a 0.002 M solution at 298 K.

the sweep-side concentrations were very low and were measured using nonspecific UV absorption at 220 nm. These were the only experiments that used this analytical technique. The higher relative permeation rates for the aromatic amino acids may result from their greater solubility (implied from the sorption results) in the membrane phase, which offsets the expected lower diffusion coefficient due to size.

FIGURE 6 shows the results of permeation experiments run at a feed-side concentration of 0.01 M. In these experiments, the pH of the feed-side was buffered to a value between 5.1 and 5.8, except for the lysine (Lys), which was run at its pI (isoelectric pH) of 9.7. Note that if the glutamic acid (Glu) is not buffered, it will have a very high initial permeability; however, in time, it solubilizes the chitosan membrane.

In general, the permeabilities are lower than those of the previous membrane (see FIGURE 5) by a factor of two. This may be due to the factors mentioned earlier, but it may also be a result of non-Fickian diffusion for amino acids through chitosan. For non-Fickian diffusion, permeability would depend on the concentration driving force

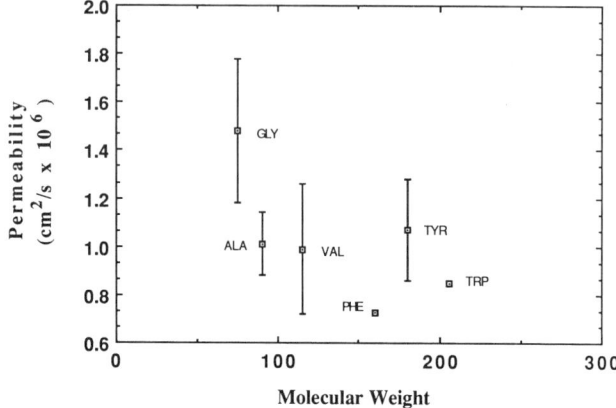

FIGURE 5. Permeability versus molecular weight for glycine (Gly), alanine (Ala), valine (Val), phenylalanine (Phe), tyrosine (Tyr), and tryptophan (Trp). The feed concentration is 0.002 M at 298 K in unbuffered DI water. All experiments were run on a single membrane (2–88) made from high molecular weight chitosan polymer.

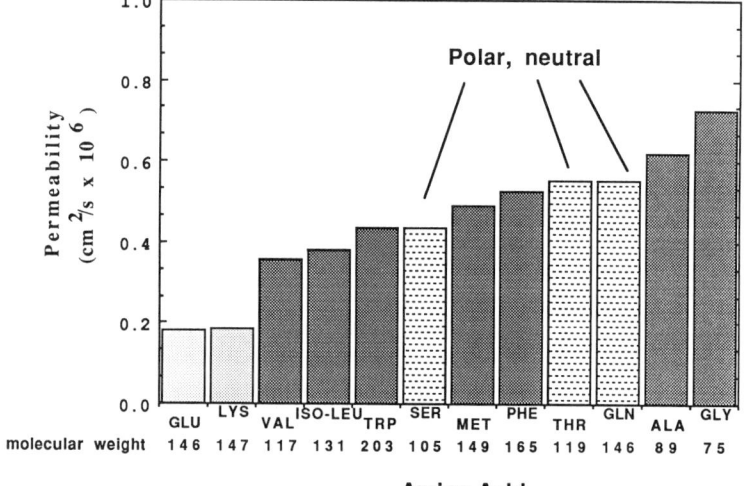

FIGURE 6. Permeability versus amino acid for glutamic acid (Glu), lysine (Lys), valine (Val), isoleucine (Iso-Leu), tryptophan (Trp), serine (Ser), methionine (Met), phenylalanine (Phe), threonine (Thr), glutamine (Gln), alanine (Ala), and glycine (Gly). The feed concentration is 0.01 M at 298 K with the pH adjusted (with phosphate buffer) to 5.13–5.8, except for lysine, which was at its pI of 9.7. All experiments were run on a single membrane (7–29) made from high molecular weight chitosan polymer.

and, in many cases, this decreases as the initial Δ(concentration) increases, such as in facilitated transport. FIGURE 7 presents permeability as a function of the concentration driving force, using a single membrane, for three amino acids. (The lines through the data points are simply interpolations to provide a visual focus for related groups of data and are not regression fits.) There seems to be a driving force effect that is stronger for the Gly and Ala than for the threonine (Thr).

Additionally, we show two sets of data for Gly. The filled squares show experiments with the driving force decreasing in sequential runs and the empty squares are for driving force increasing. These runs were made to test whether there was any

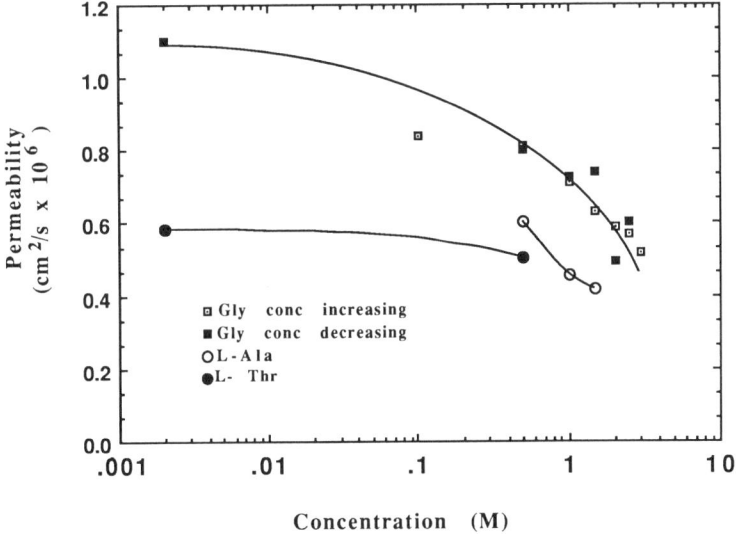

FIGURE 7. Permeability versus concentration in the feed side of glysine (Gly), alanine (Ala), and threonine (Thr). The filled squares represent a sequence of decreasing Gly concentration in subsequent experiments and the open squares represent increasing Gly concentration. The feed solutions were unbuffered and the experiments were run at 298 K on a single membrane (4-88). Curves through data are not regression fits.

irreversible, concentration-dependent sorption on the membrane that changes its permeability. The data do not support such a hypothesis.

The permeability differences (see FIGURE 6) between amino acids do not strictly follow a MW variation, but MW does seem to exert some effect, as seen in FIGURE 8. There are possibly at least two branches. However, the distribution of amino acids between those groupings does not follow any straightforward correlation. For example, serine (Ser) and Thr both contain a hydroxyl, but Thr also contains an extra methyl group (therefore, it is a bulkier molecule), yet it permeates 25% faster.

Other observations are:

- Glu, which is buffered, is in a more negatively charged form in these experiments. It probably does not interact with the amine groups of the chitosan and its permeability would be governed by diffusional resistances.
- Lys, which is at its pI and without any exceptional chemical attraction for the glucosamine, has a large aspect ratio, similar to the Glu, and they both have very similar permeabilities in these forms.
- The neutral amide of glutamic acid, glutamine (Gln), permeates three times as fast as the buffered Glu.

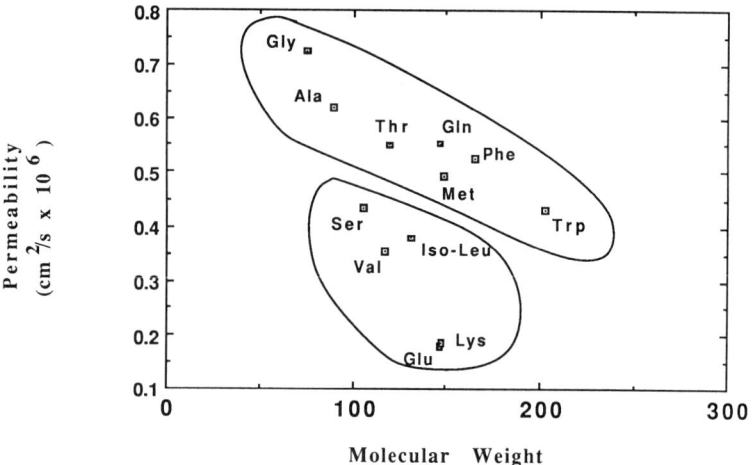

FIGURE 8. Permeability versus molecular weight for glutamic acid (Glu), lysine (Lys), valine (Val), isoleucine (Iso-Leu), tryptophan (Trp), serine (Ser), methionine (Met), phenylalanine (Phe), threonine (Thr), glutamine (Gln), alanine (Ala), and glycine (Gly). The feed concentration is 0.01 M at 298 K with the pH adjusted (with phosphate buffer) to 5.13–5.8, except for lysine, which was at its pI of 9.7. All experiments were run on a single membrane (7–29) made from high molecular weight chitosan polymer.

Some effects of pH and counterions on Gly permeation are shown in FIGURE 9. The pH and buffer effects are small with respect to experimental error. The effect of the Cu^{2+} loading is significant and is consistent with reports[17] of ligand-exchange between Gly and Cu^{2+} in chitosan. This increase in flux was a transient effect because we did not have Cu^{2+} in the feed and sweep streams. We speculate that the presence of the Cu^{2+} mediated (i.e., lowered) the interaction between the Gly molecules and the glucosamine units, which allowed for a faster diffusion rate of the Gly.

TABLE 3 lists the mixture separation results for a DI water solution containing 0.01 M of each of the indicated pairs of amino acids. The entry labeled ideal separation factor is simply the ratio of the single-component permeabilities. In the two reported

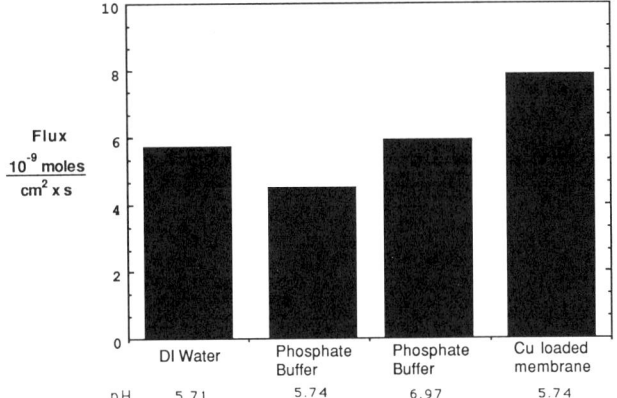

FIGURE 9. Glycine flux from 0.01 M solutions at various buffering conditions. Also shown is the flux through a membrane that was preloaded with Cu^{2+}. All experiments were run at 298 K with high molecular weight chitosan membranes.

cases, the experimentally determined mixture separation factor was higher than the ideal.

In these experiments, the amino acid flux of each component is less than the single-component result at an equivalent driving force, that is, each component is slowed by the presence of the other species. The greatest effect was on the Phe.

Protein (YADH) Membrane Permeation

We ran two experiments with chitosan membranes made by the normal casting procedure. One membrane had covalently attached cibachrome blue dye bound to it, which is known to have affinity for YADH.[18] In both experiments, we tried to permeate YADH from yeast lysate over a 24-h period. Samples taken from the sweep side were assayed for total protein as well as for the specific YADH activity. In neither of the experiments was any protein permeation detected.

FIGURE 10 shows the results obtained using the supported membrane with crosslinking. Almost immediately (within 30 min), YADH activity is detected on the sweep

TABLE 3. Amino Acid Mixture Permeation[a]

	Ideal Separation Factor[b]	Measured Separation Factor[b]
Gly/L-Ala	1.16	1.49
Gly/L-Phe	1.38	1.69

[a] This is for a DI water solution containing 0.01 M of each component.

[b] Separation factor $= \dfrac{\text{flux of } B \times \Delta(\text{driving force of } A)}{\text{flux of } A \times \Delta(\text{driving force of } B)} = B/A.$

side and, in 8–9 h, an equilibrium is reached between the sweep and the feed side. These results indicate the ability to create a gel membrane that is permeable to proteins under a concentration gradient driving force alone.

CONCLUSIONS

We have determined that chitosan films, made from commercial polymer sources, will allow permeation of amino acids with fluxes of 1 to 10 nmol/(cm^2-s) under millimolar concentration gradients only. There apparently are size, charge, and

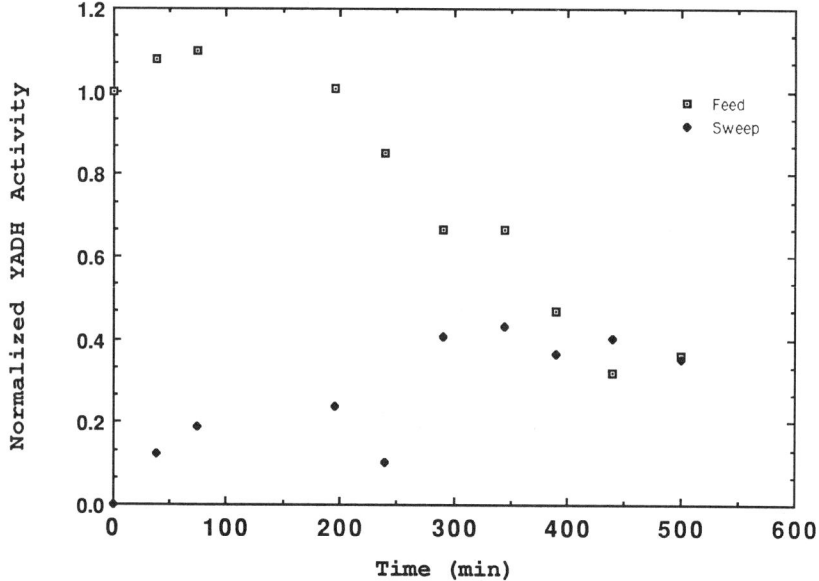

FIGURE 10. Yeast alcohol dehydrogenase (YADH) permeation through a gel chitosan supported membrane. The vertical axis plots normalized YADH activity for either the feed side or the sweep side. The concentration is normalized to the activity of the original feed sample.

molecular interactions between the amino acids and the chitosan repeat structure that can provide a basis for selective separation. Amino acid permeability is a function of concentration, which indicates that non-Fickian flux behavior and control of pH and counterions can exert an effect on the permeability.

The sorption of amino acids in chitosan does not show saturation behavior. The aromatic acids, especially Tyr and Trp, show the greatest affinity. Amino acid permeability does not follow a strict molecular weight correlation, indicating the importance of the polymer-solute interactions.

A gel chitosan membrane, made via controlled cross-linking, allowed permeation of a 141,000-dalton protein under its concentration gradient alone. This protein did not permeate through an uncross-linked chitosan film.

Further work is being performed to clarify the nature of the interactions between the amino acid residues and the chitosan polymer. We hope to extend this understanding in order to develop membranes with a greater permeability and selectivity for biological molecules in the molecular weight range of 100 to 500. These membranes may also be useful to separate other organic molecules in bioreactor streams.

REFERENCES

1. MUZZARELLI, R. A. A. & E. R. PARISER, Eds. 1978. Proceedings of the First International Conference of Chitin/Chitosan. MIT Sea Grant Report No. MITSG 78-7. Massachusetts Institute of Technology, Cambridge, Massachusetts.
2. MUZZARELLI, R. A. A. 1986. Whole cells and enzymes immobilized on chitosan (editor's report). In Chitin in Nature and Technology. R. A. A. Muzzarelli, C. Jeuniaux & G. W. Gooday, Eds. Plenum. New York.
3. KIENZLE-STERZER, C. A., D. R. SANCHEZ & C. K. RHA. 1980. Characterization of chitosan film. In Proceedings of the Eighth International Congress of Rheology, Naples, Italy (September 1–5, 1980).
4. KIENZLE-STERZER, C. A., D. R. SANCHEZ & C. K. RHA. 1982. Mechanical properties of chitosan films: effect of solvent acid. Makromol. Chem. **183:** 1353–1359.
5. MUZZARELLI, R. A. A., F. TANFANI, M. EMANUELLI & S. MARIOTTI. 1983. The characterization of N-methyl, N-ethyl, N-propyl, N-butyl, and N-hexyl chitosans, novel film forming polymers. J. Membr. Sci. **16:** 295–308.
6. MUZZARELLI, R. A. A. 1986. Filmogenic properties of chitin/chitosan (editor's report). In Chitin in Nature and Technology. R. A. A. Muzzarelli, C. Jeuniaux & G. W. Gooday, Eds. Plenum. New York.
7. MUZZARELLI, R. A. A. 1977. Chitin. Pergamon. Elmsford, New York.
8. YANG, T. & R. R. ZALL. 1984. Chitosan membranes for reverse osmosis application. J. Food Sci. **49:** 91–93.
9. CHO, I. & Y-C. LEE. 1980. A potential source of clean chitin and its film properties. Int. J. Biol. Macromol. **2:** 52.
10. RUTHERFORD, F. A. & W. A. DUNSON. 1984. The permeability of chitin films to water and solutes. In Chitin, Chitosan, and Related Enzymes. J. P. Zikakis, Ed. Academic Press. New York.
11. URAGAMI, T., F. YOSHIDA & M. SUGIHARA. 1988. Studies on syntheses and permeabilities of special polymer membranes—59. Active transport of organic ions through crosslinked chitosan membrane. Sep. Sci. Tech. **23:** 1067–1082.
12. AVERBACH, B. L. 1978. Film-forming capability of chitosan. In Proceedings of the First International Conference of Chitin/Chitosan. R. A. A. Muzzarelli & E. R. Pariser, Eds. MIT Sea Grant Report No. MITSG 78-7. Massachusetts Institute of Technology, Cambridge, Massachusetts.
13. JIN, M., S. BISCHKE & S. K. SIKDAR. 1988. Glycine permeation through Na^+, Ag^{++}, and Cs^+ forms of perfluorosulfonated ion-exchange membranes. J. Sep. Sci. Tech. **23**(nos. 14 & 15): 2293–2308.
14. MOORE, S. & W. H. STEIN. 1948. Photometric ninhydrin method for use in the chromatography of amino acids. J. Biol. Chem. **176:** 367.
15. POLTA, J. A. & D. C. JOHNSON. 1983. The direct electrochemical detection of amino acids at a platinum electrode in an alkaline chromatographic effluent. J. Liq. Chromatogr. **6**(10): 1727–1743.
16. VALLEE, B. L. & F. L. HOCH. 1955. Zinc, a component of yeast alcohol dehydrogenase. Proc. Natl. Acad. Sci. U.S.A. **41:** 327.

17. MUZZARELLI, R. A. A. 1978. Modified chitosans and their chromatographic performances. *In* Proceedings of the First International Conference of Chitin/Chitosan. R. A. A. Muzzarelli & E. R. Pariser, Eds. MIT Sea Grant Report No. MITSG 78-7. Massachusetts Institute of Technology, Cambridge, Massachusetts.
18. CLONIS, Y. D. & D. A. P. SMALL. 1987. High performance dye-ligand chromatography. *In* Reactive Dyes in Protein and Enzyme Technology. Y. D. Clonis, A. Atkinson, C. J. Bruton & C. R. Lowe, Eds. Stockton Press. London.

Diffusion of Proteins in Porous Membranes

RUTH E. BALTUS AND ZHONG LU

Department of Chemical Engineering
Clarkson University
Potsdam, New York 13676

INTRODUCTION

When the size of a solute is comparable to the pore size through which it is diffusing, reduced diffusion rates are often observed. This phenomenon is called hindered diffusion. The rate of solute transport through a membrane containing identical cylindrical pores is given by

$$N = A_p J_z = A_p D_p \frac{\Delta C_p}{\ell}, \tag{1}$$

where D_p is the intrapore diffusion coefficient based on the intrapore concentration driving force, ΔC_p. The effective diffusion coefficient, D_{eff}, is defined by considering the flux based on the bulk solution driving force, ΔC_∞:

$$N = A_p K D_p \frac{\Delta C_\infty}{\ell} = A_p D_{\text{eff}} \frac{\Delta C_\infty}{\ell}, \tag{2}$$

where K is the equilibrium partition coefficient that relates the solute concentration at each end of the pore to the bulk solution concentration. The theoretical interpretation of hindered diffusion was first presented by Renkin[1] and the problem was later reanalyzed by Anderson and Quinn[2] and Brenner and Gaydos.[3]

A general expression for the partition coefficient of spherical solutes based on statistical mechanics is

$$K = \left(\frac{C_p}{C_\infty}\right)_{\text{eq}} = \frac{\int_0^{R_0} \exp[-E(r)/kT]\, r\, dr}{\int_0^{R_0} r\, dr}, \tag{3}$$

where $E(r)$ is the interaction energy between a solute molecule at position r and the pore wall and where R_0 is the pore radius. For solutes limited to steric or "hard wall" interactions,

$$E = \infty \quad r > R_0 - A \tag{4}$$

$$E = 0 \quad r \leq R_0 - A \tag{5}$$

where A is the solute radius. The integration in equation 3 with the conditions in equations 4 and 5 yields

$$K = (1 - \lambda)^2, \tag{6}$$

where

$$\lambda = \frac{A}{R_0}. \quad (7)$$

The intrapore diffusion coefficient is a probability weighted average of $f(r)^{-1}$, where $f(r)$ is the friction coefficient of a sphere translating in a pore at position r:

$$\frac{D_p}{D_\infty} = \frac{\int_0^{R_0} \frac{f_\infty}{f(r)} \exp[-E(r)/kT] \, r \, dr}{\int_0^{R_0} r \, dr}, \quad (8)$$

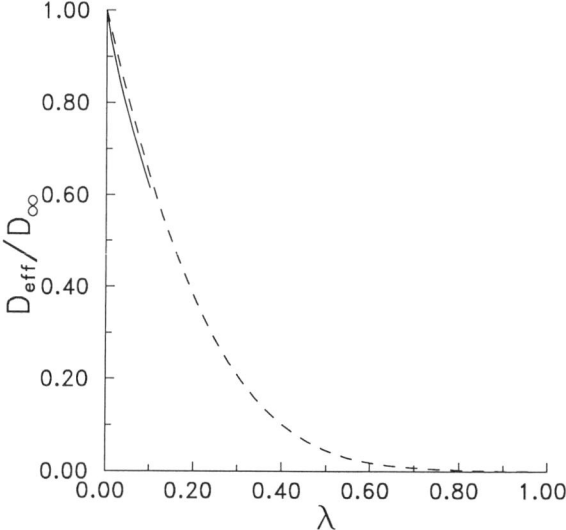

FIGURE 1. Theoretical prediction for the effective diffusion coefficient as a function of the ratio of solute to pore radius: (- - -) prediction of Renkin[1] assuming $f(r) = f(0)$; (———) prediction of Brenner and Gaydos[3] incorporating $f(r)$.

where f_∞ is the friction coefficient in bulk solution. By assuming $f(r) \approx f(0)$ and hard sphere interactions, Renkin[1] derived the following expression:

$$\frac{D_p}{D_\infty} = 1 - 2.104\lambda + 2.089\lambda^3 - 0.948\lambda^5. \quad (9)$$

The combination of equations 3 and 9 yields a prediction for D_{eff} called the Renkin equation, which is shown in FIGURE 1. FIGURE 1 also includes a prediction for D_{eff} derived by Brenner and Gaydos[3] in which the position dependence of f is included.

The hard sphere theory (equations 4 and 5) no longer applies for $E(r)$ when there are long-range forces between the solute and the pore wall. The effect of electrostatic interactions between a rigid solute and the pore wall has been investigated by Smith and Deen.[4] The electrical potential, Ψ, within the electrolyte solution in the pore is described using the linearized Poisson-Boltzmann equation:

$$\nabla^2 \Psi = \kappa^2 \Psi, \tag{10}$$

where κ^{-1} is the Debye length,

$$\kappa^{-1} = \left[\frac{F^2}{\epsilon RT} \sum_i (z_i^2 c_{i\infty}) \right]^{-1/2}. \tag{11}$$

The solution to equation 10 was obtained where constant surface charge densities on the solute surface and on the pore wall were used as boundary conditions. The interaction energy $E(r)$ is related to $\Psi(r)$ by considering the difference in the free energy of the system with the charged solute inside the charged pore and the free energies of the isolated solute and the isolated pore. Discrete values of $E(r)$ were determined and the integral in equation 3 was evaluated numerically to obtain a value of the partition coefficient K. The results of Smith and Deen[4] showed that electrostatic interactions can have a significant effect on solute partitioning. When attractive interactions are large enough to overcome steric restrictions, partition coefficient values greater than one are predicted.

EXPERIMENTAL WORK

In this study, the diffusion of two proteins, trypsin and α-chymotrypsinogen-A, was investigated. Experiments were conducted in two different buffer solutions for each protein. One buffer was at the isoelectric pH of the protein, whereas the other solution was at a lower pH. Additional salt was not added.

The porous membranes used for these experiments were prepared in our laboratory from thin ($\sim 7 \mu$m) sheets of muscovite mica by a track-etch process.[5] Uniform capillary pores having 60° rhombohedral cross sections were formed by first irradiating a mica disk with fission fragments from a ^{252}Cf source and then etching the tracks in aqueous hydrofluoric acid. The pore density in a membrane is controlled by the irradiation time and the pore size is controlled by the etching time.

All membrane parameters were determined prior to experimentation with the proteins. The pore length is equal to the membrane thickness because the source is well collimated. The thickness was determined by weighing the mica disk prior to irradiation. The pore density was known from a source calibration. The pore size was determined by measuring the flux of benzoic acid. The bulk phase diffusivity of this solute was known and its size was small relative to the pore sizes investigated. Therefore, $D_{\text{eff}} = D_\infty$ and the membrane pore size was easily calculated from a flux measurement.

The diffusion experiments were conducted using a diaphragm diffusion cell and the procedure for these experiments is shown in FIGURE 2. Because the membranes were very thin, steady-state diffusion in the pores of the membrane was assumed. The experiments were designed so that $C_H \gg C_L$ at all times and $k^* A_p \ll q$. With these simplifications, the steady-state material balance for solute in the membrane combined with the unsteady-state material balance for solute in the low-concentration chamber

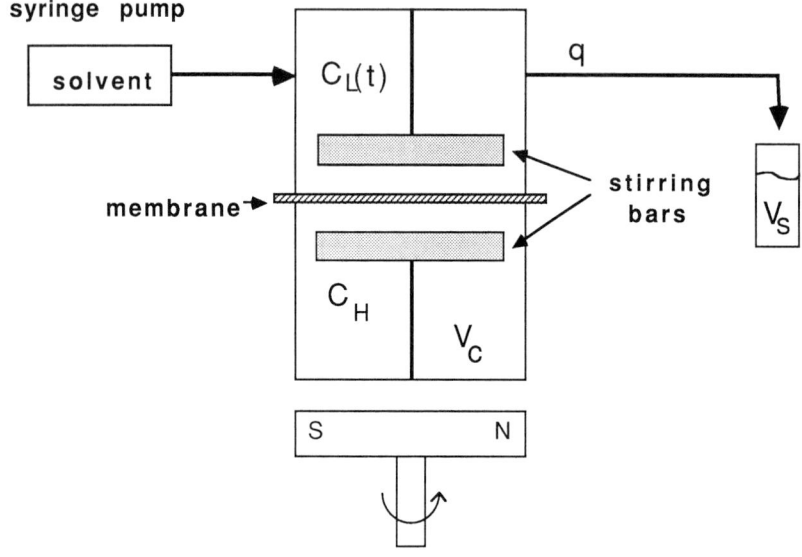

FIGURE 2. Diffusion and sampling process.

yielded

$$\frac{C_L(t)}{C_H} = \frac{k^* A_p}{q}[1 - \exp\frac{-qt}{V_c}], \quad (12)$$

where

$$A_p = n\pi R_0^2. \quad (13)$$

The measured concentration in a sample was related to $C_L(t)$ by averaging C_L over the sample collection time. The protein concentration in each sample was determined by measuring ultraviolet light absorbance. The mass-transfer coefficient, $k^* = D_{\text{eff}}/\ell$, was determined from an appropriate plot of measured concentration versus time because q and all parameters in A_p were known. The effective diffusivity was determined from k^*.

Experiments were conducted at room temperature. The solute concentration was always <2%, allowing us to neglect any concentration dependence of D_{eff}. Further experimental details can be found elsewhere.[6]

RESULTS

The results from our experiments are presented in FIGURE 3. The measured effective diffusivities were normalized using a literature value for the bulk phase diffusivity.[7] The solute radius was determined using this D_∞ value and the Stokes-Einstein equation. The following observations can be made from these results:

(1) although these two proteins have comparable molecular weights and D_∞ values, the effective diffusion coefficient for trypsin is ~4–5 times larger than that of chymotrypsinogen;
(2) the observed D_{eff} values for chymotrypsinogen are relatively insensitive to pore size;
(3) the results for both proteins are in poor agreement with the Renkin equation, with D_{eff} values greater than the values in bulk solution.

A likely explanation for the results obtained for trypsin is that an autolysis reaction was degrading this enzyme, yielding low molecular weight species in solution. Because the diffusive flux through a membrane was monitored by measuring the ultraviolet light absorbance, it was not possible to distinguish between trypsin and the smaller reaction products that would have larger diffusivities in solution. Unfortunately, we were not aware of the possibility of this reaction when these experiments were undertaken.

We can propose two explanations for the large effective diffusivities that were observed for chymotrypsinogen and it is quite likely that both were to some extent responsible for the observations. The first explanation is that attractive electrostatic forces between the diffusing protein and the pore wall were influencing the transport process. Attractive interactions of sufficient magnitude could increase the intrapore protein concentration above the bulk solution value. Therefore, the effective driving force would be increased over that based on bulk solution concentrations. The other factor that may have influenced the measured diffusivities in these experiments was the

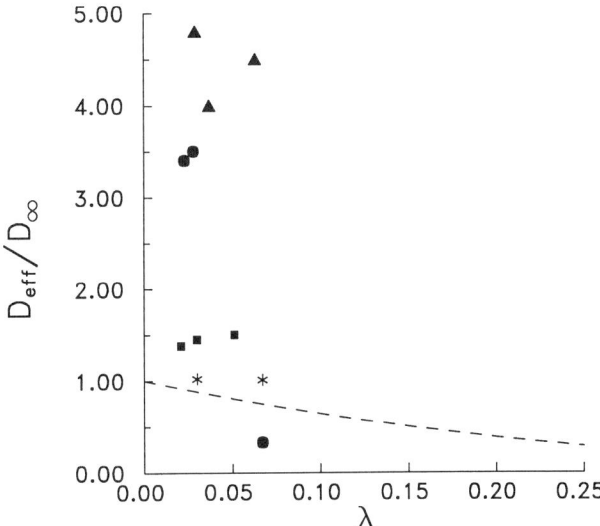

FIGURE 3. Comparison of theoretical and experimental values for the effective diffusion coefficient as a function of the ratio of solute to pore radius. Dashed line: theoretical prediction of Renkin.[1] Experimental values: (●) trypsin, pH = 7.6; (▲) trypsin, pH = 10.8 (isoelectric point); (∗) chymotrypsinogen, pH = 8.0; (■) chymotrypsinogen, pH = 9.5 (isoelectric point).

presence of impurities in the protein samples. The proteins were used as obtained from the supplier without any further purification. Impurities in the samples were probably of low molecular weight and thus would have a characteristic diffusion coefficient larger than chymotrypsinogen. It was not possible to distinguish between these low molecular weight impurities and the chymotrypsinogen in the UV absorbance measurements. Therefore, the measured diffusion coefficients were likely an average diffusion coefficient for this mixture.

We have compared our results to those predicted using the theory of Smith and Deen[4] to illustrate the effect that attractive electrostatic interactions could have on the measured diffusivities. This comparison for the chymotrypsinogen data is shown in FIGURE 4. A pore wall charge of -0.025 C/m^2 was used for these calculations. This

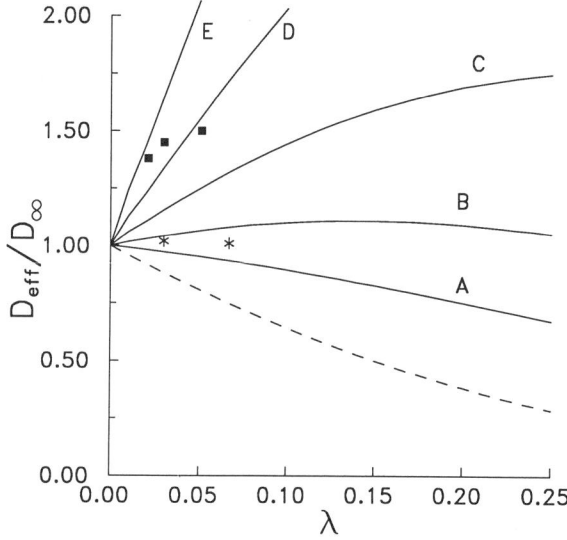

FIGURE 4. Comparison of theoretical and experimental values for the effective diffusion coefficient as a function of the ratio of solute to pore radius: (- - -) theoretical prediction of Renkin;[1] (—) theoretical predictions when electrostatic interactions are included in the partition coefficient calculations.[4] A pore surface wall charge of -0.025 C/m^2 and solute surface charge densities of 0.0075 C/m^2 (curve A), 0.010 C/m^2 (curve B), 0.0125 C/m^2 (curve C), 0.015 C/m^2 (curve D), and 0.020 C/m^2 (curve E) were used in the calculations. Data symbols are the same as in FIGURE 3.

value was based on previous results from streaming potential measurements with mica membranes.[8] The partition coefficient was determined for solutes of various surface charges. The effective diffusivities were then calculated by combining this prediction for K with the intrapore diffusivity predicted for an uncharged solute (equation 9). Although the net charge on these proteins is zero at the isoelectric point, it is possible that the surface charge may not be zero. Therefore, it is possible that electrostatic interactions were influencing these measurements, even under isoelectric conditions.

Electrostatic interactions can also provide an explanation for the insensitivity of the D_{eff} values to changes in pore size. When there are attractive forces between solute and pore wall, the decrease in accessible pore volume that occurs as pore size is decreased could be counterbalanced by an increase in the interaction energy between the protein and the pore wall.

CONCLUSIONS

An experimental investigation of the diffusion of two proteins in well-characterized porous membranes has been undertaken. The observed diffusivities for proteins diffusing in pores of radii of 100–500 Å were found to be relatively insensitive to pore size and greater than the bulk phase value. We propose several explanations for these observations. Impurities in the protein solution may be responsible for the measurement of erroneously high flux values. Attractive electrostatic interactions between the proteins and the pore wall may be causing an increase in the intrapore concentration driving force, leading to an enhancement in the rate of diffusion. Finally, protein autodegradation reactions may be yielding low molecular weight products that have a larger characteristic diffusivity than the original protein.

The phenomena discussed above may also be occurring in many of the processes where protein diffusion in small pores is important. Hence, it is important that we gain a better understanding of the parameters influencing protein diffusion in small pores. We are planning to conduct more experiments in which the influence of impurities and the possibility of autodegradation reactions are eliminated. In addition, the charge characteristics of the proteins and the pore wall will be accurately determined by titration, electrophoresis, and streaming potential measurements. Experiments will be conducted in solutions with a variety of ionic strengths and pH. Therefore, a comparison of results for systems with different Debye lengths will be possible. These experiments will allow for a quantitative comparison with theory.

REFERENCES

1. RENKIN, E. M. 1954. J. Gen. Physiol. **38**: 225.
2. ANDERSON, J. L. & J. A. QUINN. 1974. Biophys. J. **14**: 130.
3. BRENNER, H. & L. J. GAYDOS. 1977. J. Colloid Interface Sci. **58**: 312.
4. SMITH, F. G. & W. M. DEEN. 1983. J. Colloid Interface Sci. **91**: 571.
5. QUINN, J. A., J. L. ANDERSON, W. S. HO & W. J. PETZNY. 1972. Biophys. J. **12**: 990.
6. LU, Z. 1986. Protein diffusion in porous membranes. M.Sc. thesis, Clarkson University.
7. CREIGHTON, T. E. 1983. Proteins—Structure and Molecular Principles. Freeman. San Francisco.
8. WESTERMANN-CLARK, G. B. & J. L. ANDERSON. 1983. J. Electrochem. Soc. **130**: 839.

APPENDIX

Nomenclature

A = hydrodynamic radius of the solute (m)
A_p = pore area (m^2)
$c_{i\infty}$ = electrolyte concentration (mol/m^3)
C_H = solute concentration in high-concentration chamber (mol/m^3)
C_L = solute concentration in low-concentration chamber (mol/m^3)

C_p = intrapore solute concentration (mol/m³)
C_∞ = bulk phase solute concentration (mol/m³)
D_{eff} = effective diffusion coefficient (m²/s)
D_p = intrapore diffusion coefficient (m²/s)
D_∞ = bulk phase diffusion coefficient (m²/s)
E = interaction energy between solute and pore wall (J)
f = frictional coefficient for solute in pore (kg/s)
f_∞ = frictional coefficient for solute in bulk solution (kg/s)
F = Faraday's constant (9.648 × 10⁴ C/mol)
J_z = solute flux (mol/m²-s)
K = solute partition coefficient
k = Boltzmann's constant (1.38 × 10⁻²³ J/K)
k^* = mass-transfer coefficient (m/s)
ℓ = pore length (m)
n = number of pores
N = solute flux (mol/s)
q = flow rate during sample collection (m³/s)
r = radial position in the pore
R = gas constant (8.314 J/mol-K)
R_0 = pore radius (m)
t = time during diffusion experiment (s)
T = temperature (K)
V_c = half-cell volume (m³)
z_i = valence on ions in solution
ϵ = solvent dielectric permittivity (C/V-m)
κ^{-1} = Debye length (m)
λ = ratio of solute to pore radius (A/R_0)
Ψ = electrical potential of solution (V)

A Continuous Enzyme Membrane Reactor Retaining the Native Nicotinamide Cofactor NAD(H)

MICHAEL W. HOWALDT,[a,b] KLAUS D. KULBE,[c] AND HORST CHMIEL[c]

[a]*Department of Chemical Engineering*
California Institute of Technology
Pasadena, California 91125

[c]*Fraunhofer-Institut für Grenzflächen- und Bioverfahrenstechnik*
D-7000 Stuttgart, Federal Republic of Germany

INTRODUCTION

A variety of interesting products may be synthesized using NAD(H)-dependent enzyme processes. Of particular interest is the conversion of renewable and readily available substrates such as carbohydrates into higher valued products.[1] A prerequisite for an economical operation of such processes is the regeneration of the expensive coenzymes. By coupling two coenzyme-dependent enzyme reactions, the NAD(H) is continuously regenerated; that is, instead of stoichiometric amounts, only catalytic amounts of coenzyme are required. In FIGURE 1, such a process is depicted, where glucose/fructose mixtures are simultaneously converted into gluconic acid and mannitol via the enzymes, glucose-dehydrogenase (GDH, EC 1.1.1.47) and mannitol-dehydrogenase (MDH, EC 1.1.1.67).

For continuous operation, the catalysts have to be retained in the reactor. Transport limitations, as they occur when enzymes are immobilized on insoluble carriers, may be avoided by using a convective membrane reactor. In such a reactor, native enzymes are employed and are prevented from leaving the reactor by an ultrafiltration membrane. However, for economic reasons, not only the enzymes, but also the coenzymes have to be retained in the reactor. Because the molecular weights of the products and the coenzymes are often similar, a separation via ultrafiltration membranes is problematic. In the literature, several concepts are described in order to reduce the cost for NAD(H), each of which suffers from some disadvantage(s):

  (i) Separation of the coenzyme from the product stream using ion-exchange columns[2]—this approach requires additional equipment and elution media.
  (ii) Retention via utilization of the affinity between enzymes and coenzymes[3]—this concept works best when the molar coenzyme concentration is smaller than the molar enzyme concentration (based on the number of active sites per enzyme). In practice, this requirement reduces the coenzyme concentration in the reactor to low values that are, in general, smaller than the Michaelis

---

[b]Present address: Lehrstuhl für Bioprozesstechnik, Universität Stuttgart, D-7000 Stuttgart 80, Federal Republic of Germany.

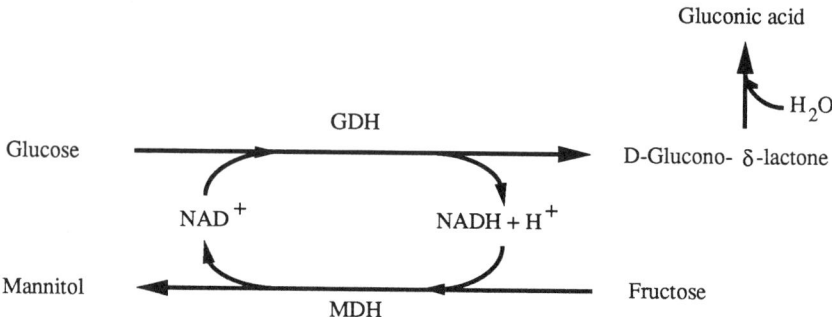

FIGURE 1. The coupled NAD(H)-dependent GDH/MDH system.

constant for the coenzyme so that the activity of the enzymes is insufficiently used.

(iii) Use of a soluble polymer-NAD(H) derivative[4,5]—due to the increased molecular weight of the coenzyme, conventional ultrafiltration membranes may be used. However, immobilization losses occur and the kinetic properties of the coenzyme derivative may differ appreciably from the properties of the native coenzyme. Some enzymes do not accept such derivatives at all.[6]

(iv) Covalent binding of the coenzyme to an enzyme[7]—losses during the immobilization procedure may occur and changes of the kinetic properties may result. In addition, breaking of the covalent bond or deactivation of either the enzyme or the coenzyme in this complex renders the whole complex inactive.

It was shown previously[6] that both NAD$^+$ and NADH are negatively charged in the pH range between pH 3 and pH 9 and may be retained in a membrane reactor using a negatively charged ultrafiltration membrane. Using this approach, the disadvantages of the above-mentioned concepts may be avoided. In this presentation, we report new results regarding the concept of electrostatic interaction between the coenzyme and the membrane as a means to prevent coenzyme losses.

MEMBRANE CHARACTERISTICS

The properties of the membrane are described elsewhere.[6] Briefly, the membrane NTR 7410x is a modification of a commercially available membrane (series 7400) produced by NITTO Electric Industrial Company, Japan. A supporting matrix from neutral polysulfone is covered by a (0.2–0.5)-μm-thin layer of sulfonated polyether sulfone with a fixed charge density of 1 mequiv/g.

KINETICS OF THE SYSTEM

Glucose-dehydrogenase from *Bacillus megaterium* is a commercially available enzyme (Merck). Mannitol-dehydrogenase from *Saccharomyces cerevisiae* was isolated[8] and characterized at the Fraunhofer-Institut für Grenzflächen- und Bioverfahrenstechnik (Stuttgart, Federal Republic of Germany). For this enzyme system, native coenzyme has to be used because the GDH does not accept a coenzyme-polymer derivative.

It was shown earlier that NADH retention by the 7410x membrane is better than NAD$^+$ retention.[6] Therefore, the performance of an enzyme membrane reactor depends on whether the coenzyme is present mostly in its oxidized or its reduced form. Here, f_c is the ratio of oxidized coenzyme to total coenzyme:

$$f_c = \frac{NAD^+}{NAD^+ + NADH}, \qquad 0 \leq f_c \leq 1, \tag{1}$$

where f_c is governed by the actual activities of both enzymes in the system under operating conditions, that is, a high activity of the NAD$^+$-reducing enzyme and a low activity of the NADH-oxidizing enzyme, leading to a low f_c. Similar kinetic properties of both enzymes in the coupled system allow the manipulation of f_c over a wide range by varying the relative activities of both enzymes in the system. If, however, the kinetic properties differ widely, then f_c is strongly influenced by the concentrations of all substances in the system. For instance, if one enzyme shows a considerably higher K_m value for the substrate than the other enzyme, then the activity of this enzyme will decrease faster with increasing conversion than the activity of the second enzyme. The same behavior occurs when the product strongly inhibits enzyme activity. In order to model the behavior of such a reaction system, the kinetics of both enzymes have to be known as a function of the concentration of all substances that are present in the reactor.

Using an extended Michaelis-Menten equation, the characteristics of the GDH and the MDH were described over a wide range of coenzyme, substrate, and product concentrations.[9] The kinetic properties of the MDH are much less favorable than those of the GDH: the K_m value for fructose is about 40 times higher than the corresponding K_m value for glucose (TABLE 1) and there is a strong product inhibition of the MDH by its product mannitol (FIGURE 2). Therefore, with increasing conversion in the GDH/MDH system, the activity of the MDH decreases much faster than the activity of the GDH. This leads to very small f_c values, which is advantageous for the retention of the total coenzyme in the membrane reactor.

TABLE 1. K_m Values for the GDH and the MDH at pH 7.5[a]

		GDH	MDH
fructose	$K_{m,S}$		152 ± 18.2
glucose	$K_{m,S}$	3.93 ± 0.210	
NADH	$K_{m,Co}$		0.286 ± 0.0266
NAD$^+$	$K_{m,Co}$	0.0817 ± 0.0029	

[a] All values are given in mmol/L.

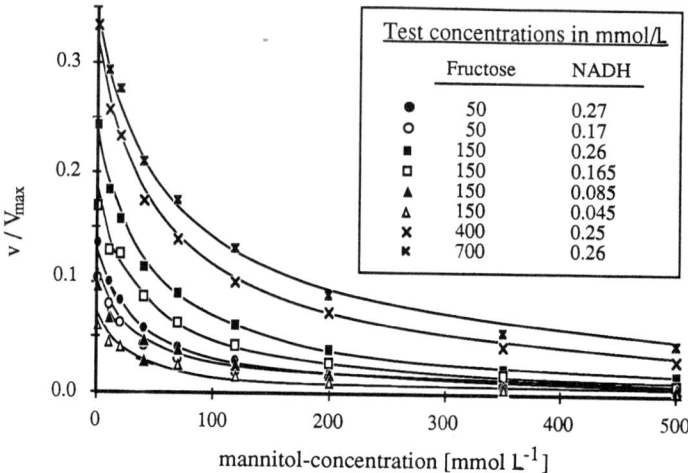

FIGURE 2. Product inhibition of the MDH by mannitol at pH 7.5 and with different fructose and NADH concentrations. Activities were determined by measuring the decrease in absorption at 340 nm. Test concentrations of fructose and NADH are given in the figure.

CSTMR EXPERIMENTS

Experimental Protocol

The experiments were carried out in continuous stirred tank membrane reactors (CSTMR) of 12.8 and 27.4 mL volume, respectively, which were stirred by a magnetic stirrer bar. A circular piece of membrane of 10-cm^2 area was supported by a porous plate; that is, the reactor was of the flat membrane configuration (FIGURE 3). Residence times were varied between 2.2 and 5.5 hours, and ideality of the reactor was assured by determining the residence time distribution using a step change in inlet concentration of a tracer dye. The feed was 50 mM TRA-buffer, containing 11 mM

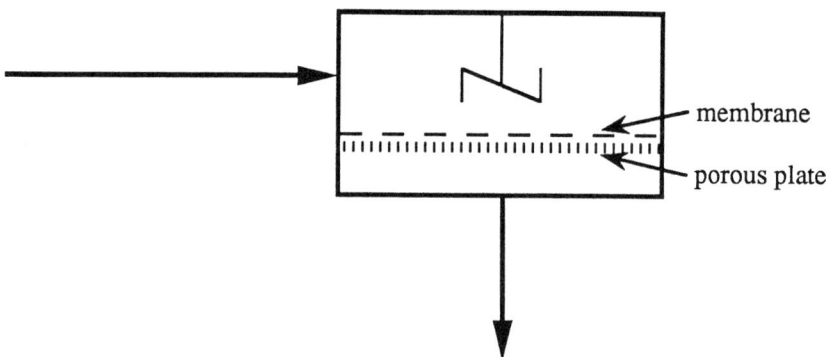

FIGURE 3. Schematic diagram of the reactor.

glucose and 10 mM fructose, 0.02% NaN_3 as a bactericide, and 10 mM NaCl in order to stabilize the GDH. The experiments were performed at pH 7.5 because, at lower pH values, the half-life time of NADH drops considerably.[9] The temperature was maintained at 25 °C. Sugar concentrations were determined via HPLC (BIOTRONIK LC 5000 sugar analyzer); gluconic acid and glucono-δ-lactone were measured using the Boehringer Mannheim Test kit (no. 428191).

In an earlier set of experiments,[6] no coenzyme was fed to the reactor and the decrease of NAD(H), which was initially present in the reactor, was monitored. However, in the experiments of the present study, coenzyme was continuously fed to the reactor at a concentration $C_{Co,f}$ in order to compensate for the imperfect retention and the deactivation of the coenzyme:

$$C_{Co,f} = C_{Co,r}(1 - R + \tau \cdot k_d), \qquad (2)$$

where the deactivation of the coenzyme was modeled as a first-order reaction with deactivation constant k_d, which was equal to $0.033\ d^{-1}$ at the conditions used in this study. $C_{Co,r}$ is the desired coenzyme concentration in the reactor. The retention was determined using equation 3:

$$R = 1 - \frac{C_{Co,ft}}{C_{Co,r}}, \qquad (3)$$

where $C_{Co,ft}$ is the coenzyme concentration in the filtrate. The samples from the reactor and the filtrate were taken simultaneously so that the effects of coenzyme deactivation could be neglected. Because the transmembrane fluxes were of the order of 10^{-6} m/s and the reactor was stirred at $5\ s^{-1}$, the effects of concentration polarization were small and the observed retentions corresponded to the true retentions.

Results

FIGURE 4 shows the results of a long-term experiment. Coenzyme was continuously fed to the reactor as NAD^+ at a concentration of $0.0234\ C_{Co,to}$, where $C_{Co,to}$ is the total concentration of coenzyme in the reactor at the start of the experiment. The factor 0.0234 was determined using both equation 2 and a numerical model described in detail elsewhere.[10] The operating conditions are given in the figure notation.

Instead of reaching a steady-state value, the conversion in the reactor passed through several maxima. As can be seen from FIGURE 5, where the enzyme activity in the reactor is given as a function of time, the decline in conversion was caused by the deactivation of the enzymes. Each arrow denotes the addition of enzymes to the reactor and this resulted in a rapid increase in conversion. The MDH activity was constantly lower than the GDH activity, which is a measure that assisted in maintaining a low f_c value in the reactor. The experimental conditions for the activity assays were as follows: 0.3 mmol/L NADH and 400 mmol/L fructose for the MDH assay; 1.5 mmol/L NAD^+ and 100 mmol/L glucose for the GDH assay. The change in absorbance at 340 nm was measured. Under these conditions, the measured activity v corresponded to 35% and to 85% of V_{max} for the MDH and for the GDH, respectively.

With increasing operating time, there was a pronounced increase of the stability of the MDH: While over the first couple of days the decrease in enzyme activity was rather fast (i.e., a half-life time of 1.5 days), at later times in the experiment the MDH was much more stable (i.e., a half-life time of 20 to 29 days). This increased stability was probably due to the protective effect of high protein concentrations on protein stability. In addition, during the start-up of the membrane reactor, the protein adsorption at the membrane was much stronger than at later times when many of the binding sites at the membrane were already saturated. In order for the GDH to be sufficiently stable, a minimum NaCl concentration in the system was required. Similar to the MDH stability, the GDH stability increased with operating time, although the increase was not as pronounced as with the MDH.

In order to determine the retentions for the coenzyme, samples were taken from the filtrate and the reactor and the retention was calculated using equation 3. The samples were analyzed for NADH by measuring the absorbance at 340 nm. The NAD^+ concentration was determined via an enzymatic assay, where NAD^+ was reduced to NADH with GDH and glucose. In this experiment, an average retention of 98.5% for the total coenzyme NAD(H) was obtained.

DISCUSSION

It was shown that high retentions for the native coenzyme NAD(H) may be obtained in a membrane reactor utilizing a negatively charged ultrafiltration membrane. The degree of retention of the total coenzyme (NAD^+ and NADH) depends on f_c. This ratio is a complex function of the kinetic properties of both enzymes (K_m values

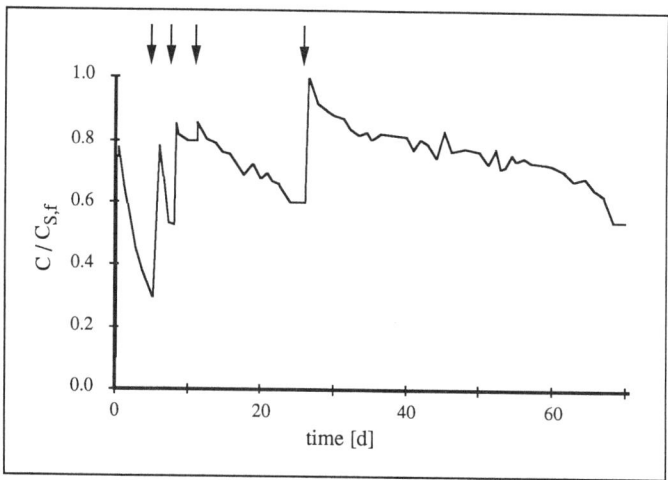

FIGURE 4. Long-term experiment in an enzyme membrane reactor. Mannitol concentration was measured in the filtrate and was normalized with respect to the fructose concentration in the feed. Substrate concentrations in the feed were as follows: NAD^+ = 0.0117 mmol/L; glucose = 11 mmol/L; fructose = 10 mmol/L. Initial concentrations in the reactor were: glucose = 11 mmol/L; fructose = 10 mmol/L; NAD^+ = 0.5 mmol/L; gluconic acid = mannitol = NADH = 0 mmol/L. τ = 3.6 h; — = mannitol.

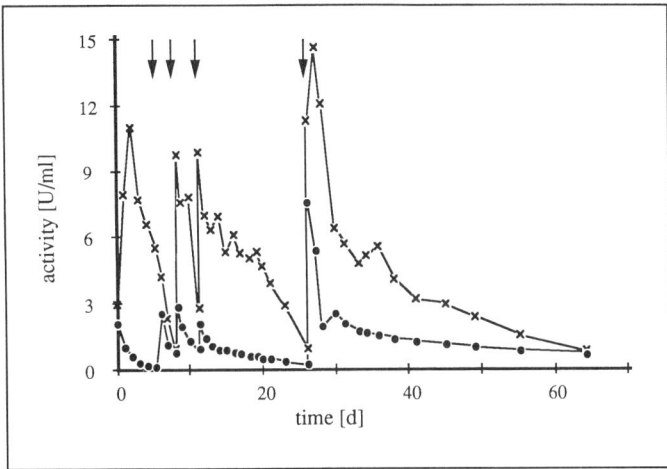

FIGURE 5. Enzyme activities in the membrane reactor. Experimental conditions were the same as in FIGURE 4. Test conditions for the activity tests are described in the text: × = GDH activity; ● = MDH activity.

for the substrate; product inhibition; enzyme stability) and of the operating conditions (absolute concentrations of substrates and products in the reactor). The efficacy of this approach, then, depends on whether a regenerating enzyme may be found that possesses favorable kinetic properties.

In the case of the GDH/MDH system, not only are the kinetic properties favorable, but there are also two products of commercial value. However, the hydrolysis of glucono-δ-lactone leads to an increase in ionic strength and a decrease in the pH of the system. As long as the ionic strength of the system is much smaller than the fixed charge density of the membrane, the Donnan potential is high and the retention is good. Increasing the ionic strength reduces the Donnan potential so that the electrostatic interaction becomes weaker and the retention for charged substances becomes smaller. We are currently investigating the GDH/MDH system utilizing higher substrate concentrations than in this study. The maximum feasible concentration is expected to be a function of the conversion, the residence time (which determines the degree of hydrolysis that takes place in the reactor), the enzyme concentration (because higher protein levels lead to increased retentions[6]), and the pretreatment of the membrane.

Because the disadvantages that are associated with other methods of retaining the coenzyme in the membrane reactor are avoided, the approach of using electrostatic interactions between the membrane and the coenzyme is a promising way of making coenzyme-dependent enzyme reactions more economical.

ACKNOWLEDGMENTS

We thank NITTO Electric Industrial Company, Japan, for providing us with the charged membranes. The technical assistance of Ute Sieglen is much appreciated.

REFERENCES

1. KULBE, K. D. & H. CHMIEL. 1988. Coenzyme-dependent carbohydrate conversions with industrial potential Ann. N.Y. Acad. Sci. **542:** 444–464.
2. FINK, D. J. & V. W. RODWELL. 1975. Kinetics of a hollow fiber dehydrogenase reactor. Biotechnol. Bioeng. **17:** 1029–1050.
3. MIYAWAKI, O., K. NAKAMURA & T. YANO. 1982. Experimental investigation of continuous NAD recycling by conjugated enzymes immobilized in ultrafiltration hollow fiber. J. Chem. Eng. Jpn. **15:** 224–228.
4. WICHMANN, R., C. WANDREY, A. F. BUECKMANN & M-R. KULA. 1981. Continuous enzymatic transformation in an enzyme membrane reactor with simultaneous NAD(H)-regeneration. Biotechnol. Bioeng. **23:** 2789–2802.
5. GU, K. F. & T. M. S. CHANG. 1988. Conversion of α-ketoglutarate into L-glutamic acid with urea as ammonium source using multienzyme systems and dextran-NAD^+ immobilized by microencapsulation within artificial cells in a bioreactor. Biotechnol. Bioeng. **32:** 363–369.
6. HOWALDT, M. W., A. GOTTLOB, K. D. KULBE & H. CHMIEL. 1988. Simultaneous conversion of glucose/fructose mixtures in a membrane reactor. Ann. N.Y. Acad. Sci. **542:** 400–405.
7. JACOBI, T. & C. WOENCKHAUS. 1987. Binary enzyme reactors with modified dehydrogenases. Proceedings 4th European Congress on Biotechnology **1:** 72–75.
8. SCHWAB, U. 1986. Zur enzymatischen Umwandlung von Glucose-Fructose-Gemischen in Gluconsäure und Mannit bzw. Sorbit. Ph.D. thesis, University of Hohenheim.
9. HOWALDT, M. W. 1988. Reaktionstechnische Untersuchungen gekoppelter coenzymabhängiger Enzymsysteme in Membranreaktoren. Ph.D. thesis, University of Stuttgart.
10. HOWALDT, M. W., K. D. KULBE & H. CHMIEL. 1989. Rejection of the native coenzyme NAD(H) in a continuous enzyme membrane reactor. Manuscript in preparation.

APPENDIX

Nomenclature

$C_{Co,f}$ = coenzyme concentration in the feed [mmol/L]
$C_{Co,ft}$ = coenzyme concentration in the filtrate [mmol/L]
$C_{Co,r}$ = coenzyme concentration in the reactor [mmol/L]
$C_{Co,to}$ = total coenzyme concentration in the reactor at the start of the experiment [mmol/L]
f_c = ratio of oxidized coenzyme to total coenzyme
GDH = glucose-dehydrogenase
k_d = deactivation constant $[d^{-1}]$
MDH = mannitol-dehydrogenase
NAD(H) = total concentration of NAD^+ and NADH
R = retention
τ = residence time $[d]$
v = activity of an enzyme [U/mL]
V_{max} = maximum activity of an enzyme [U/mL]

Covalently Attached GRGD on Polymer Surfaces Promotes Biospecific Adhesion of Mammalian Cells[a]

STEPHEN P. MASSIA AND JEFFREY A. HUBBELL[b]

Department of Chemical Engineering
University of Texas at Austin
Austin, Texas 78712-1062

INTRODUCTION

The ability of a surface to support cell adhesion is central to the biocompatibility of that material in several applications, including mammalian cell culture. Adhesion, spreading, and contraction on solid substrates are prerequisites for growth of normal anchorage-dependent cells *in vitro*.[1,2] A basic understanding of the molecular mechanisms underlying these processes has developed within the past two decades.

Fibronectin (FN) was the first cell adhesion molecule (CAM) that was shown to be involved in the adhesion of some avian and mammalian cell types to extracellular substrates.[3,4] In the *in vitro* environment, FN is available from the serum in a form known as cold-insoluble globulin (CIg). For normal attachment and spreading to occur, CIg must be adsorbed to the culture surface.[5,6]

More recently, it has been shown that FN consists of several protease-resistant domains, each of which contains specific binding sites for other extracellular molecules and for the cell surface.[7] The cell attachment activity has been localized to a tripeptide sequence, Arg-Gly-Asp (RGD), located in the cell binding domain of FN as well as in several other CAMs.[8] Substrate-bound RGD containing peptide, obtained by direct adsorption to the substrate or by cross-linking of peptide to adsorbed albumin or IgG, promotes fibroblast attachment and spreading. Furthermore, this activity is competitively inhibited by soluble RGD-containing peptides added to the medium.[8]

Affinity chromatography of cellular extracts on cell attachment–promoting FN fragments combined with specific elution utilizing synthetic RGD-containing peptides yielded a receptor with two 140-kDa subunits.[9] The mammalian FN receptor and other RGD-directed receptors are typically heterodimers of two subunits, α and β.[10] Families of these receptors consist of members with the same or very similar β subunits, whereas the α subunits are distinct and restrict the receptor's affinity to one or a few CAMs.[11] Collectively, these structurally and functionally related receptor families are known as the integrin superfamily.[12,13]

Nearly all mammalian cell adhesion to synthetic polymer surfaces is controlled by adsorbed proteins and is receptor-mediated. Immediately after a culture substrate is placed in a protein solution, proteins adsorb to the surface. If there are receptors for

---

[a] This work was supported by Grant No. CBT-881020268 from the National Science Foundation and Grant No. HL-39714 from the National Institutes of Health.

[b] To whom all correspondence should be addressed.

some of these adsorbed proteins on the cell surface and if the conformation of the adsorbed protein is not so extensively altered by adsorption as to destroy the high ligand-receptor affinity, then cell adhesion and spreading can result. If the cells are seeded on a substrate in the absence of adsorbed proteins, then the proteins on the cell surface may directly adsorb to the surface and the cell will, provided favorable conditions, secrete its own proteins toward the surface in the form of an extracellular matrix. If the substrate does not support protein adsorption or if it supports high-affinity adsorption of a protein for which there is not a cell-surface receptor, then the substrate will not support cell adhesion. Modification of commercial cell attachment surfaces by radio-frequency plasma discharge or by attaching tertiary or quaternary amines, as is commonly done, simply changes the patterns of protein adsorption. In no case does the cell actually touch the surface except through these intermediate adsorbed proteins.

The purpose of this work was to see if a surface could be designed and synthesized that would support receptor-mediated cell adhesion in the absence of any intermediate adsorbed proteins, that is, to produce a surface that was entirely self-sufficient in its support of cell adhesion and spreading. Immobilization of entire proteins, such as collagen or fibronectin, can accomplish this, but is associated with the difficulties of proteolysis, protein degradation, and protein denaturation. To circumvent this, the small, thermally stable peptide region of most CAMs, Arg-Gly-Asp (RGD), was covalently coupled to the surface of polymer films with a Gly N-terminal linker in the form of Gly-Arg-Gly-Asp (GRGD). This produced stable surfaces that were intrinsically bioadhesive, that is, the material surfaces contained groups with a high affinity for cell-surface receptors completely independent of adsorbed CAMs from the culture medium. This surface modification provides a systematic methodology for developing well-characterized substrata that may simplify optimization of a culture system.

MATERIALS AND METHODS

Culture Methods

NIH/3T3 fibroblasts (ATCC, Rockville, Maryland) were cultured in Dulbecco's modification of Eagle's minimum essential medium (DMEM) supplemented with 10% calf serum in a humidified 5% carbon dioxide atmosphere at 37 °C. Porcine aortas were obtained from sacrificed miniature swine. Endothelial cells were isolated by the method of Jaffe et al.[14] with a modification to facilitate perfusion of the luminal surface of the vessel with collagenase. Porcine aortic endothelial cells (PAE) were maintained in DMEM supplemented with 10% fetal calf serum with the same incubation conditions as above.

Surface Modification Procedure

GRGD was grafted on polymer surfaces via the glycyl terminal amine using the tresyl chloride immobilization method of Nilsson and Mosbach.[15] Two polymers were modified, poly(hydroxyethyl methacrylate) (abbreviated PHEMA) and poly(ethylene terephthalate) (abbreviated PET). The coupling method utilized activation of surface hydroxyl moieties by tresyl chloride in an organic solvent for the reaction components, but a nonsolvent for the polymer. The tresyl leaving group was then displaced in aqueous solvent by the terminal amine of the peptide.

Poly(ethylene terephthalate) (PET) has no available hydroxyl groups for tresyl chloride activation; therefore, a surface hydroxylation procedure was developed. An electrophilic aromatic substitution that adds hydroxymethyl groups to the PET films was employed (FIGURE 1). Commercially available PET films were immersed in 18.5% (v/v) formaldehyde and 1 M acetic acid for four hours at room temperature. PHEMA films did not require pretreatment because their surfaces are amply supplied with hydroxyethyl groups.

The modified PET and unmodified PHEMA films were tresyl-activated in 20 mL dry ether containing 40 μL of 2,2,2-trifluoroethanesulfonyl chloride (tresyl chloride) and 2 mL of triethylamine for 15 minutes at room temperature. Activated films were then rinsed with 0.2 M sodium bicarbonate pH 10 buffer and placed in the same buffer containing 80 ng/mL GRGD for 20 hours at room temperature to couple the peptide. FIGURE 2 graphically depicts a modified PET film with hydroxymethyl moieties and the subsequent steps involved with tresyl activation and GRGD coupling.

Cell Spreading Assay

NIH/3T3 and PAE cells were detached from culture flasks by trypsinization and were suspended in serum-free medium [DMEM with 1 mg/mL bovine serum albumin (BSA)]. The cells were washed twice by centrifugation in the BSA-containing medium and were seeded at a density of 3000 cells per cm^2 of film, unless otherwise noted, in serum-free medium; they were then incubated in the normal culture environment. Spread cells (adherent cells that have become flat in shape) were scored by morphological features such as distinct nuclei, pseudopodia, and polygonal shape (FIGURE 3). Cells were visualized at 200× magnification using Hoffman modulation contrast optics on a Leitz Fluovert inverted stage microscope. Cell growth was also assessed by determining spread cell counts at various time points for cells cultured in media supplemented with 10% calf serum. The films were not sterilized for short-term spreading assays, but were sterilized via UV irradiation for studies requiring incubation periods greater than four hours. It was assumed that this sterilization procedure did not significantly affect the peptides attached to the films because activities of the modified surfaces were not reduced to unmodified control levels.

Actin Stress Fiber Visualization

NBD Phallacidin (7-nitrobenz-2-oxa-1,3-diazolylphallacidin) (Molecular Probes, Eugene, Oregon) was employed to visualize actin stress fibers and microfilament bundles in cells attached to the modified surfaces. Samples were prepared according to the manufacturer's procedure and 1000× images were viewed utilizing the Fluovert microscope equipped with a Leitz E3 excitation filter and UV illumination.

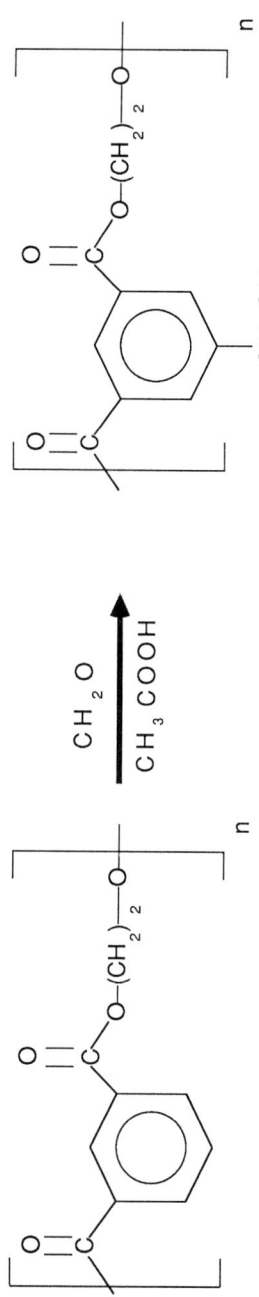

FIGURE 1. PET films were hydroxymethylated via an electrophilic aromatic substitution. The reaction was carried out at room temperature with 18.5% (v/v) formaldehyde and 1 M acetic acid.

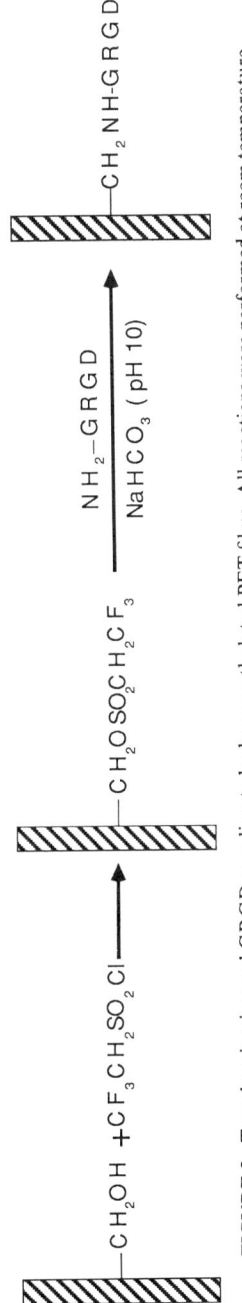

FIGURE 2. Tresyl activation and GRGD coupling to hydroxymethylated PET films. All reactions were performed at room temperature.

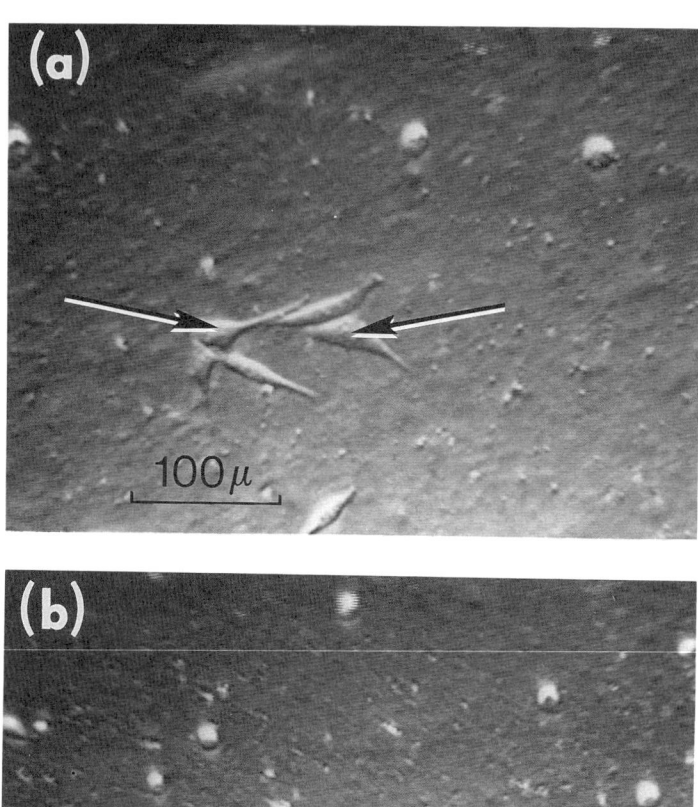

FIGURE 3. Adherent and spread 3T3 fibroblasts on (a) GRGD-coupled films and (b) untreated PET films. Arrows indicate individual cells that were scored as spread cells (200× Hoffman illumination).

Soluble Peptide Competition Studies

In the competition studies, 3T3 fibroblasts were preincubated for 30 minutes in serum-free medium containing 90 µg/mL RGDS or no peptide. The additional serine residue (S) enhances the affinity of adhesion receptors for RGD peptides that are in solution. The cells were then inoculated at a density of 3000 cells/cm^2 on GRGD-derivatized PET and cultured for three hours under normal conditions. A total count of attached cells (including cells that have not spread, but are adherent to the surface) and a differential spread cell count were determined on each film so that the percent of attached cells that are spread could be calculated and plotted (see FIGURE 7 later).

RESULTS AND DISCUSSION

The GRGD-derivatized substrates were characterized by their ability to support active adhesion of cells on these surfaces. The PET pretreatment was optimized by coupling GRGD to tresyl chloride–activated films that were hydroxylated for various time periods. Cell spreading assays using NIH/3T3 fibroblasts were performed to determine conditions that supported a maximal response. Four hours of pretreatment appeared to be optimal for maximum cell adhesion and spreading (data not shown). PHEMA films were derivatized with GRGD utilizing low concentrations of peptide (80 ng/mL), which resulted in an increase in cellular adhesion by three orders of magnitude (FIGURE 4). Comparison of GRGD-coupled PET films with pretreated untresylated films that were incubated with GRGD for the normal coupling time demonstrated that little peptide adsorbed to the film or that the adsorbed peptide was not available for the receptor-mediated adhesion response (FIGURE 5). The GRGD-modified surfaces supported much better 3T3 adhesion than the untreated PET even in the presence of serum, which is indicative of an intrinsic activity on the modified surface (FIGURE 6). The competition experiment resulted in a 75% reduction of attachment to the modified surfaces in the presence of 90 µg/mL RGDS, which further demonstrates the biospecific activity of the substrates (FIGURE 7). Gross morphology (FIGURE 3) of spread 3T3 fibroblasts in serum-free medium on the modified films appeared normal; however, microfilament bundle and stress fiber formation could not be detected under these conditions (data not shown). These results indicate, as others have shown with adsorbed RGD peptides,[16–18] that the complete adhesive response of

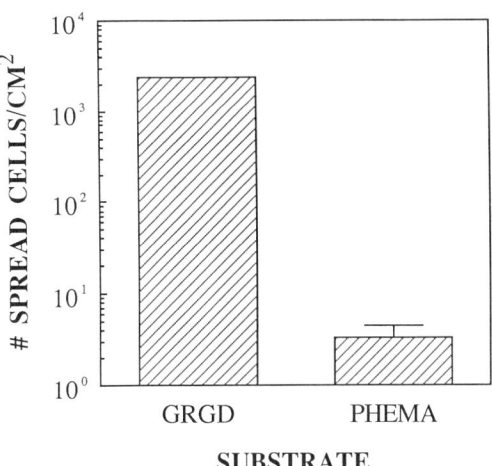

FIGURE 4. Spreading of 3T3 fibroblasts on GRGD-coupled and untreated PHEMA films in serum-free medium. Extent of cell spreading was determined three hours after seeding of the substrates.

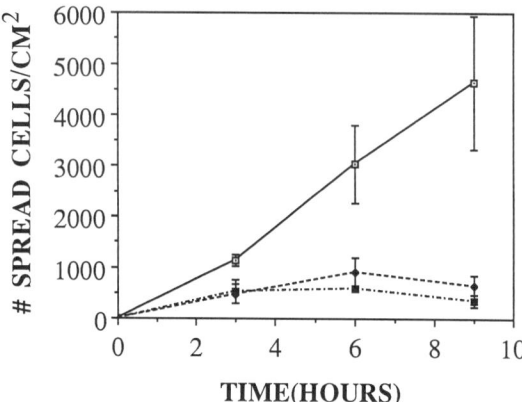

FIGURE 5. Comparison of 3T3 spreading on GRGD-coupled PET versus GRGD adsorbed to PET: GRGD coupled to PET (□); GRGD adsorbed to PET (♦); untreated PET (■).

this cell line and others is not obtained on these modified surfaces. This is not a general phenomenon, though, as some cell types including normal rat kidney fibroblasts and Nil 8, a normal hamster fibroblast cell line, have been shown to fully respond to substrates containing only RGD peptides.[18] Growth (but not attachment and spreading) on the GRGD-derivatized PET was serum dependent (data not shown) and was similar to that on unmodified PET (FIGURE 8), but the initial attachment and spreading was more rapid, as indicated by the observation that the GRGD curve leads the control curve in this figure.

Attachment and spreading (but not growth) of porcine aortic endothelial cells on the GRGD-coupled surfaces were also serum independent (data not shown), as expected, because vascular endothelial RGD-directed receptors have been

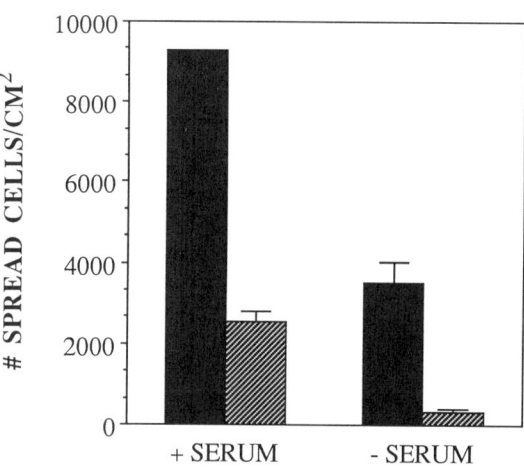

FIGURE 6. 3T3 spreading on the modified (solid bars) and unmodified (hatched bars) films in serum-free and complete medium. The extent of cell spreading was determined two hours after inoculation of each film. Cells were seeded at a density of 10,000 cells/cm^2.

FIGURE 7. Effects of soluble RGDS (90 μg/mL) on spreading of 3T3 cells on the GRGD-derivatized PET films in serum-free medium. Cells were preincubated for 30 minutes with RGDS prior to inoculation of the films. The percentage of cell spreading based on the total number of cells (both spread and rounded, but adherent cells) attached to the surface was determined three hours after inoculation: cells preincubated with RGDS (hatched bar); untreated control cells (solid bar).

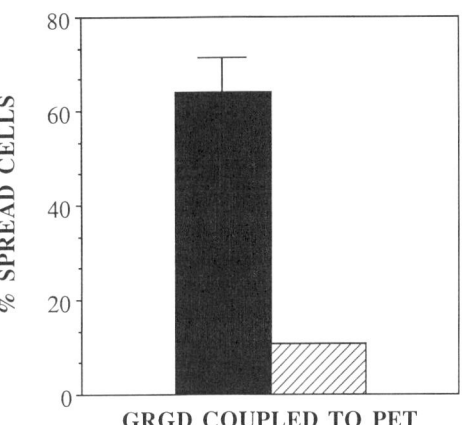

characterized.[19] PAE cell spreading in complete medium was much more extensive on the GRGD-derivatized films than on the untreated films at 4 hours, but both surfaces had confluent monolayers of cells at 24 hours (data not shown). These observations indicate that the rate of PAE cell attachment was more rapid on the modified surface.

CONCLUSIONS

Our surface modification technique is applicable to a wide variety of polymers; it was demonstrated on two materials, one with and one without native surface hydroxyl moieties. It provides a means for obtaining stable, chemically defined surfaces for use in studying cellular responses to insoluble extracellular matrix signals. It provides a means by which to decouple cell adhesion and spreading from protein adsorption. In this sense, it may be useful for those who prefer to use serum-free cell culture media (i.e., media in which purified proteins are added individually rather than introduced

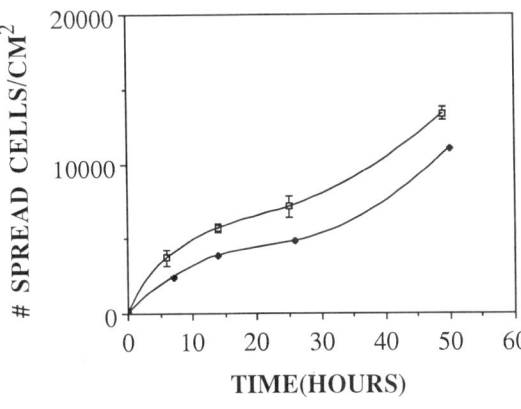

FIGURE 8. Growth kinetics of 3T3 cells on modified and unmodified PET surfaces. These studies were performed in complete medium: GRGD-derivatized PET (□); untreated PET (♦).

from serum) in that cell adhesion molecules do not need to be included. Whether these surfaces are capable of supporting cell growth at very low serum concentrations remains to be determined.

It should be understood that, in the presence of media proteins, the GRGD surface is rapidly covered by adsorbed proteins. This is not problematic, however, as our studies with albumin in the culture media (FIGURES 4, 5, and 7) indicated that the high-affinity RGD-integrin association is capable of competing favorably with the adsorbed proteins. Cell detachment may be accomplished by calcium chelation because the RGD-integrin affinity is calcium dependent.

It should also be noted that cell function is highly dependent upon the cell attachment surface.[20,21] This surface may provide a local environment that is closer to the natural one *in vivo* and hence stimulate stronger adhesion or higher productivity or both. Such measurements of surface-induced changes in cell function are currently under way.

REFERENCES

1. GRINNELL, F. 1978. Cellular adhesiveness and extracellular substrata. *In* International Review of Cytology. Volume 53. G. Bourne & J. Danielli, Eds.: 67–149. Academic Press. New York.
2. COUCHMAN, J. R., D. A. REES, M. R. GREEN & C. G. SMITH. 1982. J. Cell Biol. **93:** 402–410.
3. PEARLSTEIN, E. 1976. Nature **262:** 497–500.
4. KLEINMAN, H. K., E. B. MCGOODWIN & R. J. KLEBE. 1976. Biochem. Biophys. Res. Commun. **72:** 426–432.
5. GRINNELL, F. 1976. Exp. Cell Res. **97:** 265–274.
6. GRINNELL, F. 1976. Exp. Cell Res. **102:** 51–62.
7. HYNES, R. O. & K. M. YAMADA. 1982. J. Cell Biol. **95:** 369–377.
8. PIERSCHBACHER, M. D. & E. RUOSLAHTI. 1984. Nature **309:** 30–33.
9. PYTELA, R., M. D. PIERSCHBACHER & E. RUOSLAHTI. 1985. Cell **40:** 191–198.
10. PYTELA, R., M. D. PIERSCHBACHER & E. RUOSLAHTI. 1985. Proc. Natl. Acad. Sci. U.S.A. **82:** 5766–5770.
11. FITZGERALD, L. A. & D. R. PHILLIPS. 1985. J. Biol. Chem. **260:** 11366–11374.
12. RUOSLAHTI, E. & M. D. PIERSCHBACHER. 1987. Science **238:** 491–497.
13. HYNES, R. O. 1987. Cell **48:** 549–554.
14. JAFFE, E. A., R. L. NACHMAN, C. G. BECKER & C. R. MINICK. 1973. J. Clin. Invest. **52:** 2745–2756.
15. NILSSON, K. & K. MOSBACH. 1981. Biochem. Biophys. Res. Commun. **102:** 449–457.
16. WOODS, A., J. R. COUCHMAN, S. JOHANSSON & M. HOOK. 1986. EMBO J. **5:** 665–670.
17. STREETER, H. B. & D. A. REES. 1987. J. Cell Biol. **105:** 507–515.
18. SINGER, I. A., D. W. KAWKA, S. SCOTT, R. A. MUMFORD & M. W. LARK. 1987. J. Cell Biol. **104:** 573–584.
19. CHERESH, D. A. 1987. Proc. Natl. Acad. Sci. U.S.A. **84:** 6471–6475.
20. VARIANI, J., J. H. HADSAY, R. G. SITRIN, P. G. BRUBAKER & W. A. HILLEGAS. 1986. In Vitro **22:** 575–582.
21. AUBERT, N., G. REACH, H. SERNE & M. JOZEFOWITCZ. 1987. J. Biomed. Mater. Res. **21:** 585–602.

PART V. BIOREACTORS I (MICROBIAL SYSTEMS)

Immobilization of Growing Cells and Its Application to the Continuous Ethanol Fermentation Process

JOHN J. JOUNG AND G. P. ROYER

Amoco Technology Company
Naperville, Illinois 60566

INTRODUCTION

Cell immobilization improves the handling characteristics of industrial microorganisms and increases the volumetric productivity of a bioreactor. Early examples of immobilized cell processes include conversion of glucose to fructose, bioconversion of steroids, amino acid production, production of aminopenicillanic acid, and ethanol fermentation.[1-6]

One major problem encountered in immobilizing growing cells is the lack of expansibility of the conventional cell carriers: cell growth is restricted in the carrier, resulting in limited catalyst life as well as limited catalyst stability. Conventional carriers such as polyacrylamide gels or carrageenan gels, which have frequently been used for cell immobilization, have limited expansibility.

Recently, we[7] prepared an expansible cell carrier suitable for immobilizing growing cells. The carrier gels, which were prepared with a soluble adduct of polyethyleneimine (PEI)–alginate, permitted the expansion of the beads to over 1000% of the initial volume as the immobilized cells continued to grow in the carrier gel. The biocatalyst beads were geometrically intact, retaining stable activity for more than three months, even after the high expansion of beads had occurred. When the carrier gels were used to immobilize yeast for ethanol fermentation, a high ethanol productivity of 30–60 g/liter-gel/h with a three-month catalyst life was observed in a continuous bioreactor.

In this discussion, we report the performance of the yeast beads in a continuous recycle reactor. An ethanol process employing the immobilized yeast beads is discussed and its economics are compared to a conventional batch-fermentation process.

THEORETICAL CONSIDERATIONS

Bioreactor

The immobilized cell reactor that we studied for ethanol fermentation was a continuous recycle reactor. The reactor consisted of a cylindrical reactor vessel, yeast beads, an aeration device, a heat exchanger, a recycle pump, a feed pump, and a sump pump as shown in the schematic diagram (FIGURE 1). The fermentation medium containing glucose substrate with minimal amounts of nutrients was continuously fed to the reactor, whereas the fermentation broth was recycled across the bed of the immobilized yeast beads. The fermented broth containing the product ethanol was

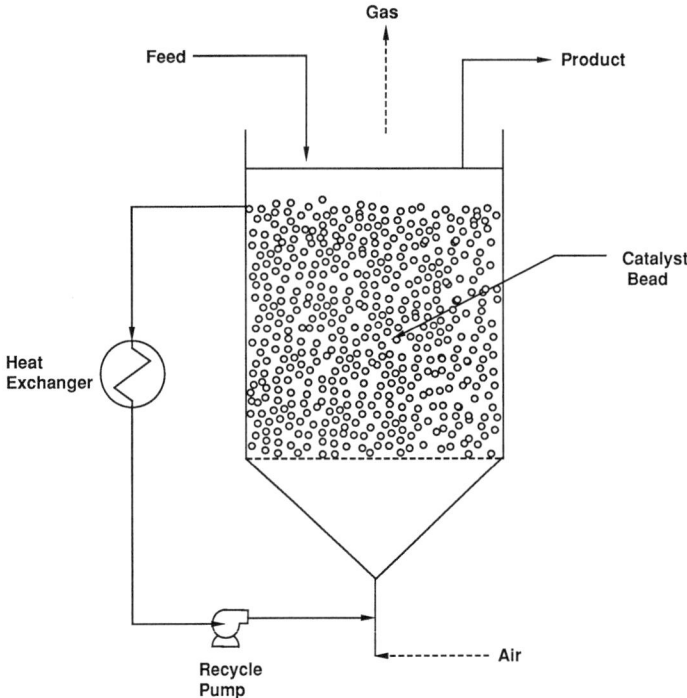

FIGURE 1. A schematic diagram of the recycle reactor with yeast beads.

continuously withdrawn from the reactor by the sump pump to maintain a constant liquid level. A low rate of aeration allowed yeast cells in the beads to continue to proliferate, maintaining the high activity of the yeast beads for ethanol fermentation. The air and carbon dioxide bubbles generated in the catalyst bed gradually ascended until they were released to the atmosphere. Heat that was dissipated in the reactor was removed by the heat exchanger in the recycle stream, thus maintaining an isothermal condition in the reactor.

Previously, a mathematical model describing the performance of the recycle reactor under a steady-state condition was proposed.[8] In this model, the fermentation kinetics followed the Monod equation and the rate-controlling step for ethanol fermentation was assumed to be the diffusion of glucose through the catalyst bead and the surrounding boundary layer. The thickness of the boundary layer was governed by the physical characteristics of the fermentation broth, the bead size, and the flow of fluid across the beads following the mass-transfer analogue of Ranz and Marshall's equation.

The solution of the mathematical model with proper boundary conditions resulted in the overall effectiveness factor (η) of the yeast catalyst in the reactor as follows:

$$\eta = \frac{\eta_0}{2\beta} [(\beta - M - 1) + \{(\beta - M - 1)^2 + 4\beta\}^{1/2}], \tag{1}$$

$$\eta_0 = \frac{3(1+\beta)}{\phi^2}\left(\frac{d\bar{c}}{d\bar{r}}\bigg|_{\bar{r}=1}\right), \qquad (2)$$

$$\phi = \left(\frac{R^2 v o}{K_m D_s}\right)^{1/2}, \qquad (3)$$

$$M = \frac{\eta_0 \phi^2}{3[1 + (1/2)\text{Nu}]}, \qquad (4)$$

$$\text{Nu} = 2.0 + 0.6(\text{Re})^{1/2}(\text{Sc})^{1/3}, \qquad (5)$$

$$\text{Re} = \frac{2RV\rho}{\mu}, \qquad (6)$$

$$\text{Sc} = \frac{\mu}{\rho D_f}. \qquad (7)$$

Thus, the effectiveness factor of the biocatalyst in the recycle reactor is expressed as a function of the Thiele modulus, the substrate saturation number, and the reactor Nusselt number.

Continuous Fermentation Process

The ethanol fermentation process produces 190-proof grain alcohol from corn starch. The capacity of the plant is 50 million (MM) gallons per year at 0.9 stream factor. The process consists of cooking and saccharification, continuous fermentation, and distillation steps.

Saccharification enzymes and yeast beads required to operate the plant are also produced in the plant. The major difference of this process from conventional batch processes is that many batch fermentors are replaced with three serially connected, continuous bioreactors that are filled with yeast beads. All other unit processes are similar to those described by R. Katzen.[9]

Referring to the schematic diagram in FIGURE 2, starch slurry of 37% solid content is adjusted to pH 6.2 with lime and is cooked in a tubular reactor with direct steam injection to the starch stream. The cooked starch is liquefied by the addition of amylase at 105 °C in the next tubular reactor. The stream temperature is reduced to 57 °C before entering the saccharification reactor. The saccharification reactor is a continuous reactor at pH 4.3 with one-day residence time. Glucoamylase is added to the

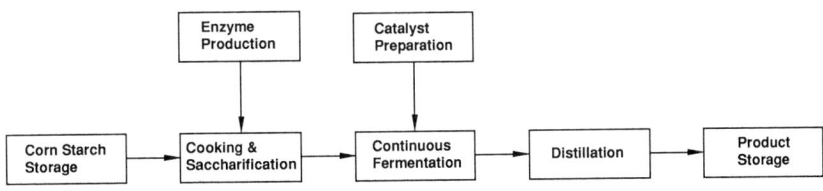

FIGURE 2. A schematic diagram of the continuous ethanol process.

reactor and the reaction continues an additional two days until 97 DE (dextrose equivalent) sugar is obtained in the holding tank. The enzymes used are crude extracts that were produced in the enzyme production section.

Moving to the ethanol fermentation section, the fermentation medium prepared with a portion of the starch hydrolysates is continuously fed to the first reactor. The effluent of the first reactor and the remaining portion of the starch hydrolysates are fed to the second reactor. The third reactor is used for scavenging the residual sugar from the fermentation broth, bringing the ethanol concentration to 7.2% prior to distillation. The fermentation is carried out at 30 °C and pH 3.7. Because of the high ethanol productivity of the bioreactor, only three 200,000-gallon reactors are required for the plant. This is in contrast to the conventional batch process, which requires at least 4,000,000 gallons of total fermentor volume. The spent yeast beads are replaced with freshly made beads as the activity of the beads decreases.

The fermentation broth is stored in a surge tank until it is fed to the distillation unit. The stripper-rectifier recovers the product as 190-proof ethanol and the bottom stream is sent to the waste treatment plant. The thermal system is integrated in order to improve energy efficiency.

The distilled product is stored in storage tanks for bulk shipping. Steam is generated by coal-fired boilers and the electricity is purchased. Plant cooling is provided by a cooling tower and water wells. Approximately 95% BOD is removed from the waste stream before it is released. More detailed information pertinent to this plant and the reference batch process is available in the technical report that was prepared by Katzen and associates.[9]

EXPERIMENTAL WORK

Yeast Beads

Yeast beads of 0.2–3.0 mm D were prepared with dehydrated baker's yeast (Universal Foods Corporation, Milwaukee, Wisconsin) in the PEI-alginate gels. The oil-phase pelletization method employed a 2-L round flask that was equipped with an overhead agitator. Equal volumes of 10–25% yeast slurry and 2% PEI-alginate adduct were mixed to homogeneity immediately before the pelletization. A 300-mL aliquot of the yeast-polymer mixture was transferred to the 2-L flask containing 600 mL of oil while a gentle agitation by the overhead agitator was provided. The composition of the oil phase was determined depending on the size of the beads desired. For example, a bead size of 0.7–1.7 mm D was obtained with the formulation consisting of 400 mL corn oil, 200 mL kerosene, and 0.5 mL sorbitan mono-oleate. As soon as the yeast-polymer mixture dispersed in the oil phase to form spheres of desired size, 7 g of finely crushed calcium chloride powder was added to the flask in order to cure the yeast beads. The cured beads were harvested after an agitation period of 20 min. The entire procedure was completed in physiologically mild conditions at room temperature.

Evaluation of the Biocatalyst

The yeast beads were subjected to a series of tests in order to determine their properties, which are listed with results in TABLE 1.

The biocatalytic activity test consisted of two parts: performance and stability. Yeast beads were prepared for a range of Thiele moduli by changing the bead diameter or yeast loading. The effectiveness of the biocatalyst was determined by measuring the ethanol productivity in the recycle reactor.

A glass column of 7 cm D and 40 cm L was used as the reactor vessel. A schematic diagram of the bioreactor is shown in FIGURE 1. The minimal medium[7] contained 10–14% glucose as the fermentation substrate. Ethanol fermentation was conducted at 25 °C and pH 3.7 with a fixed recycle rate of 4.5 L/h. The aeration rate was 0.05 vvm or less. When a steady fermentation was reached for a given feed rate and a given residual glucose concentration in the fermentation broth, small amounts of the fermentation broth and yeast beads were taken from the reactor for the determination of the catalyst activity. The feed rate, the bed volume, the concentrations of ethanol and glucose in the broth, the average bead size, and the content of viable yeast cells in the beads were determined.

TABLE 1. Selected Properties of the Yeast Beads

Property	Description
shape	sphere
size	0.7–1.7 mm D
specific gravity	1.07–1.10
bulk density	0.70–0.73 g/cm^3
water content	75–80%
cell loading	up to 85%, dry basis
bead expansion	>1000%
mechanical strength	17 psi (tensile)
strength of the gel matrix	100–200 psi
matrix, before activation	closed cell
matrix, after activation	porous
solute permeability	permeable to hydrophilic solutes
stability limitations	unstable at pH 8 or above and to metal chelating agents
storage	2% calcium chloride solution at 4 °C

Separately, the stability of the yeast beads was tested in the recycle reactor by continuously operating the bioreactor after an initial charge of the yeast beads. The catalyst activity, the physical condition of the beads, and any possible bacterial contamination were monitored periodically.

RESULTS AND DISCUSSION

Yeast Beads in PEI-Alginate Gels

The oil-phase pelletization of the yeast-polymer mixture produced spherical beads with a narrow range of size distribution for a given formulation. A photograph of the yeast beads is shown in FIGURE 3. The resilient beads were suitable for ethanol fermentation. The mild conditions employed during pelletization did not harm the yeast cells, thereby permitting a high cell viability after immobilization. The advan-

FIGURE 3. Yeast beads prepared by the oil-phase pelletization.

tages of this pelletization method also included the ability to handle viscous yeast-polymer mixtures, as well as speed, ease of scaleup, and easy selection of bead size. TABLE 1 shows selected properties of the yeast beads that were prepared with the formulation described previously in the example.

The optimum bead size for ethanol fermentation was 1.0–1.5 mm D. The smaller size beads (less than 0.5 mm D) showed a high initial activity for ethanol fermentation, but frequent inactivation of the biocatalyst by sudden cell death caused a serious problem for continuous fermentation. Beads of larger sizes (greater than 2 mm D) suffered from low activity and the formation of gas bubbles in the beads caused flotation of the beads to the surface of the fermentation broth. Because the pelletization method shown in the example resulted in beads of 0.7–1.7 mm D, no further sizing of beads by screening was necessary prior to fermentation.

The yeast beads were readily activated by contacting the beads with a fermentation medium. The minimal medium containing 0.5 g/L calcium chloride[7] was most useful for improving the performance of the yeast beads for ethanol fermentation. This medium provided a high ethanol productivity to the biocatalyst along with maintaining a sustained low cell growth of the immobilized yeast cells. This stabilized the catalytic activity of the yeast beads. Hence, a typical catalyst productivity of 30–60 g-ethanol/L-gel/h with a catalyst half-life of three months was obtained. This high activity, which is comparable to the best results obtained in different immobilization systems by others,[10,11] represents a productivity increase of the recycle reactor of an order of magnitude in comparison to the conventional batch fermentor.

The yeast beads gradually expanded to more than 1000% of the initial volume over the fermentation life span. After the yeast beads had expanded and the activity had decreased to half of the peak activity, a steady, terminal catalyst productivity of 25–30 g/L-gel/h was maintained for more than a month. FIGURE 4 shows a scanning electron micrograph of the cross section of an activated yeast bead. Yeast cells had densely populated the gels. The carrier gels resembled a sponge that allows ease of gas and liquid-phase transport in the bead.

The PEI-alginate beads also showed an apparent bacteriostatic effect. At pH 3.5–4.0, we were able to maintain the bioreactor free of bacteria for more than four months; however, severe infestation of lactic acid bacteria was a serious problem for continuous operation of the reactor in spite of the low pH condition when unmodified alginate beads were employed. The bacteriostatic effect of the PEI-alginate provides substantial operational advantages to the yeast beads because it is believed to be difficult to control bacterial contamination for many immobilized cell systems on a long-term basis.

The effectiveness factor of the yeast beads in the recycle reactor was determined with the following ranges of the reaction parameters: $(5.6 \leq \phi \leq 20.3)$, $(\beta = 4, 30, \text{and } 100)$, and $(4.5 \leq Nu \leq 6.9)$. The minor variation in Nu was due to the differences in the average bead size for the fixed recycle rate of the broth.

FIGURE 4. A scanning electron micrograph of the cross section of an activated yeast bead.

The experimentally determined effectiveness factor varied from 0.22 to 0.88 within the experimental ranges. The effectiveness factor gradually decreased as the Thiele modulus increased for given values of the substrate saturation parameter and the Nusselt number. In contrast, a higher substrate saturation value resulted in a higher effectiveness factor for given sets of the Thiele modulus and Nusselt number. FIGURE 5 compares the experimentally determined effectiveness factor to the theoretically computed results. The experimental effectiveness factor agreed well for the lower values of the substrate saturation parameter ($\beta = 4, 30$). However, the experimental values were substantially lower than the theoretical values when the substrate saturation parameter reached to 100. Perhaps, the substrate diffusion control that was assumed in the theoretical model no longer held true for such a high value of the substrate saturation parameter.

Furthermore, a strong dependence of the effectiveness factor to the substrate saturation parameter suggests that a plug-flow reactor is preferred to a back-mix reactor. However, a large-scale plug-flow reactor has operational problems and affects the stability of the yeast beads. Our choice then is the three-stage recycle reactor. The first and second reactors serve as a conversion reactor that operates at a high value of the substrate saturation parameter, whereas the last reactor scavenges glucose from the fermentation broth in order to improve the ethanol yield for the process.

A higher broth recycle rate or a higher Nusselt number improves the catalyst effectiveness by reducing the thickness of the boundary layer of the yeast beads. The optimum recycle rate should be determined after considering operational requirements such as the strength of yeast beads, reactor height, and energy cost. A reactor Nusselt number of five or more is recommended.

Process Economics

The manufacturing cost of 190-proof ethanol was estimated for the immobilized cell process in order to determine possible economic advantages of the new process over the conventional batch process. The estimation was based on a plant capacity of 50 MM gallons per year for mid-1988. All economic data that were used for the estimation of the capital cost were based on the report presented by Katzen and associates in 1978. An escalation factor of 1.8 was used in order to update the cost. All other cost bases, including raw material and utility costs, are specified in TABLE 2.

The estimated capital and manufacturing costs of the two processes are compared in TABLE 2. For a closer comparison, Katzen's process was modified so that the plant would take corn starch instead of whole corn as the raw material.

The plant construction cost of the immobilized cell process, $55.9 MM, compared favorably to that of the modified Katzen's batch process, which was $59.5 MM. The $3.6 MM or 6.1% savings of the total construction cost reflected substantial savings of the fermentor cost, which was partially offset by the additional costs of bead-processing facilities as well as more stringent requirements for the saccharification section of the immobilized cell process.

Similarly, the total manufacturing cost favored the immobilized cell process: 96.4 cents versus 101.8 cents per gallon. The total savings of 5.4 cents per gallon or 5.3% of the total cost is significant. The higher ethanol yield of the immobilized cell process (93% theoretical conversion compared to 88% of the batch process) resulted in a 4.1

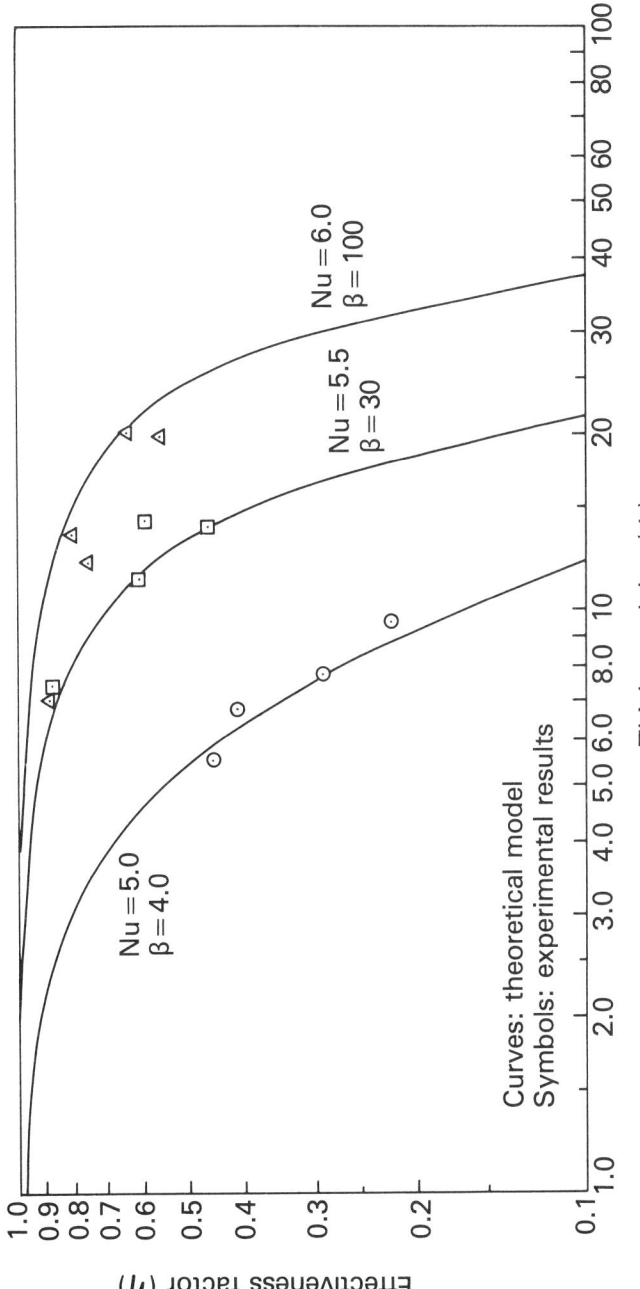

FIGURE 5. Comparison of the measured effectiveness factor to the theoretical value.

cents per gallon advantage on starch cost, but this advantage was nearly offset by the yeast credit of 3.5 cents to the batch process. Other cost items where savings were realized for the immobilized cell process included costs for chemicals and materials, labor, maintenance, and depreciation. There were no major differences in the energy cost.

The raw material cost was the major component for the total cost, which accounted for 64% of the total when corn starch was utilized for ethanol fermentation in both processes. Costs for the capital equipment, labor, and energy were about the same, sharing the major portions of the rest of the expenses.

Previous investigators cited much greater economic advantages for their respective, immobilized cell processes.[4,11] A cost reduction of 30% or more compared to the respective free-cell process was estimated in different processes or economic settings

TABLE 2. Comparative Manufacturing Cost (1988)

Product: 190-proof ethanol   *Plant Capacity:* 50 MM gallons/year

Capital Investment:

	Batch Process	Immobilized Cell Process
ISBL	$33.1 MM	$29.5 MM
OSBL	$26.4 MM	$26.4 MM
total	$59.5 MM	$55.9 MM

Manufacturing Cost:

Items	Unit Cost	Batch Process (¢/gal)	Immobilized Cell Process (¢/gal)
starch	5¢/lb	63.8	59.7
other materials	—	4.5	2.0
yeast credit	8¢/lb	(3.5)	—
coal	$40/ton	6.2	6.1
electricity	5¢/kWh	3.9	3.8
labor	$20/h	9.0	8.1
maintenance	(3% capital)	3.6	3.4
depreciation	(10% capital)	11.9	11.1
tax & insurance	(2% capital)	2.4	2.2
total	—	101.8¢/gal	96.4¢/gal

from ours. When corn starch was utilized as the starting material for ethanol fermentation, the contribution of ethanol fermentation to the overall process became less significant. Also, the economic burdens for the preparation of yeast beads and the conversion of starch to high glucose hydrolysates were substantial. At any rate, a cost savings of 5.4 cents per gallon with the immobilized cell process is still significant.

SUMMARY

The oil-phase pelletization of yeast cells in polyethyleneimine-alginate gels resulted in resilient biocatalyst beads of 0.7–1.7 mm D, which were suitable for ethanol fermentation. The bead was expansible to 1000% or more of its original volume over the span of the catalyst life as the immobilized cells continued to proliferate in the gel matrix. The yeast beads showed a high ethanol productivity of 30–60 g/L-gel/h with a three-month half-life in a minimal medium. The apparent bacteriostatic effect of PEI-alginate maintained a consistently high ethanol yield of 92–94% theoretical over the life span of the biocatalyst. The effectiveness factor of the yeast beads in a recycle reactor reached 0.88 at $\phi = 7$, $\beta = 30$, and $Nu = 5.5$.

When the yeast beads were applied to a continuous fermentation process converting corn starch to ethanol, the manufacturing cost was estimated at 96.4 cents per gallon of 190-proof ethanol. This compares favorably with the estimated cost of 101.8 cents for the conventional batch process. The 5.4 cents per gallon cost advantages or the 5.3% savings in total cost was attributed to the higher ethanol yield, the reduced capital cost requirements, and the lower labor cost.

ACKNOWLEDGMENTS

We thank Yao-En Li and Robert S. Blaskovitz for their assistance in the laboratory work.

REFERENCES

1. MOSBACH, K. & P. LASSON. 1970. Biotechnol. Bioeng. **12:** 19–27.
2. SLOWINSKI, W. & S. E. CHARM. 1973. Biotechnol. Bioeng. **15:** 973–979.
3. TOSA, T., T. SATO, T. MORI & I. CHIBATA. 1974. Appl. Microbiol. **27:** 886–889.
4. CHIBATA, I. & T. TOSA. 1976. *In* Immobilized Enzyme Principles. Volume 1. L. B. Wingard, E. Katchalski-Katzir & L. Goldstein, Eds.: 329–357. Academic Press. New York.
5. DURAND, G. & J. M. NAVARRO. 1987. Process Biochem. **September:** 14–23.
6. WADA, M., J. KATO & I. CHIBATA. 1980. Eur. J. Appl. Microbiol. Biotechnol. **10:** 275–287.
7. JOUNG, J. J., C. AKIN & G. P. ROYER. 1987. Appl. Biochem. Biotechnol. **14:** 259–275.
8. JOUNG, J. J. 1986. A mathematical analysis on the effectiveness factor of an immobilized live cell catalyst. ACRM86-16. Amoco Corporation, Naperville, Illinois.
9. KATZEN, R. 1978. Grain motor fuel alcohol technical and economic assessment study. HCP/J6639-01. National Technical Information Service, Springfield, Virginia.
10. NAGASHIMA, M., M. AZUMA, S. NOGUCHI, K. INUZUKA & H. SAMEJIMA. 1984. Biotechnol. Bioeng. **26:** 992–997.
11. LUONG, J. H. T. & M. C. TSENG. 1984. Appl. Microbiol. Biotechnol. **19:** 207–216.

APPENDIX

Nomenclature

Symbols

β = substrate saturation parameter, $\beta = c/K_m$
η = catalyst effectiveness factor
η_0 = intrinsic effectiveness factor in the absence of film resistance
ϕ = Thiele modulus, $\phi = R(voK_m^{-1}D_s^{-1})^{1/2}$
ρ = density of liquid (g/cm^3)

μ = fluid viscosity (g-cm^{-1}-s^{-1}), 1.0×10^{-2}
c = substrate concentration (g/L)
co = substrate concentration at catalyst surface (g/L)
$\bar{c}$ = dimensionless concentration, $\bar{c} = c/co$
D_f = diffusivity of the fluid (cm^2/s), 1.5×10^{-5}
D_s = diffusivity of the substrate (cm^2/s), 6.0×10^{-6}
K_m = substrate saturation constant (g/L), 4.4×10^{-1}
M = a mass-transfer modulus as defined by equation 4
m = cell concentration (g/L)
Nu = Nusselt number
r = radius (cm)
$\bar{r}$ = dimensionless radius, $\bar{r} = r/R$
R = radius of the catalyst bead (cm)
Re = Reynold number, Re = $2RV\rho\mu^{-1}$
Sc = Schmidt number, Sc = $\mu\rho^{-1}D_f^{-1}$
V = fluid velocity (cm/s)
vo = maximum catalyst activity (g-L^{-1}-s^{-1}), $vo = m \cdot vs$
vs = specific activity of yeast (g-g^{-1}-s^{-1}), 3.58×10^{-4}

Mechanisms of Oxygen Transfer Enhancement during Submerged Cultivation in Perfluorochemical-in-Water Dispersions[a]

JAMES D. McMILLAN AND DANIEL I. C. WANG

Department of Chemical Engineering
Massachusetts Institute of Technology
Cambridge, Massachusetts 02139

INTRODUCTION

Oxygen transport rates frequently limit the productivities of aerobic fermentations, and methods for improving oxygen supply are widely sought. Cultivation in perfluorochemical-in-water dispersions is one attractive method for increasing oxygen transfer capabilities. As we have shown previously, maximum oxygen transfer rates increase by over 400% (on an aqueous-phase basis) in 40% (v/v) perfluorochemical-in-water dispersions.[1]

With moderate agitation, perfluorochemicals are easily dispersed in aqueous media. Thus, cultivation in dispersions of perfluorochemicals can be performed in existing stirred tank bioreactors without vessel modification (other than installing an effluent gas condenser to recover entrained perfluorochemical). This technique also can be used with other methods for improving oxygen transfer, such as increasing the sparged gas oxygen partial pressure. Chibata *et al.*[2] were among the first to recognize the efficacy of this strategy. In 1974, a United States patent was issued covering the cultivation of aerobic microorganisms in the presence of low-viscosity, water-immiscible fluids, such as perfluorochemicals, in which oxygen is highly soluble.

The physical properties and the use of perfluorochemical fluids as oxygen transport fluids have been reviewed by Riess and Le Blanc[3] and by Lowe.[4] In addition to exhibiting high oxygen solubilities, perfluorochemicals are highly biocompatible and have other physical properties such as high density, high boiling point, and low vapor pressure that make their use in fermentation attractive. TABLES 1 and 2 present selected physical properties for perfluorochemicals similar to those used in this study. These tables also list physical properties for water and several hydrocarbons.

Whereas oxygen transfer enhancement has been clearly demonstrated during cultivation in perfluorochemical-in-water dispersions, the degree to which this technique can alleviate oxygen transfer limitations is unclear. Previous research using other

---

[a] J. D. McMillan gratefully acknowledges partial support by a United States National Science Foundation Graduate Fellowship while this research was conducted. The authors also wish to acknowledge the financial support of the Sun Company, Radnor, Pennsylvania, and of the NSF-ERC Initiative under Cooperative Agreement No. CDR-88-03014 to the MIT Biotechnology Process Engineering Center.

TABLE 1. Selected Physical Properties of Fluids (at 25 °C)

Fluid	T_{bp} (°C)	P_{vp} (torr)	ρ (g/cm³)	μ (cp)	σ_a (dynes/cm)	σ_{ow} (dynes/cm)	$S_{o/w}{}^a$
Water[b]	100	23.7	1.00	0.9	72	—	—
Perfluorochemicals[c]							
FC-40	155	3	1.87	4.1	16	52[d]	+4.0
FC-43	174	1.3	1.88	5.3	16		
FC-75	102	31	1.77	1.4	15		
Hydrocarbons							
paraffin oil[e]			0.88	109	30	47.8	−6.6
kerosene[e]			0.79	1.22	27.7	45.6	−2.1
toluene[e]	111[f]	28.2[g]	0.86	0.52	27.3	35.2	+8.7
oleic acid[e]	286[f]		0.89	20.7	31.5	15.6	+24.1
n-hexadecane[h]	287[i]		0.77	3.03[j]	27.6	53.8	−9.4
n-dodecane[h]	216[i]		0.74	1.35[i]	25.4	49.3	−2.7

[a] Calculated by equation 4.
[b] Reference 32.
[c] Reference 33.
[d] Reference 34.
[e] Reference 8 (30 °C).
[f] Reference 35.
[g] Calculated by the Antoine vapor pressure correlation (reference 36).
[h] Reference 12 (30 °C).
[i] Reference 37.
[j] Interpolation between 30 °C data (reference 12) and 20 °C data (reference 37).

"oil-in-water" systems has produced little consensus regarding the mechanisms by which a water-immiscible "oil" phase interacts to influence mass transfer. Consequently, this research was undertaken to ascertain the mechanisms controlling gas-liquid oxygen transfer in perfluorochemical-in-water dispersions. This study also provides an expanded framework for assessing the potential for oxygen transfer enhancement in other "oil-in-water" systems. The nomenclature used herein is defined in the APPENDIX at the end of this report.

LIQUID FILM CONTROL AT THE GAS-LIQUID INTERFACE

Under conditions typical for aerobic cultivation of unicellular microbes, the rate of oxygen diffusion into the liquid film adjacent the gas-liquid interface controls the overall rate of oxygen transfer.[5,6] The volumetric oxygen transfer rate, OTR, depends on both the overall volumetric mass transfer coefficient, $K_l a$, and the driving force for mass transfer. $K_l a$ is usually based on the overall concentration driving force for mass transfer between the gas and liquid phases, as shown by equation 1:

$$\text{OTR} = K_l a(c^* - c_l). \tag{1}$$

The subscript "l" on the overall mass transfer coefficient, K, indicates that K_l is essentially equal to the individual liquid film mass transfer coefficient, k_l, because of liquid film control.

In many experiments, it is the partial pressures of oxygen in the gas and liquid phases that are actually measured rather than the concentrations. The concentration-based driving force in equation 1 is converted into a partial pressure–based driving force using Henry's law. Equation 2 expresses the oxygen transfer rate using a partial pressure driving force:

$$\text{OTR} = K_l a H(p_g - p_l), \qquad (2)$$

where Henry's law is defined by

$$c_i = H_i p_g. \qquad (3)$$

In equation 2, the effect of the liquid-phase oxygen solubility on the overall concentration driving force is manifested in Henry's constant, H.

In heterogeneous oil-in-water systems, an effective Henry's constant, H_{eff}, must be used in equation 2. As discussed later, information about liquid film geometry in oil-in-water systems is inferred by examining different methods of calculating H_{eff}.

TABLE 2. Oxygen Permeability and Solubility Ratios for Selected Fluids (at 1 atmosphere O_2 and 25 °C)

Fluid	Diffusivity[a] $10^5 \times D$ (cm²/s)	Solubility (mL O_2/100 mL)	Henry's Constant[b] H (mMol/L-atm)	Henry's Ratio H/H_{AQ} (–)	Permeability Ratio[c] P/P_{AQ} (–)
Water	2.4	3.2[d]	1.3	1	1
Perfluorochemicals[d]					
FC-40	4.8[e]	37	15.1	11.6	16.4
FC-43	4.8[e]	36	14.7	11.2	15.8
FC-75	4.8[e]	51	20.9	15.9	22.5
Hydrocarbons					
paraffin oil		10.4[f]	4.2	3.2	
kerosene		11.2[f]	4.6	3.5	
toluene	5.8	21.5[f]	8.8	6.7	10.5
oleic acid	0.3	12.7[f]	5.2	4.0	1.3
hexadecane	1.8	22.0[g]	9.0	6.9	5.9
dodecane	3.0	24.0[g]	9.8	7.5	8.4

[a] Calculated by the Wilke-Chang correlation (reference 36).
[b] Calculated by equation 3.
[c] Oxygen permeabilities for each fluid as calculated by equation 5.
[d] Reference 32.
[e] Reference 34.
[f] Reference 8 (30 °C).
[g] Reference 12 (30 °C).

MECHANISMS OF OXYGEN TRANSFER ENHANCEMENT

The presence of an oil phase may influence gas-liquid mass transfer in two fundamental ways: the presence of oil may change the specific interfacial area available for gas-liquid mass transfer or it may affect the rate of solute flux across the gas-liquid interface. Changes in the gas-liquid interfacial area occur when the oil phase alters the gas-liquid interfacial tension in the system or when the dispersed oil droplets affect bubble breakup or coalescence. Alternatively, changes in the rate of solute flux across the gas-liquid interface occur when the oil phase alters either the liquid film hydrodynamics or the solute permeability.

Spreading Behavior of Oil Phase

The potential of an oil phase to affect the gas-liquid interfacial tension has been discussed by many researchers,[7-12] but no quantitative correlation for this behavior has been published. Several researchers have concluded that the spreading behavior of an oil determines the oxygen transfer characteristics of a given oil-in-water system. The spreading coefficient, $S_{o/w}$, of an oil on water is defined as the negative of the total change in free energy occurring when an oil spreads upon a unit area of water:[7]

$$S_{o/w} = \sigma_{wa} - (\sigma_{oa} + \sigma_{ow}). \tag{4}$$

As defined by equation 4, $S_{o/w}$ is a measure of the degree that an oil phase either beads up or spreads out when contacting an aqueous phase. When the change in free energy upon spreading is positive, oils bead up; when the change is negative, oils spread. In terms of the spreading coefficient, oils with a negative spreading coefficient ($S < 0$) bead up, whereas those with a positive spreading coefficient ($S > 0$) spread out.

Yoshida et al.,[8] expanding on earlier work by Eckenfelder and Barnhart[9] and by Davies,[10] found that oxygen transfer into oil-in-water systems could be explained on the basis of an oil's spreading behavior. Beading oils ($S < 0$) such as n-paraffin and kerosene impeded oxygen transfer, whereas spreading oils ($S > 0$) such as oleic acid and toluene behaved like surfactants.

Yoshida et al.[8] hypothesized that a reduction in oxygen transfer rate in the presence of beading oils resulted from decreased rates of oxygen diffusion through oil droplets within the liquid film. However, this mechanism of flux retardation is likely only to apply to highly viscous oils. As TABLES 1 and 2 show, oxygen permeabilities in low-viscosity beading hydrocarbons are generally higher than in water. For example, oxygen permeabilities in hexadecane and dodecane are sixfold and eightfold greater than in water, respectively. Therefore, we expect the presence of low-viscosity beading oils to improve liquid film permeability rather than retard it. Reduced permeabilities can occur, though, in the presence of high-viscosity oils due to low oxygen diffusivities. Oleic acid, for example, has an oxygen permeability roughly equal to that of water, in spite of its relatively high oxygen solubility. More viscous oils such as n-paraffin may actually have oxygen permeabilities below that of water and, in this case, flux retardation is anticipated.

In contrast to the hypothesis of Yoshida et al., Linek and Benes[11] concluded that the reduction in $K_l a$ in $S < 0$ oil-in-water systems is caused by decreased interfacial area, a, rather than by decreased flux. On the other hand, Hassan and Robinson[12] found that K_la for beading oils could either increase or decrease with increasing oil volume fraction, depending upon the oil used. Increases in K_la were attributed to hydrodynamic changes caused by oil droplets impinging on the liquid film surrounding the gas bubbles, but interfacial area effects were not considered. They thus hypothesized that, in $S < 0$ oil-in-water systems, K_la is affected by both hydrodynamic enhancement and flux retardation mechanisms. Consequently, they concluded that the overall behavior of K_la is determined by the relative influence of these opposing mechanisms, which presumably vary with the oil used.

In contrast to beading oils, Yoshida et al.[8] found that oils that spread at an aqueous interface ($S > 0$) behave as surfactants. At very low concentrations, a sharp decrease in K_l occurs due to the spread oil causing the local liquid slip velocity at the gas-liquid interface to decrease. However, opposing this effect on K_l, the gas-liquid interfacial tension falls as the loading of spreading oil is increased. This causes the mean bubble size to decrease. There is thus a gradual increase in specific gas-liquid interfacial area with increasing oil loading. Due to these effects on K_l and a, the overall K_la exhibits a rapid decrease followed by a gradual increase as oil loading is increased. Yoshida et al.[8] reported enhancement in K_la for spreading oils above oil loadings ranging from 1% to 10% (v/v); Linek and Benes[11] have also observed enhancement in the presence of spreading oils.

Proposed Pathways for Oxygen Transfer

Many researchers have speculated on the pathway(s) for oxygen transfer in oil-in-water systems. Whereas many oxygen transfer pathways have been described, conclusions about specific pathways remain speculative. For beading oils, the accepted pathway seems to be gas → water → oil.[11-14] In this case, the gas → water transfer occurs into the liquid film and no direct gas → oil transfer takes place. No well-accepted pathway has been proposed for spreading oils. Due to their spreading behavior, spreading oils may permit direct gas-oil contact, at least over some portion of a gas-liquid interface. Linek and Benes,[11] however, concluded that substantial direct gas-oil contact does not occur in $S > 0$ oil-in-water systems; they believe that only a trace of spreading oil at a gas-liquid interface can cause significant interfacial tension effects. Thus, whereas a trace of oil at the gas-liquid interface behaves like a surfactant, the bulk of the oil remains as dispersed droplets.

A few researchers have proposed parallel oxygen transfer from the gas phase into both aqueous and oil phases.[1,15] Linek and Benes[11] concluded that direct transfer from gas into both liquid phases takes place in water-in-oil systems (continuous oil phase) for both spreading and beading oils. More recently, Das et al.[16] observed that transfer from a gas phase into both liquid phases within a liquid film occurred. It was not clear, though, if transfer into the oil phase occurred directly from the gas phase. In contrast to other researchers, Yoshida et al.[17] concluded that direct gas → oil transfer is a significant pathway for oxygen absorption during hydrocarbon fermentation.

Hydrodynamic Enhancement

Researchers have also observed that the presence of small droplets, solids, or whole cells in aqueous systems can alter gas-liquid mass transfer rates.[18-30] A number of these studies have shown that significant changes in the interfacial area occur in the presence of dispersed particles. In addition, several studies have concluded that modest changes in liquid film hydrodynamics can take place. Experimentally, it is difficult to distinguish between the mechanism of inertial impaction hypothesized by Wise et al.[18] and by Hassan and Robinson[12] and the hydrodynamic effect proposed by Andrews et al.[26,27] Both of these mechanisms predict an increase in K_l with increased particle (solid, cell, or liquid droplet) loading. However, as Andrews et al.[27] demonstrate, this type of hydrodynamic effect plateaus above a critical particle loading.

Godbole et al.[24] showed that mass transfer behavior depends upon particle composition. Indeed, significant changes in K_l have been seen primarily when the absorbing solute has a high absorptive capacity in the particle, such as in oxygen transfer into activated carbon-in-water slurries.[21,25,28,29,31] Certainly, the effective solubility or permeability of the solute in the liquid can increase in the presence of dispersed particles onto which the solute can absorb or if the solute has a high solubility. Thus, when the particle size is less than the liquid film thickness, absorptive particles may influence mass transfer behavior in a manner independent of any hydrodynamic changes associated with the presence of the particles.[21,28] Consequently, a hydrodynamic enhancement mechanism is unnecessary for explaining increased mass transfer in the presence of particles that have a high affinity for the solute.

The literature cited above indicates that changes in $K_l a$ in the presence of inert solids are dominated by changes in a. There is no strong evidence that inert solids can cause a large hydrodynamic effect on K_l. Studies by Elstner and Onken[22] and by Alper and Ozturk,[28] for example, show that the K_l is only marginally affected by the presence of inert solids. Inert solids apparently cause minor hydrodynamic dampening due to the effect that particle loading has on the effective dispersion viscosity.[29] In our opinion, this purely hydrodynamic effect is insignificant in comparison with potential changes in interfacial area and medium permeability. Therefore, we have not considered the effect of hydrodynamic enhancement in this study. Rather, our focus is to distinguish between interfacial area–associated enhancement and flux-associated permeability enhancement. As discussed below, we expect permeability enhancement to be important in perfluorochemical-in-water dispersions.

Permeability Enhancement

Whereas the importance of oxygen solubility in oil-water systems has been suggested by many researchers, the importance of oxygen permeability has been widely overlooked. Permeability, as defined by equation 5, is a more accurate measure of the capacity to transport solute than solubility alone:

$$P_i = D_i^{1/2} c_i^*. \tag{5}$$

Defined in this manner, solute permeability in a phase is proportional to the maximum solute flux achievable in this phase.

TABLE 2 shows oxygen solubilities and diffusivities in water, perfluorochemicals, and hydrocarbons. The diffusivity of oxygen in perfluorochemicals is not known with great confidence because conflicting data are reported in the literature. Therefore, oxygen diffusivities for the perfluorochemicals in TABLE 2 are taken as twice that of water, which is an estimate supported by a recent review by Junker.[34] Using these data, Henry's constants and oxygen permeabilies are calculated by equations 3 and 5, respectively; ratios of their value relative to water are tabulated. As TABLE 2 shows, Henry's constant ratios and permeability ratios are greater for perfluorochemicals than for the other fluids listed. Significantly, this conclusion holds even if the diffusivity of oxygen in perfluorochemicals is taken to be the same as that of water. Thus, TABLE 2 shows that solubility and diffusivity both positively influence oxygen permeability in perfluorochemicals; under similar conditions, oxygen permeabilities in perfluorochemicals are higher than in hydrocarbons.

Experimental Approach

Increased liquid film permeability may account for mass transfer enhancement in perfluorochemical-in-water dispersions. Surface renewal modeling of permeability enhancement, for example, predicts a linear increase in mass flux with increasing perfluorochemical volume fraction.[1] Because maximum oxygen transfer rates on a total liquid volume basis increase linearly with increasing perfluorochemical loading, up to volume fractions of 0.2–0.3, modeling results support the conclusion that enhancement is due to increased medium permeability. However, to verify that mass transfer enhancement is caused by increased dispersion permeability requires separating the effect of interfacial area from that due to permeability.

Consequently, the following experiments were undertaken to elucidate the mechanism(s) responsible for overall enhancement. Separation of permeability effects from interfacial area behavior requires subdividing $K_l a$ into its individual components, K_l and a. Fortunately, the overall $K_l a$ is easily measured during cultivation experiments and we need only to measure either K_l or a to understand the behavior of both. Because perfluorochemical-in-water dispersions are unstable in the absence of vigorous agitation, K_l cannot be measured directly using a surface aeration technique. Consequently, we chose to measure a directly. The interfacial area in cell-free perfluorochemical-in-water dispersions was measured as a function of perfluorochemical loading. This measurement, coupled with $K_l a$ data from cultivation studies, provides the basis for a mechanistic evaluation of enhancement in perfluorochemical-in-water dispersions.

MATERIALS AND METHODS

Maximum Oxygen Transfer Rate Studies

Organism: Escherichia coli *K12 Wild Type*

This organism was obtained from the Yale University School of Medicine *E. coli* Genetic Stock Center (CGSC). It is designated as CGSC no. 4401 in their culture collection.

Maintenance of Organism

Stock cultures are maintained on agar slants at 4 °C and are transferred monthly. The complex LB agar medium used for strain propagation is: 10 g/L NaCl; 10 g/L tryptone; 5 g/L yeast extract; and 15 g/L agar.

Perfluorochemical

Fluorinert Fluid FC-40 (3M Company) was used in all experiments. FC-40 contains a mixture of perfluorochemicals, primarily trialkylamines with alkyl chain lengths between three and five, but the exact composition is proprietary. See TABLES 1 and 2 for selected physical properties of FC-40.

Fermentation Conditions

Fed-batch fermentations of *Escherichia coli* were carried out at 37 °C, 16.7 psia, and pH 7.0 in defined minimal media in perfluorochemical-in-water dispersions over perfluorochemical volume fractions ranging from 0 to 0.425. Fermentations were conducted at both 1.25-L and 11.5-L scales. The 1.25-L (working volume) fermentations were conducted using fed-batch feeding of glucose as the carbon source at an agitation rate of 500 rpm and an aeration rate of 0.8 vvm (0.5 hp/1000 L) in a 2-L (total volume) Setric 2M fermentor. The 1.25-L fermentations were begun with no perfluorochemical and then perfluorochemical was added stepwise as required (keeping the total liquid volume constant). The 11.5-L (working volume) fermentations also were conducted as fed-batch systems with glycerol as the carbon source at 800 rpm and 1.0 vvm (5 hp/1000 L) in a 14-L (total volume) Chemap LF fermentor. The 11.5-L fermentations were begun and maintained at a constant perfluorochemical volume fraction throughout the fermentation. Further details on the experimental protocol used for maximum oxygen transfer rate determinations are available in reference 1.

Interfacial Area Determination

Interfacial area experiments were performed at 37 °C in the 14-L Chemap fermentor (11.5-L working volume) using phosphate-buffered saline (PBS) solutions at the same concentration as during fermentations, that is, 6.0 g/L Na_2HPO_4 and 3.0 g/L KH_2PO_4. Sparged perfluorochemical-in-water dispersions were agitated at 800 rpm and aerated at 1.0 vvm, identical to that used during fermentation. A photographic method was used to measure air bubble size distributions over perfluorochemical volume fractions ranging from 0 to 0.15. Photographs were taken through a viewing port filled with distilled water to reduce the effect of vessel curvature. The port was positioned so that the well-mixed zone between the top and middle impellers could be photographed. The viewing port, constructed out of 0.32-cm Plexiglas held together with epoxy, measured 10.8 cm in width and 11.1 cm in height. Vessel curvature caused the depth to vary between 3.8 cm at the center and 5.7 cm at the sides. The port was mounted to the glass vessel using RTV 108 silicone sealant (GE Company). Black-and-white photographs of well-lighted dispersions were taken using a Nikon FA 35-mm camera equipped with a 55-mm macro lens and operated manually at a shutter speed of 10^{-3} seconds. Upon enlargement of photographs to 8 inches × 10 inches, a minimum of 150 bubbles per photograph were analyzed, using a calibrated Optomax Image Analyzer model VIDS III (ITI, Burlington, Massachusetts) to compile bubble size distribution data. Gas holdup, ϕ_g, was measured directly by changes in the dispersion height at the vessel wall. The specific gas-liquid interfacial area is calculated according to equation 6:

$$a = \frac{6\phi_g}{d_{sm}}, \qquad (6)$$

where d_{sm} is defined as

$$d_{sm} = \frac{\sum_{i=1}^{n} d_i^3}{\sum_{i=1}^{n} d_i^2}. \qquad (7)$$

RESULTS AND DISCUSSION

Fermentation Results

The top panel of FIGURE 1 shows dissolved oxygen and biomass profiles during two 11.5-L *Escherichia coli* fermentations. One is a control fermentation with no perfluorochemical and the other is with 35% (v/v) perfluorochemical. In the control fermentation, we observe an initial period of aerobic exponential growth until the cell density is about 4 gDCW/L. At this point, the dissolved oxygen reaches zero (below detection limit) and oxygen-limited growth begins. The ensuing period of oxygen-limited growth, between 9 and 13 hours, is characterized by a linear biomass versus time relationship.

Under oxygen-limited growth, the culture is partially anaerobic. This condition triggers fermentative energy metabolism, and end products such as lactate and acetate begin to accumulate in the medium. The buildup of acetate, in particular, is deleterious to cell growth. Unpublished shake-flask experiments show that cell growth is completely inhibited at acetate concentrations above 10 g/L. It is therefore desirable to prolong the onset of oxygen-limited growth to avoid acetate formation. By 13 hours in the control fermentation, acetate has accumulated sufficiently (~10 g/L) to inhibit further growth. At this point, the culture enters stationary phase at a final cell density of 15 gDCW/L.

Furthermore, the top panel of FIGURE 1 shows that the period of aerobic exponential growth is extended in the presence of 35% (v/v) perfluorochemical. When dissolved oxygen becomes limiting at just over 12 hours, the cell density has reached nearly 20 gDCW/L (on a per liter aqueous-phase basis). The higher biomass concentration means that acetate accumulates more rapidly during oxygen-limited growth in this fermentation. Thus, the period of oxygen-limited growth is shorter than in the control, occurring from 12 to 14 hours. This fermentation culminates at a final cell density above 30 gDCW/L, which is more than twice that achieved without perfluorochemical.

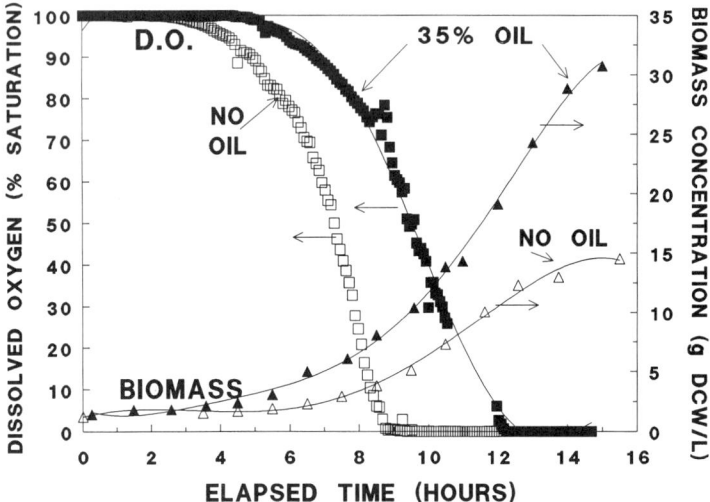

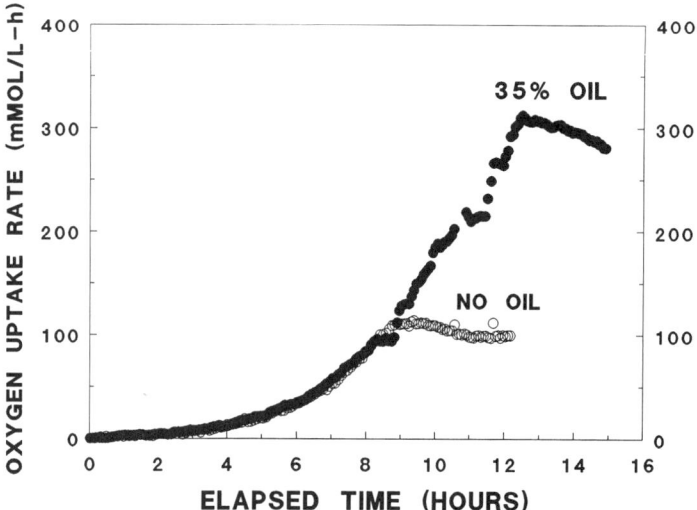

FIGURE 1. Characteristic growth kinetics of *Escherichia coli* with (solid symbols) and without (open symbols) perfluorochemical—top: dissolved oxygen (■, □) and dry cell weight (▲, △) profiles; bottom: oxygen uptake rate (●, ○) profiles.

The bottom panel of FIGURE 1 shows oxygen uptake rate profiles (expressed on per liter aqueous-phase bases) during these fermentations. In both fermentations, the oxygen uptake rate increases exponentially with elapsed time, paralleling the exponential increase in cell density during aerobic growth. However, when the oxygen uptake rate approaches the maximum oxygen transfer rate, the culture becomes oxygen-limited. During oxygen limitation, the driving force for oxygen transfer is at its maximum and the oxygen uptake rate equals the maximum oxygen transfer rate, OTR_{max}. The control fermentation attains a maximum oxygen transfer rate (OTR_{max}^0) of approximately 100 mMol O_2/L-h at nine hours. In contrast, in the presence of 35% (v/v) perfluorochemical, OTR_{max} is about 300 mMol O_2/L-h.

FIGURE 2 shows OTR_{max} as a function of perfluorochemical loading for fed-batch fermentations carried out at both 1.25-L and 11.5-L scales. The upper curve shows data from 11.5-L fermentations and the lower curve shows data from 1.25-L fermentations. On both scales, there is a linear increase in OTR_{max} on a per liter aqueous-phase basis with increasing perfluorochemical volume fraction. On a total liquid basis, these curves level off at perfluorochemical volume fractions between 0.2 and 0.3 (not shown).

Interfacial Area Results

FIGURE 3 shows gas bubble size distribution data for sparged perfluorochemical-in-phosphate buffer dispersions at perfluorochemical loadings of 0, 5, 10, and 15% (v/v). The bubble size distribution is not markedly affected by the presence of 0 to 15% perfluorochemical. The Sauter mean diameter of air bubbles in these dispersions remains constant at about 1.5 mm. Gas holdup remains constant as well, at about 0.1 m^3 gas/m^3 total liquid. FIGURE 4 shows interfacial areas calculated from size distribution and gas holdup data. No dependence on perfluorochemical loading is observed.

Discussion

Fed-batch fermentations of *Escherichia coli* show that cultivation in perfluorochemical-in-water dispersions can improve fermentor productivity through increased oxygen transfer capability. The steeper slope of the biomass versus time curve in the top

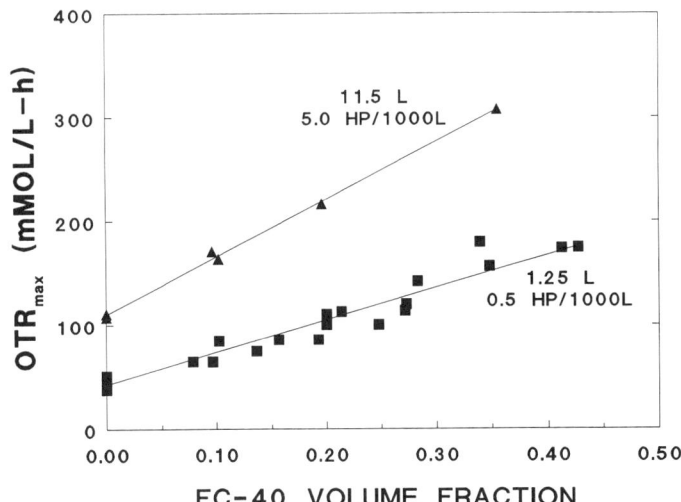

FIGURE 2. Maximum oxygen transfer rate (OTR_{max}) data during *Escherichia coli* cultivation as a function of the volume fraction of perfluorochemical. The lower curve (■) shows a summary of the data at the 1-L scale, whereas the upper curve (▲) refers to the data at the 11-L scale.

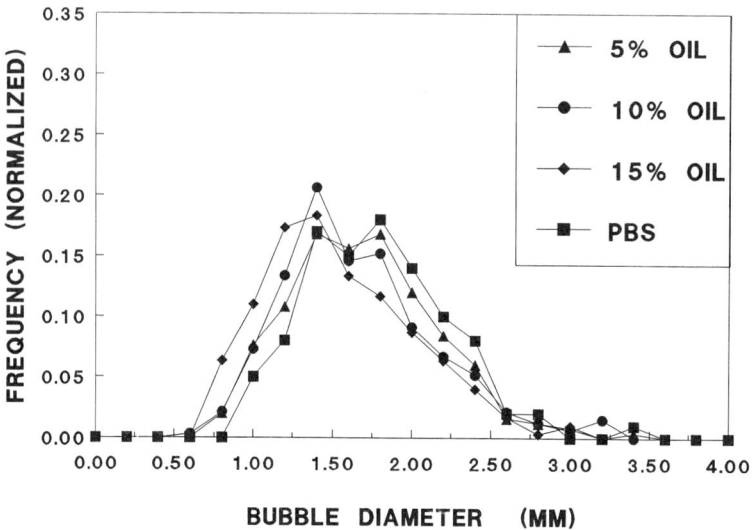

FIGURE 3. Bubble size distributions in cell-free perfluorochemical–in–phosphate buffer dispersions.

panel of FIGURE 1 shows that the specific biomass productivity improves in the presence of perfluorochemical. The greater area under this curve indicates that overall productivity also increases in perfluorochemical-in-water dispersions. The bottom panel of FIGURE 1 illustrates that increases in fermentor productivity and final cell concentration are due to higher oxygen transfer capabilities using these dispersions.

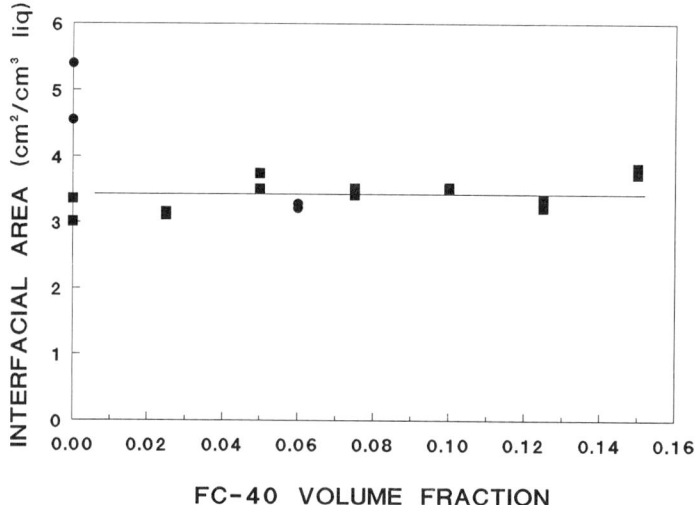

FIGURE 4. Specific gas-liquid interfacial area as a function of perfluorochemical loading.

FIGURE 2 shows that enhancement in maximum oxygen transfer rates is similar on both 1-L and 10-L scales over a tenfold difference in power input. On both scales, OTR_{max} increases linearly on an aqueous volume basis. The steeper slope of the OTR_{max} versus perfluorochemical loading curve at the higher power input level (11.5-L scale) may indicate that even better oxygen transfer enhancement is possible at higher power inputs. Significantly, when OTR_{max} is expressed on a total liquid volume basis, these curves exhibit saturation behavior above perfluorochemical loadings of 20–30% (v/v).

Interfacial area experiments show essentially no effect of FC-40 on specific gas-liquid interfacial area. However, the photographic technique only detects bubbles that are in the well-mixed zone between impellers and within 1–2 cm of the vessel wall. Thus, bubble size distribution results apply only to this local region. Nonetheless, if substantial changes were occurring in the bulk, we would also expect to see changes occurring locally. Because these are not observed, we are confident that the bulk bubble size distribution did not change significantly.

The calculated Sauter mean diameters represent local values. However, the dispersion height–based gas holdup measurement measures the bulk average holdup. In addition, the gas holdup measurement technique has a ±20% absolute error.[38] Thus, calculated interfacial area values represent ±20% estimates rather than precise bulk average values. Due to the fact that no change in the estimated interfacial area is observed, it is reasonable to conclude that no marked change in the bulk interfacial area takes place under these conditions.

As previously discussed, we would observe changes in the estimated gas-liquid interfacial area if spreading occurred. Interfacial area data therefore indicate that perfluorochemical droplets are not "wetting" or spreading at the gas-liquid interface. Consequently, we conclude that the aqueous phase is continuous at the gas-liquid interface. This conclusion is surprising at first because the spreading coefficient of FC-40 (see TABLE 1) indicates that FC-40 droplets spread at an air-aqueous interface. However, the discrepancy between spreading behavior predictions and interfacial area data can probably be explained by the spreading coefficient strictly applying only at thermodynamic equilibrium. In our experiments, perfluorochemical droplet–air bubble interactions are rapid and the time scale of these interactions may be too short for spreading to occur. As TABLE 1 shows, the spreading coefficient of FC-40 is only 4 dynes/cm. Thus, for FC-40, both the driving force for spreading and the rate of spreading are small (relative to spreading hydrocarbons). This supports the hypothesis that longer interaction times are required for FC-40 droplets to spread at the interface.

Analysis of Proper Driving Force

When the oxygen transfer rate into "oil-in-water" systems is modeled using equation 2, an effective Henry's constant, H_{eff}, must be used to quantify the effect of partial pressure on the concentration driving force. The method by which H_{eff} is calculated depends on the liquid film geometry. For example, if perfluorochemical directly contacts the gas-liquid interface in proportion to its volume fraction, then a volume fraction–based effective Henry's constant is appropriate, as shown by equation 8:

$$H_{eff} = (1 - \phi_o)H_w + \phi_o H_o. \qquad (8)$$

Alternatively, if the aqueous phase is continuous at the gas-liquid interface, then an effective Henry's constant based on the aqueous-phase solubility alone should be used, as shown by equation 9:

$$H_{\text{eff}} = H_{\text{w}}. \tag{9}$$

We can use equations 8 and 9 to test the consistency of our hypothesis that the aqueous phase is continuous at the gas-liquid interface. This is accomplished by analyzing the theoretical consistency of K_l values calculated from independent OTR_{max} fermentation data using either equation 8 or equation 9.

The assumption necessary to perform this analysis is that the cell-free interfacial area results are valid for actual fermentations. This assumption is justified because 11-L fermentations were performed at the same operating conditions (working volume, agitation rate, gas sparge rate, and phosphate buffer concentration) as those used in cell-free interfacial area experiments. If perfluorochemical significantly affected the interfacial area, we would expect to observe this behavior in both the presence and absence of cells.

If we assume that no changes in interfacial area occurred during fermentation, then permeability enhancement must be responsible for the observed increase in OTR_{max} with increasing perfluorochemical loading. In terms of equation 2, the importance of solubility on permeability enhancement is reflected by H_{eff}, whereas that of diffusivity is reflected by K_l. The behavior of H_{eff} can be approximated by equations 8 and 9. Rearranging equation 2 yields an expression for K_l:

$$K_l = \frac{OTR}{aH_{\text{eff}}(p_g - p_l)}. \tag{10}$$

At OTR_{max}, the oxygen partial pressure in the bulk liquid phase is zero. By taking the ratio of K_l in the presence of perfluorochemical to that without, K_l^0, we obtain an expression that can be analyzed from the 11-L fermentation data:

$$\frac{K_l}{K_l^0} = (1 - \phi_o) \frac{OTR_{\text{max}}}{OTR_{\text{max}}^0} \frac{H_w}{H_{\text{eff}}} \frac{p_g^0}{p_g}. \tag{11}$$

The factor of $(1 - \phi_o)$ puts equation 11 on a total liquid volume basis.

FIGURE 5 shows enhancement ratios in K_l/K_l^0 and H_{eff}/H_w plotted as functions of perfluorochemical volume fraction using H_{eff} calculated by both equations 8 and 9. As this figure shows, the enhancement in the K_l/K_l^0 ratio depends upon the method used to calculate H_{eff}. If H_{eff} is calculated using equation 8, K_l/K_l^0 decreases with increased perfluorochemical loading. If H_{eff} is identical to H_w, K_l/K_l^0 increases up to 0.2 volume fraction and then plateaus.

Mass transfer film theories predict K_l to vary as a power function of the diffusivity, D^α, where $0 < \alpha < 1$.[5,6] Because the diffusivity of oxygen in perfluorochemical is higher than in water, we expect the ratio of K_l/K_l^0 to increase (or at least remain level) with increasing perfluorochemical loading. Consequently, the K_l/K_l^0 ratios inferred by equation 11 are inconsistent with theory when H_{eff} is calculated by equation 8. Apparently, equation 8 overpredicts H_{eff}. In contrast, K_l/K_l^0 ratios are consistent with theory when H_{eff} equals H_w.

This analysis supports the conclusion that the aqueous phase is continuous at the gas-liquid interface. Hence, it appears that the low continuous aqueous-phase permeability limits the extent of enhancement that can be realized when cultivating in perfluorochemical-in-water dispersions.

CONCLUSIONS

Oxygen transfer capabilities dramatically increase when cultivation is carried out in perfluorochemical-in-water dispersions. Improved oxygen transfer capacity translates into greater fermentor productivity and higher final cell concentration. Maximum

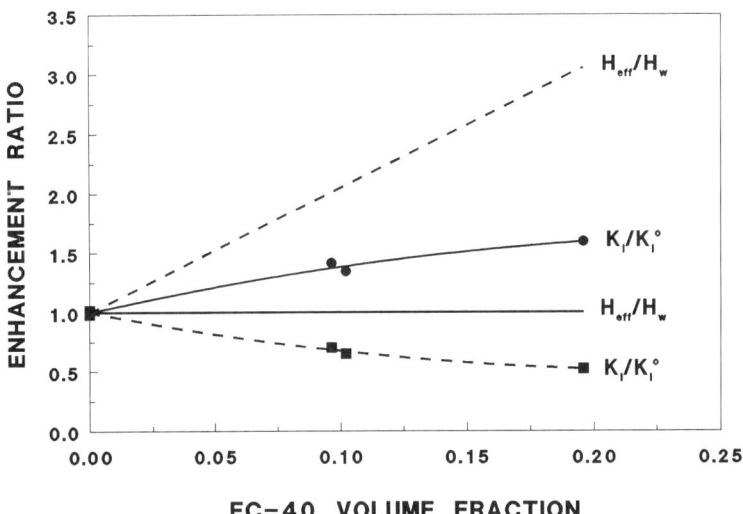

FIGURE 5. Enhancement ratios in H_{eff}/H_w and K_l/K_l^0 as functions of perfluorochemical volume fraction using H_{eff} calculated by either equation 8 (dashed lines) or equation 9 (solid lines).

volumetric gas-liquid oxygen transfer rates (OTR_{max}) increase linearly on a per liter aqueous-phase basis with increasing perfluorochemical volume fraction in both 1-L and 10-L fermentations. The ratio of increase in OTR_{max} is more pronounced at higher power inputs.

The interfacial area in cell-free perfluorochemical–in–phosphate buffer dispersions is independent of perfluorochemical loading from 0 to 15% (v/v). Extrapolation of these results to OTR_{max} fermentation data indicates that enhancement in oxygen transfer is due to increased oxygen permeability in the liquid film in the presence of perfluorochemical.

From interfacial area and OTR_{max} data, we conclude that the aqueous phase is continuous at the gas-liquid interface, despite the predicted spreading behavior of perfluorochemical. Apparently, the time scale for perfluorochemical droplet–air bubble interactions is too short to permit the spreading of droplets. Moreover, inferred K_l values are theoretically consistent when a driving force based on aqueous-phase solubility alone is used.

Based upon these conclusions, we are developing a more comprehensive model of gas-liquid mass transfer in perfluorochemical-in-water dispersions. We hope that this model will be capable of describing the leveling-off in enhancement observed at higher oil loadings, as well as predicting the influence of agitation and sparge rate on enhancement. The development and validation of this model will be the subject of a later communication.

REFERENCES

1. MCMILLAN, J. D. & D. I. C. WANG. 1987. Enhanced oxygen transfer using oil-in-water dispersions. *In* Biochemical Engineering V. Volume 506, p. 569–582. Ann. N.Y. Acad. Sci. New York.
2. CHIBATA, I., *et al.* 1974. Cultivation of aerobic microorganisms. United States patent no. 3,850,753.
3. RIESS, J. G. & M. LE BLANC. 1982. Solubility and transport phenomena in perfluorochemicals relevant to blood substitution and other biomedical applications. Pure Appl. Chem. **54**(12): 2383–2406.
4. LOWE, K. C. 1987. Perfluorocarbons as oxygen-transport fluids. Comp. Biochem. Physiol. **87A**(4): 825–838.
5. WANG, D. I. C., C. L. COONEY, A. L. DEMAIN, P. DUNNILL, A. E. HUMPHREY & M. D. LILLY. 1979. Fermentation and Enzyme Technology, p. 158 and p. 183. Wiley. New York.
6. BAILEY, J. E. & D. F. OLLIS. 1986. Biochemical Engineering Fundamentals (second edition), p. 464 and p. 474–476. McGraw–Hill. New York.
7. HIEMENZ, P. C. 1986. Principles of Colloid and Surface Chemistry (second edition), p. 314–317. Dekker. New York.
8. YOSHIDA, F., T. YAMANE & Y. MIYAMOTO. 1970. Oxygen absorption into oil-in-water emulsions: a study of hydrocarbon fermentors. Ind. Eng. Chem. Process Des. Dev. **9**(4): 570–577.
9. ECKENFELDER, W. W., JR. & E. L. BARNHART. 1961. The effect of organic substances on the transfer of oxygen from air bubbles in water. AIChE J. **7**(4): 631–634.
10. DAVIES, J. T. 1963. Mass-transfer and interfacial phenomena. Adv. Chem. Eng. **4**: 1–50.
11. LINEK, V. & P. BENES. 1976. A study of the mechanism of gas absorption into oil-water emulsions. Chem. Eng. Sci. **31**: 1037–1046.
12. HASSAN, I. T. M. & C. W. ROBINSON. 1977. Oxygen transfer in mechanically agitated aqueous systems containing dispersed hydrocarbon. Biotechnol. Bioeng. **19**: 661–682.
13. MATSUMURA, M., M. OBARA, H. YOSHITOME & J. KOBAYASHI. 1972. Oxygen equilibrium distribution and its transfer in an air-water-oil system. J. Ferment. Technol. **50**(10): 742–750.
14. YAMANE, T. & F. YOSHIDA. 1974. Comments on oxygen absorption into oil-water emulsions. J. Ferment. Technol. **52**(7): 445–450.
15. MIMURA, A., T. KAWANO & R. KODAIRA. 1969. Biochemical engineering analysis of hydrocarbon fermentation. I: Oxygen transfer in the oil-water system. J. Ferment. Technol. **47**(3): 229–236.
16. DAS, T. R., A. BANDOPADHYAY, R. PARTHASARATHY & R. KUMAR. 1985. Gas-liquid interfacial area in stirred vessels: the effect of an immiscible liquid phase. Chem. Eng. Sci. **40**(2): 209–214.
17. YOSHIDA, T., D. YOKOYAMA, K. CHEN, T. SUNOUCHI & H. TAGUCHI. 1977. Oxygen transfer in hydrocarbon fermentation by *Candida rugosa*. J. Ferment. Technol. **55**(1): 76–83.

18. WISE, D. L., D. I. C. WANG & R. I. MATELLES. 1969. Increased oxygen mass transfer rates from single bubbles in microbial systems at low Reynolds numbers. Biotechnol. Bioeng. **11**: 647–681.
19. MEHTA, V. D. & M. M. SHARMA. 1971. Mass transfer in mechanically agitated gas-liquid contactors. Chem. Eng. Sci. **26**: 461–479.
20. KURTEN, H. & P. ZEHNER. 1979. Slurry reactors. Ger. Chem. Eng. **2**: 220–227.
21. ALPER, E., B. WICHTENDAHL & W-D. DECKWER. 1980. Gas absorption mechanism in catalytic slurry reactors. Chem. Eng. Sci. **35**: 217–222.
22. ELSTNER, F. & U. ONKEN. 1981. Effect of liquid phase properties on mass transfer in gas/liquid dispersions. Ger. Chem. Eng. **4**: 84–89.
23. ALBAL, R. S., Y. T. SHAH, A. SCHUMPE & N. L. CARR. 1983. Mass transfer in multiphase agitated contactors. Chem. Eng. J. **27**: 61–80.
24. GODBOLE, S. P., A. SCHUMPE & Y. T. SHAH. 1983. Hydrodynamics and mass transfer in bubble columns: effect of solids. Chem. Eng. Commun. **24**: 235–258.
25. QUICKER, G., A. SCHUMPE & W-D. DECKWER. 1984. Gas-liquid interfacial areas in a bubble column with suspended solids. Chem. Eng. Sci. **39**(1): 179–183.
26. ANDREWS, G. F., J. P. FONTA, E. MARROTTA & P. STROEVE. 1984. The effects of cells on oxygen transfer coefficients. I: Cell accumulation around bubbles. Chem. Eng. J. **29**: B39–B46.
27. ANDREWS, G. F., J. P. FONTA, E. MARROTTA & P. STROEVE. 1984. The effects of cells on oxygen transfer coefficients. II: Analysis of enhancement mechanisms. Chem. Eng. J. **29**: B47–B55.
28. ALPER, E. & S. OZTURK. 1986. Effect of fine solid particles on gas-liquid mass transfer rate in a slurry reactor. Chem. Eng. Commun. **46**: 147–158.
29. SCHUMPE, A., A. K. SAXENA & L. K. FANG. 1987. Gas/liquid mass transfer in a slurry bubble column. Chem. Eng. Sci. **42**(7): 1787–1796.
30. OGUZ, H., A. BREHM & W-D. DECKWER. 1987. Gas/liquid mass transfer in sparged agitated slurries. Chem. Eng. Sci. **42**(7): 1815–1822.
31. CHANDRASEKARAN, K. & M. M. SHARMA. 1977. Absorption of oxygen in aqueous solutions of sodium sulfide in the presence of activated carbon as catalyst. Chem. Eng. Sci. **32**: 669–671.
32. 3M. 1985. Product manual on Fluorinert electronic liquids, p. 9 and p. 66.
33. 3M. 1987. Fluorinert electronic liquids brochure no. 98-0211-2588-9(27.5)NPI.
34. JUNKER, B. H. 1988. Monitoring and assessment of aqueous/perfluorocarbon fermentation systems. Ph.D. thesis, p. 33 and p. 324. Department of Chemical Engineering, Massachusetts Institute of Technology, Cambridge, Massachusetts.
35. MERCK INDEX (tenth edition). 1983. Merck. Rahway, New Jersey.
36. REID, R. C., J. M. PRAUSNITZ & T. K. SHERWOOD. 1977. The Properties of Gases and Liquids (third edition). McGraw–Hill. New York.
37. CRC HANDBOOK OF CHEMISTRY AND PHYSICS (60th edition). 1979. CRC Press. Boca Raton, Florida.
38. CHARPENTIER, J. C. 1982. What's new in absorption with chemical reaction? Trans. Inst. Chem. Eng. **60**: 131–156.

APPENDIX

Nomenclature

a = specific gas-liquid interfacial area [m^{-1}]

c^* = hypothetical concentration of solute in liquid phase in equilibrium with bulk gas phase [mMol/L]

c_i = concentration of solute in liquid phase i in equilibrium with bulk gas phase of solute partial pressure p_g [mMol/L]

c_1 = concentration of solute in bulk liquid [mMol/L]

d_i = diameter of the i-th gas bubble [m]

d_{sm} = Sauter mean (or volume to surface average) diameter, as defined by equation 6
D_i = diffusivity of solute in fluid i [m²/h]
DCW = dry cell weight [g/L]
D.O. = dissolved oxygen [% saturation]
H_{eff} = effective Henry's constant for perfluorochemical-water liquid mixture, as defined by equation 8 or equation 9 [mMol/L-atm]
H_i = Henry's law solubility constant for solute in phase i, as defined by equation 3 [mMol/L-atm]
H_o = Henry's law solubility constant for perfluorochemical phase, as defined by equation 3 [mMol/L-atm]
H_w = Henry's law solubility constant for aqueous phase, as defined by equation 3 [mMol/L-atm]
K_l = overall mass transfer coefficient for gas-liquid mass transfer based on overall concentration driving force [h⁻¹]
K_l^0 = overall mass transfer coefficient without perfluorochemical present [h⁻¹]
OTR = volumetric oxygen transfer rate [mMol/L-h]
OTR_{max} = maximum volumetric oxygen transfer rate [mMol/L-h]
OTR_{max}^0 = maximum volumetric oxygen transfer rate without perfluorochemical present [mMol/L-h]
OUR = volumetric oxygen uptake rate [mMol/L-h]
P_i = permeability of solute in phase i, as defined by equation 5 [mMol/cm²-s^{1/2}]
p_g = partial pressure of solute in bulk gas phase [atm]
p_g^0 = partial pressure of solute in bulk gas phase at OTR_{max}^0 [atm]
p_l = partial pressure of solute in bulk liquid phase [atm]
P_{vp} = normal vapor pressure [torr]
rpm = revolutions per minute [min⁻¹]
T_{bp} = normal boiling point [°C]
$S_{o/w}$ = spreading coefficient for oil spreading on water [dynes/cm]
vvm = volumes of gas per total volume of liquid per minute [min⁻¹]
μ = viscosity [cp]
ρ = density [g/cm³]
σ_a = surface tension against air [dynes/cm]
σ_{ow} = oil-water interfacial tension [dynes/cm]
σ_{wa} = surface tension of water against air [dynes/cm]
σ_{oa} = surface tension of oil against air [dynes/cm]
ϕ_g = gas holdup [m³ gas/m³ total liquid]
ϕ_o = oil volume fraction [m³ oil/m³ total liquid]

The Hyperthermophilic Archaebacterium, *Pyrococcus furiosus*

Development of Culturing Protocols, Perspectives on Scaleup, and Potential Applications[a]

I. I. BLUMENTALS, S. H. BROWN, R. N. SCHICHO,
A. K. SKAJA, H. R. COSTANTINO, AND R. M. KELLY

Department of Chemical Engineering
Johns Hopkins University
Baltimore, Maryland 21218

INTRODUCTION

The development of cultivation protocols for hyperthermophilic archaebacteria (i.e., bacteria capable of growth around temperatures of 100 °C) will likely be based on standard engineering and microbiological approaches as well as on new methodologies that specifically apply to these novel microorganisms. Both biological and engineering perspectives on the growth and metabolic characteristics of the expanding collection of bacteria growing at extreme temperatures have been addressed recently.[1,2] The potential application of hyperthermophilic archaebacteria and their associated biomolecules is still predicated on a clearer understanding of their metabolism and on the establishment of effective techniques for the generation of biomass at sufficient levels for protein purification.

To date, there have been relatively few bacteria that, in pure culture, have optimal growth temperatures at or above 100 °C. TABLE 1 contains the present list of these so-called hyperthermophiles; this collection will likely expand as global geothermal environments are probed further. Although these microorganisms have somewhat diverse metabolic characteristics, a common thread is their ability, at least facultatively, to metabolize elemental sulfur. Sulfur reduction at elevated temperatures may be an important bioenergetic characteristic of hyperthermophiles.

An interesting aspect of those hyperthermophiles isolated thus far is the relatively low biomass yields that have been obtained. Whereas an understanding of biological function and biocatalysis at high temperatures may lead to important insights with biotechnological application, the most clear near-term developments will likely involve enzymes isolated from hyperthermophiles. This assumes, of course, that sufficient amounts of biomass can be generated from which these enzymes can be recovered. Fortunately, in recent years, there have been hyperthermophilic bacteria isolated that have been shown to grow to still low, but significant, cell densities. The genus, *Pyrococcus*, in particular, currently contains two species, *Pyrococcus furiosus* and *Pyrococcus woesei*, both of which can be grown to cell densities in excess of 10^8

---

[a] This work was supported in part by the National Science Foundation (Grant No. CBT-8813608).

cells/mL. Further work with growth optimization will likely improve these yields to the point where the isolation of their associated enzymes for further study and application appears feasible.

The development of cultivation protocols for hyperthermophilic microorganisms such as *P. furiosus* and *P. woesei* presents some interesting problems that are not

TABLE 1. The Sulfur-metabolizing Hyperthermophiles

Organism (DSM no.)	Metabolism	Habitat	Growth Conditions	Miscellaneous
Pyrococcus furiosus (3638)	obligate heterotroph; S° respiration; fermentation?	marine solfataric mud (Italy)	70–103 °C (opt. 100 °C) pH 5.0–9.0 (opt. 7.0)	cocci 0.8–2.5 μm S° stimulates growth by 10× G+C 38%
Pyrococcus woesei (3773)	heterotroph; S° respiring	marine solfataras (Italy)	97–105 °C (opt. 100–103 °C) (pH opt. 6.0)	spherical 0.5–2 μm G+C 37.5%
Pyrodictium brockii (2708)	obligate autotroph; H_2-S° autotrophy	same as *Pyrococcus furiosus*	82–110 °C (opt. 105 °C) pH 5.0–7.0 (opt. 5.5)	disks 0.3–2.5 μm growth and sulfide production stimulated by YE G+C 62%
Pyrodictium occultum (2709)	same as *Pyrodictium brockii*	same as *Pyrodictium brockii*	same as *Pyrodictium brockii*	growth not stimulated by YE G+C 62%
Thermoproteus sp. strain Geo 3 (no DSM number)		terrestrial, solfataras *Thermoproteus* spp.	75–102 °C (opt. 100 °C)	motile
Pyrobaculum islandicum (4148)	facultative autotrophic H_2-S autotrophy	solfataric and geothermal waters	74–102 °C (opt. 100 °C)	rods (2.5 μm) strict anaerobe G+C 46%
Pyrobaculum organotrophum (4185)	obligately heterotrophic	same as *Pyrobaculum islandicum*	78–102 °C (opt. 102 °C)	3–5 μm rods

encountered when working with more conventional organisms growing at mesophilic temperatures. Probably the most significant problem is the relatively little information generated to date on the growth and metabolism of hyperthermophilies. Along these lines, development of an effective defined growth medium, establishment of an optimal growth environment, and determination of key nutritional characteristics—basic microbiological questions—have yet to be accomplished for any of the hyperthermophiles. In addition, production of large amounts of biomass presents the unfavorable prospect of very poor volumetric efficiency of fermentors with the additional problem of dealing with the hazards and corrosivity associated with high levels of biologically generated hydrogen sulfide. Thus, difficulties with cultivation of hyperthermophiles represent the key technological roadblock at present.

Despite culturing problems, work has proceeded to investigate the proteins from hyperthermophiles. Here, other problems are encountered. Whereas it appears that conventional enzyme purification methodology (i.e., precipitation, sizing, ion exchange) will be useful, high levels of purification may be difficult. For example, affinity techniques that are based on interactions with the active site will have to be carried out at elevated temperatures in the range of the enzyme's activity. Another problem in studying enzymes from these bacteria arises in assaying their stability and activity. Cofactors and coenzymes, such as NADH, have limited stability at temperatures above 100 °C. Also, if comparison to mesophilic counterparts is desired, the instability of substrates such as sugars above 100 °C presents problems in interpreting kinetic data. Buffer solutions will have to be chosen based on the stability of their pK_a with respect to temperature. The experimental systems used for enzyme activity assays at elevated temperatures will require some novel approaches.

The discussion presented here intends to illustrate some of the difficulties mentioned above, using the hyperthermophile *P. furiosus* as an example system.

The Genus, Pyrococcus

The genus, *Pyrococcus*, currently consists of two species: *Pyrococcus furiosus* and *Pyrococcus woesei*. Both of the organisms are obligate anaerobic heterotrophs and were isolated by Stetter and colleagues[3,4] from shallow geothermal marine sediments near Vulcano, Italy. Whereas both species are similar in some aspects of their growth and metabolism, several differences do exist (see TABLE 1).

P. furiosus grows in both the presence and the absence of elemental sulfur. In the presence of elemental sulfur, carbon dioxide and hydrogen sulfide are produced as growth-associated products. If sulfur is left out of the medium, hydrogen and carbon dioxide are produced with the hydrogen becoming growth-inhibiting at higher levels. Growth yields are approximately tenfold lower in the absence of sulfur, but they can be improved to above 10^8 cells/mL (as is the case with sulfur) if hydrogen is stripped from the medium as it is produced. Malik *et al.*[5] showed that growth under one atm of hydrogen (in excess of inhibiting levels in the absence of sulfur) is no different than growth under one atm of helium if elemental sulfur is added to the medium. It is not yet clear whether elemental sulfur represses the hydrogen evolution system in *P. furiosus* or if inhibition is averted by converting hydrogen to a noninhibiting product. Along these lines, Kelly and Deming[2] showed that several new bands appear on SDS-PAGE gels run on membrane-containing extracts from *P. furiosus* when grown on elemental sulfur.

The growth characteristics of *P. woesei* are even less well understood than those of *P. furiosus* and no metabolic products (gaseous or in solution) have yet been identified. This bacterium can grow on complex carbon sources such as yeast extract or bactotryptone under CO_2 and CO_2-H_2 atmospheres in the absence of elemental sulfur. However, sulfur presence is necessary for good growth in these complex media formulations when only H_2 is present in the gas phase. Also, *P. woesei* is able to grow on starch provided that H_2 and sulfur are present.[4]

MATERIALS AND METHODS

Batch Culturing Conditions

Pyrococcus furiosus (DSM 3638) was obtained from the Deutsche Sammlung von Mikroorganismen, Federal Republic of Germany. The organism was grown routinely in sealed 125-mL serum bottles placed in a 98 °C bath (New Brunswick Scientific, Edison, New Jersey). Whereas the organism can grow at temperatures at or above 100 °C, slightly lower temperatures were used in this experiment to avoid having to pressurize glass culturing vials to prevent boiling. Silicone fluid (Dow-Corning 200 fluid, Midland, Michigan) was used in the bath instead of water to avoid losses from evaporation at these high temperatures. Culturing was performed under quiescent conditions.

Growth of the organism in Chesapeake Bay sea water (CBW)-based media was done as previously described.[5] Bacteria were also grown in artificial sea water (ASW) supplemented with 0.1% yeast extract and 0.5% tryptone. The artificial sea water used is a modification of the formulation of Kester *et al*.[6] In short: equal volumes of solution A and solution B were mixed by slowly adding B to A while stirring (solution A: 47.8 g/L NaCl, 8.0 g/L Na_2SO_4, 1.4 g/L KCl, 0.4 g/L $NaHCO_3$, 0.2 g/L KBr, and 0.06 g/L H_3BO_3; solution B: 21.6 g/L $MgCl_2 \cdot 6H_2O$, 3.0 g/L $CaCl_2 \cdot 2H_2O$, and 0.05 g/L $SrCl_2 \cdot 6H_2O$). The resulting solution was supplemented with 0.25 g/L NH_4Cl, 0.14 g/L K_2HPO_4, and 1.0 g/L sodium acetate.

The defined medium consisted of ASW supplemented with amino acids (80 mg/L Glu, Gly; 60 mg/L Arg; 50 mg/mL Cys, Pro; 40 mg/L Thr, His, Ile, Lys, Leu, Asn; 30 mg/L Met, Phe, Ser, Ala, Trp; and 20 mg/L Val, Gln, Asp) and trace elements. The trace element solution was prepared according to the description for the *Methanogenium* medium, DSM no. 141 (DSM catalogue of strains, 1983). All media formulations included 30 g/L elemental sulfur (unless otherwise specified) and 1.0 mg/L resazurin (redox indicator). Anaerobic conditions were achieved by flushing the media with helium and by addition of 0.5 g/L Na_2S.

Continuous Culturing Conditions

The medium used for continuous culture experiments consisted of 0.5% tryptone and 0.1% yeast extract in CBW. Resazurin, elemental sulfur, and Na_2S were added as indicated above. The culture vessel was a five-neck round-bottom flask with a total volume of 2 L. Continuous culture experiments were generally run with a liquid working volume of 750 mL. Anaerobic conditions were maintained by continuous sparging with prepurified nitrogen at a rate of 50 mL/min. The sparging also served to mix the contents of the vessel (no additional agitation was supplied).

The temperature in the culture vessel was maintained at 98 °C using a heating mantle and a proportional temperature controller (Cole-Parmer, Chicago, Illinois).

Medium was maintained under sterile, anaerobic conditions and was added to the reactor using a peristaltic pump (Cole-Parmer). A constant reactor volume was maintained with a dip tube. Teflon PFA tubing was used between the feed reservoir and the reactor, except for a short section of silicone tubing in the pump head itself.

Media addition to the reactor was initiated during the late log phase of a batch culture. Feed rate changes were made in the direction of increasing dilution rate, and a minimum of three reactor volume changes were allowed after each adjustment for the system to reach steady state. An additional 2.0 grams of sulfur was added to the reactor after every other dilution rate increase to ensure that an excess of sulfur was always present. In general, the elemental sulfur was contained in the reactor working volume and little, if any, was carried out in the reactor effluent.

Microscopy

Bacterial growth was followed by cell enumeration using epifluorescence microscopy (EFM) on a Zeiss microscope with acridine orange stain.[7] Samples for scanning electron microscopy (SEM) were prepared by fixing the cells with glutaraldehyde (2% final concentration) and filtering an aliquot onto a 0.22-μm filter. The cells on the filter were then dehydrated by placing the sample in successively higher concentrations of ethanol. Following the dehydration, the samples were critical point–dried using liquid CO_2 and then gold-coated and mounted onto a sample stub for analysis.

Gas Analysis

Hydrogen sulfide and carbon dioxide production was measured using a Varian 3700 gas chromatograph (Varian Associates, Sunnyvale, Califorina) with a 6 foot × 1/8 inch HayeSep-N column (Alltech) and a thermal conductivity detector. A standard 286/10 microcomputer (CompuAdd, Houston, Texas) with a DAS-16 A/D interface (MetraByte, Taunton, Massachusetts) was used for data acquisition and peak integration.

RESULTS AND DISCUSSION

Growth Limitation Studies

In most microbiological systems, growth is eventually limited by either the availability of a required nutrient or the production of an inhibitory (toxic) product. A determination of how either or both of these factors relate to the growth of hyperthermophilic and extremely thermophilic archaebacteria has been somewhat elusive. For example, Jannasch et al.[8] showed that two extremely thermophilic isolates from deep sea hydrothermal vents, apparently belonging to the genus, *Desulfurococcus*, grew to densities of the order of 10^8 cells/mL while consuming only about 2% of the organic carbon available in yeast extract. Experiments with individual amino acids and peptides did not reveal a limiting substrate and the authors concluded that corespiration and coutilization of various substrates might explain the observed phenomena.

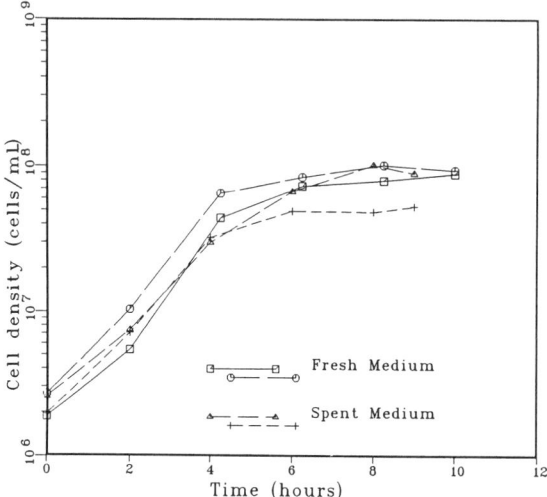

FIGURE 1. Growth of *Pyrococcus furiosus* on spent and fresh media. The spent medium was obtained by centrifuging and then filtering out cells that had reached stationary phase. Fresh sulfur was added to the spent medium prior to reinoculation. Results show duplicate growth curves for fresh and spent media experiments.

Even less is known, thus far, about the hyperthermophiles. Parameswaran *et al.*[9] determined that the chemolithotroph, *Pyrodictium brockii*, can be limited by the availability of the gaseous substrates, carbon dioxide or hydrogen, although no explanation for the cessation of growth at cell densities of the order of 10^7 was apparent. Because *P. furiosus* grows rapidly to cell densities of over 10^8 cells/mL in medium that is not necessarily optimal, further work with this species is appropriate. In this section, we describe some of our efforts in progress to identify the factor(s) involved in growth cessation in *P. furiosus* cultures.

Stoichiometric and Rate-limiting Substrates for P. furiosus

No apparent limiting substrate has yet been clearly identified for *P. furiosus*, but there has been a significant amount of effort towards this aspect. For example, FIGURE 1 shows that growth in spent medium was as good or better than growth in fresh medium. The spent medium was prepared by growing cells in ASW-based medium up to the stationary phase and then filtering out the cells using a 0.22-μm filter. This result suggests that no apparent toxic product is excreted into the medium during growth nor is there depletion of an essential nutrient.

Other experiments support the idea that growth does not stop because of nutrient availability problems. Addition of trace elements, vitamins, or extra carbon source to the complex ASW-based medium formulation did not improve biomass yields nor cell growth rates. Also, supplementing the cultures with fresh sulfur did not improve growth; it had been thought that agglomeration of sulfur after prolonged incubation at elevated temperatures might reduce its available surface area. Parallel studies on pH changes in the medium during growth and experiments in which different buffers were added to the media formulations indicate that changes in pH during growth are not responsible for growth limitations. Addition of chelating agents such as EDTA and citrate did not affect growth in any way.

All these results seem to point out that nutrient limitations are not present; however, to clearly demonstrate that this is the case, more data are needed. Efforts are being made to determine if these hyperthermophilic bacteria require the presence of less common trace elements (Se, Cr, Pb, and W) or other specific organic compounds (purines, pyrimidines, carbohydrates, vitamins, and organic acids among others) (or both). Also, the presence of a lysogenic phage is currently under investigation. Viruslike particles have been described in *Pyrococcus woesei* and it is thought that they might be responsible for sudden and rapid lysis of the cell cultures.[4] Because both *Pyrococcus* species have been isolated from the same location and they belong to the same genus, it is possible that a phage may also be present in *P. furiosus* and that its induction may lead to cell lysis and therefore to apparent low cell yields.

Defined Medium Experiments

In order to better understand cell metabolism and growth limitations, efforts to develop a defined medium were begun (see MATERIALS AND METHODS for composition). FIGURE 2 shows that *P. furiosus* is able to grow in an ASW-based medium, supplemented with amino acids and trace elements, to a cell density of 1.5×10^7 cells/mL. The cells transfer well in this medium provided that it is supplemented with 100 ppm of tryptone. The biomass yield in the defined medium is approximately ten times lower than that obtained in complex ASW-based or CBW-based media. Also,

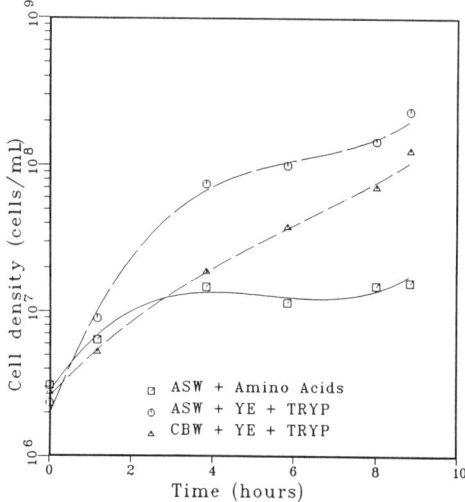

FIGURE 2. Growth of *Pyrococcus furiosus* in different media—ASW: artificial sea water; CBW: Chesapeake Bay sea water; amino acids: medium supplemented with amino acid mixture (see text for composition); and YE + TRYP: medium supplemented with 0.1% yeast extract and 0.5% tryptone.

the growth rates in the defined medium are slower in comparison to the rates observed using the complex formulations.

The above results indicate that there is still room for improvement in media formulation. So far, addition of several vitamins, chelating agents, and various buffers has not positively affected growth rates or cell yields. Similarly, supplementing the medium with glucose, sucrose, lactose, or starch had no effect on growth. This last observation is in agreement with previous reports that show no indication of carbohydrate utilization by *P. furiosus*[3] or most other hyperthermophilic species.[4,10,11]

In spite of the lower cell yields obtained with the defined medium, the use of a reproducible and determined nutrient source has been helpful in identifying key nutrients for *P. furiosus*. Among the amino acids, proline and cysteine were found to be essential for *P. furiosus* growth. The absence of either one of these amino acids in the medium results in growth rates and final cell densities that are half of those obtained when both amino acids are present. In contrast, tyrosine seems not to be relevant for the cells. These results together with ongoing efforts to identify other essential amino acids through radiolabeling experiments will permit the development of a simpler and better defined medium and will provide an insight of the cellular metabolism. Another important consideration for an effective defined medium is the identification of the critical element(s) supplied by the 100 ppm of tryptone. At this low level, it is thought that a component (or components) present at trace levels may be a key factor in improving growth yields; identifying such a component (or components) will likely be difficult. Prior work in elucidating the essential components of yeast extract for *Thermoplasma acidophilum* by Smith et al.[12] illustrates the possible complexities in ascertaining critical nutritional requirements in complex media.

Nutritional work along the lines described above represents an important, if tedious, element of efforts to better characterize and improve growth of *P. furiosus*. There have been some interesting observations that have emerged during these efforts that may be useful in further efforts with this organism and other hyperthermophiles. For example, it is interesting to note that medium composition has a profound effect on cell size. Cells are significantly smaller when grown in the defined medium compared to cells grown in the presence of tryptone and yeast extract (see FIGURE 3). Moreover, a connective material, possibly an exocellular polysaccharide, has been observed during growth of *P. furiosus* (as also seen in FIGURE 3). Attempts to isolate and characterize this material are under way.

Cell-Sulfur Interactions

It has been well established that the hyperthermophilic bacteria have either a strong dependency on or a definite requirement for sulfur to grow. However, the nature of the cell-sulfur interaction is not well understood. As mentioned earlier, Kelly and Deming[2] showed that a distinct set of proteins are induced when *P. furiosus* is grown in the presence of sulfur and that some of these proteins might be membrane-associated. So far, there is no indication that *Pyrococcus* attaches directly to the elemental sulfur in the culturing vials. This result suggests that the cells might be secreting an extracellular product involved directly in sulfur metabolism or responsible for sulfur solubilization.

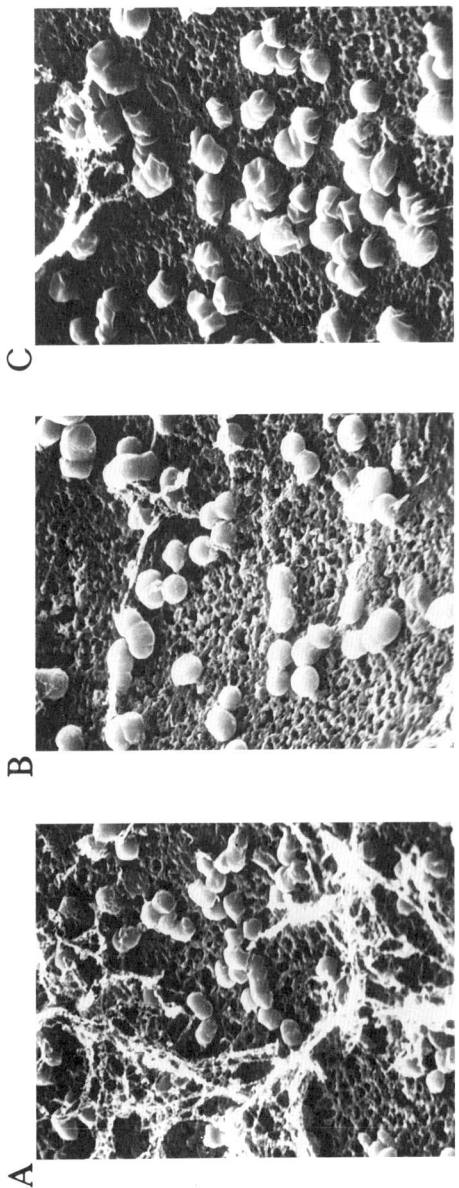

FIGURE 3. Scanning electron microscopy of *Pyrococcus furiosus*. Cells were grown in (A) ASW supplemented with yeast extract and tryptone, (B) ASW supplemented with amino acids, and (C) CBW supplemented with yeast extract and tryptone (magnification: ×9200).

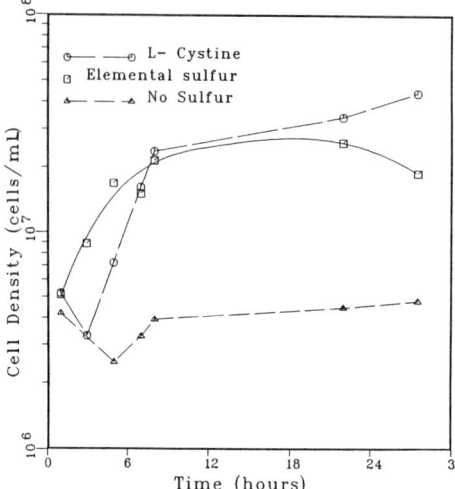

FIGURE 4. Growth of *Pyrococcus furiosus* in defined medium on different sulfur compounds. Results show that growth kinetics and extent of growth are very similar for cells grown on L-cystine and elemental sulfur, thus indicating that *P. furiosus* can effectively utilize sulfur bound in a -C-S-S-C- bond.

In order to better understand these cell-sulfur interactions, we have used model sulfur compounds to determine if any of them can be utilized as substitutes for elemental sulfur. The results obtained show that *P. furiosus* cannot reduce sulfates or sulfites, but can readily use L-cystine, which is the cysteine dimer. FIGURE 4 shows that growth rates and final cell densities are very similar for cells grown on elemental sulfur and cystine. The biotic production of H_2S (FIGURE 5) is also comparable in both cases,

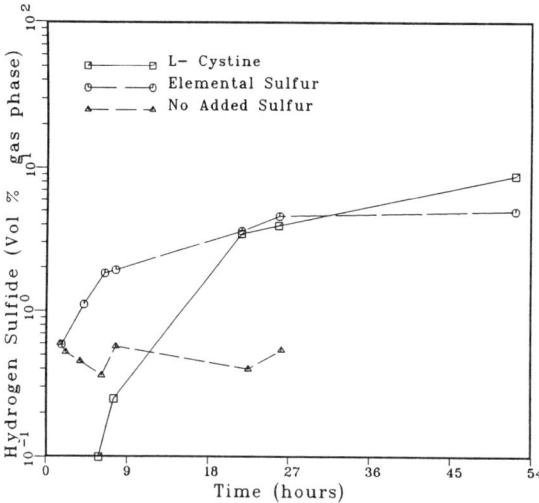

FIGURE 5. Biotic hydrogen sulfide production on different sulfur compounds. Gas production corresponds to the growth curves shown in FIGURE 4.

but the CO_2 production (FIGURE 6) is higher on cystine than it is on sulfur. At this point, it is not clear if the higher carbon dioxide production corresponds directly to the metabolization of the cysteine carbon backbone or if it is associated with a higher energy expenditure when the cells are grown on cystine. Radiolabeling experiments using ^{14}C and ^{35}S isotopes will most likely provide the answer to this problem. It is interesting to note that the above experiment indicates that *P. furiosus* can effect the breakage of carbon-sulfur bonds to further reduce the sulfur and produce H_2S. This not only gives an insight in the mechanism involved in sulfur utilization (maybe a carbon-sulfur intermediate is produced), but it also opens up the possibilities for other model compounds to serve as potential electron acceptors for *P. furiosus*.

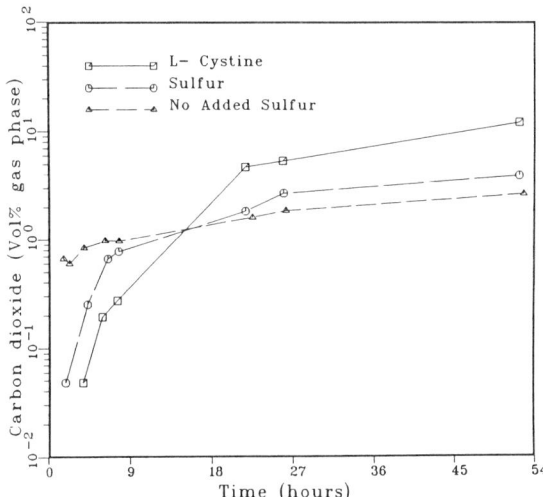

FIGURE 6. Biotic carbon dioxide production on different sulfur compounds. Gas production corresponds to the growth curves shown in FIGURE 4.

Generation of Biomass/Continuous Cultivation of P. furiosus

Two continuous culture experiments were performed with *P. furiosus* under the conditions discussed earlier. In the first, an initial dilution rate of 0.1 h^{-1} was selected and this was increased in steps to 0.3 h^{-1}. The second experiment covered a dilution rate range of 0.5 to 1.04 h^{-1}. It should be noted that a dilution rate of 1.04 h^{-1} corresponds to a doubling time of 40 minutes, which is the maximum that has been reported for *P. furiosus* in batch culture. The cell densities and specific gas production rates from these experiments are shown in FIGURE 7. Gas production data are based on gas phase measurements. The discrepancies in the cell density and gas production curves from the two experiments are most likely due to variations in media composition resulting from the use of two different CBW batches.

Two aspects of these results are particularly interesting. First, the specific CO_2 and H_2S production rates are fairly constant over a wide range of dilution rates. Additionally, the ratio of the two gas production rates is also fairly constant over this range. Although the metabolic implications of these observations are not yet clear, data over a wider range of dilution rates will likely reveal if gas production profiles are indicative of changes in this organism's energy metabolism.[13]

Perhaps the more interesting result, at least from a scaleup perspective, is that cell densities approaching batch maxima can be achieved at high dilution rates. Considering that these maximal cell densities are low in comparison with those achieved with many mesophiles, it is apparent that the volumetric efficiencies of large batch fermentors will be very low. Consequently, a prudent strategy for generating large amounts of *P. furiosus* biomass at reasonable volumetric efficiencies might involve operating a relatively small continuous reactor at high dilution rates as opposed to utilizing a large

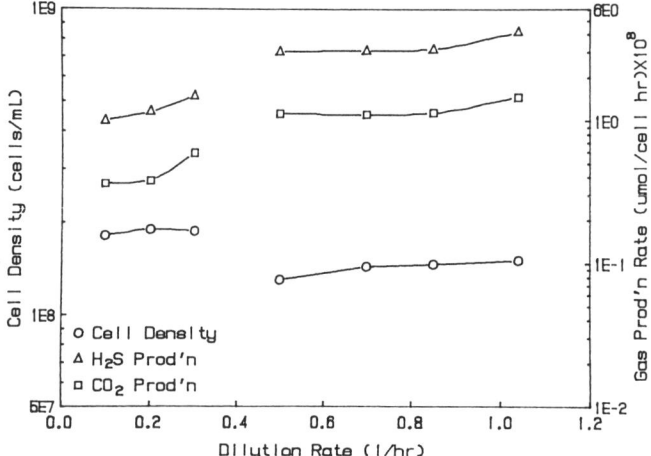

FIGURE 7. Continuous culture of *Pyrococcus furiosus* at 98 °C in a glass reactor (750-mL working volume). Sulfur was maintained in excess in the reactor. Gas production data are based on gas phase analysis only.

batch reactor. For example, a continuous reactor with a working volume of 5 L and a dilution rate of 1 h^{-1} could have a cell mass productivity equivalent to that of a batch reactor ten times its size. The obvious hazards resulting from handling hot liquid containing appreciable levels of hydrogen sulfide are greatly reduced with the smaller continuous system. Moreover, large systems typically are constructed of steel, which would be more susceptible to corrosion resulting from the combination of high temperatures, high salt concentrations, and dissolved hydrogen sulfide. Smaller systems might be constructed of glass, thus avoiding corrosivity problems. At these temperatures, sterility considerations, normally a significant problem for large-scale or long-term continuous culture, are also avoided, making this approach particularly appealing.

Potential Applications of Hyperthermophilic Archaebacteria

Opportunities related to hyperthermophilic archaebacteria in biotechnology may be either direct or indirect. Certainly, biocatalysis at elevated temperatures involving biotransformations known to be carried out by mesophiles and moderate thermophiles should be considered. There also is the potential for new types of biocatalysis involving unique pathways. Schicho et al.[14] have shown that portions of sulfur bound in coal can be reduced to sulfide by *P. furiosus* and *P. brockii*. The results from experiments involving *P. furiosus* utilization of the disulfide cystine suggest that this hyperthermophile may be able to metabolize portions of the organic sulfur in coal.

There is no doubt that the study and commercial utilization of proteins from hyperthermophiles is attractive. Work in this area is just beginning. The hydrogenase systems in *P. brockii*[15] and *P. furiosus*[16] have been characterized biochemically to some extent, although specific factors that stabilize these proteins have yet to be determined. High levels of proteolytic activity have been identified in extracts of *P. furiosus*.[17] Also, amylolytic activity has been found in extracts of *Pyrococcus woesei*.[18] The purification of the above enzymes is in progress. Given the capability to generate sufficient amounts of biomass, identification of other enzymes in hyperthermophiles should proceed rapidly.

Maybe the most profound contribution that will arise from the study of hyperthermophilic archaebacteria will be in protein engineering. This rapidly expanding field makes use of fundamental insights on protein structure and function to improve the activity/stability/purification aspects of a protein in a particular application. The forces that stabilize proteins from hyperthermophiles, if understood, could be conferred on less thermal (or otherwise) stable proteins to create a more effective catalyst. By doing so, prospects for chemical and biochemical synthesis using enzymes are considerably enhanced.

SUMMARY

From this brief discussion, it is clear that there are many obstacles to overcome before hyperthermophilic archaebacteria will be an important aspect of biotechnology. Nevertheless, the prospects are intriguing. The nature of the environments that harbor these organisms and the consequent requirements for their controlled culture suggest that chemical and biochemical engineers can play an important role in elucidating their scientific and technological aspects.

ACKNOWLEDGMENTS

R. M. Kelly expresses his appreciation to M. W. W. Adams (Department of Biochemistry, University of Georgia) and to R. J. Maier and C. B. Anfinsen (Department of Biology, Johns Hopkins University) for helpful discussions. Finally, the authors thank G. J. Olson (National Institute of Standards and Technology) for his helpful comments and analytical assistance.

REFERENCES

1. STETTER, K. O. 1986. *In* Thermophiles: General, Molecular, and Applied Microbiology. T. D. Brock, Ed. Wiley. New York.
2. KELLY, R. M. & J. W. DEMING. 1988. Biotechnol. Prog. **4:** 47–62.
3. FIALA, G. & K. O. STETTER. 1986. Arch. Microbiol. **145:** 56–60.
4. ZILLIG, W., I. HOLZ, H. P. KLENK, J. TRENT, S. WUNDERL, D. JANEKOVIC, E. IMSEL & B. HAAS. 1987. Syst. Appl. Microbiol. **9:** 62–70.
5. MALIK, B., W-W. SU, H. L. WALD, I. I. BLUMENTALS & R. M. KELLY. Biotechnol. Bioeng. **34:** 1050–1057.
6. KESTER, D. R., I. W. DUEDALL, D. N. CONNORE & R. M. PYTKOWICZ. 1967. Limnol. Oceanogr. **12:** 176–178.
7. HOBBIE, J. E., R. J. DALEY & S. JASPER. 1977. Appl. Environ. Microbiol. **33:** 1225–1228.
8. JANNASCH, H. W., C. O. WIRSEN, S. J. MOLYNEAUX & T. A. LANGWORTHY. 1988. Appl. Environ. Microbiol. **54:** 1203–1209.
9. PARAMESWARAN, A. K., W-W. SU, R. N. SCHICHO, C. N. PROVAN, B. MALIK & R. M. KELLY. 1988. Appl. Biochem. Biotechnol. **18:** 53–73.
10. STETTER, K. O., H. KONIG & E. STACKEBRANDT. 1983. Syst. Appl. Microbiol. **4:** 535–551.
11. NG, T. K. & W. F. KENEALY. 1986. *In* Thermophiles: General, Molecular, and Applied Microbiology. T. D. Brock, Ed. Wiley. New York.
12. SMITH, P. F., T. A. LANGWORTHY & M. R. SMITH. 1975. J. Bacteriol. **124:** 884–892.
13. BROWN, S. H., I. I. BLUMENTALS & R. M. KELLY. 1988. AIChE Annual Meeting, Washington, District of Columbia, abstract no. 149k.
14. SCHICHO, R. N., S. H. BROWN, G. J. OLSON, E. J. PARKS & R. M. KELLY. 1989. Fuel **68:** 1368–1375.
15. PIHL, T., R. N. SCHICHO, R. M. KELLY & R. J. MAIER. 1988. Proc. Natl. Acad. Sci. U.S.A. **86:** 138–141.
16. ADAMS, M. W. W. 1988. Personal communication. Department of Biochemistry, University of Georgia.
17. SKAJA, A. K., R. LESSICK, C. B. ANFINSEN & R. M. KELLY. 1988. AIChE Annual Meeting, Washington, District of Columbia, abstract no. 23o.
18. LADERMAN, K. & C. B. ANFINSEN. 1988. Personal communication. Department of Biology, Johns Hopkins University.

Diauxic Metabolism of *Hansenula polymorpha*

Steady- and Unsteady-State Considerations

JAMES D. BRYERS AND TIMOTHY YEH

Center for Biochemical Engineering
Duke University
Durham, North Carolina 27706

INTRODUCTION

Prior to the escalation in oil prices, interest in the physiology of *Hansenula polymorpha* was in regard to single cell protein production. Currently, *H. polymorpha* is the organism of commercial choice (Unilever, Merck) for production of a number of enzymes related to the oxidation of either methanol or glucose. Potentially, *H. polymorpha*, like a number of related yeasts, may prove an excellent host for the expression of heterologous genes. For example, a similar yeast, *Pichia pastoris*, as host organism shows high expression of a foreign gene of extracellular invertase.[1]

Most commercial microbial processes use either pure or mixed cultures growing on mixed, frequently complex, carbon and energy sources. In nature, microorganisms grow in the presence of a variety of substrates that serve either as carbon and energy sources or as nitrogen sources. However, fundamental studies on microbial growth and metabolism have been largely restricted to pure cultures supplied with a single source of carbon and energy. One exception to this is the diauxic growth behavior of microorganisms as studied extensively in batch cultures.

Many microbes employ two distinct strategies to cope with binary mixtures of carbon substrates. Under nutrient sufficient growth conditions, as in batch cultures, the dual carbon sources are utilized sequentially, with the substrate that supports the highest growth rate being metabolized first. Under substrate-limited growth conditions, as in chemostat cultures, both substrates are utilized simultaneously at low dilution rates; this is a phenomenon described as either "mixed substrate growth" or "secondary substrate metabolism".

Methylotrophic yeasts and bacteria (e.g., *H. polymorpha*, *Pichia pastoris*, *Candida boidinii*, and *Methylomonas methanica*) are capable of mixed substrate growth.[2-7] Despite the ability of *H. polymorpha* to grow diauxically on methanol and glucose, enzymes related to the oxidation of either substrate are still produced commercially in continuous steady-state culture using a single carbon and energy limiting substrate: methanol (C_1) for the alcohol-oxidizing enzymes and glucose (C_6) for the glucose-oxidizing enzymes. The work presented here indicates that the level of enzymes related to C_1 and C_6 oxidation of the mixed substrates can vary by a factor of two or three depending upon growth rate and mixture composition. The aim of our research group is to explore possibilities of enhanced enzyme production under transient conditions of growth rate and substrate mixture composition. More philosophically, our aim is to quantify the metabolic control mechanisms employed by microorganisms during prolonged perturbations. Toward these goals, this work presents (1) an unsteady-state structured model of *Hansenula polymorpha* growing diauxically on mixtures of glucose and methanol, (2) a calibration of the just-stated model to published steady-state data, and (3) techniques for studying unsteady-state metabolism and preliminary results.

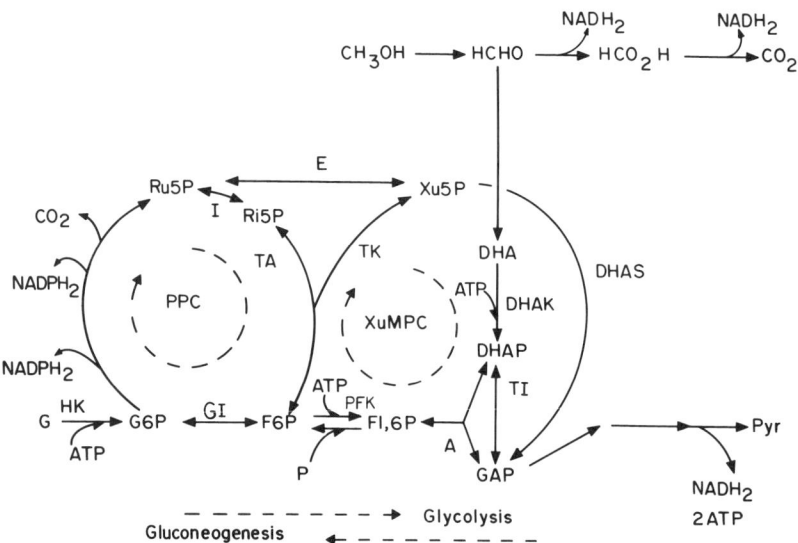

FIGURE 1. Interrelationship between metabolic pathways of glucose and methanol in methylotrophic yeasts: A = aldolase, DHAK = dihydroxyacetone kinase, DHA(P) = dihydroxyacetone (phosphate), DHAS = dihydroxyacetone synthase, E = ribulose-5-phosphate epimerase, F6P = fructose 6-bisphosphatase, G = glucose, GAP = glyceraldehyde 3-phosphate, GI = glucose-6-phosphate isomerase, G6P = glucose 6-phosphate, HK = hexokinase, I = ribose-5-phosphate isomerase, PFK = phosphofructokinase, PPC = pentose phosphate cycle, Pyr = pyruvate, Ri5P = ribose 5-phosphate, Ru5P = ribulose 5-phosphate, TA = transaldolase, TI = triosephosphate isomerase, TK = transkelolase, Xu5P = xylulose 5-phosphate, and XuMPC = xylulose monophosphate cycle. The flows of carbon through the pathways are indicated by arrows.

Diauxic Metabolism of Methylotrophic Yeasts

Hansenula polymorpha is a facultative methylotrophic yeast capable of metabolizing either methanol as the sole carbon and energy source or a variety of more complex organic compounds. The physiology of methylotrophic yeasts has been studied extensively[2-10] and it indicates that most methylotrophic yeasts share an identical pathway (FIGURE 1).

Methanol metabolism begins within specialized organelles termed peroxisomes where methanol is oxidized to formaldehyde via methanol oxidase with a concomitant production of hydrogen peroxide. Destructive effects of high hydrogen peroxide levels are negated by high activities of methanol catalase in the peroxisomes. Either (1) formaldehyde can leave the peroxisome to the cytosol where it is dissimilated to CO_2 by formaldehyde dehydrogenase and formate dehydrogenase or (2) formaldehyde and xylulose 5-phosphate can be reformed by dihydroxyacetone synthase into dihydroxyacetone and glyceraldehyde 3-phosphate; both of these products move into the cytosol to be assimilated by the xylulose monophosphate cycle (XuMPC).[10]

Glucose is metabolized by methylotrophic yeasts by channeling a portion of the entering glucose through glycolysis while the other portion first passes through the oxidative pentose phosphate cycle (PPC). In carbon-limited chemostat cultures, methylotrophic yeasts can assimilate both substrates simultaneously, thus posing a unique regulatory dilemma for such organisms. Whereas glucose is metabolized via glycolysis or the pentose phosphate cycle, thereby producing C_3 units for dissimilation and assimilation, carbon from methanol is assimilated by the XuMP cycle, which goes in the reverse direction.

Steady-State Chemostat Results

Regulation of enzyme synthesis for these interlocking pathways has been studied extensively in steady-state chemostat cultures by Egli and co-workers.[8-11]

Cultivation of *Hansenula polymorpha* in a steady-state chemostat culture with a mixture of glucose and methanol of fixed composition resulted in the observation by Egli *et al.*[12] that methanol continued to be utilized to completion at dilution rates that were higher than the μ_{max} normally found when methanol was supplied as the sole carbon and energy source (FIGURE 2).

When methanol is the sole source of carbon and energy, the maximum dilution rate observed is 0.18–0.19 h^{-1} with a biomass yield between 0.37–0.39 g dry biomass per g methanol utilized. With glucose as the sole growth-limiting substrate, the maximum dilution rate observed is 0.51 h^{-1} and the yield is a constant 0.55 g biomass per g glucose up to a dilution rate of 0.40 h^{-1}.

When methanol/glucose mixtures of a fixed composition are fed to the cells, the same pattern of utilization is observed irrespective of the C_1/C_6 ratio of the mixture. At low dilution rates, both substrates are utilized to completion (i.e., their residual concentrations in the chemostat are below the detection limit of 2 mg/L); in contrast, at high dilution rates, only glucose is metabolized by the cells and the influent methanol is not utilized and thus accumulates in the culture medium. The transition dilution rate at which the cells switch from mixed substrate growth to single substrate growth is dependent on the composition of the C_1/C_6 mixture in the feed. Results of Egli *et al.* (plotted in FIGURE 2) show that cells fed with mixtures of methanol and glucose are able to utilize methanol at dilution rates that are considerably higher than the maximum dilution rate observed for cells growing with methanol alone $[D(C_1) = 0.19$ $h^{-1}]$. The lower the ratio of methanol to glucose, the higher is the dilution rate up to which methanol is completely utilized in the presence of glucose. For all mixtures tested, the overall biomass yield is approximately additive, that is, it is the sum of $S_0(C_1)Y_{X/C_1}$ and $S_0(C_6)Y_{X/C_6}$, at dilution rates where mixed substrate growth occurs.

For all mixtures tested, the residual concentration of glucose in the culture is always below the detection limit of 2 mg/L at dilution rates below 0.46 h^{-1}. It appears from the residual glucose concentrations in FIGURE 2 that there is no influence of the

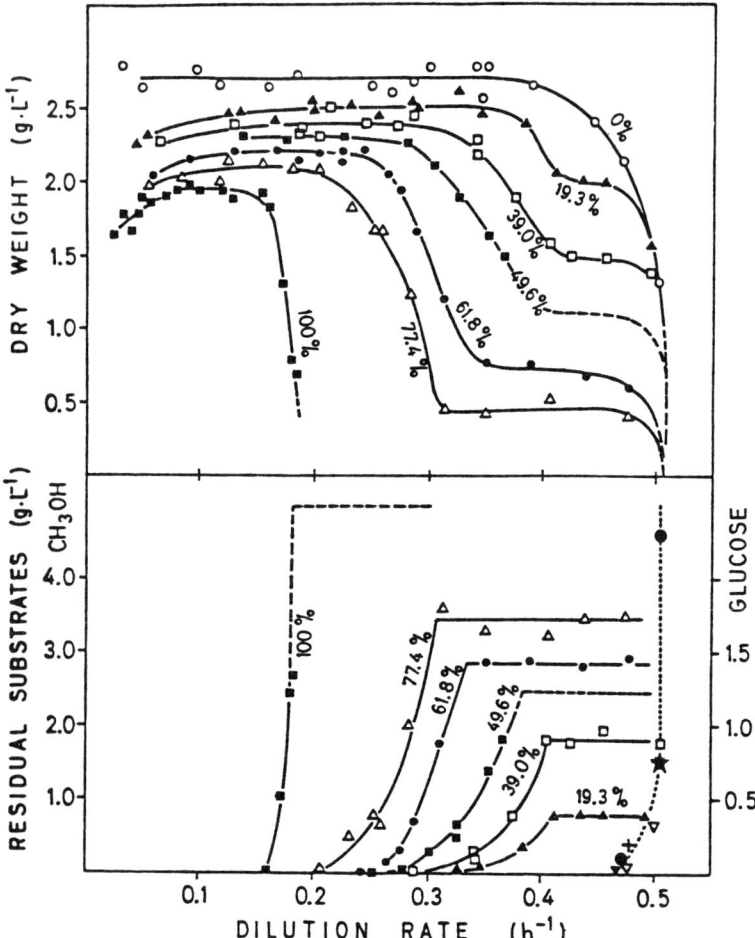

FIGURE 2. Cell dry weight formed and residual concentrations of carbon substrates during growth of *Hansenula polymorpha* CBS 4732 in a chemostat as a function of the composition of the methanol/glucose mixture supplied in the influent. The total concentration ($C_1 + C_6$) input was 5.0 g/L. Symbols for residue glucose are: (●) = 0% C_1, (▽) = 19.3% C_1, (★) = 39.0% C_1, (+) = 61.8% C_1, and (●) = 77.4% C_1.[12]

C_1/C_6 mixture composition on the maximum dilution rate with respect to glucose, which remains at 0.51 h^{-1} irrespective of the C_1/C_6 mixture used. However, a reduction in the biomass yield with respect to glucose is seen when the concentrations of unutilized methanol exceed 0.5 g/L. This effect is most pronounced in the experiment using a mixture consisting of 77.4% C_1 and 22.6% C_6, where $Y_{X/C}$ dropped from 0.55 to 0.35–0.45.

The levels of enzymes associated with the conversion of either methanol (C_1) or glucose (C_6) also depend upon the substrate mixture and the growth rate. At a fixed inlet substrate mixture of $C_1/C_6 = 61.2/38.8$ (w/w), FIGURE 3 illustrates the pertinent enzyme levels during the shift from simultaneous substrate conversion at low growth rates to exclusive C_6 conversion at high growth rates. At a fixed dilution rate of 0.15 h^{-1}, FIGURE 4 illustrates the dependence of the various enzymes on the relative proportions of C_1 and C_6 in the substrate.

MATHEMATICAL MODEL OF DIAUXIC METABOLISM

An unsteady-state "structured" mathematical model is derived here and is correlated to existing batch and steady-state continuous data for *H. polymorpha* grown on mixtures of methanol/glucose substrate.

Unstructured models describe microbial growth in terms of changes in total biomass and a single growth-limiting substrate. Such models provide good approximations to situations where microbial cell composition is essentially constant, that is, "balanced growth". Unstructured models are not capable of describing internal details of microbial growth, cell compositional changes under transient growth conditions, nor multiple reaction sequences.

Conversely, structured models can establish relationships between cell internal composition and growth, provide detail on the enzymatic regulation of cellular metabolism, and simulate multiple reaction sequences as functions of prevailing system conditions. Such models are obviously better suited to simulating enzymatic physiological control under transient growth conditions. For example, a change in the *in vitro* enzyme activity observed during a perturbation can result from an alteration in the specific enzymatic activity of the enzyme or from a change in the enzyme concentration itself (or both). The ability to take either effect into account is only possible with a structured model.

In general, a material balance over a control reaction volume in matrix notation can be written as

$$\left\{\begin{array}{c} \text{rate} \\ \text{of} \\ \text{accumulation} \end{array}\right\} = \left\{\begin{array}{c} \text{net rate} \\ \text{of} \\ \text{transport} \end{array}\right\} + \left\{\begin{array}{c} \text{rate} \\ \text{of} \\ \text{transformation} \end{array}\right\}$$

or

$$da/dt = a^\bullet + a^*, \tag{1}$$

where a is the component vector that enumerates the individual components of interest. The net rate of transport vector $a^\bullet$ would consist of a mass rate of input and output terms for each of the "m" components in the vector a. The exact form of the input-output terms depends on whether the component enters (leaves) the reaction volume as a dissolved component, as a gas, or within the cell biomass.

Any multiple reaction sequence can be represented in matrix form by summing the products of each component with its stoichiometric coefficient for each reaction, that is,

$$Y \cdot a = 0, \tag{2}$$

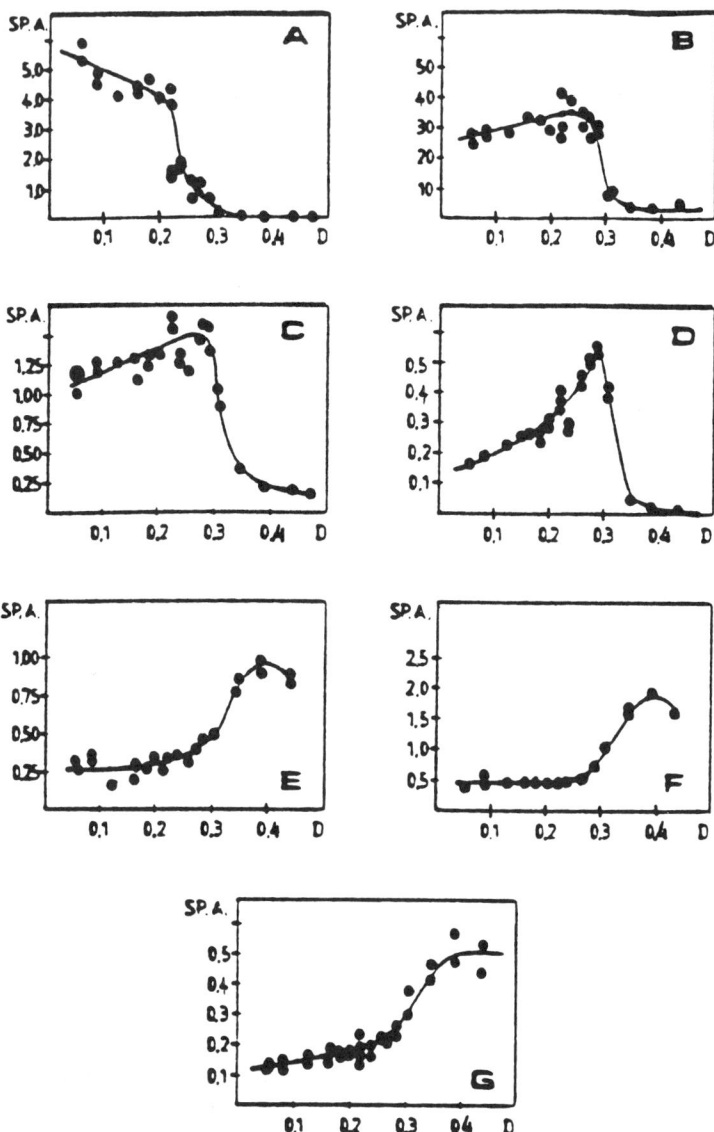

FIGURE 3. Specific activities (SP.A.) of alcohol oxidase (A), catalase (B), formaldehyde dehydrogenase (C), formate dehydrogenase (D), hexokinase (E), glucose-6-phosphate dehydrogenase (F), and 6-phosphogluconate dehydrogenase (G) in cell-free extracts of *H. polymorpha* grown in a chemostat culture with a mixture of glucose and methanol [(38.8% C_6)/(61.2% C_1) (w/w)] as the carbon source as a function of the dilution rate D.[8–10]

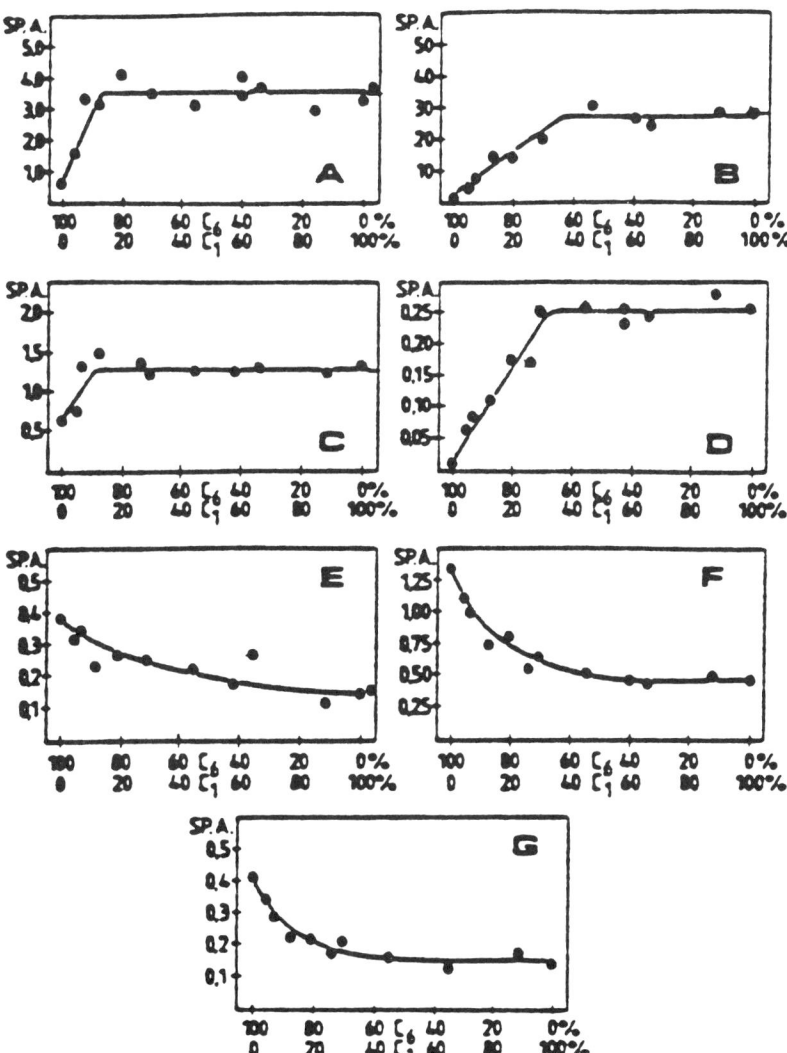

FIGURE 4. Specific activities of alcohol oxidase (A), catalase (B), formaldehyde dehydrogenase (C), formate dehydrogenase (D), hexokinase (E), glucose-6-phosphate dehydrogenase (F), and 6-phosphogluconate dehydrogenase (G) in cell-free extracts of *Hansenula polymorpha* grown in a carbon-limited chemostat culture at a fixed dilution rate of 0.15 h^{-1} as a function of the composition (given in % w/w) of the C_1/C_6 mixture supplied as the carbon and energy source in the inflowing medium.[8-10]

where

$$Y = \begin{bmatrix} \pm Y_{1,1} & & Y_{1,n} \\ \pm Y_{m,1} & \cdots & \cdot Y_{m,n} \end{bmatrix}$$

(Y = the stoichiometric coefficient matrix). For the proposed model, stoichiometric coefficients were derived on a unit mass basis as explained in Bryers and Irvine.[13] Each reaction in equation 2 has a rate, r_i, associated with it that can also be written in terms of a unit mass of the individual reaction's limiting reactant. Rates for the entire system can be summarized in a rate vector, r:

$$r = \begin{bmatrix} r_1 \\ r_2 \\ \cdot \\ \cdot \\ \cdot \\ r_n \end{bmatrix}. \quad (3)$$

Now, the transformation vector in equation 1, a^*, can be written as

$$a^* = Y^T r. \quad (4)$$

The metabolic pathway for methanol and glucose conversion by *H. polymorpha* is illustrated in FIGURE 1 and can be represented as at least 19 individual conversion steps, where each one can be further described assuming basic enzyme kinetics, that is,

$$S_{i,j} + E_i \rightleftharpoons E_i S_{i,j} \rightarrow E_i + S_{i,j+1}, \quad (5)$$

where $S_{i,j}$ = reactant j in reaction i binding with enzyme i to produce an enzyme-substrate intermediate that produces the free enzyme and the next reaction's intermediate, $S_{i,j+1}$. In the system depicted in FIGURE 1, $i = 21$ and $j = 18$. Consequently, the component vector (and thus the entire model) would become rather burdensome if all possible reaction details were included. A major objective of this project was to derive a practical simplification to the conversion sequence in FIGURE 1 while retaining sufficient detail to simulate both steady-state and transient diauxic growth dynamics. In our simplified version, we assume that oxidation and conversion of either C_1 or C_6 carbon is mediated separately by a single enzyme and that repression, induction, or activation of these two enzymes can be modeled depending on the influence of the intracellular metabolites on either enzyme activity or synthesis. A detailed discussion of this approach can be found in Papageorgakopoulou and Maier[14] and in Roels.[15]

The total rate of new biomass (X) synthesis can be written as

$$r_g = (Y_{C1} r_{C1}) + (Y_{C6} r_{C6}), \quad (6)$$

where Y_{C1} and Y_{C6} are the stoichiometric yield coefficients for methanol and glucose metabolism, respectively. The net rate of biomass production is equation 6 minus the rate of cell mass lost to maintenance or endogenous decay, that is,

$$r_{net} = r_g - k_e X. \quad (7)$$

The substrate consumption rates in equation 6, r_{C1} and r_{C6}, are modeled using the following Michaelis-Menten reaction expressions:

$$r_{C1} = \nu_{C1} E_{C1} X C_1 / (K_{C1} + C_1) \tag{8}$$

and

$$r_{C6} = \nu_{C6} E_{C6} X C_6 / (K_{C6} + C_6), \tag{9}$$

where ν_{C1} and ν_{C6} are the maximum utilization rate constants for methanol (C_1) and glucose (C_6), respectively, per unit mass of appropriate enzyme; E_{C1} and E_{C6} are the methanol- and glucose-metabolizing enzyme concentrations per unit biomass; and K_{C1} and K_{C6} are the half-saturation coefficients for each enzyme.

Material balances over a well-mixed continuous bioreactor for biomass (X) and for each substrate are summarized in FIGURE 5. Upon inspection, we note that the dynamic behavior of the enzymes, E_{C1} and E_{C6}, must also be taken into account in order to solve these material balances. Auxiliary equations for the net accumulation of enzymes within the cell are of the following form:

$$\left\{ \begin{array}{c} \text{net accumulation} \\ \text{of enzyme} \\ \text{mass fraction} \end{array} \right\} = \left\{ \begin{array}{c} \text{rate of} \\ \text{enzyme} \\ \text{synthesis} \end{array} \right\} - \left\{ \begin{array}{c} \text{rate of} \\ \text{enzyme} \\ \text{denaturation} \end{array} \right\} - \left\{ \begin{array}{c} \text{rate of enzyme} \\ \text{"dilution" within} \\ \text{cell due to growth} \end{array} \right\}.$$

The advantage of this generic interpretation of enzyme accumulation is that the induction and the repression of a particular enzyme can be modeled by the said effect on either enzyme synthesis or denaturation. For example, glucose may repress methanol-oxidizing enzymes by either decreasing the synthesis rate or increasing denaturation. Additionally, genetic control over enzyme synthesis can be easily linked to specific messenger RNA levels (mR); such an approach could be easily expanded into a complex metabolic model, as per Peretti and Bailey,[16] provided sufficient genetic information exists for the microbe. The rates of E_{C1} synthesis and denaturation are, respectively,

$$rs_{E1} = k_1 (\text{mR}) \tag{10}$$

and

$$rd_{E1} = k_{d1} E_p. \tag{11}$$

Veenhuis et al. have insinuated that glucose indirectly "represses" methanol-metabolizing enzymes by inducing the production of proteolytic enzymes, E_p, which denature the E_{C1} (equation 11).

The rates of E_{C6} synthesis and denaturation are, respectively,

$$rs_{E6} = k_6 r_{C6} / X \tag{12}$$

and

$$rd_{E6} = k_{d6} E_{C6}. \tag{13}$$

Two additional equations are required to account for the dynamic concentrations of messenger RNA (mR) coding for E_{C1} and the proteolytic enzymes, E_p. The rate of

Biomass Material Balance

$$dX/dt = (-DX) + r_{net}$$

$$r_{net} = r_g - k_e X$$

$$r_g = (Y_{C1} \cdot r_{C1}) + (Y_{C6} \cdot r_{C6})$$

Enzyme Balances

$$dE_1/dt = + (rs_{E1} - rd_{E1}) - (r_{net} E_1/X)$$

$$dE_6/dt = + (rs_{E6} - rd_{E6}) - (r_{net} E_6/X)$$

$$dE_P/dt = + (rs_{EP} - rd_{EP}) - (r_{net} E_P/X)$$

$$d[mR]/dt = + (rs - rd)_{mR} - (r_{net} [mR]/X)$$

$E_1, E_6, E_P, [mR]$ = mass protein/mass cell

Substrate Material Balances

$$dC_1/dt = D(C_1^O - C_1) - r_{C1}$$

$$dC_6/dt = D(C_6^O - C_6) - r_{C6}$$

$$r_{C1} = V_{C1} E_{C1} X C_1 / (K_{C1} + C_1)$$

$$r_{C6} = V_{C6} E_{C6} X C_6 / (K_{C6} + C_6)$$

$rs_{E1} = k_1 [mR]$ $\qquad rd_{E1} = k_{d1} \cdot EP$

$rs_{E6} = k_6 r_{C6}/X$ $\qquad rd_{E6} = k_{d6}$

$rs_{EP} = k_P C6/(K_{EP} + C6)$ $\qquad rd_{EP} = k_{dEP}$

$rs_{mR} = v_{mR} \cdot Q$ $\qquad rd_{mR} = k_{dmR} \cdot [mR]$

Model Parameters

$Y_{C1} = 0.38$ gmX/gmC1 $Y_{C6} = 0.55$ gmX/gmC6 $k_e = 0.001$ h^{-1}

$V_{C1} = 1.8 \times 10^4$ gmC1/gmE1-h $V_{C6} = 1.0 \times 10^4$ gmC6/gmE6-h

$K_{C1} = 0.04$ gmC1/ℓ $K_{C6} = 0.015$ gmC6/ℓ $k_1 = 1 \times 10^{-3}$ gmE1/gmMR-t

$k_{d1} = 2.5 \times 10^{-5}$ gmE1/gmEP-h $k_6 = 1 \times 10^{-5}$ gmE6/gmC6 $k_{d6} = 5 \times 10^{-6}$ gmE6/gmX-h

$k_P = 1 \times 10^{-3}$ gmEP/gmX-h $\qquad v_{mR} = 0.3$ h^{-1}

$k_{dEP} = 5 \times 10^{-5}$ gmEP/gmX-h

$k_{dMR} = 0.05$ h^{-1} $\quad Q =$ gmMR/gmX (See FIGURE 6)

FIGURE 5. Summary of the structured model for diauxic growth.

specific mRNA synthesis is modeled as

$$rs_{mR} = v_{mR} \cdot Q, \qquad (14)$$

where the Q function (as per references 15 and 17) relates the mass fraction of specific mRNA to a system parameter, for example, glucose or methanol concentration or overall growth rate. In this work, we elected to use the data of mRNA for methanol oxidase as a function of dilution rate as reported by Guiseppin[18] and as replicated in FIGURE 6. Denaturation of mRNA is simulated as

$$rd_{mR} = k_{dmR}[mR]. \qquad (15)$$

Analogous rate expressions for the synthesis and degradation of the proteolytic enzymes, E_p, are

$$rs_{E_p} = k_p C_6 / (K_{E_p} + C_6) \tag{16}$$

and

$$rd_{E_p} = k_{dE_p} E_p. \tag{17}$$

Auxiliary equations for E_{C1}, E_{C6}, E_{mR}, and E_p are also provided in FIGURE 5 along with the numerical constants used in all computer simulations. Due to the complexity of the model, solutions were obtained using a dynamic simulation package named Advanced Continuous Systems Language (ACSL). The model in FIGURE 5 was calibrated by comparing computer simulations to the steady-state experimental results of Egli et al.[12] in FIGURES 2–4. The solid lines superimposed upon the data within FIGURES 2–4 indicate the excellent agreement between model predictions and the parameters of biomass, methanol, and glucose concentrations.

ONGOING UNSTEADY-STATE EXPERIMENTS

The above model requires kinetic data for the synthesis and denaturation rates of key enzymes involved in C_1 and C_6 metabolism. The aforementioned constants were determined as per Bruinenberg et al.[19] by removing and concentrating cell mass from a steady-state chemostat (C_1/C_6 influent mixture = 60% C_1/40% C_6; dilution rate = 0.2 h^{-1}) and then exposing the cells to an aerated solution of glucose (concentration = 5.0 g/L). Under batch conditions, we determined both the specific activity and the absolute enzyme amount (by HPLC extraction) of methanol oxidase,

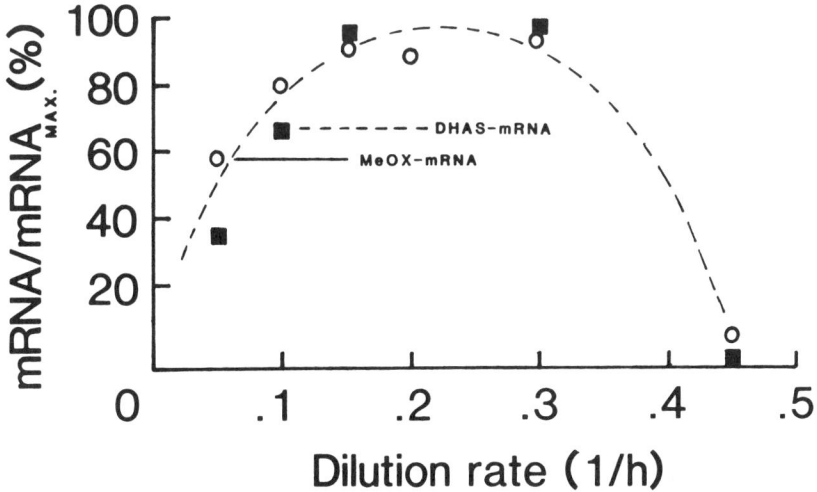

FIGURE 6. Concentration of mRNA specific for methanol oxidase as a function of growth rate.[18]

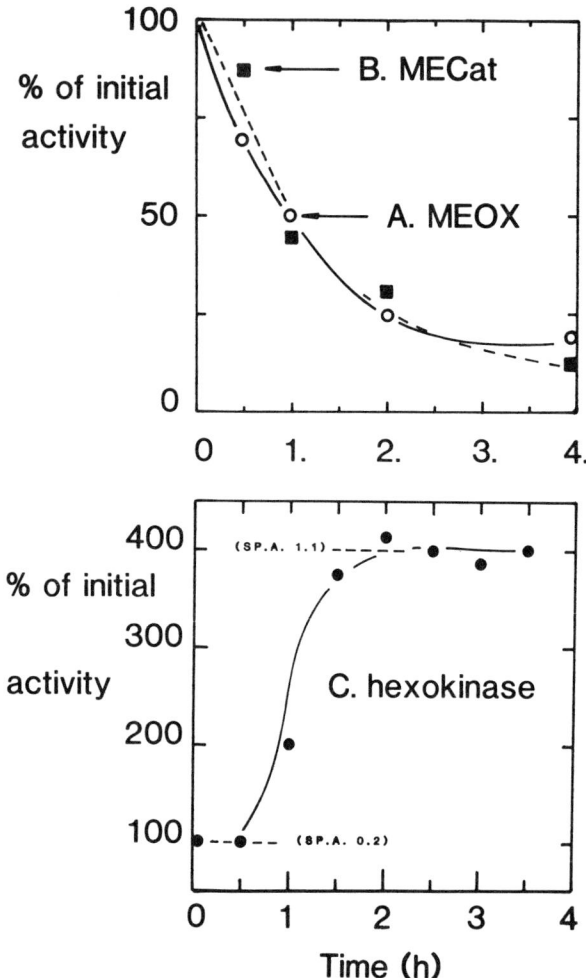

FIGURE 7. Response of enzymes in *Hansenula polymorpha* to exposure to excess glucose conditions. Cells removed from the chemostat at 60% C_1/40% C_6 at $D = 0.2$ h^{-1} and exposed to 5.0 g/L glucose: (A) methanol oxidase, (B) methanol catalase, and (C) hexokinase.

methanol catalase, and hexokinase. Results for these transient batch tests (FIGURE 7) provide denaturation kinetics for methanol oxidase and synthesis rates for hexokinase. Batch experiments like the ones above were repeated, thereby exposing cells from a chemostat (influent = 20% C_1/80% C_6, $D = 0.2$ h^{-1}) to a batch solution of excess methanol.

As from basic kinetic parameters, these batch experiments indicate that specific activities determined by Egli *et al.*[8–11] varied with dilution rate or nutrient composition due to induction/repression mechanisms and not due to allosteric control. Illustrated in FIGURE 8 are the relative characteristic response times for various metabolic processes in any cell; the rates shown in FIGURE 7 are too slow for allosteric control mechanisms. Conversely, it appears that the regulation of E_{C1} and E_{C6} enzyme levels occurs at a response time on the same order as that for the reactor system.

Consequently, it may be feasible to maintain both glucose- and methanol-metabolizing enzymes at near-maximum levels by operating the bioreactor under transient conditions, that is, by fluctuating between the extreme steady-state conditions that promote dominance of one enzyme group over another. Three prolonged

CHANGES IN THE ORGANISMS

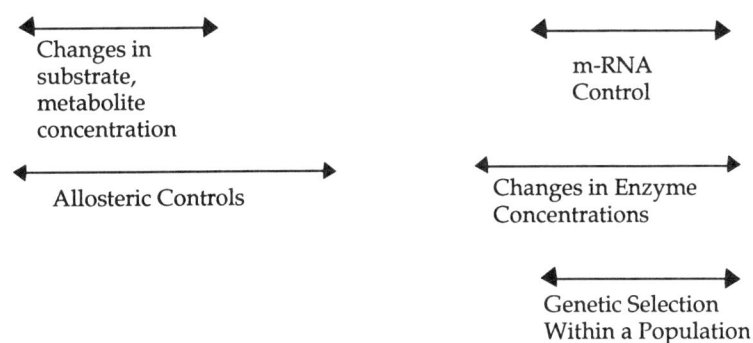

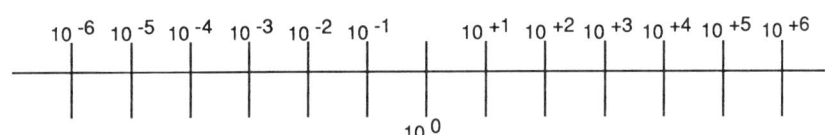

REACTOR SYSTEM CHANGES

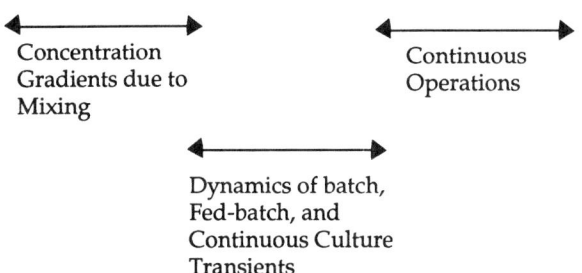

FIGURE 8. Characteristic time scales in the bioprocesses.

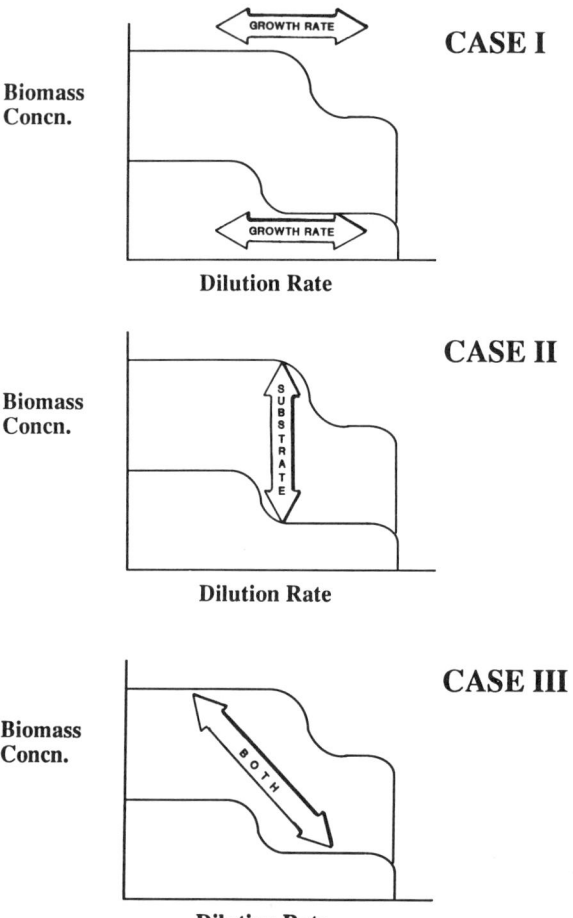

FIGURE 9. Potential perturbations that affect enzyme productivity.

perturbations are suggested, as illustrated in FIGURE 9: growth rate changes at constant influent substrate composition, influent substrate composition changes at constant growth rate, and changes in both growth rate and substrate mixtures. Currently, we are investigating (1) the feasibility of enhanced enzyme productivity by prolonged transients and (2) the metabolic control of carbon fluxes through the pathways of FIGURE 1 during such transients.

Methods Developed

Methods for the *in vitro* enzyme specific activities are detailed in Egli *et al.*[8-11] Rather than analyze for all enzymes indicated in FIGURE 1, we have selected only key enzymes critical to methanol and glucose metabolism (i.e., methanol oxidase, dihydroxyacetone synthase, hexokinase, and phosphofructokinase). Sample collection and preparation of cell-free extracts for enzyme assays are detailed elsewhere.[8-11]

To determine the effects of prolonged perturbations on cell metabolism, we are characterizing the cell cycle frequency distributions for yeast populations, employing flow cytometry to correlate individual cell DNA, RNA, total protein content, and cell size. Intracellular DNA is stained with Hoechst 33342,[20] cells are washed three times with a solution of 1 mL concentration HCl in 99 mL 70% ethanol, and then cells are resuspended in deionized water prior to flow cytometry analysis. Laser output is 0.75 W and sample fluorescence is detected at 350 nm. Cellular RNA is determined by staining with Pyronine Y;[21] laser output is 0.5 W and sample fluorescence is detected at 546 nm. DNA and RNA cell contents are determined simultaneously on the same sample by dual-laser emission techniques.[21] Protein content on a separate sample is determined by staining cells with fluorescein isothiocyanate (FITC)[22] in 0.5 M $NaHCO_3$; laser output is 0.75 W and sample fluorescence is detected at 510 nm. Due to the tenacity of the *H. polymorpha* yeast cell walls, cells were permeabilized with Triton X-100 (2% v/v) in H_2O prior to staining.

An example of the DNA frequency distribution for *H. polymorpha* during a batch growth experiment is provided in FIGURE 10. Illustrated in FIGURES 11 and 12 are DNA and total protein frequency distributions for an initial perturbation to a chemostat culture. In this latter continuous experiment, a culture at a steady-state dilution rate of 0.22 h^{-1} was shifted to a dilution rate of 0.48 h^{-1} and then back down to 0.22 h^{-1}; the inlet substrate concentration and composition, respectively, were 5 g/L total and 20% C_1/80% C_6. Prior to the shift back to a dilution rate of 0.22 h^{-1}, the elapsed time spent at $D = 0.48$ h^{-1} was two reactor residence times.

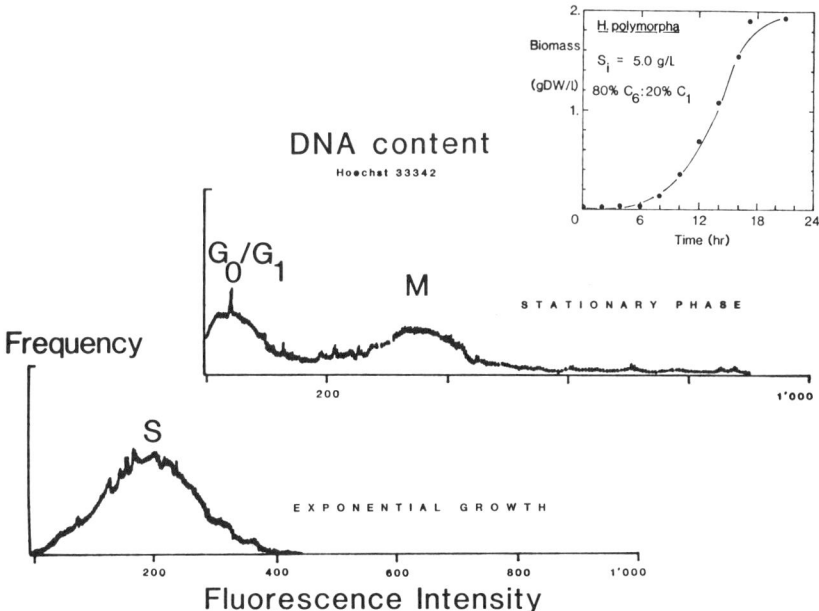

FIGURE 10. Flow cytometry results of *H. polymorpha* during batch growth.

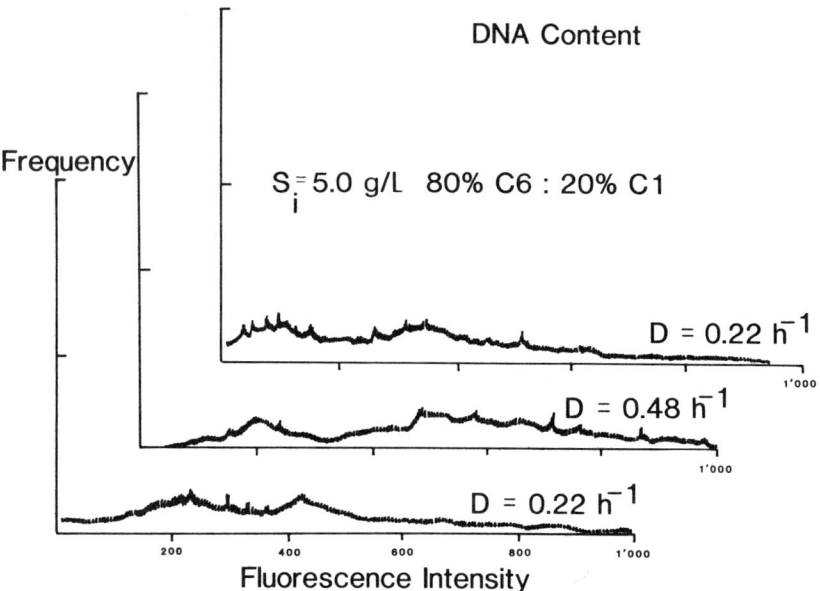

FIGURE 11. DNA content of *H. polymorpha* during one cycle of growth rate transients: $D \to 0.2\ h^{-1} \to 0.48\ h^{-1} \to 0.20\ h^{-1}$; $S_i = 5.0\ g/L$, 20% C_1/80% C_6.

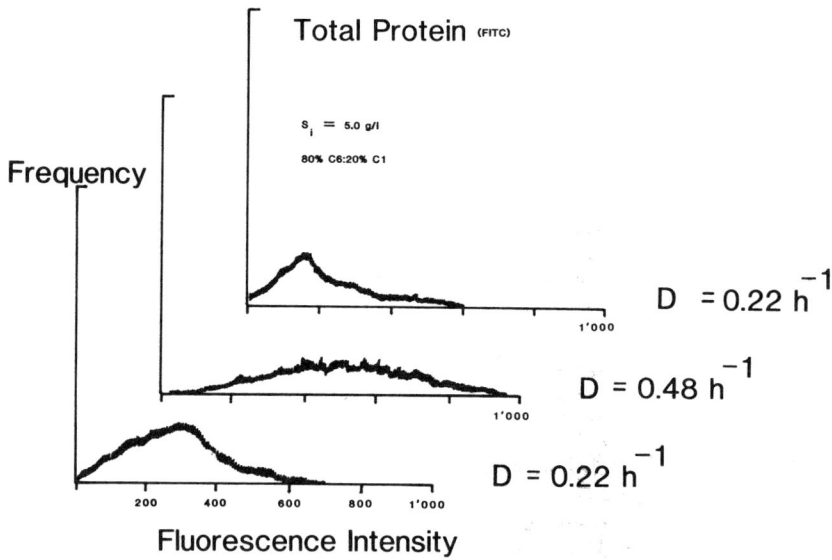

FIGURE 12. Total protein content of *H. polymorpha* during one cycle of growth rate transients: $D \to 0.2\ h^{-1} \to 0.48\ h^{-1} \to 0.20\ h^{-1}$; $S_i = 5.0\ g/L$, 20% C_1/80% C_6.

To complement the flow cytometry analysis, we are developing methods (1) to quantify the flux of both methanol- and glucose-carbon through the methylotrophic pathways and (2) to assess prolonged transient effects on these carbon fluxes. To this end, we have combined a radiolabeled tracer technique[23] with mass-labeled tracer-NMR analysis;[24] elaborate details of these adapted methods will be presented elsewhere,[25] so only a cursory overview will be given here.

Preliminary assessment of these combined tracer methods was carried out on a steady-state culture of *H. polymorpha*, with dilution rate = 0.2 h^{-1}, influent substrate = 5.0 g/L, and influent substrate composition = 60% C_1/40% C_6. Unlike the batch approach employed in den Hollander *et al.*,[24] a miniature aerated chemostat

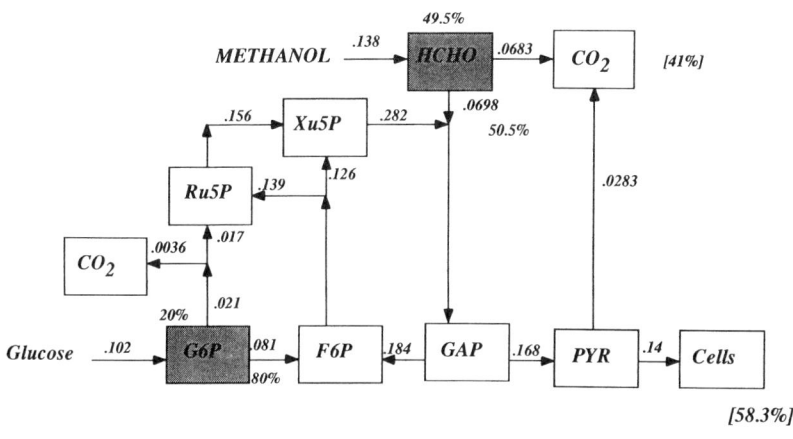

FIGURE 13. Total net carbon flux of methanol and glucose through the methylotrophic pathway at steady state: growth rate = 0.2 h^{-1}, S_i = 5.0 g/L, 60% C_1/40% C_6 inlet composition. Fluxes are determined from combination radiolabeled and mass-labeled tracer studies; see text.

was fabricated from a 10-mm-OD NMR tube. Cells from the main bioreactor were placed into the miniature chemostat and the growth conditions initiated were identical to those in the main system. Use of the miniature chemostat minimized the culture volume, as well as the amounts of tracer used, and allowed NMR diagnosis under actual growth conditions. Radiolabeled ^{14}C-glucose, ^{14}C-methanol, or both were injected, as per Chu *et al.*,[23] into the steady-state culture and the effluent cells and evolved CO_2 were collected over sufficient time to account for 99% of the tracer.

Mass-labeled tracers, ^{13}C-methanol or either 1-[^{13}C-] or 6-[^{13}C-] glucose, were also pulsed into the miniature chemostat while within the NMR. ^{13}C-NMR spectra were collected at 90.55 MHz on a Bruker WH 360 NMR spectrometer; temperature in the culture was maintained at 37 °C. NMR spectra were collected for 1-min periods with pulse intervals of 0.5 s and 45° flip angles. Broadband decoupling of the protons was employed at a power of 2 W. Extracts of cell intermediates were measured with gated decoupling to avoid nuclear Overhauser effects. The backward flow of carbon through fructose 1,6-bisphosphate was determined from the "scrambling" of the 1-[^{13}C-] and 6-[^{13}C-] glucose labels into the C_1 and C_6 positions of trehalose; this method plus radiolabeling allowed calculation of the net carbon flux through phosphofructokinase. Results of the combined radiolabeled and mass-labeled methods are provided in FIGURE 13, which indicates the net carbon mass flux for *H. polymorpha* grown in steady state ($D = 0.2 \text{ h}^{-1}$; $S_i = 5.0 \text{ g/L}$; composition = 60% C_1/40% C_6). With these encouraging results, we are currently extending these methods to the transient experiments proposed in FIGURE 9.

REFERENCES

1. TSCHOPP, J. F., P. F. BRUST, J. M. CREGG, C. A. STILLMAN & T. R. GINGERAS. 1987. Nucleic Acids Res. **15:** 3859–3876.
2. HARDER, W. & L. DIJKHUIZEN. 1982. Philos. Trans. R. Soc. London Ser. **B297:** 459.
3. MATELES, R. I., S. K. CHIAN & R. SILVER. 1967. *In* Microbial Physiology and Continuous Culture. Proc. 3rd Int. Symp. Continuous Culture Microorganisms. E. O. Powell *et al.*, Eds.: 32. H. M. Stationary Office. London.
4. VAN VERSEVELD, H. W., J. P. BOON & A. H. STOUTHAMER. 1979. Arch. Microbiol. **121:** 213.
5. GOTTSCHAL, J. C. & J. G. KUENEN. 1980. Arch. Microbiol. **126:** 33.
6. GOTTSCHAL, J. C. & J. G. KUENEN. 1980. FEMS Microbiol. Lett. **7:** 241.
7. EGGELING, L. & H. SAHM. 1981. Arch. Microbiol. **130:** 362.
8. EGLI, TH., O. KÄPPELI & A. FIECHTER. 1982. Arch. Microbiol. **131:** 1.
9. EGLI, TH., O. KÄPPELI & A. FIECHTER. 1982. Arch. Microbiol. **131:** 8.
10. EGLI, TH., N. D. LINDLEY & J. R. QUAYLE. 1983. J. Gen. Microbiol. **129:** 1269.
11. EGLI, TH., J. P. VAN DIJKEN, M. VEENHUIS, W. HARDER & A. FIECHTER. 1980. Arch. Microbiol. **124:** 115.
12. EGLI, TH., C. BOSSHARD & G. HAMER. 1986. Biotechnol. Bioeng. **28:** 1735–1741.
13. BRYERS, J. D. & R. L. IRVINE. 1987. Structured Models of Biological Processes. Bioenvironmental Systems, Vol. II. D. L. Wise, Ed. CRC Press. Boca Raton, Florida.
14. PAPAGEORGAKOPOULOU, H. & W. J. MAIER. 1984. Biotechnol. Bioeng. **26:** 275.
15. ROELS, J. A. 1983. Energetics and Kinetics in Biotechnology. Elsevier. Amsterdam/New York.
16. PERETTI, S. W. & J. E. BAILEY. 1986. Biotechnol. Bioeng. **28:** 1672.
17. YAGIL, G. & E. YAGIL. 1971. Biophys. J. **11:** 11–20.
18. GUISEPPIN, M. L. F. 1988. Optimization of methanol oxidase production by *Hansenula polymorpha*. Ph.D. dissertation. Technische Universiteit Delft. Delft, the Netherlands.
19. BRUINENBERG, P. G., M. VEENHUIS, J. P. VAN DIJKEN. J. A. DUINE & W. HARDER. 1982. FEMS Microbiol. Lett. **15:** 45–50.
20. SHAPIRO, H. M. 1981. Cytometry **2:** 143.
21. VAN DILLA, M. A., R. G. LANGLOIS, D. PINKEL, D. YAJKO & W. K. HADLEY. 1983. Science **220:** 620–621.
22. DORAN, P. M. & J. E. BAILEY. 1986. Biotechnol. Bioeng. **28:** 73–87.
23. CHU, I-M., C. M. BUSSINEAU & E. T. PAPOUTSAKIS. 1985. Biotechnol. Bioeng. **27:** 1623–1633.
24. DEN HOLLANDER, J. A., K. UGURBIL, T. R. BROWN, M. BEDNAR, C. REDFIELD & R. G. SHULMAN. 1986. Biochemistry **25:** 203–211.
25. BRYERS, J. D. & T. YEH. 1990. Diauxic carbon fluxes in the methylotrophic yeast, *H. polymorpha*: radio- and mass-labeled tracer studies. Bioprocess Eng. In press.

Nutrient Transport and Cellular Morphology in Immobilized Cell Aggregates[a]

J. D. FOWLER AND C. R. ROBERTSON

Department of Chemical Engineering
Stanford University
Stanford, California 94305

INTRODUCTION

This report summarizes a body of work concerning two different types of reactors for immobilized bacteria. In these reactors, the bacteria are retained as a dense aggregate within a physically defined region of the reactor, so the product stream leaving the reactor is cell-free. The two types of reactor employ either diffusion or convection through the cell aggregate to supply nutrients to the cells and to remove products. A schematic of such an experimental reactor is depicted in FIGURE 1.

Amongst the advantages of employing immobilized cells in reactors are improved volumetric productivity, reduced pretreatment demands on the nutrient medium, and a cell-free product stream. The motivation for a detailed investigation of the behavior of such systems arises with the failure to achieve these potential benefits, principally because of mass-transfer limitations.

A second, and not incidental, motivation for part of this work was to establish that the behavior of immobilized cells is not intrinsically different from that of cells in the more common suspension cultures; that is, cells respond to the local concentration of nutrients and products in the same manner in immobilized cell aggregates as in suspension. This is not an insignificant point, concerning as it does the determinism of cell cultures. There is a body of literature demonstrating that immobilization alters microbial metabolism. It may be contended that this can be attributed to changes in the local environment in terms of nutrients, products, or water activity.[1-3] It should be noted, however, that some effects of immobilization have been explained instead by invoking other effects such as changes in cell permeability[4,5] or rigidity of the immobilizing matrix.[6]

In this report, we shall describe how certain experimental techniques were adapted and applied to immobilized cell reactors to characterize their behavior. A unifying theme for much of this work is the morphology of the cells. Confined growing cells can exert considerable pressure (in excess of three atmospheres[7]) on their environment; one consequence of this is the distortion and ultimate collapse of neighboring cells that are starved for nutrients. The first part of this report deals with verifying the existence of

---

[a]This research was supported by funds provided by the Center for Biotechnology Research (San Francisco, California), the Defense Advanced Research Projects Agency, the Monsanto Company, and the National Science Foundation.

regions of nutrient exhaustion in diffusionally supplied reactors and with describing cell volume collapse therein. The second part concerns convectively supplied reactors, where mass-transfer limitations on cell metabolism are not apparent and where the principal mass-transfer resistance is an explicit function of cell morphology.

DIFFUSIONALLY SUPPLIED IMMOBILIZED CELL REACTORS

General Experimental Overview

The experiments to be described below were conducted with *Escherichia coli* grown anaerobically in glucose-limiting minimal media and at 37 °C unless otherwise noted. Details of the several procedures used are published elsewhere and are referenced accordingly.

In the diffusionally supplied reactor, several nutrient-carrying porous hollow fibers were threaded into the shell space of the reactor (see FIGURE 1 for a schematic). Bacteria were then inoculated into the shell space, being supplied with nutrients by diffusion across the wall thickness of the fibers carrying the nutrient stream.

Reaction-Diffusion Model—Formulation and Results

A simple model of the reaction-diffusion process was formulated to characterize the behavior of the diffusionally supplied reactor described above. The model employs the governing equations for reaction and diffusion within the cell aggregate, for flux continuity at the cell aggregate–membrane interface, and for simple diffusion within the membrane. The boundary conditions are for no flux at the shell wall and for flux at the membrane wall driven by the difference between the wall concentration and the

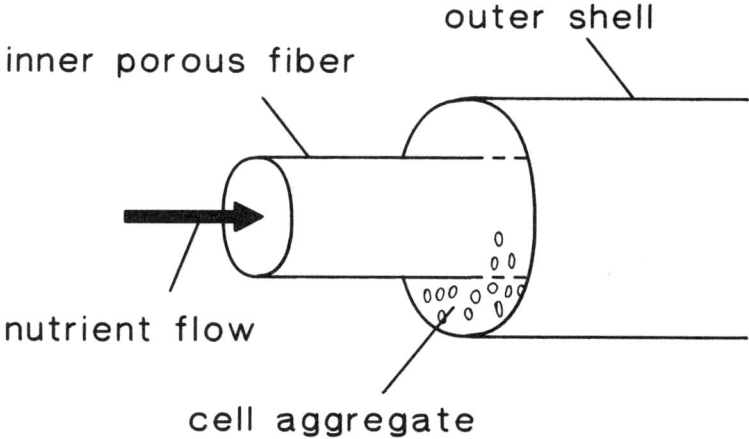

FIGURE 1. Schematic of an immobilized cell reactor.

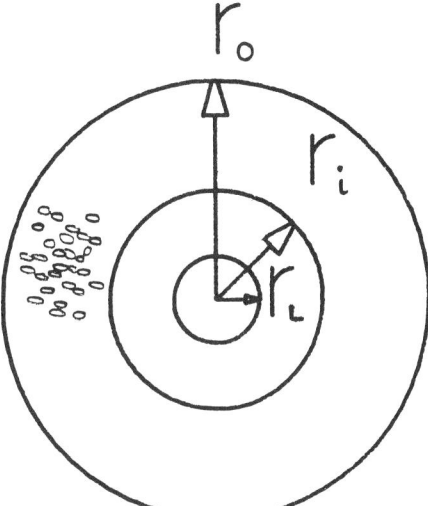

FIGURE 2. Simplified geometry of a diffusionally supplied immobilized cell reactor.

lumen mixing-cup average concentration, $\overline{C}$.[b] The governing equations are as follows:

$$D_c \frac{1}{r}\frac{d}{dr}\left(r\frac{dC}{dr}\right) = R(C), \quad r_i < r < r_o, \tag{1a}$$

$$D_m \frac{dC}{dr_-} = D_c \frac{dC}{dr_+}, \quad r = r_i, \tag{1b}$$

and

$$D_m \frac{1}{r}\frac{d}{dr}\left(r\frac{dC}{dr}\right) = 0, \quad r_L < r < r_i, \tag{1c}$$

with boundary conditions

$$k_L(\overline{C} - C) = -D_m \frac{dC}{dr}, \quad r = r_L, \tag{1d}$$

and

$$\frac{dC}{dr} = 0, \quad r = r_o. \tag{1e}$$

To apply this analysis, it was necessary to make an approximation to the geometry of the experimental system. The simplified geometry is as shown in FIGURE 2, where

---

[b] In a more complete version of this model,[8] $\overline{C}$ in boundary condition 1d is a function of axial position. With a differential mass balance in the lumen and with the neglect of axial diffusion, concentration profiles along the length of a reactor can then be obtained by numerical integration.

the exterior radius, r_o, was chosen to maintain the same surface area to volume ratio by using $r_o = (r_s^2/n)^{1/2}$.

The reaction rate was assumed to be zero order in the limiting nutrient concentration. This is a justifiable treatment for glucose, for which the substrate affinity is very high[9] and whose uptake rate may be limited by mass transfer at the cell outer membrane.[10] It also was assumed that the growth reaction ceased when the concentration of the buffering species dropped below a critical concentration, thereby accounting for the growth-inhibiting action of accumulated excreted organic acids. The reaction rate was assumed constant, which is equivalent to assuming steady state in cell concentration and metabolism. This treatment and further details of the model are discussed more completely elsewhere.[11]

The equations can be solved exactly to yield an implicit relation for the thickness of the growing region in which the glucose concentration is nonzero. In nondimensional form, this distance δ is given implicitly by

$$1 - \frac{\phi^2}{4}\{1 - \delta^2(1 - \ln \delta)\} = -\frac{\phi^2 \beta}{2}(1 - \delta^2). \tag{2}$$

All the parameters required by the model were measured independently, with the kinetic constants being measured in suspension cultures. Typical results for the model are shown in TABLE 1.

Verification of the model results lends support to two principal conclusions: (i) the metabolic behavior of immobilized cells is not intrinsically different from that of cells in suspension culture and (ii) the growth region in diffusionally supplied reactors is restricted to a narrow domain close to the cell-nutrient interface. The model was verified experimentally by transmission electron microscopy, autoradiography, and NMR spectroscopy, as discussed in the following sections.

Transmission Electron Microscopy

In order to examine the morphology of cell aggregates in a reactor, electron micrographs were taken of cross sections of the reactor. The procedure used was to perfuse the reactor with a solution of buffered gluteraldehyde, postfix with osmium tetroxide, stain with uranyl acetate, dehydrate in ethanol series, embed in low-viscosity epoxy resin (a modification of the technique of Oliveira *et al.*[12]), and poststain thin sections (60–90 nm) with lead citrate. A typical micrograph is shown in FIGURE 3.

In FIGURE 3, two distinct regions can be seen within the cell-containing shell space.

TABLE 1. Results of the Diffusion-Reaction Model

Anaerobically grown *E. coli* at 37 °C	
glucose (limiting nutrient)	3 g/L
lumen radius	135 μm
membrane thickness	190 μm
cell aggregate thickness	422 μm
flow rate	3.4 mL/h
predicted thickness of growing region = 25 μm	

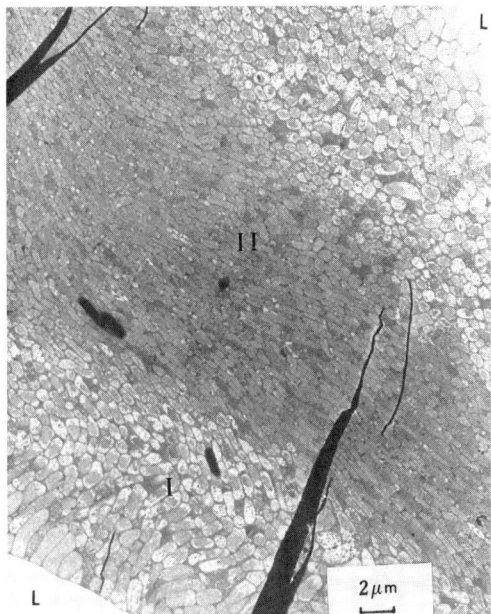

FIGURE 3. Transmission electron micrograph of a cross section through a diffusionally supplied reactor. "L" denotes the medium-carrying hollow fibers, and regions I and II of the cell aggregate are discussed in the text.

The region nearest the medium-supplying inner fibers, designated I, contains cells of normal size and ellipsoid morphology. The region further away from the inner fibers, designated II, contains cells that are greatly distorted and that have different staining characteristics from those in region I (although no significance can be attached to this per se). The size of region I is in agreement with the prediction of the model above. In order to establish the nature of the two regions, more detailed investigations (to be discussed below) were performed.

Autoradiography

Autoradiography is a technique used to visualize regions of metabolic activity. In the form used here, it relies on the fact that growing cells require sulfur for protein synthesis and that this sulfur must be taken up from the nutrient medium. In this experimental procedure, the medium feed to the reactor was switched to a pulse of "hot" medium containing ^{35}S in addition to the stable ^{32}S isotope followed by a chase of "cold" medium to elute out any unincorporated ^{35}S. The cell aggregate was then fixed and embedded as described for electron microscopy. Sections of the aggregate (2 μm thick) were glued to microscope slides and dip-coated in nuclear-track photographic emulsion. The emulsion was exposed for 1–7 days to the β-particles released by decaying ^{35}S atoms. The emulsion was then developed and fixed so that the region of sulfur incorporation and protein synthesis could be observed with a light microscope. An important experimental concern was that the pulse period be long enough to ensure

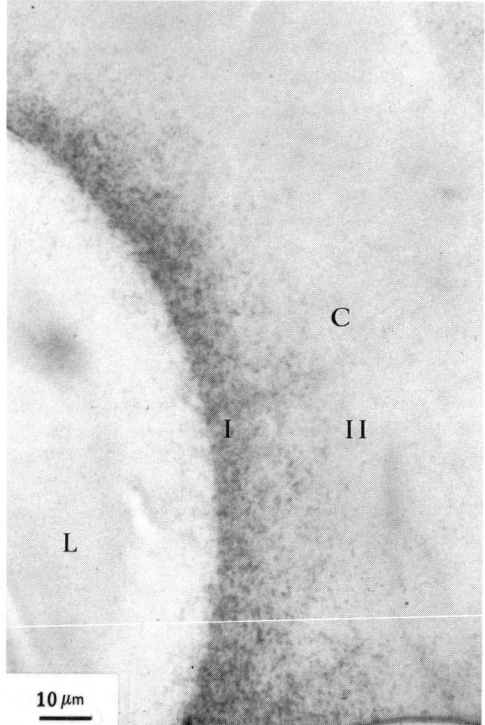

FIGURE 4. Autoradiograph of a cross section through a diffusionally supplied reactor. "L" denotes the lumen of the medium-carrying hollow fiber, "C" denotes the cell aggregate, and regions I and II are discussed in the text.

exposure of the whole aggregate to the labeling medium, that is, pulse period $\gg$ diffusion time. Further details of the procedure are described by Karel and Robertson.[13]

The results of the autoradiographic investigation are illustrated in FIGURE 4. It is apparent that the cells in region I of FIGURE 4 are actively synthesizing protein, whereas those in region II are not. Comparison with FIGURE 3 is illuminating because it is clear that there is an excellent correlation between protein synthesis and normal cellular morphology. A reasonable inference is that the cells in region II are metabolically inactive because they are starved for the limiting nutrient, glucose. Further evidence of mass-transfer limitations is provided by the NMR spectroscopy study that follows.

Nuclear Magnetic Resonance (NMR) Spectroscopy

NMR spectroscopy measures the characteristic magnetic resonances of atomic nuclei that have nonzero quantum spin. In a magnetic field, the quantum spin levels become nondegenerate and transitions between the levels can be made more likely by applying oscillations of the appropriate frequency in the applied magnetic field. The presence of nuclei that undergo transitions at given frequencies leads to resonances, or peaks, in the field strength spectrum. With suitable calibration, these peaks can be assigned to chemical species and can be used to measure their concentrations. Because the local magnetic field is affected by the local electron density and by the screening of adjacent molecules, the location of assigned peaks in the spectrum can be used to assay the local environment; again, with suitable calibration, it is possible to determine the local pH.

For this investigation, the reactor was placed within the probe of an NMR spectrometer and *E. coli* was grown *in situ* on a phosphate-buffered medium. The inorganic ^{31}P phosphate peak was used to monitor noninvasively the pH within intracellular and extracellular regions of the reactor, with a time resolution of the order of one minute. Details of the procedure are given by Briasco and Robertson.[14]

FIGURE 5 is the spectrum obtained from a reactor that was perfused with nutrient medium at 37 °C. Two distinct inorganic phosphate peaks can be seen. The first at pH 7.2 corresponds to the medium in the lumina of the inner hollow fibers. The second peak at pH < 5.5 can be assigned to the intracellular region of the cell aggregate. No peak is visible for the extracellular region because the spectra recorded by the NMR spectrometer represent spatial averages over the reactor volume and the extracellular region represents an insignificant volume fraction of the cell aggregate. Similarly, the growing region is observed to be a relatively small fraction of the reactor volume (predicted effectiveness factor was 0.34) and no peak is visible for those cells in this region. The value of the pH < 5.5 peak is too low to be attributable to growing cells, which maintain their intracellular pH at 7.8.[15] Instead, it is apparent that most of the cell aggregate contained cells unable to maintain a normal physiological intracellular pH.

FIGURE 6 is the NMR spectrum obtained from another reactor, this time cultivated at 16 °C. The reaction-diffusion model here predicted an effectiveness factor of unity,

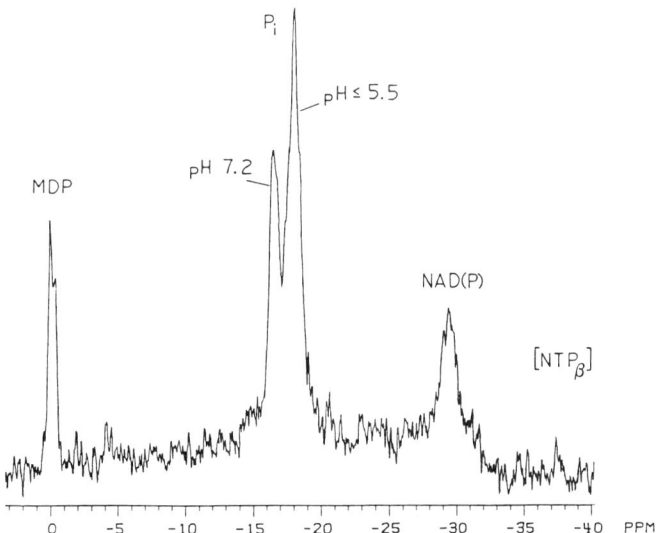

FIGURE 5. NMR spectrum of a diffusionally supplied reactor cultivating cells at 37 °C.

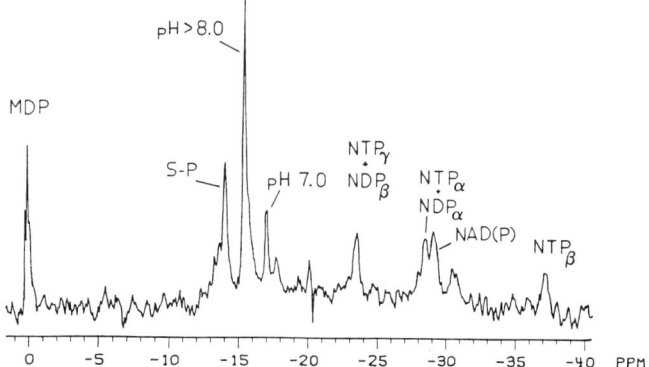

FIGURE 6. NMR spectrum of a diffusionally supplied reactor cultivating cells at 16 °C.

indicating that growing cells should have occupied the whole reactor volume. This prediction is supported by the NMR spectrum, in which, again, only two inorganic phosphate peaks are discernible. The peak at pH > 8.0 corresponds to the nutrient medium for this experiment, whereas that at pH 7.0 was assigned to energy sufficient cells. (The low value probably reflects the difficulty of obtaining accuracy in *absolute* pH values by this technique. This point is discussed by Briasco *et al.*[16])

Taken together, the results presented above confirm the predicted metabolic behavior in the growing and nutrient-starved regions of diffusionally supplied reactors. They also indicate that cells in the starved region are unable to maintain a normal intracellular pH, but may instead accumulate the organic acids that are the waste products of anaerobic catabolism. The next topic of investigation was to examine the means whereby pressure exerted by growing cells can lead to collapse of neighboring cells in a nutrient-starved region.

Characterization of Cell Collapse

For this work, cells were grown between two planar membranes held clamped between two hollow chambers. The planar geometry facilitated reactor fabrication and characterization, but was not intrinsically different from that of the hollow-fiber reactors employed above. The upper chamber was perfused with medium, allowing diffusion into the cell aggregate from the upper surface. The lower chamber could be pressurized with nitrogen gas to apply a controlled stress to the cell aggregate. A schematic of the planar reactor is given in FIGURE 7.

To determine the response of the cell aggregate to the applied stress, procedures were needed for measuring the cell volume and the osmotically active volume fraction. Estimates of the former were obtained from transmission electron micrographs of cell aggregates fixed under conditions of applied stress. The latter was measured by tracer exclusion, using [^{3}H]-methoxyinulin, [^{14}C]-glycerol, and [^{3}H]-water following the procedure reported by Stock *et al.*[17] For these experiments, the medium perfusing the upper chamber was replaced with medium lacking glucose to induce carbon and energy starvation in the cells.

A simple analysis of the volumetric response of cells to applied stress quantitatively explains the results of the tracer exclusion studies. It is assumed that the stress applied to the cells is borne solely by the internal osmotic pressure of the cells, that there is negligible tensile stress in the cell wall, and that the cell aggregate is in mechanical equilibrium. Each of these assumptions is justifiable for nutrient-starved cells, which have lost turgor. Deviations of the solution from ideality are assumed to remain constant, so the osmotic pressure within the cells is given by

$$\Pi_i = \frac{RT \sum_j \gamma_j^* n_j^*}{(V - b)}, \qquad (3)$$

where the asterisks denote quantities that remain constant, V is the total cell volume, and b is the osmotically inactive volume. Equating the difference in osmotic pressures across the cell wall with the applied stress, σ, results in the following expression for cell volume:

$$V/V_0 \approx \frac{(1 + \sigma b/\Pi_0 V_0)}{(1 + \sigma/\Pi_0)}. \qquad (4)$$

The results of this analysis are compared with experimental results in FIGURE 8. The two lines on the figure are for different values of the ratio b/V_0. These two bounds were obtained from the excluded volumes of tritiated water and of glycerol relative to that of inulin (giving $b/V_0 = 0.11$ or 0.4, respectively). It may be seen that the results confirm the validity of the description of cellular volume, as well as the ability to

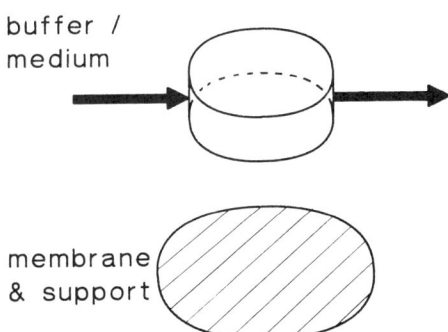

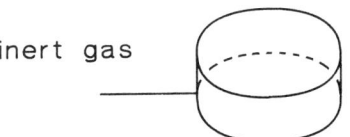

FIGURE 7. Schematic of an experimental system for applying stress to cell aggregates.

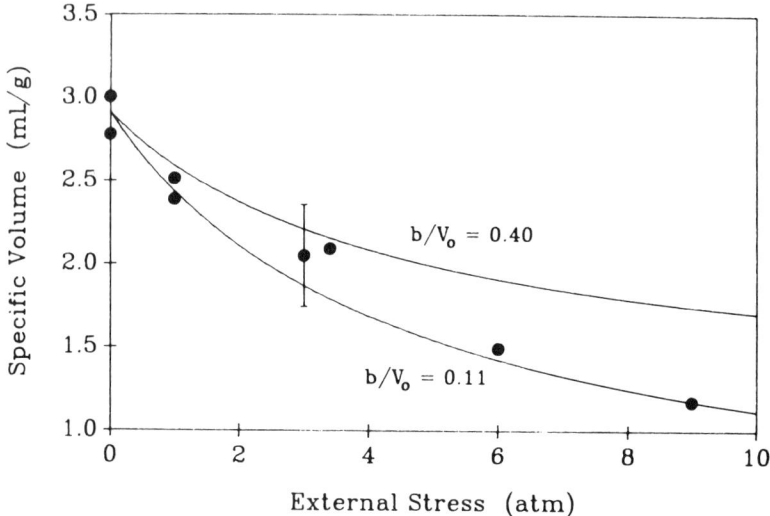

FIGURE 8. Comparison of experimental and predicted cellular volume for nutrient-starved bacteria as a function of applied hydrostatic stress.

quantitatively predict the collapse of cells in nutrient-starved regions of diffusionally supplied reactors.

CONVECTIVELY SUPPLIED IMMOBILIZED CELL REACTOR

Experimental System and Introduction

The results discussed earlier clearly show that, in a diffusionally supplied immobilized cell aggregate, cells not in proximity to the nutrient source may experience mass-transfer limitations unless growth rates are depressed, for example, through using lower temperatures. A possible means of overcoming these limitations is to employ convection through the cell aggregate. To this end, we have investigated an experimental convectively supplied immobilized cell reactor, which is depicted schematically in FIGURE 9.

In this reactor, cells were supported on a porous membrane clamped between two hollow chambers. The cells were held against the membrane by the drag force exerted by a convective current of nutrient medium through the cell aggregate and membrane. A pressure transducer was used to monitor the pressure drop required to drive controlled flow rates between the two chambers of the reactor. Again, the bacterium *Escherichia coli* was cultured anaerobically on glucose-limiting minimal media, but here at 35.0 °C. Details of the experimental procedure are available elsewhere.[18] The principal mass-transport parameter in such a system is the hydraulic permeability, L_p, which is defined by a Darcy-type equation for flow through a porous medium:

$$Q = L_p \frac{\Delta P}{\mu}. \tag{4}$$

The purpose of this investigation was to determine whether mass-transfer limitations on bacterial metabolism could be overcome in a convectively supplied reactor and to quantify the hydraulic permeability of an immobilized cell aggregate.

Transmission Electron Microscopy (TEM)

The first indication of the behavior of cells in such a system comes, as before, from an examination of their morphology using TEM. The same procedure was used in preparation of micrographs as that discussed previously. A typical micrograph is presented in FIGURE 10.

Several features of FIGURE 10 are noteworthy. First, the cells retain normal size and morphology throughout the cell aggregate, even at the surface of the supporting membrane. In this region, the cells were subjected to considerable stress by those cells above them, as the pressure drop across the thickness of the cell layer was approximately 10 psi at the time of cell fixation. (It is not possible from the micrograph alone to be certain that any cell distortion did not relax during the fixation and embedding procedure, but it is significant that no lysed or collapsed cells are visible.) Second, the density of cells within the aggregate is quite uniform, so the cell aggregate can be treated as homogeneous when considering mass-transfer phenomena. Third, the cells are highly oriented in the direction of medium flow.

The significance of the cell orientation is discussed later. The first two observations of homogeneity within the aggregate support the contention that mass-transfer limitations have been eliminated by the use of convective transport. This is examined in more detail in the following sections.

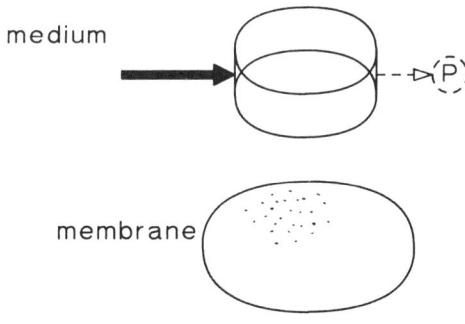

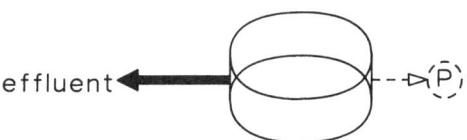

FIGURE 9. Schematic of a convectively supplied immobilized cell reactor.

Autoradiography

Autoradiography was performed on cells grown in the convectively supplied reactor using a similar procedure to that discussed above. The results of these experiments are represented by FIGURES 11 and 12, which respectively show a grazing section parallel to the membrane surface and a cross section through the thickness of the cell layer.

Image analysis of the density of silver grains in the photographic emulsion indicated that, to a first approximation, the rate of sulfur incorporation and protein synthesis was uniform throughout the cell aggregate. It is noteworthy that the cell

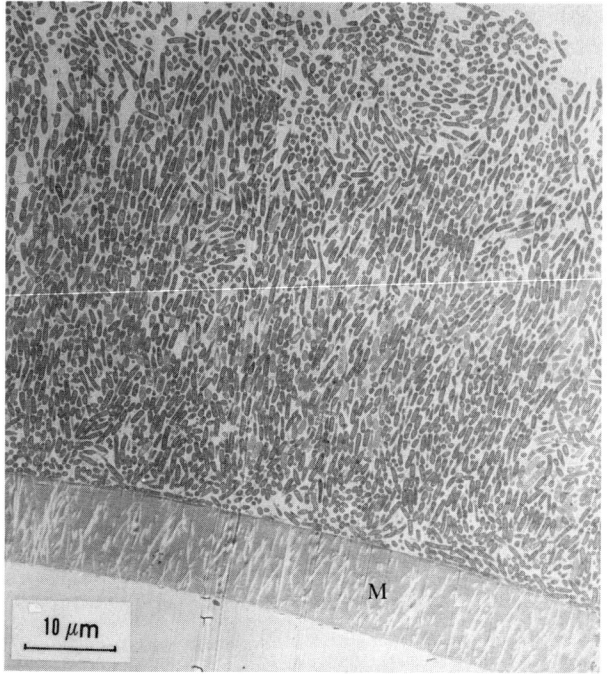

FIGURE 10. Transmission electron micrograph of a convectively supplied cell aggregate. "M" denotes the membrane supporting the cell aggregate, and the direction of flow is top to bottom.

aggregates in these experiments were up to 100 times thicker than the growing region in diffusively supplied reactors at approximately the same temperature and that no mass-transfer limitations were detectable.

Metabolic Behavior and Hydraulic Permeability

It is apparent that there are no nutrient limitations in the convectively supplied reactor because of the uniformity of cellular metabolism. From the previously dis-

FIGURE 11. Autoradiograph of a convectively supplied cell aggregate—grazing section.

100 μm

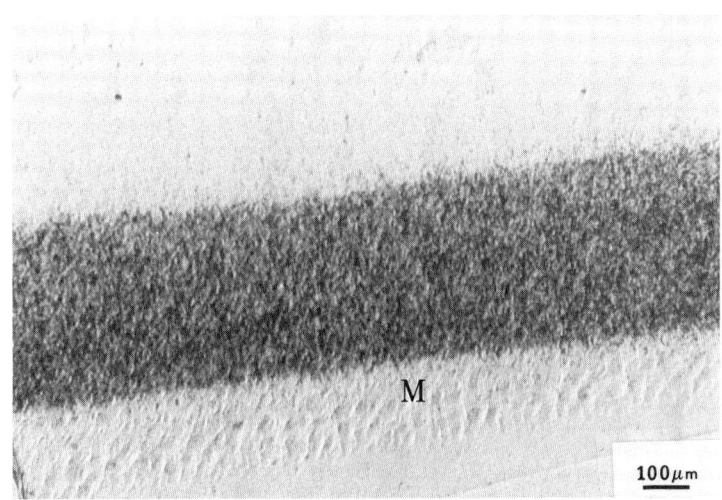

FIGURE 12. Autoradiograph of a convectively supplied cell aggregate—cross section. "M" denotes the membrane supporting the cell aggregate, and the direction of flow is top to bottom.

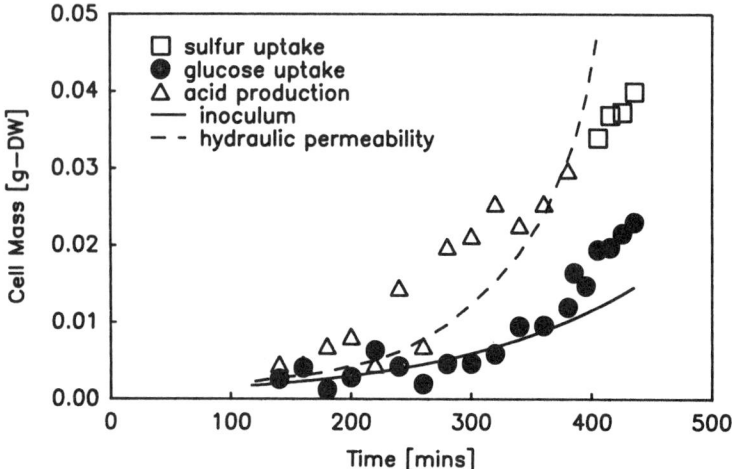

FIGURE 13. Estimates of cell mass in a convectively supplied immobilized cell reactor.

cussed work on diffusionally supplied cell aggregates, it was known that the metabolic behavior in an immobilized cell aggregate could be described by kinetic parameters measured in suspension cultures. By using suspension-culture values for the specific rates (per unit cell mass) of glucose uptake, sulfur uptake, and acid production, estimates could therefore be made of the cell mass within the convectively supplied reactor. From the experimental rates of glucose consumption, acid production, and sulfur incorporation, three independent estimates have been obtained (shown as discrete data on FIGURE 13). These values agree quantitatively with the cell mass in the reactor extrapolated from the inoculum mass by using the suspension-culture growth rate (solid line in FIGURE 13). The agreement of all these estimates of cell mass supports the conclusion that normal exponential cell growth has been achieved in the experimental reactor.

A fifth independently obtained estimate of cell mass is shown in FIGURE 13 (dotted line). This estimate was based upon measurements of the hydraulic permeability defined in equation 4. The experimental hydraulic permeability was used in the Carman-Kozeny equation below (equation 5) to calculate the thickness of the cell aggregate throughout the course of an experiment. For this purpose, it was necessary to obtain values for the void fraction and the hydraulic diameter of the cells in the aggregate. These were measured from transmission electron micrographs taken of cell aggregates at the end of experiments. Their respective values were 0.65 and 2.0 μm:

$$L_p \times L = k = \frac{D_p^2 \phi^3}{180(1 - \phi)^2}. \tag{5}$$

It may be seen in FIGURE 13 that the estimate of the cell mass derived from the hydraulic permeability is in good quantitative agreement with the other estimates. From this, it may be concluded that the hydraulic permeability of cell aggregates is, in general, well described by the Carman-Kozeny relation.

Non-Darcy Behavior

Although equation 5 applies in general, experiments show that the Darcy relation (equation 4) does not strictly describe convection through cell aggregates. Instead, the hydraulic permeability is a function of flow rate, decreasing at higher flow rates and vice versa. This fact is illustrated in FIGURE 14, which plots the experimental hydraulic resistance (reciprocal permeability) made dimensionless by an empirical curve fit through the permeability at a flow rate of 6 mL/min. In this experiment, the flow rate through the cell aggregate was periodically changed. Two points are noteworthy. First, the changes in permeability are completely reversible over a time scale of the order of one minute or less. Second, the relative change in permeability corresponds approximately to the relative change in flow rate.

The cause of the non-Darcy phenomenon described has not yet been determined and it is the topic of continuing study. Preliminary results indicate that the degree of orientation of the cells with flow is affected by the flow rate. This could explain the observed behavior if uniformly stronger orientation allowed the cell aggregate to pack more densely at higher flow rates. Alternatively, any elastic deformation of the cells may vary under the influence of the stresses exerted by their neighbors as the pressure drop across the cell aggregate changes. Of course, it is possible, if not likely, that both mechanisms may operate simultaneously. In order to investigate the morphological response of the cell aggregate in more detail, we are presently employing image analysis techniques to obtain a quantitative description of cell size, orientation, shape, and volume fraction. It must be pointed out that neither of the hypothetical mechanisms proposed here could explain the correspondence of relative changes in flow rate and hydraulic permeability other than by a coincidental combination of independent changes in void fraction, hydraulic diameter, and bed thickness.

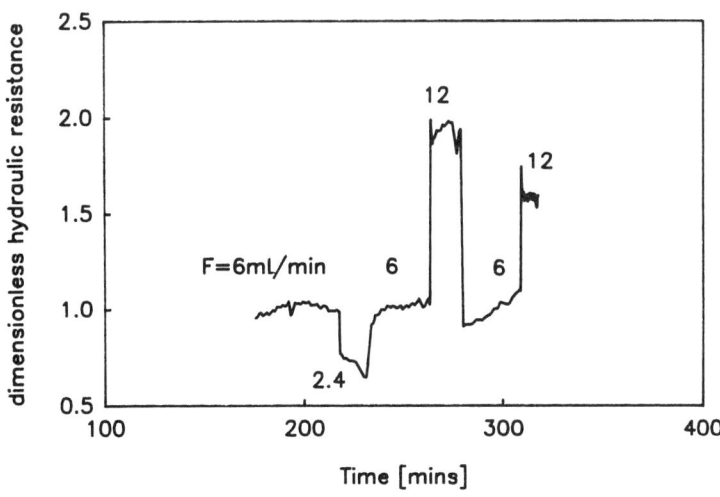

FIGURE 14. Variation of dimensionless hydraulic resistance with flow rate.

CONCLUSIONS

In summary, it is apparent that mass-transfer limitations are present throughout most of the cell aggregate in diffusionally supplied reactors. These limitations can be overcome by employing convective transport to supply nutrients and to remove products. Continuing work is necessary to further our understanding of the resistance to convection in a convectively supplied reactor. On a practical side, it should be noted that continued cell growth in either diffusively or convectively supplied cell aggregates will inevitably lead to increased mass-transfer resistances and to degradation of system performance.

A general conclusion to be drawn from this work is that immobilized cell systems are amenable to quantitative description and that, for *E. coli*, the same metabolic parameters can be employed for this purpose as for cells in suspension cultures.

REFERENCES

1. GRAJEK, W. & P. GERVAIS. 1987. Influence of water activity on the enzyme biosynthesis and enzyme activities produced by *Trichoderma viride* TS in solid-state fermentation. Enzyme Microb. Technol. **9:** 658–662.
2. HAHN-HAGERDAL, B., B. JONSSON & E. LOHMEIER-VOGEL. 1985. Shifting product formation from xylitol to ethanol in pentose fermentations using *Candida tropicalis* by adding polyethylene glycol (PEG). Appl. Microbiol. Biotechnol. **21:** 173–175.
3. HAMSTRA, R. S., M. R. MURRIS & J. TRAMPER. 1987. The influence of immobilization and reduced water activity on gaseous-alkene oxidation by *Mycobacterium* PY1 and *Xanthobacter* PY2 in a gas-solid bioreactor. Biotechnol. Bioeng. **29:** 884–891.
4. DEROSA, M., A. GAMBACORTA, E. ESPOSITO, E. DRIOLI & S. GAETA. 1982. Thermophilic microbial cells immobilized in cellulose acetate membranes. Biochimie **62:** 517–522.
5. DRIOLI, E., G. IORIO, M. DEROSA, A. GAMBACORTA & B. NICOLAUS. 1982. High-temperature immobilized-cell ultrafiltration. J. Membr. Sci. **11:** 365–370.
6. KETEL, D. H., A. C. HULST, H. GRUPPEN, H. BRETELER & J. TRAMPER. 1987. Effects of immobilization and environmental stress on growth and production of non-polar metabolites of *Tagetes minuta* cells. Enzyme Microb. Technol. **9:** 303–307.
7. STEWART, P. S. & C. R. ROBERTSON. 1989. Volume limitation of microbial growth: studies with entrapped *Escherichia coli*. Appl. Microbiol. Biotechnol. **30:** 34–40.
8. KAREL, S. F. & C. R. ROBERTSON. 1989. Cell mass synthesis and degradation by immobilized *Escherichia coli*. Biotechnol. Bioeng. **34:** 337–356.
9. KOCH, A. L. 1979. Microbial growth in low concentrations of nutrients. *In* Strategies of Microbial Life in Extreme Environments. Dahlem Koferentzen Life Sciences Report No. 13. M. Shilo, Ed.: 261–269. Verlag Chemie. Weinheim.
10. NIKAIDO, H. 1986. Transport through the outer membrane of bacteria. Methods Enzymol. **125:** 265–278.
11. KAREL, S. F. 1987. Reaction and diffusion in immobilized living bacteria: characterization with radioisotope labelling and autoradiography. Ph.D. thesis. Stanford University.
12. OLIVEIRA, L., A. BURNS, T. BISALPUTRA & K-C. YANG. 1983. The use of an ultra-low viscosity medium (VCD/HXSA) in the rapid embedding of plant cells for electron microscopy. J. Microsc. **132:** 195–202.
13. KAREL, S. F. & C. R. ROBERTSON. 1989. Autoradiographic determination of mass-transfer limitations in immobilized cell reactors. Biotechnol. Bioeng. **34:** 320–336.
14. BRIASCO, C. A. & C. R. ROBERTSON. 1988. Diffusional limitations of immobilized *E. coli* in hollow-fiber reactors: influence on ^{31}P NMR spectroscopy. Biotechnol. Bioeng. Submitted.
15. KASHKET, E. R. 1982. Stoichiometry of the H^+-ATPase of growing and resting, aerobic *Escherichia coli*. Biochemistry **22:** 5534–5538.

16. BRIASCO, C. A., D. A. ROSS & C. R. ROBERTSON. 1988. A hollow-fiber reactor design for NMR studies of microbial cells. Biotechnol. Bioeng. Submitted.
17. STOCK, J. B., B. RAUCH & S. ROSEMAN. 1977. Periplasmic space in *Salmonella typhimurium* and *Escherichia coli*. J. Biol. Chem. **252**: 7850–7861.
18. FOWLER, J. D. 1990. Convective transport in immobilized cell aggregates. Ph.D. thesis. Stanford University.

APPENDIX

Nomenclature

b = osmotically inactive volume [m^3]
C = concentration [mol/L]
$\overline{C}$ = mixing-cup average concentration [mol/L]
D_c = diffusivity of cell aggregate [m^2 s^{-1}]
D_m = diffusivity of membrane [m^2 s^{-1}]
D_p = hydraulic diameter of cells [m]
F = flow rate through cell aggregate [mL min^{-1}]
k = Darcy permeability [m]
k_L = mass-transfer coefficient [m s^{-1}]
L = thickness of cell layer [m]
L_p = hydraulic permeability [m^2]
n = number of nutrient-supplying hollow fibers in the reactor [–]
n_j = number of intracellular moles of species j [–]
Q = flow rate through porous media [m^3 s^{-1}]
r = radial distance [m]
r_i = outer wall radius of nutrient-supplying hollow fiber [m]
r_L = inner wall radius of nutrient-supplying hollow fiber [m]
r_o = outer radius of reactor [m]
r_s = shell radius of reactor [m]
R = gas constant [J mol^{-1} K^{-1}]
$R(C)$ = reaction rate [J L^{-1} s^{-1}]
T = temperature [K]
V = cell volume [m^3]
V_0 = initial cell volume [m^3]
β = reciprocal Biot number [–]
γ_j = activity coefficient of species j [–]
δ = thickness of growing region
ΔP = pressure drop [N m^2]
μ = viscosity [kg m^{-1} s^{-1}]
σ_0 = applied stress [N m^2]
Π_i = intracellular osmotic pressure [N m^2]
Π_0 = medium osmotic pressure [N m^2]
ϕ = void fraction [–]

Fermentation Development of Recombinant *Pichia pastoris* Expressing the Heterologous Gene: Bovine Lysozyme

R. A. BRIERLEY, C. BUSSINEAU,[a] R. KOSSON,
A. MELTON, AND R. S. SIEGEL

*Salk Institute Biotechnology/Industrial Associates
SIBIA
San Diego, California 92138*

INTRODUCTION

The methylotrophic yeast, *Pichia pastoris*, has been shown to be an outstanding host for high-level heterologous gene expression.[1-4] The alcohol oxidase promoter was isolated and cloned by Ellis *et al.*[5] and transformation of *Pichia pastoris* was first reported in 1985.[6] The success of the *Pichia* expression system is linked to the strong, tightly regulated alcohol oxidase (*AOX1*) promoter. The strength of the promoter is demonstrated by the observation that AOX comprises up to 30% of the soluble protein in extracts of *Pichia pastoris* grown on methanol (MeOH) in a chemostat.[7] Another key feature of the system is that high cell densities can be achieved using a simple MeOH-salts medium.[8] The strong promoter coupled with the high-cell-density fermentations have allowed production of recombinant products at high intercellular (e.g., 400 mg/L hepatitis surface antigen)[1] and extracellular (e.g., 2.5 g/L invertase and 250 mg/L bovine lysozyme)[2,4] concentrations.

The recombinant *Pichia* fermentations referenced above[1,2,4] employed an MeOH utilization mutant (Mut⁻) created by integration of the recombinant expression cassette into the host genome at the *AOX1* locus, resulting in disruption of the *AOX1* structural gene. The Mut⁻ strains were grown in a repressed, batch regime for 24 hours, followed by an MeOH-induced, fed-batch regime for 100–200 hours. The long induction period is a consequence of the slow growth rate of the Mut⁻ strain on MeOH ($\mu = 0.01$–0.04 h^{-1}). The Mut⁻ strains are still able to utilize MeOH due to the presence of the *AOX2* gene, which yields 10–20 times less AOX activity than the *AOX1* gene.[9] The original Mut⁻ protocols offer several desirable features. They are extremely easy to scale up to large volumes and they consistently give maximal levels of expression.[1] Additionally, low growth rates may be desirable for production of certain recombinant products.[1,9,10] Furthermore, the Mut⁻ strains are not as sensitive to transient high residual MeOH concentrations, which can cause as much as a 99% loss of AOX activity and cell death following certain culture perturbations of wild-type methylotrophic yeast.[7,11-13]

Whereas the original process based on the Mut⁻ strain produces a high yield of recombinant protein, the slow fermentation limits productivity. In addition, the growth

---

[a]Current address: Chiron Incorporated, Fermentation Division, Emeryville, California.

rate, the MeOH uptake rate, and the cell yield decrease throughout the fermentation; these factors limit final cell density in the fermentation.

Thus far, the development of recombinant *Pichia* fermentations has only taken advantage of a relatively rudimentary understanding of the properties of the *AOX1* promoter. Much is known of the biochemical pathway for MeOH utilization in yeast,[14-16] specifically, the characteristics of the key enzyme, alcohol oxidase (AOX).[7,16] Studies on the regulation of AOX activity in closely related methylotrophic yeasts belonging to *Pichia* or *Hansenula* genera have uncovered several regulatory mechanisms. At the transcriptional level, induction by MeOH and repression by non-C_1 carbon sources play significant roles in regulating enzyme synthesis,[11,17-21] with the latter being the dominant control. Catabolic repression may operate to a certain extent in MeOH-grown cultures as well.[9,13,21]

The repressing non-C_1 carbon sources are members of either a glucose- or an ethanol-type family of compounds and they exert their effect to varying degrees.[20] Derepression occurs either by exhaustion of a repressing substrate (in batch cultures) or by substrate-limited growth (in fed-batch or continuous culture).[11,17,18] Finally, non-C_1 compounds such as glucose and ethanol also exert posttranslational effects by inactivating the AOX enzyme and causing breakdown of the peroxisomal matrix.[16] Medium components other than carbon sources, such as trace metals and nitrogen sources, have also had effects on both AOX expression and cell yields on MeOH.[8,22,23]

By taking advantage of what is known concerning AOX regulation, more productive processes can be developed for heterologous protein expression in *Pichia pastoris* because the expression of the heterologous protein is driven by the *AOX1* promoter. This study describes two approaches that significantly improve the productivity of recombinant *Pichia* fermentations.

The first approach increased the productivity of Mut$^-$ strains by the use of a mixed-carbon feed containing various compositions of a multicarbon substrate, glycerol, in addition to MeOH. Although mixed-feed fermentations (employing glucose, glycerol, sorbitol, and formate in conjunction with MeOH) have been previously utilized with methylotrophic yeast[12,24-27] to increase cell yield and productivity, their application in high-cell-density, recombinant cultures is unique.

The second approach employed recombinant hosts that were wild-type for MeOH utilization (Mut$^+$) in which the expression cassette was integrated into a site other than the *AOX1* locus. These strains utilize MeOH with a growth rate of 0.14 h^{-1}. In both approaches, higher specific growth rates decreased the time required to reach maximal recombinant product levels. Moreover, the Mut$^+$ strains are amenable to the high-productivity, MeOH-limited, continuous process previously developed for wild-type *Pichia pastoris*.[28]

The cloning and expression of bovine lysozyme c2 in *Pichia pastoris* have been recently described.[2,3] Bovine lysozyme is a 14,000–molecular weight mammalian protein. Over 90% of the recombinant lysozyme is secreted into the medium, representing the majority of total protein in the fermentor broth. Lysozyme concentration is easily measured with a standard *Micrococcus lysodekticus* lysis assay.[29] Recombinant *Pichia pastoris* strains expressing bovine lysozyme c2 were used as a model system to develop more productive fermentation processes.

RESULTS

Mut⁻ Methanol Fed-Batch

To establish a basis for comparison, a fermentation was first carried out using the original Mut⁻ protocol in which cell mass accumulated under repressed (excess glycerol) conditions. This was followed by an inducing (MeOH fed-batch) phase during which the residual MeOH was maintained between 0.2% and 0.8%. This protocol yielded a lysozyme concentration of 250 mg/L (TABLE 1). However, the induction phase required 175 h to attain the maximum lysozyme level, resulting in a low volumetric productivity of 1.1 mg/L-h for the fermentation (TABLE 1).

This process scaled to 10 liters very easily, as described in previous reports.[1] FIGURE 1 shows a comparison of the 10-liter and the 1-liter results. The higher product levels and productivity were expected for the 10-liter fermentation because a higher initial glycerol concentration was used (7% versus 4%). The 10-liter fermentation attained 325 mg/L in 125 h, giving a volumetric productivity of 2.1 mg/L-h for the induction phase. The product yield per unit cell of 5.2 mg/g cell was the same as that obtained in the 1-liter fermentation. Completely reducing the residual MeOH concentration had little or no effect on the yield of lysozyme from the Mut⁻ strain (data not shown).

Mut⁻ Mixed-Feed Fed-Batch

The initial attempt at increasing productivity with a mixed-feed strategy made use of a 4:1 (3.6 g glycerol/h, 0.9 g MeOH/h) feed composition. The maximum lysozyme concentration obtained was 180 mg/L. Whereas the 4:1 mixed-feed process produced a lower yield of lysozyme than the original protocol, the induction period was only 39 h, resulting in a 3-fold improvement in volumetric productivity (TABLE 1). In an attempt to further increase the level of lysozyme, a 2:1 ratio of glycerol to MeOH was evaluated; this ratio was achieved by reducing the glycerol feed rate by 50%, from 3.6 to 1.8 g/h. This protocol yielded 290 mg/L in 43 h, which is a 4-fold increase in volumetric productivity with respect to the original Mut⁻ process (TABLE 1).

Because the original Mut⁻ process utilizes a residual MeOH concentration (0.2–0.8%) to sustain maximum MeOH utilization, a third mixed-feed protocol was developed in which residual MeOH accumulates. In this protocol, the MeOH feed was

TABLE 1. One-Liter Fermentation Development of Recombinant *Pichia pastoris* Expressing Bovine Lysozyme c2

	MeOH-Fed Mut⁻	Mixed-Fed Mut⁻			MeOH-Fed Mut⁺
		4:1	2:1	NL[a]	
Maximum lysozyme concentration (mg/L)	250	180	290	375	450
Time induced (hours)	175	39	43	45	42
Cell density (dry g/L)	60	82	85	103	84
Volumetric productivity (mg/L-h)	1.2	3.4	4.8	5.6	7.7
Lysozyme yield per cell (mg/g)	5.2	2.3	3.7	4.0	5.6
Maximum ethanol concentration (mg/L)	10	210	100	100	30

[a] NL = nonlimiting.

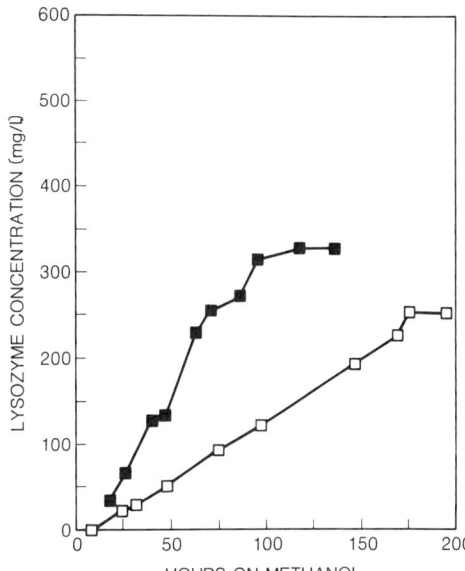

FIGURE 1. Plot showing the concentration of lysozyme with time in the induction phase of the Mut$^-$ strain grown on MeOH alone in a 2-liter fermentor (open squares) and in a 14-liter fermentor (closed squares).

increased for the 2:1 fermentation to a point where excess (nonlimiting) MeOH would begin to accumulate. With a nonlimiting (NL) MeOH feed, the cells utilized MeOH up to their maximal uptake rate and, thus, cell density increased at a maximal rate. The NL mixed-feed protocol enabled an MeOH uptake rate of 2–5 g/h. Hence, the ratio of glycerol to MeOH for the NL fermentation ranged from 1:1 to 1:3. By comparison, the maximum MeOH uptake rate for the original Mut$^-$ MeOH fed-batch fermentation was 0.5–1.0 g/h.

FIGURE 2 compares the lysozyme production in 2-liter fermentors under the four different protocols. The NL mixed-feed fermentation yielded 375 mg/L in 45 h, giving a 4.5-fold increase in volumetric productivity over the original MeOH-fed Mut$^-$ fermentation.

The mixed-feed fermentations gave significantly lower yields of product per cell than the MeOH-fed Mut$^-$ fermentation (TABLE 1). One explanation for the lower yields could be repression of the *AOX1* promoter by the metabolic by-product, ethanol (EtOH). TABLE 1 shows the maximum EtOH concentrations observed in each fermentation. The 4:1 fermentations produced twice the concentration of EtOH than did the 2:1 and NL fermentations (200 versus 100 mg/L) (TABLE 1). In other experiments, it has been observed that EtOH levels as low as 50 mg/L caused repression of the *AOX1* promoter. Thus, it is possible that the lower yields in the mixed-feed fermentations are caused by EtOH repression.

Mut$^+$ Methanol Fed-Batch

An alternative approach to increasing the productivity of recombinant *Pichia pastoris* fermentations employed the use of an Mut$^+$ strain. Because the expression

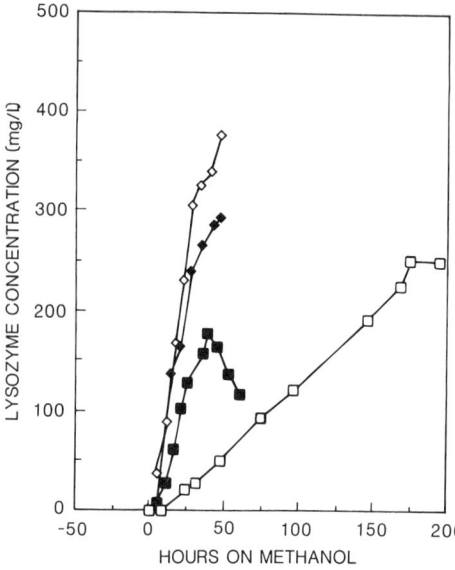

FIGURE 2. Plot showing the concentration of lysozyme with time in the induction phase in 1-liter fermentations of the Mut⁻ strain grown on MeOH alone (open squares), on a 4:1 glycerol to MeOH mixture (closed squares), on a 2:1 glycerol to MeOH mixture (closed diamonds), or on a limiting glycerol/nonlimiting (NL) MeOH mixture (open diamonds).

cassette is not integrated into the *AOX1* gene (i.e., the wild-type *AOX1* gene is maintained), the Mut⁺ strain exhibits a wild-type growth rate (0.14 h⁻¹ versus 0.035 h⁻¹ for Mut⁻ strains). Unlike the Mut⁻ strain, the Mut⁺ strain is very sensitive to changes in the residual MeOH level. For this reason, a protocol to circumvent the MeOH sensitivity of the Mut⁺ strain was developed. The key aspects of the Mut⁺ protocol include a glycerol-limited phase and a ramped MeOH feed during transition from glycerol to MeOH feeds. After the cell density is increased in a repressed, glycerol-excess batch phase, a glycerol feed is initiated at a limiting rate to further increase cell density and to allow derepression of the *AOX1* promoter. During this phase of the fermentation, AOX is synthesized at a level approximately 2–5% of that made during MeOH growth (data not shown). This glycerol fed-batch stage provides a smooth transition into the MeOH fed-batch stage.

The MeOH feed is initiated gradually with intermittent pauses to monitor for culture response. A sudden increase in the dissolved oxygen indicates that the culture is substrate-limited, that is, ready for a ramping-up of the MeOH feed rate. The feed is then increased by 10% every 30 minutes until the maximum feed rate (6 g/h) is obtained. This ramping procedure balances fast buildup of cell density and optimal expression from the *AOX1* promoter. It should be noted that the MeOH ramping is very important to the productivity of the system. In a fermentation in which a 40–60%/h ramping rate was used, lysozyme production slowed significantly during the fermentation and only 100 mg/L was produced in 25 h. Another important aspect of the MeOH ramping is that the feed should be increased at short time intervals to avoid any sudden increases in the MeOH feed rate.

The Mut⁺ protocol produced 84 g/L of dry cells and 450 mg/L of bovine lysozyme in 42 h, giving a 6.5-fold increase in the volumetric productivity over the original Mut⁻

fermentation and an approximately 40% improvement over the Mut⁻ (NL) mixed-feed fermentation (TABLE 1, FIGURE 3). The 1-liter Mut⁺ fermentation was scaled to 10 liters; a 16-fold increase in initial glycerol concentrations and MeOH feed rate was used to scale from the 1-liter to the 10-liter fermentations. FIGURE 4 shows a time course of lysozyme concentration in the 1- and the 10-liter fermentations of the Mut⁺ strains. The 10-liter fermentation reached 590 mg/L in less than 30 h. The volumetric productivity was 13 mg/L-h compared to 2.1 mg/L-h for the 10-liter Mut⁻ fermentation.

The high productivity for the Mut⁺ process (13 mg/L-h) has been reproduced several times. The 6.5-fold increase in productivity relative to the original Mut⁻ protocol is not necessarily surprising because the Mut⁺ strain exhibits a 5-fold higher growth rate than the Mut⁻ strain and almost twice the cell density can be attained on MeOH.

Mut⁺ Continuous Culture Fermentations

Continuous culture allows sustained production of maximal concentrations of lysozyme. Starting a basal salts feed at the end of a fed-batch phase initiated the continuous culture phase. The MeOH concentration in the total feed determines the cell density that can be achieved. In continuous culture, a 25–30% (w/v) MeOH feed yielded 85–120 g dry cell/L. FIGURE 5A shows the inverse correlation between lysozyme concentration and dilution rate. A dilution rate of 0.052 h⁻¹ yielded bovine lysozyme at a concentration of 300 mg/L in the cell-free broth; in contrast, a slower dilution rate of 0.045 h⁻¹ produced broth containing 440 mg/L lysozyme. When the

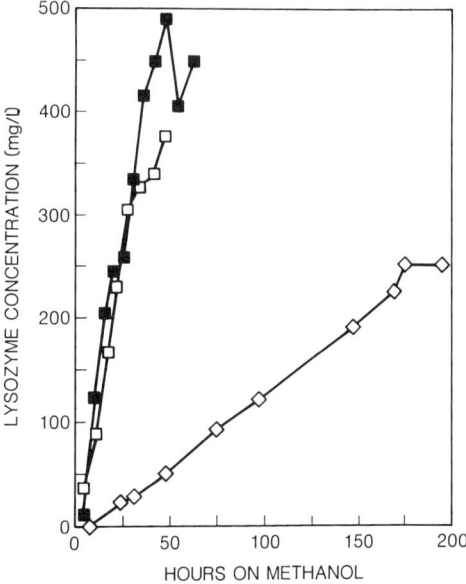

FIGURE 3. Plot showing the concentration of lysozyme with time in the induction phase in 1-liter fermentations of the Mut⁻ strain grown on MeOH alone (open diamonds) or on a limiting glycerol/nonlimiting (NL) MeOH mixture (open squares), or of the Mut⁺ strain grown under the limited MeOH protocol (closed squares).

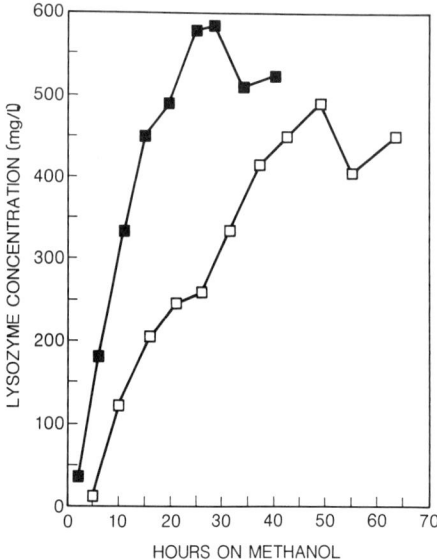

FIGURE 4. Plot showing the concentration of lysozyme with time in the induction phase of the Mut+ strain grown under the limited MeOH protocol in a 2-liter fermentor (open squares) and in a 14-liter fermentor (closed squares).

dilution rate was decreased even further ($0.035\ h^{-1}$), the lysozyme concentration reached 600 mg/L. One reactor volume of whole broth is produced every 19, 22, and 29 h, respectively, for the three dilution rates studied.

A comparison of total lysozyme production in two Mut+ continuous fermentations and in the original Mut− MeOH fed-batch fermentation is shown in FIGURE 5B. After 200 h of MeOH induction, the two Mut+ continuous runs (including 30 h in fed-batch) produced 17 and 19 g of lysozyme versus 3 g in 125 h in the original Mut− protocol. The relatively low lysozyme concentration observed early in FIGURE 5A was due to a suboptimal MeOH feed rate during the fed-batch portion of the run. Even though the volumetric productivity was lower for the fed-batch phase (5.7 mg/L-h), the productivity was high throughout the continuous phase, ranging between 12 and 15 mg/L-h. With high productivity being maintained for an extended period of time in the continuous process, the levels achieved in the fed-batch process become less critical when continuous culture is used. The time required to increase cell density in the glycerol phases of the fermentation along with the turnaround time and the inocula preparation time also become less significant with the increase in production output for the continuous process.

CONCLUSIONS

Two methods for improving the productivity of recombinant *Pichia pastoris* fermentations were examined. The first involved the use of a high-cell-density, mixed-substrate, fed-batch fermentation. The mixed-feed consisted of a repressing carbon source, glycerol, and MeOH as the inducing carbon source. Because the glycerol was fed into the fermentor at limiting amounts, repression of the *AOX1* promoter did not occur and the growth rate and MeOH uptake rate of the Mut⁻ strain increased as a result of the growth on glycerol as well as MeOH during the induction phase. The optimal NL fermentation, utilizing limiting glycerol and excess MeOH feeds (1:2 glycerol:MeOH), resulted in a 4.5-fold improvement in volumetric productivity of the production phase. Scaleup and application of mixed feeds in continuous culture are currently being addressed.

The second method involved the use of Mut⁺ strains in which MeOH utilization is similar to wild-type *Pichia pastoris*. The Mut⁺ phenotype allowed development of a process that increased the volumetric productivity of the induction stage by 6.5-fold. This improvement was demonstrated at the 1- and 10-liter scale and at cell densities in

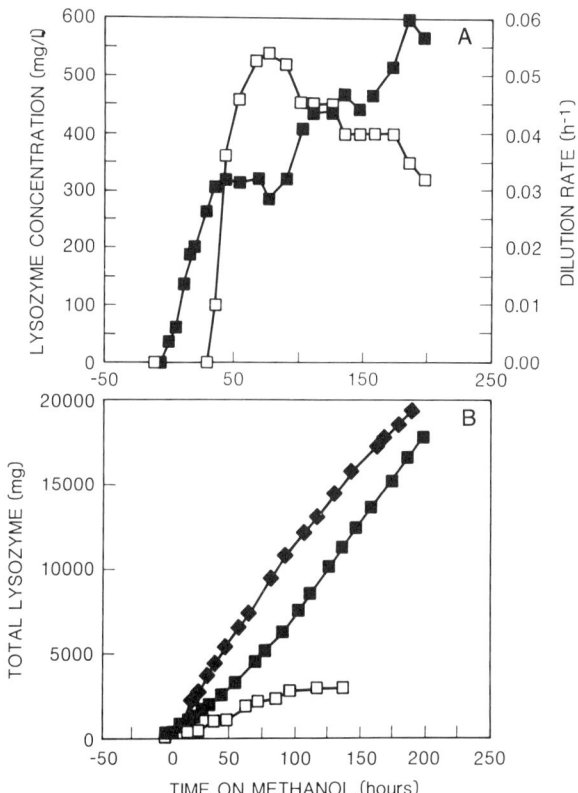

FIGURE 5. (A) Plot showing the concentration of lysozyme (closed squares) and the dilution rate (open squares) with time in the induction phase of the Mut⁺ strain grown in an 8-liter continuous culture mode using a 25% by weight MeOH feed. (B) Plot showing the total lysozyme produced with time in the induction phase of 10-liter fermentations using an Mut⁻ strain grown in the excess MeOH fed-batch fermentation (open squares), using an Mut⁺ strain grown under the optimized limited MeOH fed-batch fermentation followed by continuous culture mode (closed squares), and using an Mut⁺ strain grown under a nonoptimized limited MeOH fed-batch fermentation followed by continuous culture mode (closed diamonds).

excess of 130 g/L dry weight. The Mut⁺ strain also allowed the use of continuous culture for production of bovine lysozyme at a volumetric productivity of 12–15 mg/L-h. An 8-liter (working volume) continuous fermentation produced 19 g of bovine lysozyme in 200 h compared to the original Mut⁻ process, which produced 3 g of lysozyme in the same fermentor in 125 h.

Mut⁺ strains can also be carried out in mixed-feed fermentations to further increase the cell density and growth rate (i.e., productivity) during induction on MeOH. The MeOH-fed Mut⁺ and the mixed-fed Mut⁻ protocols have been used for several other heterologous (intercellular and secreted) products with the same favorable results that were seen with the lysozyme model system.

EXPERIMENTAL PROTOCOL

Strain

The *Pichia pastoris* strain GS115 (*his4*)[6] was used as the host strain for transformation with restriction fragments of the bovine lysozyme expression plasmid pSL12A.[2,3] This plasmid contains the *AOX1* promoter, regions from the 5' and 3' ends of the *AOX1* structural gene, the bovine lysozyme gene including the complete signal sequence, and the *HIS4* gene from *Pichia pastoris*. The Mut⁻ strain (A37)[2] was obtained by transformation with a *Bgl* II fragment of pSL12A; the Mut⁺ strain (L1)[3] was obtained by transformation with an *Sal* I fragment of pSL12A. The strains were carried on YNB without amino acids (Difco Labs, Detroit) plus 2% glucose agar plates with monthly transfers. Inocula were grown overnight at 30 °C and at 200 rpm in YNB without amino acids plus 2% glycerol plus phosphate buffer (pH 6.0).

Cell Density Determinations

Cell density was calculated by centrifuging whole fermentor broth for 10 minutes at a minimum of $4000g$ and weighing the cell pellet to determine the grams per liter of wet cells. A correlation factor of 0.25 was used to calculate the dry cell concentrations.

Ethanol and Methanol Concentration Determinations

EtOH and MeOH concentrations were determined on a Hewlett-Packard model 5890A gas chromatograph fitted with a flame ionization detector using a 6 foot by 0.25 inch glass packed column (super c9, 80/100, Alltech).

Bovine Lysozyme Assays and Calculations

Lysozyme concentrations were determined by a turbidimetric assay using lyophilized *Micrococcus lysodekticus* cells in 100 mM sodium phosphate buffer, pH 5.0.[29] A linear decrease in absorbance (at 450 nm) was measured over 3 minutes at 30 °C and was directly compared to previously purified recombinant bovine lysozyme[3] to calculate the concentrations of fermentor samples.

Concentrations (mg/L) of bovine lysozyme were determined from the cell-free broth. Actual fermentor (whole broth) concentrations were 20–30% lower than that in the cell-free broth, depending on the cell density of the reactor.

Total lysozyme (mg) was calculated from the whole broth lysozyme concentration. Volumetric productivity (mg/L-h) for the fed-batch fermentations was calculated by dividing the whole broth concentration of lysozyme by the time of induction required. Volumetric productivity for continuous culture was calculated by multiplying the whole broth lysozyme concentration by the dilution rate.

Yield of lysozyme per cell (mg/g) was calculated by plotting total lysozyme with total cells and using linear regression to determine the slope of the plot. The slope is the yield of lysozyme per cell. The correlation coefficient for the slopes was greater than 0.95.

Fermentor Start-up and General Operation

The 2-liter fermentors (L. H. Fermentation, Hayward, California; Biolafitte, LSL Biolafitte, Princeton, New Jersey) were autoclaved at a 700-mL volume containing 225 mL of $10\times$ basal salts (42 mL/L 85% phosphoric acid, 1.8 g/L calcium sulfate–$2H_2O$, 28.6 g/L potassium sulfate, 23.4 g/L magnesium sulfate–$7H_2O$, and 6.5 g/L potassium hydroxide) and 30 g glycerol. After sterilization, 3 mL of a YTM_4 trace salts solution (5.0 mL/L sulfuric acid, 65.0 g/L ferrous sulfate–$7H_2O$, 6.0 g/L cupric sulfate–$5H_2O$, 20.0 g/L zinc sulfate–$7H_2O$, 3.0 g/L manganese sulfate–H_2O, and 0.1 g/L biotin) was added and the pH was adjusted to 5.0 with the addition of concentrated ammonium hydroxide; the pH was then controlled at 5.0 with the addition of a 20% (v/v) ammonium hydroxide solution containing 0.1% (v/v) Struktol J673 antifoam (Struktol Company, Stow, Ohio) throughout the fermentation. Excessive foaming was sensed by a foam probe and was controlled by the addition of 5% (v/v) Struktol J673 antifoam. The fermentors were inoculated with a volume of 10–50 mL. Upon exhaustion of the initial glycerol charge, a glycerol feed was started as described later. The dissolved oxygen of the fermentation was maintained above 20% of air saturation by increasing the airflow rate up to 3 L/min and by increasing the agitation speed up to 1500 rpm during the fermentation.

Ten-liter fermentations (in a 14-liter Biolafitte fermentor) were started at a 5.0-liter volume containing 2.4 liters of $10\times$ basal salts and 360 g of glycerol for the Mut^+ MeOH fed-batch protocol or at an 8-liter volume containing 3.2 liters of $10\times$ basal salts and 480 g of glycerol for the Mut^- MeOH fed-batch protocol. For the Mut^+ fermentation, 29 mL of a YTM_4 trace salts solution was added after sterilization and the pH was adjusted and subsequently controlled at 5.0 with the addition of ammonia gas throughout the fermentation. For the MeOH-fed Mut^- fermentation, 40 mL of IM_1 trace salts[1] solution was added after sterilization and the pH was adjusted as above. Excessive foaming was controlled with the addition of 10% Struktol J673 antifoam. The fermentor was inoculated with a volume of 200–500 mL. Upon exhaustion of the initial glycerol charge, a feed was started as outlined later. The dissolved oxygen was maintained above 20% by increasing any or all of the following: the airflow rate up to 40 L/min, the agitation up to 1000 rpm, or the pressure of the fermentor up to 1.5 bar during the fermentation.

Mut⁻ Methanol Fed-Batch Fermentation

Upon glycerol exhaustion, an MeOH feed containing 12 mL/L YTM_4 trace salts was initiated for the 1-liter fermentation. The 10-liter fermentation utilized a pure MeOH feed with an addition of 40 mL of IM_1 trace salts[1] every two days. The MeOH feed rates were set so as to maintain a residual MeOH concentration of 0.1% to 0.8% in the fermentation broth.

Mut⁻ Mixed-Feed Fed-Batch Fermentation

After the glycerol batch phase was completed, a 50% (w/v) glycerol feed containing 12 mL/L YTM_4 trace salts was started at 5.4 mL/h. After 6 h of glycerol feeding, the glycerol feed was decreased to 3.6 mL/h and an MeOH feed containing 12 mL/L YTM_4 trace salts was initiated at 1.1 mL/h. After 5 h, the MeOH feed was increased to give a residual MeOH concentration between 0.2% and 0.8%. The fermentation is carried out for 40–50 h on the MeOH and glycerol feed.

The 2:1 fermentation was carried out as the NL fermentation except that the MeOH feed was not increased beyond 1.1 mL/h, giving a 2:1 ratio (by weight) of glycerol to MeOH throughout the fermentation.

The 4:1 protocol was also started in the same manner as the NL fermentation. During the simultaneous glycerol/MeOH feeding, the 4:1 fermentation had a glycerol feed rate that was twice that used in the 2:1 protocol, thereby giving a 4:1 ratio of glycerol to MeOH throughout the fermentation.

Mut⁺ Methanol Fed-Batch Fermentation

After glycerol exhaustion, a 50% (w/v) glycerol feed containing 12 mL/L YTM_4 trace salts was started at 12 mL/h for the 1-liter fermentation or at 200 mL/h for the 10-liter fermentation and then run for a total of 7 h. After 6 h on the glycerol feed, an MeOH feed containing 12 mL/L YTM_4 trace salts was pulsed at 1.1 mL/h (1-liter) or 11 mL/h (10-liter) fermentation for 5 minutes. When a rise in dissolved oxygen was seen following the MeOH pulse, the MeOH feed was turned back on for another 5-minute interval. The pulsing process was repeated several times until an immediate response in the dissolved oxygen was observed following the MeOH feed cessation; once this occurred, the MeOH feed was run continuously and the feed rate was increased by 20% per hour at 30-minute intervals. The MeOH feed was increased until a feed rate of 7.6 mL/h (1-liter) or 126 mL/h (10-liter) was reached. The fermentation was then carried out for 40–60 h for the 1-liter fermentation or for 25–35 h for the 10-liter fermentation.

Mut⁺ Continuous Culture Fermentation

Continuous fermentation was initiated by the addition of a 4× basal salts feed. Dilution rates between 0.03 and 0.08 h^{-1} included an MeOH feed (containing 12 mL/L of YTM_4 and IM_1 trace salts) as 20–30% (by weight) of the total feed. A weight controller was used to maintain a constant weight of the fermentor by activating a pneumatic harvest value.

ACKNOWLEDGMENTS

The authors wish to thank Mary Ellen Digan and Stephen Lair for providing the bovine lysozyme expressing strains; Pete Kellaris, Bill Craig, and Lillian Wondrack for providing purified recombinant lysozyme; and Lori Wilson for typing the manuscript. The authors are also indebted to Bill Craig, Mary Ellen Digan, Martin Gleeson, Michael Harpold, Jean Sartor, Walter Dreger, and Greg Thill for their helpful comments in the preparation of the manuscript.

REFERENCES

1. CREGG, J. M., J. F. TSCHOPP, C. STILLMAN, R. SIEGEL, M. AKONG, W. S. CRAIG, R. G. BUCKHOLZ, K. R. MADDEN, P. A. KELLARIS, G. R. DAVIS, B. L. SMILEY, J. CRUZE, R. TORREGROSSA, G. VELICELEBI & G. P. THILL. 1987. High-level expression and efficient assembly of hepatitis B surface antigen in *Pichia pastoris*. Bio/Technology **5**: 479–485.
2. DIGAN, M. E., J. TSCHOPP, L. GRINNA, S. V. LAIR, W. S. CRAIG, G. VELICELEBI, R. S. SIEGEL, G. R. DAVIS & G. P. THILL. 1988. Secretion of heterologous proteins from the methylotrophic yeast, *Pichia pastoris*. Dev. Ind. Microbiol. **29**: 59–69.
3. DIGAN, M. E., S. V. LAIR, R. A. BRIERLEY, R. S. SIEGEL, M. E. WILLIAMS, S. B. ELLIS, P. A. KELLARIS, S. A. PROVOW, W. S. CRAIG, G. VELICELEBI, M. M. HARPOLD & G. P. THILL. 1988. Production of bovine lysozyme via secretion from the yeast, *Pichia pastoris*. In preparation.
4. TSCHOPP, J. F., G. SVERLOW, R. KOSSON & L. GRINNA. 1987. High-level secretion of glycosylated invertase in the methylotrophic yeast *Pichia pastoris*. Bio/Technology **5**: 1305–1308.
5. ELLIS, S. B., P. F. BRUST, P. J. KOUTZ, A. F. WATERS, M. HARPOLD & T. R. GINGERAS. 1985. Isolation of alcohol oxidase and two other methanol regulatable genes from the yeast *Pichia pastoris*. Mol. Cell. Biol. **5**: 1111–1121.
6. CREGG, J. M., K. J. BARRINGER, A. Y. HESSLER & K. R. MADDEN. 1985. *Pichia pastoris* as a host system for transformants. Mol. Cell. Biol. **15**: 3376–3385.
7. COUDERC, R. & J. BARATTI. 1980. Oxidation of methanol by the yeast, *Pichia pastoris*. Purification and properties of the alcohol oxidase. Agric. Biol. Chem. **44**: 2279–2289.
8. SIEGEL, R. S. & R. A. BRIERLEY. 1988. The methylotrophic yeast *Pichia pastoris* can be produced in high cell density fermentations with high cell yields as a vehicle for recombinant protein production. Biotechnol. Bioeng. In press.
9. CREGG, J. M. & K. R. MADDEN. 1988. Development of the methylotrophic yeast, *Pichia pastoris*, as a host system for the production of foreign proteins. Dev. Ind. Microbiol. **29**: 23–42.
10. TSCHOPP, J. F., P. F. BRUST, J. M. CREGG, C. A. STILLMAN & T. R. GINGERAS. 1987. Expression of the lacZ gene from two methanol regulatable promoters in *Pichia pastoris*. Nucleic Acids Res. **15**: 3859–3876.
11. GLEESON, M. A. & P. E. SUDBERY. 1988. The methylotrophic yeasts. Yeast **4**: 1–5.
12. HAZEU, W. & R. A. DONKER. 1983. A continuous culture study of methanol and formate utilization by the yeast *Pichia pastoris*. Biotechnol. Lett. **5**: 399–404.
13. SWARTZ, J. R. & C. L. COONEY. 1981. Methanol inhibition in continuous cultures of *Hansenula polymorpha*. Appl. Environ. Microbiol. **41**: 1206–1213.
14. ANTHONY, C. 1982. The Biochemistry of Methylotrophs. Academic Press. New York.
15. OGATA, K., H. NISHIKAWA & M. OHSUJI. 1969. A yeast capable of utilizing methanol. Agric. Biol. Chem. **33**: 1519–1520.
16. VEENHUIS, J., J. P. VAN DIJKEN & W. HARDER. 1983. The significance of peroxisomes in

the metabolism of one-carbon compounds in yeasts. Advances in Microbial Physiology **24:** 1–82. Academic Press. New York/London.
17. EGLI, TH., J. P. VAN DIJKEN, M. VEENHUIS, W. HARDER & A. FIECHTER. 1980. Methanol metabolism in yeasts: regulation and synthesis of catabolic enzymes. Arch. Microbiol. **124:** 121–124.
18. EGGLING, L. & H. SAHM. 1978. Derepression and partial insensitivity to carbon catabolite repression of the methanol dissimilatory enzymes in *Hansenula polymorpha.* Eur. J. Appl. Microbiol. Biotechnol. **5:** 197–202.
19. ROGGENKAMP, R., Z. JANOWICZ, B. STANKOWSKI & C. P. HOLLENBERG. 1984. Biosynthesis and regulation of the peroxisomal methanol oxidase from the methylotrophic yeast *Hansenula polymorpha.* Mol. Gen. Genet. **194:** 489–493.
20. SIBIRNY, A. A., V. I. TITORENKO, B. D. EFREMOV & I. I. TOISTORKOV. 1987. Multiplicity of mechanisms of carbon catabolite repression involved in the synthesis of alcohol oxidase in the methylotrophic yeast *Pichia pinus.* Yeast **13:** 233–241.
21. VAN DIJKEN, J. P., R. OTTO & W. HARDER. 1976. Growth of *Hansenula polymorpha* in a mixed-limited chemostat. Arch. Microbiol. **111:** 137–144.
22. EGLI, TH. & A. FIECHTER. 1981. Theoretical analysis of media used in the growth of yeasts on methanol. J. Gen. Microbiol. **123:** 365–369.
23. EGLI, TH. & J. R. QUAYLE. 1986. Influence of the carbon:nitrogen ratio of the growth medium on the cellular composition and the ability of the methylotrophic yeast *Hansenula polymorpha* to utilize mixed carbon sources. J. Gen. Microbiol. **132:** 1779–1788.
24. EGLI, TH., O. KAPPELI & A. FIECHTER. 1982. Regulatory flexibility of methylotrophic yeasts in continuous culture: simultaneous assimilation of glucose and methanol at fixed dilution rates. Arch. Microbiol. **131:** 1–7.
25. EGLI, TH., C. BOSSHARD & G. HAMER. 1986. Simultaneous utilization of methanol-glucose mixtures by *Hansenula polymorpha* in chemostat: influence of dilution rate and mixture composition on utilization pattern. Biotechnol. Bioeng. **28:** 1735–1741.
26. EGGLING, L. & H. SAHM. 1981. Enhanced utilization-rate of methanol during growth on a mixed substrate. Arch. Microbiol. **130:** 362–365.
27. MUELLER, R. H., O. V. SYSOEV & W. BABEL. 1986. Use of formate gradients for improving biomass yield of *Pichia pinus* growing continually on methanol. Appl. Microbiol. Biotechnol. **25:** 238–244.
28. SHAY, L. K., H. R. HUNT & G. H. WEGNER. 1987. High-productivity fermentation process for utilizing industrial microorganisms. J. Ind. Microbiol. **2:** 79–85.
29. SHUGAR, D. 1952. Measurement of lysozyme activity and the ultraviolet inactivation of lysozyme. Biochim. Biophys. Acta **8:** 302–309.

Large-Scale Growth of *Bordetella pertussis* for Production of Extracellular Toxin

N. ANDORN,[a] J. B. KAUFMAN,[a] T. R. CLEM,[b] R. FASS,[a] AND J. SHILOACH[a,c]

[a]*Biotechnology Unit*
Laboratory of Cellular and Developmental Biology
National Institute of Diabetes
and Digestive and Kidney Diseases
National Institutes of Health
Bethesda, Maryland 20892

[b]*Biomedical Engineering and Instrumentation Branch*
Division of Research Services
National Institutes of Health
Bethesda, Maryland 20892

INTRODUCTION

The bacterium, *Bordetella pertussis*, is responsible for the upper respiratory infection known as whooping cough. The organism secretes several proteins that have been proven to play a role in its virulence.[1] One of the proteins, pertussis toxin, is an oligomeric protein that conforms to the A-B model of other bacterial toxins.[2] It enzymatically transfers the adenosine diphosphate (ADP) ribosyl moiety of NAD to certain acceptor proteins that are involved in the regulation of cyclic nucleotide metabolism.[3] Pertussis toxin is an effective immunogen in human volunteers[4] and has been suggested as the basis for the development of a new acellular vaccine.[5] To facilitate the testing of whooping cough vaccine, we have undertaken to isolate large amounts of purified pertussis toxin from the supernatant of *Bordetella pertussis* grown in submerged culture.

A major obstacle in this endeavor is the difficulty associated with the submerged growth of this microorganism. The problem lies in the tendency of the bacteria to float on the air bubbles and, as a result, to be removed from the media and stop growing. Therefore, complete elimination of foam is needed.

Bordetella pertussis is a slow-growing microorganism with a doubling time of 3.5 to 4 hours; consequently, its oxygen consumption rate is relatively low. Thus, the traditional solution for the foam problem was vortex agitation and surface aeration,[6] which eliminated the foam production. This type of operation was not a satisfactory one because only one-third of the available fermentor volume was being used and the dissolved oxygen level became limiting. Recently, we reported on the use of bottom aeration with oxygen-enriched air in an antifoam-containing media.[7] This approach

---

[c]To whom all correspondence should be addressed.

enabled us to produce up to 4 mg per liter of pertussis toxin in submerged culture and it allowed the use of the full working volume of the fermentor. The addition of relatively large amounts of antifoam to the growth media was found to cause some difficulties with later purification procedures. In order to reduce the amount of antifoam and prevent foam formation, a different type of fermentor was needed. In the following sections, we describe the use of a modified draft-tube fermentor designed for the submerged growth of *Bordetella pertussis* and for the production of its extracellular toxin while avoiding foam formation with the addition of relatively small amounts of antifoam.

MATERIALS AND METHODS

Bacterial Strain

Phase 1 *Bordetella pertussis* strain CS (LDM1 collection) was used. This strain was isolated in China from a patient and proved to be satisfactory for vaccine production in that country.

Culture Media

Modified, chemically defined, Stainer-Scholte[8] liquid medium was used throughout this work as previously described.[7] Four mL of antifoam C (containing silicone and a nonionic emulsifier; Sigma Chemical) was added per 12-L medium.

Fermentation

Lyophilized cultures were rehydrated with sterile saline and grown on Bordet-Gengou (BG) blood agar plates (15% sheep blood) for three days at 37 °C. The bacteria from two plates were used to inoculate 200 mL of medium in a 500-mL flask, which was then incubated for 20 h at 37 °C on a rotary shaker (200 rpm). Next, 100 mL of this culture was used to inoculate a 2.8-L Fernbach flask with 1.2 L of medium. The Fernbach flasks were incubated for 20 h at 37 °C. One flask was used to inoculate the 14-L bench-top fermentor and three flasks were used to inoculate the 50-L draft-tube fermentor. Both fermentors were inoculated to an initial A_{650} of 0.1–0.3.

Construction of the Draft-Tube Fermentors

A special 14-L draft-tube fermentor was built as shown in FIGURE 1. A stainless steel tube with heat exchanger was fabricated and a vaned disk impeller was installed at the bottom of the tube. In the 50-L fermentor, the baffles were removed and the draft tube was installed with an impeller at the bottom. The air jet at the bottom of both fermentors was eliminated and the air was supplied from the top (FIGURES 1 and 2). The fermentors were filled slightly above the inner tube in order to create an overflow at a mixing speed between 700 and 1000 rpm. In the 14-L fermentor, the dissolved oxygen concentration was monitored and controlled by a gas flow controller (B. Braun). Other parameters (temperature, agitation, pH, etc.) were controlled by the

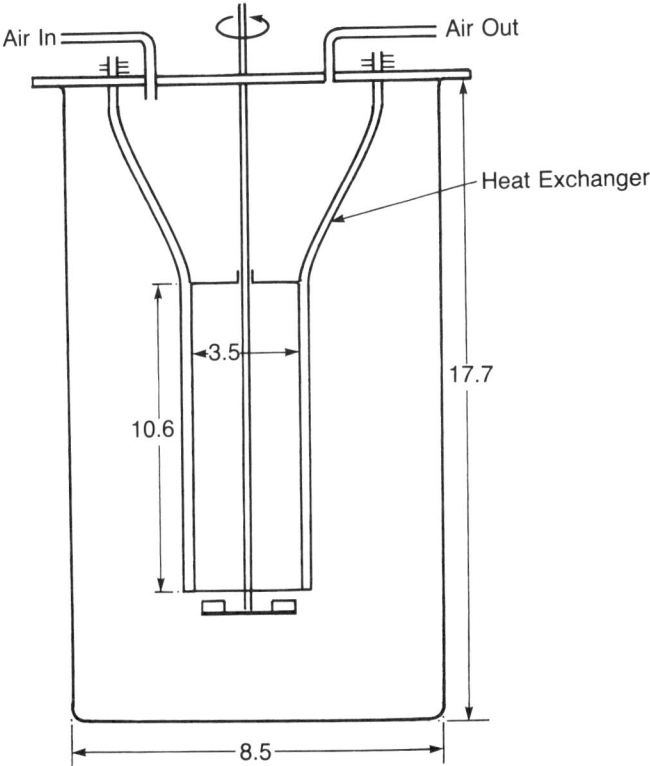

FIGURE 1. Draft-tube vessel for the bench-top fermentor. The vessel was constructed by modifying a 14-L glass Magnaferm fermentor. The conventional heat exchanger, sparger, and baffle assembly were removed and a draft-tube system was installed instead. All dimensions are in inches.

Magnaferm fermentation system (New Brunswick). The growth in the 50-L fermentor was controlled by a Gen II control system (New Brunswick) connected to a PC computer, which was used for data acquisition and on-line analysis.

Analytical Methods

The concentration of pertussis toxin was determined by measuring its ADP–ribosyl transferase activity, as previously described.[7]

The concentration of ATP in the culture at various stages of the fermentations was determined by the Firefly luciferase/luciferin method using Picozyme F reagents (Packard Instrument Company, Downers Grove, Illinois) and a Picolite luminometer (Packard). The concentration of glutamic acid in the growth medium was determined using a glutamic acid test kit (Boehringer Mannheim). Proline concentration in the culture supernatant was determined by the method of Vogel and Shimura.[9] Fluores-

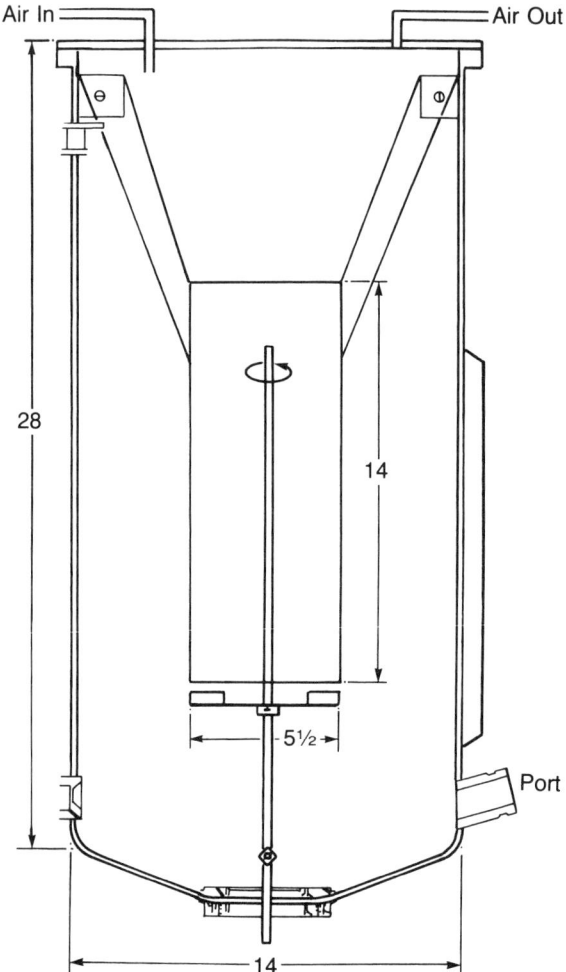

FIGURE 2. A 50-L draft-tube fermentor. The baffles and sparger of a conventional STR were replaced by the draft tube and the vaned disk impeller. All dimensions are in inches.

cence of the culture medium was continuously determined using an in-place–sterilizable fluorescence electrode (Ingold Electrodes, Switzerland).

Comparative oxygen absorption rate (OAR) determinations in the 14-L fermentor were done using the sulfite oxidation method.[10]

RESULTS

Oxygen Absorption Rate (OAR) Measurements

The OAR of the 14-L fermentor was measured with and without the draft-tube special assembly (TABLE 1). It is clear that the oxygen–liquid phase mass-transfer coefficient ($K_L a$) value obtained in the stirred tank reactor (STR) under the regular growth conditions,[7] specifically, 14.1 mM O_2/L/min, can be obtained with the draft-tube fermentor using surface aeration at an agitation speed of 700 rpm. The results indicate that a draft-tube assembly with top aeration will be able to support the initial growth of *Bordetella pertussis*. At high cell concentrations, oxygen-enriched air will be needed, as it was with the STR.[7]

Submerged Growth of Bordetella pertussis *in the Draft-Tube Fermentor*

FIGURE 3 displays typical growth parameters of the bacteria grown in Stainer-Scholte media in a 14-L draft-tube fermentor. Air was supplied from the top at a rate of 2 liters per minute until the dissolved oxygen concentration fell below 20%. At this point, the air was replaced by pure oxygen and the dissolved oxygen concentration was kept around 30%. The profile of the dissolved oxygen concentration during the course of the fermentation shows that the use of top aeration with pure oxygen did not cause extreme changes in the dissolved oxygen level. Turbidity, ATP, and NADH measurements done during the fermentation course indicate that the use of pure oxygen did not affect the growth pattern (FIGURES 3C and 4). The pattern of proline and glutamic acid consumption during the fermentation in the 14-L draft-tube fermentor (FIGURE 3A) was similar to the pattern obtained with the STR fermentation of *B. pertussis* in the same medium.[7] The top aeration with pure oxygen enabled the culture to grow at a constant rate until it entered the stationary phase at a cell concentration of 10^{10} cells/mL ($A_{650} = 4.5$, FIGURES 3 and 4).

These results are comparable to those achieved with the conventional STR.[7] A similar growth pattern was also obtained in a 50-L draft-tube fermentor as shown in the computer output given in FIGURE 4. Both fermentations were performed using only 0.3 mL of antifoam C per liter of culture. This amount of antifoam was enough to prevent foam production during the course of the fermentation and did not inhibit the growth of the bacteria nor interfere with the toxin purification.

DISCUSSION

Recently, we reported on the large-scale cultivation of *Bordetella pertussis* in submerged culture.[7] The method we reported involved growing the bacteria in a conventional STR where air supply was replaced with pure oxygen and the dissolved oxygen level was kept around 30%. This novel method gave us satisfactory results and

TABLE 1. Oxygen Absorption Rate of the Modified Draft-Tube Fermentor and a Stirred Tank Reactor[a]

Fermentor Type	Volume (L)	Agitation (rpm)	Airflow (L/min)	$K_L a$ (mM O_2/L/min)
STR	10	300	2	14.1
Draft-tube	12	700	2	18.1

[a]The air to the stirred tank reactor was supplied using a ring sparger. The air to the draft-tube fermentor was supplied from the surface. Measurements were conducted by the sulfite oxidation method in the same vessel by changing the inner assembly.

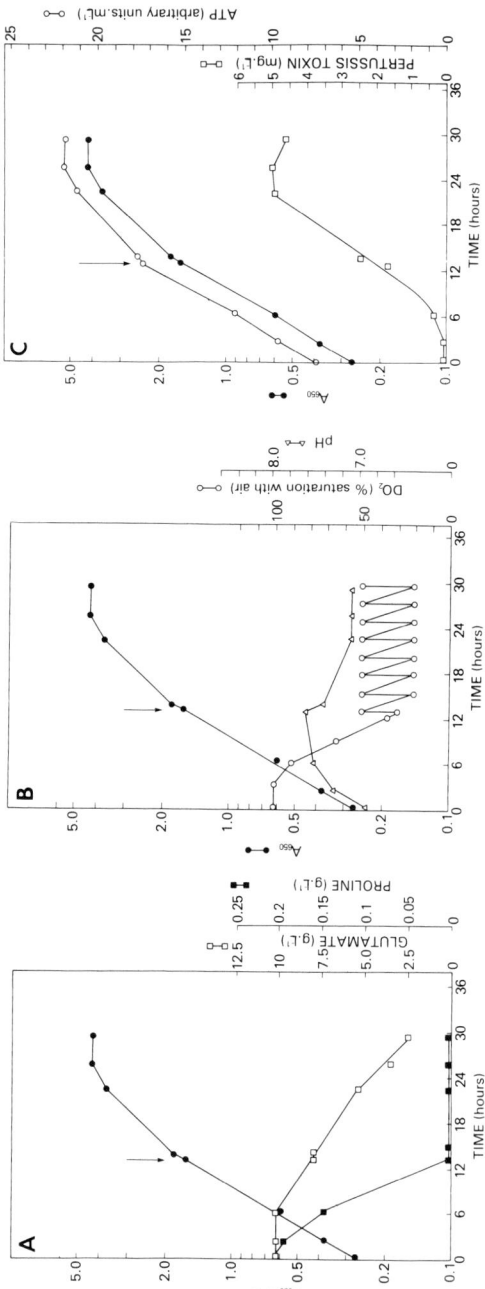

FIGURE 3. Growth of *Bordetella pertussis* in a modified bench-top draft-tube fermentor and production of pertussis toxin. (A) Amino acid consumption during the course of the fermentation. (B) Dissolved oxygen (DO_2) and pH profiles during the fermentation: Air was supplied from the top at a rate of 2 L/min until the dissolved oxygen level fell below 20% (arrow). At this point, the air was replaced with pure oxygen and the dissolved oxygen concentrations were kept around 30%. (C) ATP and pertussis toxin in the culture supernatant during the course of the fermentation.

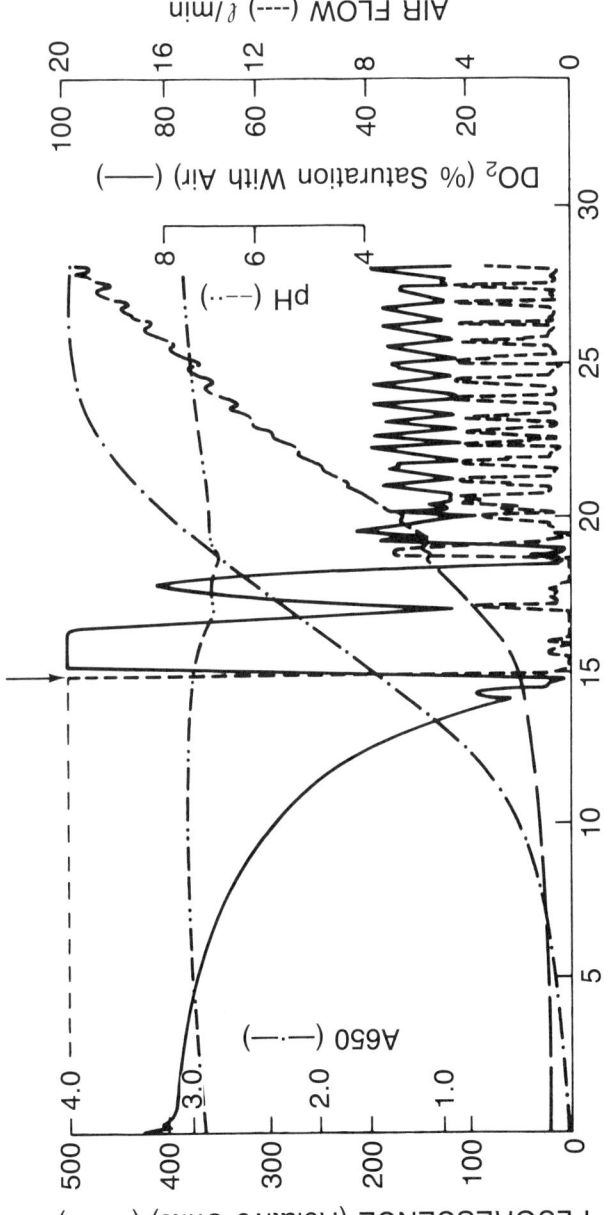

FIGURE 4. Fermentation profile of *Bordetella pertussis* in a 50-L fermentor. The bacteria were grown in the 50-L modified draft-tube fermentor. The growth parameters are seen in the computer output. The arrow indicates the time when pure oxygen was introduced into the system.

enabled us to produce enough toxin for the first stage of clinical trials to test the toxoid as an acellular vaccine.

However, the submerged growth of *Bordetella pertussis* in STR has some drawbacks: the bottom sparging of oxygen has a potential for foam production; it was difficult to control the dissolved oxygen concentration around a certain value; and we could not exceed 300 rpm or 0.2 v/v/m without major foam production, even in the presence of antifoam. Consequently, we searched for a better system. The requirements from such a system would be to eliminate completely the foam production, to use less antifoam, and to enable the controlling system to keep the dissolved oxygen level around 30%. As we demonstrated in our previous work, these were important factors in the growth of this microorganism.

To achieve these goals, we decided to explore the possibility of supplying oxygen from the top rather than from the bottom via a draft-tube system. In the regular draft tube (air-lift loop), the injected air (or gas) at the bottom of the vessel was responsible for the mixing and the circulation. These fermentors are believed to be more efficient than the STR fermentor with respect to the oxygen transfer.[11,12] In our proposed system, a vaned disk impeller was installed at the bottom of the draft tube and the sparger from the bottom of the fermentor was replaced with an air overlay. The fermentor was filled slightly above the inner tube in order to create an overflow when the mixing was at high speed between 700 and 1000 rpm.

This system was found to work in a reproducible manner. It performed as well as the STR system in terms of *B. pertussis* growth and toxin production. It was better than the conventional STR system concerning both the demand for antifoam and the ability to control the dissolved oxygen level. Most important, it did not produce foam during the course of the fermentation. These features should make it possible to further develop the fermentation process with a goal of achieving better cell density and higher toxin concentration.

REFERENCES

1. PITTMAN, M. 1979. Pertussis toxin: the cause of the harmful effects and prolonged immunity of whooping cough. A hypothesis. Rev. Infect. Dis. **1:** 401–412.
2. TAMURA, M. K., M. NOGIMORY, S. MURAI, K. ITO, T. KATADA, M. VI & S. ISHII. 1982. Subunit structure of islet-activating proteins, pertussis toxin, in conformity with an A-B model. Biochemistry **21:** 5516–5522.
3. KATADA, T. & M. VI. 1981. Islet-activating proteins. J. Biol. Chem. **256:** 8310–8317.
4. SEKURA, R. D. 1986. Safety and antigenicity of pertussis toxoid vaccine evaluated in adult volunteers. Workshop on Acellular Pertussis Vaccine, p. 135–140. National Institutes of Health. Bethesda, Maryland.
5. ROBBINS, J. B. 1984. Toward a new vaccine for pertussis. Microbiology 1984, p. 176–183. American Society for Microbiology.
6. CAMERON, J., P. ROUSSEAU & G. TREPANIER. 1985. *Bordetella pertussis*: growth in stainless steel. J. Biol. Stand. **13:** 97–100.
7. ANDORN, N., Y. L. ZHANG, R. D. SEKURA & J. SHILOACH. 1988. Large scale cultivation of *Bordetella pertussis* in submerged culture for production of pertussis toxin. Appl. Microbiol. Biotechnol. **28:** 356–360.
8. STAINER, D. W. & M. J. SCHOLTE. 1971. A simple chemically defined medium for the production of phase I *Bordetella pertussis*. J. Gen. Microbiol. **63:** 211–220.
9. VOGEL, H. G. & Y. SHIMURA. 1971. Spectrophotometric determination of lysine. *In*

Methods in Enzymology. Volume 17B. H. Tabor & C. W. Tabor, Eds.: 228–229. Academic Press. New York.
10. COOPER, C. M., G. A. FERNSTROM & S. A. MILLER. 1957. Performance of agitated gas liquid contactors. Ind. Eng. Chem. **36:** 504–509.
11. SCHIIGERE, K., U. OELS & J. LUCKE. 1977. Bubble column bioreactor. *In* Advances in Biochemical Engineering. Volume 7. T. K. Ghose, A. Fiechter & W. Blakebrough, Eds.: 1–84. Springer-Verlag. Berlin/New York.
12. BULL, D. N., R. WIATONA & T. E. STINNETT. 1983. *In* Advances in Biotechnological Processes. Volume 1. A. Mizrahi & A. L. Van Wezel, Eds.: 1–30. Alan R. Liss. New York.

PART VI. BIOREACTORS II (HIGHER EUKARYOTIC SYSTEMS)

Growth Kinetics of Free and Immobilized Insect Cell Cultures[a]

S. N. AGATHOS,[b,c] Y-H. JEONG,[b] AND K. VENKAT[b,d]

[b]Department of Chemical and Biochemical Engineering
Rutgers University
Piscataway, New Jersey 08855-0909

[c]Waksman Institute
Rutgers University
Piscataway, New Jersey 08854

[d]H. J. Heinz World Headquarters
Pittsburgh, Pennsylvania 15320

INTRODUCTION

In Vitro *Cultivation of Insect Cells in the Spotlight*

As this decade is drawing to a close, a significant change is occurring in our current thinking and evaluation of promising biotechnological production systems. A sober reassessment of the heady predictions of the late 1970s and 1980s has led to the recognition that the capabilities of genetically modified bacteria and lower eukaryotes were largely overestimated in regard to their ability to produce most commercially valuable proteins. Key obstacles have been the technical difficulties of introducing higher eukaryotic gene sequences into microbes, the incompatibility of foreign gene overexpression with host cell physiology, and—probably most crucial—the inability of the host cell to perform many of the posttranslational modifications required for the correct biological functioning of many animal proteins. Therefore, it appears increasingly likely that the majority of economically attractive (e.g., therapeutic) animal proteins must be produced on an industrial scale using animal, particularly mammalian, cells. This is providing a sizable impetus for highly visible research-and-development activities focusing on more efficient cultivation of animal cells *in vitro*.

Against this backdrop of recent concerted strides in animal cell technology, there has emerged an unexpected and burgeoning interest in, specifically, the development of improved and scaled-up systems of insect cell culture (ICC). Earlier ICC schemes were motivated by fundamental research needs based on the physiology, pathology, and ecology of insects and had been geared to more efficient systems of cultivation in view of the prospect for mass-producing viral insecticides as an environmentally sound alternative to chemical pest control.[1,2] However, the current excitement about ICC as a novel biotechnological tool is largely due to the recent development of baculovirus vectors for high-level and faithful expression of many foreign genes whose products are correctly processed posttranslationally (glycosylation, folding, phosphorylation, etc.).[3,4]

---

[a]The experimental work at Rutgers University was supported through a Rutgers University Research Council grant and through Biomedical Research Support Grant No. PHS RR 07058-21 to S. N. Agathos.

Approximately 150 biotechnological companies are reportedly in the process of evaluating or actually using insect cell lines for product manufacture or new process development[5] due to the immediate acceptance and the high perceived potential of the insect cell-baculovirus system for the synthesis of bioproducts of medical and agricultural significance.[3,4,6] Among the most visible recent applications of this system is the development of the first candidate vaccine against the AIDS virus, HIV, now in clinical trials,[7] following the successful expression of specific sheath proteins of the HIV in insect cell culture.[8-10] Many other heterologous proteins, a number of them with posttranslational processing requirements for their proper therapeutic functioning, have already been produced in the same expression system.[3,6] It is expected, then, that ICC will be the focus of further technological development, including scaleup efforts, in order to meet increasing production requirements for many desirable biologicals. This article presents an introductory overview of insect cell applications, surveys a number of biochemical engineering considerations underlying the successful development of production-scale ICC systems, and contributes in brief our recent experimental findings on the kinetic behavior of insect cells grown in suspension and in matrix-confined (immobilized) form.

BIOTECHNOLOGICAL APPLICATIONS OF INSECT CELL SYSTEMS

Insect cell systems of interest for large-scale applications are not limited only to the cell types that can serve as hosts to baculoviruses (primarily cells of lepidopteran species, i.e., moths and butterflies), but include also cells of dipteran insects that transmit medically important diseases (mosquitoes) or provide well-established model genetic systems (fruit fly, *Drosophila*). Some insect cell systems with current industrial potential are summarized briefly in TABLE 1.

The most useful group of baculoviruses, with applications in the production of viral insecticides as well as of recombinant proteins, are known as nuclear polyhedrosis viruses (NPV). Bioinsecticides approved for use in agriculture (annual crop and orchard protection) and in forestry include NPV that have as their target the following lepidopteran insects: *Heliothis zea* (cotton bollworm), *Autographa californica* (alfalfa looper), *Spodoptera frugiperda* (fall armyworm), *Trichoplusia ni* (cabbage looper), *Lymantria dispar* (gypsy moth), *Orgyia pseudotsugata* (Douglas fir tussock moth), and *Neodiprion sertifer* (European pine sawfly).[11-13] High-level heterologous gene expression has been achieved using *Autographa californica* NPV (AcNPV) as the prototype viral vector for the efficient and rapid production of biologically active proteins ranging from lymphokines (interferons, interleukins) to enzymes, viral antigens, and blood cell growth factors.[3,6] The genetically manipulated AcNPV vector has been used to transfect cell lines of *Spodoptera frugiperda in vitro* (primarily cell line Sf 9, which has emerged as something of a standard) for foreign gene expression;[3,4,6,14] however, it recently has been shown that *Trichoplusia ni* cell cultures can also be used for the same type of application.[15] An alternative, *in vivo* system for efficient heterologous protein production utilizes whole insect larvae of *Bombyx mori* (silkworm) infected with the recombinant baculovirus vector BmNPV.[16,17] With this *in vivo* system, biologically active human α-interferon and murine interleukin-3 have been produced and the potential is high for the manufacture of many other therapeutic bioproducts.[18]

The cultivation of dipteran insect cells is a modern biotechnological tool targeted at already established or potential applications that include production of the following: vaccines against arboviruses for human and veterinary uses, viral antigens for diagnos-

TABLE 1. Insect Cell Systems with Industrial Potential

Cell Type	Product	Area for Research and Development
(a) DIPTERAN CELLS		
Aedes albopictus (forest day mosquito) *Aedes aegypti* (yellow fever mosquito)	Arbovirus antigen diagnostics Vaccines (dengue, Japanese encephalitis, etc.) Viral pesticides Recombinant proteins	Large-scale cell culture (bioreactor design, media optimization, enhanced cell density and viability, cell disaggregation) Novel expression vectors, gene amplification Chromosomal integration
Drosophila melanogaster (fruit fly)	Recombinant proteins	Large-scale cell culture Novel expression vectors, gene amplification
(b) LEPIDOPTERAN CELLS		
Bombyx mori (silkworm)	Recombinant proteins (interferons, interleukins, etc.)	Baculovirus vector construction Manipulation of recombinant protein processing Large-scale larva culture (automated production, recovery)
Spodoptera frugiperda (fall armyworm) *Trichoplusia ni* (cabbage looper)	Viral insecticides Recombinant proteins (lymphokines, cell growth factors, enzymes, viral antigens, etc.)	Baculovirus vector construction Manipulation of recombinant protein processing Large-scale insect cell culture (bioreactor design, serum-free media, enhanced cell density and viability, cell adhesion)

tic purposes, and viral pesticides.[19] Mosquito cells of the species *Aedes albopictus* and *Aedes aegypti* are convenient vehicles for the *in vitro* propagation of arboviruses (= arthropod-borne viruses) that are the causative agents of serious infectious diseases, including dengue hemorrhagic fever, Rift Valley fever, Japanese encephalitis, West Nile fever, and others.[19-21] Another representative dipteran, *Drosophila melanogaster* (fruit fly), is one of the most thoroughly studied organisms with respect to gene expression and differentiation. Besides its continuing usefulness as a model system in tissue culture,[22] this insect lends itself to stable foreign gene expression both through its P transposable element that permits integration of the foreign gene into host chromosomes[23] and through the development of new vectors leading to efficient heterologous protein production by chromosomal integration of up to 10^3 gene copies per cell.[24] Stable genome integration and expression of a foreign gene has also been reported recently for the mosquitoes *Anopheles gambiae*,[25a] *Aedes triseriatus*,[25b] and *Aedes aegypti*,[25c] whereas transient heterologous gene expression following the efficient transfection of cultured mosquito (*Aedes albopictus*) cells has been recently demonstrated.[26,27]

Thus, dipteran cells, including mosquito cells in culture, present themselves as attractive potential hosts for overexpression of heterologous gene products,[28] given the constantly accruing knowledge of their biochemistry and genetics.[29] In contrast to the lepidopteran cell–baculovirus vector system, which is a lytic transformation system for foreign gene expression (i.e., the viral vector leads to the lysis and eventual death of the host cells), dipteran cells appear to be prime candidates for stable, long-term heterologous gene product formation. The industrial implication of these two contrasting systems lies in the realization that the former one is suitable for batch culture, whereas the latter may be amenable to continuous culture configurations (e.g., perfusion culture). Of course, the overall productivity of any proposed ICC-based process will depend critically not only on the absolute level of foreign gene expression possible in the candidate system, but also on the time required for this level to be attained reproducibly, including preculture time, "down" (turnaround) reactor time, and time for recovery, separation, and purification of the product stream.

Before concluding this brief survey of the current industrial potential of insect cell systems, it should be pointed out that insect cells, especially in culture, do not derive their perceived usefulness solely from the various viruses they are hosting nor from their capacities as "factories" for genetically engineered products, although these are indeed the most highly visible areas of application today. The scope of ICC applications is widening to include products exclusively derived from specific insect cell lines. Concerted research efforts may lead to the discovery of new, medically and agriculturally important substances naturally formed by insect cells in culture, as illustrated by the recent patenting in Japan of a potent nontoxic, thermostable virucidal and antibiotic 40–amino acid protein that is produced by cultured cells of the dipteran, *Sarcophaga peregrina* (flesh fly).[30]

BIOPROCESS ENGINEERING CONSIDERATIONS IN INSECT CELL CULTURE

It is clear from the foregoing section that efficient cultivation methods are a vital prerequisite in the commercial exploitation of insect cell systems. Industrial production of bioinsecticide baculoviruses still represents the largest-volume specialty biochemical manufactured from insect cells and it is carried out at present *in vivo* using whole insect larvae.[31,32] However, there are many inherent drawbacks in the *in vivo* process: it tends to be expensive, time-consuming, labor-intensive, not easily compatible with automatic process control, potentially challenging in terms of product recovery, and difficult to scale up economically. There may be some mitigating factors in manufacturing high-value-added recombinant proteins *in vivo* using the relatively large–sized larvae of *B. mori* (cheaper artificial diets, new automated technology for mass cultivation of silkworm larvae),[17,18] but, as a broadly applicable generic strategy, the *in vitro* process of ICC appears to be more suitable for large-scale production of biologicals. Within the framework of ICC techniques, the propagation of insect cells in monolayers or stationary culture may also be unsuitable for large-scale production due to the decrease in the surface-to-volume ratio upon an increase in the scale of operation, with a corresponding decrease in the profitability of the process. Thus, the most promising process development approaches are being oriented primarily towards cultivation of insect cells in suspension, using bioreactor technology and engineering principles that are currently applied successfully to microbial and mammalian cell product manufacturing.

The early pioneering development of insect tissue culture media[33–35] and of reproducibly *in vitro* propagated (= "established") cell lines or strains[34,36–38] was crucial in the evolution of standardized techniques for the growth of insect cells in laboratory-scale cultivation.[22,39–41] This knowledge was subsequently applied to schemes of more massive production of insect cells in culture (ranging in volumes from 100 mL to 10 L, but typically below 3 L), which were particularly aimed at obtaining large amounts of virus free from microbial contamination.[42–46] These developments are a testimony to the spectacular improvement of media and of aseptic insect tissue culturing techniques over the last four decades. In addition, this progress in culturing techniques has opened up the possibility of routinely producing high quantities of insect cells over a short time through bioprocess technology and engineering, irrespective of the eventual use of the insect cell mass accumulated.

Although large-scale ICC is in many respects analogous to the mass culturing of vertebrate animal (e.g., mammalian) cells on an industrial scale and, from the bioprocess engineering point of view, related to fermentation technology, there is still limited understanding and little published documentation of the behavior of insect cells growing in bioreactors. Before discussing the bioprocess engineering considerations pertaining to the efficient and scalable cultivation of insect cells in bioreactors, a comparison of insect cells with mammalian cells is in order in terms of (A) nutritional and physical requirements for growth (physiology), (B) cell propagation technology, and (C) the expression system for production of regulated biologicals (see TABLES 2A, 2B, and 2C, respectively).

Cell Lines and Media

Attempts to culture insect cells and organ tissue *in vitro* date back to the second decade of the twentieth century, yet the modern efforts to formulate convenient cultivation media and continuous cell lines are developments of the 1950s and 1960s. Wyatt[33] formulated a synthetic medium consisting of sugars, amino acids, organic acids, and inorganic salts on the basis of the chemical composition of hemolymph, that is, the equivalent of blood in insects. This medium, supplemented with heat-inactivated hemolymph, became the first rigorously designed insect cell medium capable of supporting long-term growth of insect cells *in vitro*. Grace[34] modified Wyatt's medium by adding a complement of vitamins and increasing the osmotic pressure, in recognition of the corresponding nutritional and physicochemical requirements of several types of insect cells for *in vitro* growth. It was in this medium, still in widespread use today, that Grace established the first insect cell line[34] in 1962.

Since then, more than 200 insect cell lines have been established from seven orders, with the most numerous being those from Diptera (95) and those from Lepidoptera (73).[5] These cell lines (representing approximately 75 different species of insects) are being compiled regularly by Hink.[47-49] For the establishment and serial propagation of these cells, more than 60 culture media have been formulated to date—many of them expressly designed for ICC, some originally used for vertebrate cell cultivation, some

TABLE 2A. Comparative Aspects of Cultured Insect and Mammalian Cells: Physiology

	Insect Cells	Mammalian Cells
Average diameter	10 μm	15 μm
Average doubling time	$\approx$18 h	$\approx$24 h
Optimum temperature	28 °C	37 °C
Optimum pH	6.0–7.0	7.0–7.3
Osmolarity	$\approx$320 mOsm/kg	$\approx$300 mOsm/kg
Detailed nutrition and physiology	largely unknown	progressively well understood
Media requirements		
C-energy source	yes	yes
buffer	yes	yes
organic acids	yes	no
lipids (sterols)	yes	no
defined supplements	more often	less often
serum	less often	more often
undefined supplements (yeast extract, peptone, lactalbumin)	more often	less often
growth atmosphere	0–5% CO_2	5% CO_2

supplemented or not with serum.[5,48] Furthermore, established insect cell lines are continuous in the sense that they are "immortal", although, unlike continuous mammalian cell lines, they do not appear to be transformed (= propagating in a tumorlike manner). Insect cell lines have been established from a variety of tissues, but mostly from undifferentiated ovarian or embryonic tissue. As a result, they may consist of a mixture of cell types often differing in morphology and physiology (e.g., susceptibility to virus infection),[50,51] but efforts are under way to develop clonally pure cell lines that would be useful for fundamental research and for the production of biologicals in ICC.

It should also be noted that the dramatic increase in the number of continuous cell lines established in the last decade is largely due to the development and improvement

TABLE 2B. Comparison of Cultured Insect and Mammalian Cells: Cell Propagation Technology

	Insect Cells	Mammalian Cells
Cell line maintenance	relatively easy	more difficult
Versatility of suspension/attachment	yes	no
Immortality	yes	only transformed cell lines
Contact inhibition	mild or absent	yes, except for a number of transformed and lymphoid cell lines
Detachment from substratum surface	gentle force	trypsinization
Sensitivity to changes in pH, DO, temperature, osmotic shock	relatively low	relatively high
Dependence of growth on inoculum size	yes	yes
Asepsis	required	required

of culture media, most of which are now commercially available.[5,48,52] The biochemically rational approach in formulating media for ICC is to elucidate the nutritional requirements of cultured insect cells. In addition, chemically defined media are desirable for exact metabolic studies and simpler, low-cost media are continually sought for use in practical applications of ICC, including large-scale production of biologicals.[53] Despite the progress made to date, the nutritional requirements of insect cells in culture have not been studied as extensively as for cultured mammalian cells and the available information is limited and widely scattered in the literature. Hink[5,54] and Mitsuhashi[52,55] have provided excellent summaries of the existing information on nutritional requirements of cultured insect cells.

TABLE 2C. Comparison of Cultured Insect and Mammalian Cells: Expression System for Production of Regulated Biologicals

	Insect Cells	Mammalian Cells
Gene expression level	+++	+
Ease of genetic manipulation	relatively easy (small genome)	relatively difficult (large genome)
Posttranslational modification	yes	yes
Mammalian oncogene expression	no	yes
Pathogenicity of viral vectors to humans	no	yes

Briefly, most insect cell lines utilize sugars, proteins and peptides, amino acids, organic acids, vitamins, lipids, and inorganic salts. Media typically contain mixtures of sugars that function as energy sources, but glucose and fructose are preferentially utilized by the majority of cell lines. Although amino acid metabolism varies among cell lines, a number of more widely applicable generalizations are emerging. For instance, about 14 amino acids are usually essential for the growth of insect cells as compared to 10 for mammalian cells. Nonessential amino acids for most cultured insect cells include α- and β-alanine, asparagine, aspartic acid, glutamic acid, and glycine; therefore, these can be eliminated to reduce the cost of culture media.

In modifying media composition, it is important to recognize that some amino acid(s) or other ingredient(s) may not be strictly required for insect cell growth, but may be essential for virus production and foreign gene expression. Organic acids typically included in ICC media are malic, succinic, fumaric, and α-ketoglutaric acids, although only malic acid is known to be growth enhancing if used singly. A number of vitamins, especially those of the water-soluble B vitamin group, tend to be contained in most media. Among lipids, exogenous sources of polyunsaturated fatty acids and sterols (e.g., cholesterol) are included. Sterols are considered essential media ingredients for the growth of insect cells, which, unlike vertebrate cells, cannot synthesize sterols *de novo*. Inorganic salts are important ingredients of ICC media because they contribute to maintaining the correct range of ion balance and osmotic pressure. Insect cells in culture are generally flexible to the ionic conditions of the medium and usually grow at somewhat higher osmotic pressures than mammalian cells, but the Na/K ratio may be restrictive for some cell types. Proteins and peptides are incorporated in ICC media predominantly by supplementing the defined components with complex chemically undefined substances. Among the latter, serum (primarily fetal bovine serum, FBS) is the single most widely used complex supplement. It is a source not only of proteins (fetuin and albumin fractions) and smaller peptides, but also of several free amino acids as well as vitamins, lipids, and metal ions bound to serum proteins in complexes which increase their bioavailability. Serum is the most expensive component of the majority of ICC media. In addition, it has variable composition from lot to lot, may contain cytotoxic factors, is susceptible to mycoplasma and virus contamination, and contributes to large-scale bioprocessing difficulties both upstream (e.g., foaming in bioreactor) and downstream (e.g., product separation and purification). Most significantly, serum is not necessary for insect cell growth, as demonstrated by the development of many serum substitutes, and there are continuing and successful efforts to replace serum with cheaper defined and undefined compounds. Defined proteins proposed for serum replacement include globulins, bovine serum albumin, transferrin, etc., whereas undefined substances include peptone, yeast extract, tryptose phosphate broth, and protein hydrolysates like lactalbumin hydrolysate (LH) and yeastolate (YL).

Finally, a number of miscellaneous media components with no direct nutritional role are also included in media formulations because they are thought to contribute to the survival and protection of the cells both during static maintenance and subculturing and, especially, under the intense conditions of cultivation in agitated and aerated vessels (see later). These include methylcellulose, PVP-40, Darvan®, Pluronic® polyols (thickening agents presumed to protect cells from hydrodynamic shear–induced cell damage and death and presumed to reduce cell clumping), silicones and Tweens®

(antifoam and surface-active agents), α-tocopherol acetate (antioxidant), etc. Occasionally, antibiotics (penicillin, streptomycin) are added to the media to protect against microbial contamination.

Hink and Hall[49] have summarized the most commonly used media for insect cell cultivation *in vitro*. As a broad generalization, the most frequently used media for dipteran cells, other than *Drosophila*, are Mitsuhashi and Maramorosch's MM medium[35] with or without serum (MM-SF)[56] and MM mixed 1:1 with Varma and Pudney's VP_{12} medium,[57] which is designated MM/VP_{12}. For *Drosophila* cells, M3 medium[58] or M3 mixed 1:1 with Echalier's D22 medium,[59] designated DS, are widely employed. Most cell lines from lepidopteran insects are cultured in Grace's medium[34] and its modifications (TNM-FH,[37] IPL-41,[46] TC-100,[60] BML/TC-10,[60] etc.), as well as in MM medium.[35]

The quest for media simplification and cost reduction has led to a number of low-serum, low-serum/low-protein, and serum-free media formulations. For instance, Hink and co-workers[61] developed a serum-free and serum protein–free medium for culture of *Trichoplusia ni* (cell line TN-368) cells by supplementing Grace's basal medium with LH, YL, tryptose, peptone, glucose, glutathione, oleic acid, and cholesterol (plus PVP-40 and methylcellulose as thickening agents); that is, FBS was totally replaced by chemically undefined and defined ingredients. The same cell line was cultivated in serum-free medium by Vail *et al.*[62] These workers succeeded in adapting the cells to Hink's TNM-FH medium[37] from which whole egg ultrafiltrate, bovine serum albumin, and FBS were successively deleted.

Moreover, modifications of media may contribute to a simplification of preparation and to a savings in production cost, themselves significant considerations in large-scale applications, but these same modifications may also deprive the cells of their ability to maintain differentiated functions and to produce heterologous proteins because often these functions do not exhibit the same nutritional requirements as cell proliferation.[5,63] In the example of Vail's medium development,[62] this serum-free medium supported lower cell growth, but good virus production. Röder[64] cultivated three lepidopteran cell lines in BML/TC-10 medium[60] from which the concentration of FBS was gradually decreased and replaced by egg yolk emulsion until the cells became fully adapted to the same medium now containing 1% egg yolk and no FBS. This medium's utility was demonstrated in bioreactor cultivation (up to 10-L agitated vessels) of the above cell lines in terms of both good growth and NPV baculovirus formation. Production costs of viral insecticide output were reduced by 95% thanks to this serum-free medium. In the cases of serum-free media cited here, the serum is replaced by more or less undefined components. Such "semidefined" media are generally the least expensive kinds of serum-free media because crude natural substances are used to replace costlier purified chemicals. Serum-free MM medium (MM-SF)[56] serves as a good prototype of semidefined media with the lowest cost of ingredients and the simplest preparation.

Recent work by Mitsuhashi[53] has brought about further simplification and cost reduction of MM-SF medium: the inorganic salt mix has been replaced with diluted seawater and the glucose has been replaced by table sugar. The new medium, designated MTCM-1601, consists of 25 mL seawater, 0.65 g LH, 0.5 g YL, 0.8 g table sugar of supermarket grade, and distilled water added to a total volume of 100 mL. The resulting solution has a pH of ≈6.5, so no pH adjustment is necessary. The medium is sterilized by autoclaving for 15 minutes. Mitsuhashi[53] showed that this medium,

MTCM-1601, was a step towards the ideal of a "totipotent" medium in that it could support the growth of at least 15 dipteran and lepidopteran cell lines.

In a similar manner, Koike and Sato[65] have developed a series of autoclavable insect culture media by further simplifying the salt components of serum-free MM medium (MM-SF). These media, designated no. 8, no. 10, and no. 15, supported good growth of nine lepidopteran cell lines not only in stationary cultures, but also in spinner flask (500 mL) and jar fermentor (40 L) suspension cultures.[65] Also noteworthy for its simplicity and demonstrated capacity to support growth of *Spodoptera frugiperda* cells and production of recombinant proteins is the serum-free/protein-free semidefined medium recently developed by Maiorella et al.[66] In this work, IPL-41[46] basal medium (i.e., containing only inorganic salts, sugars, amino acids, organic acids, trace elements, and vitamins, but no serum or other supplements) was supplemented with ultrafiltered yeast extract and a specially prepared stable Pluronic® polyol–lipid microemulsion containing Pluronic® F-68 and Tween 80® as a dual detergent system along with cod liver oil and cholesterol (lipids) plus α-tocopherol acetate (antioxidant). In this medium, Maiorella et al.[66] were able to reach the same cell growth rate and extent as well as the same titer of recombinant colony stimulating factor as in IPL-41 medium supplemented with 10% FBS. Moreover, this medium was successfully used in a 21-L airlift bioreactor due to the inherent suitability of its formulation for sparged cultivation (see below).

Finally, an important landmark in serum-free media formulation has been the development of a few completely chemically defined media for ICC. Among them, the often-cited CDC medium developed by Wilkie et al.[67] consists of almost 70 chemical compounds and supports the growth of cell lines from the lepidopteran *Spodoptera frugiperda* and the dipteran *Aedes aegypti* and *Aedes stephensi*. Remarkably, this serum-free/protein-free medium allowed the propagation of these three cell lines without any previous adaptation step. Although this and other chemically defined media have extremely complex compositions, they are very useful for exact biochemical studies of cultured cells.

Even though media formulations play an important role in the efficient and economical cultivation of insect cells on a large scale, the growth physiology of these cells is also affected by physicochemical requirements such as temperature, pH, gas exchange, and osmolarity. The optimal temperature range for the growth of most insect cell lines in culture is between 25 and 30 °C, that is, considerably below that for mammalian cells. For most applications of ICC, including cultivation of insect cells in bioreactors, the control of the temperature at the optimum set point (e.g., 28 °C) does not present excessive cooling requirements for metabolic heat removal. The pH required for optimal *in vitro* growth of lepidopteran cells falls in the range between 6.0 and 6.25,[68] whereas mosquito cells prefer a pH range between 6.5 and 7.0.[69] The pH in culture tends to change, usually monotonically, to more basic or more acidic values as various medium components are consumed and as various metabolic products are excreted. Thus, it is important to maintain the pH within the optimal range. This is done usually by incorporating buffers in the media, typically phosphate or bicarbonate. In the latter case, the dissociation equilibrium is stabilized under an atmosphere of 5% CO_2, especially in stationary cultures, which is a practice widely employed in vertebrate tissue culture. However, as with ionic balance in general, insect cells are known to tolerate pH changes easier than mammalian cells, possibly because the former originate from tissues with wider physiological pH fluctuations than vertebrate tissues. The osmolarity of ICC media is generally higher than that of mammalian cell media as insect cells have been shown to grow in higher osmotic pressure and also to tolerate a wider range of osmotic pressure.[54,69] Therefore, the osmolarity of mammalian cell media is set between 290–300 mOsm/L, whereas that of insect cell media can exceed 400 mOsm/L.[54,69]

Large-Scale Insect Cell Culture

The design and implementation of production-scale ICC must be based not only on an adequate understanding of nutritional and physical requirements for cell growth and product formation, but also on practical information on the behavior of insect cell lines in different culture configurations (free suspension, attached growth, matrix-confined growth), in long-term cultivation (sustained viability), and under environmental conditions inherent in scaleup (mechanical agitation, liquid medium circulation, aeration by sparging or over free liquid surface or through indirect means, etc.) plus a host of other considerations affecting both upstream and downstream bioprocessing. Although a reasonable body of such information already exists for mammalian cell cultivation,[70–73] there is a serious lack of comparable data for ICC.

Certain general insect cell properties with a direct impact on *in vitro* cultivation technology and scaleup are given in TABLE 2B and some of them have already been discussed above in connection with cell lines. Because of the mixed character of insect cell lines (originating from undifferentiated embryonic tissues or from specific organ tissues), the same cell line can be grown (usually after some adaptation) as either surface attached or suspended.[22,69,74] This versatility in growth mode, which is in contrast with most mammalian cell lines, increases the choices of bioreactor and culturing strategy for industrially attractive insect cells. In addition, the contact inhibition exhibited by many mammalian cell lines is considerably milder or absent in insect cells, which instead have a natural tendency to form aggregates (clumps) both in attached and in suspension culture.[68,69] The attachment forces between the insect cells and the solid surface for growth ("substratum") are also weaker than for mammalian cells; hence, gentle mechanical force can be used for their detachment instead of trypsin, which is the standard enzymatic approach in mammalian cell culture. Finally, both mammalian and insect cells require a minimum seeding density (inoculum size) for initiation of growth in most suspension and surface (stationary) cultures. In practice, insect cell cultivation is initiated at inoculum levels of approximately 10^5 cells/mL of medium.

The design and scaleup of bioreactor vessels for mass production of insect cells and their products hinge on the aerobic nature of the metabolism of these cells and on the constraints imposed upon available equipment to supply the oxygen required by the cells for growth, differentiation, and product formation. It should be understood that "scale" does not necessarily mean bioreactor size, but may refer to increased productivity, provided that the concentration of viable or product-forming cells (or both) is increased with an increase in scale. Indeed, there is a clear trend towards increased cell concentrations by one or two orders of magnitude for industrial mammalian cell bioreactors over and above what is typically achieved in simple batch suspension cultures.[72,75]

In order to satisfy the oxygen demand of production systems employing insect or mammalian cells, the standard approaches of microbial fermentation technology such as agitation and aeration by introduction of bubbles (sparging) are too aggressive, considering the larger size of animal cells and their lack of cell walls. The rates of oxygen mass transfer required for large-scale production of cultured insect cells are thought to result in cell damage and death due to hydrodynamic shear stress. The hydrodynamic forces frequently referred to as "shear" stem from mass transfer–assisting operations like agitation and air sparging in freely suspended culture and also medium circulation for surface grown cells (e.g., on microcarriers) and they are generally considered as the most significant bottleneck in insect cell culture scaleup. Reports on unusually high shear sensitivity of insect cells,[46,68] on susceptibility of these cells to damage and death by direct interaction with air bubbles,[45,76] and on unusually high oxygen demand of insect cells compared to mammalian cells[77,78] are often cited to substantiate the perceived need for special bioreactor designs. However, there are also emerging reports pointing to some disagreement on the high degree of shear sensitivity and on the numerical values of oxygen demand of cultured insect cells.[66,79,80] Nonetheless, because these concerns have guided most documented efforts to develop production-level ICC systems to date, our brief review of bioreactors will address these two inextricably connected factors, namely, shear stress and oxygen supply.

Growth in Suspension

There are obvious advantages to using existing bioreactors of standard design, for example, stirred tank (jar) fermentors, for insect cell growth with minimum modification. Hink's group pioneered efforts in this direction since the mid-1970s, working with *Trichoplusia ni* (TN-368) as a useful system for viral pesticide production. Hink and Strauss[43] compared insect cell growth in jar bioreactors of 400 mL and 2 L with well-established patterns of cell growth in 100-mL laboratory spinner flasks. The scaled-up systems equipped with agitators allowed the same growth extent ($\approx 3 \times 10^6$ cells/mL) as the 100-mL spinner, but at a considerably slower growth rate (6–7 days versus 3–4 days in the spinner). Another scaleup system, a 400-mL vibromixer, gave even slower growth and the cells exhibited unhealthy morphology. These phenomena were attributed to mixing-associated shear in the larger bioreactors. The same workers also addressed successfully the problem of cell clumping by supplementing the medium with 0.1% methylcellulose to increase viscosity. Furthermore, Hink and Strauss[43] compared oxygen supply between the surface-aerated spinner flask and the larger vessels and found that, by combining air sparging with surface aeration, the 400-mL and 2-L stirred reactors allowed for higher cell growth than the 100-mL spinner, which uses only surface aeration.

The oxygen requirements of *T. ni* cells were investigated by Streett and Hink,[81] who found that the oxygen uptake rate (OUR) of these cells after being infected with AcNPV baculovirus increased to twice the OUR value of exponentially growing uninfected cells. In contrast to these measurements (and to the high OUR reported by other workers[77,78]), Maiorella et al.[66] reported that there was no difference between the oxygen demand of *S. frugiperda* cells before and after infection with AcNPV baculovirus, although these values, as expected, were about three times as high as for cells in stationary phase. Clearly, more research is needed in order to establish unambiguously the magnitude of oxygen demand of uninfected and infected insect cells in suspension culture.

Further work by Hink's group[44,68] revealed that *T. ni* could grow well in 2- to 3-L jar bioreactors of standard stirred tank design, equipped with marine impellers, under sparged aeration, provided 0.1–0.3% methylcellulose (or other viscosity-enhancing agents like Darvan®) was included in the medium. Under these conditions, they reported routine attainment of final cell densities of 5×10^6 cells/mL in 5 days, with initial densities of 2×10^5 cells/mL and a corresponding doubling time of only 14 hours during exponential growth. Maintaining growth comparable with 100-mL spinners (surface aeration) in the scaled-up 2- to 3-L stirred tanks could be achieved by increasing the aeration rate from 5 cm^3/min (for the spinners) to 750 cm^3/min (after 2 days at 125 cm^3/min) and by increasing the impeller speed from 100 rpm (for the spinners) to 220 rpm. Cell damage was prevented under these conditions by increasing the methylcellulose content to 0.3% for further enhancement of medium viscosity and, by adding a silicone-based antifoam emulsion, growth rate was enhanced, possibly due to better gas mass transfer. No mechanism-based analysis of these process improvements is available at present, although it seems that a combination of effective aeration, viscous inert polymer additives, and agitation with low-shear impellers (propeller type) may enhance the rate and extent of insect cell growth in stirred bioreactors.

Hink's group[44,68] also found that dissolved oxygen (DO) levels affected the growth of insect cells in the stirred fermentors. When DO was allowed to drop freely from an initial 100% to a 2% level after 24 h, the cells stopped growing; in contrast, continuous maintenance of DO at 100% caused vacuolation and precipitate formation after the cell concentration reached a plateau. In addition, a constantly maintained DO of 15% caused cell vacuolation at 120 h and cell concentration diminished rapidly beyond this point. A 50% DO level controlled throughout the culture allowed the same growth as 100% DO and it was routinely adopted for cell growth and virus production.

Workers in Germany[45,64] succeeded in culturing lepidopteran cell lines at even higher scales by using 10-L jar fermentors. Miltenburger and David[45] were able to meet the oxygen demand of the insect cells by introducing a semipermeable silicone rubber tube system for aeration by diffusion in the 10-L vessel. In this way, the fragile insect cells were not exposed to the shear forces generated by direct interaction with swarms of air bubbles. An added advantage of this system was that foaming was avoided and thus also the usual problems of cell entrainment and flotation, which are lethal to the cells. This system was developed further by Eberhardt and Schügerl[82] in anticipation of the limitation of useful tube length for medium oxygenation upon further scale increase. An increase of inlet air pressure from 1.0 to 1.5 bar resulted in a 50% increase in cell yield of *S. frugiperda* and in considerable enhancement of growth rate. At a low agitation speed (35 rpm), mixing was inefficient and a portion of the cells settled at the bottom of the bioreactor (mass transfer limitation); on the other hand, at high agitation (75 and 100 rpm), a limit in cell yield was reached due to the adverse effects of higher shear forces. For intermediate agitation (50 rpm) and upon increasing the oxygen transfer rate (OTR) by either introducing pure oxygen in the immersed tubing or using air both through the tubing and over the free fluid surface, the final cell density was enhanced to $(4-5) \times 10^6$ cells/mL, that is, to values routinely obtained in 50-mL spinner flasks.

Quantitative assessments of hydrodynamic shear effects on insect cells cultivated in stirred or airlift bioreactors are in evidence in the recent literature. Both shear stress associated with agitation speed and that associated with direct air sparging have been addressed. Tramper and co-workers[76] assessed systematically the shear stress experienced by *S. frugiperda* cultivated in a 1-L round-bottom bioreactor equipped with a marine impeller (medium containing 0.1% methylcellulose) by varying agitation and aeration rates and by measuring the kinetics of cell growth and cell death. In two different runs, the viable cell decrease started at 220 rpm (corresponding to a shear stress of 1.5 Nm^{-2}) and at 510 rpm (corresponding to a shear stress of 3 Nm^{-2}), respectively. In the same work,[76] by employing a rotary viscometer, Tramper *et al.* confirmed that the critical shear stress at which cell viability is declining is between 1–4 Nm^{-2}. They also used a bubble column reactor (18 cm height and 3.5 cm diameter) to establish quantitatively the effect of air sparging on cell viability. From these experiments, they found that a first-order death rate constant for these insect cells was proportional to the airflow rate. On the basis of these findings and given their repeatedly unsuccessful efforts to grow the cells in an airlift reactor, Tramper *et al.*[76] recommended alternative bioreactor designs for scalable oxygen supply, such as membrane bioreactors (i.e., a separate vessel for oxygenation of the medium that is to be supplied to the cell growth chamber) and oxygen supply through semipermeable tubing, as described by Eberhardt and Schügerl.[82] Subsequent work by Tramper and co-workers[83,84] extended the findings of their bubble column experiments[76] to a rational bubble column design approach for growth of fragile insect cells. Because measurable loss of viability was seen to occur in the course of the bubble rise through the columnar reactor and in the region of bubble bursting at the liquid surface, a "killing volume" hypothesis was advanced to formulate an explicit equation for the first-order death rate constant and for the minimum specific surface area of the air bubbles to supply sufficient oxygen.[83,84] The "killing volume" represents the hypothetical volume associated with each rising air bubble during its lifetime in which (volume) all viable cells are killed.

The experimental findings of adverse effects on insect cells in directly sparged vessels have also been seen with mammalian cells in sparged systems[85,86] and have been explained in terms of the damage generated by the shear forces acting on the cells in the region of bubble disengagement from the free liquid surface.[85,86] For example, Handa-Corrigan *et al.*[86] found that higher bubble column bioreactors ensure better cell growth. The same explanations underlie also the reported failure of Koike and Sato[65] to grow *Mamestra brassicae* cell lines on their serum-free medium no. 8 in an airlift reactor (1 L), which is in contrast to their successful cultivation of the same cell lines in the same medium in a 40-L agitated jar fermentor (marine impeller, 150 rpm). Wudtke and Schügerl[79] confirmed the importance of bubble disengagement for the loss of insect cell viability in sparged cultivation systems by showing that a bubble column reactor in which the free suspension surface was covered with paraffin oil to prevent bubble bursting generated significantly less cell debris and maintained higher cell viability over a longer time in comparison to a control system.

In a significant development in commercial insect cell culture scaleup, Maiorella *et al.*[66] were able to circumvent the bad prognosis that had been established for airlift reactors with direct air sparging. These workers[66] exploited the inherent scalability of the airlift bioreactor (OTR increase with increased vessel volume)[87] by (a) controlling the airflow stream and its composition, (b) controlling the air bubble diameter at 0.5–1 cm by judicious choice of sparger orifice diameter (at this size range, bubbles do not coalesce and, unlike smaller bubbles, have lower surface tension), and (c) including in the medium Pluronic® F-68 polyol, which is known to protect mammalian[85,88] and insect[89] cultured cells from damage due to mechanical agitation as well as to air sparging. In this way, Maiorella et al.[66] were able to scale up the growth of *S. frugiperda* and the production of recombinant macrophage colony stimulating factor following infection with a modified AcNPV viral vector from a 3-L spinner flask to a 21-L airlift reactor. The protective effects of Pluronic® F-68 for insect cells against shear-induced damage due to agitation or sparging (or both) have been recently demonstrated by Murhammer and Goochee, who demonstrated the polymer's usefulness both in a 3-L agitated and sparged tank bioreactor and in a 670-mL airlift bioreactor for the cultivation of *S. frugiperda* and for the production of recombinant β-galactosidase using a modified AcNPV baculovirus vector.[89]

In conclusion, there is an important need for more rational mechanism-based explanations of the, admittedly, highly complex interactions among turbulent power dissipation, hydrodynamic shear and air bubble swarms, and their biological effects on suspended insect cells. In addition, a number of inert polymers, other than pluronic polyols, should be screened for their shear-protective properties. In these ways, the development of bioprocesses making use of insect cell suspension culture and their scaleup will be based on strategies maintaining optimal oxygen transfer coefficients and minimal values of hydrodynamic shear.

Growth in Attached and Immobilized Culture

The quest for high bioreactor productivity can be approached by achieving high cell densities with surface growth–dependent cells. Weiss and Vaughn[90] provide an extensive review of attached cell culture systems that have been proposed and used for large-scale production of baculoviruses. Most notable among them are banks of roller bottles,[42,46] with growth surfaces ranging from 490 cm^2 to 1750 cm^2 (corresponding to media volumes from 100 mL to 250 mL, respectively), rotating at 1 revolution every 8.5 min. Despite the initial labor required for this setup, remarkably reproducible growth was achieved and some of the highest specific growth rates (doubling times of 8.35 to 10.22 h) were seen for the *S. frugiperda* cell line cultures. Weiss and Vaughn[90] have also evaluated a surface growth–dependent perfusion system, the Dyna Cell Propagator or bulk culture vessel, for continuous growth of *S. frugiperda* cells. This system consists of a 1.7-L polystyrene bottle containing a spiral core film that provides 9500 cm^2 of internal growth surface for continuously perfused media. The scaleup of roller bottles, bulk culture vessels, and similar surface growth–dependent systems is limited by inherent design difficulties in maintaining consistently high cell concentrations because increases in reactor volume are not followed by commensurate increases in available growth surface (surface-to-volume ratio diminishes with scale).

This substantial bottleneck has been addressed in mammalian cell culture technology by the development of microcarriers for the propagation of anchorage-dependent cells.[91] This technology has been used recently for the large-scale production of mosquito cells and arboviruses.[19,20] Lazar and co-workers[19] used cellulose-based microgranular microcarriers for the growth of *A. aegypti* and successfully scaled up the process to a volume of 8 L. The process could be used for semicontinuous production of Sindbis virus and of West Nile virus over a 16-day period, with both virus yields higher than those obtained with conventional monolayer virus cultures. Igarashi and Srivastava[20] were able to grow *Aedes albopictus* cells on Cytodex-1 microcarriers for mass production of Japanese encephalitis virus in a 1-L spinner flask at 60 rpm. A promising recent development of attached growth ICC was reported by Shuler and co-workers,[15] who could culture an attachment-dependent strain of *Trichoplusia ni* in a packed-bed configuration using 3-mm glass beads as the packing material of their columnar bioreactor and allowing separately oxygenated medium to circulate through the biocatalyst bed. In this design, aeration is decoupled from the main bioreactor and, hence, both oxygen supply and hydrodynamic shear are not strongly dependent upon bioreactor scale. This provides for easy scaleup. The same group showed that the *T. ni* cells in this attached growth bioreactor system could reach high levels of cell density and viability that allowed the production of up to 33% of their total protein content in the form of heterologous β-galactosidase after being infected with a modified AcNPV vector.[15] Another modern approach to achieving high cell densities and thus high bioreactor productivities, while at the same time protecting shear-sensitive animal cells from detrimental mechanical stress, is cell immobilization and, especially, entrapment.[92,93] Recent results by King *et al.*[94] indicate the attainment of 8×10^7 cells/mL matrix densities when they entrapped *S. frugiperda* cells infected with a temperature-sensitive baculovirus in semipermeable microcapsules.

In this as well as most other aspects of large-scale ICC, valuable lessons from experience with mammalian cell technology are adapted to the peculiarities of insect cells in culture. The twin and interconnected factors of hydrodynamic stress and oxygen supply are likely to dominate among the bioprocess engineering considerations for production-level systems. Therefore, the designs of insect cell bioreactor systems should be based as much as possible on (a) increased basic understanding of insect cell nutrition and physiology and (b) quantitative kinetic and mass transfer descriptions of insect cells in suspension and in attached or immobilized culture. Current efforts under way to address the latter point are illustrated by work linking mechanistically physiological effects, such as cell damage and death, to hydrodynamics of turbulent flow.[83,86,95,96] The commitment to such fundamental studies should not only bring about satisfactory explanations for empirically found ways that circumvent the undesirable physiological effects in animal cell reactors (e.g., use of thickening agents, modifications of oxygenation systems, etc.), but it should also result in fresh approaches to optimal design, operation, and performance of many bioprocessing systems.

EXPERIMENTAL WORK

Our interest in ICC has been focused on the growth of *Aedes albopictus* (mosquito) cells for the efficient mass production of arboviruses. In our preliminary experiments, we studied the growth kinetics of *A. albopictus* both in stationary culture (in petri or multiwell dishes) and in suspension culture (in spinner flasks and in agitated/aerated bioreactors) as a first step in establishing the physical requirements for the efficient propagation of these insect cells in large scale. Our studies additionally examined the feasibility of high cell densities and semicontinuous cultivation by immobilizing *A.*

albopictus cells in open-structured porous microspheres. Here, we present a brief account of our kinetic results, which are described in more detail elsewhere.[97]

MATERIALS AND METHODS

Cells and Medium

The C7-10 cells used in this study were derived from the *Aedes albopictus* line of Singh[36] and they are a subclone of line LT C-7 described by Sarver and Stollar.[98] For the cell culture in dishes, E-5 medium was used.[98,99] This is Eagle's minimal medium (MEM)[100] supplemented with nonessential amino acids, glutamine, 5% heat-inactivated fetal bovine serum, sodium bicarbonate, penicillin, and streptomycin.

For the culture in suspension and on microspheres, cells were cultured in Joklik's modified Eagle's minimal medium supplemented with nonessential amino acids, 5% heat-inactivated fetal bovine serum, and sodium bicarbonate.[99]

Experimental Methods

First, monolayer culture: Cells were cultured as monolayers in 35-mm petri dishes containing 2 mL of E-5 medium at 28 °C under a 5% CO_2 incubator.[27] The inoculum level was 1×10^5 cells/mL of medium. The cell number was counted in a Coulter counter.[27,29]

Second, suspension culture in the spinner reactor: Cells were cultured in 50-mL and 100-mL spinner flasks in Joklik's medium at 28 °C and 5% CO_2 in a CO_2 incubator. The inoculum level was 2×10^5 cells/mL of medium. The cell number was counted in a Coulter counter after dispersing the sample. The cell viability was tested using 0.4% trypan blue.

Third, microsphere culture in the spinner reactor: CF-IMMO® microspheres[93] from Verax Corporation (Lebanon, New Hampshire) were used as a support material. The microsphere concentration was 5 mg/L and the inoculum level was 2×10^5 cells/mL of medium. The cell number was counted after treating the microspheres with collagenase.[97]

Fourth, protein synthesis rate test: The rate of protein synthesis was evaluated by measuring the rate of ^{3}H-leucine incorporation into cells. Acid-precipitable radioactivity was measured in a scintillation counter.

Fifth, large-scale suspension culture: A 4-L modified NBS jar fermentor (Model MMF-05, New Brunswick Scientific, Edison, New Jersey) was equipped with a standard marine impeller placed close to the bottom of the vessel and a turbine impeller placed slightly below the free liquid surface. A working volume of 1 L (giving a liquid height–to–tank diameter ratio of ~0.4) was used at an agitation speed of 60 rpm at 28 °C. A 5% CO_2-in-air mixture was provided into the headspace of the reactor for surface aeration.

RESULTS

Petri-Dish Culture

FIGURE 1 shows the growth kinetics of *Aedes albopictus* cells in dishes. The doubling time estimated in exponential phase was 17 h, which shows that these cells are among the fastest propagating animal cells. This is a clear advantage for insect cells as a host system. Considerable clumping of cells was observed after four days of culturing. It was established that clumping phenomena start when the cell number becomes $(2.5-3.0) \times 10^6$ per mL of medium, and the amount of cells that form clumps increases as the overall cell number increases.

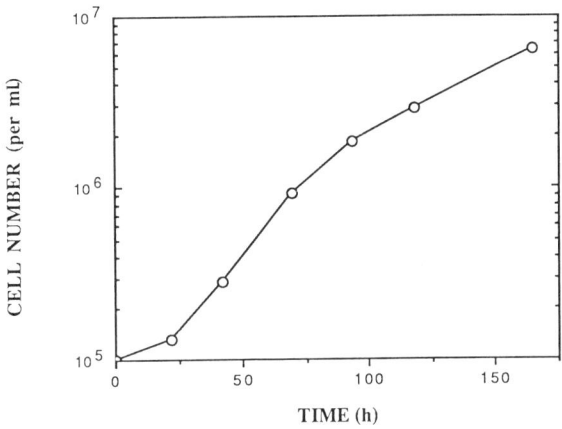

FIGURE 1. Cell growth in dishes incubated in a CO_2 incubator at 28 °C (E-5 medium). Inoculum level: 1×10^5 cells/mL.

Rate of Protein Synthesis

Clumping phenomena may affect the cell growth in the following two directions. When cells form clumps, the physical distance between cells decreases and this makes it easy to exchange necessary materials such as growth factors among cells. This ease of communication can be a positive contribution of clumping to cell growth. However, if cells form clumps, a contact inhibition signal may become stronger so that the cells do not synthesize protein and do not grow any more or cells may be subjected to internal mass transfer limitation. Investigating whether clumping phenomena contribute to cell growth positively or negatively is very important in order to decide whether clumping should be avoided or enhanced by some means in the case of large-scale cell culture. Incorporation of ^{3}H-L-leucine into cells and counting radioactivity of the cells provides a preliminary way to test the apparent efficiency of protein synthesis in a cell population subject to cell aggregation. FIGURE 2 shows that the rate of protein synthesis decreases as cell number increases.

It was observed that fully propagated cells consist of a clumped portion and an unclumped portion and that the clumped portion increases as cell number increases. If it is assumed that both the viability and the rate of protein synthesis of unclumped cells remain constant and those of clumped cells are a function of the number of cells in the system, then the drop in observed protein synthesis may reflect an adverse effect of cell clumping on protein synthesis due to an intrinsic decrease in protein synthesis rate or a drop in cell viability (or both).

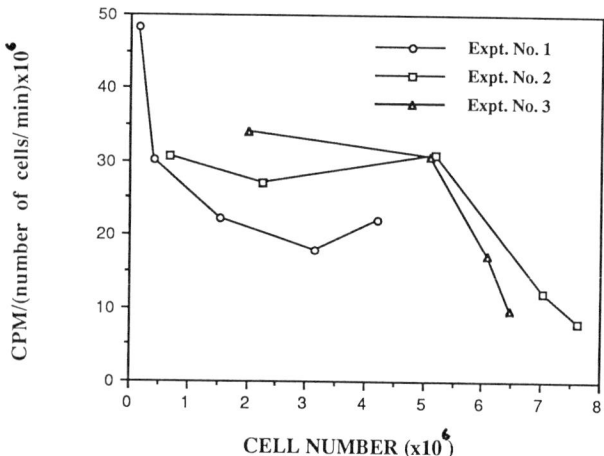

FIGURE 2. Rate of protein synthesis per unit of cells as a function of cell number. Data points were obtained from duplicate dishes. Linearity of ^{3}H-leucine uptake was good for up to 90 minutes.

The Effect of Agitation in Suspension Culture

FIGURE 3 shows the growth kinetics in suspension culture (spinner) at different agitation speeds. Up to 100 rpm, both the maximum attainable cell number and the specific growth rate increase as agitation speed increases. However, if the agitation speed increases further, these values decrease. Thus, an optimum agitation speed exists and would be 100 rpm in the case of a spinner reactor system (FIGURE 4). For spinner reactors of larger volumes, this optimum may be different.

In the case of 60 rpm, agitation is not enough to maintain a uniform distribution of cells in the spinner; hence, a large clump grew at the bottom of the spinner. In the case of 80 rpm, there was still a large cell aggregate plus several smaller clumps as well near the bottom. The size of the clumps was observed to decrease as the agitation speed increased further and there were only a few clumps that could be easily observed by the

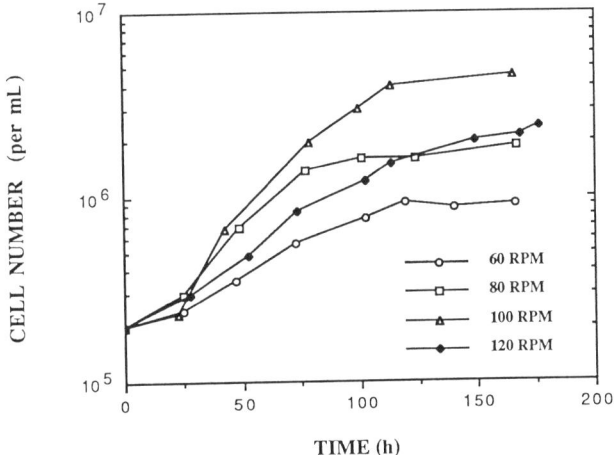

FIGURE 3. Effect of rpm in suspension culture (28 °C, in a spinner reactor at 5% CO_2).

naked eye in the case of 120 rpm. Therefore, as agitation speed increases, the cells become well distributed and less prone to form clumps, perhaps because clumping may require sufficient contact time for two separate cells or clumps.

FIGURE 5 represents the average viability of cells during the culture. As expected, this behavior of viability according to the agitation speed has the same trend as the growth kinetics of the cells. In the case of 60 rpm, the one large cell aggregate is

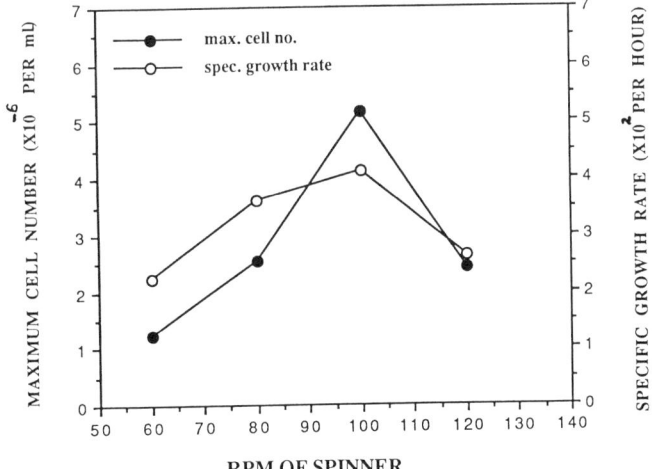

FIGURE 4. Effect of agitation speed on the maximum cell number and the specific growth rate in suspension culture.

obviously subject to internal mass transfer inside the clump and thus viability is low. It was repeatedly observed that cells at the center of large clumps were dead (results not shown). This also agrees with the result of the test of protein synthesis, that is, clumping phenomena may decrease the apparent efficiency of protein synthesis by decreasing either the cell viability or the specific protein synthesis rate. However, as rpm increases, the size of the clumps decreases so that the effectiveness factor due to internal mass transfer increases and the viability of the cells increases because of sufficient nutrient supply, as is well known in fungal cell aggregates.[101]

As rpm increases further to 120 rpm, the viability decreases due to shear. This shear vulnerability at higher rpm can be confirmed from the results of an initial shear sensitivity test (FIGURE 6). During the first few hours of the cultures, cells do not grow due to their adaptation to the new environment and do not even form clumps because of the low number of cells. Consequently, the change in viability during this period is only

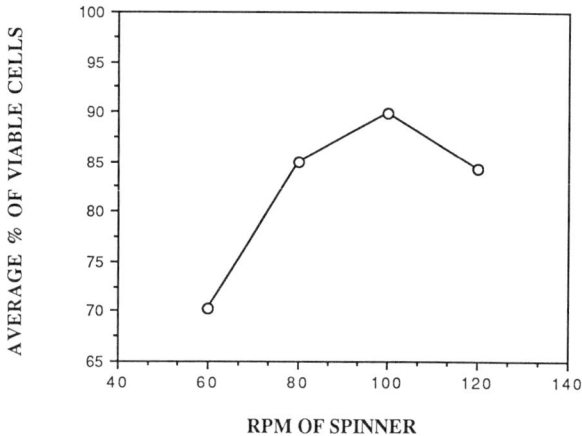

FIGURE 5. Average cell viability during the whole period of suspension culture.

due to shear forces. This figure shows that the viability decreases more steeply as rpm increases.

In summarizing all the previous results, the following can be inferred. At low rpm, maldistribution of cells in the reactor causes a large size of clumps and a low external mass transfer coefficient. The large size of clumps causes an internal mass transfer limitation that apparently results in decreased cell viability and decreased external mass transfer area. The small external mass transfer area, the low external mass transfer coefficient, and the internal mass transfer limitation result in a low growth rate at low rpm. As rpm increases, the distribution of cells in the reactor improves and the external mass transfer coefficient increases. This results in an enhancement of both external and internal mass transfer and in an increase of growth rate. However, as rpm increases, the decrease of cell viability due to shear forces becomes dominant; thus, the growth rate starts to decrease again. Therefore, an optimum rpm exists in compromising between mass transfer and shear sensitivity.

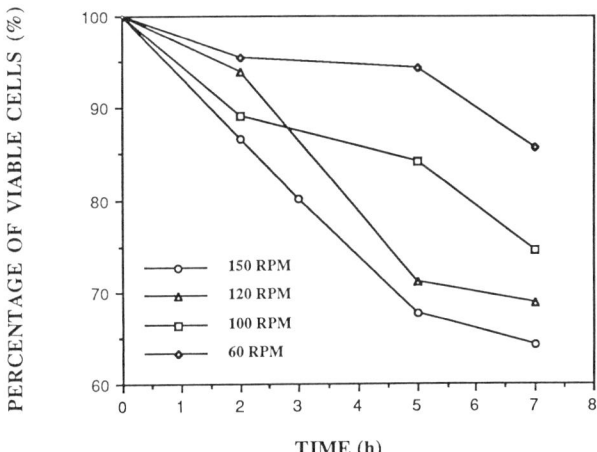

FIGURE 6. Initial cell viability in a spinner (trypan blue test). Level of cells seeded: 2×10^5 cells/mL.

Immobilized Insect Cell Culture

FIGURE 7 shows the growth kinetics of insect cells grown in open-pore, collagen-based microsphere culture. The relatively high number of cells in the beads compared with that in the medium shows that adsorption or entrapment characteristics of the collagen-based beads are excellent for these cells because a maximum cell number of 2×10^7/mL of bead was obtained. FIGURE 8 shows a comparison of a microsphere

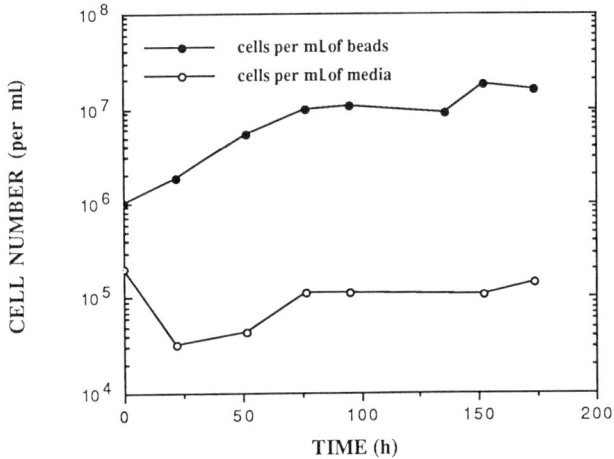

FIGURE 7. Growth of cells in microsphere culture (CF-IMMO®) (28 °C, in a spinner reactor at 5% CO_2).

culture and a suspension culture in terms of the number of cells per unit volume of medium. Almost the same growth kinetics were observed. Given the fact that the collagen beads were only added at ≈40% (v/v) in the medium and that the SEM showed less than a full cell load in the microspheres,[97] much higher cell densities per volume of bioreactor are possible.

Large-Scale Suspension Culture

We failed several times to grow the insect cells in an agitated reactor equipped with a direct sparging system. We also experienced failures with a bubble column and a tapered airlift reactor aerated directly through sparging into the medium. The direct sparging may create shear at air/liquid interfaces.[76,86] However, providing a 5% CO_2-in-air mixture into the headspace as well as surface aeration led to successful scaleup, and the maximum cell number obtained was 3.6×10^6 cells/mL of media according to FIGURE 9.

CONCLUSIONS

- Mosquito cell culture is a promising system for large-scale production of biologicals because of its high propagation rate and feasibility of growing in suspension culture.
- Clumping is not efficient for cell growth and this should be avoided if large numbers of cells are required.

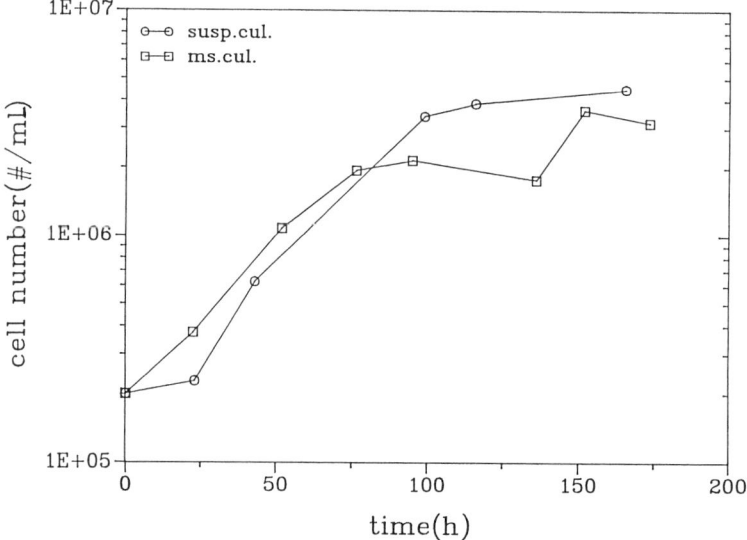

FIGURE 8. Comparison of suspension and microsphere culture (28 °C, in a spinner reactor at 5% CO_2 incubation, 100 rpm).

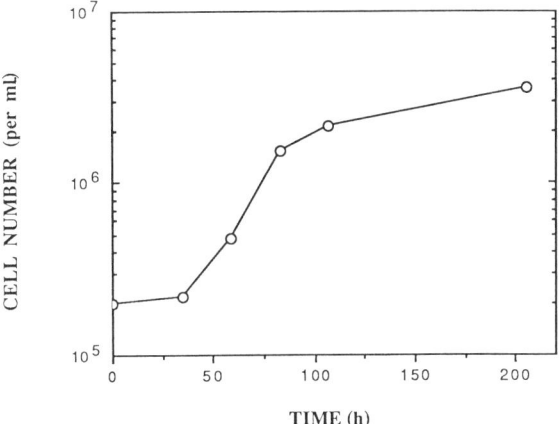

FIGURE 9. Insect cell growth in suspension culture (agitated, surface-aerated scaleup reactor). Working volume: 1 L. Agitation speed: 60 rpm (marine impeller).

- In suspension culture, mass transfer limitation governs the growth rate at low agitation speed and shear sensitivity governs at high agitation speed. Between them, there exists an optimum agitation speed.
- Collagen-based open structure microspheres prove to be a good support for immobilized growth of mosquito cells and they may be used in a large-scale fluidized reactor system.
- The large-scale suspension culture developed allows the reactor to "escape" from the CO_2 incubator and provides the possibility of further scaleup.

ACKNOWLEDGMENTS

The use of the facilities of V. Stollar's laboratory at the University of Medicine and Dentistry of New Jersey is appreciated. The gift of collagen microspheres from Verax Corporation (Lebanon, New Hampshire) is gratefully acknowledged. The technical assistance and suggestions of A. M. Fallon, formerly of the University of Medicine and Dentistry of New Jersey, are acknowledged with gratitude.

REFERENCES

1. BURGES, H. D. 1981. Microbial Control of Pests and Plant Diseases 1970–1980, p. 949. Academic Press. New York/London.
2. MARTIGNONI, M. E. 1984. *In* Chemical and Biological Controls in Forestry. W. Y. Garner & J. Harvey, Eds.: 55–67. Amer. Chem. Soc. Washington, District of Columbia.
3. LUCKOW, V. E. & M. D. SUMMERS. 1988. Bio/Technology **6:** 48–55.
4. MILLER, L. K. 1988. Annu. Rev. Microbiol. **42:** 177–199.
5. HINK, W. F. 1988. Presented at the American Institute of Chemical Engineers (AIChE) Annual Meeting, Washington, District of Columbia, November 27 to December 2, 1988 (abstract no. 26A).

6. FRASER, M. J. 1989. In Vitro **25**: 225–235.
7. VAN BRUNT, J. 1987. Bio/Technology **5**: 1118–1121.
8. MADISEN, L., B. TRAVIS, S-L. HU & A. F. PURCHIO. 1987. Virology **158**: 248–250.
9. COCHRAN, M., B. ERICKSON, J. KNELL & G. SMITH. 1987. *In* Vaccines 1987, p. 384–388. Cold Spring Harbor Laboratories. Cold Spring Harbor, New York.
10. HU, S-L., S. G. KOSOWSKI & K. F. SCHAAF. 1987. J. Virol. **61**: 3617–3620.
11. PODGWAITE, J. D. 1985. *In* Viral Insecticides for Biological Control. K. Maramorosch & K. E. Sherman, Eds.: 775–797. Academic Press. New York.
12. YOUNG, S. Y. & W. C. YEARIAN. 1986. *In* The Biology of Baculoviruses, Vol. II. R. R. Granados & B. A. Federici, Eds.: 157–179. CRC Press. Boca Raton, Florida.
13. BOHMFALK, G. T. 1986. *In* Reference 12, p. 223–235.
14. SUMMERS, M. D. & G. E. SMITH. 1987. A manual of methods for baculovirus vectors and insect cell culture procedures. Texas Agricultural Experiment Station Bulletin No. 1555.
15. SHULER, M. L., T. CHO, T. WICKHAM, O. OGONAH, M. KOOL, D. A. HAMMER, R. R. GRANADOS & H. A. WOOD. 1990. This volume.
16. MAEDA, S., T. KAWAI, M. OBINATA, H. FUJIWARA, T. HORIUCHI, Y. SAEKI & M. FURUSAWA. 1985. Nature **315**: 592–594.
17. MIYAJIMA, A., J. SCHREURS, K. OTSU, A. KONDO, K. ARAI & S. MAEDA. 1987. Gene **58**: 273–281.
18. MAEDA, S. 1989. Annu. Rev. Entomol. **34**: 351–372.
19. LAZAR, A., L. SILBERSTEIN, S. REUVENY & A. MIZRAHI. 1987. Dev. Biol. Stand. **66**: 315–323.
20. IGARASHI, A. & A. K. SRIVASTAVA. 1988. *In* Invertebrate and Fish Tissue Culture. Y. Kuroda, E. Kurstak & K. Maramorosch, Eds.: 137–139. Springer-Verlag. Berlin/New York.
21. ECKELS, K. H., D. R. DUBOIS, P. L. SUMMERS & P. K. RUSSEL. 1989. *In* Invertebrate Cell System Applications, Vol. II. J. Mitsuhashi, Ed.: 141–146. CRC Press. Boca Raton, Florida.
22. LENGYEL, J., A. SPRADLING & S. PENMAN. 1975. *In* Methods in Cell Biology **10**: 195–208. Academic Press. New York.
23. RUBIN, G. M. & A. C. SPRADLING. 1982. Science **218**: 348–353.
24. VAN DER STRATEN, A., H. JOHANSEN, R. SWEET & M. ROSENBERG. 1988. *In* Invertebrate and Fish Tissue Culture. Y. Kuroda, E. Kurstak & M. Maramorosch, Eds.: 131–134. Springer-Verlag. Berlin/New York.
25. (a) MILLER, L. H., R. K. SAKAI, P. ROMANS, R. W. GWADZ, P. KANTOFF & H. G. COON. 1987. Science **237**: 779–781; (b) McGRANE, V., J. O. CARLSON, B. R. MILLER & B. J. BEETY. 1988. Am. J. Trop. Med. Hyg. **39**: 502–510; (c) MORRIS, A. C., P. EGGLESTON & J. M. COMPTON. 1989. Med. Vet. Entomol. **3**: 1–7.
26. DURBIN, J. E. & A. M. FALLON. 1985. Gene **36**: 173–178.
27. FALLON, A. M. 1986. Exp. Cell Res. **166**: 535–542.
28. FALLON, A. M. & J. H. WILLIS. 1985. Trends Biotechnol. **3**: 217–218.
29. FALLON, A. M. & V. STOLLAR. 1987. Adv. Cell Culture **5**: 97–137.
30. SANWA CHEM. RES. INSTITUTE. 1989. Japanese patent no. JO 1096–197.
31. SHAPIRO, M. 1986. *In* The Biology of Baculoviruses, Vol. II. R. R. Granados & B. A. Federici, Eds.: 31–61. CRC Press. Boca Raton, Florida.
32. SHIEH, T. R. & G. T. BOHMFALK. 1980. Biotechnol. Bioeng. **22**: 1357–1375.
33. WYATT, S. S. 1956. J. Gen. Physiol. **39**: 841–852.
34. GRACE, T. D. C. 1962. Nature **195**: 788–789.
35. MITSUHASHI, J. & K. MARAMOROSCH. 1964. Contrib. Boyce Thompson Inst. **22**: 435–460.
36. SINGH, K. R. P. 1967. Curr. Sci. **36**: 506–508.
37. HINK, W. F. 1970. Nature **226**: 466–467.
38. VAUGHN, J. L., R. H. GOODWIN, G. J. TOMPKINS & P. McCAULEY. 1977. In Vitro **13**: 213–217.
39. SPRADLING, A., R. H. SINGER, J. LENGYEL & S. PENMAN. 1975. *In* Methods in Cell Biology **10**: 185–194. Academic Press. New York.
40. GRANADOS, R. R. 1976. Adv. Virus Res. **20**: 189–236.

41. MARKS, E. P. 1980. Annu. Rev. Entomol. **25:** 73–101.
42. VAUGHN, J. L. 1976. J. Invertebr. Pathol. **28:** 233–237.
43. HINK, W. F. & E. M. STRAUSS. 1976. *In* Invertebrate Tissue Culture. E. Kurstak & K. Maramorosch, Eds.: 297–300. Academic Press. New York.
44. HINK, W. F. & E. M. STRAUSS. 1980. *In* Invertebrate Systems in Vitro. E. Kurstak, K. Maramorosch & A. Dubendorfer, Eds.: 27–33. Elsevier/North-Holland. Amsterdam/New York.
45. MILTENBURGER, H. G. & P. DAVID. 1980. Dev. Biol. Stand. **46:** 183–186.
46. WEISS, S. A., G. C. SMITH, S. S. KALTER & J. L. VAUGHN. 1981. In Vitro **17:** 495–502.
47. HINK, W. F. 1980. *In* Invertebrate Systems in Vitro. E. Kurstak, K. Maramorosch & A. Dubendorfer, Eds.: 553–578. Elsevier/North-Holland. Amsterdam/New York.
48. HINK, W. F. & D. R. BEZANSON. 1985. *In* Techniques in the Life Sciences, Vol. C1—Techniques in Setting Up and Maintenance of Tissue and Cell Cultures. E. Kurstak, Ed.: C111/1–C111/30. Elsevier. Amsterdam/New York.
49. HINK, W. F. & R. L. HALL. 1989. *In* Invertebrate Cell System Applications, Vol. II. J. Mitsuhashi, Ed.: 269–293. CRC Press. Boca Raton, Florida.
50. GRANADOS, R. R., A. C. G. DERKSEN & D. G. DWYER. 1986. Virology **151:** 472–476.
51. CORSARO, B. G. & M. J. FRASER. 1987. In Vitro **23:** 855–862.
52. MITSUHASHI, J. 1982. *In* Advances in Cell Culture. Volume 2. K. Maramorosch, Ed.: 133–196. Academic Press. New York.
53. MITSUHASHI, J. 1988. *In* Invertebrate and Fish Tissue Culture. Y. Kuroda, E. Kurstak & K. Maramorosch, Eds.: 15–18. Springer-Verlag. Berlin/New York.
54. HINK, W. F., D. A. RALPH & K. H. JOPLIN. 1985. *In* Comprehensive Insect Physiology, Biochemistry, and Pharmacology—Vol. 10, Biochemistry. G. A. Kerkut & L. I. Gilbert, Eds.: 547–570. Pergamon. Elmsford, New York.
55. MITSUHASHI, J. 1989. *In* Invertebrate Cell System Applications, Vol. I. J. Mitsuhashi, Ed.: 3–20. CRC Press. Boca Raton, Florida.
56. MITSUHASHI, J. 1982. Appl. Entomol. Zool. **17:** 575–581.
57. VARMA, M. G. R. & M. PUDNEY. 1969. J. Med. Entomol. **6:** 432–439.
58. SHIELDS, G. & J. H. SANG. 1977. Drosophila Inf. Serv. **52:** 161.
59. ECHALIER, G. 1976. *In* Invertebrate Tissue Culture. E. Kurstak & K. Maramorosch, Eds.: 131–150. Academic Press. New York.
60. GARDINER, G. R. & H. STOCKDALE. 1975. J. Invertebr. Pathol. **25:** 363–370.
61. HINK, W. F., E. M. STRAUSS & D. E. LYNN. 1977. In Vitro **13:** 177.
62. VAIL, P. V., D. L. JAY & C. L. ROMINE. 1976. J. Invertebr. Pathol. **28:** 263–267.
63. GOODWIN, R. H. & J. R. ADAMS. 1980. *In* Invertebrate Systems in Vitro. E. Kurstak, K. Maramorosch & A. Dubendorfer, Eds.: 493–498. Elsevier/North-Holland. Amsterdam/New York.
64. RÖDER, A. 1982. Naturwissenschaften **69:** 92–93.
65. KOIKE, M. & K. SATO. 1988. *In* Invertebrate and Fish Tissue Culture. Y. Kuroda, E. Kurstak & K. Maramorosch, Eds.: 7–10. Springer-Verlag. Berlin/New York.
66. MAIORELLA, B., D. INLOW, A. SHAUGER & D. HARANO. 1988. Bio/Technology **6:** 1406–1410.
67. WILKIE, G. E. I., H. STOCKDALE & S. J. PIRT. 1980. Dev. Biol. Stand. **46:** 29–37.
68. HINK, W. F. 1982. *In* Microbial and Viral Pesticides. E. Kurstak, Ed.: 493–506. Dekker. New York.
69. KURTTI, T. J. & U. G. MUNDERLOH. 1984. Adv. Cell Culture **3:** 259–302.
70. TOLBERT, W. R. & J. FEDER, Eds. 1985. Large-Scale Mammalian Cell Culture. Academic Press. New York.
71. ARATHOON, W. R. & J. R. BIRCH. 1986. Science **232:** 1390–1395.
72. LYDERSEN, B. K., Ed. 1987. Large-Scale Cell Culture Technology. Hanser Pub. Munich/New York.
73. SEAVER, S. S., Ed. 1987. Commercial Production of Monoclonal Antibodies. Dekker. New York.
74. GOODWIN, R. H. 1975. In Vitro **11:** 369–378.
75. TYO, M. A. & R. E. SPIER. 1987. Enzyme Microb. Technol. **9:** 514–520.

76. TRAMPER, J., J. B. WILLIAMS, D. JOUSTRA & J. M. VLAK. 1986. Enzyme Microb. Technol. **8:** 33-36.
77. STOCKDALE, H. & G. R. GARDINER. 1976. *In* Invertebrate Tissue Culture. E. Kurstak & K. Maramorosch, Eds.: 267-296. Academic Press. New York.
78. WEISS, S. A., T. ORR, G. C. SMITH, S. S. KALTER, J. L. VAUGHN & E. M. DOUGHERTY. 1982. Biotechnol. Bioeng. **24:** 1145-1154.
79. WUDTKE, M. & K. SCHÜGERL. 1987. *In* Modern Approaches to Animal Cell Technology. R. E. Spier & J. B. Griffiths, Eds.: 297-315. Butterworths. London.
80. CHALMERS, J., Y. BAE, R. BRODKEY & W. F. HINK. 1988. Presented at the American Institute of Chemical Engineers (AIChE) Annual Meeting, Washington, District of Columbia, November 27 to December 2, 1988 (abstract no. H100).
81. STREETT, D. A. & W. F. HINK. 1978. J. Invertebr. Pathol. **32:** 112-113.
82. EBERHARDT, U. & K. SCHÜGERL. 1987. Dev. Biol. Stand. **66:** 325-330.
83. TRAMPER, J., D. SMIT, J. STRAATMAN & J. M. VLAK. 1988. Bioprocess Eng. **3:** 37-41.
84. TRAMPER, J., E. J. VAN DEN END, C. D. DE GOOIJER, R. KOMPIER, F. L. J. VAN LIER, M. USMANY & J. M. VLAK. 1990. This volume.
85. HANDA, A., A. N. EMERY & R. E. SPIER. 1987. Dev. Biol. Stand. **66:** 241-252.
86. HANDA-CORRIGAN, A., A. N. EMERY & R. E. SPIER. 1989. Enzyme Microb. Technol. **11:** 230-235.
87. LAMBERT, K. J., R. BORASTON, P. W. THOMPSON & J. R. BIRCH. 1987. Dev. Ind. Microbiol. **27:** 101-106.
88. MIZRAHI, A. 1975. J. Clin. Microbiol. **2:** 11-13.
89. MURHAMMER, D. W. & C. F. GOOCHEE. 1988. Bio/Technology **6:** 1411-1418.
90. WEISS, S. A. & J. L. VAUGHN. 1986. *In* The Biology of Baculoviruses, Vol. II. R. R. Granados & B. A. Federici, Eds.: 63-87. CRC Press. Boca Raton, Florida.
91. FLEISCHAKER, R. 1987. *In* Large-Scale Cell Culture Technology. B. K. Lydersen, Ed.: 59-79. Hanser Pub. Munich/New York.
92. NILSSON, K. 1987. Trends Biotechnol. **5:** 73-78.
93. DEAN, R. C., S. B. KARKARE, N. G. RAY, P. W. RUNSTADLER & K. VENKATASUBRAMANIAN. 1987. Ann. N.Y. Acad. Sci. **506:** 129-146.
94. KING, G. A., A. J. DAUGULIS, P. FAULKNER & M. F. A. GOOSEN. 1988. Biotechnol. Lett. **10:** 683-688.
95. CHERRY, R. S. & E. T. PAPOUTSAKIS. 1986. Bioprocess Eng. **1:** 29-41.
96. CROUGHAN, M. S., J. F. HAMEL & D. I. C. WANG. 1987. Biotechnol. Bioeng. **29:** 130-141.
97. AGATHOS, S. N., Y-H. JEONG, A. M. FALLON & K. VENKAT. 1990. Biotechnol. Bioeng. Submitted.
98. SARVER, N. & V. STOLLAR. 1977. Virology **80:** 390-400.
99. MALINOSKI, F. & V. STOLLAR. 1981. Antiviral Res. **1:** 287-299.
100. EAGLE, H. 1959. Science **130:** 432-437.
101. METZ, B. & N. W. F. KOSSEN. 1977. Biotechnol. Bioeng. **19:** 781-799.

Bioreactor Development for Production of Viral Pesticides or Heterologous Proteins in Insect Cell Cultures[a]

M. L. SHULER,[b] T. CHO,[b] T. WICKHAM,[b] O. OGONAH,[b]
M. KOOL,[b] D. A. HAMMER,[b] R. R. GRANADOS,[c]
AND H. A. WOOD[c]

[b]*School of Chemical Engineering*
Cornell University
Ithaca, New York 14853

[c]*Boyce Thompson Institute for Plant Research*
Cornell University
Ithaca, New York 14850

INTRODUCTION

Why Insect Cell Cultures?

One of the most startling recent developments in biotechnology has been the rapid acceptance of insect cell cultures and the baculovirus expression vector system for commercial-scale eukaryotic protein production. In the summer of 1988, MicroGenSys received the first FDA approval for clinical tests of a potential AIDS vaccine.[1,2] They used a baculovirus expression system to produce an HIV envelope protein, gp160, in insect cells. The MicroGenSys product gained FDA approval first "... because its preparation stimulates a strong immune response in animals ...".[1] It is noteworthy that many other companies working on HIV vaccines (e.g., Repligen uses *E. coli*, Chiron uses *S. cerevisiae*, Genentech uses mammalian cells) did not achieve FDA approval as quickly even though they have been using more traditional expression systems to produce other HIV envelope proteins.[1]

Earlier attempts at large-scale insect cell tissue cultures were motivated by the possibility of "factory" production of viral insecticides. Insect viruses are attractive as insecticides because they are specific for a limited range of insects and nonpathogenic to vertebrates. In fact, four baculoviruses have been registered by the United States EPA. Attempts at commercial production of viruses currently involve diseased insects, but such production methods are difficult to apply on large scale because they require special pathogen-free, insect-growing facilities and high labor and operating costs.[3] Additionally, the product is impure, containing contaminating microorganisms, insect cuticle, and proteins. These contaminants can be highly allergenic to man. Thus, insect cell culture for virus production has been viewed as a potentially viable alternative,

---

[a]The work on the bioreactors was supported, in part, by NSF Grant Nos. ECE 8700739 and EET 8807089. The development of mathematical models was supported by a fellowship to T. Wickham through Cornell's Biotechnology Institute.

although the high cost of serum-containing medium has been a barrier. Recent developments with serum-free medium[4,5] may result in greatly reduced medium costs.

Significant commercial potential exists for the insect cell–baculovirus system. However, this system has received very little attention from biochemical engineers. The purpose of this article is twofold. The first is to summarize important characteristics of the baculovirus system and aspects of bioreactor technology that have been or could be applied to the insect-baculovirus system. The second goal is to report preliminary results on attempts to construct an appropriate bioreactor system for these cells and to develop mathematical models to describe baculovirus–insect cell systems.

STATUS OF THE INSECT CELL–BACULOVIRUS SYSTEM

Biological Aspects

The most extensively studied baculovirus is the *Autographa californica* nuclear polyhedrosis virus (AcMNPV) (see reviews by Kelly[6] and Miller *et al.*[7]). The AcMNPV has a genome of approximately 130 kilobases of double-stranded, circular DNA. Infection of cultured cells begins with the adsorption of infectious nonoccluded virus (enveloped nucleocapsids) to the plasma membrane and with the entry into the cell by either membrane fusion or endocytosis. Baculovirus replication is biphasic and occurs in the nucleus. The AcMNPV has a broad host range and can infect a variety of lepidopteran species (e.g., fall armyworm).

During the early phase, infectious nonoccluded virus is produced by budding of nucleocapsids through the plasma membrane. Late in the replication cycle, the nucleocapsids become enveloped within the nucleus and are embedded within a paracrystalline protein matrix referred to as occlusion bodies (OB). The OB matrix protein (or polyhedrin protein) is a single major protein with a molecular weight of about 28 kilodaltons. At cell death, the OB protein constitutes 40% or more of the total protein mass in the cell. Accordingly, the OB protein gene is one of the most highly expressed viral gene products described. Lysis of infected cultured cells usually commences within 36 hours postinfection and is complete by 72 hours postinfection.[8,9] For commercial use as an insecticide, the occluded form is preferred. The occluded form is relatively noninfectious in cell culture, but is highly infectious to larval hosts.

The polyhedrin gene has been mapped and sequenced. Deletion of the gene has been shown not to interfere with early infectious virus production and thus not to interfere with virus propagation in tissue culture. In cell cultures, the polyhedrin-minus mutants produce plaques that are easily discernible from the wild-type OB-plus plaques.

Therefore, baculoviruses and insect cell tissue culture have been suggested to be attractive alternatives to mammalian cell culture expression vector systems. A baculovirus–insect cell culture system avoids many of the problems inherent in protein production with cloned genes in mammalian cell culture; namely, very high level expression can be obtained—much higher than most mammalian cell systems. Furthermore, the use of transformed cells with viruses that might possibly have adverse effects on humans can be avoided.[10] Insect cell lines are continuous and are apparently nontransformed, although their growth characteristics may change with time. Insect cell systems also offer an advantage over prokaryotic systems by providing a wide variety of posttranslational modifications. Although glycosylation occurs in yeast, this posttranslational modification is done differently than in the normal (i.e., mammalian) host. The potential for high-level expression of cloned eukaryotic genes with potentially correct posttranslational modifications has recently been recognized and has induced a resurgence of commercial interest in insect cell tissue culture.

The AcMNPV is well suited as an expression vector for cloned eukaryotic genes. The virus is nonpathogenic to vertebrates and, because of its bacillus shape, it can encapsulate viral genomes with large pieces of foreign DNA. In addition, the OB gene provides the following properties: (1) it is a nonessential gene that can be replaced with foreign DNA; (2) it has a very strong promoter; and (3) it provides a convenient phenotypic marker for selection of recombinant viruses.

Several strategies have been used in constructing AcMNPV expression vectors. In each case, the polyhedrin gene has been cloned and then modified to interrupt gene expression, and foreign DNA has been inserted. The foreign DNA can be inserted in frame to produce a fusion protein. Alternatively, the 3' end of the open reading frame could be deleted and the foreign gene with its own initiation and termination codons could be inserted. Insect cells are then cotransfected with both the viral DNA and the plasmid containing the modified polyhedrin gene. Through recombination and allelic transplacement, the foreign gene is inserted into the viral DNA. The recombinant lacks a functional polyhedrin gene and therefore produces infectious viruses that produce foci lacking OB. These are easily discernible from the OB containing wild-type infection foci.

Luckow and Summers[11] have written an excellent review on baculovirus expression vectors including a table that summarizes the reported work on the baculovirus expression system. The results to data show that ". . . the very abundant expressions of recombinant proteins are antigenically, immunogenically, and functionally similar to their authentic counterparts."[11] Experience with expression of over 35 proteins has demonstrated correct targeting of proteins, including secretion, phosphorylation, several types of glycosylation, and disulfide formation. However, the glycosylations are not completely faithful to mammalian oligosaccharide processing, at least in *Spodoptera frugiperda* (fall armyworm). Different insect cell lines may exhibit differences in their ability to glycosylate mammalian proteins. It is also clear that much work remains to be done to elucidate the molecular mechanisms of such processes.

Some important examples include pioneering work by Smith *et al.*,[12] who obtained high levels (5×10^6 u) of biologically active human interferon, and Pennock *et al.*,[13] who obtained high expression levels of *Escherichia coli* β-galactosidase gene in insect cells. Since then, there have been reports of expression of human c-myc protein,[14] human interleukin-2,[15] influenza virus hemagglutinin,[16,17] the S-coded genes of lymphocytic choriomeningitis arenavirus,[18] the N and GPC proteins of lymphocytic choriomeningitis arenavirus,[19] the env gene of the AIDS virus,[20] and the large and small T antigens of Simian Virus 40.[21] The interesting aspect of this last report is that a baculovirus vector can express intron-containing genes. In addition, human interferon has been expressed in insect larvae.[22]

Spodoptera frugiperda tissue cultures have been used in the *in vitro* studies mentioned above. However, *Trichoplusia ni* (cabbage looper) tissue culture cells may also be suitable. The unique properties of the AcMNPV expression vector system make it of considerable commercial interest, particularly with products that require posttranslational modifications for biological activity.

Thus, an insect cell–baculovirus system is a plausible and potentially attractive system for producing both viral insecticides and proteins from cloned eukaryotic genes. To utilize this potential requires a much better fundamental understanding of the dependence of cellular and viral growth dynamics, as well as protein processing, on easily controllable environmental variables.

Insect cell cultures grow best at slightly acidic pH (6.2 to 6.3) and at temperatures of 27 to 30 °C, and they display moderate tolerance for a wide range of osmotic pressures.[23] Cell lines from the same donor can grow either in suspension or on surfaces.[24] Oxygen requirements are readily satisfied for cells growing as a monolayer in 25- or 75-cm^2 plastic culture flasks. Higher oxygen levels may be required for the growth of differentiated cells such as epidermal and fatbody cells.[25-27]

Media costs have been an important bottleneck to the commercialization of insect cell culture for insecticide production. Proteins in sera may also complicate protein recovery. The nutritional requirements in insect cell culture may be even more complex than for mammalian cell culture. For example, unlike cells of vertebrates, insects are unable to synthesize sterols from squalene and must be supplied such components directly. Sera may transport sterols and essential fatty acids and may play a role in cell physiology different from that in mammalian cell culture. Moreover, Inlow et al.[5] have suggested that serum may play a role in shear protection and may protect cells from trace toxic compounds.

Currently, fetal bovine serum is used at levels of 5–15% of the total volume. Vaughn et al.[4] have recently reported the use of serum replacements for the production of baculoviruses in culture. Inlow et al.[5] report a serum-free medium that supports both good growth and high expression levels for CSF-1. The recent progress on serum-free medium makes the large-scale use of insect cell systems much more economically attractive.

Insect cells appear to exhibit a form of contact inhibition. Host and viral DNA replication are dramatically reduced in high density cultures, which, of course, greatly reduces viral infectivity and replication. This inhibition appears to occur in both attached and suspension systems. Inhibition involves direct cell-to-cell contact rather than diffusible chemical species.[28]

Large-Scale Culture

Although the potential for large-scale systems has been considered for many years, almost all work has been in vessels less than 10 L (see references 29–33). Inlow et al.[5] report good results in a 20-L airlift system. As with mammalian cells, the simultaneous maintenance of adequate mixing to provide (i) a homogeneous environment, (ii) desirable levels of gaseous components, and (iii) low levels of shear has proved exceedingly difficult in large volume systems (>10 L) due to the interactive nature of agitation and gas transfer in most traditional stirred or sparged reactors (see Thilly[34] for a review of mammalian cell technology).

Insect cells may be even more shear-sensitive than mammalian cells. High shear stresses will limit cell viability and growth rates. For instance, Tramper et al.[35] have indicated that a shear stress of 1 N/m^2 causes damage to the insect cell line *S. frugiperda*. This value is below the shear stress that endothelial cells can withstand without cell damage (8.5 N/m^2).[36] However, FS-4 fibroblasts grown in microcarrier cultures show significant cell death at shear stresses above 1.6 N/m^2 and the death rate increases with increasing shear rates.[37] Therefore, insect cells appear to be at least as sensitive to fluid shear as mammalian cells.

Because cell physiology can be modulated by a cell's environment, it is anticipated that fluid shear stresses and adhesive interactions will influence the efficiency of the insect cell as a host for production of protein from recombinant DNA. There is

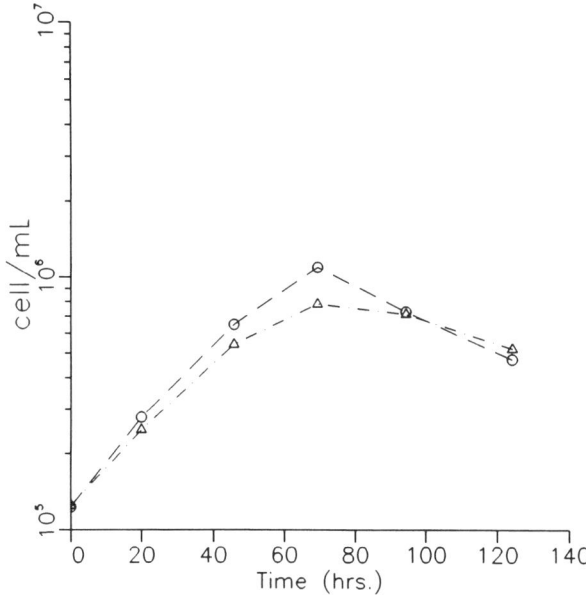

FIGURE 1a. Growth curves for *Trichoplusia ni* (*Tn* 368) in suspension cultures. Symbols (O) and (△) denote two separate experiments.

evidence obtained from mammalian cell culture that shear stresses could modulate protein synthesis. Shear stress affects not only embryonic kidney cell morphology, but increasing stress also decreases the cell's ability to produce the enzyme urokinase.[38] Endothelial cells elongate and orient in the direction of flow, and the degree of elongation is shear stress dependent.[36] Contrary to the results seen with kidney cells, shear stresses increase the production of prostacyclin from cultured endothelium.[39]

Evidence also exists to suggest adhesiveness will modulate cellular processes that are central to normal insect cell function. Folkman and Moscona[40] showed that rates of DNA synthesis, a necessary process for viral proliferation, depend on cell shape and adhesive interactions between the cell and the substrate. Recently, Ingber and Folkman[41] showed that rates of DNA synthesis for bovine endothelial cells increase exponentially

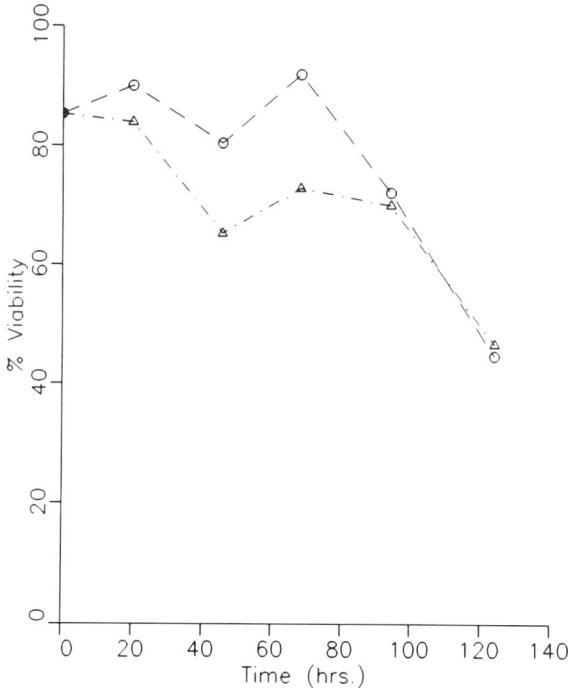

FIGURE 1b. Viability curves for *Tn* 368. Note that at the onset of stationary phase there is a marked decrease in cell viability. Symbols (O) and (△) denote two separate experiments and correspond to the symbols shown in FIGURE 1a.

with increases in cell-surface contact area. The contact area was altered by manipulation of the chemical composition of the matrix to which the cells were bound.

With specific reference to insect cell systems, Bilimoria and Carpenter[42] have selected for variants of *T. ni* that are more adhesive than the parent strain. The more adhesive strains proved to be more sensitive to infection by NPV, suggesting that the changes in morphology and intracellular structure afforded by increased adhesion increase infectivity.

Thus, it seems likely that cell physiological responses such as protein and DNA synthesis necessary for viral synthesis, viral assembly, and protein production will depend on shear stress and cell adhesion. In addition, we expect the overall viral exocytosis rate, which depends on intracellular viral transport, virus–plasma membrane fusion, and viral exocytosis, to be affected by local shear stresses on the cell surface.

In addition to the effects of fluid shear on insect behavior, low shear tolerance presents significant limitations on the ability to supply oxygen to insect cells. Oxygen requirements are relatively high (15 to 45 μg O_2/mg cell-h)[43] and these requirements nearly double during viral infection.[44] Also, differentiated cells, which might be desirable host cells, have elevated oxygen demand. Systematic studies on gas-transfer effects on insect cell–baculovirus systems are still needed.

The ability to simultaneously supply oxygen and control shear is a severe constraint on scaleup. Based on the work of Tramper et al.,[35] it is apparent that gas sparging for airlift fermentors will not be practical. Shear, mixing, and $k_L a$ do not scale proportionally in the normal stirred fermentor.[45] Although the addition of high viscosity agents such as methylcellulose (400 cP at levels of 0.1–0.3%) may help mitigate some of the shear effects, the mechanism by which this occurs is not understood. Murhammer and Gooche[46] and Inlow et al.[5] have both reported protection from liquid shear by using Pluronic F-68 as well as the resulting ability to increase reactor scale. Miltenburger and David[29] used an immersed silicone rubber tube for oxygenation. In a large-scale reactor, the length of tubing that can be used is limited because gas flow will be plug flow and the concentration of oxygen will decrease to a point where the concentration gradient becomes zero.

In any mechanically mixed suspension system, achieving good mixing and homogeneous environmental conditions throughout the reactor becomes very difficult at large scales because shear is determined by the tip speed of the impeller. If nonhomogeneous regions exist, scaleup from screening trials and bench studies is difficult because many of the cells are exposed periodically to nonoptimal growth environments. Hence, the population-averaged behavior may differ markedly from that predicted at the bench scale.

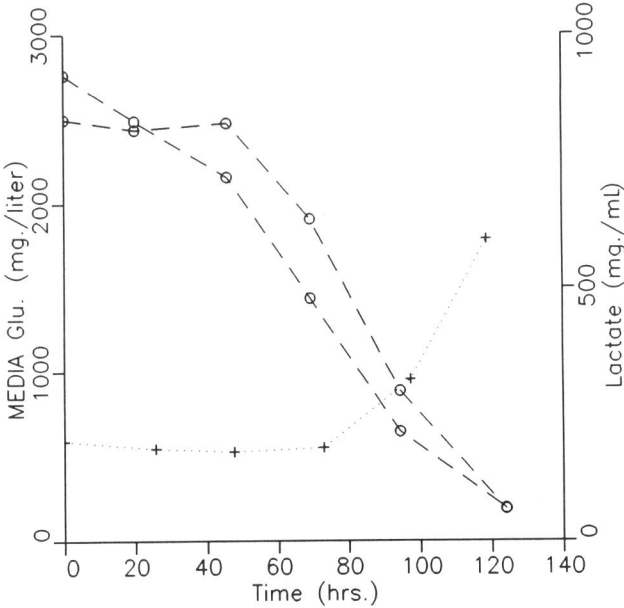

FIGURE 1c. Glucose utilization and lactate production curves for *Tn* 368 in suspension cultures. Note that the sharp decrease in glucose concentration in media and the increase in lactate concentration coincided with the onset of stationary phase. This increase in glucose uptake rate and the coincident increase in lactate concentration indicate a switch from aerobic to anaerobic metabolism.

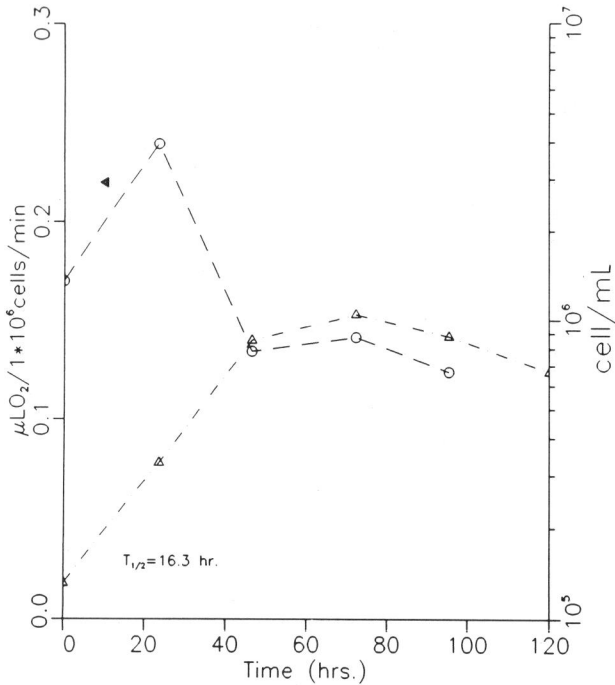

FIGURE 1d. In a separate experiment, rates of O_2 consumption and cell growth were measured simultaneously. The rate of oxygen consumption is scaled to the cell density to measure O_2 consumption on a per cell basis. O_2 consumption (○) is greatest during the initial stages of growth and it levels out at later stages when the cell density (△) is constant.

To circumvent these shear or oxygenation problems, several novel reactor concepts have been suggested for animal cell suspension cultures. These methods usually involve some form of cell entrapment (microencapsulation, bead entrapment, attachment to a ceramic matrix, and use of hollow-fiber devices; see various chapters in Lydersen[47]). Often, gas transfer takes place in a device (e.g., "oxygenator" or "gas permeator") separate from the main reactor, requiring very high fluid recirculation rates. Although such an approach places an upper limit on reactor size, the upper size will suffice for the production of most therapeutic proteins. Many of these devices work for both attachment dependent and independent cell lines, although none are ideal for all cell lines. In each of these devices, diffusional limitations with respect to the immobilized cells are a potential problem.

These nonhomogeneities and problems also exist for surface-dependent cell growth reactors. Insect cells can be adapted to grow in either mode. With surface growth, it is difficult to obtain sufficient surface area per unit reactor volume to have a high cell density. One common method to increase surface to volume area is the use of microcarriers. However, problems of shear, O_2 transfer, and heterogeneity exist. In fact, Croughan et al.[37] write that "... animal cells in microcarriers are especially susceptible to damage from excessive agitation." Microcarrier agglomeration can be a further complication.

Other bioreactor systems for attachment-dependent animal cells have been advocated. The most straightforward approach is the use of multiple roller bottles. However, roller bottles provide low surface/volume ratios (S/V) and would be very labor-intensive. House and colleagues[48] have advocated a bulk culture vessel using a coiled sheet of polystyrene film and gas sparging. Taylor et al.[49] have found such devices to be unsafe due to contamination and formation of dangerous aerosols. A titanium disk unit—an analogue of the well-known rotating biological disk (RBD) reactor commonly used in waste treatment—has been sold commercially for animal cell culture.[50] This unit provides relatively high S/V and decouples aeration and shear from a strong scale-dependence. A major limitation on size is inducing cell attachment, which requires the unit to be in the vertical position while the operating position is horizontal. Monsanto's (Invitron) perfusion reactor may overcome many of these problems (at least at culture volumes of 16 L), although the filtering requirements to maintain high medium recirculation rates are significant and probably limit the ultimate size of the reactor (see page 186 of reference 34).

All of these systems present potential problems in rapidly and effectively converting data from the T-flasks and spinner flasks used by biological scientists into large-scale reactors with predictable performance. We believe that the potential of the insect cell–baculovirus system can be better exploited if "scalable" bioreactors and

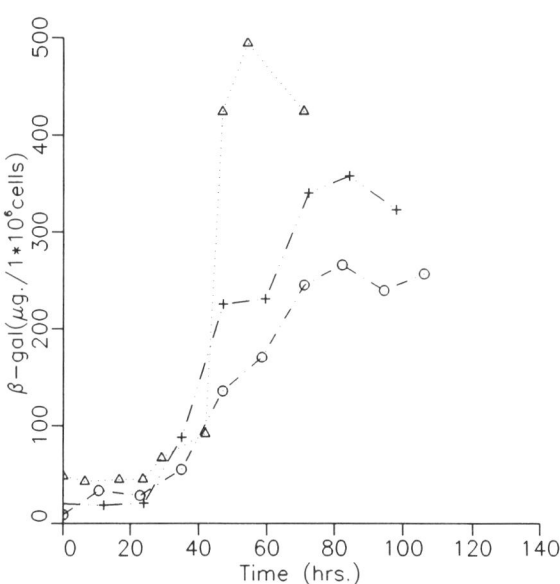

FIGURE 2. β-Galactosidase production curves for SF9 (+, O) and *T. ni* 368 ($\triangle$) cells in a spinner flask plotted as a function of time after viral infection. In the experiment denoted "+", viability was 31% at the experiment's end; in "O", viability at the experiment's end was greater than 80% (1 mg β-gal = 500 units of activity).

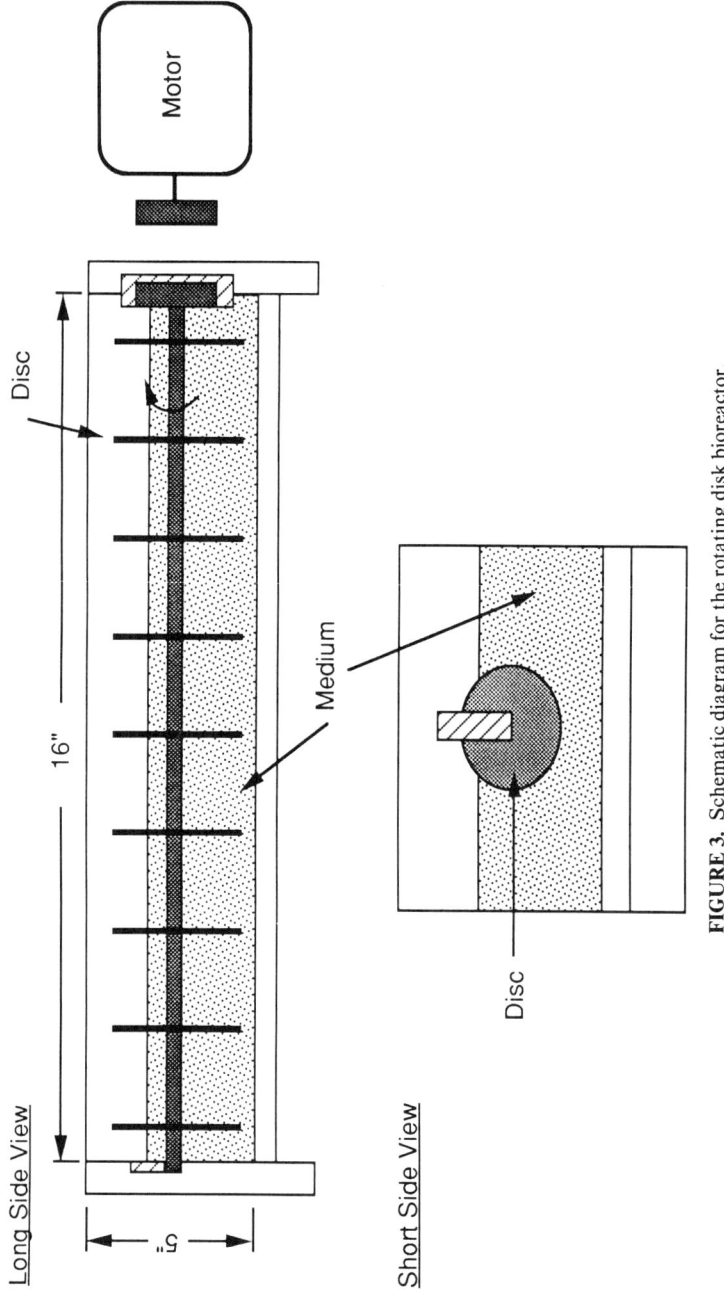

FIGURE 3. Schematic diagram for the rotating disk bioreactor.

predictive mathematical models can be developed. By "scalable", we mean a reactor system where the microenvironment (e.g., nutrients, shear, dissolved oxygen, etc.) experienced by a cell is essentially independent of reactor size. Our initial attempts to develop such systems and models are described next.

EXPERIMENTAL RESULTS

The APPENDIX at the end of this report describes the basic experimental techniques utilized in this work. The experimental work so far has had two objectives: (1)

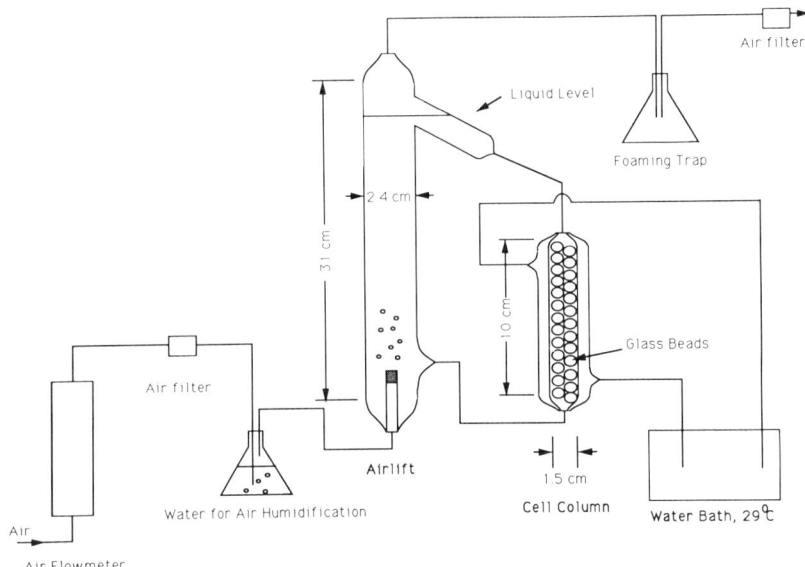

FIGURE 4. Schematic diagram of the packed-bed insect cell bioreactor and airlift for oxygen supplementation.

comparison of *T. ni* and *S. frugiperda* cells as hosts for production of proteins from recombinant DNA and (2) development of bioreactors for attachment-dependent strains of *T. ni*.

Suspension Cell Cultures

Preliminary experiments in 50-mL spinner flasks established that stirring at 60 rpm gave good growth with minimal cell lysis. Lower stirring rates (30 rpm) did not maintain cells in suspension and cell aggregation became a problem. At higher speeds

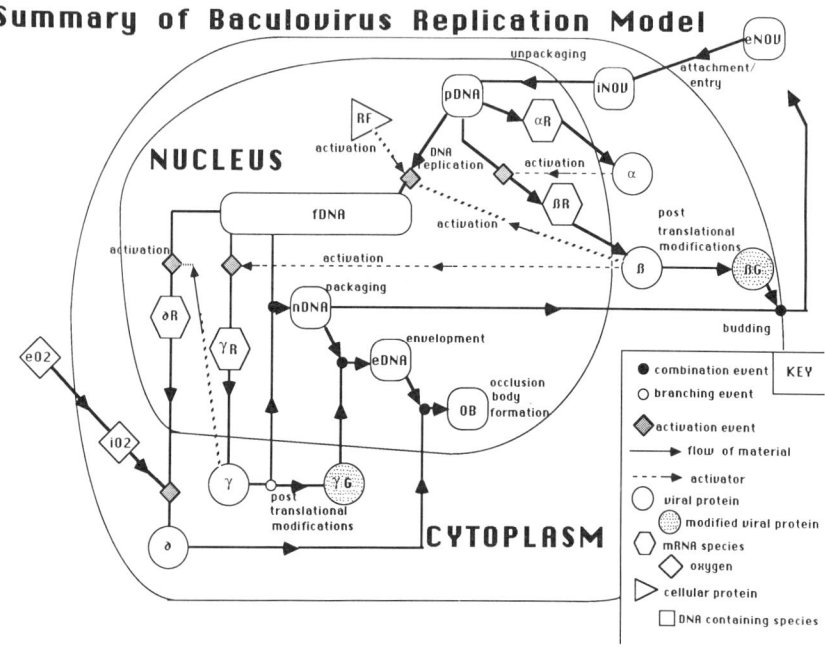

FIGURE 5a. Summarization of a baculovirus replication model developed from literature data. Baculovirus replication is temporally controlled by a cascade induction of the phases of protein synthesis.[54,55] The general scheme of the model follows the phases of protein synthesis, α, β, γ, and δ, delineated by Kelly and Lescott.[54] The events in each phase are presented in FIGURES 5b–e with the prominent populations and flows in bold type. With the possible exception of polyhedrin mRNA (δR), the syntheses of the other mRNA populations (αR, βR, and γR) do not appear to be limiting steps in viral replication. They are included to clarify the mechanisms of replication and to demonstrate that most of the temporal regulation of viral replication is transcriptionally controlled.[56,57]

(i.e., 120 rpm), extensive amounts of cell lysis became apparent. Consequently, all further experiments were conducted at 60 rpm.

Experiments were done with *T. ni* (*Tn* 368) cells. These cells showed significant alterations as they became adapted to suspension culture. Over 16 passages, the yield increased from approximately 7×10^5 to 1.3×10^6 cells/mL. Initially, the cells were largely spherical in shape and did not exhibit any of the projections exhibited by siblings grown as loosely attached cells in T-flasks. By the 16th passage, the cells again exhibited tentaclelike projections. The cellular volume of adapted cells as determined electronically (Coulter Counter) was 2.6×10^3 μm^3 (diameter of ca. 17 μm) for exponentially growing cells.

FIGURES 1a–d summarize the response of growth characteristics of suspension-adapted cells to batch culture in terms of pH variations, dissolved oxygen consumption, glucose consumption, and lactic acid formation. Note that cell concentration (FIGURE 1a) reaches a maximum of 10^6 cells/mL at about 70 hours, after which there is a slight drop in concentration. Cell viability in the same experiment (FIGURE 1b) remained high (>80% in one experiment, >65% in the other) until 70 hours, after which viability also decreased. FIGURE 1c shows the glucose concentration as a function of time, along with a typical curve for the generation of lactate for suspension culture. After 50 hours, glucose is consumed rapidly and, after 70 hours, lactate is rapidly produced. Finally, in FIGURE 1d, oxygen consumption is measured against cell growth in a separate experiment. As cell concentration drops off slightly after 70 hours, O_2 consumption plateaus. Therefore, these results indicate that suspension-adapted Tn 368 cells undergo a transition from aerobic to anaerobic growth after 70 hours in culture, resulting in a decrease in cell concentration, viability, and the consumption of glucose, as well as an increase in the generation of lactate. Doubling time is 19 ± 2 h.

The cell line from *S. frugiperda* known as SF9 has been used almost exclusively for production of proteins from recombinant DNA. As our laboratory has more experience with *T. ni* cultures, it was of interest to compare these two strains for production of β-galactosidase from a baculovirus vector. The results are given in FIGURE 2. Production of β-galactosidase was more rapid in Tn 368 cells. The level of production (as measured by β-galactosidase activity) in Tn 368 cells was about twice that for SF9 cells. It should be recognized, though, that the Tn 368 cells have a volume about two

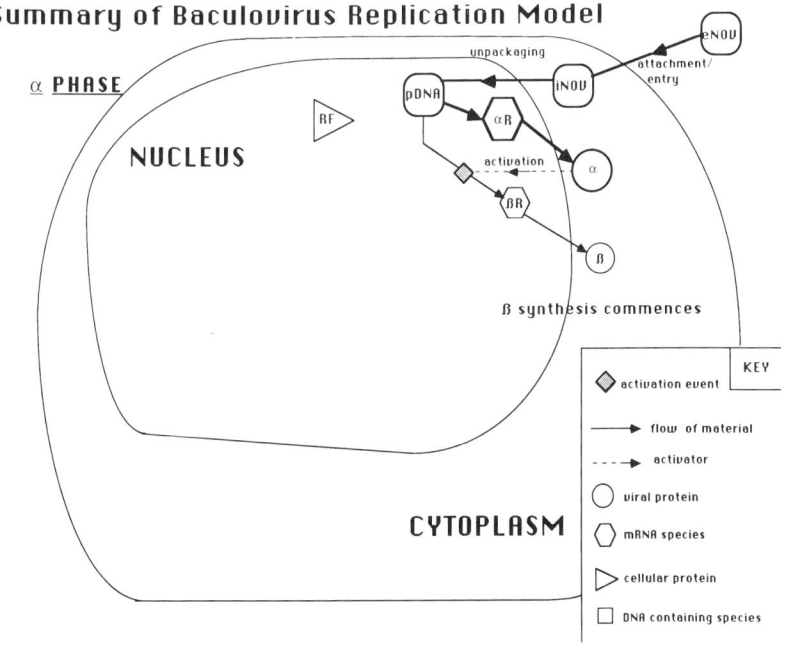

FIGURE 5b. Infection begins with the extracellular virus particle (eNOV) attaching and entering the cell by way of either endocytosis of the whole virus or fusion of the virus envelope with the cell membrane (iNOV).[9,58] Virus particles that have lost their envelope (either by fusion at the cell surface[58,59] or by fusion of the virus envelope with endocytic vacuole, stimulated by the lowered pH in this compartment[60]) are then unpackaged to release their copy of viral DNA (pDNA). Once the viral DNA is free in the nucleus, α proteins (α) are immediately synthesized by host cell machinery.[61]

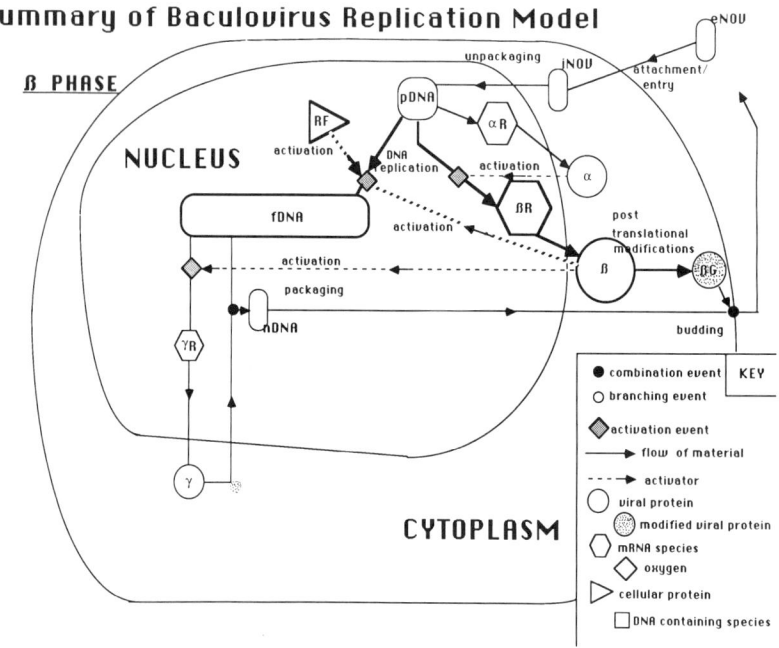

FIGURE 5c. The β phase is activated by one or more α proteins (α). Proteins in the β phase (β) are involved in: (1) DNA replication by a virus-induced DNA polymerase,[62–64] (2) activation of the γ phase, possibly by a virus-induced RNA polymerase,[61,65] and (3) the structural glycoproteins (βG) that make up the nonoccluded form of the virus.[60,66] One or more cell growth factors have been incorporated into the model that account for the observed coupling of baculovirus replication to the growth state of the cell.[28,67]

times that of SF9 so that the specific yields in both cultures are comparable (about 30% of the total protein is β-galactosidase). However, the much faster rate of production (between 40 and 50 h postinfection) in Tn 368 is intriguing and perhaps could be exploited in the correct bioreactor configuration.

BIOREACTOR STUDIES WITH ATTACHED CELLS

The ability to separate aeration and mixing is fostered in reactors where cells either self-attach or are immobilized. Hence, development of a "scalable" reactor might be more easily achieved with attached cell lines. This factor coupled with the report[42] that attached cell lines might increase viral infectivity motivated us to explore some reactor configurations for attachment-dependent variants of *T. ni* cells.

The following materials were screened for their ability to foster growth of *T. ni* cells: hydrophilic and hydrophobic polypropylene, polyethylene, polycarbonate, polystyrene, and boro-alumino-silicate [Pyrex] glass. Of these materials, boro-alumino-silicate [Pyrex] glass and polystyrene gave the best results in terms of growth rate, retention of viability, and ultimate extent of surface coverage.

One reactor system to exploit the growth of attachment-dependent cells is the rotating biological disk system (RBD). Although such systems are best known for their role in waste treatment, they have been exploited for animal cells.[50] The reactor system we used is depicted in FIGURE 3. The disk is about 40% submerged in medium. As the disk rotates, cells are alternately exposed to liquid nutrients and gas exchange through a thin liquid film as the disk rotates through the gas phase. The system can be easily characterized. Shear increases as a function of radius. Such a device facilitates experiments to determine shear effects on cells as well as effects of different materials while all other factors are maintained constant.

After several trials, it became apparent that this reactor system would not satisfy many of our needs. Even at the lowest rotational rate (0.5 rpm) attainable with the system, cell growth was abnormal. Toward the edge of the disk, cells formed irregular mucoid colonies. Cells became enmeshed in a polymer matrix apparently of their own synthesis; cells enmeshed in this mucoid polymer were clumped and did not exhibit the typical morphology of healthy cells. However, cells near the axis (within 2 mm) grew

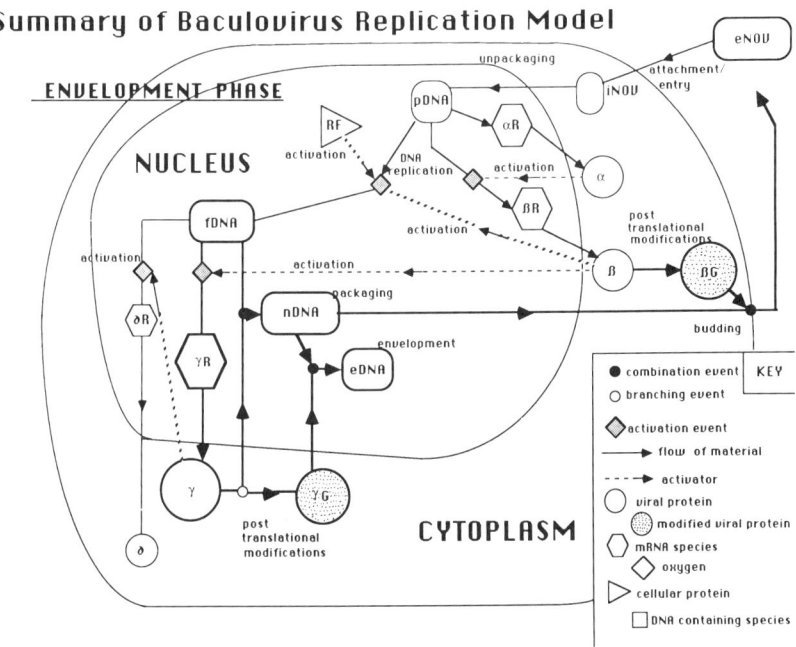

FIGURE 5d. The formation of viral structures such as nucleocapsids (nDNA), nonoccluded virus (eNOV), and occluded virus (eDNA) is a discrete process.[59] Although structural proteins are synthesized in the two previous phases, the γ-phase proteins (γ) and glycoproteins (γG) are limiting in the formation of complete nucleocapsids (nDNA) and occluded virus particles (eDNA), respectively. Synthesis of γ and δ polypeptides is dependent on DNA replication.[54,68] Nucleocapsid formation and budding can commence upon the induction of the γ phase. Budding rate is curtailed and the occluded form (eDNA) prevails when a sufficient population of posttranslationally modified γ proteins (γG) becomes available for envelopment of nucleocapsids (nDNA) within the nucleus.[69,70]

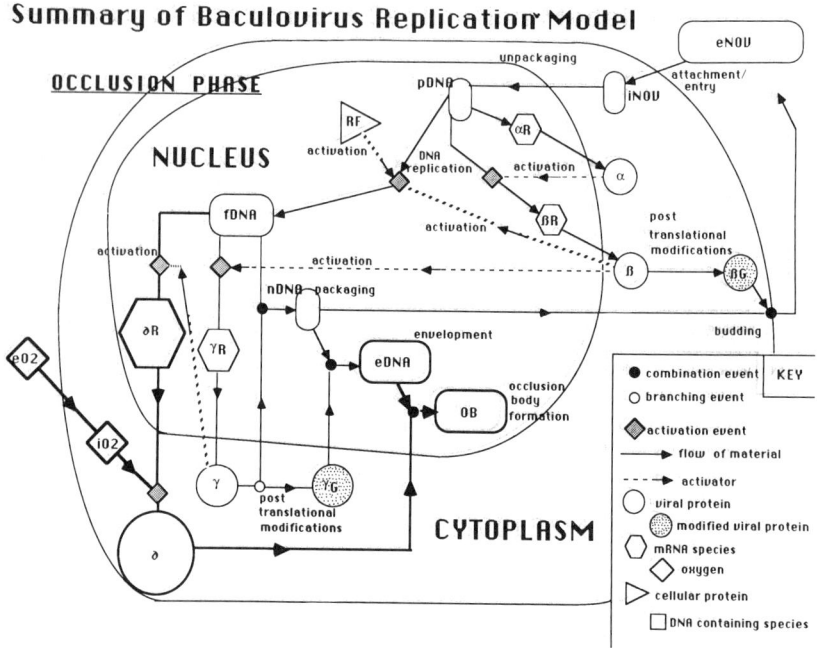

FIGURE 5e. The final δ phase is initiated by one or more γ proteins. Polyhedrin protein (δ) is hyperexpressed in this phase and is responsible for the occlusion of enveloped virus particles (eDNA) within the nucleus.[71,72] A doubling of oxygen (eO2 and iO2) consumption is noted during this phase[44] and is probably a result of the increased energy requirements of polyhedrin synthesis.

normally. The cells clearly showed a great sensitivity to shear and/or repeated movement through the gas/liquid interface.

Some protection might be provided if cells were growing inside a porous disk. Furthermore, such a configuration would have the practical advantage of greatly increasing surface area over a smooth disk. A porous glass disk (kindly supplied by Manville Corporation) was tested. Cell attachment and growth were poor. The glass fibers in the porous disk had a diameter of 10 μm. The radius of curvature of these fibers may be too large to allow good cell attachment and spreading.

An alternative to the porous disk is to wrap glass wool about the spindle and rotate the glass fibers through the liquid nutrient. Cells that do not attach initially might be enmeshed in the moving fiber bed. Such a reactor system can be simulated by stuffing a roller bottle with glass fibers. Such a system also did not give good results. Cell attachment and growth were negligible.

As an alternative system, we constructed a packed-bed reactor (FIGURE 4). Nonporous glass beads of 3-mm diameter were used. A separate unit was used for gas exchange and the oxygenated medium recirculated through the packed-bed reactor. A fluid recirculation rate of 90 mL/h was calculated to be sufficient to meet the oxygen requirements of a complete monolayer of cells.

In initial experiments, a peristaltic pump was used to assist in recirculation. In such a system, precipitates formed and growth was negligible. A separate experiment established that the pumping of the medium led to the formation of the precipitate. The medium with precipitate would not support cell growth in T-flasks, whereas unpumped medium supported normal growth. Why the action of the peristaltic pump should have this effect is not clear to us.

If, however, the aeration reservoir is also used as an airlift pump, then the bead reactor leads to excellent growth. Using *Tn* 5B1-4, which is a strongly attached variant of *T. ni*, a doubling time of 16 h can be obtained. This is comparable with the doubling time found in experiments in T-flasks. Cells grew to a monolayer on the upper halves of the beads because the reactor was inoculated from the top.

When cells were infected with the genetically modified virus at an MOI of 3 (multiplicity of infection in terms of the number of plaque-forming units per cell), significant production of β-galactosidase resulted. Most importantly, the cells remained attached to the bead surfaces after infection. Surprisingly, better than 70% of the cells appeared viable 70 hours postinfection (as determined by trypan blue exclusion). Very similar results were also obtained with an MOI of 0.1. Detachment of

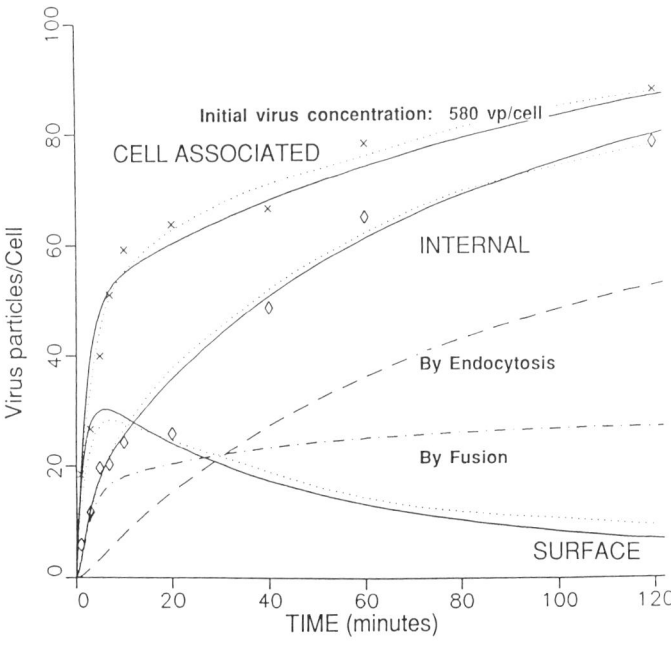

FIGURE 6. Comparison of the predictions for a model of attachment and internalization with data from Wang and Kelly.[58] The solid curves represent predictions of the model, the dotted-line curves represent a quadratic curve fit to the data, and the symbols represent the data. In this simulation, endocytosis is assumed to be independent of free surface receptor composition, whereas fusion does depend (linearly) on free surface receptor composition. The agreement between the model and the experiment is reasonably good.[73]

infected cells is often observed in T-flasks and such detachment would be deleterious to bioreactor design. Although the cause of retention of attachment is not known, it is important to note that the bead reactor provided better gas transfer and higher nutrient levels in the postinfection than the traditional T-flask would.

With an MOI of 3 at 70 h postinfection, intracellular β-galactosidase activity was 89 units/10^6 cells, whereas the 130 mL of extracellular medium had 0.5 units of activity/mL. Tn 5B1-4 is smaller than Tn 368 and has 0.53 mg of total protein per 10^6 cells (as measured by the Bio-Rad assay method based on the Bradford method).[51] The reactor had about 6×10^6 cells at the end of the experiment. Assuming that 500 units of activity is 1 mg of pure galactosidase,[52] then 35% of the total protein in the culture is β-galactosidase. This high level of production is significantly greater than reported in the literature for other insect cell–baculovirus systems making β-galactosidase (see references 13 and 53). However, because different vectors and conditions were used, a direct comparison is difficult. Nonetheless, the bead reactor system clearly shows promise as a system for production of proteins from recombinant DNA.

MATHEMATICAL MODELS

Good, dynamically accurate mathematical models would be important tools to further bioreactor design and development of operating strategies for such reactors. Sufficient information is available in the literature to make such an effort possible. This information is summarized in FIGURES 5a–e, which also depict our projection on how such a model might be structured.

The first part of this modeling attempt is to better understand how the virus enters the cell. This model is discussed in more detail elsewhere.[73] The model considers two possible modes of entry into the cell—adsorptive endocytosis and internalization by fusion. In the baculovirus–insect cell system, endocytosis yields productive infective particles, whereas fusion is a dead-end route. A model incorporating receptor-mediated attachment followed by internalization through fusion or endocytosis makes predictions consistent with available kinetic measurements and microscopic observations. FIGURE 6 compares model predictions of virus uptake to the experimental measurements of Wang and Kelly.[58] The agreement is reasonable. Moreover, Wang and Kelly[58] using electron microscopy showed mostly fusion at 10 min postinfection, which is consistent with the model. Furthermore, the predicted rate of infectious entry (presumably endocytosis) is similar in magnitude and shape (i.e., roughly linear) to that reported by Volkman and Goldsmith.[60]

In addition to being able to make reasonable predictions of virus entry, the model has given us insight into the influence of receptors on virus attachment and internalization. We have experimentally confirmed that receptors are involved in the process of virus entry.[73] Further experimental work will be required to test the model over a wide range of virus concentrations.

Such models will provide us with guidance on optimal strategies for introducing virus into the culture. Further refinements may even lead to insights on factors such as the "passage" effect (reduction of infectivity of virus after repeated use).

SUMMARY

The insect cell–baculovirus expression system has significant potential for producing proteins requiring some degree of posttranslational modification. *T. ni* cells appear to be as good a host as *S. frugiperda* cells for heterologous protein production as demonstrated by production of β-galactosidase. Attachment-dependent cells of *T. ni* can be effectively cultured in a packed-bed reactor using glass beads. When cells in such a reactor were infected, they produced 35% of the total protein as β-galactosidase. No cell detachment was observed even 70 h postinfection. A model of viral entry has been proposed and tested.

REFERENCES

1. BARNES, D. M. 1987. Science **237**: 973.
2. VON BRUNT, J. 1987. Bio/Technology **5**: 1118.
3. KHACHATOURIANS, G. G. 1986. Trends Biotechnol. **4**: 120–124.
4. VAUGHN, J. L., J. R. ADAMS, E. M. DOUHERTY & J. T. MCCLINTOCK. 1987. Abstracts, p. 207. Virology Congress, Edmonton, Alberta, Canada, August 8–9, 1987.
5. INLOW, D., D. HARANO & B. MAIORELLA. 1987. ACS National Meeting, New Orleans, Louisiana, August 30 to September 4, 1987.
6. KELLY, D. C. 1982. J. Gen. Virol. **63**: 1–13.
7. MILLER, L. K., A. J. LINGG & L. A. BULLA, JR. 1983. Science **219**: 715–721.
8. FAULKNER, P. 1981. *In* Pathogenesis of Invertebrate Microbial Diseases. E. W. Davidson, Ed.: 5–38. Alanheld, Osmum & Co. Totowa, New Jersey.
9. GRANADOS, R. R. & M. NAUGHTON. 1980. *In* Invertebrate Tissue Culture—Applications in Medicine, Biology, and Agriculture. E. Kurstak & K. Maramorosch, Eds.: 379–389. Academic Press. New York.
10. RAMABHADRAN, T. V. 1987. Trends Biotechnol. **5**: 175.
11. LUCKOW, V. A. & M. D. SUMMERS. 1988. Bio/Technology **6**: 47–55.
12. SMITH, G. E., M. D. SUMMERS & M. J. FRASER. 1983. Mol. Cell Biol. **3**: 2156–2165.
13. PENNOCK, G. D., C. SHOEMAKER & L. K. MILLER. 1984. Mol. Cell Biol. **4**: 399–406.
14. MIYAMOTO, C., G. E. SMITH, J. FARREL-TOWT, R. CHIZZONITE, M. D. SUMMERS & G. JU. 1985. Mol. Cell Biol. **5**: 2860–2865.
15. SMITH, G. E., G. JU, B. L. ERICSON, J. MOSCHERA, H-W. LAHM, R. CHIZZONITE & M. D. SUMMERS. 1984. Proc. Natl. Acad. Sci. U.S.A. **82**: 8404–8408.
16. KURODA, K., C. HAUSER, R. ROTT, H-D. KLENK & W. DOERFLER. 1986. EMBO J. **5**: 1359–1365.
17. POSSEE, R. D. 1986. Virus Res. **5**: 43–59.
18. MATSUURA, Y., R. D. POSSEE & D. H. L. BISHOP. 1986. J. Gen. Virol. **67**: 1515–1529.
19. MATSUURA, Y., R. D. POSSEE, H. A. OVERTON & H. L. BISHOP. 1987. J. Gen. Virol. **68**: 1233–1250.
20. HU, S-L., S. G. KOSOWSKI & K. F. SCHAAFL. 1987. J. Virol. **61**: 3617–3620.
21. JEANG, K-T., M. HOLMGREN-KONIG & G. KHOURY. 1987. J. Virol. **61**: 1761–1764.
22. MAEDA, S., T. KAWAI, M. OBINATA, H. FUIFIWARA, T. HORIOCHI, Y. SACKI, Y. SATO & M. FUTUSAWA. 1985. Nature (London) **315**: 592–594.
23. HINK, W. F., D. A. RALPH & K. H. JOPLIN. 1981. *In* Comprehensive Insect Physiology, Biochemistry, and Pharmacology, Volume 10 (no. 16). G. A. Kerkut & L. I. Gilbert, Eds.: 547–570. Pergamon. Elmsford, New York.
24. GOODWIN, R. H. 1975. In Vitro **11**: 369–378.
25. PHILLIPE, C. 1982. J. Insect Physiol. **28**: 257–265.
26. RIDDFORD, L. M., A. T. CURTIS & K. KIGUCHI. 1979. TCA Man. **5**: 975–985.
27. ESCHALIER, G. 1980. *In* Invertebrate Systems in Vitro. E. Kurstak, K. Maramorosch & A. Dubendorfer, Eds.: 582–592. Elsevier. Amsterdam/New York.
28. WOOD, H. A., L. B. JOHNSTON & J. P. BURAND. 1982. Virology **119**: 245–254.

29. MILTENBURGER, H. G. & P. DAVID. 1980. Dev. Biol. Stand. **46:** 181.
30. WEISS, S. A., G. C. SMITH, S. S. KALTER & J. L. VAUGHN. 1981. In Vitro **17:** 495.
31. STOCKDALE, H. & R. A. J. PRISTON. 1981. *In* Microbial Control of Pests and Plant Diseases 1970–1980, Volume 16. H. D. Burger, Ed.: 313–328. Academic Press. New York.
32. TRAMPER, J. & J. M. VLAK. 1986. Ann. N.Y. Acad. Sci. **469:** 279–288.
33. HINK, W. F. 1982. *In* Microbial and Viral Pesticides. E. Kurstak, Ed.: 493. Dekker. New York.
34. THILLY, W. G. 1986. Mammalian Cell Technology. Butterworths. London.
35. TRAMPER, J., J. B. WILLIAMS, D. JOUSTRA & J. M. VLAK. 1986. Enzyme Microb. Technol. **8:** 33.
36. LEVESQUE, M. J. & R. M. NEREM. 1985. J. Biomech. Eng. **107:** 341.
37. CROUGHAN, M. S., J-F. HAMEL & D. I. C. WANG. 1987. Biotechnol. Bioeng. **29:** 130–141.
38. STATHOPOULOS, N. A. & J. D. HELLUMS. 1985. Biotechnol. Bioeng. **XXVII:** 1021.
39. FRANGOS, J. A., S. G. ESKIN, L. V. MCINTIRE & C. L. IVES. 1985. Science **227:** 1477.
40. FOLKMAN, J. & A. MOSCONA. 1978. Nature **273:** 345.
41. INGBER, D. E. & J. FOLKMAN. 1987. Twenty-seventh Annual Meeting at the American Society for Cell Biology, St. Louis, Missouri. J. Cell Biol. **105.**
42. BILIMORIA, S. L. & W. M. CARPENTER. 1983. In Vitro **19:** 870.
43. STOCKDALE, K. & G. R. GARDINER. 1976. *In* Invertebrate Tissue Culture—Applications in Medicine, Biology, and Agriculture. E. Kurstak & K. Maramorosch, Eds.: 267. Academic Press. New York.
44. STREETT, D. A. & W. F. HINK. 1978. J. Invertebr. Pathol. **32:** 112.
45. OLDSHUE, J. Y. 1966. Biotechnol. Bioeng. **8:** 3.
46. MURHAMMER, D. W. & C. F. GOOCHE. 1987. ACS National Meeting, New Orleans, Louisiana, August 30 to September 4, 1987.
47. LYDERSEN, B. K., Ed. 1987. Large Scale Cell Culture Technology. Hanser Pub. New York.
48. HOUSE, W., M. SHEARER & N. G. MAROUDAS. 1972. Exp. Cell Res. **71:** 293.
49. TAYLOR, W. G., V. J. EVANS, C. H. FOX, R. F. CAMALIER & K. SANFORD. 1975. Biotechnol. Bioeng. **17:** 1847.
50. WHITAKER, A. M. 1972. Tissue and Cell Culture. Williams & Wilkins. Baltimore, Maryland.
51. BRADFORD, M. M. 1976. Anal. Biochem. **72:** 248–254.
52. KENNELL, D. & H. RIEZMAN. 1977. Transcription and translation initiation frequencies of the *Escherichia coli* lac operon. Mol. Biol. **114:** 1–21.
53. CARBORELL, L. F., M. J. KLOWDEN & L. K. MILLER. 1985. Baculovirus-mediated expression of bacterial genes in dipteran and mammalian cells. J. Virol. **56:** 153–160.
54. KELLY, D. C. & T. LESCOTT. 1981. Microbiologica **4:** 35.
55. WOOD, H. A. 1980. Virology **102:** 21–27.
56. ESCHE, H., H. LUBBERT, B. SIEGMAN & W. DOERFLER. 1982. Eur. Mol. Biol. Organ. J. **1:** 1829.
57. FRIESEN, P. D. & L. K. MILLER. 1985. J. Virol. **54:** 392.
58. WANG, X. & D. C. KELLY. 1985. J. Gen. Virol. **66:** 541–550.
59. BASSEMIR, U., H. G. MILTENBURGER & P. DAVID. 1983. Cell Tissue Res. **228:** 587.
60. VOLKMAN, L. E. & P. A. GOLDSMITH. 1985. Virology **143:** 185.
61. FUCHS, L. Y., M. S. WOODS & R. F. WEAVER. 1983. J. Virol. **48:** 641.
62. GORDON, J. D. & E. B. CARSTENS. 1984. Virology **138:** 69.
63. FLORE, P. H., J. P. BURAND, R. R. GETTIG & H. A. WOOD. 1987. J. Gen. Virol. **68:** 2025.
64. WANG, X. & D. C. KELLY. 1983. J. Gen. Virol. **64:** 2229.
65. GRULA, M. A., P. L. BULLER & R. F. WEAVER. 1981. J. Virol. **38:** 916.
66. STILES, B. & H. A. WOOD. 1983. Virology **131:** 230–241.
67. VOLKMAN, L. E. & M. D. SUMMERS. 1975. J. Virol. **16:** 1630.
68. DOBOS, P. & M. A. COCHRAN. 1980. Virology **103:** 446.
69. VOLKMAN, L. E., M. D. SUMMERS & C-H. HSIEH. 1976. J. Virol. **19:** 820.
70. KELLY, D. C. 1981. J. Gen. Virol. **52:** 209.
71. ROHEL, D. Z., M. A. COCHRAN & P. FAULKNER. 1983. Virology **124:** 357.
72. KNUDSON, D. L. & T. W. TINSLEY. 1974. J. Virol. **14:** 934.

73. WICKHAM, T. T., R. R. GRANADOS, D. HAMMER & M. L. SHULER. 1988. Paper presented at the 1988 AIChE Annual Meeting, Washington, District of Columbia.
74. GOODWIN, R. H. 1976. TCA Man. 3: no. 1.

APPENDIX

Materials and Methods

Suspension culture cells were grown in spinner flasks (Belco Glass Company, New Jersey). These flasks were modified slightly. The screw caps were removed from the sample ports and silicone closures (Belco Glass) were used instead. This allowed for gas exchange between the reactor and the environment while allowing the reactor to remain free from contamination.

Suspension Culture Routine Maintenance

For routine maintenance, cells were subcultured every three days. Cell densities were determined (Coulter Counter) and cell viabilities were determined (hemocytometer). Cell suspension was then concentrated by centrifugation ($1000g \times 5$ min), resuspended in fresh media, and transferred to sterile spinner flasks at a seeding density of approximately 1×10^5. Flasks were then put in a 25 °C incubator.

Virus Inoculation

Cell suspensions were concentrated by centrifugation ($1000g$ for 5 min) and the pellet was resuspended in virus inoculum. Fresh media was added until the cell density was about 5×10^6. Suspension was then incubated at 28 °C for 1 hour without agitation. Cells were then resuspended in fresh media supplemented with 100 μg/mL gentamicin sulfate and the density was adjusted to the desired value (usually about 5×10^5). Suspension was transferred to spinner flasks in 28 °C incubators. Samples were taken every 12 hours. Cells were pelleted, resuspended in NTM, and stored at -20 °C. Supernatant was stored at -20 °C. Assays were performed later.

Cell Lines, Recombinant Viruses, and Medium

The cell line used for suspension cell culture was *Trichoplusia ni* 368 derived from ovaries of cabbage looper worm. For attachment of cells, the cell line *Trichoplusia ni* 5B1-4, derived from embryonic eggs of cabbage looper worm, has been used.

The virus we used was *Autographa californica* nuclear polyhedrosis virus recombinant 246, kindly supplied by M. Cochran (MicroGenSys, Incorporated, West Haven, Connecticut). In this recombinant virus, the coding sequence for polyhedrin is replaced by the β-galactosidase gene from *Escherichia coli*.

The cells were cultured in Grace's insect medium GTC-100 (Gibco Laboratories)[74] supplemented with 0.26% tryptose broth, 4.2 mM $NaHCO_3$, 10% fetal (Gibco Laboratories, Grand Island, New York) bovine serum, and 100 ng/mL gentamicin sulfate.

Counting Cell Number and Viability Test

The cells were counted with a ZB1 Coulter Counter (Coulter Electronics, Incorporated, Hialeah, Florida). The viable cell count was determined by counting the cells that were impermeable to trypan blue (0.05%) or erythrocin (0.4%) in a hemocytometer chamber.

β-Galactosidase and Protein Assays

Cells suspended in 0.1 g M NTM-buffer (0.01 M Tris-HCl at pH 7.6, 0.1 g NaCl, 0.1 M Mg-acetate, 0.01 M β-mercaptoethanol, and 0.001 M Na_2-EDTA) were disrupted using the French press.

The cell debris was pelleted at $200\,g$ for 5 min and the supernatant was assayed for β-galactosidase activity using the O-nitrophenyl β-D-galactopyranoside (ONPG) assay. The supernatant was diluted appropriately (from 1:50 to 1:500) in Z-buffer (0.06 M Na_2HPO_4, 0.04 M NaH_2PO_4, 0.01 M KCl, 0.001 M $MgSO_4$, and 0.05 M β-mercaptoethanol, pH 7.6) so that a faint yellow color developed 10 min after the ONPG addition. The reaction was stopped by adding 1 M Na_2CO_3 and the absorbance was measured with a Beckman Acta M-VI Spectrometer at 420 nm. Specific activity was defined as nanomoles of ONPG cleaved per minute per milligram of protein. In this report, micromoles are used instead of nanomoles.

Protein was quantitated by a protein assay (Bio-Rad Laboratories, Richmond, California) based on the Bradford method.[51]

Screening of Material for Cell Attachment and Growth

The materials that were tested for cell attachment were hydrophilic and hydrophobic polypropylene of Celgard[R] microporous membrane and Porex[R] technologies, polyethylene of Porex[R] technologies, polycarbonate of Rohm & Haas, polystyrene of Dow Chemical, and Corning and Fisher and boro-alumino-silicate (Pyrex[R]) glass. The circular form of each material was placed inside the 25-mL beaker and the 5-mL cell suspension of Tn 5B1-4 was added to each of them. The percentage of attached cells to each material was calculated based on the ratio of the number of attached cells after five hours to that in the initial cell suspension. After the materials were screened for their efficient attachment, the materials were tested for the growth and viability of cells. The growth of cells attached on the surface of each screened material in petri dish was monitored under a Nikon Diaphot inverted phase contrast light microscope, the number of cells was counted in the Coulter Counter, and viability was checked.

Rotating Biological Disk Reactor

The configuration and dimensions of the system that we have used are depicted in FIGURE 3. The top of the reactor is covered with hydrophobic polypropylene membrane for gas exchange. The silicone rubber adhesive sealant (General Electric) was applied between the membrane and the frame of the reactor to prevent the leakage of liquid medium and contamination. The reactor was sterilized for 20 minutes at 15 ebs and 121 °C in autoclave. The 990 mL of TC-100 liquid medium, the 110 mL of cell suspension of *Tn* 5B1-4, and the 11 mL of gentamicin solution were well mixed in a separate, sterilized flask and were transferred to the reactor through a hole on the side of the reactor in a laminar hood. Then, the reactor was positioned vertically for five hours to allow the cells to settle by gravity force and to attach and spread on the surface of the disk. Following this, the reactor was again positioned horizontally and placed inside the 28 °C incubator, and the disks were rotated at 0.5 or 1 rpm. The rotation speed of the disks is controlled by the magnetically coupled electronic motor.

The sterilized deionized distilled water was added intermittently to compensate for the water evaporated through the top hydrophobic polypropylene membrane. After three days of operation, the reactor was dismantled and the disks were examined under the inverted phase contrast microscope. The polystyrene and glass disks were tested. Because the polystyrene is not autoclavable, the polystyrene disks were heat-treated for about two days at 80 °C in an oven, inserted into the spindle through their center holes, and then fitted into the inside of the reactor.

Packed-Bed Reactor

The general arrangement of the equipment is shown in FIGURE 4. The column that holds the beads contains about 700 3-mm-diameter borosilicate (Pyrex) glass beads (Corning Glass Works, New York) with an area available for cell growth of about 100 cm^2 (calculated from only the upper halves of the glass beads). At the bottom of the column, a 5-micron nylon membrane is placed to prevent free cells from circulating through the system.

A medium reservoir that is used as an airlift pump is connected to the cell column. When air is passed through the airlift, the medium circulates through the packed bed via the medium reservoir. The rate at which the medium circulates depends upon both the geometry of the airlift and the rate at which gas flows through it. The gas flow is monitored and controlled by an airflow meter (Cole-Parmer Instrument Company, Chicago, Illinois). It also is passed into and out from the equipment via air filters (Balston Filter Products, Lexington, Massachusetts) so as to prevent the contamination of the medium with exogenous microorganisms.

Cell Growth in the Packed-Bed Reactor

The column with the beads was first isolated from the system by closing a clamp under the column. A suspension of 8×10^5 *Tn* 5B1-4 cells was made up in 8 mL of GTC-100 medium.

After 5 hours, 120 mL of GTC medium supplemented with 0.1% antifoam C [a silicon polymer (Sigma, St. Louis, Missouri)] was added to the medium reservoir and the circulation of the medium around the system was initiated by passing air at 30 mL/min through the airlift. The circulation was continued for a further 67 h, when the cells were either harvested or infected with AcMNPV recombinant virus. Cells were harvested from the glass beads by first draining away the spent medium and then transferring the beads to a petri dish containing 20 mL of warm trypsin (4 mg/mL). After incubating this mixture in an incubator at 28 °C for 25 min, the cells were dispersed by gently expressing them from a pipette several times. The cells were counted with the Coulter Counter and the viable cell count was determined by trypan blue exclusion.

Infection with Virus in the Packed-Bed Reactor

The virus was added to the system with a multiplicity of infection (MOI) varying from 0.1 to 3. The circulation of the medium was continued for a further 72 h. Samples were removed at 24 and 48 h and were assayed for β-galactosidase activity.

After this 72 h, cells were harvested from the glass beads and the viable cell count was determined by trypan blue exclusion. For β-galactosidase activity and protein assays, the cells were pelleted at $200g$ for 5 min, washed with phosphate-buffered saline, and pelleted again at $200g$ for 5 min. Finally, cells were suspended in 0.1 g M NTM-buffer (0.01 M Tris-HCl at pH 7.6, 0.1 g M NaCl, 0.1 M Mg-acetate, 0.01 M β-mercaptoethanol, and 0.001 M Na_2-EDTA).

Production of Baculovirus in a Continuous Insect-Cell Culture

Bioreactor Design, Operation, and Modeling

J. TRAMPER,[a] E. J. van den END,[a] C. D. de GOOIJER,[a]
R. KOMPIER,[a] F. L. J. van LIER,[a] M. USMANY,[b]
AND J. M. VLAK[b]

[a]Department of Food Science
Food and Bioengineering Group
[b]Department of Virology
Agricultural University
6700 EV Wageningen, the Netherlands

INTRODUCTION

Baculoviruses are attractive biological control agents of insect pests.[1] In addition, they are exploited as expression vectors of eukaryotic and prokaryotic genes coding for proteins of medical, pharmaceutical, and veterinary importance,[2] including vaccines. Insect-cell cultures form a relatively new, but useful means for the production of baculoviruses and products thereof.[3] Mass culturing of insect cells is a key factor in the exploitation of the baculovirus expression vector system and is important for the production of baculovirus as an alternative to insect larvae.

The most severe technological constraint to large-volume cultivation of insect cells, and animal cells in general, is the supply of sufficient oxygen without aggressive sparging and stirring. This situation is essentially due to the sensitivity of insect and most animal cells to shear and other hydrodynamic forces. With the cost of labor being a limiting factor in the operation of small vessels for cell culture, the technology most suitable for scaleup is the one that involves large vessels based on the airlift principle and equipped with automatic process control.[4,5] It is our aim to quantify the shear sensitivity of insect cells, to develop a continuously operated baculovirus production system, and to model the virus replication process in the bioreactor.

The continuous baculovirus production system that we study consists of an insect-cell-producing bioreactor connected to one or more bioreactors in series where baculoviruses or recombinant proteins are produced (FIGURE 1). As we have shown[6] that mechanical stirring is redundant from the point of view of oxygen transfer (the shear sensitivity of the cells only allows relatively gentle stirring), we study in particular bubble columns and airlift-type bioreactors. Results obtained with a two-step system using *Spodoptera frugiperda* cells and *Autographa californica* nuclear polyhedrosis virus (AcNPV) are described in this report.

BIOREACTOR DESIGN

The fragility of insect and animal cells is such that even sparging of air into a culture to supply oxygen may cause a larger cell death rate than cell growth rate. A

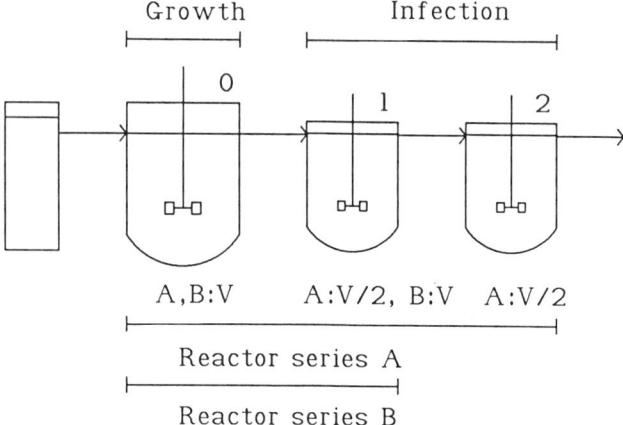

FIGURE 1. Experimental reactor configurations.

model, relating cell death to aeration by sparging, has been developed and experimentally validated.[7-10] The model is based on the following two assumptions:

(i) cell death as a result of air sparging is a first-order process:

$$C_t = C_0 e^{-k_d t}, \qquad (1)$$

where C_t and C_0 are the viable cell concentrations (cells/m^3) at time $t = t$ (s) and $t = 0$, respectively, and k_d is the first-order death-rate constant (1/s);

(ii) associated with each air bubble is a hypothetical killing volume X (m^3) in which all viable cells are killed:

$$V \frac{dC_t}{dt} = NC_t X = \frac{6FC_t X}{\pi d_b^3}, \qquad (2)$$

where V is the liquid volume of the reactor (m^3), N is the number of generated air bubbles (1/s), F is the airflow rate (m^3/s), and d_b is the diameter of the air bubbles (m).

Separation of the variables and integration between $C_t = C_0$ and $C_t = C_t$ and between $t = 0$ and $t = t$, with X constant in time, yields

$$C_t = C_0 \exp\left(-\frac{6FX}{\pi d_b^3 V} t\right). \qquad (3)$$

Combination of equations 1 and 3 leads to

$$k_d = \frac{6FX}{\pi d_b^3 V} = \frac{24FX}{\pi^2 d_b^3 D^2 H}, \qquad (4)$$

with D being the diameter of the reactor (m) and H being the height of the liquid volume in the reactor (m). In order to test the validity of equation 4, samples from a continuous culture were aerated in a bubble-column or airlift reactor in such a way that, in a period of 4 h, a significant decrease in viability could be measured. FIGURE 2 summarizes the results obtained at constant airflow F. It shows that k_d is proportional to F; in other words, X must be independent of F. Similarly, X was found to be independent of D and H for the bubble column. In contrast, k_d was independent of d_b and therefore X must be proportional to the air-bubble volume.[9] In equation 4, X can thus be replaced by a dimensionless hypothetical killing volume X':

$$X' = \frac{6X}{\pi d_b^3}, \tag{5}$$

which converts equation 4 into

$$k_d = \frac{4FX'}{\pi D^2 H}. \tag{6}$$

If compared to the well-known oxygen-transfer-rate equation, it is evident that H in particular is the parameter to manipulate in order to make the growth rate larger than the death rate in bubble columns. The dependence of k_d on H in airlift reactors was

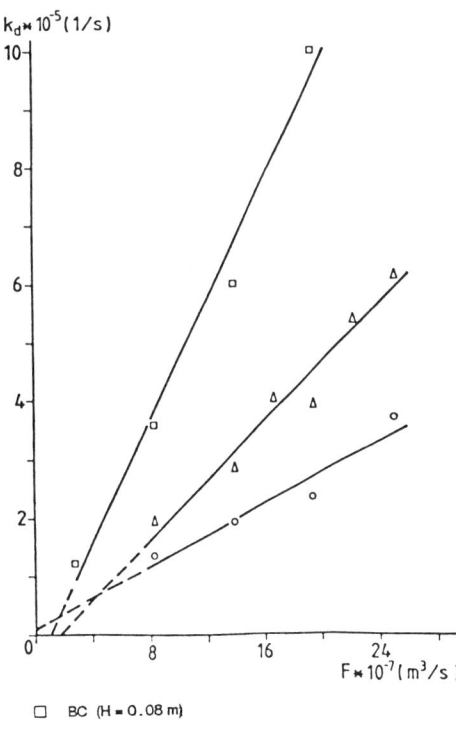

FIGURE 2. k_d as a function of F.

□ BC (H = 0.08 m)
△ ALR (H = 0.09 m)
○ ALR (H = 0.17 m)

found to deviate from that in bubble columns (FIGURE 3). Probably a change in liquid circulation pattern occurs, but more research is needed to further elucidate this phenomenon.

CONTINUOUS BACULOVIRUS PRODUCTION

The development of an efficient, preferably continuous, insect-cell culture system is necessary to obtain large quantities of baculovirus or baculovirus-derived recombinant proteins. Baculoviruses have a unique, biphasic replication cycle[11,12] as illustrated in FIGURE 4. In contrast to larvae that are infected with polyhedra, that is, protein capsules containing virus particles, insect cells can only be infected by the nonoccluded virus (NOV) form.

These NOVs are circulating in the hemolymph of infected insects. In order to initiate an infection in an insect-cell culture, hemolymph-derived NOVs of infected larvae are required. In a continuous production system, however, it is possible to take advantage of the fact that each infected cell eventually produces about 200 NOVs 12–14 hours after infection so that only hemolymph-derived NOVs are needed to initiate the infection cycle. After 24 hours, the infected cells start to produce polyhedra up to a maximum of about 100 after 48 hours. Then, the cells start to disintegrate due to cell lysis. The best time to harvest the infected cells is thus about 48 hours after infection because a simple centrifugation step can be used for convenient collection of cells (still intact) with polyhedra.

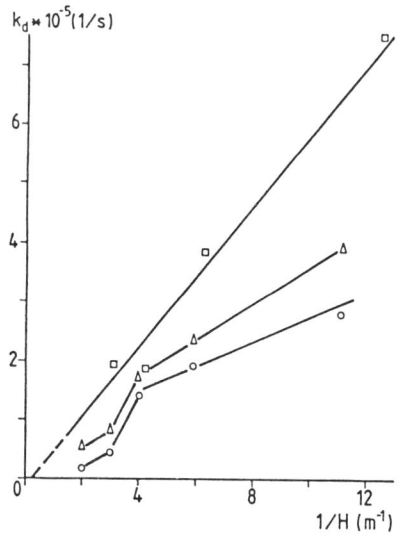

FIGURE 3. k_d as a function of $1/H$.

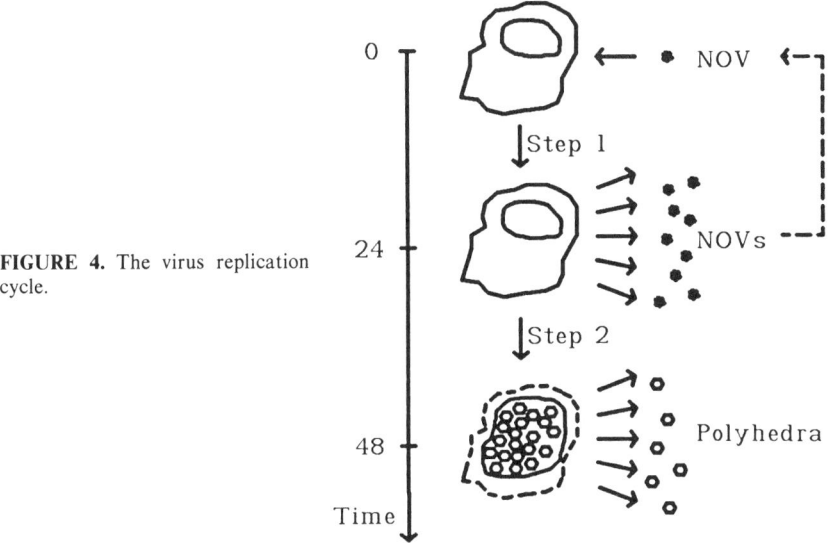

FIGURE 4. The virus replication cycle.

An important phenomenon in the infection process is the so-called "passage" effect.[13,14] Prolonged serial passage of virus in cultured insect cells results in the production of baculovirus variants with atypical polyhedrin morphogenesis and with lower capacity for causing infection. Especially in a lengthy continuous process, it is necessary to take into account this passage effect. This effect becomes significant when the number of passages surpasses 10 and is very severe above passage no. 25. By means of a computer model, we have calculated that, after about a month, the passage effect in the infection vessel (residence time of 48 h) in a continuous system becomes significant.[15] In order to test this, we have operated several continuous baculovirus production experiments.[16,17] FIGURE 5 shows some results of such a run. From day 7 to day 30, the process was in a rather stable condition. After day 30 postinoculation, the percentage of infected cells and the average yield of 25 polyhedra per infected cell started to decrease. After a run time of 56 days, polyhedra were rarely found and, when present, they showed an atypical morphology with defective occlusions upon electron microscopic inspection. This all strongly indicates the occurrence of the passage effect. More details can be found in references 16 and 17.

BACULOVIRUS PRODUCTION MODEL

For a quantitative description of the infection process, which is needed for reactor design purposes, it is important to describe the effect of crucial reactor-operating parameters such as pH, temperature, medium, and virus replication. However, the basic understanding of such a system is still poor and experimental data are scarce. We have therefore started with a model to describe the continuous viral infection process, abstracting from all possible influences imposed by reactor-operating parameters, but using the well-known concept of mass balances.[18] Regarding the viable, noninfected, insect cells and initial NOVs as substrates and regarding both the resulting NOVs and polyhedra (P) as products, the following reaction-rate equation can be written:

$$\text{insect cell} + \text{NOV} \rightarrow 200 \text{ NOV} + \text{P}. \tag{7}$$

By means of a mass balance for the viable, noninfected, insect cells over each reactor in a cascade such as depicted in FIGURE 1, it can be derived that

$$k_r = \frac{C_{j-1} + (k_g \tau_j - 1) C_j}{C_j \tau_j}. \tag{8}$$

This assumes that cell growth is a first-order process (k_g, 1/s) and is constant both in time and place and that the NOVs are, in steady state, available in excess so that equation 7 can be described as a first-order reaction (k_r, 1/s) with respect to the viable, noninfected cell concentration C_j in the j-th reactor ($j = 1, 2, 3, \ldots$) with residence time τ_j. For the special case B in FIGURE 1 with $\tau_0 = \tau_1$, and thus $k_g = 1/\tau_0$ in both vessels, equation 8 reduces to

$$k_r = C_0/(C_1 \tau_1). \tag{9}$$

The validity of these equations was experimentally tested using the results obtained from two continuous runs—one in a configuration like A (reference 17) and one like B (reference 16) in FIGURE 1. Experimental results and calculated values are listed in TABLE 1. A detailed discussion can be found in reference 18.

Most relevant to note is that the calculated reactor-rate constants, k_r, for the independent runs are close to each other such as they should be. This k_r was used to calculate the viable cell concentration in infection reactor no. 2 of configuration A. TABLE 1 shows that the measured value is rather well predicted. Taking into account cell lysis and the fact that an infected cell is not recognized as such during the first 24

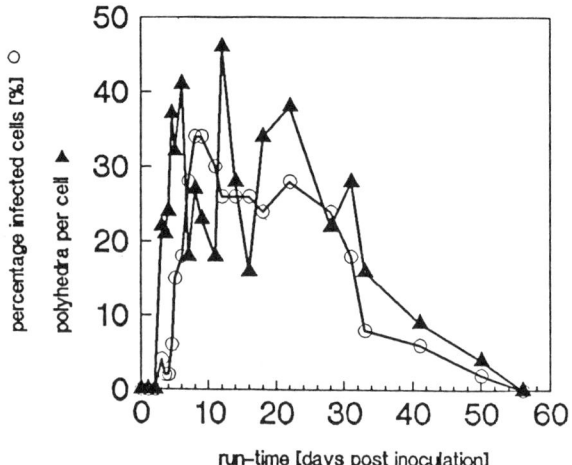

FIGURE 5. Results of a continuous baculovirus production run.

TABLE 1. Results with Two Reactor Configurations[a]

Parameters		Dimension	A (2)	B (1)
C_0	cell concentration at inlet of reactor no. 1	cells · cm^{-3}	8.1×10^5	7.8×10^5
C_1	viable cell concentration in reactor no. 1	cells · cm^{-3}	4.9×10^5	4.2×10^5
τ_1	residence time in reactor no. 1	h	30	60
I_1	infected fraction in reactor no. 1	% of cells	31	55
k_r	reaction rate constant	s^{-1}	1.07×10^{-5}	0.86×10^{-5}
C_{2p}	predicted viable cell concentration in reactor no. 2	cells · cm^{-3}	3.0×10^5	–
C_{2m}	measured viable cell concentration in reactor no. 2	cells · cm^{-3}	2.1×10^5	–
I_2	infected fraction in reactor no. 2	% of cells	59	–

[a] A and B are referred to in the text. Numbers between parentheses denote the number of continuous-operated infection vessels.

hours postinfection, it can be concluded that the viral infection process can be described by a first-order reaction rate.

CONCLUSIONS

The killing-volume hypothesis yields useful design equations relating cell growth, cell death by aeration, and oxygen supply in bubble columns. These equations have been experimentally validated on lab scale. Airlift-type growth vessels need to be further studied in this respect. The so-called passage effect resulting in virus degeneration limits the useful productivity time of a continuous system to about one month. The viral infection process in such continuous systems can be described rather well by a first-order reaction rate with respect to viable cell concentration.

REFERENCES

1. MARTIGNONI, M. E. 1984. Baculovirus: an attractive biological alternative. *In* Chemical and Biological Controls in Forestry. W. Y. Garner & J. Harvey, Eds.: 55–67. Amer. Chem. Soc. Washington, District of Columbia.
2. LUCKOW, V. A. & M. D. SUMMERS. 1988. Trends in the development of baculovirus expression vectors. Bio/Technology **6**: 47–55.
3. WEISS, S. A. & J. L. VAUGHN. 1986. Cell culture methods for large-scale propagation of baculoviruses. *In* The Biology of Baculoviruses, Volume II. R. R. Granados & B. A. Federici, Eds.: 63–87. CRC Press. Boca Raton, Florida.
4. CROUGHAN, M. S., J. F. HAMEL & D. I. C. WANG. 1987. Hydrodynamic effects on animal cells grown in microcarrier cultures. Biotechnol. Bioeng. **29**: 130–141.
5. KATINGER, H. W. D. & W. SCHEIRER. 1982. Status and development of animal cell technology using suspension culture techniques. Acta Biotechnol. **2**: 3–41.
6. TRAMPER, J., J. B. WILLIAMS, D. JOUSTRA & J. M. VLAK. 1986. Shear sensitivity of insect cells in suspension. Enzyme Microb. Technol. **8**: 33–36.
7. TRAMPER, J., D. JOUSTRA & J. M. VLAK. 1987. Bioreactor design for growth of shear-sensitive insect cells. *In* Plant and Animal Cell Cultures: Process Possibilities. C. Webb & F. Mavituna, Eds.: 125–136. Ellis Horwood. Chichester, United Kingdom.

8. TRAMPER, J. & J. M. VLAK. 1987. Design parameters for cultivation of insect cells in bubble columns. *In* Rheologie und Mechanische Beanspruchung Biologischer Systeme, p. 189–201. GVC–VDI–Gesellschaft Verfahrenstechnik und Chemieingenieurwesen. Dusseldorf.
9. TRAMPER, J., D. SMIT, J. STRAATMAN & J. M. VLAK. 1988. Bubble-column design for growth of fragile insect cells. Bioprocess Eng. **3:** 37–41.
10. TRAMPER, J. & J. M. VLAK. 1988. Bioreactor design for growth of shear-sensitive mammalian and insect cells. *In* Upstream Processes: Equipment and Techniques. Advances in Biotechnological Processes, Volume 7. A. Mizrahi, Ed.: 199–228. Alan R. Liss. New York.
11. FAULKNER, P. 1981. Baculoviruses. *In* Pathogenesis of Invertebrate Microbial Diseases. E. A. Davidson, Ed.: 3–37. Alanheld, Osmun & Co. Totowa, New Jersey.
12. KELLY, D. C. 1982. Baculovirus replication. J. Gen. Virol. **63:** 1–13.
13. MACKINNON, E. A., J. F. HENDERSON, D. B. STOLTZ & P. FAULKNER. 1974. Morphogenesis of nuclear polyhedrosis virus under conditions of prolonged passage *in vitro.* J. Ultrastruct. Res. **49:** 419–435.
14. POTTER, K. N., P. FAULKNER & E. A. MACKINNON. 1976. Strain selection during serial undiluted passage of *Trichoplusia ni* nuclear polyhedrosis virus. J. Virol. **18:** 1040–1050.
15. TRAMPER, J. & J. M. VLAK. 1986. Some engineering and economic aspects of continuous cultivation of insect cells for the production of baculoviruses. Ann. N.Y. Acad. Sci. **469:** 279–288.
16. KOMPIER, R., J. TRAMPER & J. M. VLAK. 1988. A continuous process for the production of baculovirus using insect-cell cultures. Biotechnol. Lett. **10:** 849–854.
17. VAN LIER, F. L. J., E. J. VAN DEN END, C. D. DE GOOIJER, J. M. VLAK & J. TRAMPER. 1989. Continuous production of baculovirus in a cascade of insect-cell reactors. Appl. Microbiol. Biotechnol. Accepted.
18. DE GOOIJER, C. D., F. L. J. VAN LIER, E. J. VAN DEN END, J. TRAMPER & J. M. VLAK. 1989. A model for baculovirus production with continuous insect-cell cultures. Appl. Microbiol. Biotechnol. **30:** 497–501.

Effects of Microcarriers and Serum on Local Hydrodynamics within an Airlift Column[a]

G. T. JONES, L. E. ERICKSON, AND L. A. GLASGOW

Department of Chemical Engineering
Kansas State University
Manhattan, Kansas 66506

INTRODUCTION

The airlift fermentor has been shown to be a viable bioreactor for large-scale suspension culturing of animal,[1-4] plant,[5] and insect cells,[6] although the shear sensitivity of these cells is a major concern. The airlift fermentor may potentially serve as a bioreactor for the culturing of anchorage-dependent cells (ADC) grown upon microcarriers as it is recognized as having a low shear environment. Unfortunately, understanding is limited in regard to how the interaction of turbulence with laden microcarriers promotes the loss of viability. High serum concentrations are commonly believed to provide protection against damaging shear stresses that occur within a fermentor, but it is unclear how this increased protection is provided. The purpose of the present research is to gain an understanding of the interaction of microcarriers and the turbulence present within an airlift fermentor and to establish the effect serum has upon the hydrodynamics.

Mixing within an airlift loop reactor is accomplished by sparging air into the bottom of the column and allowing degassed liquid at the top of the column to return to the bottom (see FIGURE 1). Whereas sparger aeration is the simplest method of providing oxygen and agitation to a fermentor, there are indications that mammalian and insect cells are susceptible to damage in the presence of bubbles.[7-9] At present, it is unclear whether damage to the cell is occurring due to turbulent shear, gas-liquid interfacial effects, or a combination of these phenomena. Some researchers in this area have suggested that the loss of viability is due to turbulent shear encountered in a fermentor.[10,11]

Handa *et al.*[8] have investigated gas-liquid interfacial effects on hybridoma viability within a bubble column. Not only was the rate of agitation provided by the sparged gas found to be important, but bubble size and the addition of surface active polymers also affected cell viability. Serum proteins will increase the effective viscosity of the media, affecting bubble size and rise velocity in addition to the size of dissipative eddies and other characteristics of the turbulent flow. Polymers such as carboxymethyl cellulose, hydroxyethyl starch, and pluronic polyols (copolymers of polypropylene oxide and polyethylene oxide) have also been used to protect cells from mechanical damage due to agitation and aeration.[12]

---

[a]This work was partially supported by National Science Foundation Grant No. CBT-8619943.

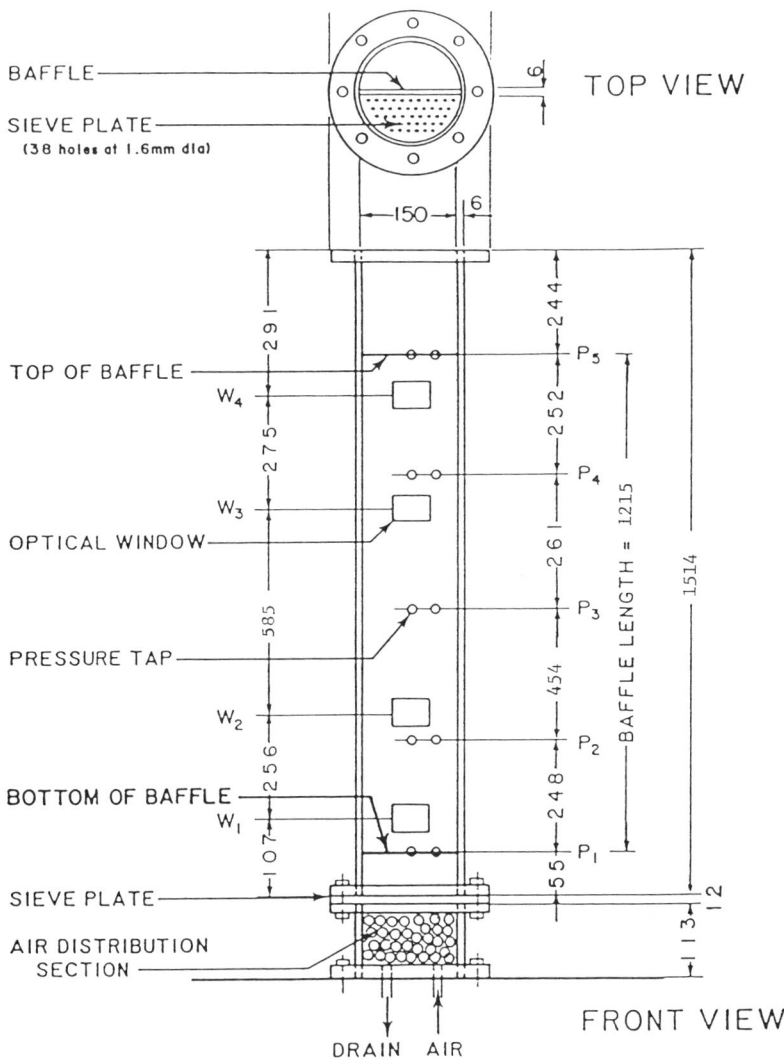

FIGURE 1. The acrylic plastic airlift column used in the experiments (all dimensions are in millimeters).

CELL CULTURE HYDRODYNAMICS

Cell culture in an airlift reactor must be considered a three-phase system. Turbulence within the reactor is a result of bubble passage with cells or microcarriers interacting with the turbulent eddies that have been generated. Integral size eddies of the turbulent bubble wake may be assumed to be proportional to the bubble diameter, which can be estimated at low gas rates by equating surface tension and buoyancy forces:[13]

$$d_B = \left(\frac{6\sigma d}{g(\rho_L - \rho_G)}\right)^{1/3},\qquad(1)$$

where d_B is the bubble diameter, σ is the surface tension, d is the orifice diameter, g is the gravitational constant, and ρ_L and ρ_G are the liquid and gas densities, respectively. The bubble rise velocity within a reactor is dependent upon interactions between the bubble size and the liquid properties and can be predicted by the method of Mendelson:[14]

$$v_B = \left(\frac{\sigma}{\rho_L r_e} + g r_e\right)^{0.5},\qquad(2)$$

where v_B is the bubble rise velocity in quiescent liquid and r_e is the equivalent radius of the bubble. Thus, both bubble rise velocity and diameter are strongly affected by the media composition and they, in turn, influence the turbulence and shear stresses encountered by cells within the reactor.

The turbulent dissipation rate must be recognized as a function of position and time. The Reynold's decomposition[15] is commonly used in the study of turbulence based upon the Navier-Stokes equations. For a statistically stationary flow, the local instantaneous velocity u_i is decomposed into a mean flow component U_i and a velocity fluctuation u_i' about the mean. Typically, U_i is a time-averaged quantity and $\overline{u_i'}$ is the standard deviation about the mean, which is more commonly known as the root-mean-square (RMS) velocity fluctuation. The local time-averaged turbulent energy dissipation rate is defined as

$$\epsilon = 2\nu \overline{s_{ij} s_{ij}},\qquad(3)$$

which is equal to

$$\frac{\nu}{2}\overline{\left(\frac{\partial u_i'}{\partial x_j} + \frac{\partial u_j'}{\partial x_i}\right)^2},\qquad(4)$$

where ν is the kinematic viscosity of the fluid, s_{ij} is the fluctuating rate of strain, and the i and j subscripts refer to the Eulerian coordinate axes.[15] Accurate measurement of ϵ through use of equation 4 is difficult. Therefore, in most work, estimates are made of local time-averaged values of ϵ using either Taylor's large-scale inviscid estimate,

$$\epsilon = A \overline{u_i'^3}/\ell,\qquad(5)$$

or by integration of experimental one-dimensional spectra. The constant, A, is to be determined and ℓ is the integral length scale that is characteristic of the system being studied.

Croughan et al.[10] have shown that there is a correlation between the loss of cell viability and the size of dissipative eddies in a stirred tank. The characteristic length scale for dissipative eddies is[15]

$$\eta = \left(\frac{\nu^3}{\epsilon}\right)^{1/4},\qquad(6)$$

where η is the size of the dissipative eddy, ν is the kinematic viscosity of the fluid, and ϵ is the local instantaneous dissipation rate of turbulent energy. This relation holds only for the case of local isotropy, that is, when the small-scale structure shows no preferred orientation.

Dissipative eddies smaller than individual entities within the flow will act upon the surface of the entity to result in a shear stress or a dynamic pressure force (or both) at the surface. The result of the entity-eddy interaction may be the deformation and rupture or increased permeability of the cell membrane. With the increased size of a microcarrier, dissipative eddies can more easily interact with a cell located on the microcarrier surface. Additionally, consideration must be given to bead-bead interactions, to rotation of the beads generated by interaction with turbulent eddies, and to bead-bead coagulation occurring due to extracellular excretions.

MATERIALS AND METHODS

The airlift column used in these studies of the local instantaneous velocity was a 151-cm-tall and 15-cm-diameter acrylic plastic cylinder with a vertical baffle along its axial length as shown in FIGURE 1. Four optically flat viewing ports along its length allow the use of backscatter laser-Doppler velocimetry (LDV) to determine local liquid velocities. Polymer microspheres from Duke Scientific (cat. no. 434) with diameters ranging from 100 to 500 μm were used as model microcarriers in the fermentor. Blood was collected from freshly slaughtered cattle and allowed to coagulate at 4 °C overnight. The serum was isolated from the clot and then centrifuged for one hour at 800g and subsequently frozen at -10 °C until use. The measuring apparatus consisted of a TSI 9100-5 model laser-Doppler velocimeter using a 350-mm front lens and a Nicolet 4094A digital recording oscilloscope. Aluminum powder was used to provide seed particles for LDV measurements.

Each recorded data set of 15,872 points was transferred to a Zenith 158 computer via an RS-232 interface and analyzed using an algorithm written in BASIC. The program was designed to detect and evaluate the characteristic frequency of suspected Doppler bursts within the data set. Four concentrations of microcarriers in water were studied: 0 g/L, 5 g/L, 10 g/L, and 15 g/L; 5% serum concentration and 5 g/L of microcarriers suspended in a 5% serum solution were also investigated.

Moreover, studies were undertaken in which flow visualization was used to examine characteristics of the flow associated with bubble movement on the upflow side of the airlift fermentor. A square three-liter acrylic plastic column was constructed to provide optically flat surfaces for macrophotography. The interior corners were filleted and air was introduced through 15 holes of 1.6 mm in diameter on one side of the dividing baffle. Glass microspheres of 10 μm in diameter were used to seed the fluid. Focused white light was used for illumination of a planar segment perpendicular to the film plane with a depth of approximately 2 to 3 mm. An Olympus OM-2 camera equipped with a 50-mm macro lens was used to record the images; the aperture was f2 to provide sufficient light and to minimize the depth of field. Kodak TX Professional film was used for all exposures and was developed using HD-110 to provide an ASA of approximately 2000. Exposures were obtained using shutter speeds of $1/30$ and $1/15$ s. Water, 5% serum, and 10% serum concentrations were investigated.

RESULTS AND DISCUSSION

FIGURE 2 displays a cumulative probability distribution for velocity for a 5% serum solution at 25 °C on the upflow side of the 1.5-m-tall airlift column for a superficial gas velocity of 1.3 cm/s. The fractional height is in relation to the entire column height. From this figure, it can be seen that the greatest variation occurs in the tail regions, which have a significant effect on estimates of the higher moments. It must be noted, though, that the results obtained are for approximately 150 to 400 observed values of velocity for each point represented. Boerner et al.[16] have indicated that 1200 "good" Doppler bursts are desirable for reproducible values of the mean velocity at low gas holdups within a bubble column. This presents problems with our experimental setup

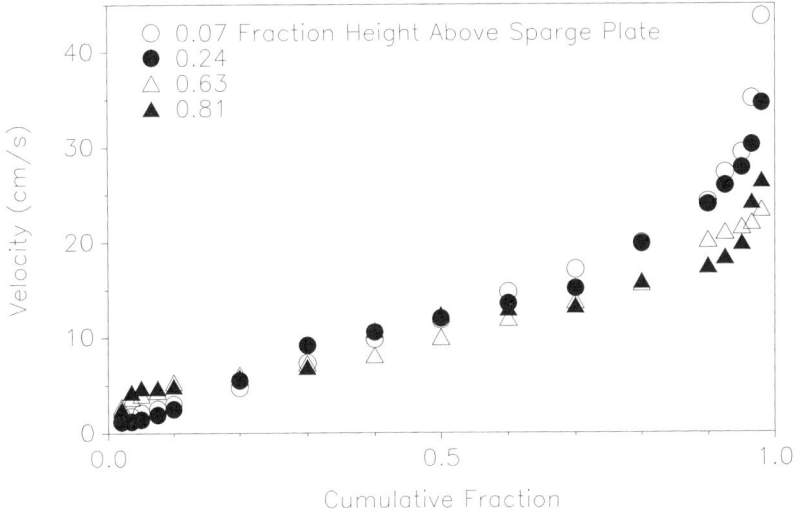

FIGURE 2. Cumulative frequency distribution for the velocity occurring for a 5% serum concentration with a superficial gas rate of 1.3 cm/s on the upflow side.

in that an inordinate amount of time would be required to make measurements at each window and condition. Additionally, the measurement of the RMS velocity is of more interest in evaluating the local time-averaged dissipation rate based upon Taylor's inviscid estimate. As the RMS velocity is the second statistical moment, this implies that significantly more than 1200 Doppler bursts are required for reproducible RMS velocities. However, a lower number of observations will allow order-of-magnitude estimates.

FIGURE 3 displays the mean liquid velocity on the upflow side of the column for a superficial velocity of 1.3 cm/s as a function of the fractional height above the sparge plate. Three systems are represented: air-water, air–5% serum solution, and air–5 g/L microcarriers–5% serum. It can be seen for the water and the 5 g/L microcarriers–5%

serum systems that significant variations exist in the values observed at each window. Nevertheless, the mean values of velocity for the upflow side obtained by averaging the observations for each window are approximately equal in both cases. This was unexpected; we had assumed that the two systems containing serum would have similar results. The mean value obtained for the 5% serum system is a significant departure from the other mean values. Consideration of the means and medians observed for the 5% serum system indicates that the means are approximately 10% greater than the observed medians, implying that the upper tail region of the cumulative probability plot has a substantial impact on the estimation of the mean velocity and hence also on the RMS velocity.

FIGURE 4 displays the RMS liquid velocities based on the ensemble standard

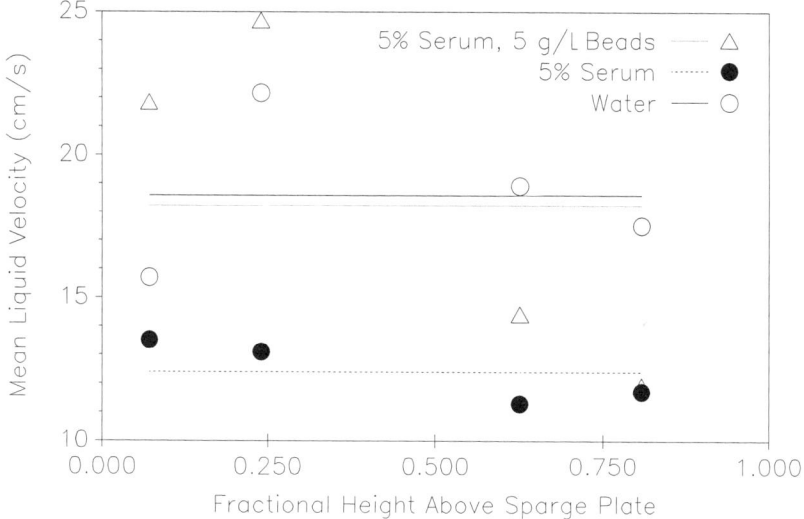

FIGURE 3. Mean velocity of the continuous phase on the upflow side of the column for a specific gas rate of 1.3 cm/s for water, 5% serum, and 5 g/L microcarriers–5% serum systems.

deviation in relation to the fractional height above the sparge plate. The lines indicate the averaged values of the RMS velocities for each specific system. Again, significant variations in the detected RMS velocities must be noted for the water and the 5 g/L microcarriers–5% serum systems. Consideration of $\overline{u_i'^3}$ alone shows a variation from ≈ 300 cm^3/s^3 to ≈ 2200 cm^3/s^3 for the 5% serum and the 5 g/L microcarriers–5% serum systems, respectively. However, the values of A and ℓ are necessary for use of Taylor's estimate of the dissipation rate. Typically, A is on the order of one to two and ℓ is roughly comparable to the Sauter mean bubble diameter.

FIGURE 5 shows the mean liquid velocities for a superficial gas velocity of 2.6 cm/s on the upflow side of the column based upon an ensemble average in relation to the fractional height above the sparge plate. The lines denote the averaged values of the

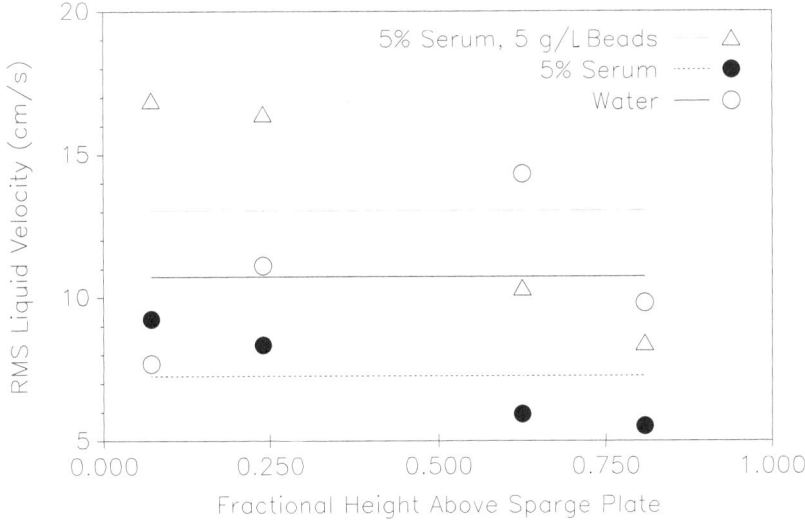

FIGURE 4. RMS velocity of the continuous phase on the upflow side of the column for a specific gas rate of 1.3 cm/s for water, 5% serum, and 5 g/L microcarriers–5% serum systems.

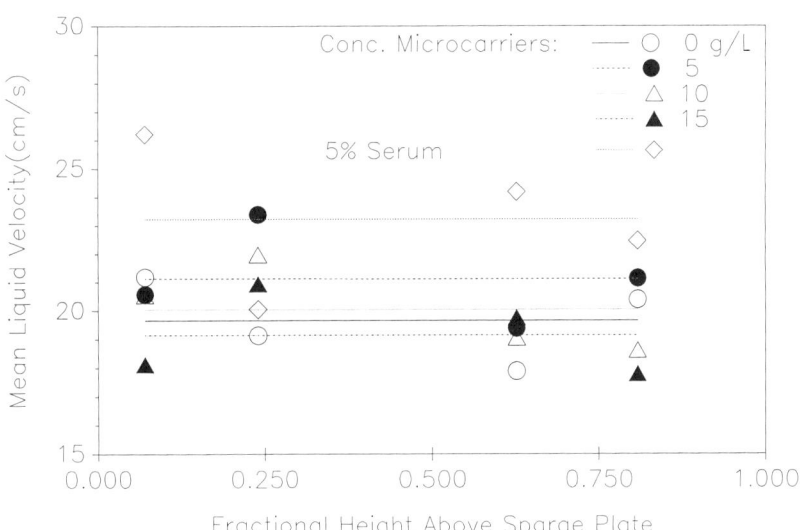

FIGURE 5. Mean velocity of the continuous phase on the upflow side of the column for a specific gas rate of 2.6 cm/s for a 5% serum solution and for an aqueous solution with different concentrations of microcarriers present.

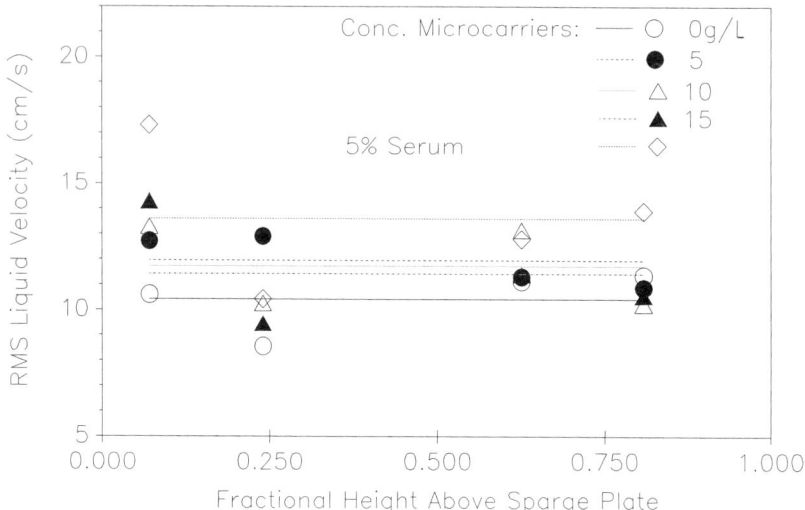

FIGURE 6. RMS velocity of the continuous phase on the upflow side of the column for a specific gas rate of 2.6 cm/s for a 5% serum solution and for an aqueous solution with different concentrations of microcarriers present.

mean velocities for the different microcarrier concentrations in water and for the 5% serum concentration. As the particle concentration increases from 5 g/L to 15 g/L, the average velocity decreases. The mean velocity for the case of 5% serum is greater than for any other reported condition. Analysis of variance performed on these data indicates that no significant differences exist between the mean velocities when all are considered together.

FIGURE 6 displays the RMS velocities based upon the ensemble standard deviation in relation to the fractional height above the sparge plate. The lines denote the averaged values of the RMS velocities for each system investigated. From this figure, it can be seen that the addition of microcarriers and the presence of 5% serum may increase the RMS velocity; however, an analysis of variance indicates again that the difference is not significant. The use of Taylor's inviscid estimate of ϵ would show that the addition of microcarriers or serum has increased the overall dissipation of energy on the upflow side, assuming of course that A and ℓ are the same for all concentrations of microcarriers and serum.

FIGURES 7a–c are typical photographs obtained in a three-liter airlift column built specifically for photographic work. The exposures displayed are obtained for water with a shutter speed of $1/15$ s at an aperture setting of $f2$ and a magnification factor of 4.7. Many small vortices are evident generally at the edges of the wake structure. In the photographs, a wake or bubble pathway had a width of 2 to 3 cm for bubbles with diameters of 4 to 8 mm. Vortical structures had a typical scale of about 0.5 cm and the higher-velocity regions of the central wake had a decidedly sinuous motion. The maximum velocities observed in the wake area were on the order of 15 to 28 cm/s. These higher velocities were generally found within a few centimeters of the moving bubble. The smallest-scale eddies identifiable in the photographs had scales of approximately 0.1–0.2 mm; if this is representative of the Kolmogorov microscale, then the local dissipation rate should be on the order of 10 to 40 cm^2/s^3. However, estimates made from Taylor's inviscid approximation suggest dissipation rates one to two orders of magnitude greater than these values. Observed velocity gradients tend to support the former because, assuming isotropicity, it is possible to obtain estimates of somewhat less than 100 cm^2/s^3. Photographic work done with serum concentrations commonly seen in cell culture resulted in very poor images, particularly with the 10% serum

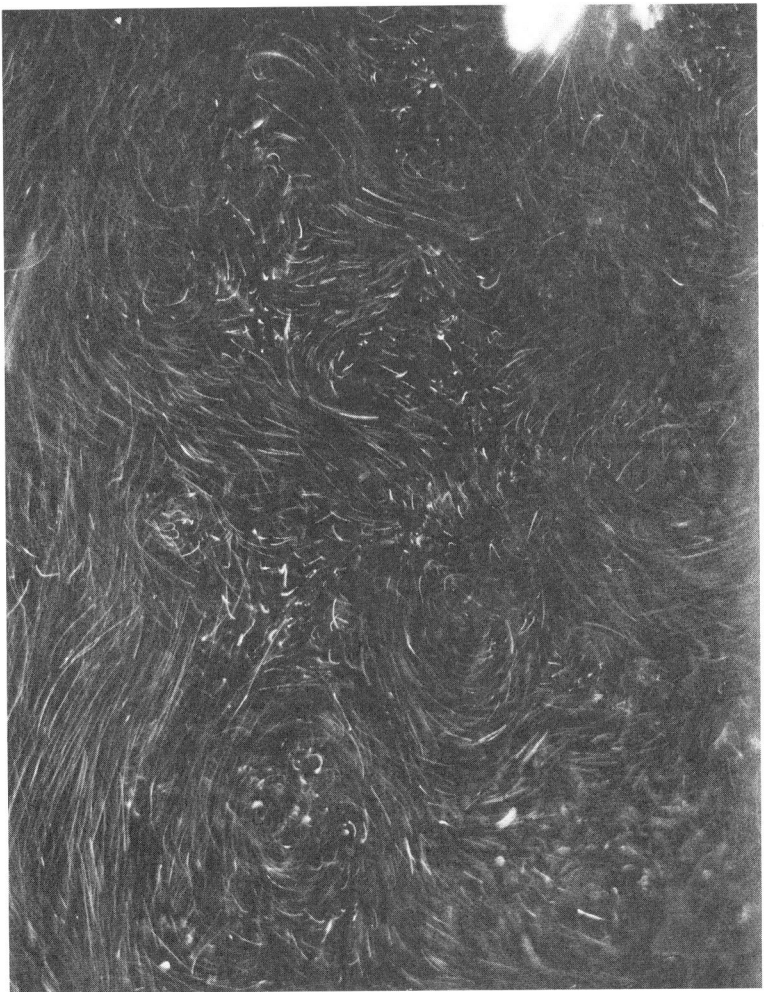

FIGURE 7a. Photograph of the wake structure trailing a bubble in an air-water system within an airlift column. The image was obtained at a shutter speed of $1/15$ s and with the ASA pushed to approximately 2000.

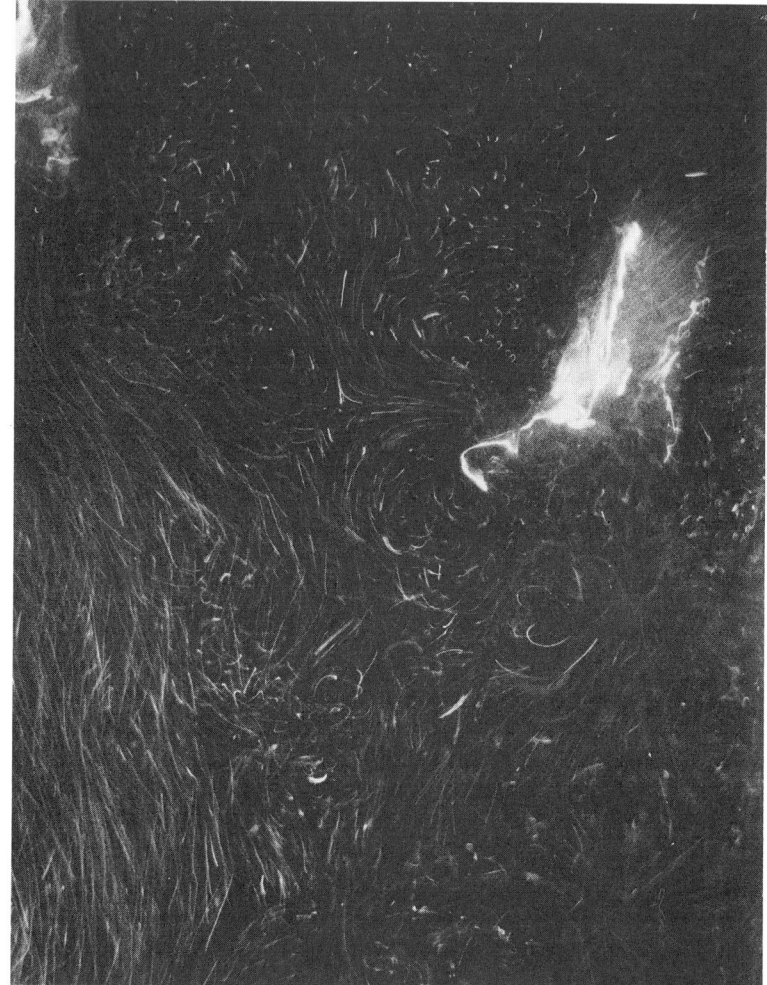

FIGURE 7b. Same as caption to FIGURE 7a.

concentration in which little could be observed. The principal problem with this approach is light dispersion resulting from light scattering centers in the serum that caused a poor planar segment definition.

CONCLUSIONS

The presence of microcarriers within the airlift reactor may promote an increase in the turbulent dissipation of energy based upon Taylor's inviscid approximation. Data obtained for effects of serum concentration upon hydrodynamics within the reactor are inconclusive. A larger number of observations is necessary to more accurately evaluate the mean and RMS liquid velocities. Measurements of the Sauter mean bubble diameter within a typical culture media are necessary to (1) evaluate the effect of microcarriers on bubble diameter and (2) establish an estimate for the integral length scale that is characteristic of a system. Hydrodynamic modeling of the interaction of microcarriers and turbulence is needed to better understand the magnitude and frequency of shear stresses occurring on the surface of a microcarrier.

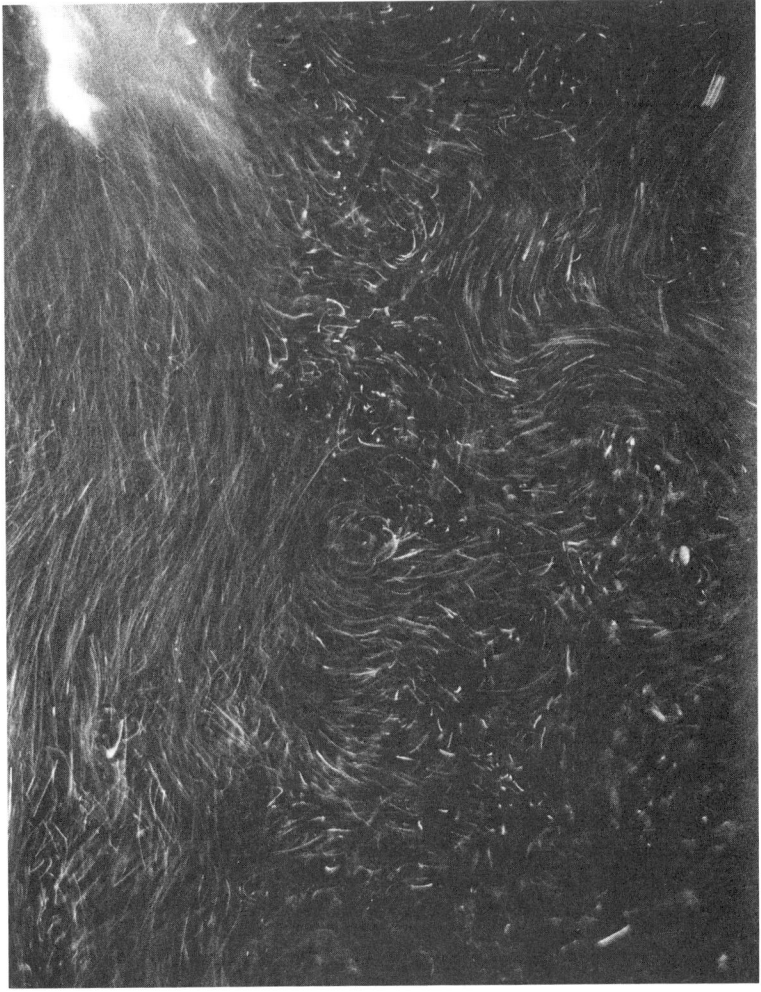

FIGURE 7c. Same as caption to FIGURE 7a.

REFERENCES

1. KATINGER, H. W-D. & W. SCHEIRER. 1979. Dev. Biol. Stand. **42:** 111.
2. BIRCH, J. R., P. W. THOMPSON, K. LAMBERT & R. BORASTON. 1985. *In* Large-Scale Mammalian Cell Culture. J. Feder & W. R. Tolbert, Eds. Academic Press. New York.
3. WOOD, L. A. & P. W. THOMPSON. 1987. Appl. Biochem. Biotechnol. **15:** 131.
4. CORTESSIS, G. P. & C. M. PROBY. 1987. Biopharm. Manuf. **1:** 30.
5. HOHL, U., B. UPMEIER & W. BARZ. 1988. Appl. Microbiol. Biotechnol. **28:** 319.
6. INLOW, D., D. HARANO & B. MAIORELLA. 1986. National ACS Meeting, New Orleans, August 30 to September 5, 1986. Microbial and Biochemical Technology Division— Paper no. 70.
7. KILBURN, D. G. & F. C. WEBB. 1968. Biotechnol. Bioeng. **10:** 801.
8. HANDA, A., A. N. EMERY & R. E. SPIER. 1987. Dev. Biol. Stand. **66:** 241.
9. TRAMPER, J., D. SMIT, J. STRAATMAN & J. M. VLAK. 1988. Bioprocess Eng. **3:** 37.
10. CROUGHAN, M. S., J-F. HAMEL & D. I. C. WANG. 1987. Biotechnol. Bioeng. **29:** 130.
11. CHERRY, R. S. & E. T. PAPOUTSAKIS. 1986. Bioprocess Eng. **1:** 29.
12. MIZRAHI, A. 1984. Dev. Biol. Stand. **55:** 93.
13. MOO-YOUNG, M. & H. W. BLANCH. 1981. *In* Advances in Biochemical Engineering, Vol. 19. A. Fiechter, Ed.: 1. Springer-Verlag. New York/Berlin.
14. MENDELSON, H. D. 1967. AIChE J. **13**(2): 250.
15. TENNEKES, H. & J. L. LUMLEY. 1972. *In* A First Course in Turbulence. MIT Press. Cambridge, Massachusetts.
16. BOERNER, T., W. W. MARTIN & H. J. LEUTHEUSSER. 1984. Chem. Eng. Commun. **28:** 29.

Continuous Cell Cultures in Fluidized-Bed Bioreactors

Cultivation of Hybridomas and Recombinant Chinese Hamster Ovary Cells Immobilized in Collagen Microspheres

N. G. RAY, A. S. TUNG, E. G. HAYMAN,
J. N. VOURNAKIS, AND P. W. RUNSTADLER, JR.

Verax Corporation
Lebanon, New Hampshire 03766

INTRODUCTION

Production of therapeutics and diagnostics on a commercial scale using mammalian cell culture technology has received much attention during the last decade.[1-3] Several systems have been proposed for culturing mammalian cells *in vitro*.[4-8] However, none of these systems have successfully addressed all the critical issues with respect to large-scale, low-cost production of proteins; namely, scaleup, product quality, productivity, long-term steady-state operation in low-cost/serum-free medium, and use of a common system for both anchorage-dependent and anchorage-independent cell lines.

A fluidized-bed bioreactor system has been developed with an objective to meet all the criteria discussed above.[9,10] A schematic of the system is presented in FIGURE 1. The basic element of this system is spherical, spongelike, microspheres made of natural bovine collagen. Collagen fiber constitutes the fundamental framework of the mammalian cell extracellular matrix *in vivo* and collagen is a natural substrate for cell adhesion *in vitro*. The microspheres, approximately 500 microns in diameter, are weighted to achieve the high specific gravity (typically 1.6 and higher) necessary to remain suspended in high-velocity (order of 75 cm/min), upward-flowing, fluidizing culture liquid. The leaflike, highly porous morphology of the microspheres provides a large number of potential binding sites for attachment and proliferation of cells. The interconnecting network of the matrix pore channels and the fluid dynamics of the culture liquid around individual flexible microspheres in the fluidized bed are the potential factors to enhance intramatrix transport of nutrients and removal of secreted products and metabolites.

In this article, we present experimental results from hybridoma and genetically engineered Chinese Hamster Ovary (CHO) cell cultures in fluidized-bed bioreactor systems. Mouse hybridoma cells were cultured in a 150-mL fluidized-bed bioreactor (Verax System 10) in order to study cell growth and immunoglobulin (Ig) production in serum-free medium. Effects of dilution rate on culture performance, namely, population growth in the collagen microspheres, Ig productivity, glucose consumption, and lactate and ammonia production, were studied at four different dilution rates under steady-state conditions. Bioreactor productivity was found to have increased

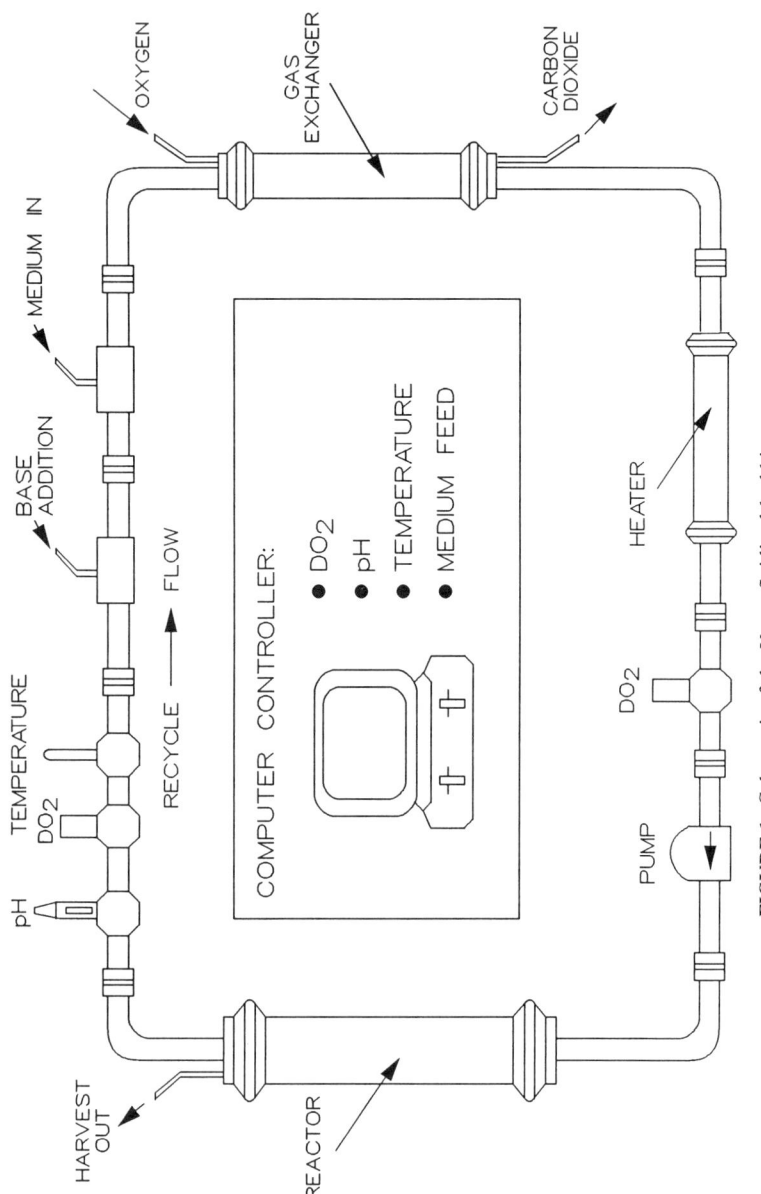

FIGURE 1. Schematic of the Verax fluidized-bed bioreactor systems.

greatly with an increase in dilution rate. In addition, a 1.6-L fluidized-bed bioreactor (Verax System 200) was used to culture a recombinant CHO cell line producing a cardiovascular therapeutic. The effects of dissolved oxygen on culture metabolic activities and productivity were investigated in this experiment.

MATERIALS AND METHODS

Bioreactor System

As shown in the schematic diagram in FIGURE 1, the bioreactor system consists of an external recycle loop through which culture liquid is continuously circulated to fluidize the collagen microspheres in the reactor vessel. The liquid, separated from the microspheres near the top of the vessel, enters the recycle loop, which contains a membrane gas exchanger, the measuring electrodes (temperature, dissolved oxygen, and pH), a heating element, a recycle pump, a flow meter, and a pressure transducer (in System 200). Independent of the recycle loop, fresh medium is pumped in and the harvest, containing product and spent liquor, is removed continuously. The culture parameters—dissolved oxygen, pH, and temperature—are controlled at the desired levels by a computer. Medium and harvest are kept at 4 °C in their respective containers.

The System 10 bioreactor has a maximum capacity of 150-mL fluidized-bed volume. The total system volume, including the recycle pump, the gas exchanger, and the recycle loop, is 750 mL. The total system volume for the System 200 is 4.5 L. Its maximum fluidized-bed volume is 1.6 L. The dissolved oxygen (D.O.) electrode in the System 10 is located at the inlet of the gas exchanger. The System 200 has two D.O. probes, one at the inlet and the other at the outlet of the gas exchanger. Oxygen transfer rate is continuously calculated and displayed by the System 200 computer. Dissolved oxygen is controlled at the inlet of the gas exchanger, where the concentration level in the recycle liquid is the lowest. In the System 10 run, D.O. was controlled at 75 ± 10 mmHg equilibrium oxygen partial pressure. The System 200 was operated at different D.O. levels to study their effects on the culture. Culture temperature and pH were controlled at 37 ± 0.2 °C and 7.2 ± 0.05, respectively. A liquid sample was removed daily to check pH against a standardized pH electrode. The recycle rates for the System 10 and the System 200 were 200 ± 10 mL/min (superficial velocity of 70 ± 4 cm/min in the fluidized-bed column) and 850 ± 50 mL/min (superficial velocity of 75 ± 5 cm/min), respectively.

Cells and Media

Hybridoma Cells

A mouse hybridoma cell line was used in this study. The cells secrete immunoglobulin G (IgG) antibody. A stock culture is maintained in high glucose containing Dulbecco's Modified Eagle's (DME) medium (Irvine Scientific), supplemented with 10% (v/v) fetal bovine serum (FBS). Inoculum was prepared from the stock culture in tissue-culture flasks in serum-free medium. The bioreactor (System 10) was inoculated while the microspheres were fluidized. The inoculum size was 1.4×10^8 viable cells. The system was operated with 100-mL fluidized-bed volume at 25% solids.

Serum-free medium was used during the entire run. The base medium consisted of a 3:1 mixture of high glucose DME medium and Ham's F-12 (Irvine Scientific). Glucose and glutamine concentrations in the feed medium were 3.2 and 0.46 g/L, respectively.

Chinese Hamster Ovary Cells

The genetically engineered CHO cells used in this study produce a cardiovascular therapeutic that is secreted in the culture liquid. The stock culture was maintained in DME (high glucose) medium, supplemented with 10% FBS. The fluidized-bed volume was 1.2 L with a solids fraction of 25% in the bed. On day no. 27, approximately two-thirds of the populated microspheres were transferred to inoculate a Verax System 2000 (24-L fluidized bed) bioreactor for a production run. The experiment reported here was carried out in the System 200 with the remaining microspheres.

The start-up medium contained 2% FBS. On day no. 9, feed medium was switched to a serum-free formulation (base medium of a 1:1 mixture of high glucose DME medium and Ham's F-12). The final glucose and glutamine levels in the medium were 3.0 g/L and 0.58 g/L, respectively. The System was inoculated with 1.1×10^9 viable cells.

Analyses

Cell concentration was determined by a Coulter Counter (Model 2M, Coulter Electronics). Matrix cell counts were performed after the populated microspheres were treated with collagenase and trypsin at 37 °C to free the cells. Cell viability was determined by the dye (erythrosine B) exclusion technique. Glucose was measured by an enzyme electrode analyzer (YSI Model 27). Lactate and ammonia levels were measured using enzymatic kits from Boehringer Mannheim and Sigma. The products were assayed by enzyme-linked immunosorbent assay (ELISA).[11]

RESULTS AND DISCUSSION

Hybridoma Cell Line

FIGURE 2 illustrates the time course profile of glucose concentration and the dilution rate (D) schedule during the entire run. D is calculated on the basis of the fluidized-bed volume, which at steady state contained approximately 90% of the total cell population in the bioreactor system. At the highest dilution rate investigated, that is, 0.8 h^{-1}, the medium perfusion rate for the 100-mL fluidized bed was 80 mL/h. As shown in the figure, D increased as the glucose concentration dropped after an initial lag period. D was then held constant at 0.32 h^{-1} until a steady state was achieved. A steady state was assumed when the free cell density, glucose, and monoclonal antibody levels in the harvest liquid remained statistically unchanged over a period of three days after at least four residence times had passed. In a similar manner, we have investigated steady-state performance at dilution rates of 0.48, 0.64, and 0.80 h^{-1}. Interestingly, the lowest steady-state culture glucose concentration was observed at the highest dilution rate. Direct cell counts in the matrix on day nos. 22 and 33 did not show any significant difference in cell density. The average was 2.7×10^8 cells/mL collagen.

FIGURE 3 presents the lactate concentration profile. There were slight increases in lactate levels with increasing dilution rates. The steady-state ammonia levels at all the dilution rates investigated were not significantly different from each other and re-

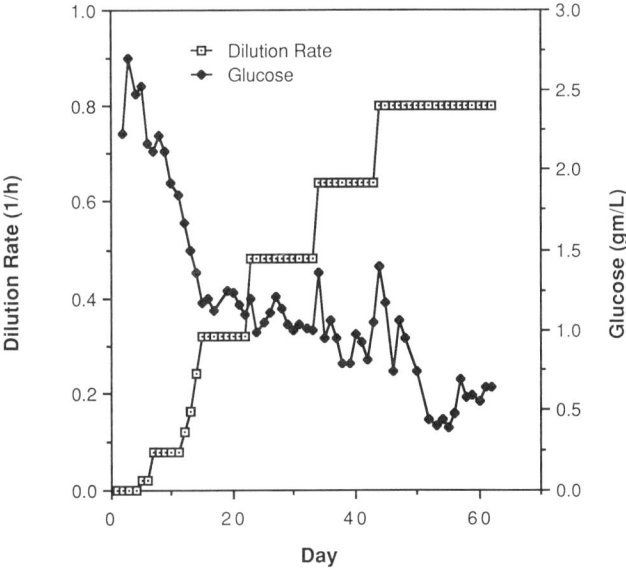

FIGURE 2. Continuous hybridoma culture in a fluidized-bed bioreactor: dilution rate and glucose concentration profiless Medium glucose concentration: 3.2 g/L.

mained between 25–30 mg/L. FIGURE 4 depicts time course profiles of the glucose consumption rate (GCR) and the lactate production rate (LPR). Both rates increased with an increase in dilution rate and the LPR-to-GCR ratio remained virtually unchanged, between 0.7–0.75. In contrast, this ratio was found to decrease with an increase in dilution rate in suspension hybridoma cultures.[12] FIGURE 5 shows the volumetric productivity in mg IgG/liter fluidized bed/h. The steady-state results for volumetric productivity and IgG concentration as a function of dilution rate are presented in FIGURE 6. The increase in reactor productivity at high dilution rates could result from an increase in cell density in the matrix or from an increase in specific productivity (or both). As stated before, the matrix cell density did not increase

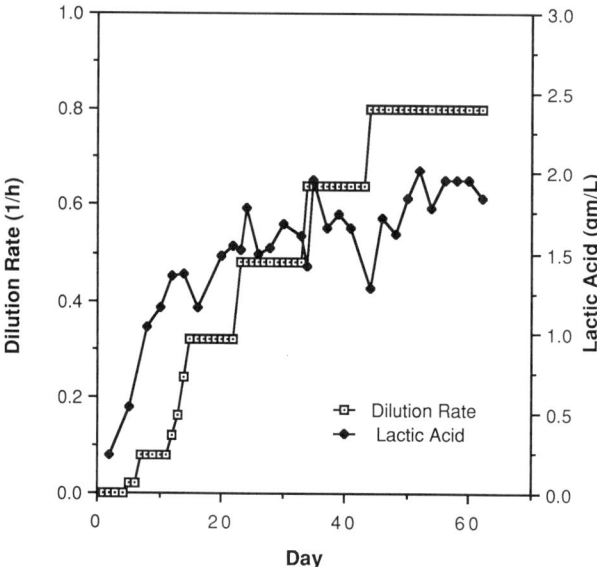

FIGURE 3. Continuous hybridoma culture in a fluidized-bed bioreactor: dilution rate and lactic acid concentration profiles.

between day nos. 22–33 of the run, although the dilution rate was raised on day no. 22 from 0.32 to 0.48 h^{-1}. It is quite possible that the available matrix spaces were occupied by the high cell number observed on day no. 22 (ca. 3×10^8 cells/mL collagen), thereby limiting any further increase in cell density in the microspheres. At that point, division of an immobilized cell would result in the release of a progeny to the culture liquid. It is therefore likely that the observed increase in glucose consumption rate in FIGURE 4 was due to an increase in the population specific growth rate, and the observed increase in volumetric productivity, shown in FIGURE 5, was the result of increased specific productivity of the immobilized population.

Available information from continuous cell culture experiments indicates that specific productivity can be profoundly affected by dilution rate. Avgerinos[13] has observed an increase in specific productivity with an increase in dilution rate, whereas Miller and co-workers[14] have shown an inverse relationship between specific antibody productivity and dilution rate. Ray et al.[12] have observed the existence of an optimum dilution rate with respect to the specific productivity in continuous suspension hybridoma cultures. The population specific growth rate of the immobilized cells in the microspheres loaded at maximum capacity can be estimated with reasonable accuracy from the cell output rates. The average steady-state cell counts per milliliter of harvest liquid at all the dilution rates investigated varied between $(1.45–1.70) \times 10^6$ cells/mL and the cell output rates were 4.6×10^7 ($D = 0.32$ h^{-1}), 7.5×10^7 ($D = 0.48$ h^{-1}), 1.1×10^8 ($D = 0.64$ h^{-1}), and 1.3×10^8 ($D = 0.80$ h^{-1}) per hour from the bioreactor. Therefore, if we assume that the microspheres were fully populated (constant density) at the above dilution rates, as indicated by some data, then the relative increase in the population specific growth rate in the microspheres would be approximately 2.9-fold at $D = 0.8$ h^{-1} compared to $D = 0.32$ h^{-1}. It should be noted that cell growth in suspension (ca. 10% of the total population was suspended) at the dilution rates investigated was not considered to be very significantly different because of the relatively constant suspended cell density and liquid environment, including pH, dissolved oxygen, and nutrient and by-product levels (e.g., glucose, lactate, and ammonia).

Chinese Hamster Ovary Cell Line

As discussed earlier, the System 200 was inoculated in 2% FBS supplemented medium. On day no. 9, serum was removed from the medium and, on day no. 27, two-thirds of the microspheres were transferred to inoculate the System 2000. The System 200 run was continued to study the effects of dissolved oxygen concentration on culture performance.

Dissolved oxygen (D.O.) concentration in the culture liquid was varied and controlled at different levels before the gas exchanger in the recycle loop, causing the reactor bottom inlet (gas exchanger outlet) oxygen concentration (monitored by a second D.O. probe) to change correspondingly. Obviously, concentration gradients did exist inside the microspheres. However, it is believed that the intramatrix transport is significantly enhanced by the highly porous morphology of the matrix and the fluid dynamic characteristics around the flexible microspheres in the fluidized bed. FIGURE

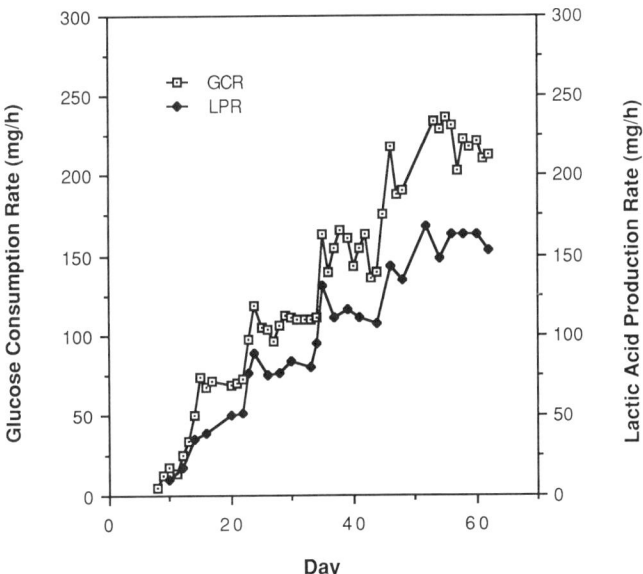

FIGURE 4. Continuous hybridoma culture in a fluidized-bed bioreactor: time course profiles of glucose consumption and lactic acid production rates.

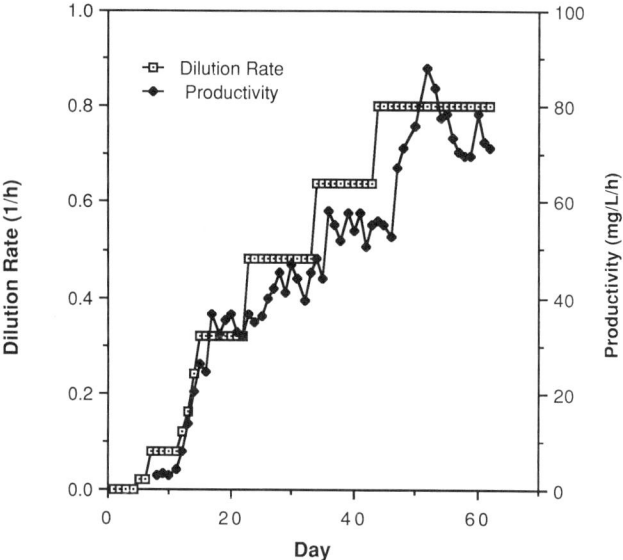

FIGURE 5. Continuous hybridoma culture in a fluidized-bed bioreactor: dilution rate and volumetric productivity during the course of the run.

7 shows D.O. levels at the inlet and at the outlet of the gas exchanger, as well as matrix-associated cell counts. Cells continued to grow in the microspheres after serum was removed from the medium. The maximum density achieved was approximately 3×10^8/mL collagen.

The concentration difference of D.O. across the gas exchanger at the controlled gas exchanger inlet concentration was primarily due to oxygen uptake by the cell population in the fluidized bed because approximately 98% of the total cells in the entire system were contained in the bed. The liquid cell concentration ranged from $(3-5) \times 10^5$ cells/mL. FIGURE 8 presents the medium feed rate (MFR) and the glucose and lactate concentration profiles during the entire run. MFR was reduced from 20 to 9 L/day after two-thirds of the populated microspheres were removed on day no. 27. The oxygen and glucose uptake rates and the lactate production rate are shown in FIGURE 9. The oxygen uptake rate (OUR) dropped from 25 to 8 mmoles/h after partial removal of the microspheres. Another drop in OUR on day no. 30 was due to an accidental withdrawal of some microspheres while a daily liquid sample was being taken.

A substantial decline in OUR from 11.7 to 7.1 mmoles/h was observed when D.O. at the gas exchanger inlet was reduced from 200 to 25 mmHg. This decline was very rapid—within two hours after the D.O. set point was stepped down—and corresponded very closely to the D.O. profiles, indicating a reduction in respiration rate due to low oxygen concentration in the culture liquid. Reduction in OUR is not considered to be the result from the death of cells because (a) no significant cell death was observed with this and other CHO cell lines within two hours of imposing an oxygen-starved condition in the culture and (b) after the initial decline, there was no significant decrease in OUR over a period of four days. Also, the original respiration rate (at high D.O.) was restored rather quickly as oxygen concentration was raised from 25 → 50 → 100 mmHg. At 100 mmHg D.O. at the inlet to the gas exchanger, OUR even exceeded the values observed at 200 mmHg. Because the matrix cell density remained essentially constant, the higher OUR may be attributed to reduced glucose levels in the culture liquid. It has been reported that the respiration rate is enhanced by low glucose levels in the culture.[12,15] It should be noted that no further increase in OUR was observed by raising D.O. from 100 to 175 mmHg.

Although oxygen uptake was significantly affected between day no. 36 and day no. 51, glucose consumption rate remained virtually constant. Also, at low oxygen levels, it appeared that cells incorporated more glucose carbon to lactate. As shown in FIGURE 9, the lactate production rate increased as D.O. concentration was reduced, and vice versa.

Glucose is metabolized via the glycolytic pathway to produce pyruvate, which can then be reduced to lactate.[16] Pyruvate can also be oxidized through the TCA cycle to produce carbon dioxide and water. Additionally, glutamine is catabolized to produce energy.[17,18] Although glutamine metabolism can yield lactate, only a small fraction of the total lactate may result from glutamine in the presence of glucose.[19] Lactate is considered to be the major end product of glucose metabolism in cultured mammalian cell systems.[20] The lactate yield on glucose (g lactate/g glucose) increased from 0.39 to 0.59 at 25 mmHg D.O. and then declined to 0.37 as the D.O. was raised to 100 mmHg at the gas exchanger inlet.

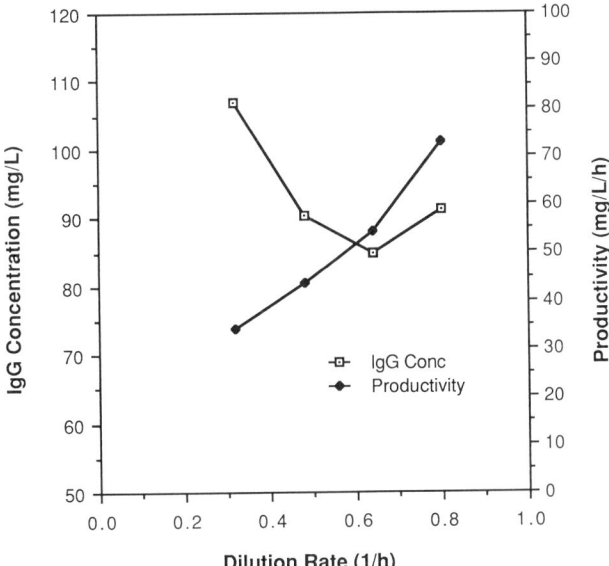

FIGURE 6. Continuous hybridoma culture in a fluidized-bed bioreactor: effect of dilution rate on steady-state immunoglobulin concentration and volumetric productivity.

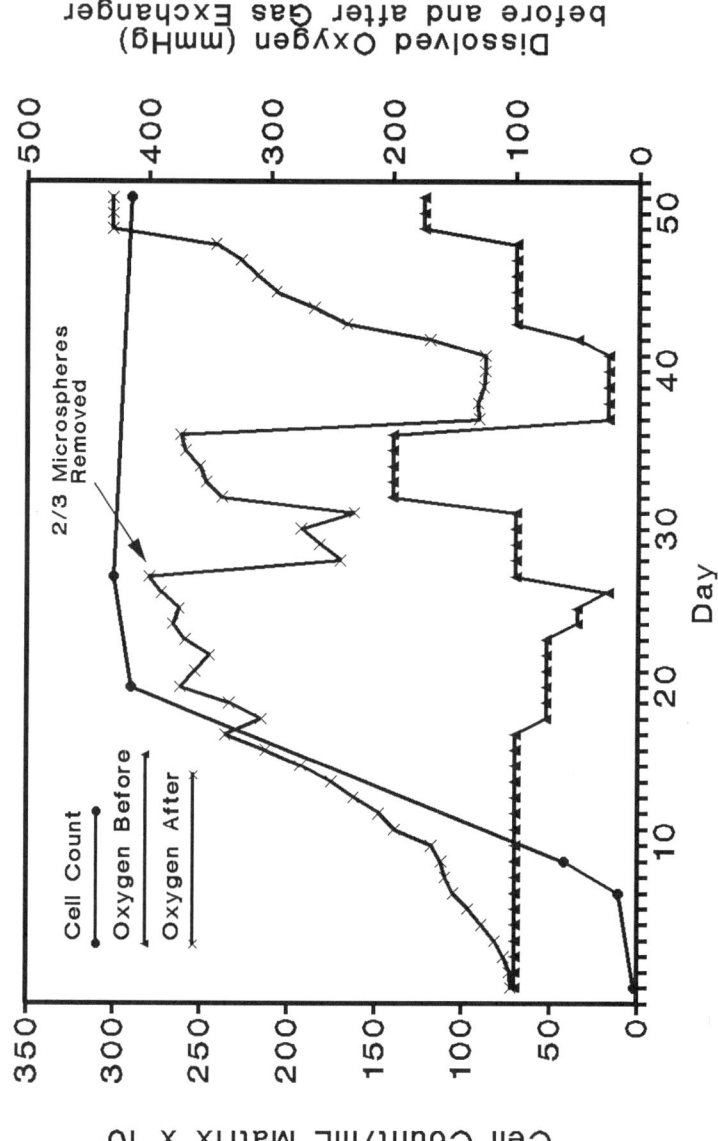

FIGURE 7. Continuous culture of Chinese hamster ovary (CHO) cells in a fluidized-bed bioreactor: matrix cell density and dissolved oxygen at the inlet and outlet of the gas exchanger.

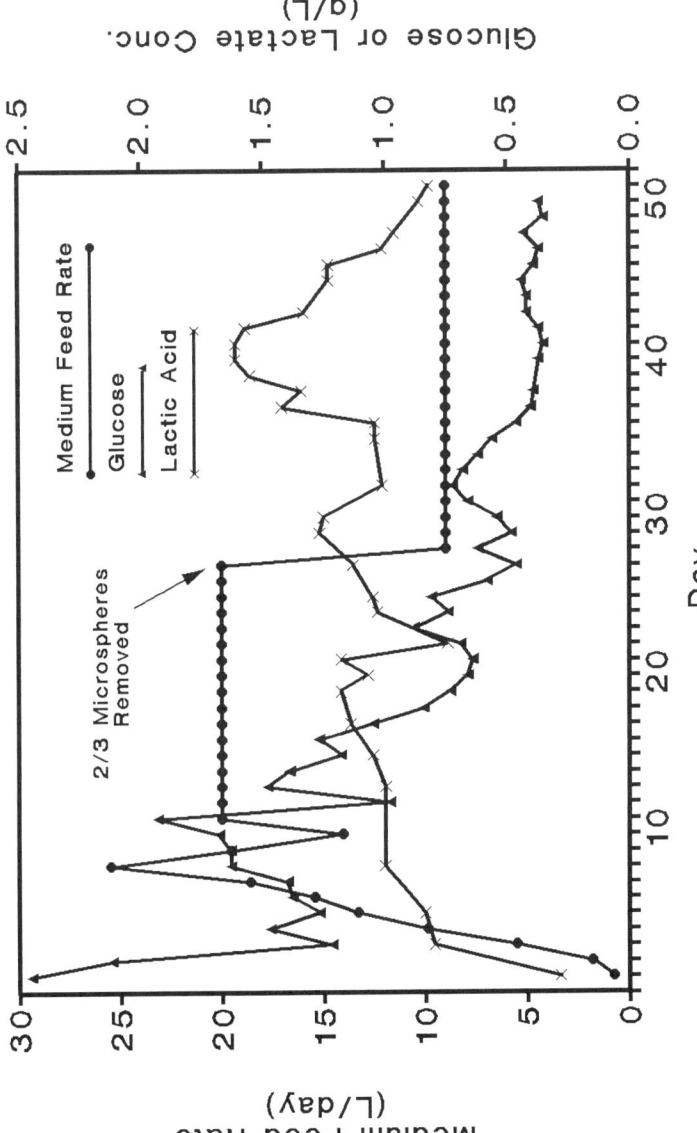

FIGURE 8. Continuous culture of CHO cells in a fluidized-bed bioreactor: medium feed rate, and glucose and lactate profiles during the course of the run. Medium glucose concentration: 3.0 g/L.

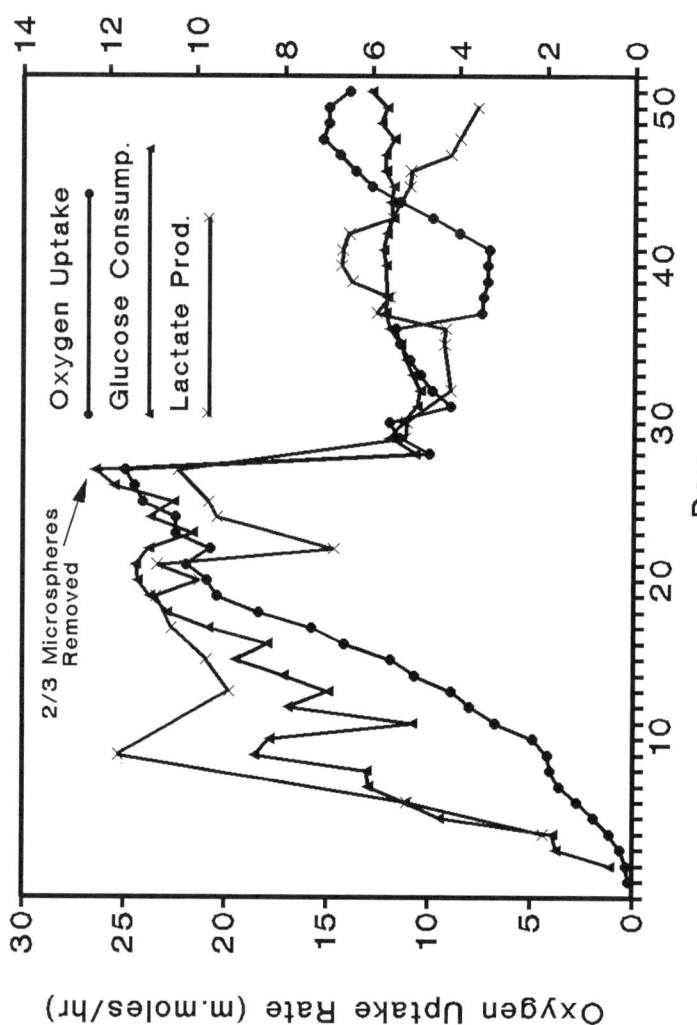

FIGURE 9. Continuous culture of CHO cells in a fluidized-bed bioreactor: time course profiles of the oxygen and glucose uptake rate and the lactate production rate.

FIGURE 10 depicts the product concentration profile and the oxygen to glucose uptake ratio (γ). A good correlation exists between them. On the one hand, γ remained low in serum-supplemented medium; however, in serum-free medium, γ started to increase and the value went up from approximately 0.5 to about 2.0 on day no. 21 and remained reasonably constant until day no. 36. The seemingly altered metabolic activities may be attributed to the following:

(a) removal of serum from the medium,
(b) decrease in glucose concentration (FIGURE 8), and
(c) overall increase in oxygen concentration in the fluidized bed (FIGURE 7).

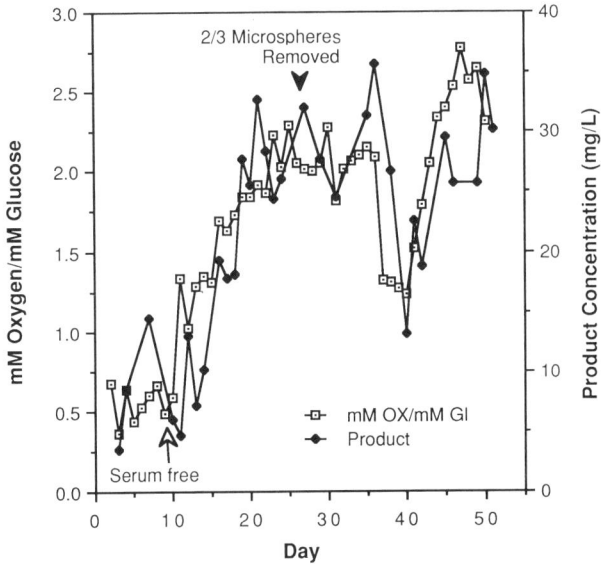

FIGURE 10. Continuous culture of CHO cells in a fluidized-bed bioreactor: product concentration and oxygen to glucose uptake ratio (mole oxygen/mole glucose).

A trend similar to the γ profile was also observed with respect to product concentration. As the dissolved oxygen concentration dropped on day no. 36 at the gas exchanger inlet, there was a sharp decline in both γ and product concentration and the values remained depressed at the low D.O. levels. When D.O. was raised, both the values went up simultaneously.

An analysis of the data suggests that the energy (ATP) production in the bioreactor was decreased significantly at the low D.O. levels (25 mmHg at the gas exchanger inlet—i.e., reactor outlet). Six moles ATP is generated for each mole oxygen utilized, whereas only one mole ATP is generated per mole lactate derived from glucose.[21] Because lactate is the end product primarily of glucose metabolism, reasonably accurate energy production rates can be determined from oxygen uptake and lactate production rates. ATP production rates were calculated to be 73.9 and 49.6 mmoles/h at the gas exchanger inlet D.O. levels of 200 and 25 mmHg, respectively. Because a decline in productivity was observed at low D.O. concentration (25 mmHg), it is quite possible that the energy transformation steps became rate-limiting for product formation. High ATP production rate and productivity were restored when D.O. levels were raised.

CONCLUSIONS

A murine hybridoma cell line and a recombinant CHO cell line were cultured in continuous fluidized-bed bioreactor systems developed by Verax. The first study investigated the effects of dilution rate on immobilized hybridoma culture performance with respect to cell growth, glucose metabolism, production of metabolites, and product (IgG) formation in serum-free medium. Cells grew to high density in the collagen microspheres. Steady-state data were obtained at four different dilution rates. IgG concentration was maximum at the lowest dilution rate investigated; however, at the higher dilution rates, the concentration differences were not significant. Glucose consumption rate and IgG productivity increased significantly with an increase in dilution rate. These increases were attributed to the higher specific growth rate and specific productivity of the immobilized cells.

The second study investigated the effects of dissolved oxygen on oxygen and glucose utilization, production of metabolic by-products, and secreted cloned gene product. A CHO cell line was cultured in a 1.6-L fluidized-bed bioreactor. Significant cell growth and improvement in productivity were observed in serum-free medium. Dissolved oxygen concentration in the culture liquid was varied to study its effect on culture performance. The concentration of dissolved oxygen in the range of 100–200 mmHg at the gas exchanger inlet, where the oxygen levels in the liquid were minimum, had insignificant effects on the culture. However, profound effects were observed when the concentration was reduced to 25 mmHg. Oxygen uptake was decreased and lactate production was enhanced considerably, but glucose uptake remained unchanged. Productivity was depressed significantly. It is concluded that production of ATP at the low D.O. levels was rate-limiting with respect to productivity. In addition, it has been found that the oxygen to glucose uptake ratio and the product formation followed a similar pattern throughout the entire run. We conclude that the oxygen/glucose ratio is an important parameter in optimizing CHO cell cultures.

ACKNOWLEDGMENTS

We wish to thank T. Moodie, P. Corriveau, M. Day, S. Campero, J. Sample, S. Warner, R. Fager, and H. Landon for technical assistance. We also are thankful to J. Coull for preparing the manuscript.

REFERENCES

1. FEDER, J. & W. R. TOLBERT. 1983. Sci. Am. **246**(1): 36.
2. BIRCH, J. R., K. LAMBERT, P. W. THOMPSON, A. C. KENNEY & L. A. WOOD. 1987. Large-Scale Cell Culture Technology. B. K. Lydersen, Ed. Hanser Pub. Munich/Vienna/New York.

3. DEAN, R. C., JR., S. B. KARKARE, N. G. RAY, P. W. RUNSTADLER, JR. & K. VENKATA-SUBRAMANIAN. 1987. Ann. N.Y. Acad. Sci. **506:** 129.
4. ALTSHULER, G., D. M. DZIEWULSKI, J. A. SOWEK & G. BELFORT. 1986. Biotechnol. Bioeng. **28:** 646.
5. REUVENY, S. 1985. Adv. Cell Culture **4:** 213.
6. POSILLICO, E. G. 1986. Biotechnology **4:** 194.
7. ARATHOON, W. R. & J. R. BIRCH. 1986. Science **232:** 1390.
8. LYDERSEN, B. K. 1987. Large-Scale Cell Culture Technology. B. K. Lydersen, Ed. Hanser Pub. Munich/Vienna/New York.
9. TUNG, A. S., JV. G. SAMPLE, T. A. BROWN, N. G. RAY, E. G. HAYMAN & P. W. RUNSTADLER, JR. 1988. Biopharm. Manuf. **1**(2): 50.
10. RUNSTADLER, P. W., JR., A. S. TUNG, E. G. HAYMAN, N. G. RAY, JV. G. SAMPLE & D. E. DELUCIA. Large-Scale Mammalian Cell Culture Technology. A. S. Lubiniecki, Ed. In press.
11. ENGVALL, E. 1980. Methods Enzymol. **70:** 419.
12. RAY, N. G., S. B. KARKARE & P. W. RUNSTADLER, JR. 1989. Biotechnol. Bioeng. **33:** 724.
13. AVGERINOS, G. C. 1988. Paper presented at the Engineering Foundation Conference on Cell Culture Engineering, Palm Coast, Florida, January 31 to February 5, 1988.
14. MILLER, W. M., H. W. BLANCH & C. R. WILKE. 1988. Biotechnol. Bioeng. **32:** 947.
15. GLACKEN, M. W., R. J. FLEISCHAKER & A. J. SINSKEY. 1986. Biotechnol. Bioeng. **28:** 1376.
16. PAUL, J. 1965. Cell and Tissue Culture (third edition), p. 42. Williams & Wilkins. Baltimore, Maryland.
17. REITZER, L. J., B. M. WICE & D. KENNELL. 1979. J. Biol. Chem. **254:** 2669.
18. DONNELLY, M. & I. E. SCHEFFLER. 1976. J. Cell. Physiol. **89:** 39.
19. ZIELKE, H. R., C. M. SUMBILLA, D. A. SEVDALIAN, R. L. HAWKINS & P. T. OZAND. 1980. J. Cell. Physiol. **104:** 433.
20. LANKS, K. W. & P. W. LI. 1988. J. Cell. Physiol. **135:** 151.
21. WHITE, A., P. HANDLER & E. L. SMITH. 1973. Principles of Biochemistry (fifth edition). McGraw–Hill. New York.

PART VII. BIOSENSORS AND ANALYTICAL METHODS IN BIOPROCESSING

Nuclear Magnetic Resonance Methods for Observing the Intracellular Environment of Mammalian Cells[a]

ERIK J. FERNANDEZ, ANTHONY MANCUSO,
MARILEE K. MURPHY, HARVEY W. BLANCH,
AND DOUGLAS S. CLARK

Department of Chemical Engineering
University of California
Berkeley, California 94720

INTRODUCTION

The behavior of highly concentrated mammalian cells in bioreactors is difficult to analyze because of the complex transport and kinetic parameters governing growth. Traditionally, the concentrations of several important nutrients and products outside the cell have been measured by conventional analytical methods. This strategy results in useful information about net production and uptake rates as well as product yields, but provides no direct information about intracellular kinetics.

Over the past several years, NMR spectroscopy has been used by many investigators to measure intracellular concentrations as well as transport and reaction rates.[1,2] ^{31}P NMR has been widely used to monitor levels of high-energy phosphorylated compounds. In addition, NMR spectroscopy is sensitive to the location of a labeled atom in a molecule and experiments involving ^{13}C and ^{15}N have provided valuable information about metabolic pathways and reaction rates. Moreover, isotope labeling studies can be performed *in vivo* and tedious and invasive isolations of labeled products are not necessary. Because NMR measurements are noninvasive, cells can be monitored during growth and in response to changes in the extracellular environment. As a result of these unique attributes, NMR spectroscopy has developed into a useful, quantitative tool for examining the metabolism of both prokaryotic and eukaryotic cells.

We are interested in using NMR spectroscopy to study the primary metabolism of hybridoma cells and to relate primary metabolic parameters to antibody formation. FIGURE 1 shows some of the salient features of hybridoma metabolism relevant to this work. Like other transformed cells, hybridomas metabolize large amounts of glucose and convert much of it to lactic acid rather than utilize it in the TCA cycle.[3] As an alternate energy source, hybridomas may metabolize glutamine, converting it to

[a]Financial support for this work was provided by the Dow Chemical Company, Merck & Company, Incorporated, the Xoma Corporation, the NSF Presidential Young Investigator Award of D. S. Clark, and the University of California Biotechnology Research and Education Program.

glutamate and α-ketoglutarate; excess amino groups from catabolized glutamine are excreted as either ammonia or alanine.[4] ^{13}C NMR spectroscopy is a valuable tool to quantitatively examine the metabolic fate of glucose and glutamine. For example, a ^{13}C atom in the C-1 position of D-glucose could be followed as it is metabolized through the glycolytic pathway and the TCA cycle, leading to the labeled intermediates shown in

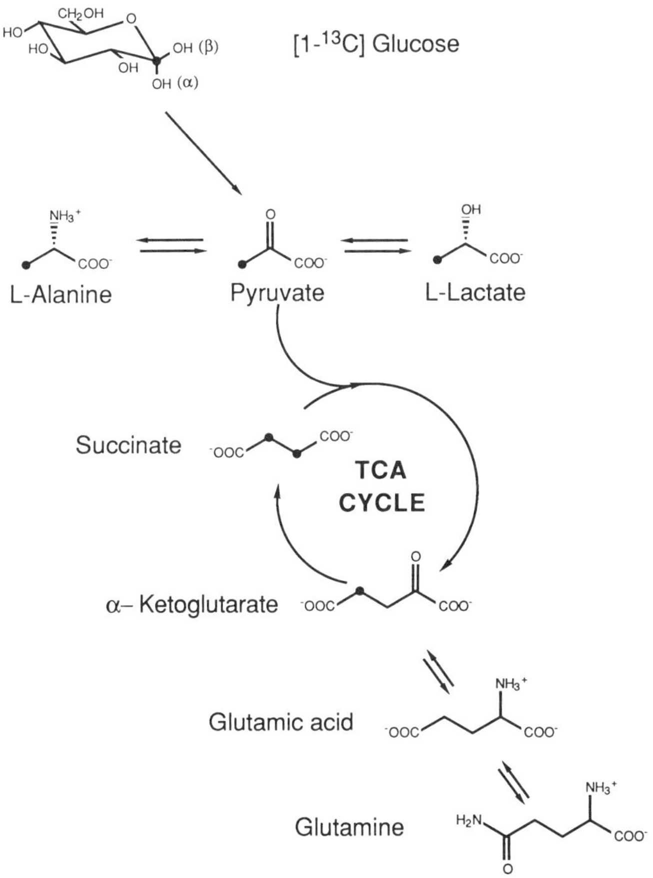

FIGURE 1. Important metabolic conversions in hybridomas. Heavy dots indicate the expected positions of ^{13}C labels derived from the metabolism of [1-^{13}C]D-glucose. For example, lactate formed from [1-^{13}C]D-glucose via glycolysis will be labeled in the C-3 position.

FIGURE 1. In addition, the energy-related processes in hybridomas may be monitored under various environmental conditions via ^{31}P NMR measurements of high-energy phosphate compounds.

The initial objective of the present research was to develop NMR spectroscopy as a tool for observing the metabolism of mammalian cells in bioreactors. The metabolism of hybridoma cells (University of California, San Francisco, cell line AB2-143.2) immobilized in a simple membrane reactor was examined. Difficulties resulting from mass-transfer limitations and the low sensitivity of the measurements were apparent, so an improved reactor as well as home-built NMR sample probes were constructed to overcome these problems.

DETECTION OF INTRACELLULAR METABOLITES

Initial experiments were designed to ascertain which of the key metabolites could be monitored in relatively short time periods. The NMR measurements were performed on hybridoma cells entrapped in dialysis tubing.[5] Cells were cultured in Iscove's modified Dulbecco's medium containing 6 mM glutamine supplemented with 1× MEM nonessential amino acids and 10% fetal bovine serum. A short segment of dialysis tubing (approximately 3 cm long, 50-kDa cutoff) was loaded with approximately 10^8 cells resuspended in fresh medium after centrifugation (1:1 volume ratio, cell pellet to fresh medium). The tubing was sealed and loaded aseptically into a 10-mm NMR sample tube with a modified top as shown in FIGURE 2A. The modified top included ports for continuous medium recirculation. A thermocouple was also included to monitor the temperature of the incoming medium. Dissolved oxygen and pH were controlled manually by equilibrating the medium with a gas mixture of CO_2 and air as shown in FIGURE 2C. A second modified NMR tube (FIGURE 2B) was used to determine the amounts of labeled compounds outside the cells.

The NMR measurements were performed using a Bruker Instruments AM-500 spectrometer (125.8 MHz for ^{13}C, 202.5 MHz for ^{31}P) to monitor the cells in the dialysis tubing. A second home-built spectrometer (45.3 MHz for ^{13}C) was used to monitor the medium circulating in the second NMR tube. Spectral acquisition conditions are described in the figure captions.

FIGURE 3 shows representative spectra collected from two ^{13}C-labeling experiments. FIGURE 3A is the ^{13}C NMR spectrum of the circulating medium during an experiment in which the cells were maintained in 12 mM glucose. FIGURE 3B is a spectrum of the NMR tube containing cells during the same experiment. The spectrum obtained from cells four hours after a step change in glucose concentration (2 mM to 10 mM) is shown in FIGURE 3C. Resonances were observed for glucose C-1, lactate C-3, and alanine C-3. The buffer (HEPES) used in these experiments also appeared in the spectrum (1.1% of all carbon in nature is ^{13}C) and served as a convenient reference for quantitation. In addition, there are two resonances that remain unidentified. Each of the spectra in FIGURE 3 represents 42 minutes of signal averaging. Thus, it was possible to monitor changes in concentration that occurred over a time scale of a few hours.

During the step-change experiment, the cells were loaded into the reactor with 2 mM [1-^{13}C]D-glucose in the circulating medium. About one hour after the cells were loaded into the spectrometer, sufficient labeled glucose was added to bring the concentration in the recirculating medium to 10 mM. As shown in FIGURE 4, the concentration of lactate in the reactor increased rapidly after the step change; then, a more gradual increase was observed. Alanine, which appeared immediately under conditions of constant high glucose concentration (12 mM), did not appear until about

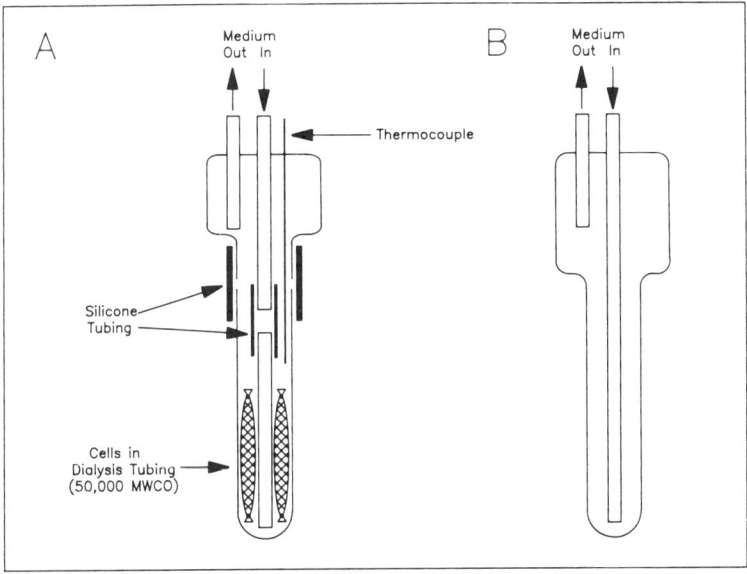

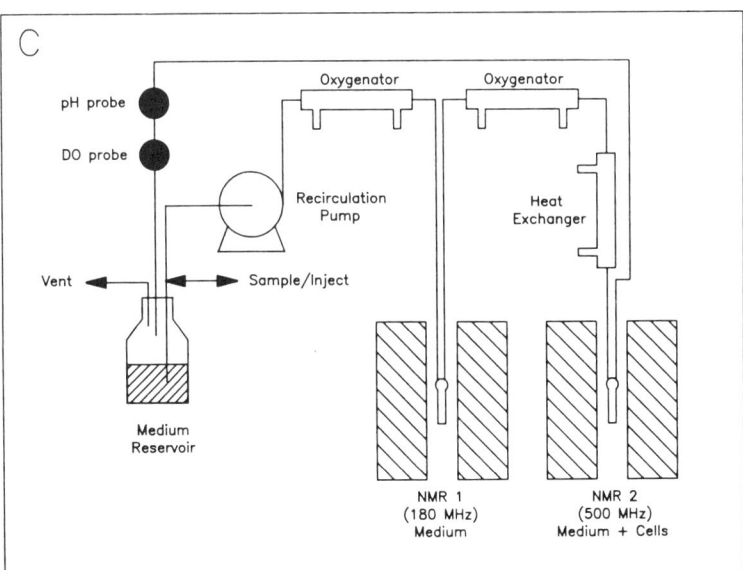

FIGURE 2. Glass sample tubes and reactor loop used in dialysis-tube experiments: (A) modified 10-mm NMR sample tube containing cells and medium, (B) modified 10-mm NMR tube containing only medium, and (C) schematic of reactor loop. Reproduced from reference 5.

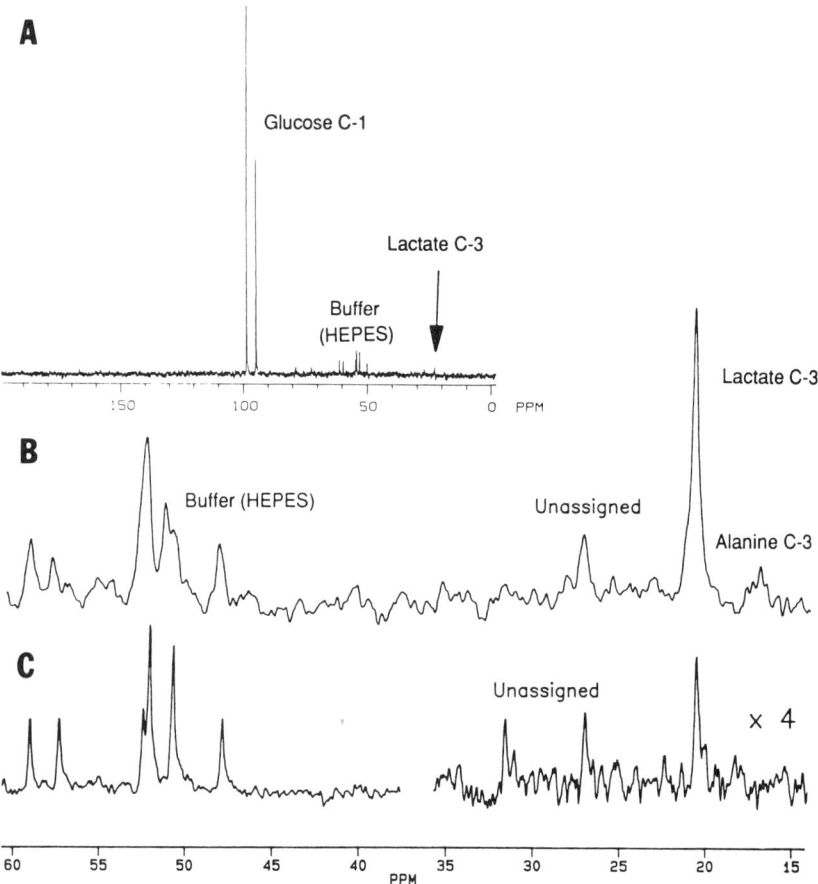

FIGURE 3. ^{13}C NMR spectra of hydridoma cells in a membrane reactor system following introduction of [1-^{13}C]D-glucose: (A) spectrum of perfusing medium two hours after introduction of 12 mM [1-^{13}C]D-glucose, (B) cells + medium two hours after being exposed to 12 mM [1-^{13}C]D-glucose, and (C) cells + medium four hours after a step change from 2 to 10 mM [1-^{13}C]D-glucose. The vertical scale of the region between 14 and 35 ppm has been expanded by a factor of four. In part (A), 1000 scans were collected using 60° pulses, a 2.5-s repetition time, 16K data points, and 3-Hz line broadening. For parts (B) and (C), 1000 scans were collected using 55° pulses, a 2.5-s repetition time, 16K data points, and 15-Hz line broadening. Reproduced from reference 5.

five hours after the step change. Resonances for two other unidentified ^{13}C-labeled compounds appeared immediately and remained relatively constant over the course of the experiment. These results illustrate how NMR spectroscopy may be used to simultaneously monitor several intracellular concentrations on-line throughout a dynamic experiment.

NMR measurements are quantitative because the spectral peak areas are directly proportional to the amount of each compound present. In most experiments, NMR spectra are obtained by repetitively collecting a signal from the sample after it has been excited by a radio-frequency pulse, with each compound requiring a different time to relax to equilibrium. If all compounds are given sufficient time for complete relaxation, the spectra acquired are termed "fully relaxed". Higher signal-to-noise can be achieved by using radio-frequency pulses separated by shorter times; however, the resulting relationship between peak area and concentration will be different for each compound.

To maximize the signal-to-noise ratio, our spectra were not acquired under fully relaxed conditions, resulting in different proportionality constants for each resonance. To correct for these differences, control spectra were acquired under the same conditions on samples prepared with known concentrations of lactate, alanine, and HEPES dissolved in medium. It was assumed that the relaxation behavior for each compound was the same in the prepared solution as in the cells. This assumption was necessary because no intracellular relaxation data for lactate and alanine are presently available.

For lactate and alanine, the concentrations measured in the reactor were converted to intracellular concentrations by subtracting the contribution of these compounds measured in the medium (either by on-line measurements in the second NMR spectrometer or by off-line NMR measurements on a medium sample). TABLE 1 shows intracellular and extracellular concentrations of lactate and alanine for experiments

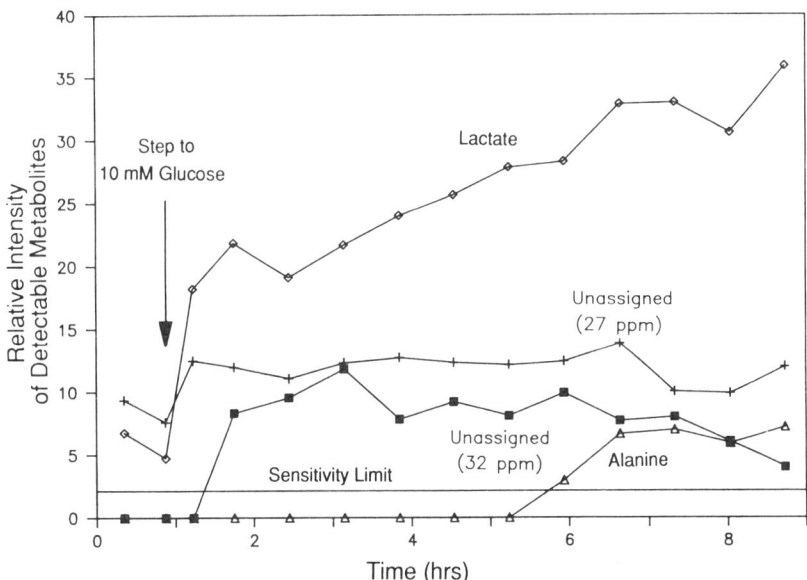

FIGURE 4. Time dependence of ^{13}C peak intensities during the [1-^{13}C]D-glucose step-change experiment. Each NMR spectrum represents 42 minutes of signal averaging and the time of each plotted point corresponds to the midpoint of the acquisition interval. Reproduced from reference 5.

where the concentration of glucose in the medium was held constant at 12 mM and for the step-change experiment outlined above. Whereas the intracellular concentrations of alanine in the two experiments were not dramatically different, the intracellular concentration of lactate was much lower in the step-change experiment. With regard to the surprisingly high concentration of lactate measured in the 12-mM experiment, this value probably reflects considerable metabolic stress and is not expected to represent typical lactate levels in hybridoma cells.

These data illustrate that NMR can provide quantitative information that could not be readily obtained in any other way. Unfortunately, two substantial limitations of NMR spectroscopy were apparent in this application: low sensitivity and restrictive sample-probe geometry. To overcome the first limitation, a high cell concentration is desirable; however, the small size of commercial probes places severe constraints on the dimensions and complexity of the reactor. In the case just described, the simple dialysis-tube reactor was unable to maintain cells in a viable state for more than about 10 hours. Hence, we constructed an alternative reactor/probe system capable of culturing hybridoma cells at high densities for extended periods.

TABLE 1. Concentrations of ^{13}C-Labeled Metabolites[a]

Experiment	Intracellular (mM)		Extracellular (mM)	
	[Lactate]	[Alanine]	[Lactate]	[Alanine]
12 mM ($t = 1$ h)	38 ± 3	3.2 ± 3.2	0.6 ± 0.2	0.3 ± 0.2
2–10 mM ($t = 7.3$ h)	5.4 ± 1.6	1.8 ± 1.6	0.51 ± 0.01	<0.1

[a]In the 12-mM experiment, the medium contribution was determined from simultaneous NMR observations of the medium; in the step-change experiments (2–10 mM), the extracellular contribution was estimated by analyzing the spent medium at the end of the experiment ($t = 9$ hours).

MEASUREMENTS OF HYBRIDOMA CELLS IN A HOLLOW-FIBER BIOREACTOR

To achieve high cell concentrations and maintain good mass-transfer characteristics, a hollow-fiber bioreactor (HFBR) was selected. The HFBR employed here was 10 cm long with an outer diameter (o.d.) of 2.74 cm. The unit contained 1350 380-μm o.d. mixed cellulose ester hollow fibers of 80% porosity with 0.2-μm pores and 100-μm-thick walls (Microgon, Incorporated, Laguna Hills, California).

To accommodate the HFBR and additional apparatus (temperature sensors and heat exchanger) inside the magnet of the NMR spectrometer, we designed and built custom NMR probes. The most important component of the probe is a radio-frequency coil designed to act as an antenna for irradiating the sample and for detecting the signal emitted by the sample. Three different coil geometries were evaluated for this application: the multiturn solenoid, the loop-gap resonator, and the slotted tube resonator. These three coils are shown in FIGURE 5. The solenoid is generally considered to provide the highest sensitivity, although this advantage is less pronounced in the case of conductive samples.[6]

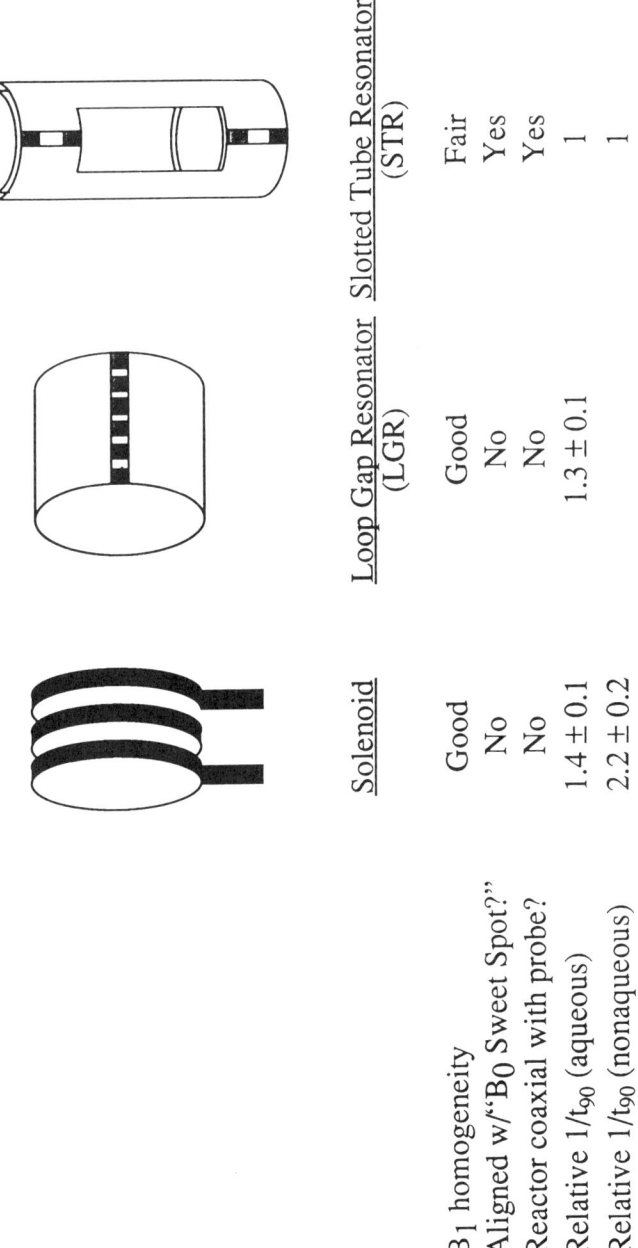

	Solenoid	Loop Gap Resonator (LGR)	Slotted Tube Resonator (STR)
B_1 homogeneity	Good	Good	Fair
Aligned w/"B_0 Sweet Spot?"	No	No	Yes
Reactor coaxial with probe?	No	No	Yes
Relative $1/t_{90}$ (aqueous)	1.4 ± 0.1	1.3 ± 0.1	1
Relative $1/t_{90}$ (nonaqueous)	2.2 ± 0.2		1

FIGURE 5. Comparison of coil designs evaluated for these studies: (A) multiturn solenoid, (B) loop-gap resonator, and (C) slotted tube resonator.

The sensitivities of the three coils were compared under the conductivity conditions employed here. The details of this comparison are presented elsewhere.[7] Coils of each type, of equal volume, were constructed for use at 73 MHz. Although the estimated signal-to-noise for a nonconductive sample was over a factor of two higher for the solenoid than for a slotted tube resonator, the sensitivity advantage was only 30–40% for samples of physiological conductivity. More important, the vertical orientation of

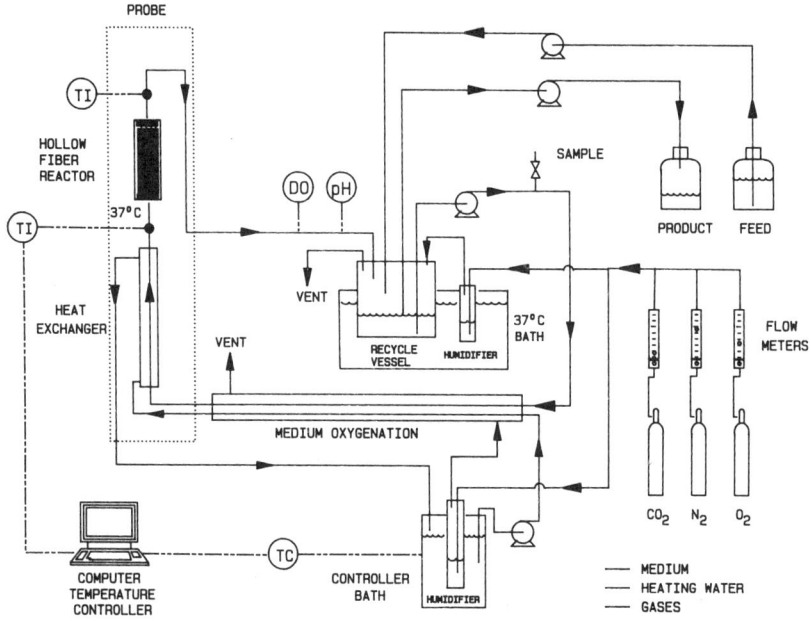

FIGURE 6. Schematic of the HFBR reactor apparatus. Medium was circulated at 120 mL/minute. Dissolved oxygen and pH were measured downstream of the reactor outlet. Dissolved oxygen and pH control was accomplished manually by exposing the medium to a mixture of carbon dioxide, nitrogen, and oxygen across silicone tubing. The headspace of the recirculation vessel was purged with this same gas mixture. Temperature was maintained at 37 ± 0.05 °C with a thermistor mounted at the inlet of the reactor (TI), along with a proportional-integral control scheme implemented on an HP-85 computer that controlled a 125-watt resistive heater in the controller bath. Fresh and spent media were added and removed at the same rate using a peristaltic pump.

the slotted tube resonator can accommodate a larger sample. This orientation also exploits a more homogeneous region of the static magnetic field; field homogeneity can strongly impact spectral resolution and signal-to-noise. Based on these factors, the slotted tube resonator was selected for use in subsequent experiments.

A schematic of the combined reactor/probe apparatus is shown in FIGURE 6. To minimize axial gradients of nutrients and waste products, the medium was circulated at a high rate (120 mL/min). The medium was passed through a section of silicone tubing ($1/8''$ i.d., $1/32''$ wall, Cole Parmer) to effect gas transfer with a mixture of nitrogen, oxygen, and carbon dioxide. The composition of this mixture was manually adjusted to control dissolved oxygen concentration and pH. The medium then entered the sample probe and was warmed to 37 °C by a glass heat exchanger. A thermistor at the inlet of the reactor, a computer, and a resistive heating element maintained the temperature of the medium entering the reactor at 37 ± 0.05 °C. Continuous operation was achieved by adding fresh medium to, and removing spent medium from, the recirculation vessel. Dissolved oxygen and pH were monitored on-line using dissolved oxygen and pH probes (Instrument Laboratories model IL530 and Ingold model 465) placed downstream from the reactor. The entire volume of the recycle vessel, reactor, heat exchanger, and connecting tubing (hereafter referred to as the "recycle loop") was 175 mL.

Results from the two apparatus are compared in FIGURE 7. The top diagram is a representative ^{31}P NMR spectrum of the dialysis membrane reactor acquired in 30 minutes. In contrast, a similar spectrum of hybridoma cells in the HFBR (bottom diagram) was obtained in 80 seconds. Thus, even though the NMR sample probe was home-built, the larger number of cells (approximately 6×10^8 cells in the sensitive region of the NMR coil) supported by the HFBR resulted in over a 20-fold decrease in the data acquisition time.

Three NMR experiments have been performed with the HFBR. ^{31}P NMR was used to measure the concentration of nucleoside triphosphates (NTP) for various concentrations of oxygen, yielding information about metabolism and mass-transfer limitations of this nutrient. Phosphorylated metabolites were also monitored during a glucose starvation-spike sequence. Finally, ^{13}C NMR was employed in the analysis of spent medium collected after a feed change from 100% ^{13}C-labeled glucose to unlabeled glucose. Each of these experiments was performed with Iscove's medium containing 15 mM glucose and 5 mM glutamine. The medium was also supplemented with 5% fetal bovine serum and $1 \times$ MEM nonessential amino acids.

EFFECTS OF OXYGEN CONCENTRATION ON NTP LEVEL

Because hybridomas are believed to derive a large portion of their NTP from oxidative phosphorylation, the concentration of NTP should be a useful indicator of the cells' access to oxygen.[5] Therefore, the ability of NMR spectroscopy to measure NTP levels on short time scales was exploited to examine the energy state of the cells for various degrees of oxygenation. In this study, the concentrations of glucose and glutamine in the circulating medium were 2.5 mM and 3.0 mM, respectively. Prior to each experiment, the dissolved oxygen concentration at the outlet of the reactor was 0.15 mM. The concentration of phosphorylated metabolites was monitored continuously with ^{31}P NMR as the oxygen concentration in the equilibrating gas mixture was changed to some lower level. After about 30 minutes, the outlet oxygen concentration was returned to 0.15 mM and the concentration of NTP was allowed to return to its original value. This procedure was repeated for six different lowered oxygen concentrations. The concentrations of NTP and the levels of dissolved oxygen during one such experiment are presented in FIGURE 8. The NTP concentration during the period of

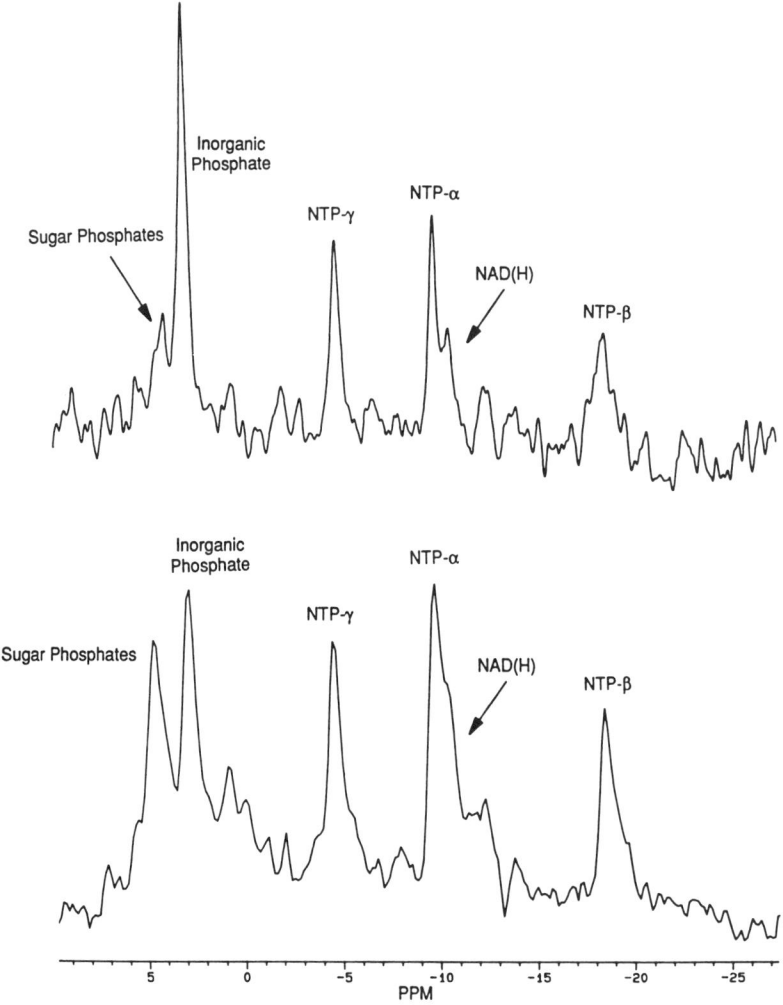

FIGURE 7. Comparison of ^{31}P NMR spectra of hybridoma cells immobilized in (top) dialysis tubing (1800 scans, 60° pulses, 1.0-s repetition time, 30-Hz line broadening) and (bottom) the HFBR (50 scans, 70° pulses, 1.6-s repetition time). The bottom spectrum was obtained on a Nalorac Quest 4300 spectrometer using selective presaturation of a broad resonance centered at 13 ppm with the DANTE pulse sequence[14] using 2000 1° pulses separated by 0.15-ms delays.

lowered oxygen concentration is plotted against the reactor-outlet oxygen concentration in FIGURE 9.

From the response of the cells, we can draw two conclusions about hybridoma energy metabolism under these conditions. FIGURE 8 shows that the concentration of NTP in the cells responds quickly to the changes in oxygen concentration. This result is consistent with the hypothesis that the cells obtain a large portion of their energy from oxidative phosphorylation, as opposed to other energy-producing pathways such as glycolysis. Moreover, whereas the concentration of oxygen was lowered by about a factor of two in the experiment summarized in FIGURE 8, the concentration of NTP dropped by only about 25%. This difference suggests that other pathways contribute significantly to energy production in the cell and/or the cells modify their metabolism in response to the reduced availability of oxygen.

The results shown in FIGURE 9 also reveal the importance of mass-transfer limitations in the reactor. From measurements made in this laboratory on this same cell line in continuous suspension culture, changes in glucose, glutamine, and oxygen uptake rates (which should relate directly to changes in NTP formation) are not observed until the cells are exposed to less than 0.02 mM oxygen (10% of air saturation).[8] However, FIGURE 9 shows that the NTP level declines at a much higher

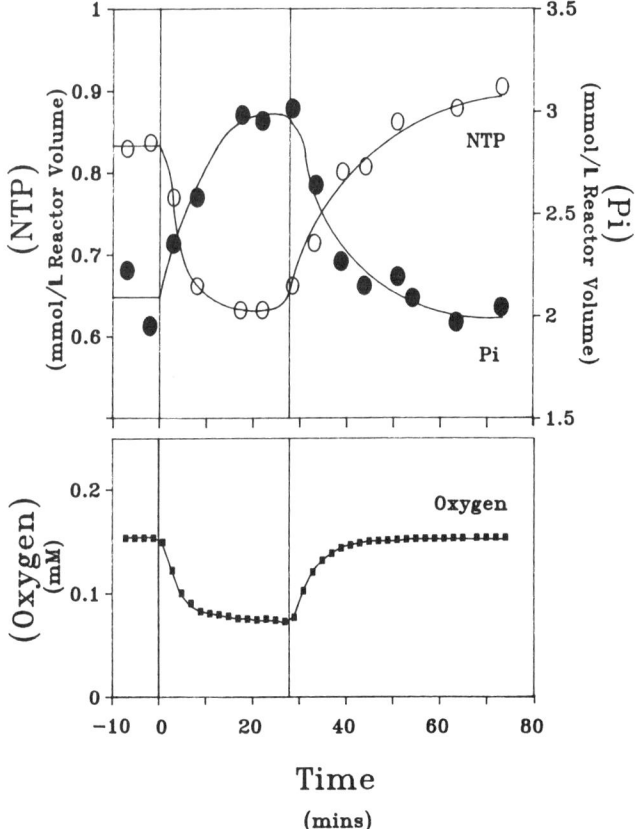

FIGURE 8. Effect of dissolved oxygen on the NTP level. Concentrations were determined by comparison with the resonance intensity of a known amount of methylene diphosphonic acid in a capillary mounted on the outside of the reactor. Concentrations are expressed in mmol/(liter reactor volume).

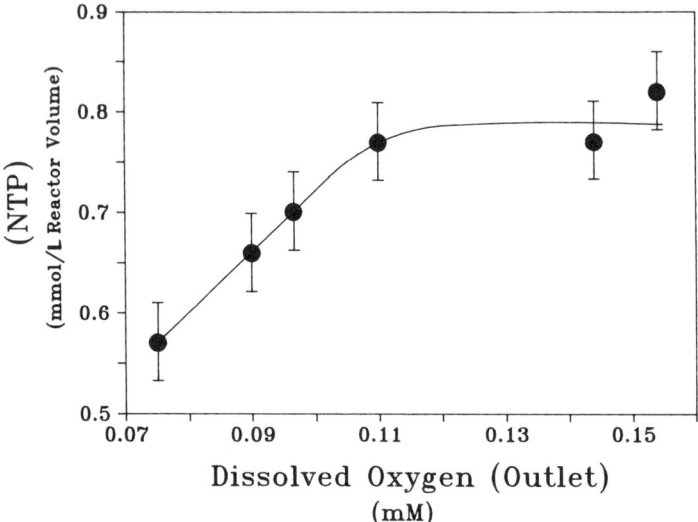

FIGURE 9. Concentration of NTP measured as a function of dissolved oxygen concentration at the reactor outlet. The error bars correspond to the concentration represented by the amplitude of the spectral noise.

oxygen concentration, suggesting that oxygen consumption becomes mass-transfer-limited below an outlet concentration of 0.11 mM. The concentration of NTP is expected to be well regulated within the cell; however, the immediate drop in NTP shown in FIGURE 8 indicates that regulatory processes are inadequate to maintain NTP levels during large changes in extracellular oxygen concentration.

The importance of oxygen mass-transfer limitations can be estimated independently from the observable modulus, Φ:[9]

$$\Phi^2 = \frac{Q_0}{D_{O_2} c_0} \left(\frac{V}{A}\right)^2,$$

where

Q_0 = zero-order consumption rate of oxygen (mmol/L-s),
D_{O_2} = diffusivity of oxygen in cell mass (cm^2/s)
c_0 = concentration of oxygen in the lumen (mmol/L),
V = volume of cell mass (cm^3), and
A = surface area for mass transfer to the cell mass (cm^2).

In general, at sufficiently low nutrient concentrations, mass-transfer limitations will become dominant. Using (i) the maximum oxygen consumption rate observed for this cell line in continuous culture, (ii) a diffusivity of oxygen equal to half of that in water,[10] (iii) the operating concentration of oxygen in the lumen (0.15 mM), and (iv) the geometric properties of the HFBR, a modulus of 0.64 was obtained. This value is consistent with the interpretation that mass transfer may have been limiting at oxygen concentrations below 0.11 mM.

In future experiments, we plan to reduce V/A by decreasing the fiber diameter and increasing the fiber density in the reactor. Of course, current technological limitations in fiber manufacture and potting set a lower limit on fiber diameter and an upper limit on the packing density. The modulus can also be lowered by reducing the concentration of cells, thus lowering the volumetric reaction rate, Q_0. This could be accomplished most simply by performing NMR measurements closer to the time of reactor inoculation while the cell concentration is lower.

DYNAMICS OF PHOSPHATE METABOLISM DURING THE GLUCOSE STARVATION-SPIKE SEQUENCE

In a second study designed to observe metabolism under dynamic conditions, the cells were allowed to deplete the available glucose and then were subjected to a spike addition of glucose. As in the oxygen limitation study, the initial feed concentration of glucose was 15 mM, resulting in a 2.5 mM steady-state glucose concentration in the recirculating medium. At the beginning of the experiment, the concentration of glucose in the feed was dropped to 10 mM for 20 hours and then to zero. Then, 2.5 hours later, the concentration of glucose in the medium dropped to less than 0.1 mM and a spike of glucose was added to bring the concentration of glucose in the recirculating medium to 3.5 mM (FIGURE 10). The normal 15 mM glucose feed medium was restarted 1.5 hours after the spike. During the course of the experiment, ^{31}P NMR spectra of the reactor were collected continuously and samples were collected for off-line analysis.

FIGURE 10 shows the response of the cells based on conventional measurements of glucose and antibody concentrations in samples of recirculating medium, along with NMR measurements of inorganic phosphate and NTP levels in the reactor. (No substantial changes in the amplitudes of other observed phosphorylated metabolite resonances were noted.) During the course of the experiment, the antibody concentration in the recycle loop underwent significant changes, indicating that the production rate changed as well.

The bottom half of FIGURE 10 reveals changes that occurred inside the cell. Notably, the NTP level was not significantly affected during the experiment, even though the glucose level changed dramatically. This behavior suggests that glutamine was the major energy source and/or the cells rapidly altered their metabolism (e.g., by increasing glutamine consumption) to maintain a constant NTP level. Indirect calculations by previous investigators have suggested that glutamine can be responsible for large amounts of energy in hybridoma cells.[11]

Whereas the effect of glucose concentration on the NTP level was small, inorganic phosphate levels in the reactor were greatly affected. The inorganic phosphate concentration presented in FIGURE 10 is the sum of intracellular and extracellular concentrations and was observed to drop by about 25% after the spike addition. This large consumption of phosphate could have been due to rapid phosphorylation of the glucose entering the cell. The sugar phosphate peak did not change significantly in area, but did vary in shape, perhaps because the relative amounts of different sugar phosphates were

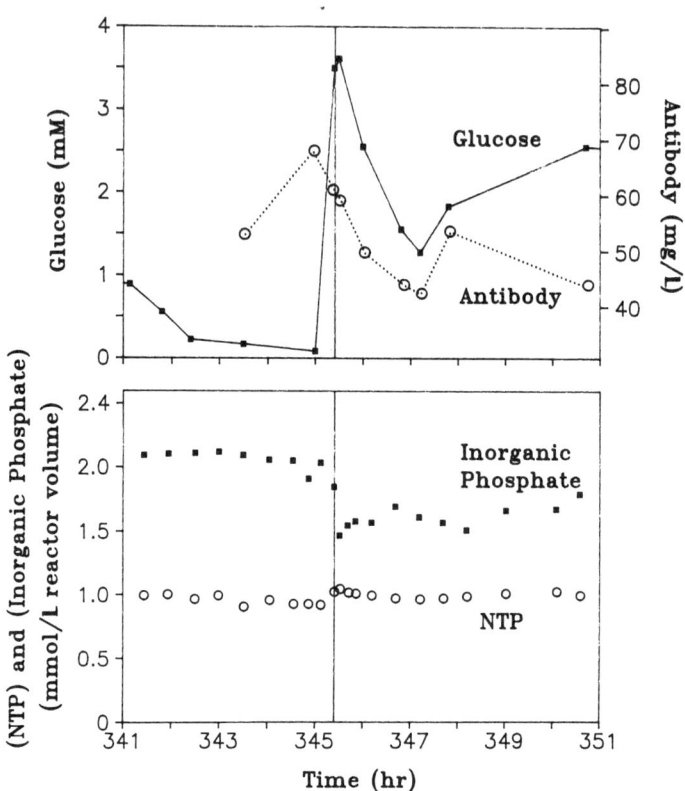

FIGURE 10. Top: Glucose and antibody concentrations in the recirculating medium. Bottom: Concentrations of nucleoside triphosphates (NTP) and inorganic phosphate (P_i) in the reactor. Phosphorylated metabolite concentrations are expressed as mmol/(liter reactor volume).

changing. With improved spectral resolution, we may be able to determine whether a change in glucose phosphate concentration accompanies such a change in P_i. Therefore, although substantial changes were observed in both the conventional and the NMR data, the relationship between glucose uptake, phosphorylated metabolite levels, and antibody production remains to be determined.

^{13}C NMR ANALYSIS OF LACTATE FORMATION

The greatest potential for obtaining general metabolic information from NMR measurements lies in ^{13}C NMR spectroscopy. This method can provide information about a much broader set of metabolic intermediates than ^{31}P NMR, and the narrow line widths and the large chemical shift range of ^{13}C provide much higher spectral resolution. Consequently, our primary objective is to study carbon metabolism by ^{13}C NMR. Only recently did we complete the construction of ^{13}C NMR probes suitable for *in vivo* studies of hybridoma metabolism in bioreactors. However, in a previous application of ^{13}C NMR, the fate of ^{13}C-labeled glucose was followed in off-line NMR measurements of recirculated medium.

At the beginning of the experiment, the feed medium was changed from Iscove's medium containing 15 mM natural glucose to Iscove's medium containing 15 mM 100% [1-^{13}C]D-glucose. The ^{13}C-labeled medium was supplied to the reactor for approximately 24 hours. This period corresponded to three residence times of the recycle loop (based on the volumetric feed rate); therefore, most of the ^{12}C-glucose in the system should have been consumed or washed out.

After the 24-hour exposure to labeled glucose, a sample of medium was taken for off-line analysis. Using NMR and conventional methods, the concentrations of total and ^{13}C-labeled lactate in the sample were determined. TABLE 2 shows the concentrations and yields of lactate from glucose based on the different methods.

The conventional enzymatic assay gave 1.8 for the total yield of lactate from glucose, which is close to the maximum theoretical value of 2.0 for complete conversion of glucose to lactate. The same value was obtained from many other samples taken before the ^{13}C-labeled glucose was added. This high yield of lactate is typical of transformed cells in general and is documented for this cell line in particular.[12]

Because NMR detects only ^{13}C atoms, only the lactate derived from ^{13}C-labeled and naturally abundant ^{13}C glucose is observed. In addition, if glucose is converted to lactate only via direct glycolysis, then a culture fed [1-^{13}C]D-glucose should produce ^{13}C-labeled lactate labeled only in the C-3 position (see FIGURE 1). Because each

TABLE 2. Comparison of Conventional and NMR Measurements of Lactate and Glucose Concentrations and Yield of Lactate from Glucose ($Y_{lac,glc}$) after Exposure to [1-^{13}C]D-Glucose[a]

Method	Total Lactate (mM)	^{13}C Lactate (mM)	Glucose (mM)	$Y_{lac,glc}$ Total	^{13}C
enzymatic assay	22.7		2.4	1.8	
^{13}C NMR		7.6	2.4		1.2
^{1}H NMR	21.9	6.9		1.8	1.1

[a]Enzymatic assays of glucose and lactate were performed with a glucose analyzer (Instrumentation Laboratory model 919) and with an enzymatic assay for lactate based on lactate dehydrogenase (Sigma Chemical). Approximately 10 mL of medium was freeze-dried and redissolved in 1.0 mL of D$_2$O for NMR analysis. Fully relaxed ^{1}H and ^{13}C NMR spectra were collected at 400.10 and 100.60 MHz, respectively, on a Bruker AM-400 spectrometer. ^{13}C NMR spectra were collected using inverse gated decoupling to avoid nuclear Overhauser effects. Concentrations were determined by comparing peak integrals (determined using the Bruker software) of lactate and glucose with those of HEPES, which had a concentration of 23.2 mM in the original sample. The total lactate concentrations were determined using both the C-2 and the C-3 ^{1}H NMR resonances and were within 5% of one another. ^{13}C and ^{1}H NMR spectra of each sample were collected twice and all the resultant concentrations were reproduced to within 5%. The amounts of ^{13}C-labeled lactate were determined from the areas of satellite C-3 ^{1}H resonances due to the ^{13}C label at C-3. Yields of ^{13}C-labeled lactate from glucose were multiplied by two because only one of the two lactate molecules derived from [1-^{13}C]D-glucose will contain ^{13}C. The glucose concentration could not be determined from the ^{1}H NMR spectrum because its resonances were not well resolved.

glucose molecule contains a single ^{13}C label and is split to produce two lactate molecules, only half the lactate formed will be ^{13}C-labeled. The calculated ^{13}C-labeled lactate yields in TABLE 2 take this fact into account. The concentration of ^{13}C-labeled lactate due to naturally abundant sources was neglected.

In addition to total lactate, the concentration of ^{13}C-labeled lactate could be determined from ^1H data because ^{13}C in a compound can be detected via bonded protons. ^{13}C atoms split a proton peak in the ^1H spectrum, resulting in separate peaks due to ^{12}C- and ^{13}C-containing material. As shown in TABLE 2, the concentration of total lactate determined by ^1H NMR was very similar to that obtained from enzymatic assay and the amount of labeled lactate determined by ^1H NMR was similar to that obtained from ^{13}C NMR.

The NMR results show that large amounts (approximately 30%) of lactate were formed by processes other than direct glycolysis. Two possible sources of unlabeled lactate are the pentose phosphate pathway and glutaminolysis. Glucose that enters the pentose phosphate pathway has the C-1 carbon removed early in the pathway. Any pentoses returning to the latter part of the glycolytic pathway could end up as lactate, but would not contain the ^{13}C label. In addition to the pentose phosphate pathway, lactate can be formed from glutamine via the TCA cycle and from malic decarboxylases that can divert the TCA-cycle carbon back to pyruvate. Lactate production from glutamine has been previously postulated.[13] By utilizing other labeled forms of glucose (e.g., [2-^{13}C]D-glucose) and labeled glutamine, we should be able to distinguish how much each of these two pathways participate in lactate production. Interestingly, even under conditions of high glucose concentration (3 mM in the recycle loop), large amounts of lactate are produced by routes other than direct glycolysis.

CONCLUSIONS

The experiments described in this report illustrate how NMR can provide unique information about a cell culture. The immobilized cells were monitored noninvasively on time scales as short as 80 seconds. This capability provided information about oxygen metabolism and mass-transfer effects, as well as glucose metabolism. The measurements were made *in situ* on a single sample, eliminating the uncertainties involved with sample-to-sample variation. However, an important remaining variable is the homogeneity of the cell sample. Whereas high cell densities are important because of the relatively low sensitivity of NMR spectroscopy, adequate mass transfer must be achieved to ensure high cell viability at the low concentrations of nutrients where critical changes in metabolism occur. We hope to improve the situation by further increasing the amount of surface area available for mass transfer in the hollow-fiber reactor.

There are several intriguing metabolic questions that could be addressed with the combination of NMR and conventional methods outlined above: How much do different pathways contribute to waste product formation under various nutrient conditions? What are the kinetics of waste product formation? How should the cells be fed to meet their energy requirements while minimizing the production of detrimental waste products? How do the concentrations of intracellular compounds correlate with the yield of monoclonal antibody? These are among the questions currently being considered in this laboratory.

ACKNOWLEDGMENT

We wish to thank Microgon, Incorporated, for constructing the hollow-fiber reactors.

REFERENCES

1. FERNANDEZ, E. J. & D. S. CLARK. 1987. Enzyme Microb. Technol. **9:** 259–271.
2. AVISON, M. J., H. P. HETHERINGTON & R. G. SHULMAN. 1986. Annu. Rev. Biophys. Chem. **15:** 377–402.
3. EIGENBRODT, E. P., P. FISTER & M. REINACHER. 1985. New perspectives on carbohydrate metabolism in tumor cells. *In* Regulation of Carbohydrate Metabolism, Volume 1. R. Beitner, Ed.: 141–179. CRC Press. Boca Raton, Florida.
4. MILLER, W. M. 1987. A kinetic analysis of hybridoma growth and metabolism. Ph.D. thesis. University of California, Berkeley, California.
5. FERNANDEZ, E. J., A. MANCUSO & D. S. CLARK. 1988. Biotechnol. Prog. **4**(3): 173–183.
6. GADIAN, D. G. 1982. Nuclear Magnetic Resonance and Its Applications to Living Systems. Oxford University Press. London/New York.
7. MURPHY, M. K., E. J. FERNANDEZ & D. S. CLARK. 1989. Magn. Reson. Med. **12:** 382–389.
8. MILLER, W. M., C. R. WILKE & H. W. BLANCH. 1987. J. Cell. Physiol. **132:** 524–530.
9. BAILEY, J. E. & D. F. OLLIS. 1986. Biochemical Engineering Fundamentals (second edition). McGraw-Hill. New York.
10. GROTE, J., R. SÜSSKIND & P. VAUPEL. 1977. Pflügers Arch. **372:** 37–42.
11. MILLER, W. M., C. R. WILKE & H. W. BLANCH. 1988. Bioprocess Eng. **3:** 103–111.
12. MILLER, W. M., C. R. WILKE & H. W. BLANCH. 1988. Bioprocess Eng. **3:** 113–122.
13. GLACKEN, M. W., R. J. FLEISCHAKER & A. J. SINSKEY. 1986. Biotechnol. Bioeng. **28:** 1376–1389.
14. MORRIS, G. A. & R. FREEMAN. 1978. J. Magn. Reson. **29:** 433–462.

Scanning Tunneling Microscopy and Atomic Force Microscopy of Biological Surfaces[a]

JOSEPH A. N. ZASADZINSKI[b] AND PAUL K. HANSMA[c]

[b]Department of Chemical and Nuclear Engineering
[c]Department of Physics
University of California, Santa Barbara
Santa Barbara, California 93106

INTRODUCTION

The first applications of microscopy to biology must have occurred shortly after the invention of simple magnifying lenses in the fifteenth century—imagine the amazement of our Renaissance ancestors as they had their first magnified view of the world. The development of the optical microscope by van Leeuwenhoek in 1660 allowed the investigation of the micron-sized world of single cells, bacteria, algae, etc. After only five short years, Robert Hooke published the first scientific drawings of insects, crystals, and feathers using the new optical microscope. Since that time, the optical microscope has played, and continues to play, an enormous role in biological studies and has produced dramatic changes in our understanding—Pasteur's germ theory of disease and Fleming's discovery of penicillin are two prime examples of fundamental breakthroughs made possible by the optical microscope. Still, the submicron world was out of reach because optical resolution is limited by the wavelength of light to about 0.2 microns.

The invention of the electron microscope by Ruska and Knoll in the 1930s pushed this resolution limit to atomic dimensions by using high-energy (short-wavelength) electrons rather than photons as the imaging medium. However, soft, hydrated biological materials are basically incompatible with the high vacuum and radiation environment of the electron microscope without substantial alteration. Hence, the real resolving power of the electron microscope was unavailable to biologists until sophisticated sample preparation techniques, namely, plastic embedments, ultrathin sectioning, and freeze-fracture replication, were developed in the 1950s. Since that time, the findings from electron microscopy have greatly influenced modern biology. The electron microscope has revealed much of what is known of viral structure, shapes, and sizes. The electron microscope has also permitted the visualization of cellular organelles such as mitochondria, plastids, and Golgi apparatus, along with the details of their internal structure. Moreover, it has been invaluable in studies of membrane

---

[a]Financial support to J. A. N. Zasadzinski was provided by the donors of the Petroleum Research Fund, by the Du Pont Company, by the Exxon Education Foundation, and by National Science Foundation Presidential Young Investigator Award No. CBT 86-57444. P. K. Hansma was supported by NSF Grant No. DMR86-13486.

structure, cell walls, and protein and enzyme processes and location. However, even today, electron image resolution is limited more by deficiencies in preparing properly preserved and lifelike samples rather than by any inherent limitations of the electron microscope.

Still, in the over 300 years of optical and 50 years of electron microscopy development, nothing could have prepared us for the incredible imaging power and simplicity of the scanning tunneling microscope (STM).[1] The STM has the greatest proven resolution of any imaging technique; for ideal samples, the lateral resolution is about 1 Å and the vertical resolution is less than 0.01 Å. The operating principle of the STM is surprisingly simple. A metal tip is brought close enough to the surface to be imaged so that, at a convenient operating voltage (2 mV–2 V), electrons begin to tunnel between the tip and the surface. This electron tunneling is a quantum effect that is exponentially dependent on the separation between the tip and the surface. As a result, the tunneling current decreases by an order of magnitude for every 1-Å increase in separation. The tip is scanned over the surface while the tunneling current is measured. A feedback network changes the height of the tip to keep the tunneling current constant in the so-called "constant current mode" or the current is monitored at constant tip height in the "constant height mode". In the more commonly used constant current mode, if the current can be kept constant to 2%, then the gap between the surface and the tip will remain constant to within 0.01 Å. An image consists of a map of the tip height versus lateral position. The result is a three-dimensional image of the scanned surface with atomic resolution (see FIGURE 1). Conventional transmission electron microscopes (TEM) can only give two-dimensional projections and, because the high-energy electrons penetrate deep into matter, the TEM has limited applications to surface investigations. A three-dimensional image with molecular resolution is only possible with the scanning tunneling microscope (STM).

In the six or so years since its invention by Gerd Binnig and Heinrich Rohrer,[1] the STM has provided the first ever three-dimensional images of atoms on the surfaces of metals and semiconductors. The STM has been used to settle many important technical questions including the 7-by-7 atomic reconstruction of the silicon surface, the surface structure of superconductors, the adsorption of gases onto the surfaces of catalytic metals, spectroscopy of single atoms, and even direct observations of electrochemical reactions in electrolyte solutions.[2] The STM can even give atomic resolution images under water and other fluids.[3] The immense importance of these discoveries was confirmed by the Nobel prize in physics in 1986, only four years after the first STM paper was published. Modern STMs have been so simplified that they are commercially available from a number of vendors and are even being used as a quality-control check for thin-film magnetic recording heads, for stampers used to manufacture compact disks, and for the tips of certain cutting tools.[4]

However, there are still surprisingly few images of biological surfaces in the literature[5-10] and even fewer images with sufficient resolution that actually contribute new information on biological structures.[11,12] This is because the STM can only be used to image rigid, conductive surfaces such as metals, silicon, graphite, etc.; the STM is virtually useless for imaging materials that are fluid, nonconductive, or both. Even so, the incredible resolution and three-dimensional imaging capability of the STM makes it imperative to develop a way of applying the STM to the investigation of the relationship between structure and function in biomaterials, polymers, and colloids.

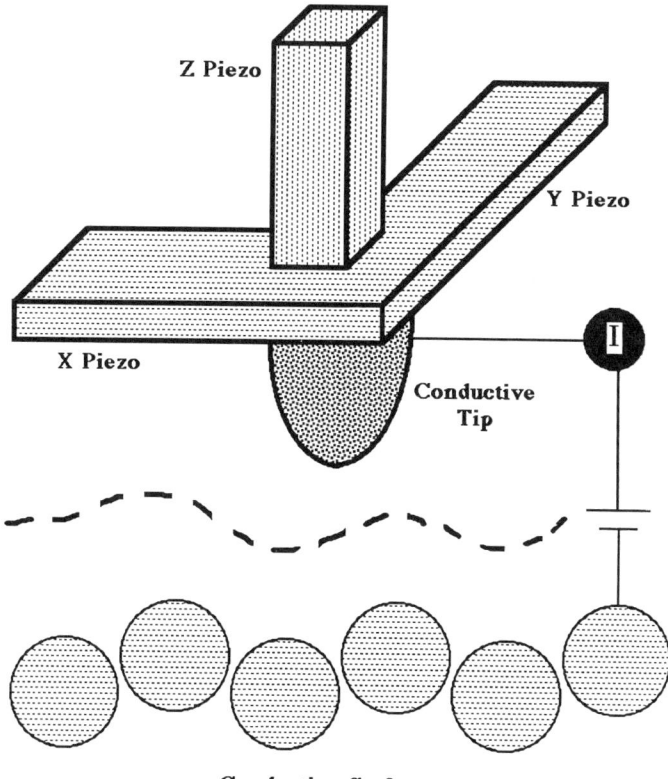

FIGURE 1. Operating principles of the scanning tunneling microscope. A metal tip mounted on a piezoelectric crystal is brought close enough to the surface to be imaged to initiate a tunneling current. As the metal tip is scanned across the surface in the x direction, a feedback loop adjusts the height of the tip, z, to keep the tunneling current constant. An image consists of a series of scans of z versus x displaced by a constant Δy. The image is actually a tracing of contours of constant electron density at a particular energy determined by the bias voltage. This reduces to a simple topograph for surfaces of constant composition.

Like samples for the electron microscope, biological samples for the STM must be modified to make them conductive and rigid enough to be imaged at high resolution. For the STM to work properly, about 10^{10} electrons/second (1 nanoamp) must flow in a 0.1-nm^2 area. This sort of conductivity is not found in most biomaterials. Specialized preparation techniques need to be developed before the ultimate resolution of the STM can be applied to biological systems. In our favor, the STM has only been around for 6 years as compared to the over 300 years of optical microscopy and 50 plus years of electron microscopy. As the potential benefits of atomic resolution images of proteins, enzymes, membranes, etc., are so vast and important, investigators worldwide are engaged in active research to make biosurfaces compatible with the STM. Others, including the developers of the STM,[13] have developed a new type of hybrid microscope, specifically, the atomic force microscope (AFM) that can image nonconductive surfaces (FIGURE 2). The AFM measures the deflection of a diamond tip suspended by a weak spring as it is dragged across a surface.[4] The AFM was designed with biological surfaces in mind; however, at the present state of development, only fairly rugged surfaces can be imaged.[4,14] As the AFM continues to be improved and our experience with biosurface sample preparation for the STM increases, the possibility of imaging biological materials at the molecular level is a realistic possibility.

EXPERIMENTAL METHODS

There are two difficulties to overcome before biological and other nonconductors can be routinely imaged with the STM. Most biological materials are extremely poor

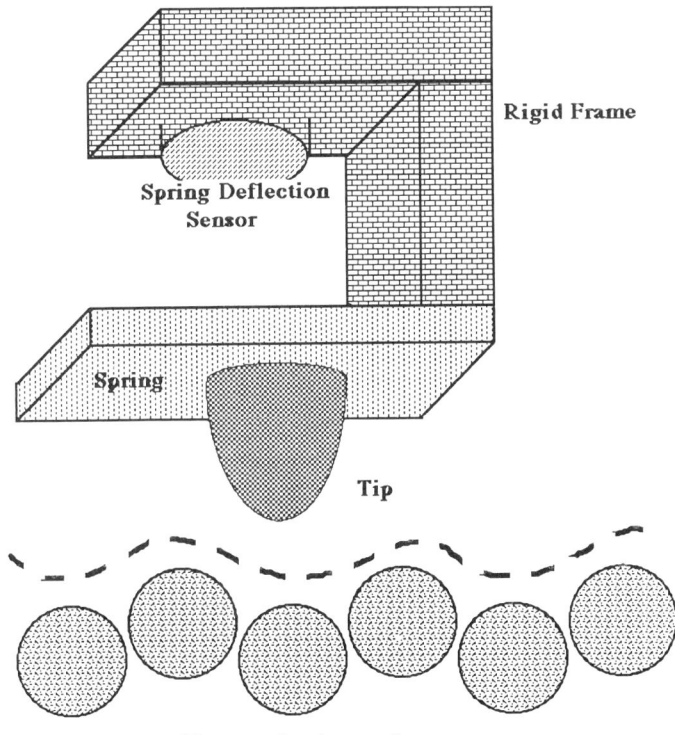

FIGURE 2. Schematic view of the force sensor for an atomic force microscope (AFM). The essential features are a probe, a spring, and a method of measuring the deflection of the spring. The spring deflection, in practice, has been measured using electron tunneling to the back of the spring, by optical interferometry between the back of the spring and a reference plate, or by deflection of a laser light beam bounced off a mirror on the back of the spring. The probe follows a path that is an accurate topograph of the sample surface without any need for a conductive sample.

conductors of electricity—too poor for the tunneling current to propagate sufficiently so that it can be measured. The second limitation is that biomaterials and polymers are relatively "squishy" in comparison to the crystalline solids that are compatible with the STM. At room temperature, membranes, proteins, and macromolecules such as DNA are dynamic structures. Brownian and other thermal fluctuations can be large in comparison to the molecular features of interest. Both of these limitations must be overcome by appropriate sample modifications that make minimal changes in the specimen microstructure while providing sufficient conductivity and rigidity for high resolution imaging.

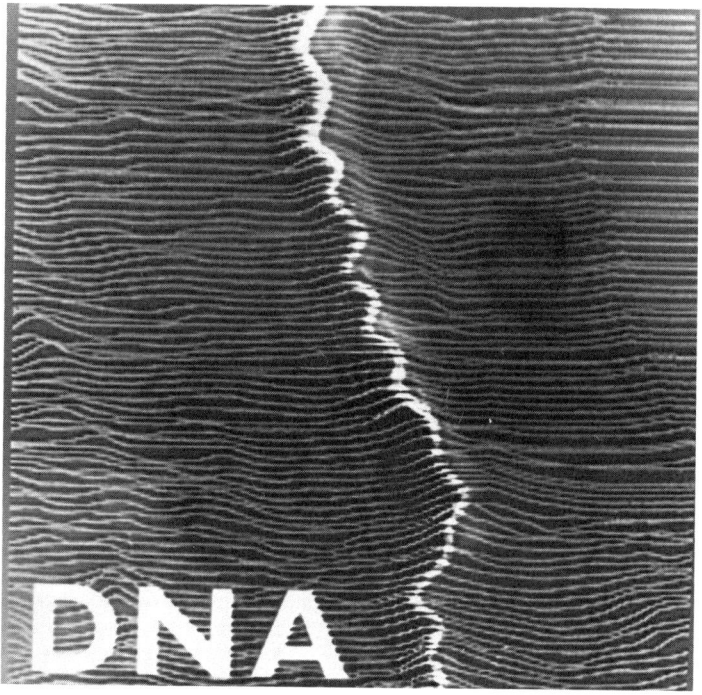

FIGURE 3. Direct STM image of DNA dried down on a carbon substrate. The image has been turned upside down because a long mound is easier to visualize than a long trough. The periodicity of the zigzag is about 4 nm, which is in rough agreement with the helical periodicity of DNA. (Figure by Binnig, used with permission.[15])

Direct Imaging of Biomaterials

The first attempts at STM imaging of biomaterials involved mounting dried specimens of DNA[15] onto conductive surfaces and then attempting to image the surfaces directly with the STM (see FIGURE 3). The idea behind this simple procedure was that the biological material would behave like an insulating gap between the metal tip and the underlying conductive surface. If the layer of biomaterial was thin enough, tunneling could occur through the biomaterial, just like through the vacuum that normally exists between the tip and the specimen in the STM. In fact, Giaevar's experiments with thin insulating oxides between metal strips were some of the first demonstrations that electrons did tunnel.[16] More recently, Foster, Frommer, and Arnett[17] have imaged single organic molecules pinned to a graphite surface with the STM with atomic resolution. Lang et al. have used the STM to image Langmuir-Blodgett films of cadmium arachidate and dimyristoylphosphatidic acid on graphite with molecular resolution.[18] Unfortunately, although it is possible to detect the presence of larger biological macromolecules such as DNA, the images obtained are fairly low resolution and often unreproducible. It is unclear what the mechanism of electron transport is through thicker layers of biological material. FIGURE 3 shows the image of DNA on a carbon substrate.[15] The tunneling tip of the STM, which started about 5 nm from the sample surface, actually dipped toward the DNA molecule rather than rising up over it. The image was turned upside down because a long mound is easier to visualize than a long trough. The DNA appeared to have a roughly periodic zigzag length of 4 nm, which is about what is expected from the twisting periodicity of the DNA double helix. Although many investigators have been involved in direct imaging of biomaterials with the STM, most have abandoned their attempts due to this sort of lack of resolution and reproducibility.

Metal-coated Samples

To overcome the lack of conductivity of DNA and other biological macromolecules, Amrein et al.[11] applied a 1-nm layer of platinum-iridium-carbon alloy to complexes of DNA and recA protein[19] adsorbed and then freeze-dried onto a platinum-carbon surface. Slow freeze-drying is a standard electron microscopy method of preparing dehydrated biological specimens without the structural alterations that accompany simple room-temperature drying from the liquid.[20] The samples were freeze-dried for three hours and the Pt-Ir-C films were applied by rotary shadowing at an elevation angle of 60° in a standard high-vacuum freeze-fracture apparatus (Balzers BAF 400T). The samples were imaged in the constant current mode with a bias voltage of 0.8 volts, tip negative, with mechanically sharpened Ir-Pt tips or Au tips that were chemically etched in concentrated HCl.

FIGURE 4 shows the processed STM images of a segment of the freeze-dried recA-DNA complex or, in fact, the conductive metal coating on the recA-DNA complex. FIGURE 4A shows the three-dimensional representation with simulated shading. FIGURE 4B is a top view in gray-tone representation. The height is increased from bright to dark. The striations, which are due to the right-handed single helix of the complex, are indicated by arrows. FIGURE 4C is a top view with simulated shading of the segment, with the filament structures enhanced and the background corrugation suppressed. Many of the striations in top view show a tripartite structure that indicates that one helical turn of the complex contains six such parts, which corresponds to the established number of recA monomers per turn.[19] The width of one of the complexes is about 12 nm and the height is about 7 nm, indicating some compression in the vertical direction. The bar in the figure is 10 nm. In comparison to the uncoated DNA in

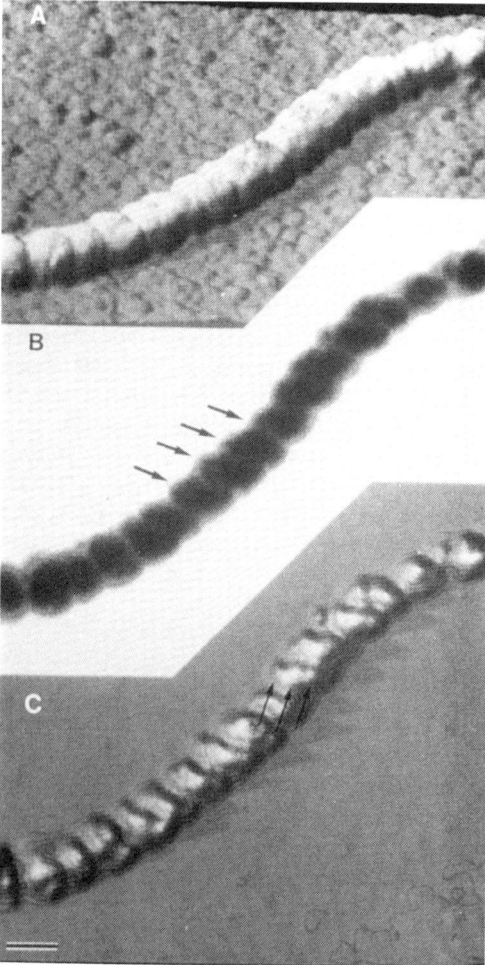

FIGURE 4. Processed STM image of a segment of the freeze-dried recA-DNA complex coated with a Pt-Ir-C film: (A) Three-dimensional representation with simulated shading. (B) Top view in gray-tone representation. The striations, caused by the right-handed single helix, are indicated by arrows. (C) A top view with the filament structures enhanced to demonstrate the tripartite structure of one striation as indicated by the arrows. (Figure from Amrein et al.,[11] used with permission).

FIGURE 3, the image detail in the metal-coated DNA complex is greatly improved. Clearly, resolution on the scale of biological macromolecules is possible with metal-coated specimens.

Freeze Fracture

For membranes and more fluid biomaterials that cannot be simply adsorbed or freeze-dried onto a substrate, a more elaborate "fixation" procedure for STM is necessary. The best general fixation technique for fluid materials is freeze-fracture replication. Freeze-fracture replication involves (1) quick-freezing the sample to kinetically trap fluid structures, (2) fracturing the sample under vacuum at cryogenic temperatures to expose internal structure, and (3) replicating the fracture surface with an evaporated layer of platinum and carbon. In this process, the fluid and nonconductive material is replaced by a rigid, conductive surface replica that is an almost ideal specimen for STM imaging.

For freeze-fracture STM, the initial quick-freezing step is the most important to achieve minimal disruption of the structure. If the cooling rate is too slow, the material can reorganize into some new structure, not always the one we wish to study. To obtain the best results, the specimen must be frozen at such a rate that the molecules do not have sufficient time to rearrange themselves. In practice, this means a cooling rate of about 10,000 °C/s. To achieve these rates, we sandwich a (10–50)-μm-thick layer of sample liquid between two 100-μm-thick copper plates in an enclosed, temperature-controlled oven (± 0.1 °C) and then rapidly cool the sandwich by placing it between two opposed high velocity jets of liquid propane cooled by liquid nitrogen to <100 K. This cooling rate is sufficient to freeze water in a vitreous state and to preclude rearrangement of dispersed macromolecules or biomembranes. After freezing, the sample sandwich is fractured at cryogenic temperatures (100 K) in a high-vacuum ($<10^{-7}$ torr) freeze-etch apparatus. The complementary fracture surfaces are then immediately coated with a 1.5-nm-thick shadowing film of platinum-carbon followed by a 15-nm-thick layer of carbon at normal incidence to strengthen the shadowing film. The samples and replicas are brought to ambient temperature and pressure, and any remaining sample is cleaned from the replica films. Further details of the freeze-fracture procedure are given in Zasadzinski and Bailey.[21]

The metal surface films form near-exact replicas of the fracture surface and are ideal specimens for both transmission electron and scanning tunneling microscopy as they are rigid and highly resistant to radiation or electric field damage, have virtually zero vapor pressure at room temperature, and are highly conductive. FIGURE 5A is an STM image of a freeze-fracture replica of the P'_β or "ripple" phase of a naturally occurring phospholipid liquid crystal, dimyristoylphosphatidylcholine (DMPC) dispersed in water.[12] The replicas were oriented so that the side directly contacting the sample surface was imaged by the STM. The characteristic three-dimensional ripples of the P'_β phase are readily apparent as regular mounds spaced about 130 Å apart. What is surprising is the fine-scale features such as the smaller periodicity perpendicular to the ripples as shown in FIGURE 6A. These undulations are only 5 to 10 Å in height and are 30 Å in lateral separation. These are the highest resolution features ever seen on a biomembrane surface; they may show how the molecules that make up the biomembrane are organized.

The resolution of a freeze-fracture replica appears to be far superior when imaged by the STM than by the TEM. (Compare FIGURES 5A and 6A to FIGURES 5B and 6B.) One explanation for this is that the STM only images the side of the replica film in direct contact with the original sample surface. Presumably, this side of the replica matches the original fracture surface most exactly. There is less influence of platinum migration, recrystallization, etc., on the images because the STM is only influenced by the topography of the replica surface. In the TEM, the image is formed by scattering variations caused by differences in the replica thickness. The electron beam samples the entire replica and any variations or imperfections in the replica film caused by local crystallization, preferential attachment, surface migration, etc., are imaged. This is best illustrated by the drawing in FIGURE 7. Each ripple, on average, is coated by several metal grains, held together by the carbon film. The entire shape of the

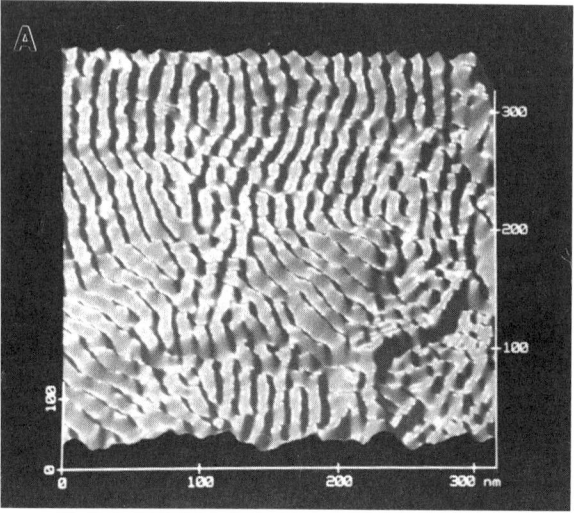

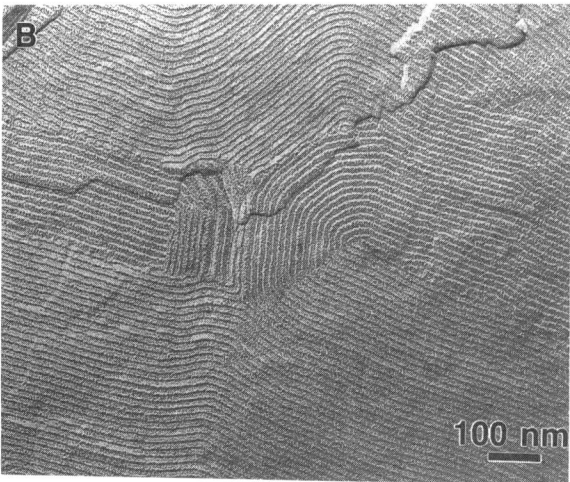

FIGURE 5. (A) Three-dimensional scanning tunneling micrographs of the platinum-carbon replica of the ripple phase of dimyristoylphosphatidylcholine-water liquid crystalline membranes. (B) TEM image of the platinum-carbon replica of DMPC-water membranes. The two-dimensional projection gives an impression of the third dimension, but no quantitative information.

crystallites determines the image in the TEM; hence, the resolution is limited by how much platinum crystallization and migration has influenced the local replica mass-thickness.[22] As long as the replica is a continuous layer of platinum, though, the STM does not care about the thickness or shape of the crystallites or the thickness of the carbon backing film. In fact, better resolution is obtained for thicker replicas than thinner ones, which is in contradiction to what is normally found for TEM images.[23] If the side of the crystallites contacting the surface retains the surface profile, then resolution is enhanced as seen in FIGURES 5 and 6. The influence of deposition parameters, the choice of metal composition, and the type of deposition used all influence how well the surface can be replicated. To really understand the details of the

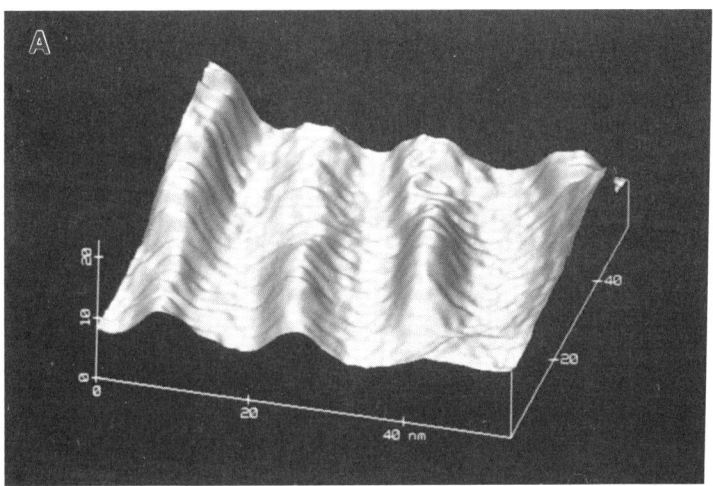

FIGURE 6. (A) High magnification view of the ripple phase of DMPC-water. The ripple amplitude of ~45 Å and the asymmetric ripple contour are well defined and easily visible. (B) High magnification TEM view of the ripple phase. At this magnification, the image is dominated by the metal grains making up the replica. Most of the three-dimensional information is lost. The resolution is clearly inferior to the STM image in part A.

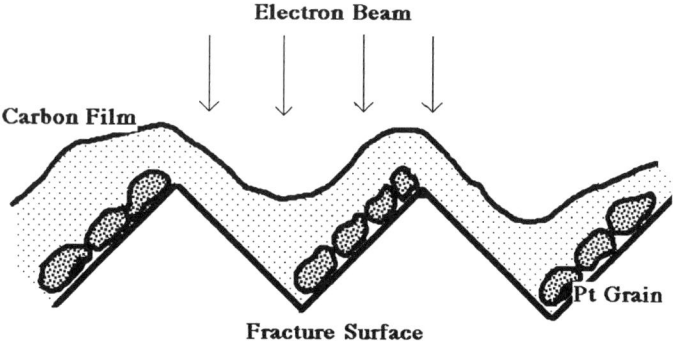
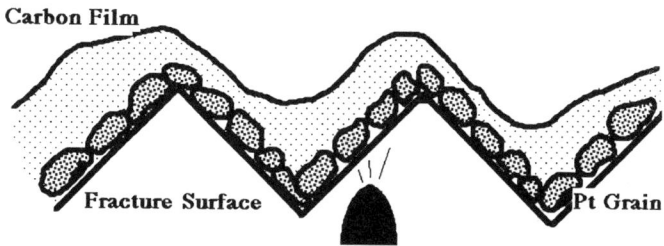

FIGURE 7. Artist's conception of a freeze-fracture replica of a ripple phase such as those in FIGURES 5 and 6. Each ripple is coated by, at most, a few metal grains, which are then coated by a thicker layer of carbon (drawing not to scale). The metal grains have developed a crystalline structure due to the migration of the metal to sites of low energy. This migration distorts the electron microscope image, the contrast of which depends on the simple relationship between metal thickness and surface contour. The STM images only the surface layer of the replica that was in direct contact with the original fracture surface. The migration and crystallization of the metal film has much less influence on the final images; hence, the resolution is improved.

surface features revealed in FIGURE 5A, we need to better understand the replication process and the way in which the STM produces an image. However, these results show that, in addition to direct, three-dimensional images, the STM appears to give better resolution for the same replica.

Atomic Force Microscopy

In the three years since its invention, the AFM has been used to obtain the first ever atomic resolution images of nonconductors. The atomic force microscope, as shown schematically in FIGURE 2, works along the same lines as a record player. A probe—here, a small diamond fragment—is pressed against the surface by a weak spring with a constant tracking force of about 10^{-8} newtons.[14] As the probe is dragged across the surface, the spring is deflected by the contours of the sample surface. This deflection is then measured by detecting the tunneling current between the spring, which was made from a gold-coated tungsten wire, and a platinum-coated razor blade that acted as the reference surface for tunneling. The height of the sample surface relative to the reference surface is then adjusted by changing the voltage to a piezoelectric crystal on which the sample was mounted just as in the STM.

FIGURE 8A shows the processed image of the surface of a crystalline flake of DL-leucine obtained with the AFM. The white dots in the computer-filtered image represent the topological peaks of methyl groups at the end of individual DL-leucine molecules. In the plane of the image, the leucine molecules are spaced about 0.53 nm, which is in good agreement with X-ray diffraction measurements from the bulk.[14] In FIGURE 8B, the molecular graphics program CHEMX was used to generate the packed crystalline structure from published crystallographic data. The array of methyl groups shown in FIGURE 8B was rotated to an orientation similar to that in FIGURE 8A. The close correspondence between the surface image obtained by the AFM (FIGURE 8A) and the bulk projection obtained by X-ray diffraction (FIGURE 8B) shows that the surface of this amino-acid crystal is a simple termination of the bulk; there is no evidence of reconstruction. Unfortunately, most biological molecules could not withstand even such a small force as the approximately 6×10^{-9} N used to record this image.

| 1 nm

FIGURE 8A. Computer-processed gray-scale image of the DL-leucine surface. The AFM was operated using a tunneling junction that sensed the deflection of the spring. The white spots represent the methyl groups at the terminal end of the DL-leucine molecule that project furthest from the surface. The methyl groups are spaced about 0.53 nm apart.

DISCUSSION AND SUGGESTIONS FOR FURTHER WORK

Neither the STM nor the AFM will provide routine images of untreated biomaterials without significant improvements in both instruments. However, both instruments are much more compatible with hydrated and dynamic surfaces than the electron microscope, so less sample alteration is necessary. Both the STM and the AFM can operate at room temperature and pressure and even under fluids, including water. Both instruments are much less destructive of samples than the electron microscope because

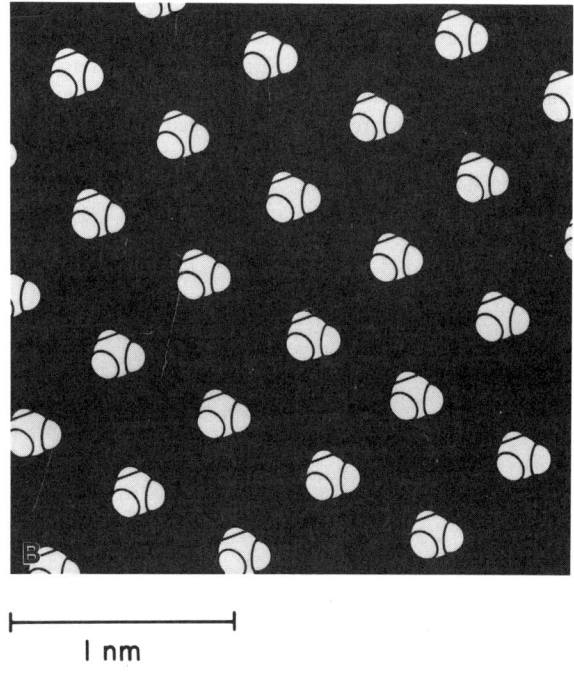

FIGURE 8B. Array of methyl groups spaced according to X-ray data and rotated to an orientation similar to that in FIGURE 8A. Drawn with the program CHEMX (Chemical Design Limited, Oxford, United Kingdom).

the energy input, a few millivolts versus hundreds of kilovolts, is much less. The energy in the electron beam of a conventional transmission electron microscope is sufficient to break most organic chemical bonds and usually causes gross structural rearrangements on the time scales required for imaging. The higher the magnification of a transmission electron microscope, the higher the electron flux required for imaging. Also, electron micrographs are two-dimensional projections, as opposed to the three-dimensional topographs provided by the STM and the AFM. Hence, even at their present level of development, the STM and the AFM can provide a wealth of new information on biological structure. The commercial availability of the STM at relatively modest prices will also spur new workers to join in the hunt. Still, the scientist who hopes for molecular and eventually atomic resolution images of biological materials faces many challenges before this goal can be reached. However, it is important to remember that scanning probe microscopy is still in the early stages of development.

In the near future, refinements in metal coating and freeze-fracture techniques will undoubtably lead to higher resolution images, although atomic resolution seems impossible. The goal of the metal coating process is to reproduce the fractured or freeze-dried surface as accurately as possible with an amorphous, well-defined, electrically conductive layer that can easily be imaged by the STM. At present, resolution is limited by imperfections in the metal films caused by interactions between the metal vapor and the specimen surface. If the randomly arriving atoms of metal vapor were trapped at the point of their first contact, a spatial resolution close to the atomic dimensions of the metal could be obtained. Such resolution has not yet been achieved because the condensing metal atoms migrate along the surface while dissipating their thermal kinetic energy until they are eventually trapped, either by strong binding sites on the specimen surface or by previously condensed atoms, crystallites, or grains of the evaporation material. During continued deposition of metal, these first nucleation sites grow, forming stable, immobile centers that eventually merge with neighboring centers to form a continuous film. Therefore, the main problem in enhancing faithful replication is reducing the surface mobility of the condensing metal atoms while inhibiting the formation and growth of crystal nuclei.

A reduction in surface recrystallization and migration and a consequent improvement in resolution might be obtained by the following:

(1) lowering the fracture surface temperature to reduce the lateral mobility of condensing atoms,
(2) using higher melting point metals that are less mobile at the fracture surface temperature,
(3) evaporation of more than one metal or alloy, or metals with carbon, to inhibit crystal nucleation,
(4) reducing the deposition rate to decrease the likelihood of forming stable aggregates.

In addition to these deposition parameters, topological and chemical features of the fracture surface can also have an influence on the extent of lateral mobility of condensing atoms. This "preferential nucleation" or "decoration" often occurs at crystal lattice steps or at surface discontinuities where the fracture surface presents a sharp edge. Decoration is much less of a factor in the STM as the excess metal accumulated at the nucleation site is not visible to the STM, whereas it is visible to the TEM.

One of the future goals of research in this area is to qualitatively and, more importantly, quantitatively determine the optimal metal film composition that provides the highest resolution STM images. A number of investigators have examined various metals, alloys, and mixtures for use as replicating films for the TEM, although no real consensus exists.[23-26] Amrein et al.[11] have used a mixture of platinum (Pt), iridium (Ir), and carbon (C) for scanning tunneling microscopy. They claimed (although no direct evidence was presented) that the Pt-Ir-C films were more stable and had smaller grain sizes than Pt-C or Ir-C films and that the tunneling current was less noisy, resulting in better resolution. Akahori et al.[25] have examined Pt, tungsten (W), tantalum (Ta), and molybdenum (Mo) films produced by electron beam evaporation and ion sputtering with electron microscopy and electron diffraction. The W and Ta films had the finest grain size as evidenced by TEM images and by electron diffraction patterns, but no differences were observed between films deposited by ion sputtering as opposed to electron beam heating. Unfortunately, the W and Ta films were very sensitive to chemical degradation during cleaning and mounting the replicas for TEM examination. Jaklevic et al.[26] used gold films to provide conductive surfaces for the STM and optimized the metal thickness to provide a smooth continuous layer. There are still no objective criteria for choosing one metal film over another; it is possible that the best film for one surface might not be the best for all surfaces.

Atomic resolution will probably be the result of new atomic force microscopes that can operate at much lower forces combined with more reliable force sensors. A certain level of physical rigidity is required for imaging with the AFM. Forces in the 10^{-10}-N range or less may be necessary to image fragile biological structures. It may be possible to use the AFM at low temperatures to image rapidly frozen biological surfaces to take advantage of the increased rigidity inherent in frozen materials. It is important to remember that atomic force microscopes are really only now approaching some degree of reliability in only a few research labs. A commercially available AFM is still a few years off. The biggest problems with the AFM are: (1) reproducibly measuring the small deflections of the springs, (2) producing springs with a high resonant frequency (to minimize the effect of vibrations) that are robust and large enough to locate with the measuring device, and (3) reproducibly making sharp tips and keeping them clean. Once these problems are substantially solved, the additional problems of sample preparation can be explored so as to make best use of the resolving power of the instrument. Still, the evolution of the AFM has been much more rapid than that of the STM, in part because of the experience gained in building and refining the STM. The evolution is driven, however, by the importance and need for creating instruments that can image nonconducting surfaces with atomic resolution.

ACKNOWLEDGMENTS

We thank O. Marti, V. Elings, and C. E. Bracker for helpful comments and discussions.

NOTES AND REFERENCES

1. BINNIG, G., H. ROHRER, CH. GERBER & E. WEIBEL. 1982. Surface studies by scanning tunneling microscopy. Phys. Rev. Lett. **49:** 57–61.
2. HANSMA, P. K. & J. TERSOFF. 1987. Scanning tunneling microscopy. J. Appl. Phys. **61:** R1–R23.
3. SONNENFELD, R. & P. K. HANSMA. 1987. Atomic resolution microscopy in water. Science **232:** 211–213.
4. HANSMA, P. K., V. B. ELINGS, O. MARTI & C. E. BRACKER. 1988. Scanning tunneling microscopy and atomic force microscopy: some applications to biology and technology. Science **242:** 209–216.

5. BINNIG, G. & H. ROHRER. 1983. In Trends in Physics. J. Janta & J. Panatoflicek, Eds.: 38. European Physical Society. Petit-Lancy, Switzerland.
6. BARÓ, A. M., R. MIRANDA, J. ALAMAN, B. GARCIA, G. BINNIG, H. ROHRER, CH. GERBER & J. L. CARRASCOSA. 1985. Determination of surface topography of biological specimens at high resolution by scanning tunneling microscopy. Nature (London) **315**: 253–254.
7. TRAVAGLINI, G., H. ROHRER, M. AMREIN & H. GROSS. 1987. Scanning tunneling microscopy on biological matter. Surf. Sci. **181**: 380–390.
8. STEMMER, A. O., R. REICHELT, A. ENGEL, J. P. ROSENBUSCH, M. RINGGER, H. R. HIDBER & H. J. GÜNTHERODT. 1987. Surf. Sci. **181**: 394–402.
9. DAHN, D. C., M. O. WATANABE, B. L. BLACKFORD, M. H. JERICHO & T. J. BEVERIDGE. 1988. Scanning tunneling microscopy imaging of biological structures. J. Vac. Sci. Technol. **A6**: 548–552.
10. LINDSAY, S. M. & B. BARRIS. 1988. Imaging deoxyribose nucleic acid molecules on a metal surface under water by scanning tunneling microscopy. J. Vac. Sci. Technol. **A6**: 544–547.
11. AMREIN, M., A. STASIAK, H. GROSS, E. STOLL & G. TRAVAGLINI. 1988. Scanning tunneling microscopy of recA-DNA complexes coated with a conducting film. Science **240**: 514–516.
12. ZASADZINSKI, J. A. N., J. SCHNEIR, J. GURLEY, V. ELINGS & P. K. HANSMA. 1988. Scanning tunneling microscopy of replicas of biomembranes. Science **239**: 1014–1016.
13. BINNIG, G., C. F. QUATE & CH. GERBER. 1986. Atomic force microscope. Phys. Rev. Lett. **56**: 930–933.
14. GOULD, S., O. MARTI, B. DRAKE, L. HELLEMANS, C. E. BRACKER, P. K. HANSMA, N. L. KEDER, M. M. EDDY & G. D. STUCKEY. 1988. Molecular resolution images of amino acid crystals with the atomic force microscope. Nature **332**: 332–334.
15. Figure by G. Binnig in reference 2.
16. GIAEVAR, I. 1974. Electron tunneling and superconductivity. Rev. Mod. Phys. **4**: 245–250.
17. FOSTER, J., J. E. FROMMER & P. C. ARNETT. 1988. Molecular manipulation using a tunneling microscope. Nature (London) **321**: 324–326.
18. LANG, C. A., J. K. H. HÖRBER, T. W. HÄNSCH, W. M. HECKL & H. MÖHWALD. 1988. Scanning tunneling microscopy of Langmuir-Blodgett phloems on graphite. J. Vac. Sci. Technol. **A6**: 368–370.
19. EGLEMAN, E. H. & A. STASIAK. 1986. Structure of helical recA-DNA complexes. J. Mol. Biol. **191**: 677–697.
20. GROSS, H. 1987. In Cryotechniques in Biological Electron Microscopy. R. A. Steinbrecht & K. Zierold, Eds: 205–234. Springer-Verlag. Berlin/New York.
21. ZASADZINSKI, J. A. N. & S. BAILEY. 1989. Applications of freeze-fracture replication to problems in materials and colloid science. J. Electron Microsc. Technol. **13**: 309–334.
22. MISELL, D. L. 1978. Practical Methods in Electron Microscopy. Volume 7. A. M. Glauert, Ed. North-Holland. Amsterdam.
23. ROBARDS, A. W. & V. B. SLEYTR. 1985. Practical Methods in Electron Microscopy. Volume 10. A. M. Glauert, Ed. Elsevier. Amsterdam/New York.
24. GROSS, H., T. MÜLLER, I. WILDHABER & H. WINKLER. 1985. High resolution metal replication quantified by image processing of periodic test specimens. Ultramicroscopy **16**: 287–304.
25. AKAHORI, H., Y. NAKAJIMA, K. TERASAWA, H. ISHII & I. NONAKA. 1986. High resolution freeze replica by means of high melting point metal shadowing. J. Electron Microsc. **35**: 202–207.
26. JAKLEVIC, R. C., L. ELIE, W. SHEN & J. T. CHEN. 1988. Application of the scanning tunneling microscope to insulating surfaces. J. Vac. Sci. Technol. **A6**: 448–453.

Uses of Fluorescence Sensors

For the Monitoring of Immobilized Cell Culture Fluorescence and as Optical Biosensors[a]

K-D. ANDERS,[b] W. MÜLLER,[b] T. SCHEPER,[b]
AND A. F. BÜCKMANN[c]

[b]Institut für Technische Chemie
Universität Hannover
D-3000 Hannover 1, Federal Republic of Germany

[c]Gesellschaft für Biotechnologische Forschung
D-3000 Braunschweig, Federal Republic of Germany

INTRODUCTION

During recent years, the use of miniaturized fluorescence probes for biotechnological processes has become increasingly important.[1-5] On-line and *in situ* monitoring of NAD(P)H-dependent fluorescence has primarily been used to obtain information about the metabolic state of cells or to estimate biomass concentration. Most studies have been performed on suspended cell cultures.

Two different applications of fiber optic–based fluorescence probes are presented in this report: the use for the monitoring of immobilized cells as well as its use as an optrode for an optical biosensor system.

IMMOBILIZED CELL EXPERIMENTS

FIGURE 1 shows the design of the measuring system for calcium alginate–immobilized cell experiments. Fluorescence was measured in the small, light-proof metal reactor, which was used in a packed-bed configuration for the immobilized cell experiments. This reactor was connected to the fermentor via a recirculation loop; the medium was pumped through the system at high flow rates.[5,6] An Ingold Fluorosensor (Ingold Meßtechnik, Switzerland) was interfaced to the metal reactor via a sensor port (FIGURE 1). *S. cerevisiae* was used in the experiments.

SUSPENDED AND IMMOBILIZED CELL CULTIVATIONS

The influences of immobilization on the cells were studied by monitoring the variations in the NADH-dependent culture fluorescence in suspended and immobilized cell cultivations.

---

[a]This project was supported in part by the DFG and the BMFT.

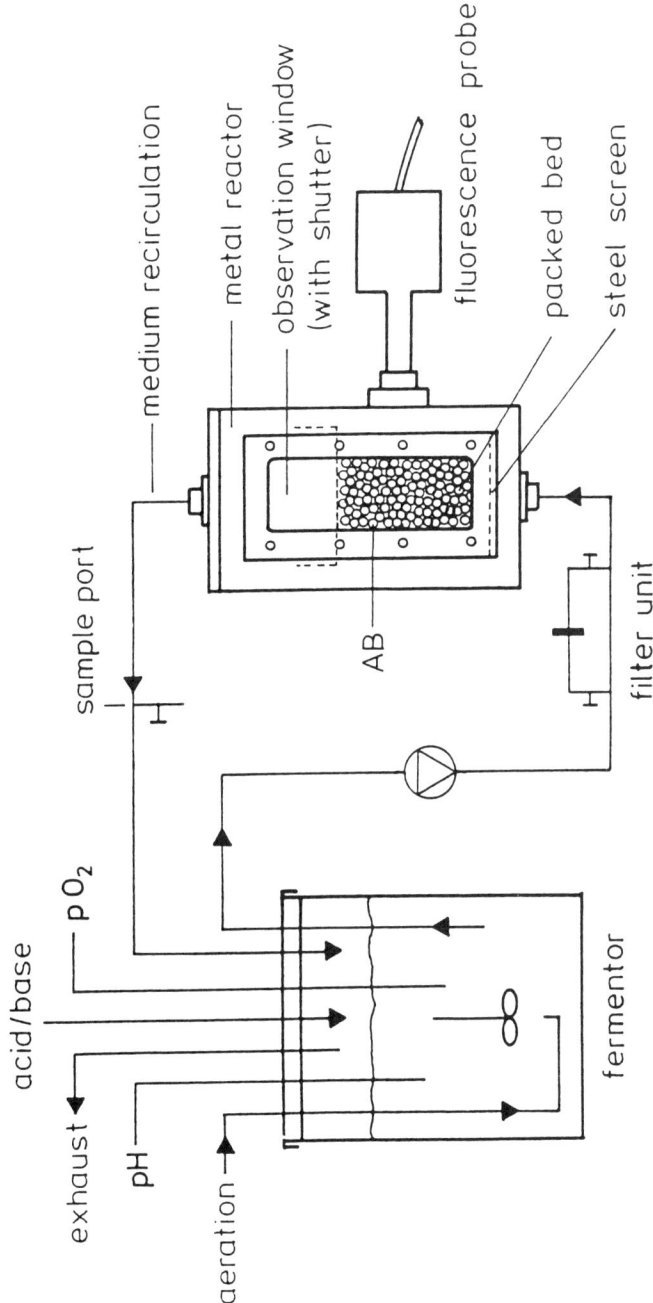

FIGURE 1. Experimental setup (AB = alginate beads).

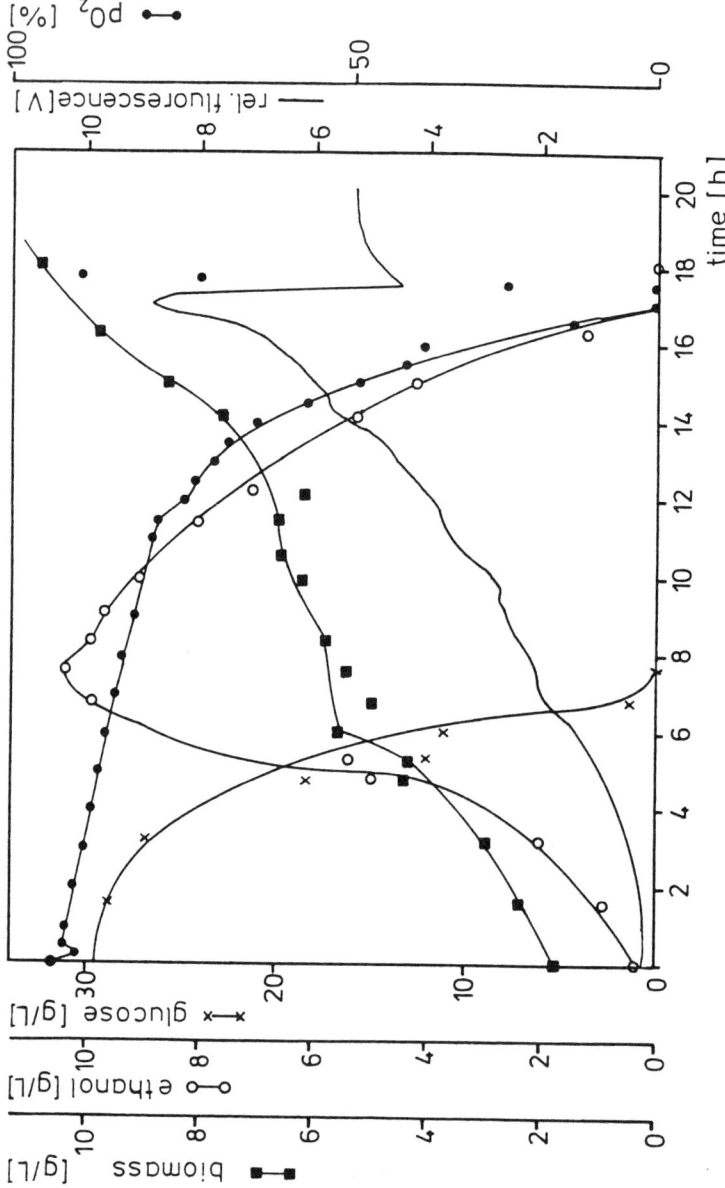

FIGURE 2. Suspended cell cultivation (defined medium with 3% glucose, according to Schatzmann[8]).

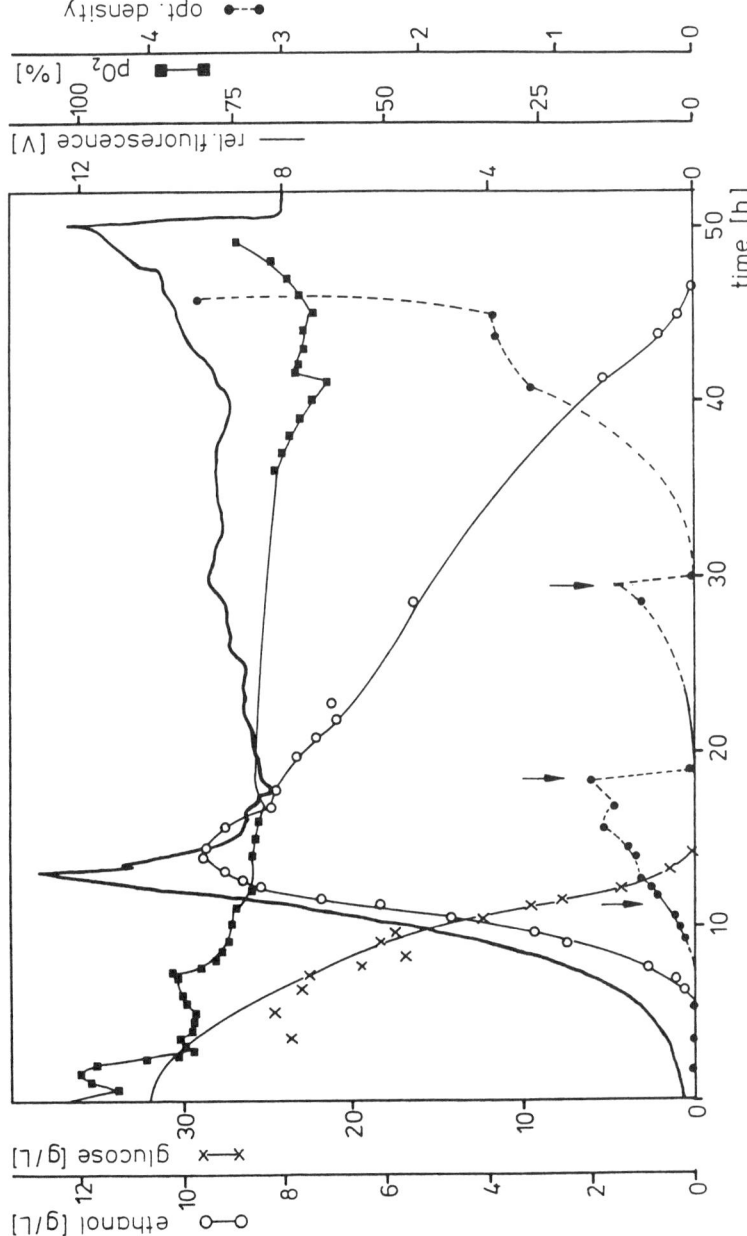

FIGURE 3. Immobilized cell cultivation (defined medium with 3% glucose, according to Schatzmann[8]).

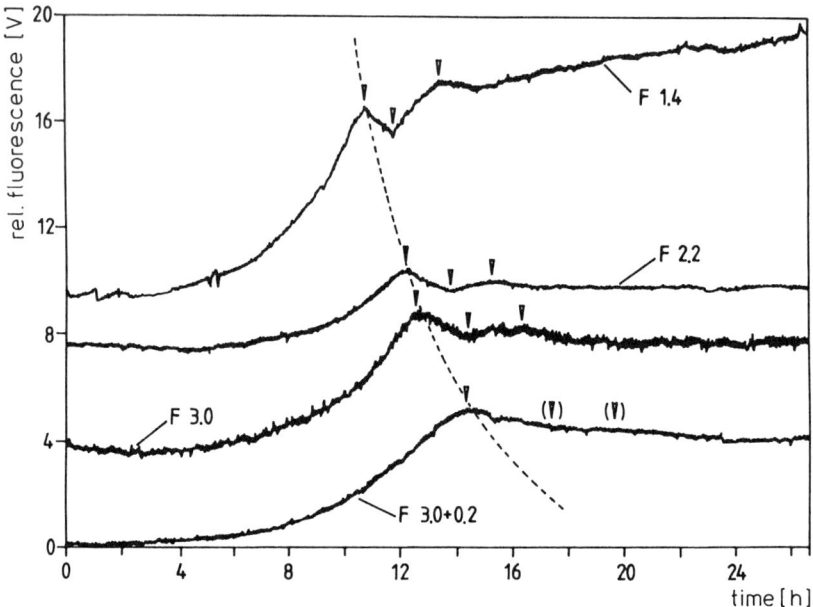

FIGURE 4. Variation of bead diameter (arrows indicate corresponding maxima, the dashed line connects maxima at fermentation times when glucose is consumed completely, and bead diameters are indicated by the symbol F). Alginate beads were prepared with defined diameter according to Vorlop and Klein.[9]

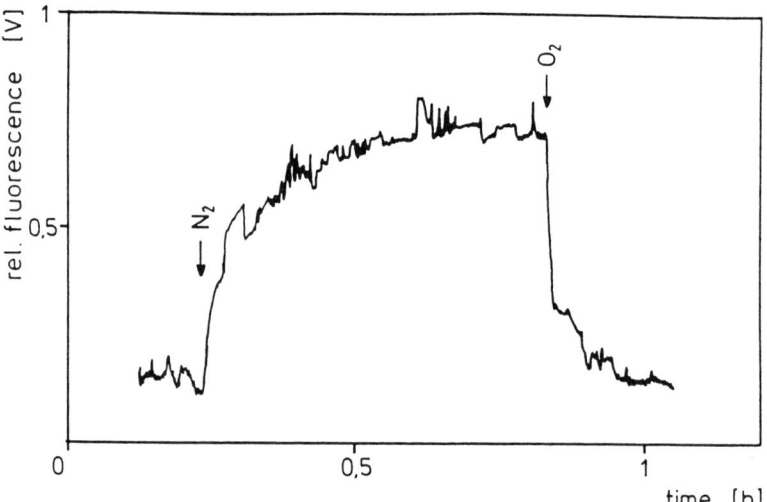

FIGURE 5. Aerobic/anaerobic transition experiments.

FIGURE 2 shows the results for an aerobic batch fermentation of *S. cerevisiae* and, in FIGURE 3, those from a cultivation with immobilized cells are presented. The background fluorescence signal was small and constant during all of these experiments. In the immobilized cell experiment, the effect of the suspended cells was negligible during the first 45 hours (OD data are given in FIGURE 3). The first parts of both cultivations were similar, except that the immobilized cells needed about six hours longer to consume all the glucose. One of the biggest differences is the fluorescence peak during the immobilized cell experiment that occurred shortly before the diauxic lag phase. Small undulations can be observed in the suspended cell experiment at this time. The much steeper signal decrease in the immobilized cell experiment may have been caused by oxygen-transfer limitations through the alginate beads. These limita-

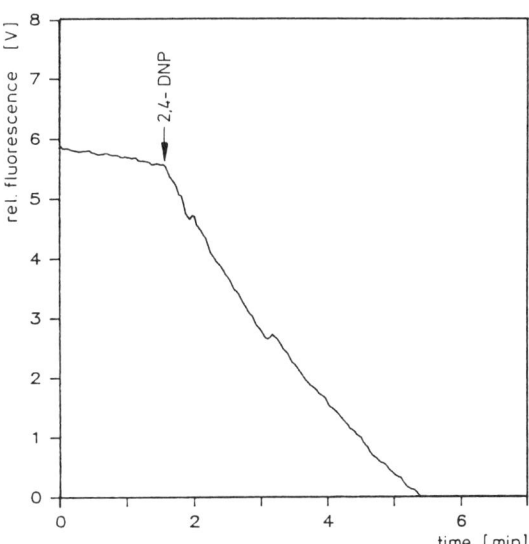

FIGURE 6. Pulse experiment with 2,4-dinitrophenol (addition of toxin is indicated by an arrow).

tions lower the ethanol consumption rate. The NADH pool is now lower than during the suspended cell cultivations. Although the ethanol consumption rate was slow, the end of the cultivation was also indicated by a steep signal decrease, just as in the suspended cell experiments.

These effects are more obvious in experiments with different bead diameters. The fluorescence data from these experiments are shown in FIGURE 4. Whereas the mean bead diameter for the studies in FIGURE 3 was 3.5 mm, the diameter was varied in the range from 1.4 to 3.0 mm for these experiments. In addition, 3.0-mm beads were coated with a 0.2-mm cell-free alginate layer to prevent the outgrowth of cells. All cultivation conditions were the same during these experiments (e.g., inoculum yeast concentration and cell mass at the end of all experiments). Only the surface viewed with the probe could not be kept constant. This results in slightly different offsets, slopes, and absolute maximum fluorescence intensities. The different maxima as a function of fermentation time are not affected. All curves in FIGURE 4 are offset from one another in order to make the differences more obvious. The results show that the diffusion limitation caused by the alginate barrier becomes larger with increasing bead diameter.

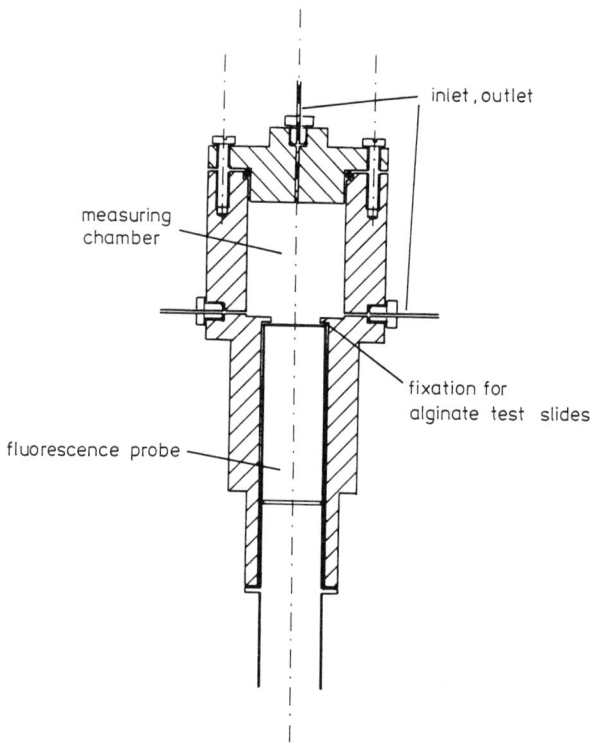

FIGURE 7. Flow-through cell for monitoring the dynamic behavior of immobilized cells to defined substrate or toxin pulses.

PULSE EXPERIMENTS

FIGURE 5 shows some aerobic/anaerobic transition experiments with immobilized cells in the measurement reactor. When the gas flow is switched from air to nitrogen, the NADH level increases because the coenzyme can no longer be oxidized through oxidative phosphorylation. When the reactor is aerated again, the NADH level decreases immediately.

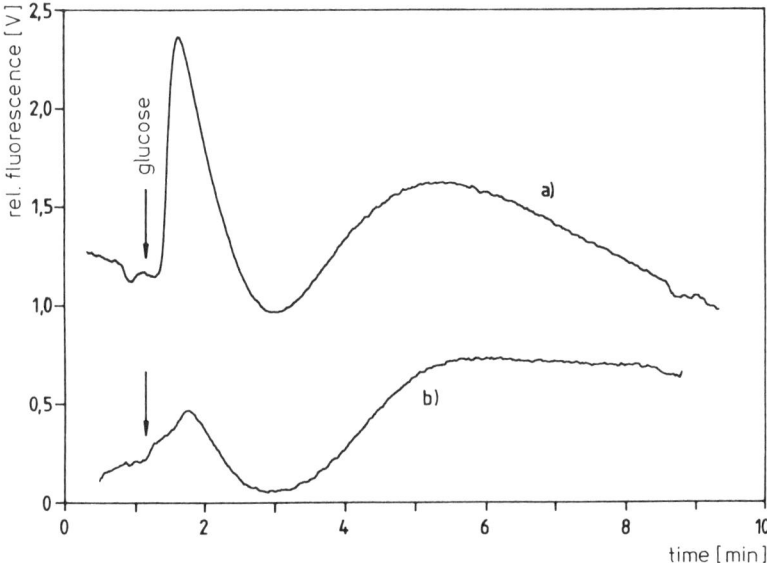

FIGURE 8. Pulse experiment with cyanide: (a) NADH after a glucose pulse; (b) cyanide was pulsed before this NADH signal to a glucose pulse was monitored.

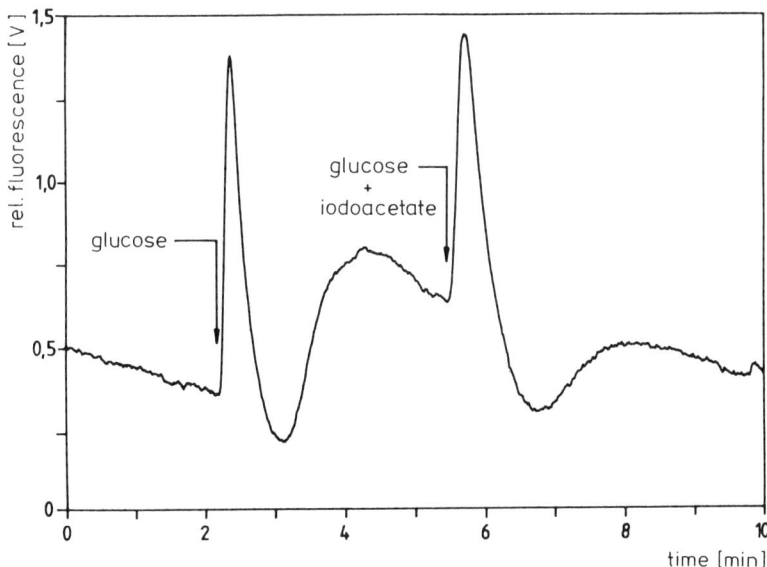

FIGURE 9. Pulse experiment with iodoacetate (glucose pulse prior to glucose/iodoacetate mixture pulse).

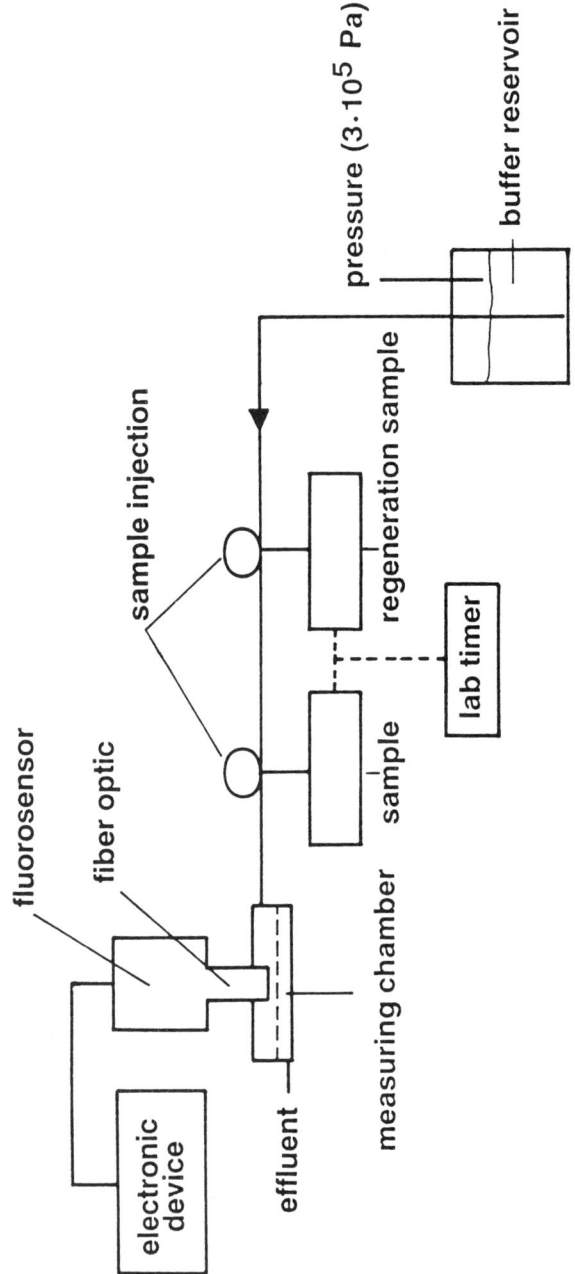

FIGURE 10. The analysis system for the optical biosensor.

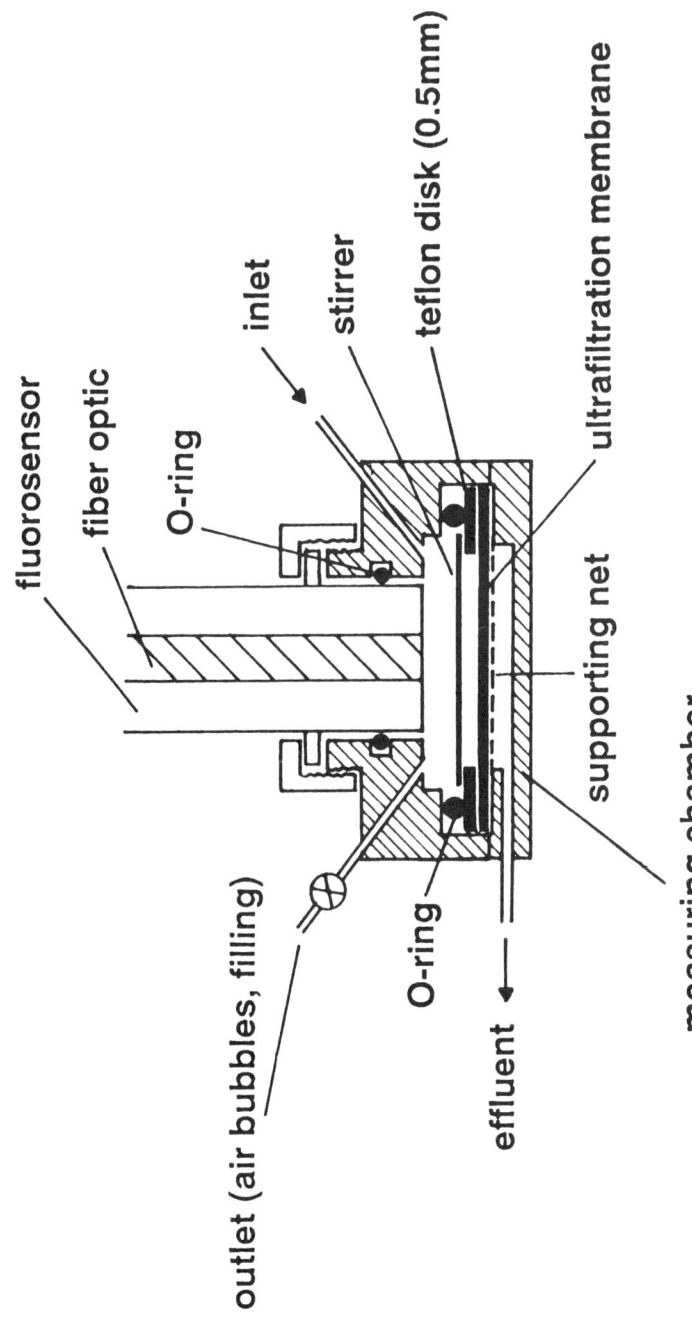

FIGURE 11. The measuring cell.

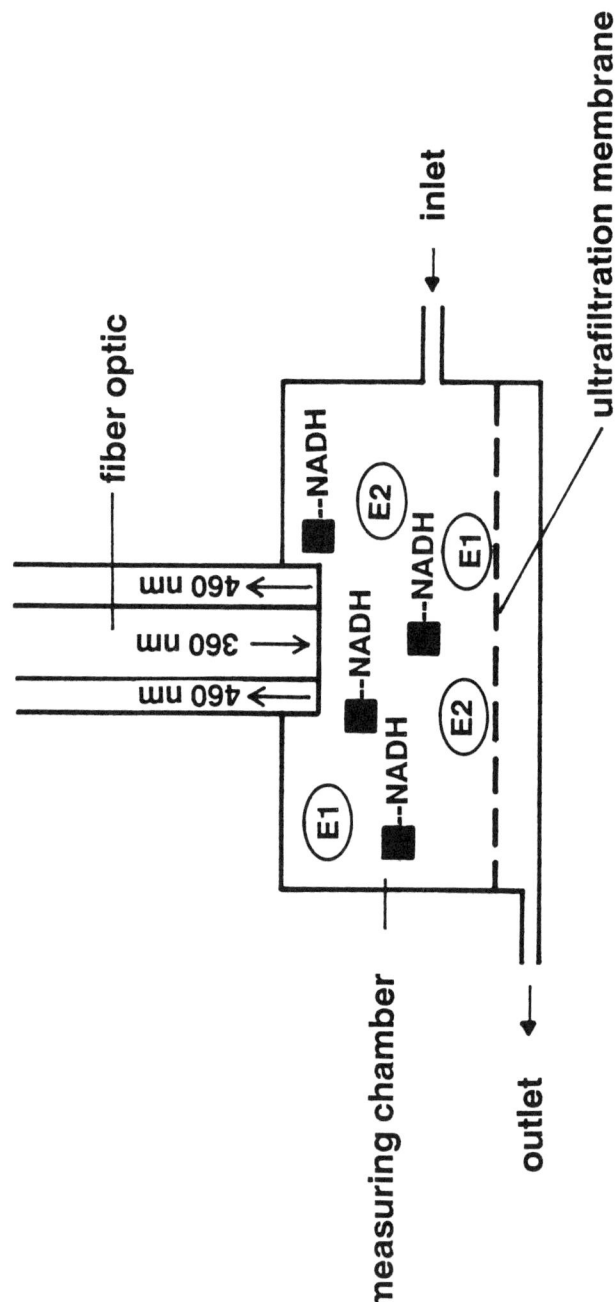

FIGURE 12. The principle of analysis (filled squares indicate macromolecular weight NADH).

FIGURE 6 shows a toxin-pulse experiment with 2,4-dinitrophenol, which was injected into the medium flow. This substance blocks the phosphorylation of ADP to ATP, but does not interrupt the respiratory chain. As a result, NADH oxidation is uncontrolled.

Large amounts of cell-loaded alginate beads are needed for the toxin-pulse experiments in the measurement reactor. Thin alginate layers loaded with cells or small volumes of alginate beads can be used in the flow-through cell shown in FIGURE 7. These layers are prepared on thin glass slides and can be fixed in front of the observation window. The NADH-dependent fluorescence of these test cells is monitored as substrate-free, oxygen-saturated medium is pumped continuously through the

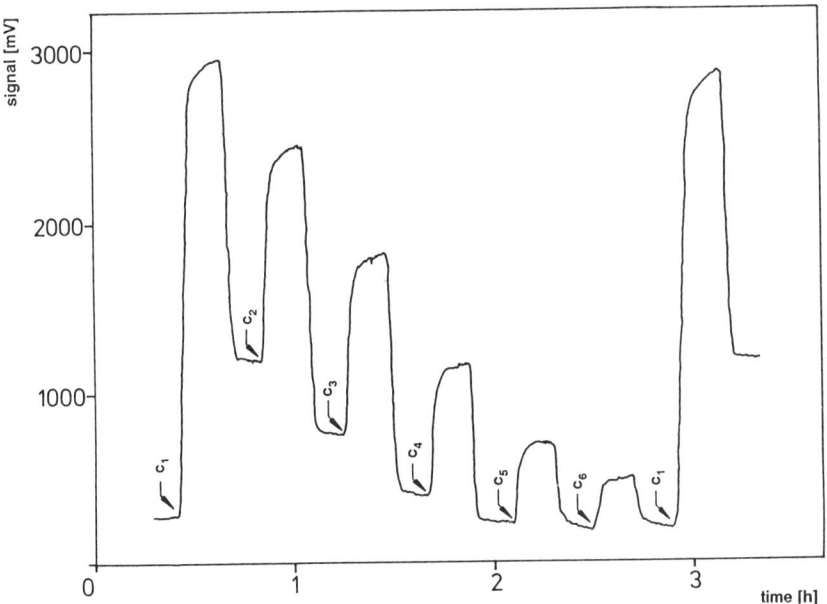

FIGURE 13. Lactate analysis ($c_1 = 5$ g/L; $c_2 = 2.5$ g/L; $c_3 = 1.25$ g/L; $c_4 = 0.625$ g/L; $c_5 = 0.3125$ g/L; $c_6 = 0.156$ g/L).

device. Samples can be injected into the buffer stream. The immobilized cells are in a starved state when substrate-free medium is pumped through the system. The contact time of glucose pulsed to the carrier stream is in the range of approximately 25 s.

The dynamic answer of these starved cells to glucose pulses is shown in FIGURE 8a. Substrate was pulsed to the carrier flow every 20 minutes. Two peaks can be observed: the first peak represents the changes in the cytoplasmic NADH pool and the second one represents those in the mitochondrial pool. The oscillatory behavior of yeast cells to glucose pulses is described by several authors. During our experiments, no changes in the dissolved oxygen concentrations were observed. We suggest that one

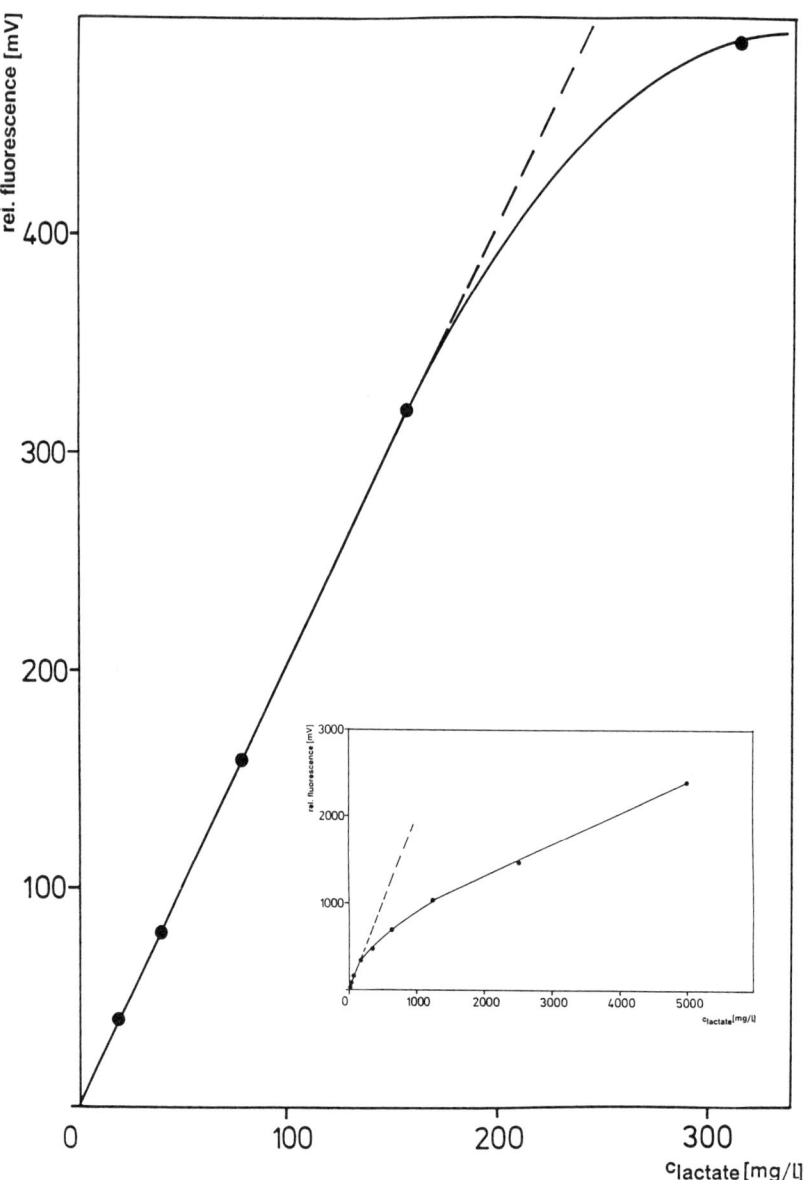

FIGURE 14. Lactate calibration curve (given for two different concentration ranges).

peak represents cytoplasmic NADH and the second one represents mitochondrial by the following experiments. When cyanide is injected before the second glucose pulse (FIGURE 8b), the fluorescence signal is completely different. Cyanide blocks the electron transfer in the respiratory chain, so NADH accumulates to a constant value. The respiratory chain is located inside the mitochondria, thereby causing the accumulative effect shown by the second peak. Cytoplasmic enzymes affected by cyanide lower the intensity of the first NADH peak.

Another toxin-pulse experiment is shown in FIGURE 9. A mixture of glucose and iodoacetate was injected after the first glucose pulse. The toxic effect of iodoacetate, which blocks several enzymes in glycolysis, is shown by the different fluorescence signal of the test cells.

OPTICAL BIOSENSOR

A new type of optical biosensor was developed, using an Ingold Fluorosensor, to monitor coenzyme-dependent enzymatic reactions.[7] The whole system is shown in FIGURE 10, the measuring chamber is shown in FIGURE 11, and the principle is shown in FIGURE 12. The enzymatic reaction occurs in a solid membrane reactor in which the enzymes, lactate dehydrogenase (LDH) and glutamic pyruvic transaminase (GPT), as well as PEG macromolecular NADH (MW = 20,000),[6] are entrapped by an ultrafiltration membrane (cutoff: 5000 daltons). The changes in the NADH content during the enzymatic reaction are monitored via the fiber optic of the fluorescence probe. The system is operated continuously at high pressure (3×10^5 Pa) using flow injection analysis. The buffer is pumped into the upper compartment of the measuring chamber. A special magnetic stirring device is used. The enzymatic reaction takes place in this compartment and all small reaction products leave the cell through the ultrafiltration membrane. The flow rate during the experiments was 0.18 mL/min. A 50 mM potassium phosphate buffer (pH = 8.5) was used in which the samples (0.3 mL) were injected. The coenzyme in the sensor had to be regenerated after each sample. The

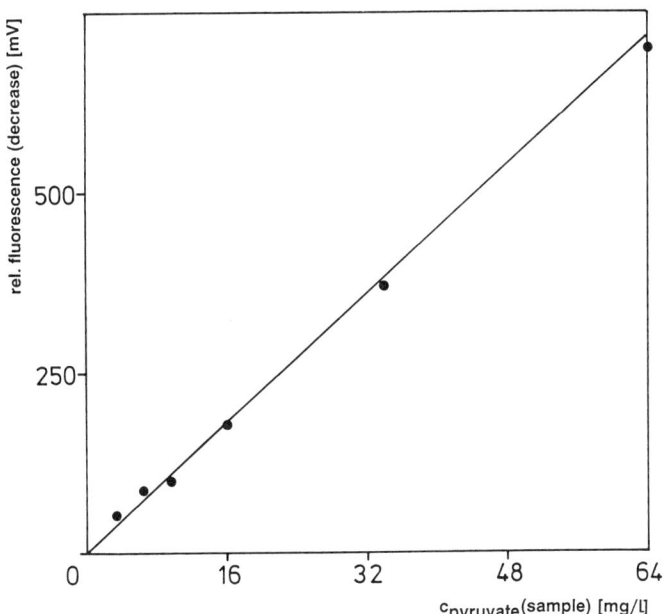

FIGURE 15. Pyruvate calibration curve (linear range).

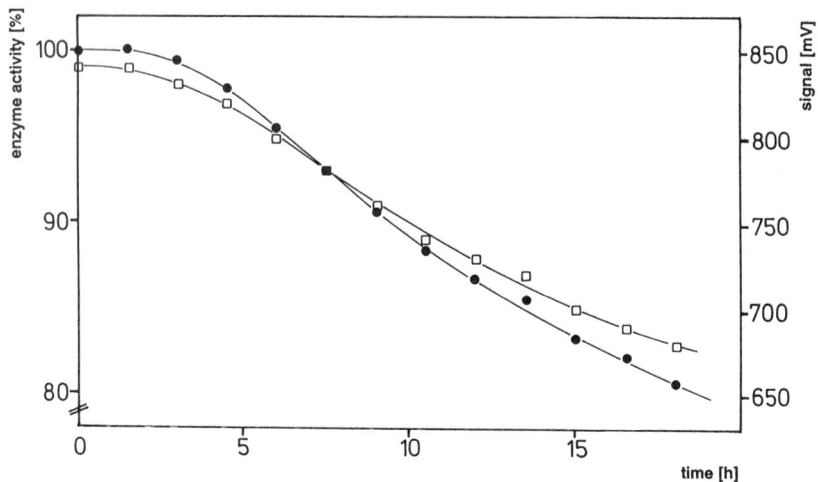

FIGURE 16. Long-term stability (●: enzyme activity; □: signal height).

regeneration solution was injected after each test sample or when the sensor capacity was exhausted.

The system was tested for the analysis of lactate and pyruvate. The principle is shown below:

$$\text{lactate} + \text{NAD}^+ \xrightleftharpoons{\text{LDH}} \text{pyruvate} + \text{NADH} + \text{H}^+$$

$$\text{pyruvate} + \text{L-glutamic acid} \xrightarrow{\text{GPT}} \text{L-alanine} + \alpha\text{-ketoglutamic acid.}$$

THE ANALYSIS

Lactate and pyruvate can be analyzed with this system. When lactate is to be determined, both the LDH and the GPT reaction are necessary. The equilibrium constant of the lactate/pyruvate reaction is very low, so the GPT reaction is needed to consume the produced pyruvate. Glutamic acid must be present in the lactate samples as substrate for the GPT reaction. During the LDH reaction, NADH is produced, which is monitored by the sensor. In order to regenerate the sensor, glutamic acid–free pyruvate samples are injected. Thus, NADH is oxidized while pyruvate is converted to lactate due to the equilibrium conditions—that is why the whole system can also be run for pyruvate analysis.

In FIGURE 13, the analysis of different lactate concentrations (every 20 minutes) is shown. A constant pyruvate concentration was injected for regeneration.

The calibration curve for lactate analysis is shown in FIGURE 14 and that for pyruvate analysis is presented in FIGURE 15. Long-term tests showed that the activity losses are low (at 30 °C). This activity loss and the effect on the signal height are plotted in FIGURE 16. The half-life of the enzyme is 60 hours, which is comparable to the native enzymes at 25 °C.

CONCLUSIONS

The results presented show that fiber optic fluorescence probes offer a large variety of applications. One application is the monitoring of culture fluorescence, especially of immobilized cells, to study cell metabolism and to detect toxic substances in the medium flow. The other application is an optical biosensor in which coenzyme-dependent enzymatic reactions are monitored for sensitive analysis. Here, the solid membrane reactor technique using molecular increased PEG coenzymes is coupled with the fiber optical fluorescence probe.

ACKNOWLEDGMENT

We gratefully acknowledge the advice of K-D. Vorlop on the immobilization procedures.

REFERENCES

1. ARMIGER, W. B., J. F. FORRO, L. M. MONTALVO & J. F. LEE. 1986. The interpretation of on-line process measurements of intracellular NADH in fermentation processes. Chem. Eng. Commun. **45:** 197.
2. BEYELER, W., A. EINSELE & A. FIECHTER. 1981. On-line measurements of culture fluorescence: method and application. Eur. J. Appl. Microbiol. Biotechnol. **13:** 10.
3. ZABRISKIE, D. W. & A. E. HUMPHREY. 1978. Estimation of fermentation biomass concentration by measuring culture fluorescence. Appl. Eur. Microbiol. Biotechnol. **35:** 337.
4. SCHEPER, T., T. LORENZ, W. SCHMIDT & K. SCHÜGERL. 1987. On-line measurement of culture fluorescence for process monitoring and control of biotechnological processes. Ann. N.Y. Acad. Sci. **506:** 431.
5. REARDON, K. F., T. SCHEPER & J. E. BAILEY. 1987. *In-situ* fluorescence monitoring of immobilized *Clostridium acetobutylicum*. Biotechnol. Lett. **8**(11): 817–822.
6. REARDON, K. F., T. SCHEPER & J. E. BAILEY. 1987. Metabolic pathway rates and culture fluorescence in batch fermentations of *C. acetobutylicum*. Biotechnol. Prog. **3:** 153.
7. BÜCKMANN, A. F. 1987. A new synthesis of coenzymatically active water soluble macromolecular NAD^+ and $NADP^+$ derivatives. Biocatalysis **1:** 1783–1786.
8. SCHATZMANN, H. 1975. Anaerobes Wachtums von *Saccharomyces cerevisiae*. Ph.D. thesis no. 5504. ETH Zürich, Switzerland.
9. VORLOP, K-D. & J. KLEIN. 1981. New developments in the field of cell immobilization—formation of biocatalysts by ionotropic gelation.

PART VIII. BIOPROCESS OPTIMIZATION AND CONTROL

Multivariable Control of Continuous and Fed-Batch Bioreactors[a]

DANIEL WEI,[b] SATISH J. PARULEKAR, AND
WILLIAM A. WEIGAND[c]

*Department of Chemical Engineering
Illinois Institute of Technology
Chicago, Illinois 60616*

INTRODUCTION

In recent years, there has been considerable interest in developing process control strategies for multivariable control problems, that is, problems where multiple process variables are to be controlled and where multiple variables can be manipulated. For large-scale commercial processes such as cell mass and primary metabolite production processes as well as for laboratory-scale operations for modeling microbial growth, there is a strong incentive to develop efficient control schemes that would enable rapid start-up and stabilization of the stationary states in continuous bioreactors and the desired states in fed-batch bioreactors, given the slow dynamics usually associated with microbial growth and the risk of contamination that accompanies it. The design and operation of multivariable control schemes is more difficult compared to the design of single-input/single-output control schemes due to interactions among various manipulated and controlled variables. In designing a multivariable control system, one of the key considerations is the determination of the proper pairing of controlled and manipulated variables to minimize interactions. Previous attempts at analyzing the multivariable control of fermentation processes have included the application of static analysis to determine the controller pairing for a baker's yeast fermentation[1] and the proposal of a control scheme with dilution rate and substrate feed concentration as manipulated variables to control cell mass concentration.[2] The latter multivariable control scheme has been shown to be superior to single-variable control schemes.[2]

Large disturbances in the manipulated variables occur during switchover from batch operation to fed-batch/continuous operation and during step changes in dilution rate and/or feed concentration of the limiting substrate to estimate the kinetics of these processes at the laboratory scale. In such situations, it is the dynamic interactions rather than the steady-state interactions that must be considered in the determination of the controller configuration. In this report, we have investigated this problem using both static and dynamic analyses. As examples, we have considered continuous and fed-batch fermentation processes where the specific rates for key processes such as cell

---

[a]Partial support of this research was provided by the Amoco Foundation.
[b]Present address: UOP Incorporated, Riverside, Illinois 60546.
[c]Present address: Department of Chemical and Nuclear Engineering, University of Maryland, College Park, Maryland 20742.

growth, substrate consumption, and metabolite production are dependent solely on the concentration of the limiting substrate. The effects of dynamic interactions are illustrated by numerical simulations and experimental results.

THEORETICAL ANALYSIS

Problem Formulation

Because we consider here bioprocesses whose specific kinetics can be expressed in terms of the concentration of the limiting substrate (S), the material balances for cell mass and limiting substrate are sufficient to describe the dynamics of the bioreactor when operated in either continuous mode or fed-batch mode. These are

$$\frac{dX}{dt} = \mu(S)X - DX \tag{1}$$

and

$$\frac{dS}{dt} = D(S_F - S) - \sigma(S)X, \tag{2}$$

where X, D, S_F, μ, and σ denote cell mass concentration, dilution rate, substrate feed concentration, specific cell growth rate, and specific substrate consumption rate, respectively. The dilution rate D is the ratio of the volumetric feed rate F to the bioreactor volume V. The bioreactor volume is time-invariant in a continuous operation; in contrast, for a fed-batch operation, it varies with time as

$$\frac{dV}{dt} = F, \qquad F = DV. \tag{3}$$

Because V does not appear explicitly in equations 1 and 2, the conclusions reached by analyzing these equations are applicable to both continuous and fed-batch operations of bioreactors. The two controlled variables here are the concentrations of cell mass and limiting substrate in the bioreactor and the two manipulated variables are dilution rate and substrate feed concentration, with the objective of the multivariable control scheme being to attain a desired stationary state in continuous operation or a quasi-steady state in fed-batch operation as early as possible. In a quasi-steady state of a fed-batch operation, the concentrations of cell mass and limiting substrate become time-invariant, although the reactor volume increases with time. To determine the controller configuration that accomplishes this objective, it is necessary to linearize equations 1 and 2 around the desired stationary or quasi-stationary state. The linearized version of the process model is obtained by subtracting the material balances for cell mass and limiting substrate at the desired state from those for a transient situation (viz., equations 1 and 2):

$$\frac{dx_1}{dt} = \mu'_o X_o x_2 - X_o m_1 \tag{4}$$

and

$$\frac{dx_2}{dt} = -\sigma_o x_1 - (X_o\sigma'_o + D_o)x_2 + (S_{Fo} - S_o)m_1 + D_o m_2, \qquad (5)$$

where

$$x_1 = X - X_o, \qquad x_2 = S - S_o, \qquad m_1 = D - D_o, \qquad m_2 = S_F - S_{Fo}. \qquad (6)$$

In the equations above, the subscript "o" refers to conditions at the desired (reference) stationary state or quasi-stationary state and the prime denotes the derivative with respect to the concentration of the limiting substrate.

Static Analysis

The asymptotic relations among the deviations in the manipulated variables (m_1 and m_2) and the deviations in the controlled variables (x_1 and x_2) as the bioreactor state approaches the desired stationary or pseudostationary state are obtained from the time-invariant form of equations 4 and 5 as

$$\begin{bmatrix} x_1 \\ x_2 \end{bmatrix} = \begin{bmatrix} e_{11} & e_{12} \\ e_{21} & e_{22} \end{bmatrix} \begin{bmatrix} m_1 \\ m_2 \end{bmatrix}, \qquad (7)$$

where

$$\left. \begin{array}{l} e_{11} = [(S_{Fo} - S_o)\mu'_o - X_o\sigma'_o - D_o]/(\sigma_o\mu'_o), \\ e_{12} = \dfrac{D_o}{\sigma_o}, \qquad e_{21} = \dfrac{1}{\mu'_o}, \qquad e_{22} = 0 \end{array} \right\} \qquad (8)$$

The coefficients, e_{ij} ($i,j = 1,2$), are the so-called steady-state gains. Because $e_{22} = 0$ for the class of fermentation processes under consideration, the substrate concentration in the bioreactor is not influenced by its value in the feed and we have a one-way interaction. The control strategy should therefore be to use the dilution rate to control substrate concentration and to use the substrate feed concentration to control cell concentration. Boyle[1] reached the same conclusions.

The steady-state and dynamic gains can be more conveniently determined by considering the open-loop transfer function of the process to be controlled. The following relations among the two controlled variables and the two manipulated variables are obtained in the Laplace domain by taking the Laplace transform of the linearized process model:

$$\begin{bmatrix} \bar{x}_1 \\ \bar{x}_2 \end{bmatrix} = \begin{bmatrix} G_{11} & G_{12} \\ G_{21} & G_{22} \end{bmatrix} \begin{bmatrix} \bar{m}_1 \\ \bar{m}_2 \end{bmatrix}; \qquad G_{ij}(s) = \frac{g_{ij}(s)}{\Delta}, \qquad i,j = 1,2, \qquad (9)$$

where

$$g_{11} = -X_o[s + X_o\sigma'_o + D_o - (S_{Fo} - S_o)\mu'_o], \qquad g_{12} = X_o D_o \mu'_o, \qquad (10a)$$

$$g_{21} = (S_{Fo} - S_o)(s + D_o), \qquad g_{22} = D_o s, \qquad (10b)$$

$$\Delta = (s + \alpha_1)(s + \alpha_2), \tag{10c}$$

$$\alpha_1 = [X_o\sigma'_o + D_o + \sqrt{\beta}]/2, \tag{10d}$$

$$\alpha_2 = [X_o\sigma'_o + D_o - \sqrt{\beta}]/2, \tag{10e}$$

and

$$\beta = [X_o\sigma'_o + D_o]^2 - 4D_o(S_{Fo} - S_o)\mu'_o. \tag{10f}$$

The coefficient matrix $\underline{G}(s)$ [$\underline{G}(s) \equiv \{G_{ij}(s), i = 1,2, j = 1,2\}$] in equations 9 and 10 is the transfer function of the process under consideration. The dynamics of the fermentation process described by equations 1 and 2 is decided by the zeros of the function Δ, namely, $s = -\alpha_1$ and $-\alpha_2$.

The most widely used measure of the degree of steady-state interaction is Bristol's relative gain array,[3] whose elements are defined as the ratio of the steady-state open-loop response to the steady-state closed-loop response when a particular manipulated variable is adjusted. The elements of the Bristol array (λ_{ij}) are related to the elements of the steady-state process transfer function as shown below:[3,4]

$$\lambda_{ij} = [\underline{G}(0)]_{ij}\{[\underline{G}^T(0)]^{-1}\}_{ij}, \quad \underline{\lambda} = \underline{G}(0)\{\underline{G}^T(0)\}^{-1}, \tag{11a}$$

$$\lim_{s \to 0} G_{ij}(s) = G_{ij}(0) = e_{ij}, \quad i, j = 1,2. \tag{11b}$$

The sum of the elements in any row or any column in the Bristol array is unity.

For a system involving two controlled variables and two manipulated variables, $\lambda_{11} = \lambda_{22}$ and $\lambda_{12} = \lambda_{21}$. The identification of the Bristol array therefore requires that only a pair of diagonal and off-diagonal elements (say, λ_{11} and λ_{12}) be calculated. If the diagonal elements are larger than the off-diagonal elements, then the recommended pairing is x_1 (X) with m_1 (D) and x_2 (S) with m_2 (S_F). Otherwise, the recommended pairing is x_1 with m_2 and x_2 with m_1. The diagonal element of the relative gain array, λ_{11}, in this case is expressed in terms of steady-state process gains as

$$\lambda_{11} = \frac{e_{11}e_{22}}{e_{11}e_{22} - e_{12}e_{21}}, \quad \lambda_{12} = 1 - \lambda_{11}. \tag{12}$$

For the fermentation processes under consideration, $\lambda_{11} = 0$ and $\lambda_{12} = 1$. Because the diagonal elements are smaller than the off-diagonal ones, D should be used to control S and S_F should be used to control X. Notice that the controller pairing recommended by the Bristol relative gain array is consistent with the conclusions reached by considering steady-state gains.

Dynamic Analysis

A number of procedures based on the process dynamics are available in the literature for determining the multivariable controller configuration. These include methods based on average dynamic gain array,[5] relative dynamic gain array,[6] singular value analysis,[7] and direct Nyquist array.[8] The analysis presented here is based on the method of average dynamic gain array due to Gagnepain and Seborg[5] because, unlike other methods, this method does not require a complete dynamic model for controller pairing to be determined. Furthermore, an approximate determination of the average dynamic gains requires only a few experiments.

The average dynamic gains D_{ij} are defined as the ratio of the average change in a controlled variable x_i to the average change in a manipulated variable m_j. The average changes in the controlled variables x_i can be determined from the process dynamics over a time interval θ; a recommended value for this is the largest time constant in the open-loop transfer function for the process.[5] For a unit step change in the manipulated variable m_j, the average dynamic gain D_{ij} for the controlled variable x_i is defined as

$$D_{ij} = \frac{1}{\theta} \int_0^\theta x_i(t)\, dt. \tag{13}$$

To calculate the average dynamic gains, we consider step changes in each of the manipulated variables and obtain expressions for the Laplace transforms of the deviations in controlled variables $\bar{x}_1$ and $\bar{x}_2$. Inversion of these furnishes the profiles of the deviations in controlled variables, which can be used to calculate the average dynamic gains. The calculation of the dynamic gains requires knowledge of the variations in control variables x_1 and x_2 with time. The process dynamics is decided by the roots of the characteristic equation for the open-loop transfer function (poles of the open-loop transfer function, $-\alpha_1$ and $-\alpha_2$). Equations 10d and 10e indicate that α_1 and α_2 are complex conjugates if β is negative. The largest time constant θ is thus defined as

$$\theta = \begin{cases} 1/\alpha_2 & \text{if} \quad \beta > 0 \\ 1/\text{Re}(\alpha_2) & \text{if} \quad \beta < 0, \end{cases} \tag{14}$$

where $\text{Re}(\alpha_2)$ denotes the real part of α_2. The relations among the elements of the relative dynamic array and the average dynamic gains are similar to the relations among the elements of the Bristol relative gain array and the steady-state gains. The relations are

$$\phi_{ij} = [\underline{D}]_{ij} [\{\underline{D}^T\}^{-1}]_{ij}, \quad i,j = 1,2; \quad \underline{\phi} = \underline{D}\{\underline{D}^T\}^{-1}. \tag{15}$$

The sum of the elements in any row or any column in the relative dynamic array is unity.

For a system involving two controlled variables and two manipulated variables, $\phi_{11} = \phi_{22}$ and $\phi_{12} = \phi_{21}$. The identification of the relative dynamic array therefore requires that only a diagonal element and an off-diagonal element (say, ϕ_{11} and ϕ_{12}) be calculated. If the diagonal elements are larger than the off-diagonal elements, then the recommended pairing is x_1 (X) with m_1 (D) and x_2 (S) with m_2 (S_F). Otherwise, the recommended pairing is X with S_F and S with D. The diagonal element of the relative dynamic array, ϕ_{11}, can be expressed in terms of the average dynamic gains as

$$\phi_{11} = \frac{D_{11}D_{22}}{D_{11}D_{22} - D_{12}D_{21}}, \quad \phi_{12} = 1 - \phi_{11}. \tag{16}$$

In special situations of very low and very high dilution rates, the relative dynamic array can be identified analytically without placing any restrictions on the exact forms of the specific kinetics of cell growth and substrate consumption. At dilution rates approaching the maximum specific cell growth rate (very high dilution rates), the relative dynamic gains approach the values of

$$\lim_{D \to \mu_{max}} \phi_{11} = 1 - e^{-1} \simeq 0.63, \qquad \lim_{D \to \mu_{max}} \phi_{12} = e^{-1} \simeq 0.37. \qquad (17)$$

The recommended pairings at very high dilution rates are thus cell mass concentration with dilution rate and substrate concentration with that in the feed. It must be noticed that these pairings are different from those suggested by the Bristol relative gain array method.

The controller pairings at very low dilution rates are dependent on the nature of the specific substrate consumption kinetics at such dilution rates. The process behavior in this situation is classified into two types for convenience: (a) $\sigma_o \to 0$ as $\mu_o \to 0$ and (b) $\sigma_o \not\to 0$ as $\mu_o \to 0$. In general, very low specific cell growth rates are encountered at low substrate concentrations. (In the case of substrate inhibition kinetics, large substrate concentrations lead to very low specific cell growth rates. The operation at such large concentrations is not attractive as far as productivity is concerned. Such situations are therefore not considered here.) From the material balance for the limiting substrate (equation 2), the following are evident at the desired steady state or quasi-steady state for $\mu_o \to 0$: (a) $X_o \not\to 0$ if $\sigma_o \to 0$ and (b) $X_o \to 0$ if $\sigma_o \not\to 0$.

When both the specific cell growth rate and the specific substrate consumption rate vanish at very low substrate concentration (case a), the interaction measure, ϕ_{11}, can be expressed as

$$\phi_{11} = -\frac{\sigma_o^2}{\mu_o \sigma_o'} Y_o', \qquad (18a)$$

with the cell mass yield (Y) being defined as

$$Y = \frac{\mu}{\sigma}. \qquad (18b)$$

Because Y_o' must be finite for $S_o \to 0$, it can be deduced from equation 18a that ϕ_{11} tends to zero as the substrate concentration becomes negligible. In the present case, both the specific cell growth rate and the specific substrate consumption rate increase with increasing substrate concentration at very low values of the same ($\mu_o' > 0$, $\sigma_o' > 0$ for $S \to 0$). Hence, at very low substrate concentrations, ϕ_{11} is (i) positive if the biomass yield Y decreases with increasing S or (ii) negative if Y increases with increasing S. In the first situation, the controller pairing suggested by the relative dynamic array method is to pair x_1 (X) with m_2 (S_F) and x_2 (S) with m_1 (D) because the diagonal elements of the relative dynamic array ($\phi_{11} = \phi_{22}$) are much smaller than the off-diagonal elements ($\phi_{12} = \phi_{21}$). The controller pairings suggested by the average dynamic gain array method in this situation are the same as those suggested by the relative gain array method. In the second situation, ϕ_{11} and ϕ_{22} are negative, signifying negative interaction. This implies that changing the control variables in the closed loop has just the opposite effect from changing these in the open loop.[4] The controller pairings in the event of negative interaction cannot be decided in a simple manner using methods such as the average dynamic gain array method.[5]

When both the specific cell growth rate and the cell mass concentration vanish at very low substrate concentration (case b), equation 10f can be rearranged as

$$\frac{\beta}{D_0^2} = \psi^2 - \frac{4(S_{Fo} - S_o)}{D_o}\mu_0', \qquad \psi = 1 + \frac{(S_{Fo} - S_o)}{\sigma_o}\sigma_0'. \tag{19}$$

It is evident at very low substrate concentrations ($S_o \to 0$, $D_o \to 0$) that β is negative (α_1 and α_2 are complex conjugates), with the largest time constant θ (defined in equation 14) being

$$\theta = \frac{2}{\psi D_o} \tag{20}$$

provided $\psi > 0$. The expression for the dynamic interaction measure, ϕ_{11}, in this case has the form,

$$\phi_{11} = p/(p - q), \tag{21a}$$

where

$$p = [\eta - e^{-1}\{\phi \sin(\rho) + \eta \cos(\rho)\}] [\{X_o\sigma_0' + D_o - (S_{Fo} - S_o)\mu_0'\}$$
$$\times \{-\eta D_o(S_{Fo} - S_o)\mu_0' + \phi\eta e^{-1}(\eta \sin(\rho) - \phi \cos(\rho)) + \phi^2\eta\}$$
$$+ \{\phi(X_o\sigma_0' + D_o - (S_{Fo} - S_o)\mu_0')$$
$$- D_o(S_{Fo} - S_o)\mu_0'\}\{\phi\eta - \phi^2 e^{-1}\sin(\rho) - \phi\eta e^{-1}\cos(\rho)\}],$$

$$q = [\eta D_o(S_{Fo} - S_o)\mu_0' - 2\phi^2\eta - \phi(\eta^2 - \phi^2)e^{-1}\sin(\rho) + 2\phi^2\eta e^{-1}\cos(\rho)]$$
$$\times [\eta D_o(S_{Fo} - S_o)\mu_0' + \phi\{(S_{Fo} - S_o)\mu_0' - \phi\}$$
$$\times \{\eta - \phi e^{-1}\sin(\rho) - \eta e^{-1}\cos(\rho)\} - \phi\eta\{\phi - \phi e^{-1}\cos(\rho) + \eta e^{-1}\sin(\rho)\}],$$

$$\phi = \tfrac{1}{2}[X_o\sigma_0' + D_o], \qquad \eta = \tfrac{1}{2}\sqrt{-\beta}, \qquad \text{and} \qquad \rho = \eta/\phi. \tag{21b}$$

The complex nature of the above expressions does not permit estimation of ϕ_{11} without specifying the exact forms of cell growth and substrate consumption kinetics. In general, the controller pairings for intermediate dilution rates must be decided by numerical evaluation of the relative dynamic array.

RESULTS OF NUMERICAL SIMULATIONS

The first example considered for numerical simulations is a microbial process where the cell growth kinetics is described by the Monod equation and the substrate consumption rate is linearly proportional to the cell growth rate, that is,

$$\mu = \frac{\mu_m S}{K_s + S}, \qquad \sigma = \frac{\mu}{Y}. \tag{22}$$

This example belongs to case a ($\sigma \rightarrow 0$ as $\mu \rightarrow 0$) discussed earlier. The parameters used for numerical simulations are[9]

$$\mu_m = 1.0 \text{ h}^{-1}, \quad K_s = 0.2 \text{ g/L}, \quad Y = 0.5 \frac{\text{g cell mass}}{\text{g limiting substrate}}. \quad (23)$$

The profile of ϕ_{11} for this example (see FIGURE 1) indicates that the controller pairing below a critical dilution rate D_1 ($\phi_{11} = 0.5$ at $D = D_1$) should be X with S_F and S with D because ϕ_{11} ($=\phi_{22}$) $< \phi_{12}$ ($=\phi_{21}$). Above the critical dilution rate, X should be paired with D and S with S_F because $\phi_{11} > \phi_{12}$. The most significant difference in the average dynamic gains at low and high dilution rates was observed for the gain between substrate concentration in the bioreactor (S) and that in the feed (S_F). For example,

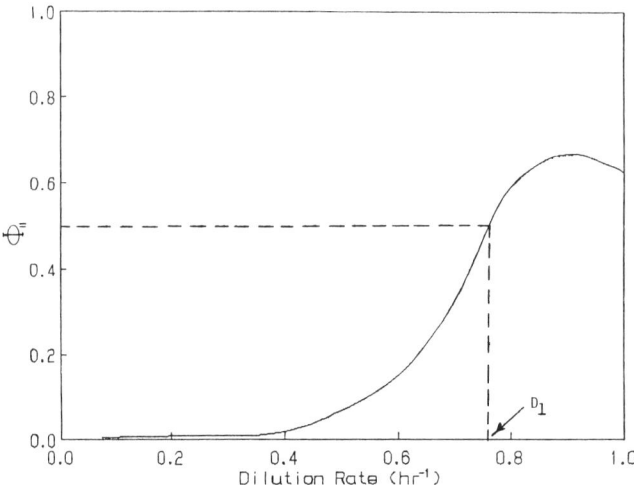

FIGURE 1. A plot of the relative dynamic array element, ϕ_{11}, versus the dilution rate for Monod growth kinetics with constant biomass-to-substrate yield.

D_{22} values for the dilution rates of 0.1 and 0.9 h^{-1} were 0.002 and 0.47, respectively. At low dilution rates, substrate concentration in the bioreactor will be insensitive to variation in substrate feed concentration.

Further illustration of this is provided by the responses of substrate concentration in the bioreactor to a step change in substrate feed concentration in a continuous culture operating at a steady state (see FIGURE 2). Over the wide range of dilution rates considered, it must be observed that the eventual substrate concentration is the same as that before introduction of the step change in S_F. This is in agreement with the conclusions reached earlier via static analysis. The profiles in FIGURE 2 also indicate that substrate concentration in the bioreactor at low dilution rates is insensitive to variation in substrate feed concentration at all times. The controller pairings suggested by dynamic analysis and static analysis are therefore identical at low dilution rates.

Unlike the situation at large times, substrate concentration in the bioreactor is altered significantly for the first few hours after the step change in (and is thus very sensitive to) substrate feed concentration at high dilution rates. The suggestions of the average dynamic gain array method, which is based on process transients during the first few hours, for controller pairings are consequently different than those based on the relative gain array method.

Further verification of the predictions of the average dynamic gain array method was obtained from closed-loop simulations. The closed-loop responses of the process subject to step changes in set points were obtained using two proportional controllers for two different controller pairings. For each controller pairing, the controllers were

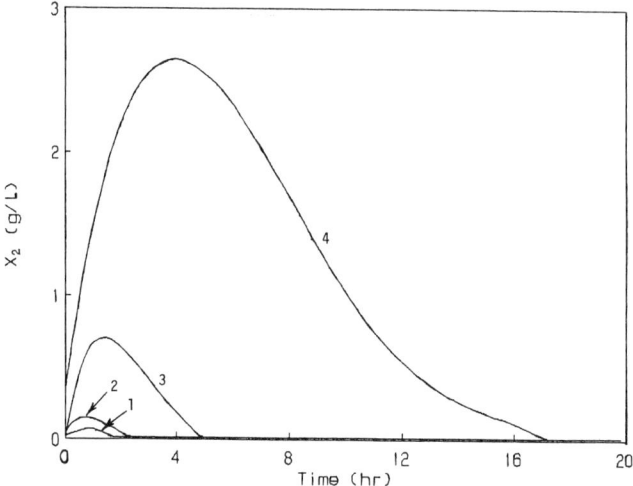

FIGURE 2. Responses of substrate concentration to a fixed step change in substrate feed concentration at various dilution rates for Monod growth kinetics with constant biomass-to-substrate yield. Curve nos. 1, 2, 3, and 4 represent simulated responses for dilution rates (h^{-1}) of 0.1, 0.5, 0.7, and 0.9, respectively.

tuned so as to minimize the sum of the integral square errors (ISE) for step changes in R_1 and R_2 (set points for cell mass concentration and substrate concentration, respectively):

$$\text{ISE} = \int_0^\infty [\{e_1(t)\}^2 + \{e_2(t)\}^2]\, dt, \quad e_1 = R_1 - X, \quad e_2 = R_2 - S. \quad (24)$$

Results of simulations for such step changes in the set points in the low and high dilution rate regions are summarized in TABLE 1. The use of controller pairing suggested by the average dynamic gain array method results in smaller ISE values, which indicates better performance of the controller.

The second example considered for numerical simulations is a microbial process where the cell growth kinetics is described by the Monod equation and significant

TABLE 1. Closed-Loop Responses to Step Changes in the Manipulated Variables of a Microbial Growth Process Described by Monod Kinetics with Constant Cell Mass–to–Substrate Yield

D (h^{-1}) Step Change		S_F (g/L) Step Change		Controller	
From	To	From	To	Pairing	ISE
0.1	0.3	8.0	12.0	X with S_F, S with D	3.279
				S with S_F, X with D	7.158
0.7	0.9	8.0	12.0	X with S_F, S with D	5.422
				S with S_F, X with D	2.211

substrate consumption occurs for cell maintenance. The process kinetics is described by equations 22 and 23 with the exception that

$$\sigma = \frac{\mu}{Y} + m \qquad (25)$$

with $m = 0.1$ h^{-1}. This example belongs to case b ($\sigma \not\to 0$ as $\mu \to 0$) discussed earlier. A profile of the dynamic interaction measure, ϕ_{11}, shown in FIGURE 3 indicates that ϕ_{11} is negative at low dilution rates ($D < D_2$). The determination of the proper controller configuration gains increased importance when ϕ_{11} is negative. The performance of a closed-loop process (process with feedback control) with incorrect controller pairing may in fact be worse than that of the same process without any feedback control. The summary of closed-loop simulations in TABLE 2 indicates that better performance of the closed-loop control system can result (i.e., ISE can be minimized) if variables with

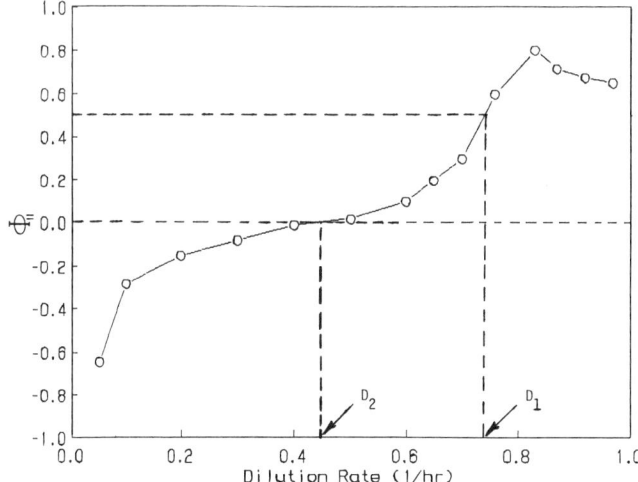

FIGURE 3. A plot of the average dynamic array element, ϕ_{11}, versus the dilution rate for Monod growth kinetics with maintenance requirement.

TABLE 2. Closed-Loop Responses to Step Changes in the Manipulated Variables of a Microbial Growth Process Described by Monod Kinetics with Maintenance Requirement

D (h^{-1}) Step Change		S_F (g/L) Step Change		Controller	
From	To	From	To	Pairing	ISE
0.1	0.3	8.0	12.0	X with S_F, S with D	5.960
				S with S_F, X with D	≥9.186[a]
0.7	0.9	8.0	12.0	X with S_F, S with D	5.393
				S with S_F, X with D	2.273

[a]The minimum ISE (9.186) is obtained when the proportional gains for both controllers approach zero (no feedback control action).

positive interaction measures are paired together (i.e., pair X with S_F and S with D). The closed-loop simulations also revealed that operation with incorrect controller pairing is restricted to a narrow range of proportional gain for the dilution rate controller in order to obviate cell washout. This problem does not arise when the controllers are paired properly. The interaction measures, ϕ_{11} and ϕ_{12}, are positive and less than unity in the interval of $D_2 < D < \mu_m$ (see FIGURE 3). The controller pairing suggested by the average dynamic gain array method for $D_2 < D < D_1$ and deduced from the closed-loop simulations for $0 < D < D_2$ is to pair X with S_F and S with D. For dilution rates in excess of D_1 ($\phi_{11} = 0.5$ at $D = D_1$), the suggested controller pairing is to pair X with D and S with S_F (because $\phi_{11} > \phi_{12}$). The results for closed-loop simulations in TABLE 2 at a high dilution rate ($D = 0.7$ h^{-1}) confirm that this is the proper controller pairing.

EXPERIMENTAL RESULTS

The controller pairings suggested by the average dynamic gain array method at low and high dilution rates were verified by open-loop experiments involving continuous cultures of a wild-type strain, namely, *Bacillus subtilis* TN106. Additional fed-batch and continuous culture experiments were conducted with *B. subtilis* TN106 and *B. subtilis* TN106[pAT5] to examine the efficacy of feedback control with proper pairing.

Materials and Methods

Bacterial strains used in this study are *B. subtilis* wild-type strain TN106 and recombinant strain TN106[pAT5]. The recombinant strain consists of the structural gene for a thermostable α-amylase from the donor species *B. stearothermophilus* CU21 carried on the recombinant plasmid pAT5 with the α-amylase negative *B. subtilis* TN106 as the host.[10] The nutrient medium, containing glucose as the limiting substrate, contained (per liter)[11] glucose (variable), 5.0 g $(NH_4)_2SO_4$, 1.0 g sodium citrate · $2H_2O$, 0.5 g $MgSO_4 · 7H_2O$, 0.14 g $FeCl_3 · 6H_2O$, 0.1 g $CaCl_2 · 2H_2O$, 0.01 g $MnSO_4 · 4H_2O$, 0.01 g $FeSO_4 · 7H_2O$, 3.0 g K_2HPO_4, 1.5 g KH_2PO_4, 2.0 g arginine,

0.5 g yeast extract, and 2 mL trace metal solution. In experiments involving the recombinant strain, 10 mg/L kanamycin was added to the medium to keep the number of plasmid-free cells very low.

A schematic diagram of the bioreactor with capability to control concentrations of cell mass and limiting substrate in the bioreactor by manipulation of dilution rate and substrate feed concentration is provided in FIGURE 4. A 2-L BioFlo Bench Top Chemostat (New Brunswick Scientific) was used for continuous culture experiments, whereas a 16-L Microgen bioreactor (New Brunswick Scientific) was used for fed-batch experiments. The pH and temperature of the culture were controlled at 6.8 and 37 °C, respectively.[11] The dissolved oxygen (DO) level was measured by a DO analyzer (New Brunswick Scientific, Model DO-40) and was maintained above 50% saturation in all experiments. The glucose concentration in the culture was measured by a YSI Model 27 Glucose Analyzer. The cell mass concentration in the culture was determined by measuring the optical density of the culture samples and by using a predetermined correlation between optical density and dry cell mass concentration. The information on cell mass and substrate concentrations, which were measured every 20 minutes, was fed to an IBM AT-compatible personal computer. The manipulated variables, that is, dilution rate and substrate feed concentration, were identified by the computer using the algorithm for PI controllers in velocity form.[12] Two feed tanks, one containing water and the other containing medium with glucose at 4 g/L, were used in the experimental setup. The manipulation of dilution rate and substrate feed concentration was accomplished by regulating the speeds of the two feed pumps by the personal computer.

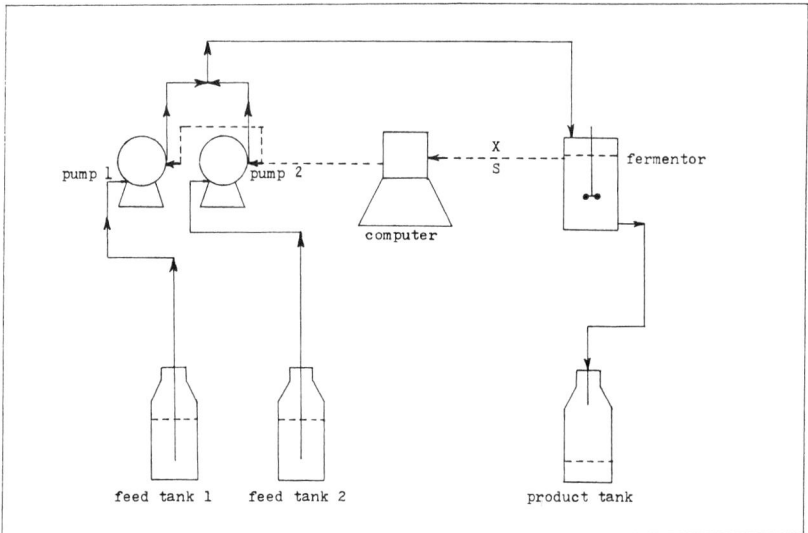

FIGURE 4. Schematic diagram of the experimental setup with capability to control concentrations of cell mass and limiting substrate in the bioreactor by manipulation of dilution rate and substrate feed concentration.

Open-Loop Experiments

These experiments were conducted with the wild-type strain *B. subtilis* TN106. The profiles of steady-state concentrations of cell mass and limiting substrate are presented in FIGURE 5. The changes in both concentrations are not significant for dilution rates below 0.7 h^{-1}. Because the steady-state cell mass concentration remains nonzero as the substrate concentration drops to zero [$\sigma_o \rightarrow 0$ and $\mu_o \rightarrow 0$ as $S_o \rightarrow 0$], the behavior exhibited here belongs to case a discussed earlier. It must be remembered for case a that the recommendations of the average dynamic gain array method are to use dilution rate to control substrate concentration and to use substrate feed concentration to control cell mass concentration at low dilution rates and to reverse the controller

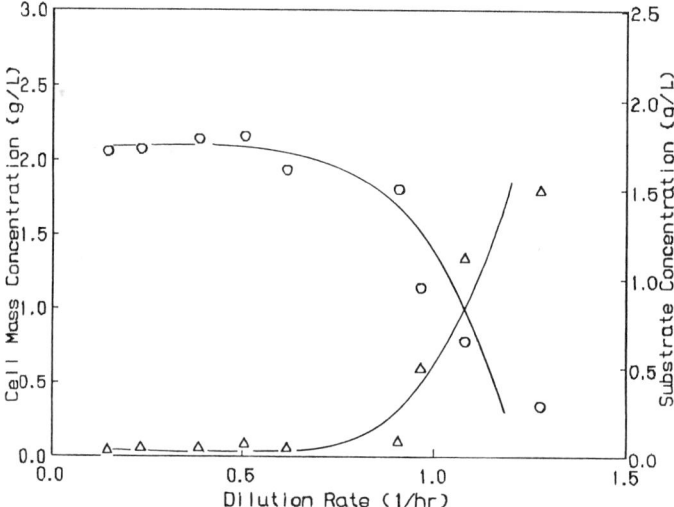

FIGURE 5. Steady-state concentration profiles for cell mass and limiting substrate obtained from continuous culture experiments using *B. subtilis* TN106 ($S_F = 2.5$ g/L). Symbols ○ and △ represent experimental data for cell mass concentration and substrate concentration, respectively.

pairing at high dilution rates. The relative gain array method, based on static analysis, suggested that S should be paired with D and X with S_F irrespective of the magnitude of D.

The response of a continuous culture operating at the stationary state with D and S_F being 0.58 h^{-1} and 1.6 g/L, respectively, to a step change in S_F is presented in FIGURE 6. At this low dilution rate, the cell mass concentration responds rapidly to such a change, whereas the substrate concentration is nearly invariant. The continuous culture initially at a similar stationary state does not respond significantly when subjected to a small step change in dilution rate (see FIGURE 7). The results in FIGURES 6 and 7 indicate that the desired controller pairing is X with S_F and S with D. In fact,

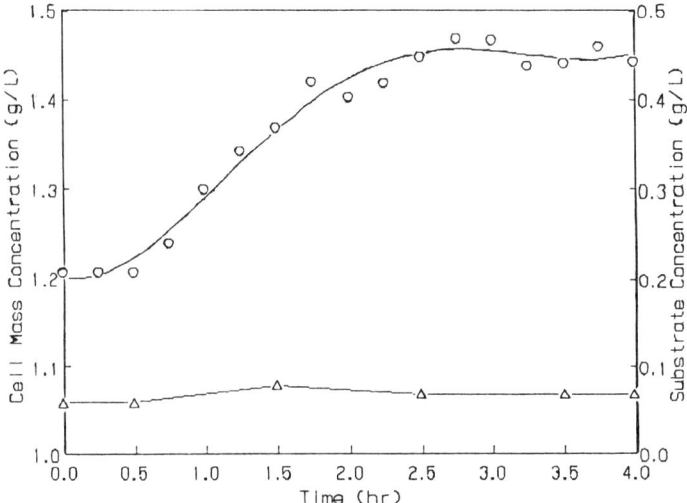

FIGURE 6. Response of continuous culture of *B. subtilis* TN106, initially at the stationary state corresponding to $D = 0.58$ h^{-1} and $S_F = 1.6$ g/L, to a step change in S_F (from 1.6 g/L to 2.1 g/L). Symbols ○ and △ represent experimental data for cell mass concentration and substrate concentration, respectively.

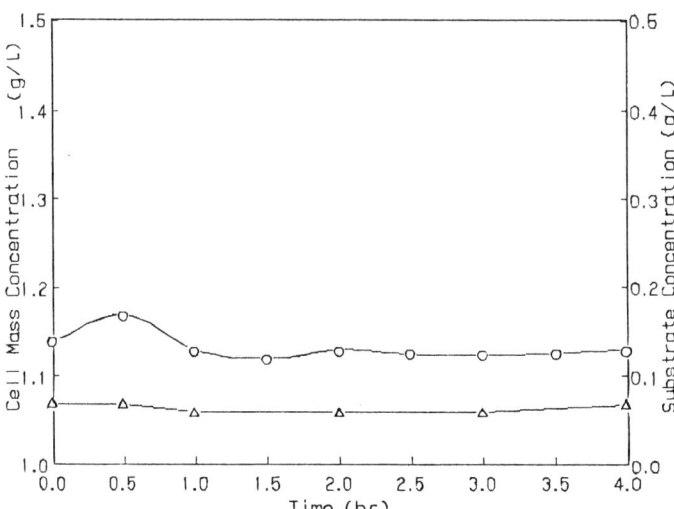

FIGURE 7. Response of continuous culture of *B. subtilis* TN106, initially at the stationary state corresponding to $D = 0.61$ h^{-1} and $S_F = 1.6$ g/L, to a step change in D (from 0.61 h^{-1} to 0.70 h^{-1}). Symbols ○ and △ represent experimental data for cell mass concentration and substrate concentration, respectively.

only one controller is needed to control X by manipulation of S_F because, at such low D, S is insensitive to variations in D and S_F.

The response of a continuous culture operating at the stationary state with D and S_F being 1.08 h^{-1} and 1.82 g/L, respectively, to a step change in S_F is shown in FIGURE 8. The cell mass concentration responds very slowly to such a step change, whereas the response of substrate concentration is rapid and substantial. The results imply that S should be paired with S_F and X with D at this high dilution rate. Eventually, S returns to its value before the introduction of the step change, whereas there is a small increase in X as the bioreactor approaches a new stationary state. These results suggest that X should be paired with S_F and S with D at large times (stationary state operation), which is consistent with the outcome of the static analysis. The results in FIGURE 8 were used to obtain approximate estimates of the average dynamic gains, D_{12} and D_{22} (defined in equation 13), by assigning θ (unknown) values over the range of 0.5–8.0 h. When a continuous culture operating at the same initial stationary state is subjected to a step change in D, both S and X change rapidly and substantially (see FIGURE 9). The results from this experiment were used to obtain approximate estimates of D_{11} and D_{21}, with θ being assigned values between 0.5 and 8.0 h. The interaction measure, ϕ_{11} (calculated using equation 16), remained above 0.8 for this range of θ. Because $\phi_{11} > \phi_{12}$ for the results presented in FIGURES 8 and 9, the average dynamic gain array method suggests that the proper controller pairing should be X with D and S with S_F, as deduced earlier from the results in FIGURE 8 alone.

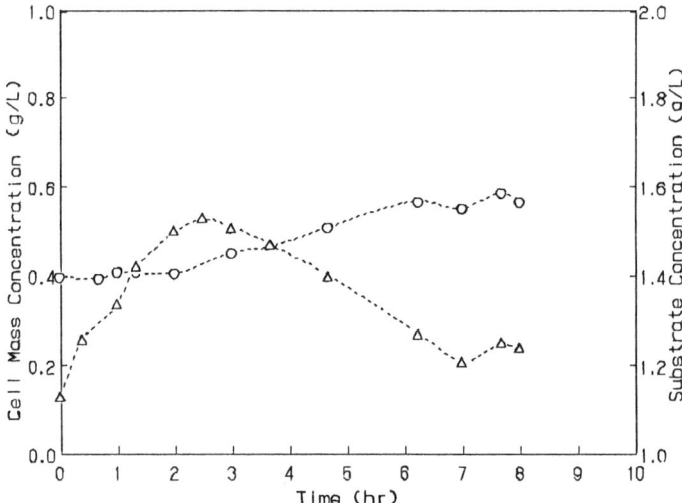

FIGURE 8. Response of continuous culture of *B. subtilis* TN106, initially at the stationary state corresponding to $D = 1.08$ h^{-1} and $S_F = 1.82$ g/L, to a step change in S_F (from 1.82 g/L to 2.33 g/L). Symbols ○ and △ represent experimental data for cell mass concentration and substrate concentration, respectively.

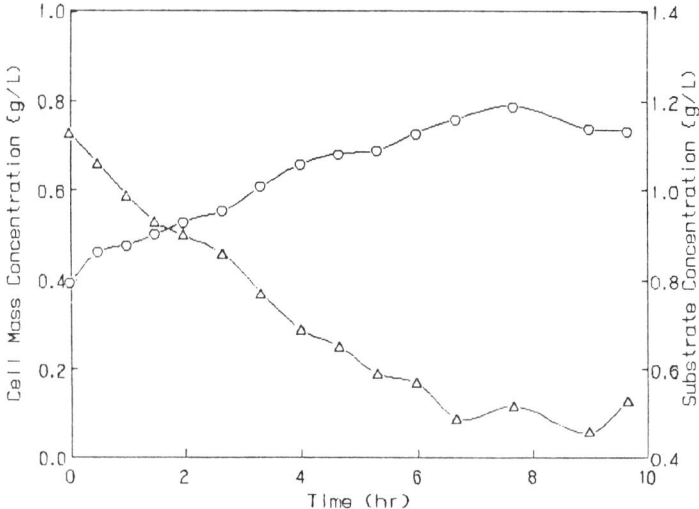

FIGURE 9. Response of continuous culture of *B. subtilis* TN106, initially at the stationary state corresponding to $D = 1.08$ h^{-1} and $S_F = 1.82$ g/L, to a step change in D (from 1.08 h^{-1} to 0.97 h^{-1}). Symbols ○ and △ represent experimental data for cell mass concentration and substrate concentration, respectively.

Closed-Loop Experiments

About 30 h is required to attain the desired stationary state for a continuous culture of *B. subtilis* TN106 after transition from batch to continuous operation with $D = 0.6$ h^{-1} (see FIGURE 10). In a similar experiment, the time required to reach stationary state after transition from batch to continuous operation was reduced to about 4 h when a feedback controller with proper controller pairing was used (see FIGURE 11). The advantage of using a feedback controller with proper pairing of controlled and manipulated variables in attaining the desired bioreactor state quickly is clearly evident from these results. In a continuous culture experiment at high dilution rate ($D = 1.28$ h^{-1}), stationary state was attained in about 7 h after transition to continuous operation using feedback controllers with controller pairing applicable at high dilution rates (see FIGURE 12).

The results to be presented next for fed-batch experiments demonstrate that the importance of feedback control of bioreactors with correct controller pairing cannot be understated. These experiments employed the recombinant strain *B. subtilis* TN106[pAT5]. With dilution rate being fixed, each semibatch operation was an exponential fed-batch operation. The cell mass and substrate concentration profiles in FIGURE 13 indicate that attainment of quasi-steady state in an uncontrolled fed-batch operation is not possible before the bioreactor volume reaches its maximum permissible value. In another experiment under identical conditions, the use of a feedback controller with proper pairing (X with S_F and S with D) enabled establishment of a quasi-steady state (see FIGURE 14), as evidenced by the near invariance of the

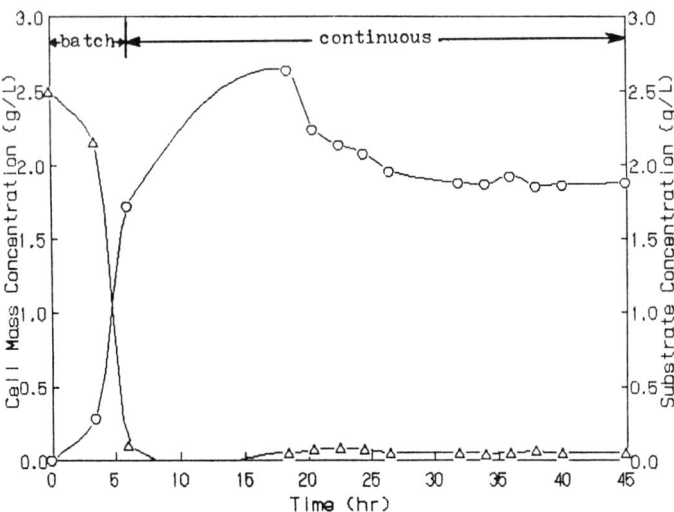

FIGURE 10. Concentration profiles in an open-loop continuous culture experiment with *B. subtilis* TN106 with $D = 0.6$ h^{-1} and $S_F = 2.5$ g/L in continuous operation. Time zero denotes the start of the batch operation preceding the continuous operation. Symbols O and △ represent experimental data for cell mass concentration and substrate concentration, respectively.

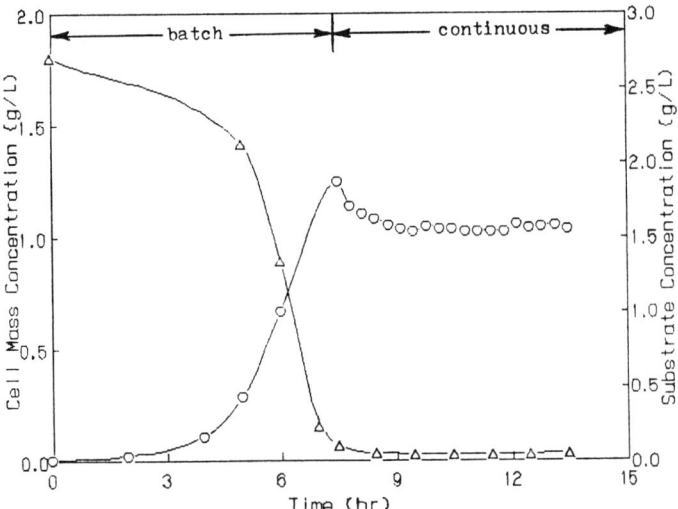

FIGURE 11. Concentration profiles in a continuous culture experiment with *B. subtilis* TN106 employing feedback control (with proper controller pairing) during continuous operation ($D = 0.61$ h^{-1} and $S_F = 1.45$ g/L). Time zero denotes the start of the batch operation preceding the continuous operation. Symbols O and △ represent experimental data for cell mass concentration and substrate concentration, respectively.

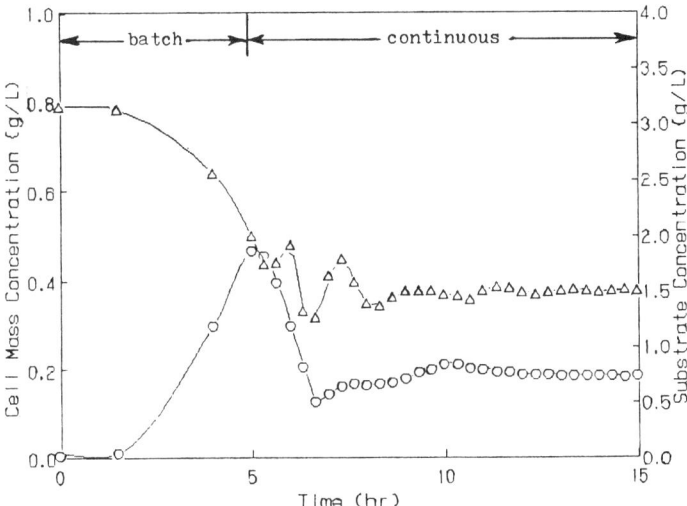

FIGURE 12. Concentration profiles in a continuous culture experiment with *B. subtilis* TN106 employing feedback control (with proper controller pairing) during continuous operation ($D = 1.28$ h^{-1} and $S_F = 2.1$ g/L). Time zero denotes the start of the batch operation preceding the continuous operation. Symbols ○ and △ represent experimental data for cell mass concentration and substrate concentration, respectively.

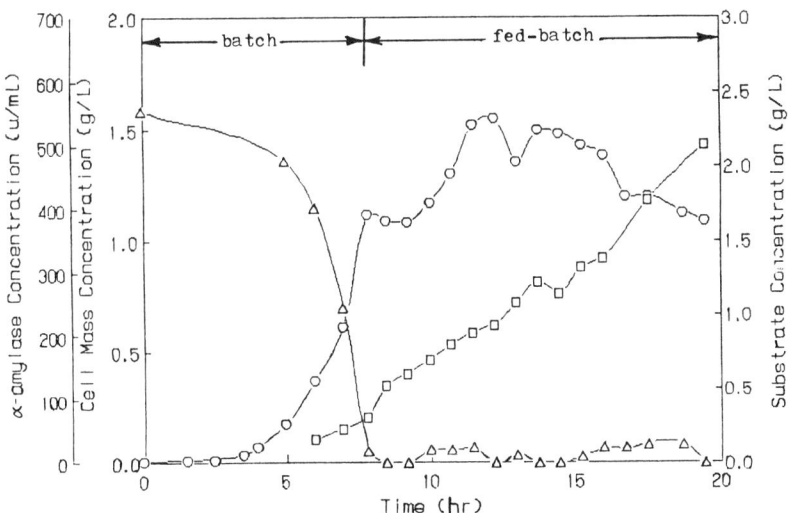

FIGURE 13. Concentration profiles in an open-loop fed-batch experiment with *B. subtilis* TN106[pAT5] with $D = 0.1$ h^{-1} and $S_F = 2.58$ g/L in fed-batch operation. Time zero denotes the start of the batch operation preceding the fed-batch operation. Symbols ○, △, and □ represent experimental data for cell mass concentration, substrate concentration, and α-amylase activity, respectively.

concentrations of cell mass and limiting substrate with time 12 h after the start of the experiment.

CONCLUSIONS

Controller configurations were determined in this study for continuous and fed-batch operations of fermentation processes where the specific kinetics of the key rate processes can be described solely in terms of the concentration of the limiting substrate. The static analysis predicted a one-way interaction at all permissible dilution rates. The control strategy based on the steady-state gains and the Bristol relative gain array

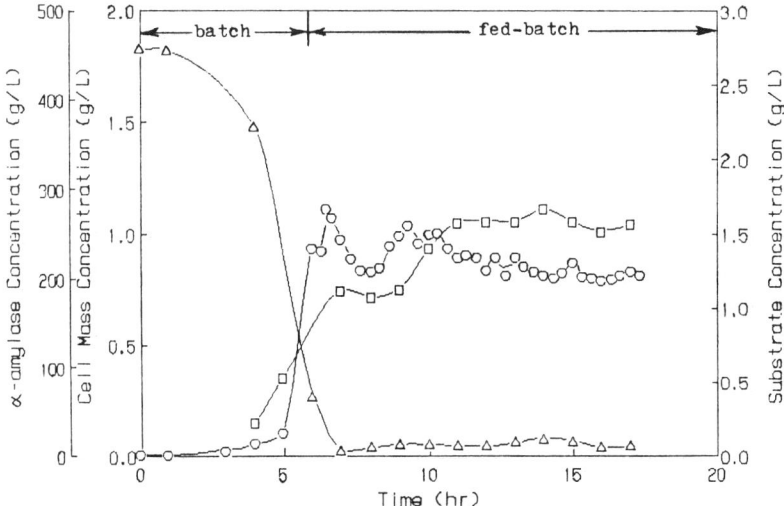

FIGURE 14. Concentration profiles in a fed-batch experiment with *B. subtilis* TN106[pAT5] employing feedback control (with proper controller pairing) during fed-batch operation ($D = 0.1$ h^{-1} and $S_F = 1.81$ g/L). Time zero denotes the start of the batch operation preceding the fed-batch operation. Symbols O, △, and □ represent experimental data for cell mass concentration, substrate concentration, and α-amylase activity, respectively.

was therefore shown to be the use of dilution rate to control substrate concentration and the use of substrate feed concentration to control cell mass concentration. In the case of large disturbances in manipulated variables, the controller configuration must be based on dynamic interactions among manipulated and controlled variables rather than on the eventual static interactions. The dynamic analysis based on the average dynamic gain array method indicated that the conclusions reached regarding controller pairing based on static interactions are not valid at all dilution rates—in particular, at high dilution rates. For certain relations between the specific rates for cell growth and substrate consumption, some of the dynamic interaction measures were negative at very low dilution rates, implying difficulty in control of the processes. These situations were studied further by numerical simulations for specific examples of kinetics and the importance of proper controller pairing for attaining the desired state was demonstrated. The general conclusions based on the average dynamic gain array and relative gain array methods were verified via closed-loop simulations for specific examples of kinetics of fermentations. Continuous culture experiments with *B. subtilis* TN106 demonstrated that only a few open-loop experiments are necessary for deducing the controller pairing. Closed-loop continuous culture experiments with this bacterium and fed-batch experiments with *B. subtilis* TN106[pAT5] established that feedback control with proper controller pairing ensures attainment of and significantly reduces the time required to reach a steady state in continuous cultures and a quasi-steady state in fed-batch cultures.

REFERENCES

1. BOYLE, T. J. 1979. Biotechnol. Bioeng. Symp. **9:** 349–358.
2. AGRAWAL, P. & H. C. LIM. 1984. *In* Advances in Biochemical Engineering/Biotechnology, Volume 30. T. K. Ghose, A. Fiechter & N. Blakebrough, Eds.: 61–90.
3. BRISTOL, E. H. 1966. IEEE Trans. Autom. Control **AC-11:** 133.
4. RAY, W. H. 1981. Advanced Process Control, p. 66–71. McGraw–Hill. New York.
5. GAGNEPAIN, J. P. & D. SEBORG. 1982. Ind. Eng. Chem. Process Des. Dev. **21:** 5.
6. TUNG, L. S. & T. F. EDGAR. 1981. AIChE J. **27:** 690–693.
7. LAU, H., J. ALVAREZ & K. F. JENSEN. 1985. AIChE J. **31:** 427–439.
8. JENSEN, N., D. C. FISHER & S. L. SHAH. 1986. AIChE J. **32:** 959–970.
9. HERBERT, D. 1961. Soc. Chem. Ind. Monogr. **12:** 247.
10. AIBA, S., K. KITAI & T. IMANAKA. 1983. Appl. Environ. Microbiol. **46:** 1059–1065.
11. WEI, D., S. J. PARULEKAR, B. C. STARK & W. A. WEIGAND. 1989. Biotechnol. Bioeng. **33:** 1010–1020.
12. STEPHANOPOULOS, G. 1984. Chemical Process Control, p. 634–637. Prentice–Hall. Englewood Cliffs, New Jersey.

APPENDIX

Nomenclature

D dilution rate (h^{-1})
$\overline{D}$ average dynamic gain array
$\overline{D}_{ij}$ elements of the average dynamic gain array (equation 13) ($i,j = 1,2$)
D_1 dilution rate at which $\phi_{11} = 0.5$ (h^{-1})
D_2 dilution rate ($D_2 > 0$) at which $\phi_{11} = 0$ (h^{-1})
e_1, e_2 errors (equation 24)
e_{ij} steady-state process gains (equation 8) ($i,j = 1,2$)
F volumetric feed rate (L/h)
g_{ij} defined in equation 10 ($i,j = 1,2$)
G_{ij} elements of the open-loop transfer function matrix (equation 9) ($i,j = 1,2$)
ISE   sum of the integral square errors (equation 24)
K_s saturation constant (g/L)
m maintenance coefficient (h^{-1})
m_1, m_2 deviations in dilution rate and substrate feed concentration from their respective values at the desired state (g/L)

p, q defined in equation 21b
S substrate concentration (g/L)
S_F substrate feed concentration (g/L)
s Laplace transform variable
t time (h)
V bioreactor volume (L)
X cell mass concentration (g/L)
x_1, x_2 deviations in concentrations of cell mass and limiting substrate from their respective values at the desired state (g/L)
Y cell mass–to–substrate yield (equation 18b)

Greek Letters

$\alpha_1, \alpha_2, \beta, \Delta$ defined in equation 10
$\overline{\lambda}$ Bristol relative gain array
λ_{ij} elements of the Bristol relative gain array ($i,j = 1,2$)
μ specific cell growth rate (h^{-1})
μ_m maximum specific cell growth rate (h^{-1})
η, ϕ defined in equation 21b
$\overline{\phi}$ relative dynamic gain array
ϕ_{ij} elements of the relative dynamic gain array ($i,j = 1,2$)
ψ defined in equation 19
ρ defined in equation 21b
σ specific substrate consumption rate (h^{-1})
θ largest time constant in the open-loop transfer function (h)

Subscript

o desired stationary or quasi-stationary state

Superscript

' differentiation with respect to substrate concentration

Intelligent Purification of Monoclonal Antibodies

P. W. THOMPSON,[a] A. C. KENNEY,[a] P. MOULDING,[a]
AND D. WORMALD[b]

[a]*Oros Instruments Limited*
Slough, Berks SL1 4LJ, England

[b]*Oros Instruments Incorporated*
Cambridge, Massachusetts 02142

INTRODUCTION

The development of biotechnology has fallen into a number of identifiable phases.

Initially, the emphasis was on the molecular biology of microbial and eukaryotic cells in order to construct new strains and to produce valuable biochemical products in the laboratory. Little consideration was given at this time to suitable methods of culturing these cells with the result that the products were generated as very dilute and highly impure solutions or even insoluble particles.

This then led to attempts to isolate and purify those products from crude samples such as fermentation and tissue culture broths using traditional methods. These methods were often unsuitable for handling dilute products, and typical isolation and purification protocols often involved numerous steps.

The next stage was to scale up these processes from procedures suitable for the laboratory bench to reliable and economic industrial-scale processes. The difficulty of scaling up the procedures developed at the laboratory bench stimulated efforts to improve product concentrations, reduce levels of impurities from fermentation, reduce the number of purification steps, and increase the recoveries from those steps. Methods involving mechanical handling of the product (such as batch centrifugation) gave way to more continuous processes already used in the chemical industry.

Latterly, efforts have been made to apply recent advances in the computer and microprocessor fields to automate those processes. These automation methods began as simple remote actuation of plant items and steadily advanced into more interactive sequence and parametric control based on preset procedures.

Developments in the downstream processing of products from biotechnology have tended to lag behind the upstream fermentation processes such that, whereas many fermentations are now operating efficiently and automatically at a large scale, the associated downstream processes are still based on labor-intensive "stretched" laboratory processes.

PURIFICATION OF MONOCLONAL ANTIBODIES

Pretreatment

Monoclonal antibodies are produced by two main methods. First, *in vivo* culture in rodents results in an ascitic fluid usually rich in lipids and contaminating immunoglobulins that may interfere with subsequent purification unless they are removed. Lipids may be removed by centrifugation,[1] adsorption,[2] or filtration,[3] although this last method can result in significant loss of product. Second, *in vitro* methods rely on the culture of hybridoma cells either in free suspension or immobilized on a surface or within an inert matrix. The products of *in vitro* methods are usually more dilute than ascites and may contain cellular debris. Clarification by centrifugation or filtration and concentration by ultrafiltration[4] may therefore be required.

Subsequent to feedstock pretreatment, there are a number of possible purification strategies that may be employed.

Precipitation

Monoclonal antibodies may be selectively precipitated using precipitants such as ammonium sulfate[5] or polyethylene glycol.[2] There are a number of disadvantages to this method: the yield of antibody is usually low (approximately 60%) and the dewatering of the resultant precipitates may involve further losses and potential exposure of the product to the atmosphere where manual removal from tubular centrifuges is necessary.

Chromatography

The principal chromatographic methods used for antibody purification are ion exchange chromatography and affinity chromatography. Both of these methods may be operated at either low or high pressure.

Ion exchange chromatography is possible with a range of resins, both anionic and cationic. Yields of greater than 90% may be achieved at purity levels of greater than 80%. The main practical difficulty of using ion exchange methods is that the initial methods development is likely to be extensive due to the spread of isoelectric points found in monoclonal antibodies. The more acidic antibodies may be difficult to purify on anion exchangers.[6]

Affinity chromatography has the advantage of purifying a species of protein quite specifically. For antibodies, the affinity ligand most commonly employed is Protein A, which will bind to virtually all IgG antibodies.[4] A number of potential disadvantages of Protein A affinity chromatography have inhibited its use. The expense of the Protein A ligand can be difficult to justify without adequate knowledge of the reusability of the matrix. Protein A column lifetimes in excess of 100 cycles have now been recorded.[7] Some IgG subclasses have been difficult to purify on Protein A. However, the recent development of buffers that promote binding of antibody to Protein A has overcome this problem.

The use of immunopurification methods based on immobilized antibodies may assist with some of the more intractable antibodies such as IgM antibodies by using an anti-IgM affinity ligand.

APPLICATION OF KNOWLEDGE-BASED SOFTWARE TO PROCESS CONTROL

Most available expert system software packages are "shells" consisting of a knowledge base and an inference engine. The knowledge base must be filled with "rules" by which a certain field of expertise is described. This approach has been extensively used for applications such as medical diagnosis and fault detection.

However, real-time process control using a knowledge base to provide control parameters as opposed to diagnostic information is limited to a few systems suitable for very large rule bases. Control systems for unit operations usually require only a few rules.

Most applications to date rely on the use of expertise to arrive at a conclusion on the basis of information available. However, in the laboratory, the required information is very often not available and has to be determined by performing experiments. The expertise lies in knowing what experiments to perform in order to obtain the information in the shortest possible time and with the least amount of effort. In addition, it may be necessary to modify experiments in the light of results obtained.

There may also be uncertainty about what to do next and hence a theoretical approach is employed; that is, the results obtained in the previous experiment were most likely caused by phenomenon X and the next experiment should test that theory. This approach can combine the use of expertise with a simple form of learning ability such that, in a sequence of operations of experiments, the next operation uses information obtained in a previous operation.

The following section discusses some of the knowledge that may be applied to the purification of monoclonal antibodies by Protein A affinity chromatography.

AREAS OF EXPERTISE IN MONOCLONAL ANTIBODY PURIFICATION

Existing methods of protein purification rely on their performance by skilled operators familiar with the variances that occur in the feedstocks to be processed. When presented with a new antibody to be purified, the biochemist has to first decide on a potential method of purification. For many, Protein A affinity chromatography is the first choice method most likely to succeed. Once this decision is made, there are a number of questions that need to be answered:

(a) What is the capacity of the Protein A matrix for this particular antibody? Antibody binding capacities can vary between 1 and 10 mg/mL of Protein A matrix.
(b) What is the highest pH at which the antibody can be eluted from the column? Traditional use of Protein A chromatography has eluted the antibody at low pH's (typically, pH 3), which can degrade the biological activity of the antibody. It is very seldom necessary to elute at such a low pH; a pH range of 4 to 7 is more

typical. Clearly, it is more beneficial to the antibody's activity to elute as close to neutral pH as possible.

(c) What is the concentration of the antibody in the feedstock? It is notoriously difficult to determine a meaningful antibody concentration in a crude antibody feedstock. ELISA methods are only 20% accurate and suffer from the problem of not necessarily identifying the same section of the molecule that attaches to Protein A. It is possible that ELISA methods can detect antibody that is in fact partially degraded.

(d) Are there any contaminating antibodies in the feedstock? Ascites may contain mouse antibodies; tissue culture supernatants may contain antibody contaminants derived from serum added to the culture medium; some monoclonal antibodies are not truly monoclonal and may contain additional heavy or light chains.

Answering these questions can consume days or even weeks of valuable time.

Once the purification methods have been devised, it may be necessary at some point to purify a larger amount of the feedstock. Unless the antibody has been produced by a different method, the information derived during the methods development phase is likely to remain valid. The one exception is concentration, which is likely to vary from batch to batch.

The current costs of Protein A matrices tend to inhibit the use of large columns. A multicycle operation of smaller columns will therefore be necessary to purify larger amounts of antibody feedstock. Such a sequence of operations makes it possible for some learning about the purification to take place. The questions to be answered by such an exercise are discussed below:

(a) Will the column capacity remain unimpaired throughout the run? Column fouling, mechanical damage to the matrix by abrasion, and leakage of Protein A from the matrix can all lead to a gradual deterioration in the capacity of the column and may potentially lead to loss of antibody.

(b) Will the column start to become fouled especially if high lipid levels exist in the feedstock? Column fouling may also lead to an increase in the pressure drop across the column.

(c) Will the elution pH remain unchanged? Changes in column characteristics can occasionally result in minor elution variations.

(d) How will contaminating antibodies be separated from the required antibody? Coelution of contaminating antibodies will lead to a loss of activity and low purity.

(e) Can multicycle purifications be left to run unattended?

AN AUTOMATED CHROMATOGRAPHY SYSTEM

An automated chromatography purification system has been devised[8] to achieve the following objectives:

(a) automatic operation of all valves and pumps,
(b) fail-safe error detection,

(c) automatic sensor calibration,
(d) automatic methods development for unknown feedstocks,
(e) on-line optimization of performance during multicycle purification of known feedstocks,
(f) automatic determination of feedstock concentration,
(g) hygienic operation,
(h) monoclonal antibody output averaging 2 g/day.

These objectives are designed to generate the capability to purify monoclonal antibodies (and potentially other proteins if alternative chemistries are employed) in a reliable, reproducible, and economic manner and to permit largely unattended purifications, thus leaving the human expert to work on more difficult or rewarding

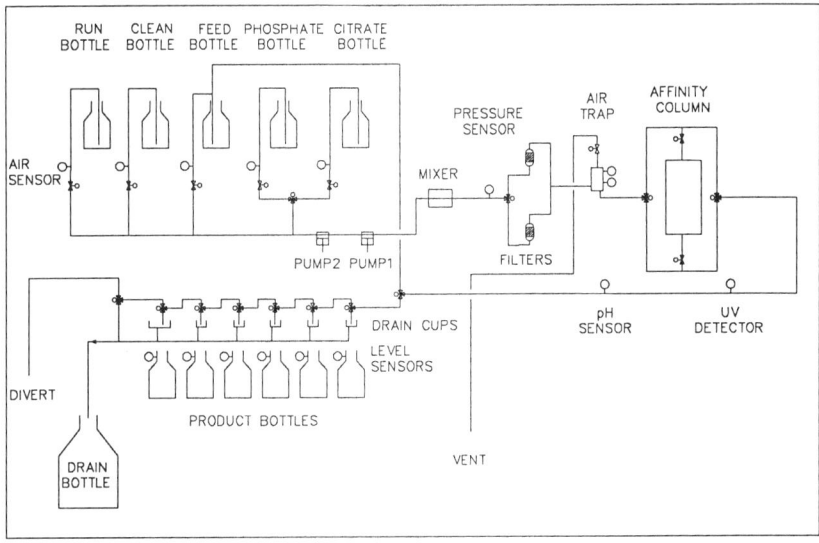

FIGURE 1. Flow diagram of an automated chromatography system.

problems. The areas of expertise described earlier have been incorporated into the software of the controlling computer.

The system hardware is illustrated in the form of a schematic flow diagram in FIGURE 1 and consists of the following features:

(a) supply bottles containing column equilibration buffer, pH control buffers, cleaning solution, and feedstock;
(b) air detectors for each supply bottle to prevent excessive ingress of air into the system;
(c) two variable speed pumps to control flow rate and to respond to high pressure conditions;

(d) an in-line static mixer to blend pH control buffers;
(e) a pressure transducer to detect high or low pressure conditions;
(f) duplex liquid filters to reduce particulate fouling of the column—one filter is in use while the other is on standby; swapover occurs on high pressure conditions;
(g) an automatic self-venting air eliminator to prevent air reaching the column;
(h) a bidirectional Protein A column permitting both upward and downward flow;
(i) a UV absorbance spectrophotometer for detection of protein concentrations in a flowing stream; the UV lamp is also monitored by a current sensor to warn of failure;
(j) an in-line pH electrode to monitor elution characteristics and to warn of any pH that is inconsistent with what is expected;
(k) product bottles for separation of individual antibodies or to collect cycles separately;
(l) product bottle level sensors to prevent overflow conditions;
(m) valves incorporating microswitch feedback to warn of valve failure and prevent incorrect fluid routing.

The whole system is controlled by an IBM-compatible personal computer incorporating the necessary input/output devices.

METHODS DEVELOPMENT

When purifying an antibody for the first time, the operator is prompted by the computer to enter the antibody name, the batch number, the volume of feedstock available, and the concentration of the antibody in the feedstock if known. After this, the system equilibrates the column with a high salt equilibration buffer to generate the correct environmental conditions for optimum antibody binding before loading the feedstock onto the column. After washing away the impurities left in free solution using the run buffer again, the column is eluted with a pH gradient constructed from a variable blend of citrate and phosphate buffers formulated to generate a linear gradient. This "characterization" experiment is designed to elucidate the separation parameters needed for that antibody. These include the capacity of the column for that antibody, the number of separable peaks resulting from the pH gradient, and the optimum elution strategy for the antibody peak of interest. For any given concentration of that antibody, the capacity is used to ensure that the column will be loaded with sufficient feedstock to utilize the available capacity without giving rise to breakthrough of the antibody. In this way, yield and efficiency are optimized. The system ensures that the elution pH is chosen to be as high as possible and hence maintains the biological activity of the antibody while allowing efficient elution from the column. The separation parameters are stored on the computer's hard disk and can be recalled for later use.

If there are occluded peaks or shoulders appearing during elution, the computer can recommend that another cycle be run and thus sets a shallower gradient to attempt to improve the resolution of each peak.

At the end of the characterization run, the system is fully cleaned to prevent cross-contamination of a subsequent feedstock.

Due to the automatic nature of the system, no operator involvement is required beyond the initial data entry except to fill any supply bottles or to change blocked filters as prompted by the computer. A typical characterization run takes about four hours to complete depending on the volume of feedstock to be loaded.

FIGURE 2 illustrates a characterization run as described earlier performed on an ascites feedstock containing mouse IgG3 as the principal antibody. The chromatogram shows the elution pH gradient and the absorbance at 280 nm. The monoclonal antibody is shown to be eluted at a different pH from that of the nonspecific IgG contaminants from the ascites fluid.

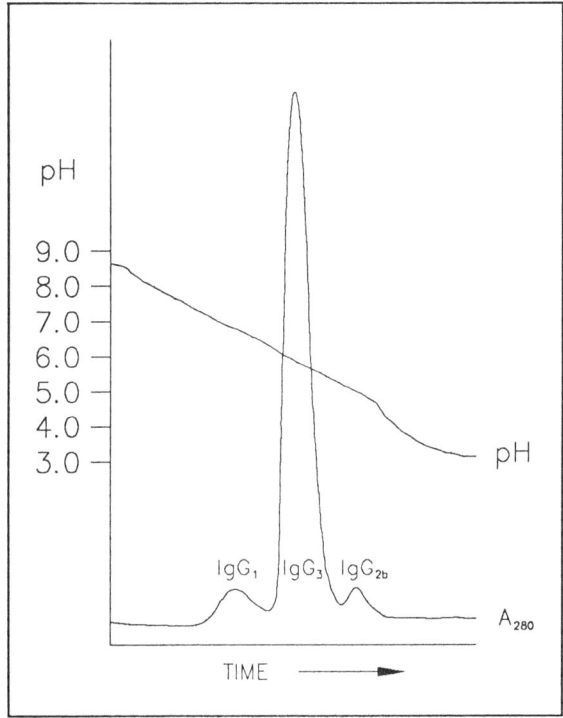

FIGURE 2. Characterization of mouse IgG3 from ascites.

FIGURE 3 illustrates another purification performed on a tissue culture concentrate containing a mouse IgG1. This demonstrates a situation more typical of tissue culture–derived antibodies where there is little or no contaminating antibody present.

MULTICYCLE PURIFICATIONS

To enable multicycle purifications to take place, the computer requires that the data for the feedstock to be purified be already on file. These data are then recovered and can be applied to control the purification about to take place.

The initial data entry again prompts for batch number, feedstock volume, and probable antibody concentration. The computer then prompts the operator to identify which peaks from the original characterization chromatogram are to be collected and which are to be discarded. This information is utilized during elution of the antibody from the column.

After equilibration of the column, sufficient feedstock is loaded onto the column so as not to exceed the capacity for that particular antibody as determined during characterization. The column is then thoroughly washed. If more than one antibody component exists, then a multistep elution is performed at different pH values whereby unwanted peaks are eluted and discarded, whereas wanted peaks are eluted and collected as defined at the start of the run. The system ensures that all peaks are eluted from the column to prevent carryover to a subsequent run, which would cause both a loss of capacity of the column and possible contamination of the subsequent purified antibody.

This multistep elution is illustrated in FIGURE 4 using an ascites feedstock containing a mouse IgG3 in which a minor contaminant had been identified during characterization. The low-level contaminant was eluted at pH 8.9 and was discarded to drain prior to further reduction of the pH to 7.5, at which point the main peak appeared and was collected. This particular example used insufficient feedstock for a multicycle purification.

At the completion of each purification cycle of a multicycle run, the results for each peak were subjected to analysis and comparison with the results achieved during the characterization purification. Depending on the degree of diversity of the yield or peak shape and on what events have already occurred, such as column cleaning sequences

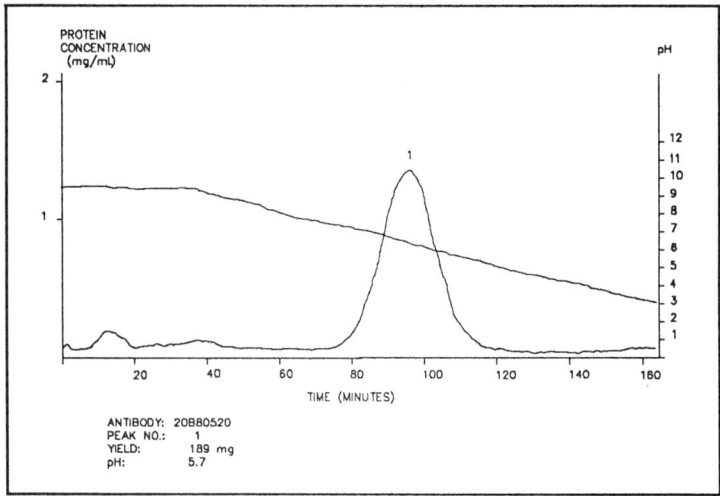

FIGURE 3. Characterization of a mouse IgG1 (20B8) from concentrated tissue culture supernatant.

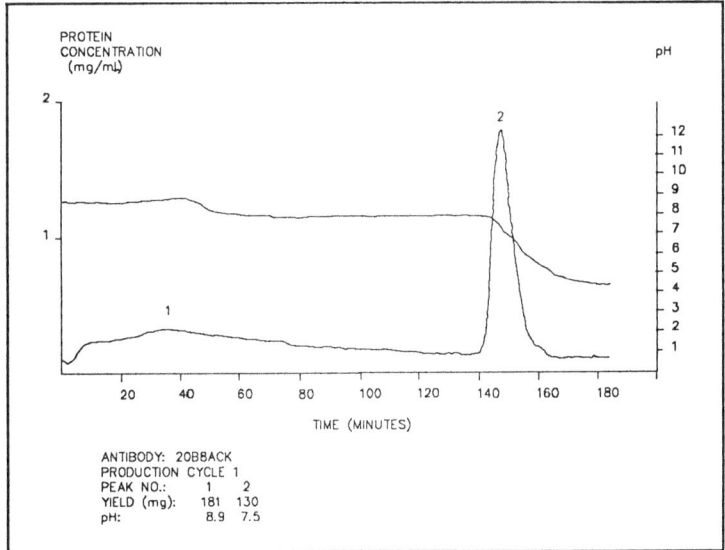

FIGURE 4. Multistep purification of a mouse IgG3.

and number of cycles achieved, the program will arrive at certain conclusions via a decision tree made up of a hierarchical consideration of the issues discussed earlier in areas of expertise.

The example shown in FIGURES 5 through 7 is a three-cycle purification of a mouse IgG1 derived from a concentrated tissue culture supernatant. The run demonstrates that the yield at the completion of the first cycle (FIGURE 5) was greater than that expected from the characterization run (FIGURE 3) such that the column may have been overloaded, thereby leading to a potential loss of product.

The program made the assumption that the antibody concentration input for the feedstock was in error and so, for the second cycle (FIGURE 6), a much reduced volume of feedstock was loaded onto the column to ensure that no overloading could take place. The subsequent elution generated a small peak from which a more accurate antibody concentration in the feedstock could be calculated.

Armed with this information, the third cycle (FIGURE 7) loaded the column up to near capacity (allowing a safety margin) and the subsequent elution indicated that the yield of antibody was now as expected.

CONCLUSIONS

The concept of incorporating both expertise and some limited learning capability into a control system for chromatographic purification of monoclonal antibodies has been realized in the form of a fully integrated antibody purifier. Whereas the rule base

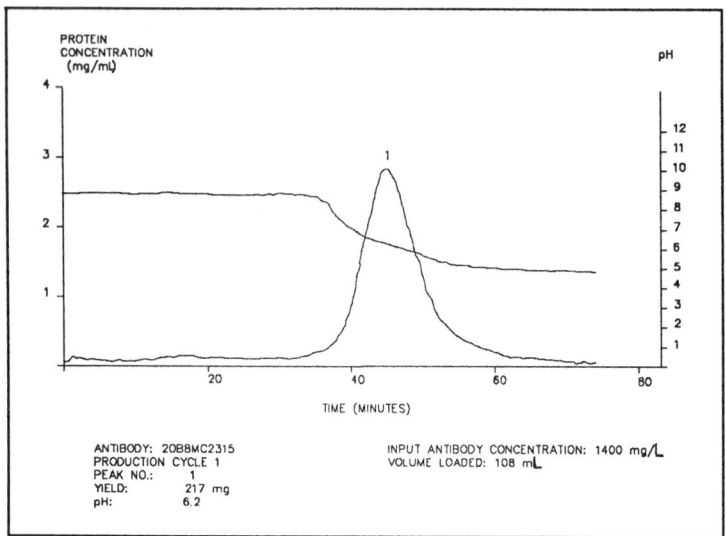

FIGURE 5. Production cycle 1 of 20B80520 antibody purification.

is still relatively small, the objective of operating the system in the absence of human interference has been achieved with a process that hitherto has required constant operator attention to correct errors arising from natural biochemical variances.

Whereas the system to date has been applied to the purification of antibodies, the structure of the controlling software programs lends itself to purification of other

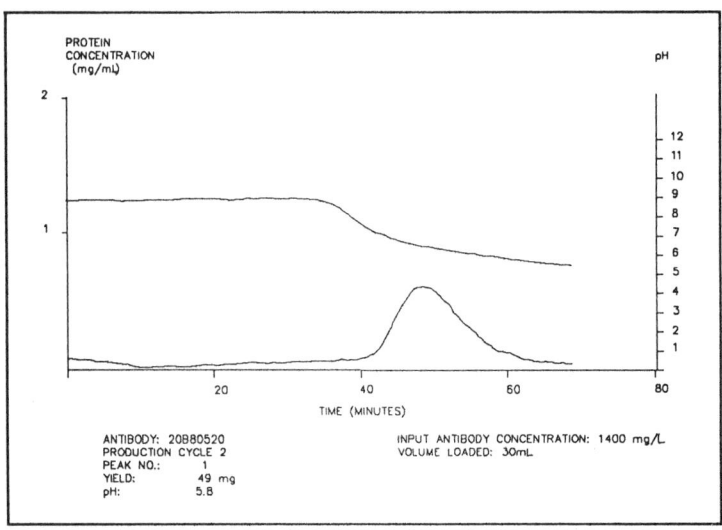

FIGURE 6. Production cycle 2 of 20B80520 antibody purification.

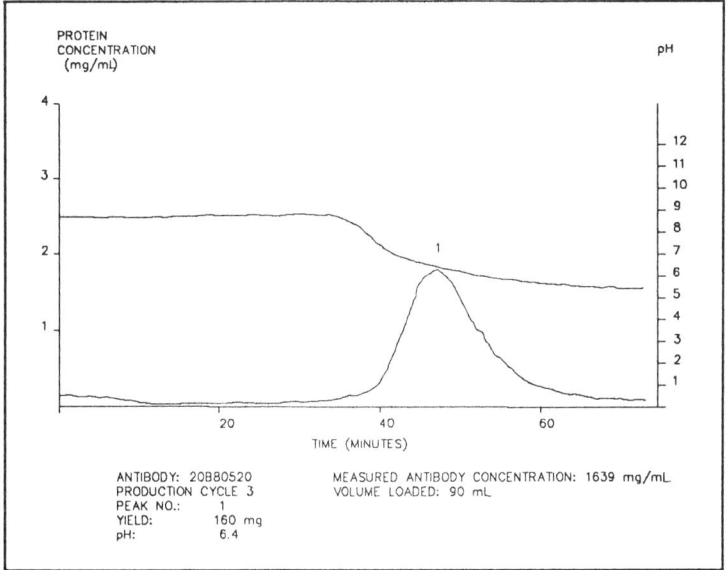

FIGURE 7. Production cycle 3 of 20B80520 antibody purification.

biochemicals by making relatively simple changes to program variables. Future developments will concentrate on enhancing the ease in which these changes may be made by the operator in order to allow him/her to develop a suite of configuration files, each specific to the purification of a particular molecule or range of molecules.

REFERENCES

1. ANNUNZIATO, M. E. & D. J. MARCIANI. 1987. Genet. Anal. Technol. **4:** 1.
2. NEOH, S. H., C. GORDON, A. POTTER & H. ZOLA. 1986. J. Immunol. Methods **91:** 231.
3. ROSS, A. H., D. HERLYN & H. KOPROWSKI. 1987. J. Immunol. Methods **102:** 227.
4. BIRCH, J. R., K. LAMBERT, P. W. THOMPSON, A. C. KENNEY & L. A. WOOD. 1987. Large Scale Mammalian Cell Culture Technology. B. K. Lydersen, Ed. Springer-Verlag. Berlin/ New York.
5. BURCHIEL, S. W., J. R. BILLMAN & F. ALBER. 1984. J. Immunol. Methods **69:** 33.
6. BOONEKAMP, P. M. & R. POMP. 1986. Sci. Tools **33:** 5.
7. KENNEY, A. C. & H. A. CHASE. 1987. J. Chem. Technol. Biotechnol. **39:** 173.
8. KENNEY, A. C. & D. J. WORMALD. 1988 (June). Am. Lab. **20:** 82.

L-Aspartic Acid Production Using Immobilized *E. coli* Cells in a Packed-Bed Reactor

Design of Reactor and Its Optimal Operation

HIROYASU SEKO, SHINOBU TAKEUCHI,
KAZUYOSHI YAJIMA, MASARU SENUMA,
AND TETSUYA TOSA

Research Laboratory of Applied Biochemistry
Tanabe Seiyaku Company, Limited
Osaka 532, Japan

Immobilized enzymes and microbial cells, which are called immobilized biocatalysts, have made it easy to produce many useful biochemicals. As new techniques using immobilized biocatalysts have been developed one after another, design and operation procedures for succeeding in commercial plants have become of major importance nowadays from the standpoint of economic aspect.

In 1969, our efforts arrived on the first commercial process using immobilized enzymes, that is, the optical resolution of D,L–amino acids using immobilized aminoacylase.[1] In 1973, one extension led to the commercial production of L-aspartic acid from fumaric acid and ammonia with aspartase activity of *Escherichia coli* cells immobilized with polyacrylamide gel.[2] This technique has simplified the unit operation and has resulted in a high yield of L-aspartic acid. Further improvements have been performed in the areas of immobilization of microbial cells and of reactor design and operation. In the former, Nishida *et al.*[3] in 1978 developed an excellent technique using κ-carrageenan as the immobilizing matrix for *E. coli* cells. Umemura *et al.*[4] obtained high selectivity cells in which the existence of by-products was almost completely eliminated. From the standpoint of reactor design and operation, Takamatsu *et al.*[5,6] rearranged the kinetics of L-aspartic acid production as a Michaelis-Menten–type form and subsequently measured the stabilization of aspartase activity of *E. coli* cells immobilized with polyacrylamide gel. Furui[7] proposed a packed-bed reactor equipped with a horizontal heat-exchange tube bundle to carry out the exothermic L-aspartic acid production successfully.

In particular, the aspartase reaction is exothermic, resulting in an uneven decay of immobilized cells in the axial direction of the reactor. This uneven decay is the major problem that reduces the life span of immobilized cells.

This report is concerned with the reactor design and operation for the exothermic L-aspartic acid production using *E. coli* cells immobilized with κ-carrageenan:

$$\text{HOOC—CH=CH—COOH} + \text{NH}_3 \underset{\text{aspartase}}{\rightleftharpoons} \text{HOOC—CH(NH}_2\text{)—CH}_2\text{—COOH} - \Delta H.$$

fumaric acid   ammonia   L-aspartic acid

Both the reaction kinetics accompanying a decay of immobilized cells and the reactor model of a multistage continuous stirred-tank type were derived experimentally. By combining the reaction kinetics and the reactor model, a simulation model was developed in a form capable of calculating the temperature and concentration profiles in the axial direction of the reactor. From numerical computations of the simulation model, a packed-bed reactor equipped with four horizontal heat-exchange tube bundles was proposed as an ideal reactor. Furthermore, an optimal temperature policy to use the immobilized cells as long as possible was performed. The result was also compared with an isothermal operation.

EQUIPMENT AND METHODS

Two different sizes of packed-bed reactors were used to confirm the reactor model that we presented. One of them was a pilot reactor and its sketch is shown in FIGURE 1. The reactor consisted of a reactor column and a horizontal heat-exchange tube bundle. The column had a bed height of 2.0 m and, into this, 1.32 m^3 of *E. coli* cells (EAPc-7 derived from ATCC 11303) immobilized with κ-carrageenan[4] was packed.

The other reactor was a small-scale glass reactor equipped with an outer jacket in which cooling water was circulated to maintain the immobilized cells at a constant temperature. The reactor had a bed height of 0.145 m and had thermocouples and samplers at three selected points in the axial direction of the bed. The amount of immobilized cells used here was 5.0×10^{-5} m^3.

Experiments and analyses were conducted in the same manner as described by Furui[7] and Takamatsu *et al.*[5] The reaction rate and stability of the immobilized cells were measured according to the experimental methods described elsewhere.[5,6] Several important values of operating variables and physical properties used in the pilot reactor are listed in the APPENDIX at the end of this report.

REACTION KINETICS

Reaction Rate

Takamatsu *et al.*[5] presented equation 1 as the reaction rate of L-aspartic acid production using *E. coli* cells immobilized with polyacrylamide gel:

$$r = R_0(yK_1 - 1 + y)/(K^*K_1/S_0 + yK_1 - 1 + y). \tag{1}$$

To examine whether equation 1 can be applied to the reaction kinetics of *E. coli* cells immobilized with κ-carrageenan or not, the reaction rates were measured in the temperature range of 10 to 45 °C. Applying the nonlinear regression technique and the analysis of variance, we recognized that equation 1 was also useful for describing the reaction kinetics of *E. coli* cells immobilized with κ-carrageenan and that the parameters in equation 1 were expressed as follows:

$$K^* = 694 \text{ mol/m}^3, \tag{2}$$

$$K_1 = (9.780 \times 10^{-6})S_0 \exp(24{,}007/RT), \tag{3}$$

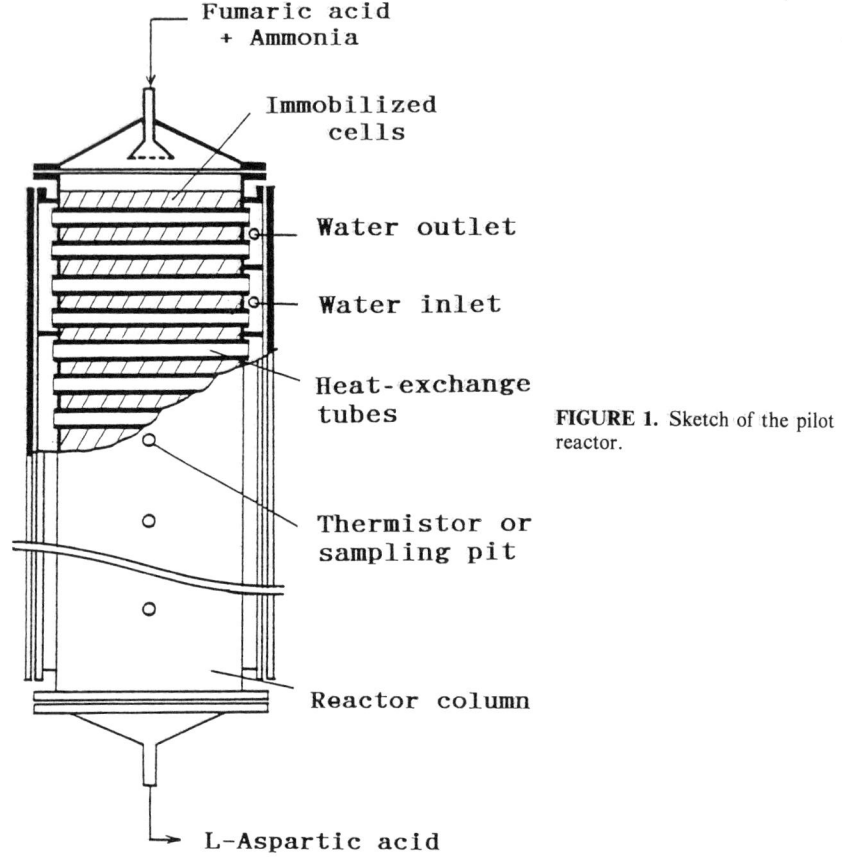

FIGURE 1. Sketch of the pilot reactor.

$$R_0 = (5.475 \times 10^4)\exp(-40{,}231/RT) \ (T: 283\text{--}298 \text{ K}), \tag{4}$$

$$R_0 = (2.203 \times 10^3)\exp(-32{,}263/RT) \ (T: 298\text{--}318 \text{ K}). \tag{5}$$

The activation energies of R_0 were 40,231 J/mol below 293 K and 32,263 J/mol above 293 K. This characteristic is very similar to the result of Rollan et al.[8]

Stability of Immobilized Cells

Because aspartase activity decreased as time increased, we introduced an activity factor (defined by equation 6) in order to show the stability of immobilized cells:

$$\phi = r/r_0, \tag{6}$$

where r_0 is the initial reaction rate at $t = 0$.

FIGURE 2 shows the plots of activity factors against reaction time on semilogarithmic ordinates. The linearity shown in FIGURE 2 suggests that the activity factor is expressed as[9]

$$\phi = \exp(-\alpha t), \quad (7)$$

where α is a decay constant and could be rearranged to the Arrhenius-type form as follows:

$$\alpha = (4.686 \times 10^6)\exp(-50{,}380/RT). \quad (8)$$

The decay of the immobilized cells was a little higher than that of polyacrylamide gel.[6] This would be attributed to the characteristics of solid supports.

FORMULATION OF THE SIMULATION MODEL

Because Furui[7] suggested that the superficial liquid flow in immobilized cells deviated a little from an ideal plug flow, we made the following simplifying assump-

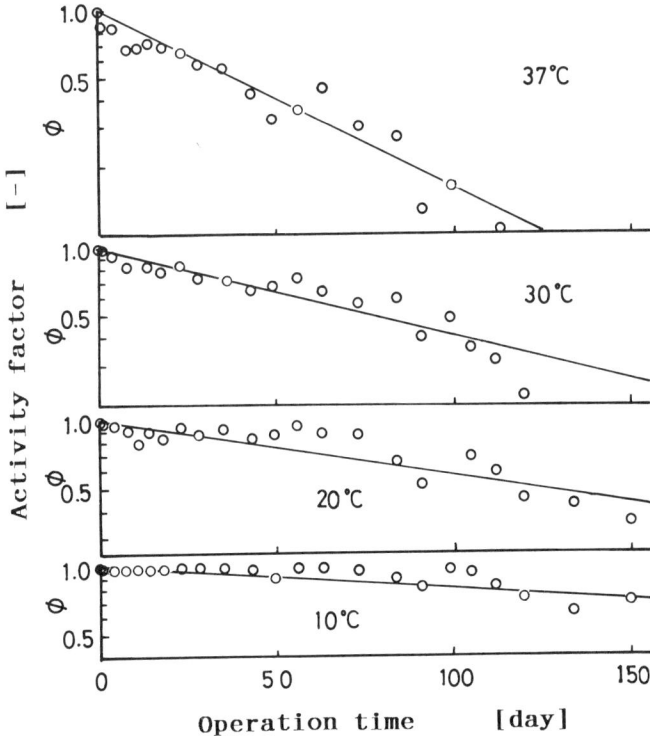

FIGURE 2. Stability of immobilized cells.

tions to develop a simulation model that could predict the concentration and temperature profiles along the bed height:

(1) the aspartase reaction proceeds in a pseudosteady state because the substrate passes the bed so quickly relative to the decay of the immobilized cells;
(2) the concentration and temperature profiles in the radial direction of the bed are negligibly small;
(3) the mass and heat transfer resistances between the immobilized cells and the substrate are infinity;
(4) internal diffusion effects of the substrate into immobilized cells can be ignored;[5]
(5) the cooling water is a perfectly mixed state in a heat-exchange bundle.

A sketch of the simulation model is illustrated in FIGURE 3; that is, the reactor consists of small N stages and each stage is perfectly mixed. In this model, mass and heat balances on the i-th stage can be expressed by equations 9 and 10, respectively:

$$FS_0(y_{i-1} - y_i) = rV_i\rho_s \qquad (9)$$

and

$$FC_f\rho_f(T_{i-1} - T_i) = UA_i(\Delta T)_{lm} + rV_i\rho_s\Delta H. \qquad (10)$$

Here, $(\Delta T)_{lm}$ indicates the logarithmic mean temperature difference between the temperatures of substrate and of cooling water on the i-th stage. It is expressed by

$$(\Delta T)_{lm} = (T_{ci} - T_{coi})/\ln\{(T_i - T_{coi})/(T_i - T_{ci})\}. \qquad (11)$$

On the other hand, the number of stages was calculated from equation 12:[10]

$$N = (Lu_0/2E) + 1. \qquad (12)$$

To examine the validity of the simulation model that combines the reactor model and the reaction kinetics, we measured the concentration and temperature profiles along the bed height using two different sizes of packed-bed reactors and then compared the observed data with the calculated data solved by the following procedure. As is obvious from FIGURE 3, the values of y_1 and T_1 on the first stage were calculated from equations 9–11 under the given operating values of y_0, T_0, and T_{c1}. Although the value of T_{co1} in equation 11 was unknown, it was estimated from the amounts of heat transfer of cooling water by using the Newton method. The obtained values of y_1 and T_1 were used as the operating values on the second stage. This procedure was successively repeated until the final N-th stage. The profiles are as follows:

(A) Concentration profiles for the small-scale reactor—FIGURE 4 shows the axial concentration profiles of fumaric acid for two different flow rates of substrate. As the number of stages, N, was equal to 4, the calculated curves

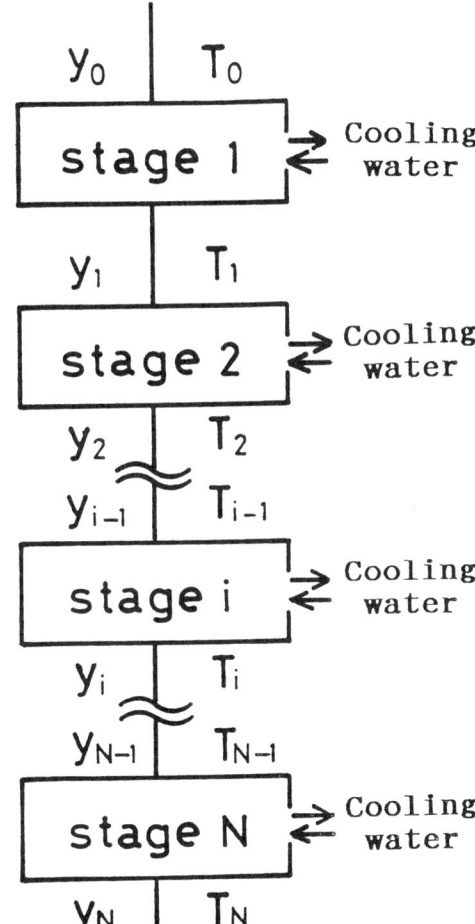

FIGURE 3. Sketch of the simulation model.

had the shape of stairs. It can be seen that the actual axial concentrations predicted from these stairs are in fair agreement with the observed ones.

(B) Temperature profiles for the pilot reactor—FIGURE 5 shows the calculated axial temperature profiles under various flow rates of substrate. In this case, as the number of stages equaled 33, the plots of the calculated values became almost smooth lines. It can be seen that the peak of maximum temperature moves to the bottom of the reactor with an increase in substrate flow rate and that the observed temperatures agree very well with the calculated ones.

These agreements in FIGURES 4 and 5 show that the reactor packed with immobilized cells can be reasonably regarded as a multistage continuous stirred-tank reactor.

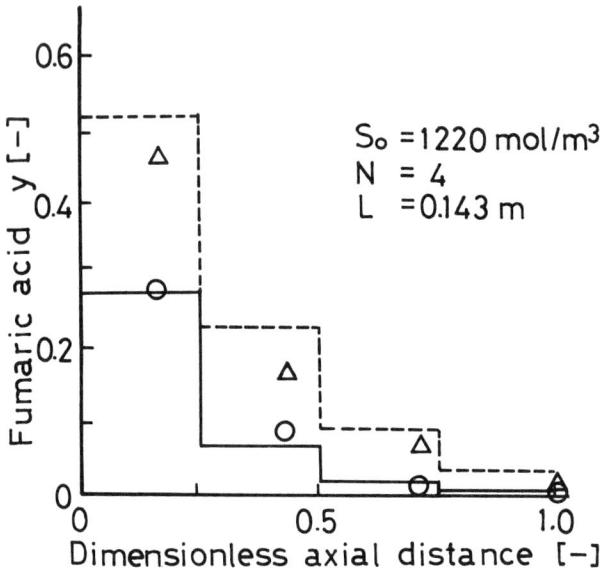

FIGURE 4. Concentration profiles for the small-scale glass reactor:

Observed	T_c [°C]	F [m^3/s]	Calculated
○	37.0	1.52×10^{-8}	—
△	10.0	0.85×10^{-8}	---

CONTROL DESIGN

Control-Plan Construction

For the control design of commercial plants, a performance index involving state variables has often been used to perform an economic optimization. Prior to the control-plan construction to use the immobilized cells effectively, attention needs to be devoted to the dot-dash line in FIGURE 5. The immobilized cells at the maximum temperature would reduce their activity rapidly, whereas the decay of immobilized cells at a lower temperature would be slow relative to that of a higher temperature. This would result in an uneven distribution of activity in the axial direction. If the heat-exchange tubes are divided into several parts and if the cooling water of different temperatures is fed into each part, then the decay of immobilized cells would be reduced.

From the numerical computations of the simulation model, optimal cooling patterns were analyzed for the pilot reactor, some imaginary reactors, and an expected final reactor. Consequently, the calculation results suggested that the heat-exchange tubes should be divided into four parts in every case (as shown in FIGURE 6) because dividing the heat-exchange tubes into more than five parts was only slightly effective for the life span of the immobilized cells.

Therefore, we wish to choose the optimal control policy in such a way that the summation of the decrease in the activity factor on the i-th stage with respect to a definite time will take the minimum value by using the four control variables of T_{c1}, T_{c2}, T_{c3}, and T_{c4}; that is,

$$P = \sum_{i=1}^{N} (\Delta\phi_i/\Delta t) \rightarrow \text{minimum} \ (i = 1–33). \tag{13}$$

This equation means a control design for the use of immobilized cells as long as possible.

Based on the practical operation, inequality constraints (equation 14) were placed on the control variables:

$$10 \ °C \leq T_{cj} \leq 45 \ °C \quad (j = 1–4). \tag{14}$$

A desired value for the outlet concentration of fumaric acid was also restricted by equation 15:

$$y_N \leq 0.01. \tag{15}$$

This value is an absolutely necessary condition to accomplish a high quality product.

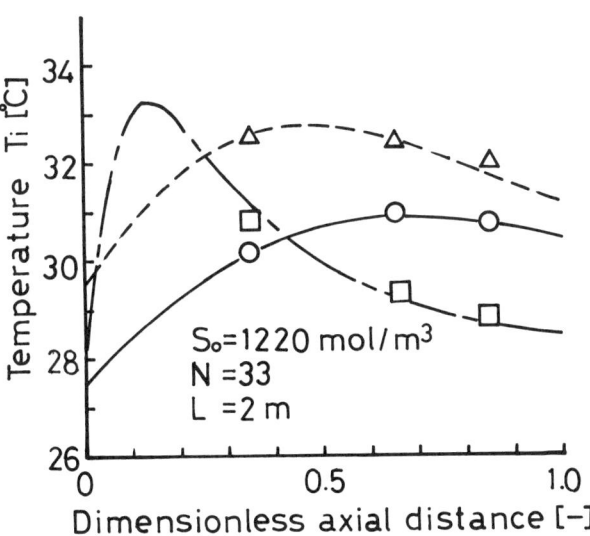

FIGURE 5. Temperature profiles for the pilot reactor:

Observed	T_c [°C]	F [m³/s]	Calculated
○	27.5	2.78×10^{-4}	—
△	29.5	1.94×10^{-4}	---
□	28.1	1.39×10^{-4}	· · ·

Optimization of the Control Design

The optimization was carried out as follows: At first, under four arbitrary temperatures of cooling water, T_{c1}, T_{c2}, T_{c3}, and T_{c4}, the concentration and temperature profiles were calculated from equations 9–11 using the initial activity of immobilized cells corresponding to $t = 0$ (days). Following this, the decrease in activity factor with respect to a definite time, that is, $(\Delta\phi_i/\Delta t)$, was calculated. This calculation was repeated until there was no change in the four temperatures of cooling water, which minimized equation 13 according to the Sequential Simplex method.[11]

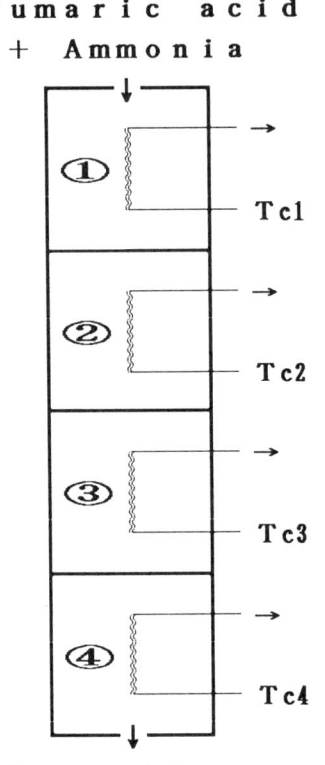

FIGURE 6. Sketch of an ideal reactor derived from the simulation model.

When equation 13 is minimized, the activity factor on the i-th stage should be shown by $\phi_i - \Delta\phi_i$ after the time of Δt. These procedures were continued as long as the necessary condition of equation 15 was satisfied.

The optimal control actions for the pilot reactor are given in FIGURE 7. When the activity of immobilized cells is high, control actions are within the lower limit of equation 14. As time increases, it is seen that adequate control actions are given for the four cooling waters because of the successive decrease in activity. In this connection,

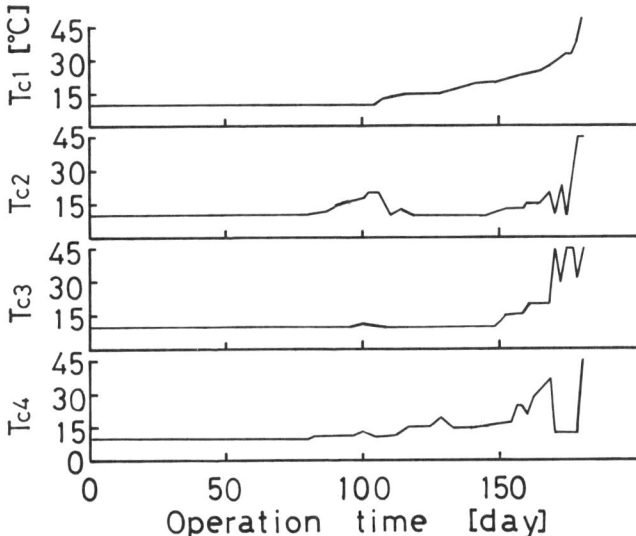

FIGURE 7. Optimal control actions for the pilot reactor.

the axial distributions of the activity factor are shown in FIGURE 8 as a function of the operation time. It can be seen that the decay is almost even with axial distance, although there are uneven curves without control.

Let us introduce the mean activity factor (defined by equation 16) to evaluate the operating conditions:

$$\bar{\phi} = \int_0^1 \phi_i \, dZ. \tag{16}$$

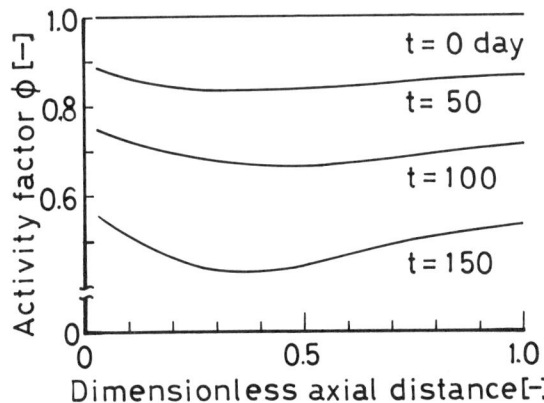

FIGURE 8. Axial distributions of the activity factor as a function of the operation time.

FIGURE 9 shows the variation in $\bar{\phi}$ under the optimization described above in comparison with $\bar{\phi}$ under an isothermal operation in which cooling water at 20 °C was fed into the four tube bundles at all times. Under the optimization, the operative use of immobilized cells was possible for 175 days and thus the mean activity factor was 0.21. In contrast, under the isothermal operation, the use of immobilized cells was only 80 days to keep the necessary condition of equation 15 and hence the final mean activity factor was 0.5. These curves mean that the half-life period[12] of immobilized cells is not always an index of the process operation.

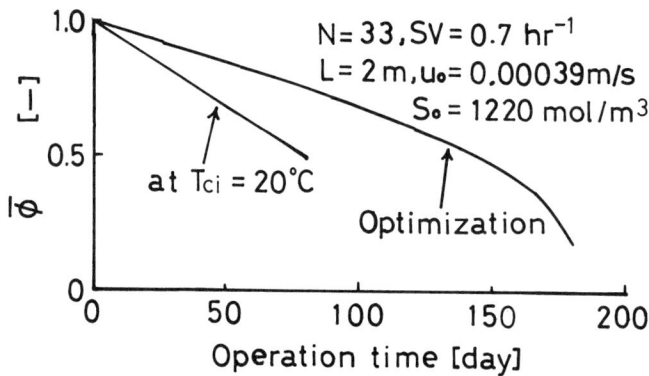

FIGURE 9. Variation in the mean activity factor under optimization and isothermal operation.

CONCLUSIONS

The following conclusions have been obtained:

(1) a reactor packed with immobilized cells was modeled by a multistage continuous stirred-tank reactor;
(2) a simulation model for the L-aspartic acid production was developed in a form able to predict the temperature and concentration profiles along the bed height;
(3) a packed-bed reactor equipped with four horizontal heat-exchange tube bundles was the most suitable for the production of L-aspartic acid;
(4) an optimal temperature policy to employ the immobilized cells as long as possible was performed;
(5) the half-life period of immobilized cells is not always an index of process operations.

ACKNOWLEDGMENTS

We are grateful to I. Chibata (president of our company), S. Saito (managing director of research and development headquarters), and S. Takamatsu and M. Furui of our laboratory for their helpful discussions and suggestions throughout this work.

REFERENCES

1. Tosa, T., T. Mori, N. Fuse & I. Chibata. 1966. Studies on continuous enzyme reactions. Enzymologia **31**: 214–224.
2. Chibata, I., T. Tosa & T. Sato. 1974. Immobilized aspartase-containing microbial cells. Appl. Microbiol. **27**: 878–885.
3. Nishida, Y., T. Sato, T. Tosa & I. Chibata. 1979. Immobilization of *Escherichia coli* cells having aspartase activity with carrageenan and locust gum. Enzyme Microb. Technol. **1**: 95–99.
4. Umemura, I., S. Takamatsu, T. Sato, T. Tosa & I. Chibata. 1984. Improvements of production of L-aspartic acid using immobilized microbial cells. Appl. Microbiol. Biotechnol. **20**: 291–295.
5. Takamatsu, S., M. Ueba & K. Yamashita. 1984. A stabilization of aspartase activity of immobilized *Escherichia coli* cells by temperature-raising operation. J. Chem. Eng. Jpn. **17**: 647–649.
6. Takamatsu, S., K. Yamashita & A. Sumi. 1980. Kinetics of production of L-aspartic acid by aspartase of immobilized *Escherichia coli* cells. J. Ferment. Technol. **58**: 129–133.
7. Furui, M. 1985. Heat-exchange column with horizontal tubes for immobilized cell reactions with generation of heat. J. Ferment. Technol. **63**: 371–375.
8. Rollan, G. C., M. C. Manca de Narda, A. A. Pesce de Ruiz Holgano & G. Oliver. 1985. Aspartate metabolism in *Lactobacillus murinus CNRS* 313. J. Gen. Appl. Microbiol. **31**: 403–409.
9. Froment, G. F. & K. B. Bischoff. 1962. Kinetic data and product distributions from fixed bed catalytic reactors subject to catalyst fouling. Chem. Eng. Sci. **17**: 105–114.
10. Kramers, H. & G. Alberda. 1953. Frequency-response analysis of continuous-flow systems. Chem. Eng. Sci. **2**: 173–181.
11. Olsson, D. M. 1974. A Sequential Simplex program for solving minimization problems. J. Qual. Technol. **6**: 53–57.
12. Wood, L. L. & G. J. Calton. 1984. A novel method of immobilization and its use in aspartic acid production. Bio/Technology **1984**: 1081–1084.

APPENDIX

Nomenclature

A_i heat-exchange surface ($\Sigma i = 10.38$), m^2
C_i specific heat of fluid $(1000 - 0.106 S_0) \times 4.1868$, J/kg-K
E axial diffusion coefficient (0.000012), m^2/s
F substrate flow rate, m^3/s
F_{ci} flow rate of cooling water on the i-th stage ($\Sigma i = 0.000257$), m^3/s
N number of stages
K^* apparent Michaelis constant, mol/m^3
K_1 modified equilibrium constant
L length of immobilized cells, m
R gas constant (8.31434), J/mol-K
R_0 maximum reaction rate, mol/kg-s
r reaction rate, mol/kg-s
S_0 initial concentration of fumaric acid (1220), mol/m^3
T temperature, K
T_i outlet temperature of substrate on the i-th stage, °C
T_{ci} inlet temperature of cooling water on the i-th stage, °C
T_{coi} outlet temperature of cooling water on the i-th stage, °C
t operation time, day

U overall heat transfer coefficient (200), W/m²-K
u_0 superficial fluid velocity, m/s
V_i volume of immobilized cells at the i-th stage ($\Sigma i = 1.32$), m³
y fractional concentration of fumaric acid
Z dimensionless axial distance (i/N)
α decay constant, 1/day
ΔH heat of reaction ($-25{,}120$), J/mol
Δt calculation step of operation time (1), day
ρ_f fluid density (1056), kg/m³
ρ_s density of immobilized cells (517), kg/m³

Modeling and Control of the Biocatalytic Conversion of Hydantoins to D–Amino Acids

L. TRANCHINO[a] AND F. MELLE[b]

Eniricerche S.p.A.
00015 Monterotondo (Roma), Italy

INTRODUCTION

D–Amino acids are known precursors of fine chemicals (β-lactam antibiotics, peptide hormones, pyrethroids, and others). They can be prepared on an industrial scale by the classical chemical resolution method (fractional crystallization of their salts with chiral amines or acids[1,2]), by chemoenzymatic processes (e.g., microbial transformation of 5-substituted DL-hydantoins followed by nitrous acid treatment;[3] stereospecific hydrolysis of amino acid amides followed by chemical hydrolysis[4]), or by a purely enzymatic process (e.g., direct microbial transformation of 5-substituted DL-hydantoins to D-amino acids[5,6]).

The direct enzymatic process developed in our laboratories,[6] utilized on an industrial scale by Recordati, involves a sequence of reactions performed by a single biocatalyst (*A. radiobacter* resting cells) expressing both D-hydantoinase and *N*-carbamoyl–D–amino acid amido hydrolase activities[6] (FIGURE 1).

In the present work, the mathematical simulation of the process was developed by focusing on the interaction between enzymatic reactions and mass-transfer processes through the cell membrane. The final goal of the work was to utilize the mathematical model for defining optimal process control strategies. The validity of the mathematical model was confirmed by an experimental check of the conversion of DL-5-isopropylhydantoin to D-valine. However, the approach that was utilized also holds for other D–amino acids.

EXPERIMENTAL APPROACH

Materials

DL-5-Isopropylhydantoin was prepared from DL-valine according to reference 7. All the chemicals utilized were reagent grade.

The microbial biomass was prepared by fermenting a strain of *A. radiobacter* NRRL B 11291 according to reference 6.

---

[a]Present address: Consorzio per il Trasferimento delle Biotecnologie S.p.A., Via Sardegna 38, 00187 Roma, Italy.
[b]Present address: Snamprogetti Biotecnologie S.p.A., Viale De Gasperi 16, 20097 S. Donato Milanese, Italy.

Analytical Methods

The hydantoinase and the N-carbamoyl–D–amino acid amido hydrolase activities of the microbial biomass were measured according to the procedure described in reference 6 using DL-5-isopropylhydantoin as the substrate.

The analysis of the DL-5-isopropylhydantoin hydrolysis mixture was carried out by HPLC (Varian Model 5060) according to reference 8.

The optical activity was measured with a Perkin-Elmer 241 MC polarimeter.

Hydrolysis of DL-5-Isopropylhydantoin

The hydrolysis was carried out in a stirred 2-L reactor (working volume of 1.5 L) at 40 °C and with a pH maintained at 7.8 by 2 N NaOH addition. An amount of 30 g (dry wt) of *A. radiobacter* cells was utilized for each experiment. The substrate DL-5-

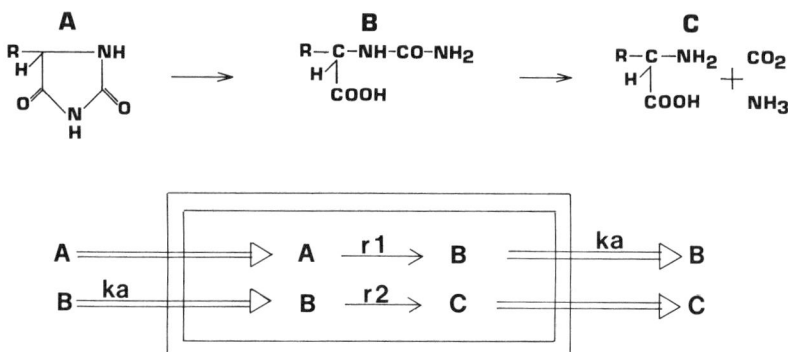

FIGURE 1. Schematic description of the biocatalytic system.

isopropylhydantoin was fed to the reactor either batchwise (addition at time = 0) or in the fed-batch mode (starting concentration = 0) at a constant rate (3.45 g/h) or at a modulated rate (controlled by a microrotocell operated by a step motor). The gaseous products (CO_2 and NH_3) were discharged from the reactor through a water seal in order to maintain the reactor at a pressure slightly higher than 1 atm.

MATHEMATICAL MODEL

System Description

The process consists of a sequence of steps (FIGURE 1):

(a) transport of the substrate A (hydantoin) through the cell membrane,
(b) enzymatic conversion of A to the intermediate B (N-carbamoyl–D–amino acid)

at the rate R_1 (hydantoinase activity), assumed to have a Michaelis-Menten kinetic expression with competitive and noncompetitive inhibition and inactivation coefficients,

(c) enzymatic conversion of B to the product C (D–amino acid) at the rate R_2 (N-carbamoyl–D–amino acid amido hydrolase), assumed to have the same kinetic expression as R_1,

(d) transport of the product C from the inside to the outside of the cell,

(e) transport of the intermediate B from the inside to the outside of the cell due to the accumulation of B into the cell at the beginning of the process and from the outside to the inside when the concentration gradient is reversed [rate $J = K_a(C_{BI} - C_{BE})$].

This last effect (e) is quite severe because the first reaction rate (V_{MA}) is normally much higher than the second one (V_{MB}) at the beginning of the process and the ratio, R_1/R_2, tends to increase with time because of the higher sensitivity to inactivation and inhibition of the second enzyme. Hence, B accumulates into the cell and is excreted outside. However, as the concentration of the substrate A decreases with time, a reverse situation is attained where $R_1 < R_2$. The concentration of B into the cell decreases and B diffuses back from the outside to the inside where it is converted to the final product C.

Whereas the permeability of the cell membrane to the intermediate (B) plays an important role in the process, substrate (A) and product (C) permeabilities are very high and we therefore assumed identical concentrations inside and outside the cell for A and C in the mathematical model.

Typical kinetic profiles are shown in FIGURE 2 for batch processes.

Mathematical Description

In TABLE 1, the differential equations describing the system are listed (see APPENDIX). They are based on the mass balance of each component and on the above-mentioned hypothesis for the transport and conversion rate equations.

All the equation parameters were adimensionalized to define generalized solutions for two consecutive enzymatic reactions coupled to a mass-transfer process (TABLE 2). Time was adimensionalized with reference to the activity of the first enzyme:

$$t = TV_{MA}/K_{MA}.$$

The activity of the second enzyme and the mass-transfer rate coefficient of the intermediate B through the cell membrane were also referred to the activity of the first enzyme in order to define the two key parameters of the process:

$$D = K_a(W/V_t)K_{MA}/V_{MA},$$

$$TR = V_{MA}/V_{MB}.$$

The system in TABLE 2 was integrated by a numerical method (subroutine DGEAR from the IMSL library running on an IBM 4331 computer).

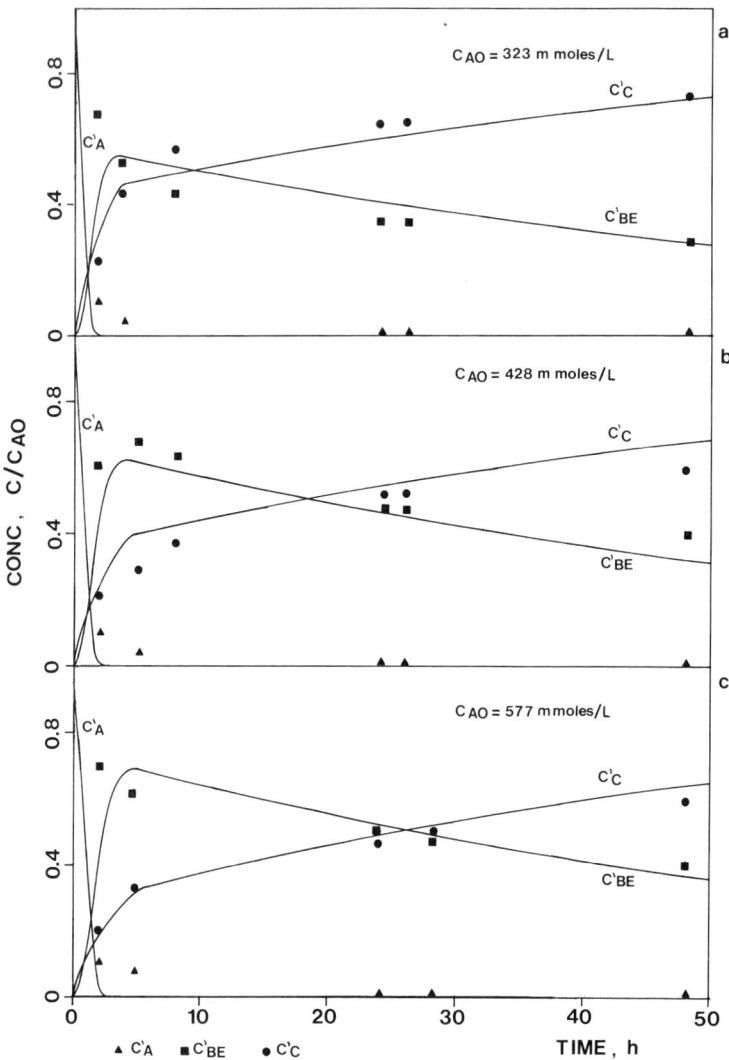

FIGURE 2. Batch conversion of DL-5-isopropylhydantoin: experimental results and theoretical curves of A, B, and C concentrations in the reaction medium as functions of time—$C_{A0} = 0.323$, 0.428, and 0.577 mol/L in parts a, b, and c, respectively. All the other parameter values are indicated in TABLE 4.

Sensitivity to the Parameters D and TR

The dynamics of the process were calculated for a set of D and TR values, with all other parameters being constant (TABLE 3).

The effect of the cell membrane permeability [expressed in terms of its ratio (D) to the first enzymatic reaction rate] is shown in FIGURE 3 in adimensional terms. The following comments can be made:

(i) The concentration of the intermediate B is roughly the same inside and outside the cell if D is high enough ($D > 0.3$, FIGURE 3a). The rate-limiting step of the process is the second enzymatic reaction and the process is quite fast, attaining a 90% conversion in an adimensional time of about 140.

(ii) When the permeability of the cell wall decreases ($D < 0.03$, FIGURE 3b), the concentration of B inside the cell becomes very high in the early phase of the process; B is excreted from the cell and, subsequently, it must diffuse back into the cell to complete the conversion. Then, the process becomes much slower (diffusion limited): 90% conversion is attained in an adimensional time of about 200 (for $D = 0.03$) or >900 (for $D = 0.003$) under the conditions indicated in TABLE 3.

(iii) When the permeability of the cell wall becomes extremely low ($D = 0.0003$, FIGURE 3c), the process again becomes very fast as B barely diffuses outside the cell; a 90% conversion is attained in an adimensional time of about 140 under the conditions indicated in TABLE 3.

The system behavior is summarized in FIGURE 4, where the inverse of the adimensional time ($1/t_X$) required to get any specified conversion (X) is plotted versus the cell wall permeability D. This parameter ($1/t_X$) is unaffected by D when the conversion is low enough ($X < 50\%$) or when the D values are very low or very high. For intermediate D values, $1/t_X$ becomes quite low and sensitive to D. The function $1/t_X = f(D)$ shows a minimum at a value of D that is lower when the conversion is higher.

TABLE 1. Mathematical Description of the System: Dimensional Parameters

$$V_t \cdot \frac{dC_A}{dT} = -W \cdot R_1 + F_v(T)$$

$$\alpha \cdot V_t \cdot \frac{dC_{BI}}{dT} = W \cdot (R_1 - R_2) - W \cdot K_a \cdot (C_{BI} - C_{BE})$$

$$(1 - \alpha) \cdot V_t \cdot \frac{dC_{BE}}{dT} = W \cdot K_a \cdot (C_{BI} - C_{BE})$$

$$\int_0^T F_v(T) \cdot dT = V_t \cdot [C_A + \alpha \cdot C_{BI} + (1 - \alpha) \cdot C_{BE} + C_C]$$

When $T = 0$, $C_A = C_{A0}$ $C_{BI} = C_{BE} = C_C = 0$ $C_I = C_C$

$$R_1 = \frac{V_{MA} \cdot C_A \cdot e^{-K_1 \cdot T}}{[K_{MA} \cdot (1 + C_I/K_{1c}) + C_A] \cdot (1 + C_I/K_{1nc})}$$

$$R_2 = \frac{V_{MB} \cdot C_B \cdot e^{-K_2 \cdot T}}{[K_{MB} \cdot (1 + C_I/K_{2c}) + C_B] \cdot (1 + C_I/K_{2nc})}$$

In the batch operation:

$$F_v(T) = 0$$

TABLE 2. Mathematical Description of the System: Adimensional Parameters

$$\frac{dc_A}{dt} = -r_1 + f_v(t)$$

$$\alpha \cdot \frac{dc_{BI}}{dt} = (r_1 - r_2) - D \cdot (c_{BI} - c_{BE})$$

$$(1 - \alpha) \cdot \frac{dc_{BE}}{dt} = D \cdot (c_{BI} - c_{BE})$$

$$\int_0^t f_v(t) \cdot dt = c_A + \alpha \cdot c_{BI} + (1 - \alpha) \cdot c_{BE} + c_C$$

When $\quad t = 0, \quad c_A = c_{A0} \quad c_{BI} = c_{BE} = c_C = 0 \quad c_I = c_C$

$$r_1 = \frac{c_A \cdot e^{-k_1 \cdot t}}{(1 + ED1 \cdot c_I + c_A) \cdot (1 + c_I \cdot ED2)}$$

$$r_2 = \frac{(c_B/TR) \cdot e^{-k_2 \cdot t}}{[(1 + ED3 \cdot c_I)/ME + c_A] \cdot (1 + c_I \cdot ED4)}$$

In the batch operation:

$$f_v(t) = 0$$

In other words, the worst situation occurs when a high conversion is required and when the cell wall has a low permeability: diffusion becomes the rate-limiting step, particularly in the final part of the process.

The sensitivity of the process dynamics to the ratio of the two enzymatic reaction rates (TR) is shown in FIGURE 5 at an intermediate cell wall permeability value ($D = 0.03$). The effect of excretion and subsequent diffusion into the cell of the intermediate B is enhanced by the increase of the ratio TR, with a consequent slowing down of the process ($TR = 6$ in FIGURE 5c). The effect disappears when the two enzymatic reaction rates become equal ($TR = 1$ in FIGURE 5a). Under these conditions, the process becomes much faster than in all the previously mentioned cases.

TABLE 3. Parameters Utilized for the Process Simulation Studies

$$\alpha = 0.05$$

$$ED1 = ED2 = ED3 = ED4 = 0$$

$$ME = 50$$

$$K_1 = K_2 = 0$$

When $\quad t = 0, \quad C_{BI} = C_{BE} = C_C = C_I = 0$

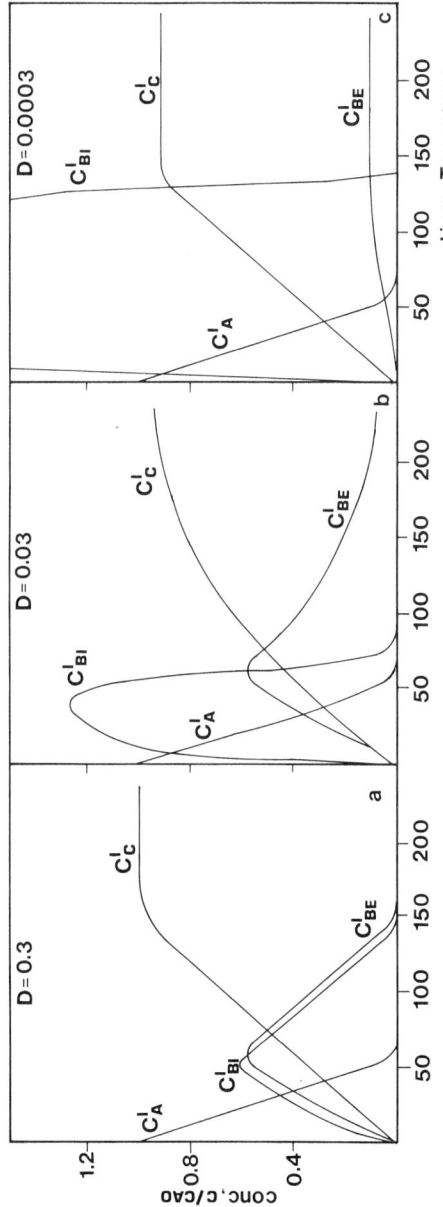

FIGURE 3. Effect of the cell membrane permeability (D) on the dynamics of a batch biocatalytic process: $D = 0.3$, 0.03, and 0.0003 in parts a, b, and c, respectively, with a constant ratio of the two enzymatic reactions ($TR = 3$). All the other parameter values are indicated in TABLE 3.

Sensitivity to the Substrate Flow Rate

The aforementioned characteristics of the system, mainly, when $TR > 1$ and D is in the intermediate range (0.001–0.01), suggest that a fed-batch mode of operation should be better than a batch one. If the substrate A is fed at a rate comparable to the second enzymatic reaction rate (R_2), then the excretion of B is minimized and the slow phase of reassimilation of B is reduced, making the whole process faster. Thus, the feeding rate—or, in other words, the adimensional time (t') required to feed a certain amount of substrate at a constant rate—becomes a key parameter of the process. In

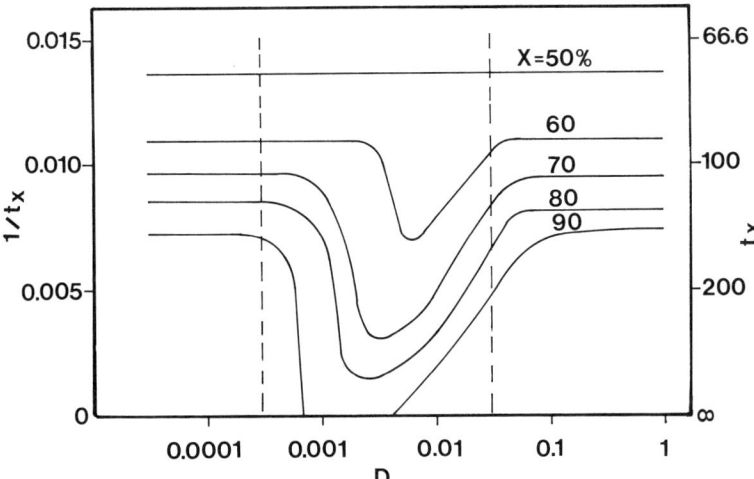

FIGURE 4. Effect of the cell membrane permeability (D) on the time (t_X) required for getting any specific conversion (X) in a batch process at $TR = 3$. All the other parameter values are indicated in TABLE 3.

FIGURE 6, the effect of t' on the dynamics of the process is shown:

(i) At very high substrate flow rates ($t' = 36$), the behavior of the fed-batch system is not very different from the batch one, indicating excretion of B and a slow conversion rate.

(ii) At intermediate substrate flow rates ($t' = 144$), production and consumption of B are quite well balanced, the diffusion of B outside the cells is negligible, and the overall process is quite fast.

(iii) At low substrate flow rates ($t' = 252$), the diffusion of B outside the cells is still negligible, but the process is slowed down because it is limited by a very low substrate feeding rate. In other words, the catalytic activity of the microbial biomass is not fully exploited under these conditions.

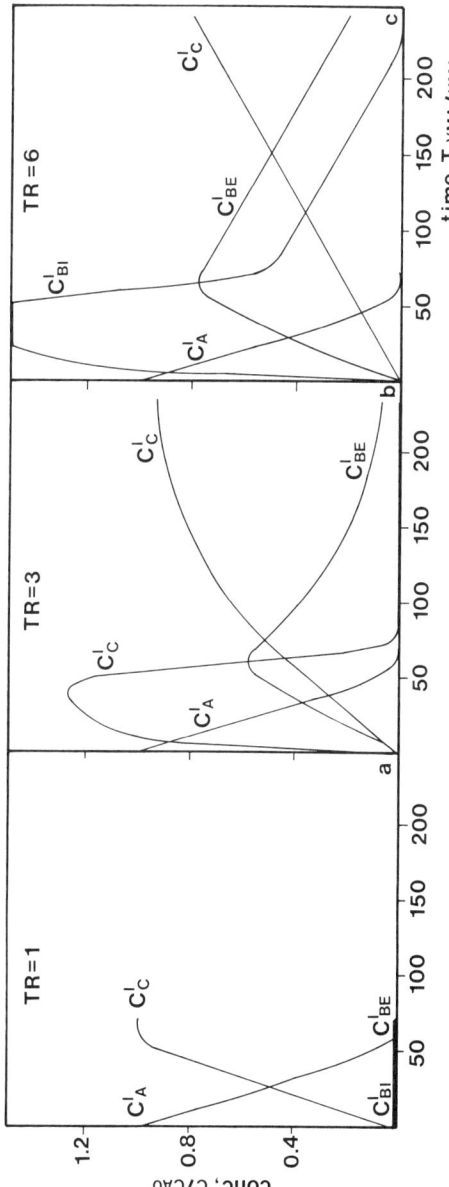

FIGURE 5. Effect of the ratio (TR) of the two enzymatic reactions on the dynamics of a batch biocatalytic process: $TR = 1$, 3, and 6 in parts a, b, and c, respectively, with a constant D value ($D = 0.03$). All the other parameter values are indicated in TABLE 3.

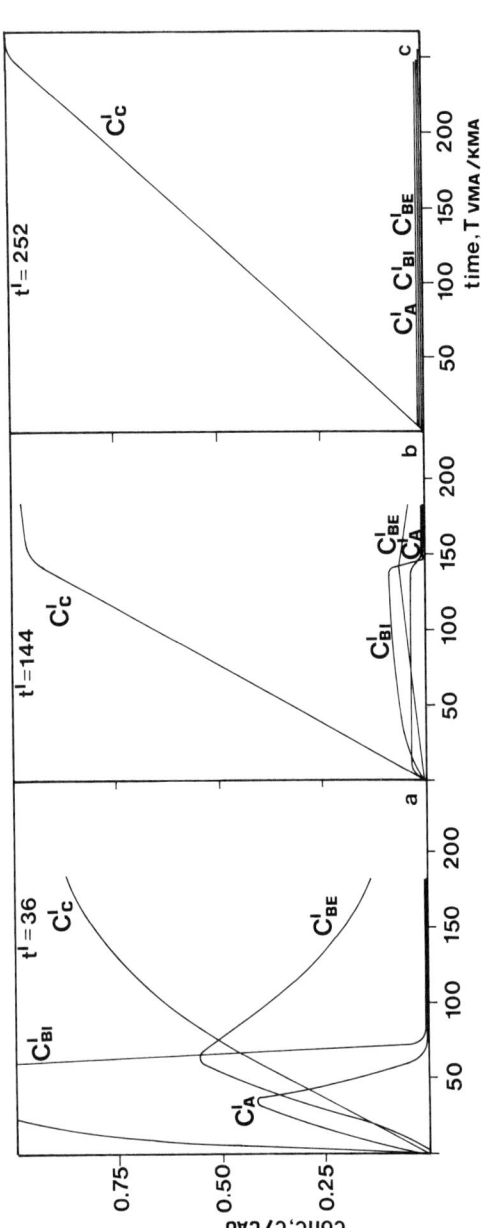

FIGURE 6. Effect of the substrate feeding rate (expressed in adimensional terms as the time t^I required to feed 1 mol/L of substrate) on the dynamics of a fed-batch biocatalytic process: $TR = 3$ and $D = 0.03$; all the other parameter values are indicated in TABLE 3.

RESULTS AND DISCUSSION

Experiments were carried out to evaluate the adequacy of the mathematical model, to determine the best values of the parameters, and to verify the conclusions of the process simulation, especially for the more efficient fed-batch operation mode.

TABLE 4. DL-5-Isopropylhydantoin Conversion to D-Valine: Operative Conditions and Model Parameters Utilized to Fit the Experimental Data

C_{A0} = 0.113, 0.323, 0.428, 0.577 mol/L
When $t = 0$, $C_{BI} = C_{BE} = C_C = 0$
W/V_t = 20 g/L
V_t = 1.5 L
V_{MA} = 0.36 mol/h/L
V_{MB} = 0.12 mol/h/L
K_{MA} = 0.05 mol/L
K_{MB} = 0.001 mol/L
$K_{1nc} = K_{1c} = K_{2c} = 0$
K_{2nc} = 0.04 mol/L
K_1 = 0.036 h^{-1}
K_2 = 0.36 h^{-1}
α = 0.02
$K_a W/V_t$ = 0.0144 h^{-1}
F_v = 0, batch mode
    = 3.45 g/h, fed-batch mode with constant feeding rate
    = $R_2 W$, fed-batch mode with modulated feeding rate, experimentally simulated by discontinuously decreasing feeding rates (see FIGURE 8)

Batch Experiments

In FIGURE 2, experimental results are compared to the theoretical curves calculated from the set of parameters indicated in TABLE 4. The three situations in FIGURE 2 refer to different substrate concentrations (C_{A0}), with the biocatalyst concentration being identical ($W = 20$ g/L). All the available data obtained under different operating conditions, feeding modes (batch or fed-batch), and catalyst concentrations, including data not shown, have been compared to the theoretical curves using the same set of parameters. The fitting, although not perfect, can be considered to be adequate as most of the parameters have been measured independently.

When C_{A0} is low (0.323 mol/L), a conversion of about 75% is obtained after 48 hours of reaction. After the initial fast phase, the rate-limiting step becomes the

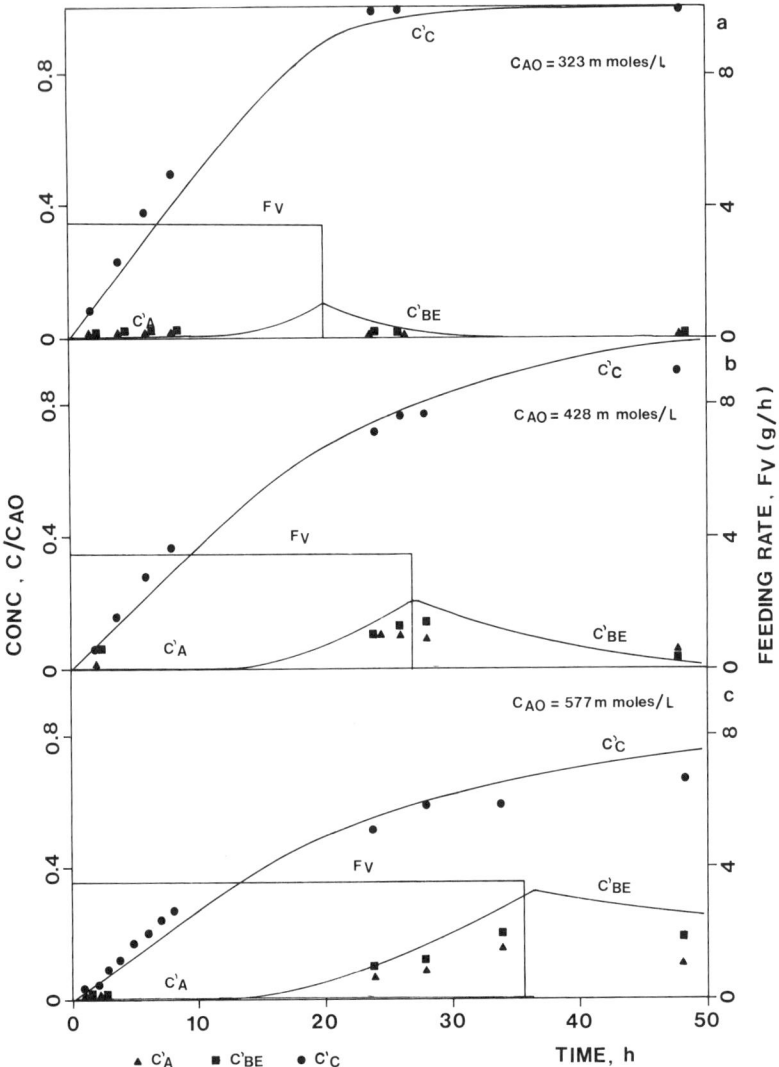

FIGURE 7. Fed-batch conversion of DL-5-isopropylhydantoin under constant feeding rate: experimental results and theoretical curves of A, B, and C concentrations in the reaction medium as functions of time—$C_{A0} = 0.323$, 0.428, and 0.577 mol/L in parts a, b, and c, respectively. All the other parameter values are indicated in TABLE 4.

diffusion of B from the outside to the inside of the cells. Then, both B and C concentrations change very slowly with time (FIGURE 2a).

If C_{A0} is higher (0.428 mol/L), the effect of excretion of B is enhanced and the final conversion attained becomes lower (65% after 48 h, FIGURE 2b).

A further increase in C_{A0} produces a slight decrease in the conversion after 48 h

(FIGURE 2c). In addition to the factors D and TR discussed above, the inactivation rates of the two enzymes (K_1 and especially K_2) play an important role for the process dynamics.

Fed-Batch Experiments with Constant Feeding Rate

In FIGURE 7, experimental results of fed-batch conversions are compared to the theoretical curves calculated from the TABLE 4 parameters. The three situations in FIGURE 7 refer to different final substrate amounts fed to the reactor at the same rate ($F_v = 3.45$ g/h hydantoin), but with increasing feeding times. The total substrate concentrations to be converted (C_{A0}) are the same as in FIGURES 2a–c; in this way, a comparison can be made between the batch and the fed-batch modes of operation. The improvement in the final conversion attained in the fed-batch mode, predicted by the model, is evident in all the cases.

Fed-Batch Experiments with Modulated Feeding Rate

Further improvements could be obtained if the substrate feeding rate was modulated in order to fit at any time the second enzyme reaction rate ($F_v = R_2W$), thereby avoiding both excretion of B (too fast feeding) and underutilization of the biocatalyst (too slow feeding). A direct experimental confirmation of this conclusion was not possible because a sensor for the reaction progress was not available. However, we simulated the process control by feeding, in a series of experiments, the substrate at predetermined feeding rates calculated according to the model predictions. FIGURE 8 shows the actual feeding rates (discontinuous periodical reduction of F_v), adopted in each experiment as functions of time, as well as the kinetics of amino acid production compared to the theoretically predicted curves.

Improved performances of the modulated fed-batch operation mode with respect to the constant feeding rate and the batch modes resulted both from the experimental results and from the theoretical predictions (FIGURES 2, 7, and 8).

CONCLUSIONS

The biocatalytic conversion of DL-5-isopropylhydantoin to D-valine by using *A. radiobacter* cells seems to be adequately simulated by a general mathematical model of two consecutive enzymatic reactions coupled with the mass-transfer phenomena of substrate, product, and intermediate through the cell membrane. The actual situation is much more complex, with more than two enzymes involved (in reference 8, the expression of a hydantoin racemase activity in *A. radiobacter* cells was demonstrated) and probably with more complex inhibition and inactivation patterns.

Nevertheless, the most important features of the system have been simulated:

    (i) The process dynamics dependence from the permeability of the cell membrane (D)—This dependence accounts for the observed variability of the

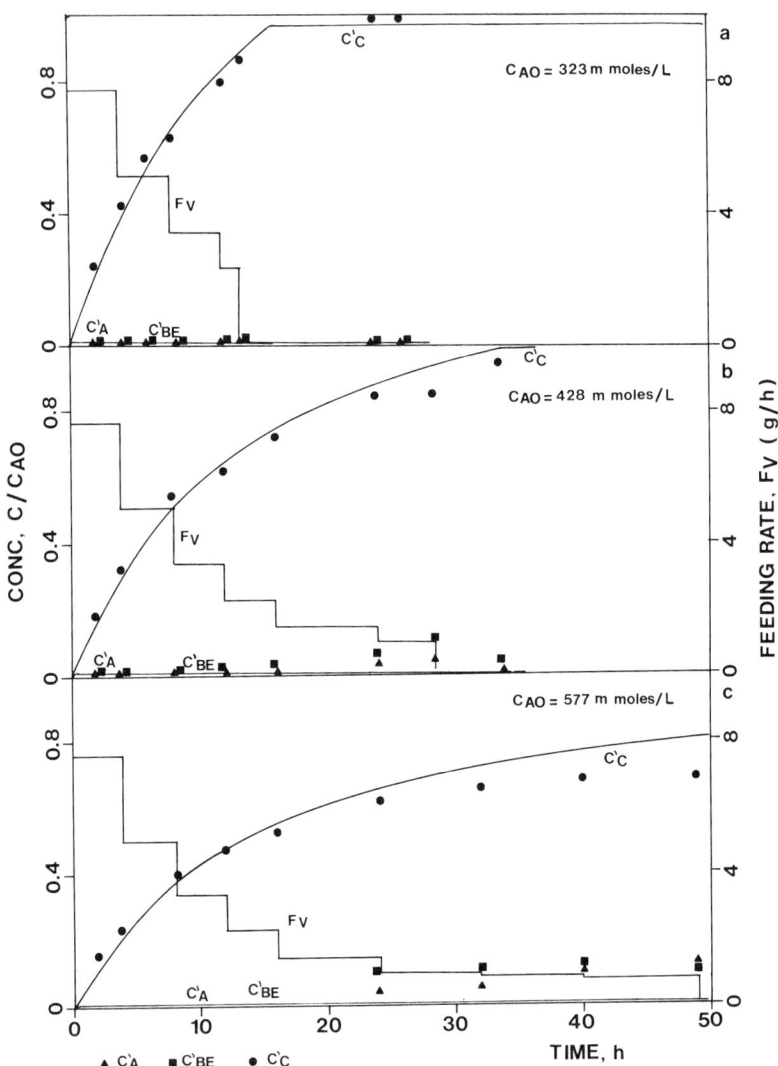

FIGURE 8. Fed-batch conversion of DL-5-isopropylhydantoin under modulated feeding rate: experimental results [obtained using predetermined discontinuously decreased feeding rates (F_v)] and theoretical curves of A, B, and C concentrations in the reaction medium as functions of time—C_{A0} = 0.323, 0.428, and 0.577 mol/L in parts a, b, and c, respectively (in the last experiment, the feeding was stopped at 0.500 mol/L as the system was not converting any more hydantoin). All the other parameter values are indicated in TABLE 4.

final conversion obtained with different samples of biocatalyst that show the same initial activity (FIGURE 3).

(ii) The process performances improvement by using a fed-batch mode of operation, rather than a batch one, and particularly by adapting the substrate feeding to the second enzyme reaction rate—The practical application of the latter operation mode is possible as the production of CO_2 or NH_3 (produced stoichiometrically with the D-amino acid) can be monitored by an appropriate probe and can drive the substrate feeding rate.

A control strategy, as proposed, seems to be appropriate for improving the utilization of the biocatalyst and for facing unavoidable problems of slight changes in biocatalyst characteristics (ratio of enzymatic activities, cell membrane permeability, etc.).

REFERENCES

1. HOLDREGE, C. T. 1974. United States patent no. 3,796,748.
2. YAMADA, S., C. HONGO & I. CHIBATA. 1978. Agric. Biol. Chem. **42**: 1521.
3. YAMADA, H., S. TAKAHASHI, Y. KII & H. KUMAGAI. 1978. J. Ferment. Technol. **56**: 484.
4. BOESTEN, W. H. J. & L. R. M. MEIJER-HOFFMAN. 1978. United States patent no. 4,080,259.
5. NAKAMORI, S., K. YOKOZEKI, K. MITSUGI, S. EGUCHI & I. IWAGAMI. 1980. United States patent no. 4,211,840.
6. OLIVIERI, R., E. FASCETTI, L. ANGELINI & L. DEGEN. 1981. Biotechnol. Bioeng. **23**: 2173.
7. SUZUKI, T., K. IGARASHI, K. HASE & K. TUZIMURA. 1973. Agric. Biol. Chem. **37**: 411.
8. BATTILOTTI, M. & U. BARBERINI. 1988. J. Mol. Catal. **43**: 343.

APPENDIX

Symbols

C_A, C_{BI}, C_{BE}, C_C, C_I	hydantoin, N-carbamoyl amino acid inside and outside the cell, amino acid, and generic inhibitor concentrations [mol/L], respectively
C_{A0}	initial concentration of hydantoin
T	time [h]
V_t	total volume of the reaction medium [L]
W	catalyst amount [g]
α	ratio of the cells volume to the total volume
R_1, R_2	enzymatic reaction rates [mol/h/g]
F_v	substrate feeding rate [mol/h]
V_{MA}, V_{MB}	maximum enzymatic reaction rates [mol/h/L]
K_{MA}, K_{MB}	Michaelis-Menten kinetic constants [mol/L]
K_{1c}, K_{2c}	competitive inhibition constants [mol/L]
K_{1nc}, K_{2nc}	noncompetitive inhibition constants [mol/L]
K_a	transport coefficient of the intermediate through the cell membrane [L/g-h]
K_1, K_2	thermal inactivation coefficients [h^{-1}]
$X = 100(C_{A0} - C_A)/C_A$	conversion [%]

Definitions of Adimensional Terms

$c = C/K_{MA}$
$r = R/V_{MA}$
$f_v = F_v/V_{MA}$
$t = TV_{MA}/K_{MA}$
$k_1 = K_1 K_{MA}/V_{MA}$
$k_2 = K_2 K_{MA}/V_{MA}$
$D = (K_a W/V_t) K_{MA}/V_{MA}$
$TR = V_{MA}/V_{MB}$
$ME = K_{MA}/K_{MB}$
$ED1 = K_{MA}/K_{1c}$
$ED2 = K_{MA}/K_{1nc}$
$ED3 = K_{MA}/K_{2c}$
$ED4 = K_{MA}/K_{2nc}$
$C' = C/C_{A0}$

An Expert System for Cultivating Operations

H. ASAMA,[a] T. NAGAMUNE,[a] M. HIRATA,[b]
A. HIRATA,[b] AND I. ENDO[a]

[a]Chemical Engineering Laboratory
RIKEN (The Institute of Physical and Chemical Research)
Wako-shi, Saitama 351-01, Japan

[b]Department of Chemical Engineering
School of Science and Engineering
Waseda University
Ohkubo 3-4-1, Shinjuku-ku, Tokyo 160, Japan

INTRODUCTION

Bioprocess automation has been required in order to shorten the development period of bioproducts, to improve reliability and flexibility of production processes, to improve a product's quality, to save manpower, and to enhance total productivity. Automation technologies in bioprocesses, though, are far behind those in other fields such as the machine industry, the electronic industry, etc. In other words, bioprocess is still controlled by operators who are experts of separating biocatalysts or of cultivation. This means that complicated and diverse bioprocesses are demanded in their operations. Thus, they are based on the operator's empirical and/or heuristic decisions utilizing knowledge stored through past experiments and experiences. Therefore, it is essential to implement the operator's knowledge on experiences and know-how in order to realize a fully automated production system in bioindustries.

On the other hand, in accordance with the progress in the computer science and information processing technologies, knowledge-based systems have been developed in various fields. Even in biotechnology, although being abstract in concept, research works of AI applications for bioprocess design, fermentation control, etc., have been reported.[1,2]

Our objective is to develop new technologies to automate cultivating operations carried out in the stage of research and development of bioproducts. We developed the BIOACS (BIO Advanced Control System[3]), which enables on-line monitoring of various state variables and optimal control based on physiological data of a microorganism, which have been stored in a bio-data base. Moreover, in order to enhance the function of the BIOACS, we are developing expert systems[4] as a supervisory system of the BIOACS by implementing knowledge necessary for bioprocesses. In this report, we discuss a concept of an autonomous and fully automated cultivation system, as well as functions of the expert system (ESCO: Expert System for Cultivating Operations), which support decision making in cultivating operations totally. Then, methods of knowledge representation and the inferring process are proposed, and an implemented prototype system is presented.

AN ADVANCED CULTIVATION SYSTEM

The BIOACS

The ESCO is situated as a supervisor of the BIOACS, which we have developed already. Characteristic features of the BIOACS are summarized as follows:

(1) on-line measurement of cell mass concentration, substrate concentration, and product concentration;
(2) on-line monitoring of physiological activities of microorganisms, which are represented by various specific rates;
(3) optimal control of bioprocesses, which is executed on the basis of the characteristic properties of the bioprocess parameters saved in a data base.

These functions are achieved by the development of an on-line turbidisensor,[5] an on-line sampling unit of cell-free culture medium,[6] accurate estimation of the specific rates via a Kalman filter,[7] and dynamic responses of the specific rates to fed-batch operations.[8]

However, from the supervisory viewpoint, an autonomous and fully automated cultivation system has to guarantee not only advanced control of the cultivation process, but also on-line and real-time diagnosis and dynamic process planning in order to cope with unexpected disturbances to the process, unknown changes of cultivating conditions, unestimable performances of the microorganisms, or any other troubles such as mechanical defects or misoperations. In this sense, the BIOACS is the best system to be upgraded to a more advanced system because it provides an on-line turbidisensor and an on-line sampling unit and it enables us to recognize the dynamic state of the process in real time.

The cultivation process is composed of various operations such as experimental design, installation of sensors, culture medium preparation, sterilization, inoculation of cells, cleaning of the fermentor, and maintenance operations. These operations are, in nature, not always procedural, but require empirical knowledge with heuristic inference. In addition, we first have to determine the cultivating conditions or parameters on process control before the experiment run because these are associated with lots of unknown or ambiguous factors. Therefore, decisions in cultivating operations made by operators are rather trial-and-error than decisive. Moreover, it is difficult to construct a definite model of cultivation with respect to the biochemical reaction. Taking account of these discussions, we present an advanced cultivation system.

Concept of an Advanced Cultivation System

A concept that we propose here is shown in FIGURE 1.[9] The system consists of the following main component systems:

(1) **BIOTRON**, an intelligent fermentation factory, which characterizes microorganisms and generates experimental data. BIOTRON is integrated with robots in order to automate cultivating operations. The BIOACS can be regarded as a primitive system of BIOTRON.

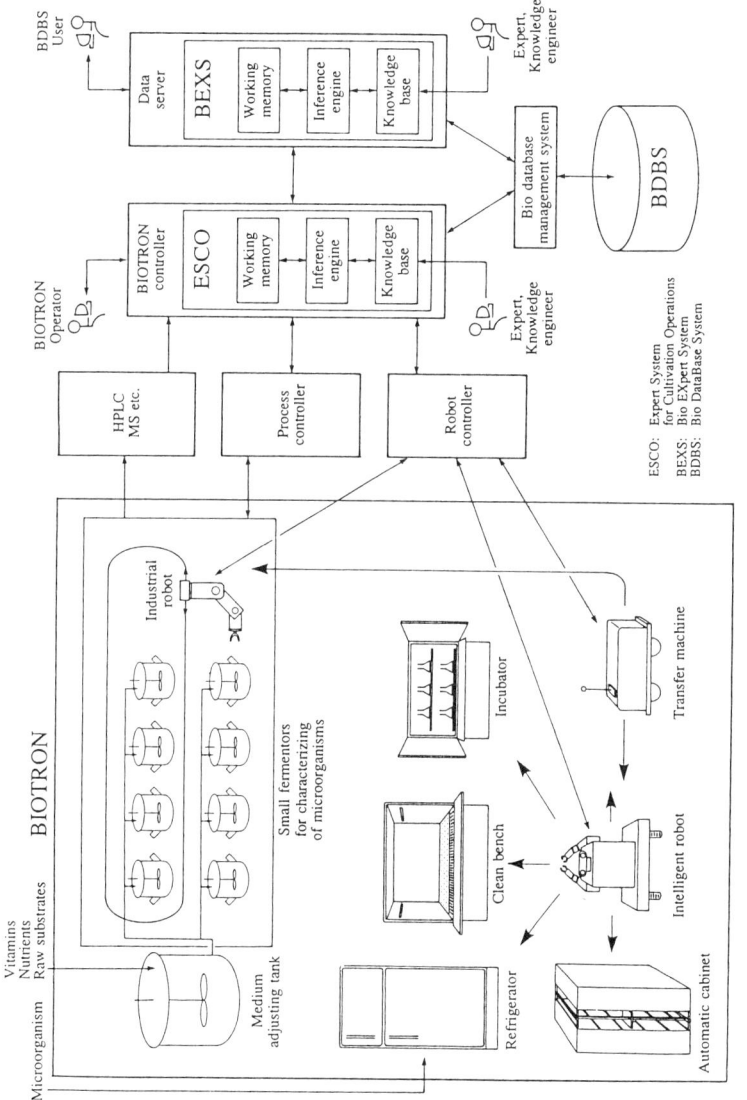

FIGURE 1. Concept of an advanced cultivation system.

(2) **BDBS** (**B**io-**D**ata **B**ase **S**ystem), storage of experimental data representing various physiological characteristics and performance of microorganisms in cultivation processes.

(3) **ESCO** (**E**xpert **S**ystem for **C**ultivation **O**perations), an expert system necessary for controlling BIOTRON. ESCO not only controls cultivation processes in BIOTRON optimally, but it also diagnoses the bioreactor system and the cultivation processes and designs the experimental fermentation processes.

(4) **BEXS** (**B**io **EX**pert **S**ystem), another expert system necessary for intelligent data service to general users. BEXS supports process planning, production design, and scheduling by utilizing characteristic data stored in BDBS.

In this report, we focus on the ESCO.

The ESCO: Expert System for Cultivating Operations

The mechanism of biochemical reaction is so complicated that knowledge and know-how of human operators play a very important part in decision making in cultivating operations. Hence, the ESCO is expected to apply knowledge engineering to various problems from upstream to downstream in the operations. Furthermore, in the stage of research and development of bioproducts, one of the main concerns is to find the optimal fermentation conditions. However, this problem is usually solved by numerous experiments and mathematical modeling. The ESCO should be provided with the following functions:

(1) *Diagnosis of Troubles*: In the case where troubles result from any misoperations or mechanical defects, the system is required to be able to discover them, to specify their root causes, and to instruct countermeasures for repairing or recovering. It is also expected to diagnose contamination by miscellaneous microorganisms.

(2) *Intelligent Process Control*: In the case where unexpected disturbances to the process occur or where variation of cultivating conditions cannot be estimated beforehand, the system is required to be able to modify the controlled process parameters while the process is running and to continue the cultivation process. It is proper that the function for the optimal control of fed-batch operations should also be included.

(3) *Dynamic Experimental Design*: In the case where experimental processes are obliged to be designed based on ambiguous or unreliable information or when data representing the performance of microorganisms are not acquired sufficiently previous to the process run, the system is required to store the observed data and to redesign the process dynamically by taking account of the data acquired in real time.

(4) *Simulation of Operations*: In order to realize these functions, it is indispensable for the system to refer to the data base on past experiments. In addition, when detecting troubles, sensed data in the cultivation process should be monitored by comparison with results from a simulator that estimates the performance of microorganisms referring to the data base.

A DIAGNOSING SYSTEM

Diagnosis of Fermentation System

Here, we discuss the methodology of knowledge representation and the inference process with respect to the function of diagnosing the bioreactor system. As mentioned above, in the case of diagnosis of fermentation processes, we should take account of not only recovery of troubles by detecting any mechanical defects or misoperations, but we also should place emphasis on intelligent control by monitoring unsuitable cultivating conditions and on data base construction by acquiring data of unknown performances of the microorganism. Thus, the system diagnosing the bioreactor system should have the function to recognize the state of the bioreactor system and processes, which is considered to be intelligent control or dynamic experiment planning. In other words, this system has a learning function in the narrow sense that it accumulates all data automatically in the data base so that it can be applicable to other cultivating operations.

In developing a prototype system, the BIOACS is assumed to be a diagnosed object. As the BIOACS has a typical structure as a bioreactor system, this system can be applied to any general bioreactor system. However, we have the necessary conditions that physiological activities can be monitored and that the system is integrated with the data base, which both have already been realized in the BIOACS. FIGURE 2 shows the flowchart of the BIOACS.

The diagnosing system should be provided with the following functions:

(1) efficient reasoning with an expert's heuristics,
(2) deep reasoning on the basis of the structure of the bioreactor system (BIOACS),
(3) evaluation of the effects caused by trouble,
(4) utilization of available characteristic data.

Furthermore, the system requires the following items:

(1) extensibility and maintainability,
(2) generality to be applied to any bioreactor system,
(3) generous framework for redundant knowledge,
(4) reliability of diagnosed results,
(5) combination with sensors and other software resources.

From the above-mentioned requirements, it is concluded that the system is required to be highly flexible by apt modulization of knowledge bases.

Knowledge Representation

In designing the configuration of knowledge bases, information or knowledge as shown below should be provided in order to diagnose a bioreactor system and cultivation processes:

(1) structural model of the bioreactor system,
(2) states or conditions of the bioreactor system and cultivation processes,
(3) heuristic logics extracted from diagnosing experts,

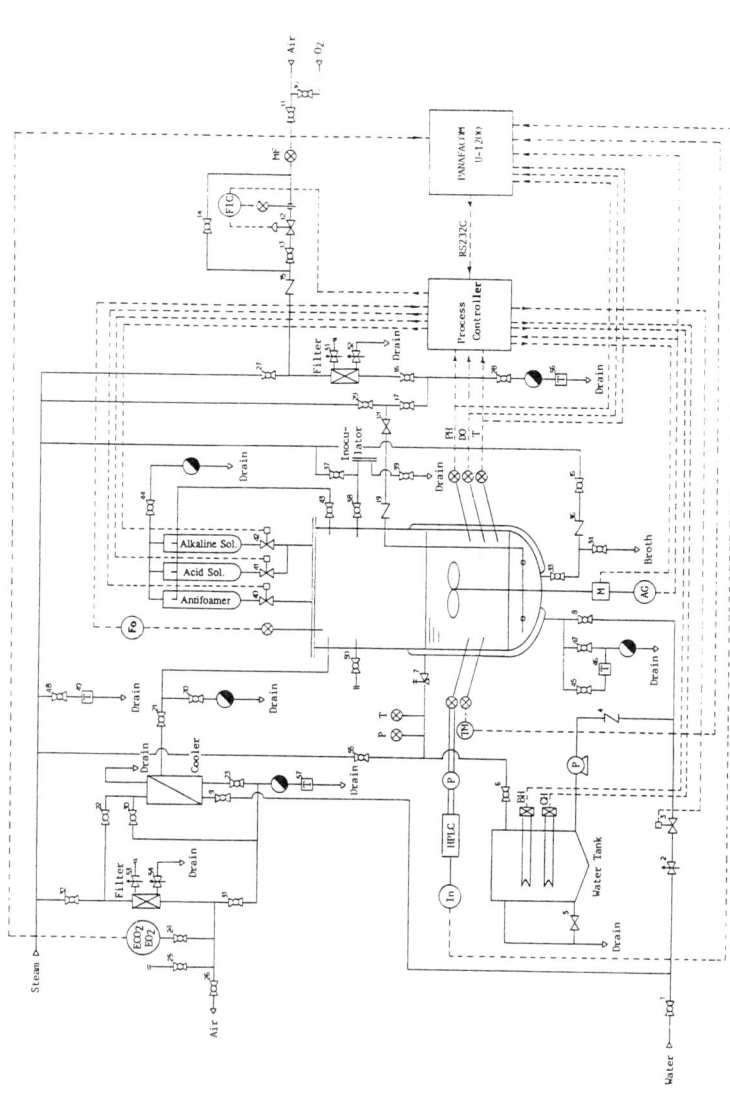

FIGURE 2. Flowchart of BIOACS. Abbreviations: BH = base heater; CH = control heater; AG = agitation speed meter; DO = dissolved oxygen concentration; HPLC = high performance liquid chromatograph; TM = turbidity meter; In = integrator; Fo = foam sensor; MF = mass flow meter.

(4) causality of events of troubles,
(5) data on characteristic properties of bioprocess parameters and physiological activities.

Taking into account these discussions and requirements mentioned in the previous section, we adopted the schemes of knowledge representation and the configuration of knowledge bases as shown in FIGURE 3.

Logics to diagnose the bioreactor system and cultivation processes are represented by production rules and they are modulized into several knowledge sources, each corresponding to the monitoring or controlling parameters such as temperature, pH, DO, etc. Knowledge on the structure of the bioreactor system composed of various equipment and knowledge on the states of the processes are represented by a network of frames, which is necessary for the application of deep reasoning. Moreover, knowledge on the causality of events of troubles is also represented by linkages of frames.

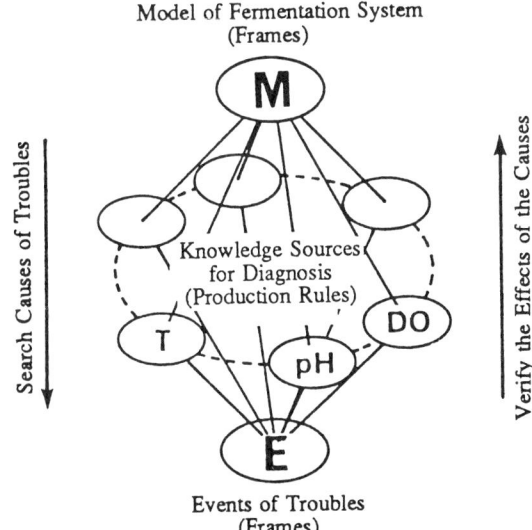

FIGURE 3. Configuration of knowledge bases.

Inferring Process

The inferring process for diagnosis is shown as follows:

(1) specification of a diagnosing item that is one of the monitoring or controlling parameters,
(2) entry of an observed unusual phenomenon, which is a trigger of diagnosis,
(3) activation of knowledge sources for diagnosing possible hypotheses, along with judgement effects with an operator's interaction,

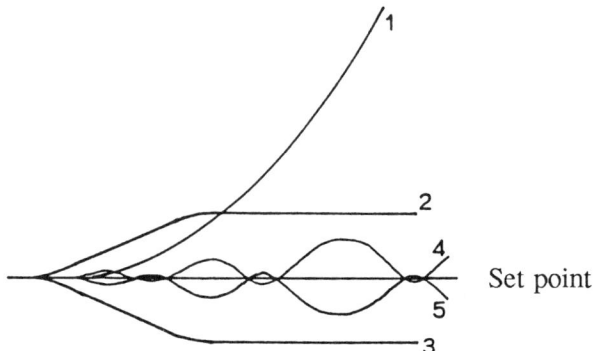

FIGURE 4. Typical troubled patterns of the time course (triggers of diagnosis).

(4) output of all the causal relations from the inferred cause until the observed event,
(5) output of the countermeasure.

The diagnosing item is specified as a typical troubled pattern against time, which is shown in FIGURE 4. The system infers to search for causes of troubles by activating knowledge sources that orderly refer to the structural model in order to obtain information on the state of the bioreactor system. When the system succeeds in searching a root cause, it estimates the effects of the cause on other parameters by tracing the causality. The system compares the real states with the estimated effects and verifies the diagnosed results. The inferred process and the assumed hypotheses are recorded on the blackboard and a simple explanation function can be achieved by displaying the contents on the blackboard before the system stops.

Prototype System

Based on the above-mentioned discussion, we prototyped an expert system for diagnosing the BIOACS with respect to temperature control, pH control, and DO

```
RULE# <1>
IF
   (LESSP (@FGET 'PANEL-AG 'SV '(0 N)) *AGSVMIN)
THEN
 1 PROPOSE
   CHANGE TYPE:  ADD
   EVENT NAME :  PRINT-CONCLUSION
   LEVEL      :  DIAGNOSTIC-STATE
   ATTRIBUTES :  DIAGNOSTIC-ELEMENTS =   'AG-FAULT
                 <CF>  0.9
                 IDENTIFY             =  'AG-SV-LOW
```

FIGURE 5. An example of production rules for diagnosis.

```
(@DEFRAME
    'FERMENTOR-A
    '(CLASS (¥VALUE (INDIVIDUAL)))
    '(AKO (¥LINK (FERMENTOR)))
    '(BROTH (¥IF-NEEDED
                ((#INPUT-HIGH-LOW
                    "Input a level of broth. (HIGH/LOW)"))))
    '(AGITATOR (¥LINK (AGITATOR-A)))
    '(THERMOMETER (¥LINK (THERMOMETER-A))))
```

FIGURE 6. An example of frames representing equipment.

control. We implemented a prototype system with AI tool *Eshell* in UTILISP on a main frame computer, namely, Fujitsu M780.

An example of production rules for diagnosis, an example of frames representing equipment, an example of frames representing events causality, and an example of frames representing root causes are shown in FIGURES 5, 6, 7, and 8, respectively. An executed example of the diagnosing system is shown in FIGURE 9.

The diagnosing system is a stand-alone system in the present stage. Therefore, the reasoning starts with a manual entry of an observed troubled pattern and the human operator should answer interactively to the questions on states of the bioreactor system according to the system prompt. A data base for characteristic data on the fermentation can be prepared, but it has not been implemented yet in the prototype system.

Concerning the specification of the expert system for diagnosing the BIOACS with regard to the control of temperature, pH, DO, and agitation speed, the number of rules for diagnosis is 141 and the number of frames is 70 for a structural model and 140 for relations between events.

CONCLUSIONS

We presented a concept of an advanced cultivation system where robots, expert systems, and a data base system are introduced. We discussed functions for the ESCO, focusing first on the diagnosis of a bioreactor system and cultivation processes. Then, taking account of requirements for the diagnosing system, we discussed the knowledge representation and the inferring process. Finally, based on these discussions, we

```
(@DEFRAME
    'ACID-ADDITION-INSUF
    '(CLASS (¥VALUE (GENERIC)))
    '(SUPERC (¥LINK (ACID-ADDITION-FAULT)))
    '(SUBC (¥LINK (ACID-CONC-INSUF) (ACID-TIMER-UNSUIT-CLOSE)))
    '(MESSAGE
        (¥VALUE ("Therefore, acid cannot be added sufficiently. "))))
```

FIGURE 7. An example of frames representing events causality.

```
(@DEFRAME
  'V42-AO-LONG
  '(CLASS (¥VALUE (GENERIC)))
  '(SUPERC (¥LINK (ALKALINE-TIMER-UNSUIT-OPEN)))
  '(MESSAGE (¥VALUE ("Open-time of VALVE-42 (AO) is set too long. ")))
  '(COUNTERMEASURE (¥VALUE ("Set shorter time to AO. "))))
```

FIGURE 8. An example of frames representing root causes.

```
INPUT DIAGNOSTIC ITEM   (1. TEMP /2. pH /3. DO) > 2
INPUT TRIGGER > 1
Input a state of PANEL-SWITCH NC for pH. (ON/OFF) > OFF
Is the solution for pH control added to broth? (YES/NO) > NO
Is there acid solution within the acid pot? (YES/NO) > YES
Input a state of PANEL-SWITCH BSW. (AUTO/OFF/MANUAL) > AUTO
Input a state of VALVE-43. (OPEN/CLOSE) > OPEN
Turn the switches BSW to MANUAL and ASW to OFF. Is acid added to
broth? (After tested, turn both switches to AUTO.)   (YES/NO) > NO


        *** RESULTS OF DIAGNOSIS ***
There is a trouble in VALVE-41.
VALVE-41 is closed.
Therefore, acid can not be added.
Because acid isn't added to broth functionally,
pH can not be controlled.

        *** COUNTERMEASURE ***
Repair VALVE-41.


 ** 異常診断 の CURRENT ノード **
NODE NAME <診断項目>
  PROTO: 異常診断
 ** DIAGNOSTIC-STATE の CURRENT ノード **
NODE NAME <DIAGNOSTIC-STATE1>
  PROTO: DIAGNOSTIC-STATE
  DIAGNOSTIC-ELEMENTS = ((PH-CONTROL-MODE-UNSUIT -1.0))
NODE NAME <DIAGNOSTIC-STATE2>
  PROTO: DIAGNOSTIC-STATE
  DIAGNOSTIC-ELEMENTS = ((ACID-ADDITION-FAULT 1.0))
  IDENTIFY            = (BSW-V41-TROUBLE-CLOSE)
```

FIGURE 9. An executed example of the diagnosing system.

prototyped an expert system for diagnosing cultivation processes and verified its efficiency.

SUMMARY

A new concept of a fully advanced cultivation system aiming at bioprocess automation is presented, which is composed of BIOTRON (a robotized cultivation factory), ESCO (Expert System for Cultivating Operations), BEXS (Bio EXpert System), and BDBS (Bio-Data Base System). Especially functions required of ESCO such as diagnosis of trouble, intelligent process control, and dynamic experimental design are discussed in this report. In designing an expert system for diagnosing a bioreactor system and cultivation processes, information and knowledge necessary for diagnosis are clarified and a new methodology for knowledge representation and a strategy of inferring processes are proposed, based on functional requirements and system requirements. Finally, specification of a developed prototype system is reported and its efficiency is verified.

ACKNOWLEDGMENT

We wish to acknowledge Noriko Koizumi at Fuji Facom Company, Limited, for her kind advice in the implementation of a prototype system using *Eshell*.

REFERENCES

1. STEPHANOPOULOS, G. & G. STEPHANOPOULOS. 1986. Trends Biotechnol. **32:** 241–249.
2. O'CONNOR, G. M. & C. L. COONEY. 1986. ACS Meeting, Anaheim, MBTD paper no. 85.
3. ENDO, I. & T. NAGAMUNE. 1987. Bioprocess Eng. **2:** 111–114.
4. ENDO, I., H. ASAMA & T. NAGAMUNE. 1989. *In* Bioproducts and Bioprocesses. A. Fiechter, H. Okada & R. D. Tanner, Eds.: 337–346. Springer-Verlag. Berlin/New York.
5. NAGAMUNE, T., *et al.* 1985. Japan patent laid-open publication no. 57-201, 954; United States patent no. 4,561,799.
6. ENDO, I., *et al.* 1985. Japan patent laid-open publication no. 57-68, 781; United States patent no. 4,501,161.
7. ENDO, I., T. NAGAMUNE & I. INOUE. 1983. Ann. N.Y. Acad. Sci. **431:** 228–230.
8. NAGAMUNE, T., I. ENDO & I. INOUE. 1984. Operation charts for a successive fed-batch fermentation of alcohol. Kagaku Kogaku Ronbunshu **10:** 506–512.
9. ASAMA, H., *et al.* 1988. Japan-Finland Symposium on Automation Technologies and AI Applications for Bioprocesses, p. 1–3.

Frequency Response Analysis of Naphthalene Biotransformation Activity[a]

JAMES W. BLACKBURN

Center for Environmental Biotechnology
and
Department of Chemical Engineering
University of Tennessee
Knoxville, Tennessee 37996

BACKGROUND

A view of mixed culture microbial systems as "complex" and "unpredictable" might be easily accepted by many scientists and engineers. From an historical perspective, many past approaches to studying microbial processes have been limited to pure culture systems or to the behavior of mixed culture systems assessed with the use of nonspecific parameters.[1,2] These simplifying approaches have often arisen from the uncertainties in system structure and kinetics inherent in mixed culture microbial systems. Microbial system processes operate at the macro or reactor, ecological, cellular, and molecular levels, all with significant interactions and with differing characteristic times. From an engineering perspective, the system contains numerous feedforward, feedback, cascade, switching (on-off or bang-bang), and other interactions with characteristic times varying from nanoseconds to many years.[3-6]

The analysis of an operating system as a "black box" is possible as capabilities for on-line monitoring improve. With this approach, that is, a *microbial systems analysis* approach, the system is allowed to achieve an equilibrium state and is then perturbed by the controlled change of a potentially critical "input" variable. The resulting changes in measured "output" variables can be recorded and analyzed with the appropriate mathematical tools and important conclusions may be derived regarding the system's structure and robustness without a priori deterministic understanding.[7]

Much of the frequency response work in the physical and engineering science area has been with linear systems that are nonadaptive; that is, systems whose component interactions are described by linear differential equations and are time-invariant. These systems permit extrapolation of a single frequency response study with high confidence.[8,9]

Processes and interactions for regulation and control of biological systems at the ecological, cellular, and molecular level allow an organism to adapt to changing conditions by continuously adjusting a complex network of nested regulatory loops.[3,5,6] In such systems, the descriptive equations may not only be nonlinear, but may even be

---

[a]This work has been supported in part by the Gas Research Institute and by the University of Tennessee's Center for Environmental Biotechnology and Waste Management Research and Education Institute.

discontinuous. Furthermore, the system may be time-variant and adaptive based on its present and past history.

Bode proposed linear approximations as a way to approximate the mathematical model of experimentally derived response data measured from electrical networks.[10] With no prior constraint for the system to be linear, he used this approach merely as a transform to approximate system performance in the frequency domain. With sufficient data, even the most nonlinear, continuous systems should yield to linearization for approximation if time-invariant. For time-variant or adaptive systems, an empirically derived Bode model measured at an instant in time should represent a "snapshot" against which time variance—or, in the case of biological systems, adaptability—can be determined through time.

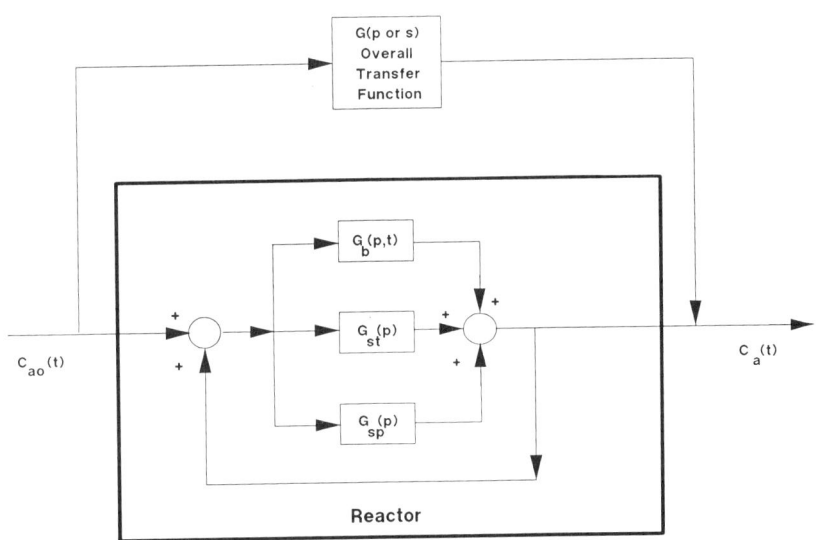

FIGURE 1. System transfer function block diagram.

SYSTEM PROCESS AND MODEL HYPOTHESES

The mixed microbial naphthalene treatment system is presented in a block diagram/transfer function format in FIGURE 1. With this representation, the naphthalene entering the system in the feed stream, $C_{ao}(t)$, is added to the naphthalene in the reactor liquid, $C_a(t)$, and is processed by three naphthalene removal mechanisms. The first removal process is the uptake of naphthalene from the reactor liquid and the subsequent biotransformation of the compound to some metabolite, $G_b(p,t)$. Independent variables p and t respectively refer to the function's dependence on various time derivatives and on the time-variant nature of the function itself. Naphthalene is also removed from the reactor liquid by mass transfer into the air stream being bubbled into the reactor, $G_{st}(p)$. Finally, naphthalene can sorb on the biomass or on other solids and/or surfaces present in the system, $G_{sp}(p)$. Stripping and sorption may both occur fast enough to assume they are equilibrium processes. The reactor is assumed to be completely mixed and residence time–related dynamics are assumed to be slow.

The nature of $G_b(p,t)$ can be explored a posteriori by selecting various mathematical expressions and observing whether the behaviors of the calculated rate parameters and constants meet expectations for that type of expression. For example, the calculated rate parameter for a single first-order biotransformation rate expression should be constant for varying substrate concentrations and, for Michaelis-Menten or Monod-type kinetics, the biotransformation rate should approximate first order at low substrate concentrations and zero order at concentrations above the saturation concentration. Inconsistent experimental results require consideration of more complex mechanisms.

Because the purpose here was to show the feasibility of the approach and not to elucidate the reaction mechanism, a single first-order mechanism was the tested hypothesis for the model. The calculation of the first-order biotransformation rate parameter was based upon an unsteady-state, difference equation–based naphthalene material balance in the reactor and used several measured system variables. The model (equation 1) may be found in TABLE 1 and a more comprehensive discussion is presented elsewhere.[11]

A method referred to as "the empirical transfer function estimate" or ETFE was used to resolve the output signal from the noise and to calculate the necessary parameters for frequency response analysis.[7] A computer algorithm named *spa* (a component of the PC-Matlab System Identification Toolbox, The Math Works) employed the discrete fast Fourier Transform to analyze both the input and output time series and to form the ETFE as the ratio of the power spectrum of the output over the input. The phase angle, the residual signal, and the noise spectrum were also calculated. In this work, the ETFE for a given perturbation frequency represented an average output-input magnitude ratio with the noise removed. For all perturbation intervals except for the lowest two frequencies, at least one-third of the early data was discarded to minimize the influence of initial transients.

EXPERIMENTAL PERTURBATION PROTOCOL

Experimental details for this work may also be found elsewhere.[11] The simultaneous operation of two feed metering pumps with sinusoidal control signals resulted in a feed relatively constant in total flow and composition, except for a sinusoid in naphthalene concentration ranging from near zero to about 10 mg/L. The upper limit was influenced gradually by an experimental difficulty and may represent a disturbance to be discussed later. A perturbation protocol was developed that employed alternating sinusoidal perturbation intervals and relaxation intervals at constant feed concentration. A summary of this protocol is shown in TABLE 2.

The shortest perturbation period tested was 1 hour. This upper frequency was limited by the minimum time required for an off-gas sample to be taken and analyzed by the dedicated gas chromatograph. This minimum sample interval was 10 minutes and no less than six samples per cycle were taken.

BIOTRANSFORMATION AND MINERALIZATION ACTIVITY

For each relaxation and perturbation period, average values of the biotransformation rate parameter, $k1b$, are presented in FIGURE 2. The average $k1b$ ranges from a minimum of 0.05 h^{-1} (during the relaxation interval before the 6-h perturbation) to 19.6 h^{-1} (during the 64-h perturbation). This represents a nearly 400-fold variation in the average rate parameter over the frequency ranges studied.

The calculated time series for the first-order biotransformation rate parameter, $k1b$, is presented in FIGURE 3. Negative values of $k1b$ in the final three perturbation intervals (4-, 2-, and 1-hour perturbation periods) arose in the calculation because of the unsteady-state accumulation terms. These terms describe the accumulation and depletion of naphthalene in the reactor liquid and the overall reactor biomass based on the change in calculated reactor liquid concentrations during a sampling interval. Nonzero values for these terms manifest additional dynamics otherwise unaccounted for in $G_b(p,t)$, $G_{st}(p)$, and $G_{sp}(p)$.

TABLE 1. Important Models and Equations

Overall Naphthalene Material Balance (Equation 1)

$$V_s \frac{dC_a}{dt} + V_s \frac{dC_a^b}{dt} = Q_f C_{ao} - (Q_f + Q_n - Q_w)C_a - Q_w X K_{ba} C_a$$

$$- (Q_f + Q_n - Q_w) X_e K_{ba} C_a - V_r(k1b)C_a - Q_{air} C_{air}$$

Difference Equation Calculation of Biotransformation Rate Parameter (Equation 2)

$$k1b_{(t)} = Q_{f(t)} \frac{C_{ao(t)}}{V_r C_{a(t)}} - \frac{1}{\text{RHRT}} - \frac{Q_w X_{(t)} K_{ba}}{V_r} - \frac{(Q_{f(t)} + Q_n - Q_{w(t)}) X_{e(t)} K_{ba}}{V_r}$$

$$- \frac{Q_{air(t)} C_{air(t)}}{V_r C_{a(t)}} - \frac{\text{SHRT}}{\text{RHRT}} \left(\frac{C_{a(t)} - C_{a(t-1)}}{C_{a(t)} \Delta t} \right) - \frac{\text{SHRT}}{\text{RHRT}} K_{ba} X_{(t)} \left(\frac{C_{a(t)} - C_{a(t-1)}}{C_{a(t)} \Delta t} \right)$$

Definition of Biomass–Aqueous Phase Partition Coefficient (Equation 3)

$$K_{ba} = \frac{f_L K_{ow}}{\rho_L} = \frac{C_a^b}{C_a \rho_B}$$

Applicable Overall Transfer Functions

Second-order lag plus lead plus dead time (equation 4)

$$G(s) = \frac{(\tau s + 1)(e^{-\tau_0 s})}{(s^2 + 2\xi s + 1)}$$

First-order lag plus dead time (equation 5)

$$G(s) = \frac{e^{-\tau_0 s}}{(\tau s + 1)}$$

TABLE 1. (Continued)

Nomenclature

Time, t, is shown as a subscript when used in conjunction with a sampled or calculated variable. The sampling interval is Δt.

C_{ao} = naphthalene concentration in the reactor feed (MV^{-1})
C_a = naphthalene concentration in the reactor liquid (MV^{-1})
C_a^b = naphthalene concentration in the reactor biomass (MV^{-1})
C_{air} = naphthalene concentration in the reactor off-gas (MV^{-1})
Q_f = feed flow rate (VT^{-1})
Q_n = nutrient flow rate (VT^{-1})
Q_w = sludge waste rate (VT^{-1})
Q_{air} = airflow rate to the reactor (VT^{-1})
X = reactor suspended solids concentration (MV^{-1})
X_e = effluent solids concentration (MV^{-1})
V_s = hydraulic volume of the reactor and clarifier (V)
V_r = hydraulic volume of the reactor (V)
$k1b$ = first-order biotransformation rate parameter (T^{-1})
RHRT = $V_r/(Q_f + Q_n)$, reactor hydraulic residence time (T)
SHRT = $V_s/(Q_f + Q_n)$, system hydraulic residence time (T)
f_L = fraction of lipidlike cell mass (dry basis) (MM^{-1})
K_{ow} = naphthalene octanol-water partition coefficient (MV^{-1}/MV^{-1})
ρ_L = average density of cellular lipids (MV^{-1})
ρ_B = average density of the biomass (MV^{-1})
$G(s)$ or $G(p)$ = overall feed-to-reactor liquid concentration transfer function—Laplace or time (differential) domain, respectively
τ = first-order system characteristic time constant (T)
τ_0 = dead time (T)
ξ = second-order system damping ratio

TABLE 2. Experimental Perturbation Protocol

Time Interval (serial day of year)	Experimental Interval Type[a]	Perturbation Frequency (cycles/h)	Perturbation Period (h)
~220–254.5	R		
254.5–270.3	P	0.016	64
270.3–279.5	R		
279.5–283.5	P	0.042	24
283.5–286.5	R		
286.5–290.5	P	0.083	12
290.5–293.5	R		
293.5–297.5	P	0.125	8
297.5–300.4	R		
300.4–303.8	P	0.167	6
303.8–307.3	R		
307.3–310.8	P	0.25	4
310.8–314.5	R		
314.5–318.7	P	0.5	2
318.7–321.3	R		
321.3–326.2	P	1.0	1

[a]R = relaxation; P = perturbation.

The first-order biotransformation rate hypothesis can be tested by examining the calculated rate parameter, $k1b$, as a function of calculated reactor liquid concentration (FIGURE 3) and this leads to a highly counterintuitive result. The $k1b$ increased dramatically as the liquid concentration approached zero. This result disagrees with the classical view of a single first-order rate constant that is expected to be constant for various liquid concentrations. Use of Michaelis-Menten or Monod kinetics also fails to resolve this problem unless surprisingly low saturation constants are employed.

Close examination of the high-frequency time series showed that these variations were very rapid, often just a few minutes. Whereas second-order or changing microbial population effects could be considered as an explanation of kinetics at low or medium

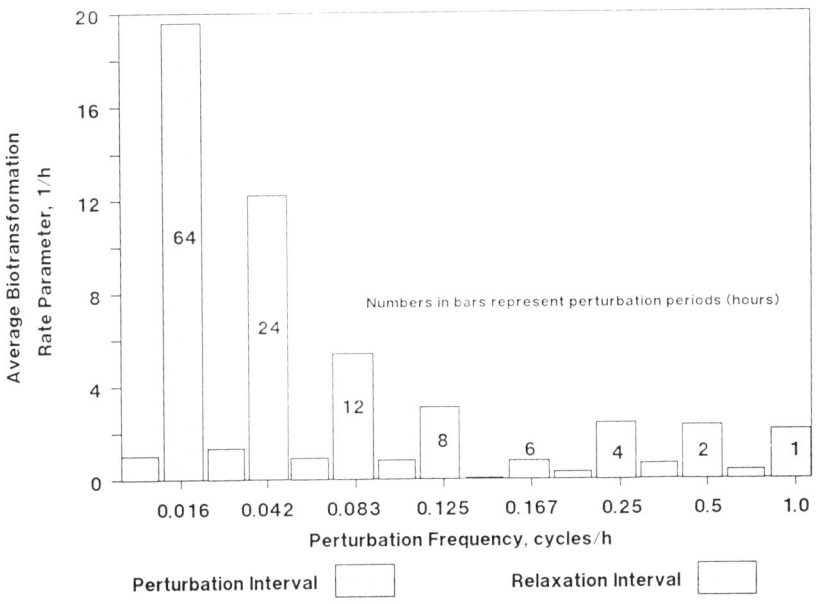

FIGURE 2. Average calculated first-order naphthalene rate parameters for all experimental relaxation and perturbation intervals.

frequencies, this is an inadequate explanation for the high-frequency data because 100-fold biotransformation kinetic changes were seen over time frames too short to lead to similar population changes. This was supported by various types of microbiological measurements.[11]

Because naphthalene biotransformation was not well described by the first-order model or other simple model forms, one must look to more complex explanations. These explanations could be that (i) processes in addition to the three assumed are operating, (ii) the abiotic process expressions are incorrect in form or structure, or (iii) biotransformation is more complex than the simple, single first-order rate process assumed. For

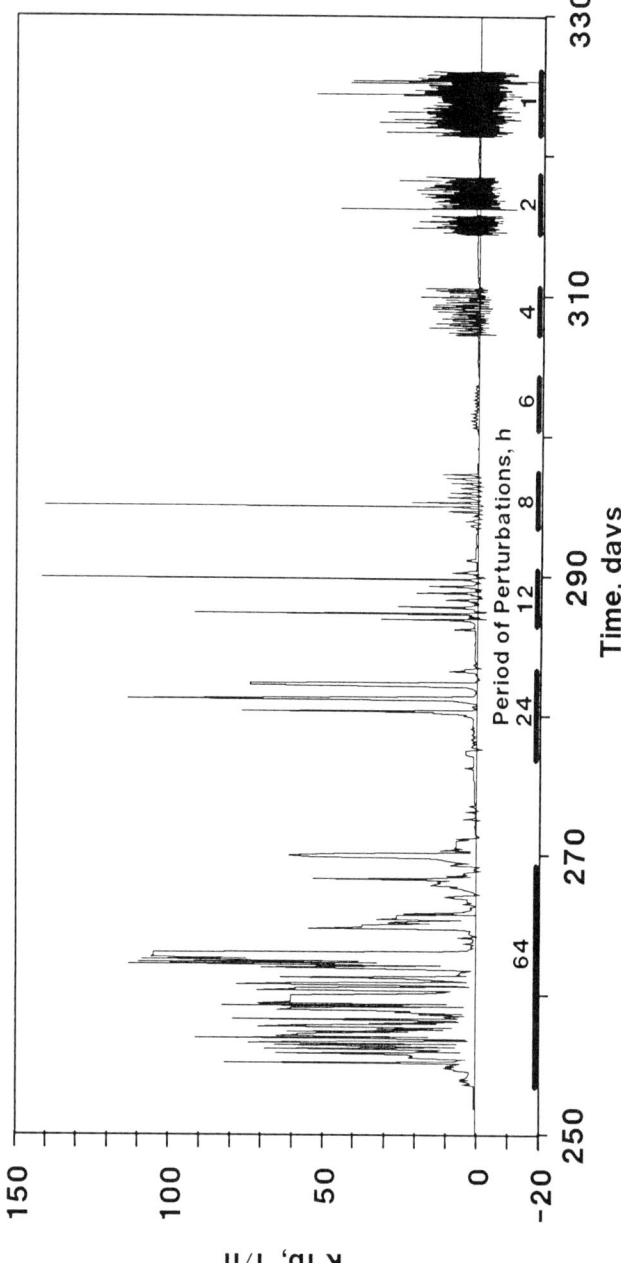

FIGURE 3. Calculated time-series, first-order naphthalene biotransformation rate parameters for all experimental intervals.

naphthalene biotransformation in this reactor system, additional processes to those assumed are unlikely.[2,12]

Looking to abiotic removal processes, one could invoke a highly nonlinear stripping process—one that departed strongly from Henry's law and led to stripping rate constants that were orders of magnitude greater at low concentrations. This behavior would require a change in the thermodynamic properties of the solution (e.g., rapidly varying surface-active agent) or some rapidly changing naphthalene chemical adduct. These conjectures are unsatisfying in light of the rapid variations seen in activity (a few minutes at high frequencies) when the reactor's liquid residence time averaged about 47 hours, which is the half-life for a conserved liquid-soluble surface-active constituent.

The alternating sign on $k1b$ suggests a sorption-desorption process. One possibility is a partitioning process on the inside walls of the 0.25-OD reactor off-gas sampling tube on the path to the automatic sample valve and GC. This contingency is being evaluated in present experimental work. Another hypothesis is that the total biomass behaved as a sorption-desorption source and that the magnitude of this process was greater than the passive sorption process already included in the calculation of $k1b$ (equation 3). This effect would require that biomass-to-liquid partition coefficients be several orders of magnitude higher than those previously determined.[2]

Assuming there is no significant sampling problem, what kind of enzymatic uptake/conversion mechanism could be switched on and off to give the activity observed? Suppose naphthalene biotransformation is a two-step process, each with independent, but possibly linked regulation. The first process could be a protein- or enzyme-linked uptake from liquid solution, possibly to aid the organism(s) in scavenging naphthalene at low environmental concentrations. Once taken in by the cell, a second enzyme (or, in naphthalene's case, a series of enzymes) proceeds to biotransform naphthalene in an independently regulated fashion.

A reversible enzymatic initial uptake process would also permit reversible desorption of naphthalene from the cells as the reactor liquid concentration fell. This would appear in the $k1b$ calculation as a negative $k1b$. Behavior such as this has not been reported in the scientific literature for these types of compounds and therefore would require conclusive study.

Other possibilities might include an ecological format where two competing organisms or guilds alternate dominance based on the naphthalene concentration. One guild could be very active at low concentrations, but inhibited at higher concentrations. The problem here is that a naphthalene storage mechanism is still needed to satisfy the negative $k1b$ behavior and that the guilds must be capable of rapid oscillation if the effect is population-based.

Supporting the notion of a reversible uptake and cellular storage mechanism, delays or dead times in naphthalene carbon flux processing have been experimentally confirmed. Radiolabeled ^{14}C-naphthalene carbon was introduced as a pulse at the beginning of each perturbation interval. The radiolabeled carbon consistently required over three days to clear the system as $^{14}CO_2$. These results for three perturbation experiments are presented in FIGURE 4 superimposed over a concurrent time line. The mineralization rate kinetics during the perturbation intervals were on the order of 30 times slower than the average calculated $k1b$ biotransformation rate. The $^{14}CO_2$ wave front's leading edge, as shown on FIGURE 4, was quite sharp (peak width is estimated to

588 ANNALS NEW YORK ACADEMY OF SCIENCES

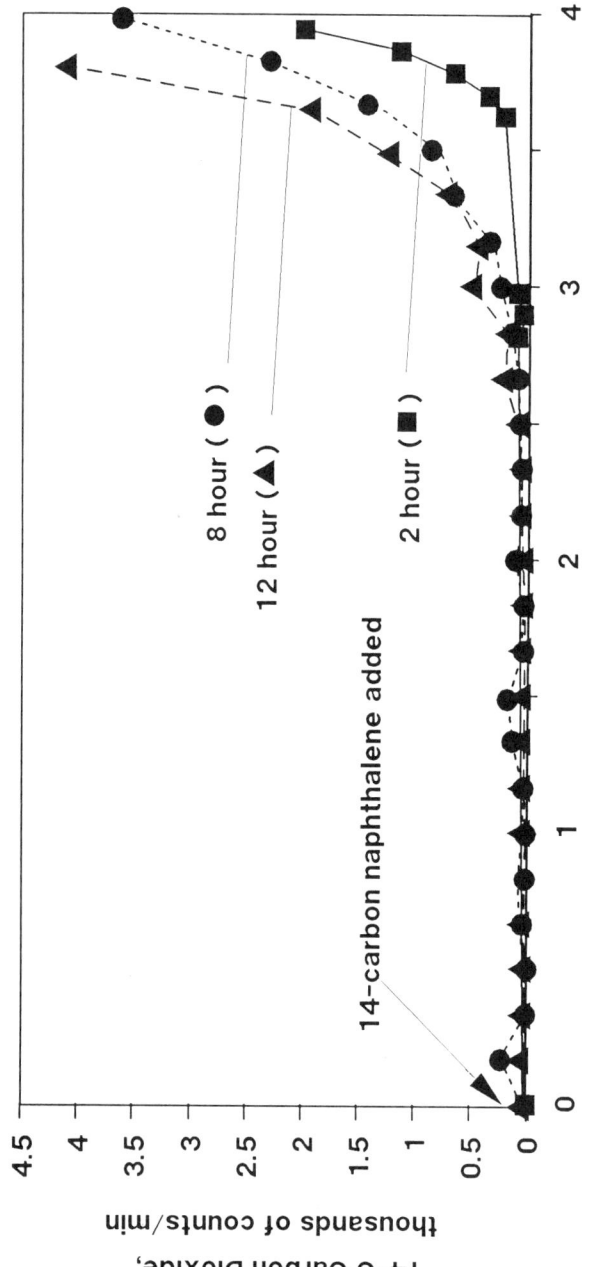

FIGURE 4. Evolution of $^{14}CO_2$ from ^{14}C-naphthalene over three experimental perturbation intervals (2-, 8-, and 12-hour periods) superimposed on a concurrent time axis.

be less than 24 hours). No radiolabel was found in the off-gas at the next sampling point (relaxation period, 16 hours later).

One known naphthalene carbon transportation delay arose from the time lag between the feed leaving the pump heads and entering the reactor. At the lowest frequency, this time (because of large bore tubing) ranged over 30 minutes. Large bore feed tubing was replaced by capillary bore tubing after the first perturbation interval and resulted in a typical dead time of about 10 minutes.

In studies of passive organic substrate uptake using nonviable, activated sludge as the sorbent, equilibrium uptake required less than 15 minutes and often less than 5 minutes.[2] Passive uptake is expected to be fast and is not the source of the three-day delay.

Another possible explanation is that the $^{14}CO_2$ formed in naphthalene catabolism was stored intracellularly or extracellularly as CO_2, bicarbonate, or carbonate. The CO_2 in the reactor that was radioactive was a very small fraction of the nonradioactive naphthalene-related CO_2 and even a smaller fraction of the nonnaphthalene-related CO_2. Naphthalene carbon never exceeded 0.5% of the dissolved organic carbon in the feed. Whereas CO_2-bicarbonate-carbonate was a system at dynamic equilibrium, it was dominated in this experiment by the nonnaphthalene CO_2 and unperturbed by the minor naphthalene oscillations. Furthermore, if the delay arose from the CO_2 chemical equilibria and if this subsystem were dominated by nonnaphthalene carbon, then the $^{14}CO_2$ waveform would be expected to have a broad bandwidth, beginning near the ^{14}C-naphthalene injection point.

If all uptake and cell processing was very fast and without delays, then the $^{14}CO_2$ peak should follow ^{14}C-naphthalene injection as follows: (1) in minutes (gas mean residence time of about 3 minutes) if the storage moiety was gaseous CO_2, (2) at about 2 days with a broad bandwidth (reactor mean liquid residence time about 47 hours) if a soluble liquid moiety was involved, or (3) at about 5 days with a broad bandwidth (mean cell residence time) if the moiety was a settleable solid. None of these options fit the observations particularly well.

The reversible active naphthalene uptake and cellular storage hypothesis with the uncoupled enzymatic naphthalene biotransformation hypothesis appears to best fit the observed data and merits closer experimental testing.

FREQUENCY RESPONSE SYSTEM ANALYSIS

The time-series input and output (example shown in FIGURE 5) were analyzed using the empirical transfer function estimate (ETFE) approach noted earlier and were plotted in a Bode-type frequency response format (FIGURE 6). The ETFEs at the highest frequencies demonstrate an approximately linear (log-log slope of -1) behavior. Nonminimal phase behavior is seen in the phase angle (still dropping at $-180°$).

On its face, the Bode diagram might suggest an appropriate overall transfer function model, $G(s)$, incorporating a dead time delay, a first-order lead, and a first- or second-order (underdamped) lag (equation 4). Unfortunately, low-frequency ETFEs may have been understated because of operation below the analytical detection threshold (-1 $\mu g/L$, liquid phase) during the valleys in the input feed sinusoid. Also, as noted earlier, the low-frequency experiments had too few cycles to allow removal of the

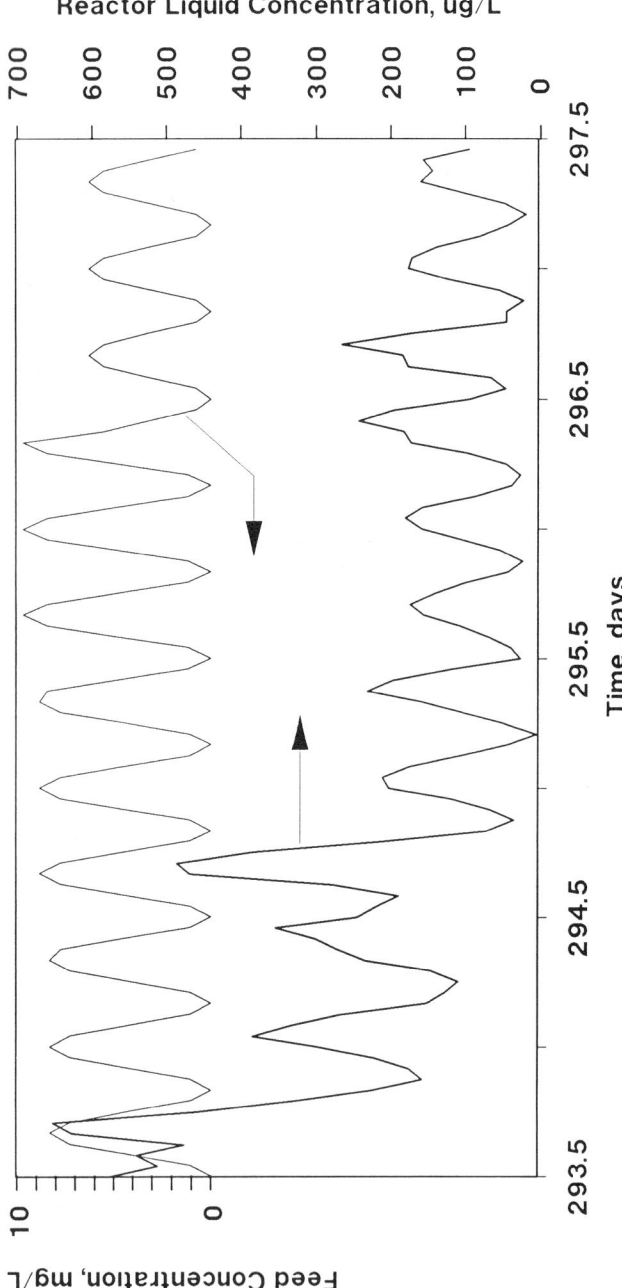

FIGURE 5. Time-series input (feed concentration) and output (reactor liquid concentration) for the 8-hour-period perturbation interval (0.125 cycles/h).

effects of initial transients. Moreover, the apparent lead at the frequency of 0.167 cycles/h (6-h period) coincided with disturbances in the naphthalene feed reservoir, further complicating interpretation. For these reasons, until further collaborative experiments can be implemented, higher-order models for $G(s)$ must be rejected in favor of the simpler first-order lag plus dead time model (equation 5).

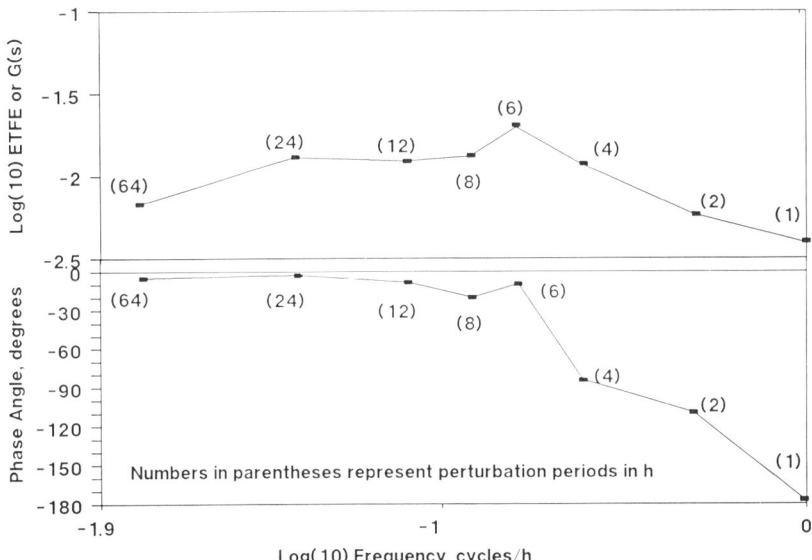

FIGURE 6. Bode frequency response diagram for the naphthalene feed–to–reactor liquid concentration transfer function.

CONCLUSIONS

Frequency response studies, supported by simultaneous measurement of potentially important system variables, are feasible and can lead to the ability to probe the nature and importance of otherwise hidden process dynamics in the operating, mixed-microbial system. Coupled with other measured and calculated variables, the critical dynamic processes can be identified and further studied.

Naphthalene uptake and biotransformation behavior was represented with a first-order plus dead time model. Some evidence exists for a higher-order model, but this must be corroborated in further work. Interesting naphthalene carbon flux dynamics suggest significant enzyme- or protein-related delays and complex reversible naphthalene sorption behavior worthy of further experimental study.

ACKNOWLEDGMENTS

I thank Terry Donaldson, Gary Sayler, Charles Moore, Paul Bienkowski, and Greg Reed. I also thank Henry King, Rod Bunn, Phil DiGrazia, Paul Dunbar, William Eng, and Jeff Bettis for their assistance in system operation, sampling, and especially the provision of the microbiological data.

REFERENCES

1. BLACKBURN, J. W. 1987. Environ. Prog. **6:** 217–223.
2. BLACKBURN, J. W., W. L. TROXLER, K. N. TRUONG, R. P. ZINK, S. C. MECKSTROTH, J. R. FLORANCE, A. GROEN, G. S. SAYLER, R. A. MINEAR, O. YAGI & A. BREEN. 1985. Organic chemical fate prediction in activated sludge treatment processes. Report no. PB 85-247647. National Technical Information Service. Springfield, Virginia.
3. BARFORD, J. P., N. B. PAMMENT & R. J. HALL. 1982. *In* Microbial Population Dynamics. M. J. Bazin, Ed. CRC Press. Boca Raton, Florida.
4. BAZIN, M. J. 1981. *In* Mixed Culture Fermentations. M. E. Bushell & J. H. Slater, Eds. Academic Press. London/New York.
5. BLACKBURN, J. W. 1989. *In* Biological Treatment of Agricultural Wastewater. M. Huntley, Ed. CRC Press. Boca Raton, Florida.
6. PROKOP, A. 1982. Int. J. Gen. Syst. **8:** 7–31.
7. LJUNG, L. 1987. System Identification: Theory for the User. Prentice–Hall. Englewood Cliffs, New Jersey.
8. MURRILL, P. W. 1967. Automatic Control of Processes. Intern. Textbook. Scranton, Pennsylvania.
9. STEPHANOPOULOS, G. 1984. Chemical Process Control—An Introduction to Theory and Practice. Prentice–Hall. Englewood Cliffs, New Jersey.
10. BODE, H. W. 1945. Network Analysis and Feedback Amplifier Design. Van Nostrand. Princeton, New Jersey.
11. BLACKBURN, J. W. 1989. Hazardous Waste and Hazardous Materials **6**(2): 173–193.
12. BLACKBURN, J. W., R. K. JAIN & G. S. SAYLER. 1987. Environ. Sci. Technol. **21:** 884–890.

Analysis of Performance Limitations in Immobilized Cell Fermentors

C. WEBB, G. A. DERVAKOS, AND J. F. DEAN

Department of Chemical Engineering
University of Manchester Institute of Science and Technology
Manchester M60 1QD, United Kingdom

INTRODUCTION

As immobilized cell fermentation research enters a second decade, little progress is apparent in the a priori prediction of the reactivity of immobilized cell particles containing viable microorganisms. This is largely due to the fact that the transport parameters of the relevant systems, such as diffusivity values or external mass-transfer coefficients, are usually known only to within an order of magnitude and the data describing the inherent kinetics of immobilized cells are not generally available. A more serious limitation, however, stems from a subtle assumption behind all of the relevant models, namely, that the reaction behavior of immobilized cell particles used in fermentation is directly analogous to that of heterogeneous catalysis. This, though, has to be verified for the immobilized cell case. For instance, viable cells are apt to be distributed nonuniformly throughout the particle, creating intraparticle spatial variations of the diffusivity with consequent implications for the intrinsic rate of reaction. Furthermore, the importance of high product concentrations in most biochemical processes necessitates the use of complex rate laws incorporating substrate and product inhibitions; this, in turn, leads to the need to consider more than one diffusing species inside the particle. Moreover, the open pore structure of some carriers used to immobilize cells may mean that diffusion is not the dominant mechanism for nutrient transport throughout the aggregate. In our opinion, the key to progress in modeling such systems is a disciplined approach to the use of assumptions, particularly of those related to growth, distribution, and physiology of the immobilized cells.

THEORETICAL ANALYSIS

Before describing the model for immobilized cell fermentation, it is necessary to identify the underlying assumptions behind it:

(a) the aggregate has a simple geometry that can be characterized by a symmetry parameter, n ($n = 0$ for a slab, 1 for a cylinder, and 2 for a sphere),
(b) the aggregate is isothermal,
(c) diffusion is the only mass-transfer mechanism inside the aggregate,
(d) diffusivity values are constant throughout the aggregate,
(e) a cell distribution function, $f_x\{z\}$, describes the spatial variation of the cell concentration, C_x, throughout the catalytic carrier,

(f) solutes are not adsorbed onto the carrier surface,
(g) the aggregate is at steady state.

Although the model is potentially applicable to any biological process involving the use of immobilized cells, just one application, namely, ethanol production, will be considered here. Using the above assumptions, a differential mass balance inside the particle can be written in terms of dimensionless quantities:

$$(\pm)\frac{\delta_j}{\sigma_j}z^{-n}\frac{d}{dz}\left(z^n\frac{dw_j}{dz}\right) = \phi^2 f_x\{z\}g_s\{w_s, w_{p1}\}, \quad (1)$$

where the subscript j refers to the diffusing species ($j = s$, $p1$, and $p2$ for substrate, ethanol, and carbon dioxide, respectively). The left-hand side of the equation is positive for substrate(s) and negative for product(s), and

$$w_j = \frac{C_j}{C_j|_b}, \quad \delta_j = \frac{D_j}{D_s}, \quad \text{and} \quad \sigma_j = \frac{Y_{js}C_s|_b}{C_j|_b}, \quad (1a)$$

where Y_{js} is the stoichiometric coefficient of j over s (i.e., for $j = s$, $Y_{js} = 1$). Moreover, ϕ is the Thiele modulus for a biocatalyst with uniform activity. Assuming that a modified Michaelis-Menten-type rate equation can be used to describe the biological activity of immobilized cells and that the cells retain their intrinsic kinetic parameters, the Thiele modulus can be defined as

$$\phi = L\left(\frac{C_{x,\text{ave}}\nu_{\max}}{D_s k_m}\right)^{0.5}, \quad (1b)$$

where L is a characteristic length (e.g., particle radius for a sphere), $\nu_{\max}$ is the maximum specific substrate uptake rate, and k_m is the Michaelis-Menten saturation constant.

The boundary conditions for equation 1 are

$$\frac{dw_j}{dz} = 0 \text{ at } z = 0 \quad \text{and} \quad \frac{dw_j}{dz} = \text{Sh}_j(1 - w_j) \text{ at } z = 1, \quad (2)$$

where Sh_j, the modified Sherwood number for species j, is given by $L(k_\ell)_j/D_j$ and k_ℓ is the liquid-phase mass-transfer coefficient.

The set of three differential equations (in equation 1) for substrate and products can be reduced to a single equation by, first, omitting the equation for CO_2 (because CO_2 concentration is assumed not to affect the reaction rate) and, second, by substituting the following relationship for w_{p1} into the rate term in the substrate balance:[1]

$$w_{p1} = \frac{\sigma_{p1}}{\delta_{p1}}w_s\big|_{z=1} + 1 + \frac{\sigma_{p1}\text{Sh}_s}{\delta_{p1}\text{Sh}_{p1}}(1 - w_s\big|_{z=1}) - \frac{\sigma_{p1}}{\delta_{p1}}w_s. \quad (3)$$

An equation similar to equation 3 can be written for CO_2 ($p2$) provided that the local CO_2 concentration does not exceed the saturation value for solubility at the local conditions. Thus, although CO_2 concentration is not used in calculating the overall reaction rate, a CO_2 concentration profile could be calculated and used to establish whether or not bubble formation would occur inside the immobilized cell matrix.[2]

The effectiveness factor, η, which characterizes the influence of external and internal mass-transfer effects on the biological reaction rate, can be defined as

$$\eta = \frac{\int_0^1 (n+1) f_x\{z\} z^n g_s\{w_s, w_{p1}\} dz}{g_s(1,1)}. \tag{4}$$

NUMERICAL SOLUTION

The differential equation (equation 1) with its associated boundary conditions (equation 2) is nonlinear and has no general analytical solution. It must therefore be solved by either an algebraic approximation or a numerical technique. Of the alternatives available, the orthogonal collocation method introduced by Villadsen and Stewart[3] appeared best suited for the solution of equation 1, having been used previously for generalizing immobilized enzyme reactions.[4]

RESULTS AND DISCUSSION

The lack of data describing the inherent or even the intrinsic kinetics of immobilized cells, combined with the fact that several other important parameters of the relevant systems are only known to within an order of magnitude, makes detailed modeling of such systems a very difficult task. One way around this problem is to use values of the various parameters that provide a comprehensive "worst" or "best" case. In the example discussed below, a set of assumptions was formulated that provided a worst case for ethanol production.

The system chosen for the study was ethanol production using immobilized *Zymomonas mobilis*, a bacterium that is not only much more active than yeast strains (a typical value of ν_{max} for *Z. mobilis*[5] is 11.3 h^{-1} compared to 1.4 h^{-1} for *Saccharomyces cerevisiae*[6]), but also, because of its geometrical characteristics, can be packed much more efficiently; this, in turn, can lead to higher cell packing densities and hence higher overall rates of reaction. The higher the rate of reaction, the greater is the requirement for fast diffusion of nutrients into and products out of the immobilized cell aggregate, thereby making the *Z. mobilis* case the worst one for diffusional limitations. Furthermore, the cells were considered to be homogeneously distributed throughout the aggregate ($f_x\{z\} = 1$); in practice, though, a spatial distribution of cells is usually observed, with the cells preferring to grow near the aggregate surface. The assumption of a uniform distribution should yield a lower effectiveness factor because fewer cells are exposed to the higher substrate and lower product concentrations in the uniform case than in the nonuniform case. This would, however, not be the case if the cells were exhibiting a corelike distribution, being packed more densely at the center, and it is also not valid when diffusivities exhibit a strong dependence on the biomass loading.

The effectiveness of an immobilized cell aggregate depends on the concentration of substrates and products in the bulk liquid. When substrates are supplied at relatively low concentrations, cells inside the particle are likely to suffer from substrate limitation or may even be out of reach of the diffusing substrate. On the other hand, substrates supplied at inhibitory concentrations might be consumed faster inside the particle than at the surface because of the substrate concentration dropping below inhibitory levels. Therefore, the use of relatively low bulk substrate concentrations in the solution to equation 1 provides a "worst case" assumption not only because of possible substrate limitation inside the aggregate, but also due to the elimination of possibly advantageous diffusional limitations brought about by substrate inhibition. The influence of bulk ethanol concentration on the effectiveness factor is not generally acknowledged in the literature. Product inhibition effects might be expected to be higher towards the center of the particle due to the concentration gradient and, therefore, the incorporation of a product inhibition term in the rate equation strengthens the "worst case" claim. Finally, the use of external mass-transfer coefficients obtained from empirical correlations applicable in two-phase systems (which do not take into account the beneficial effect of gas production on film mass transfer) is also likely to provide an overestimation of diffusional limitations.

In solving equation 1 for this worst case, it was observed over a wide range of practical conditions that particles of 1-mm diameter or less were always free from

TABLE 1. Reported Reactor Efficiencies for Ethanol Production by Immobilized Cells

Carrier	Size	Microorganism	Substrate	Reactor[a]	Efficiency	Reference
Ca-alginate	1 mm	*Kluyv. marxianus*	Jer. artichokes	PB	25%	7
Ca-alginate	fibers	*Z. mobilis*	glucose	PB	38%	8
Ca-alginate	1 mm	*Z. mobilis*	glucose	PB	52%	9
k-carrageenan	2–3 mm	*Z. mobilis*	fructose	Tower	73%	10
k-carrageenan	2–3 mm	*Z. mobilis*	fructose	ST	90%	11
Ca-alginate	3.34 mm	*Sacch. cerevisiae*	glucose	PBR	113%	12

[a]PB = packed bed, PBR = packed bed with recycle, and ST = stirred tank.

mass-transfer limitations, that is, $\eta \approx 1$. This, however, contrasts markedly with the typically low reactor efficiencies reported in the literature for simple packed-bed reactors (TABLE 1). It therefore would appear that a major factor contributing to the low overall efficiency of the immobilized cell systems used for ethanol production is not the low effectiveness of the biocatalytic particles employed, but rather the poor design of the bioreactors with respect to the transport of nutrients and products to and from the cell aggregates (i.e., through the bulk liquid). Apart from exhibiting poor mixing characteristics due to low liquid velocities, the most commonly used type of reactor, the simple packed bed, does not provide good degassing conditions for carbon dioxide, which thus becomes responsible for dead spaces in the reactor, for channeling, and even for matrix disruption. By comparison, the very few reactor efficiencies reported for relatively well mixed systems are high (TABLE 1).

Another possible reason for the discrepancy might be that it is unreasonable to assume, as is usually done, that the immobilized cell aggregate is an effective continuum. Cells immobilized in the "macropores" of a support matrix usually, in fact, form microcolonies that have another level of structure consisting of individual cells with tiny "micropores" between them. Diffusivity values in these "micropores" are likely to be much less than effective diffusivities in the support matrix, which in turn necessitates the use of a "two-phase" diffusion term in the model.[13] The usefulness of this approach is limited, though, because of the introduction of additional parameters, which are practically immeasurable, such as diameter and distribution of the microcolonies throughout the particle.

REFERENCES

1. DERVAKOS, G. A. Thesis in progress.
2. KROWELL, P. G. & N. W. F. KOSSEN. 1980. Biotechnol. Bioeng. **22**: 681.
3. VILLADSEN, J. & W. E. STEWART. 1967. Chem. Eng. Sci. **22**: 1483.
4. RAMACHADRAN, P. A. 1975. Biotechnol. Bioeng. **27**: 211.
5. LEE, J. H., M. L. SKOTNICKI & P. L. ROGERS. 1982. Biotechnol. Lett. **4**: 615.
6. TYAGI, R. D. & T. K. GHOSE. 1982. Biotechnol. Bioeng. **24**: 781.
7. MARGARITIS, A. & P. BAJPAI. 1982. Biotechnol. Bioeng. **24**: 1483.
8. GROTE, W., K. J. LEE & P. L. ROGERS. 1980. Biotechnol. Lett. **2**: 481.
9. MARGARITIS, A., P. K. BAJPAI & J. B. WALLACE. 1981. Biotechnol. Lett. **3**: 613.
10. JAIN, V. K., I. TORAN-DIAZ & J. BARATTI. 1985. Biotechnol. Bioeng. **27**: 723.
11. JAIN, V. K., I. TORAN-DIAZ & J. BARATTI. 1985. Biotechnol. Bioeng. **27**: 273.
12. CHIEN, N. K. & S. S. SOFER. 1986. Enzyme Microb. Technol. **7**: 538.
13. SALMON, P. M. & C. R. ROBERTSON. 1987. J. Theor. Biol. **125**: 325.

APPENDIX

Nomenclature

C_j	local concentration of species j (mol/L^3)
D_j	effective diffusivity of species j in the aggregate (L^2/T)
$f_x\{z\}$	dimensionless cell distribution function ($C_x/C_{x,\text{ave}}$)
g_s	dimensionless rate expression for the substrate uptake rate
k_i	substrate inhibition constant (extended M-M kinetics) (mol/L^3)
$(k_\ell)_j$	liquid-film mass-transfer coefficient for species j (L/T)
k_m	substrate saturation constant (M-M kinetics) (mol/L^3)
L	characteristic length (L)
m	exponent in the product inhibition expression
n	symmetry parameter ($= 0, 1,$ or 2)
Sh$_j$	modified Sherwood number for species j $[(k_\ell)_j L/D_j]$
w_j	dimensionless local concentration of species j ($C_j/C_j|_b$)
Y_{js}	yield coefficient (moles of j per moles of s)
z	dimensionless distance

Greek Letters

δ_j	relative diffusivity of species j to substrate
η	effectiveness factor
ν	specific substrate uptake rate (T^{-1})
σ_j	local stoichiometric coefficient for species j
ϕ	Thiele modulus

Subscripts

ave	average
b	bulk
j	species j
max	maximum
$p1$	ethanol
$p2$	carbon dioxide
s	substrate
x	cells

PART IX. MIXING AND SCALEUP OF BIOLOGICAL REACTORS

Development and Scaleup of a High-Rate Biogas Process for Treatment of Organically Polluted Effluents

A. AIVASIDIS AND C. WANDREY

Institute for Biotechnology of the Jülich Research Center
D-5170 Jülich, Federal Republic of Germany

INTRODUCTION

Due to the high concentration of industrial production in localized small urban areas in recent years, highly polluted effluents are produced. Modern waste management should therefore consider these localized situations in a special way. Conventional aerobic wastewater treatment (activated sludge process) becomes more and more problematic due to the considerable energy requirements for oxygen supply. The main goal of the so-called high-rate biogas process was thus to provide a technical solution for treatment of organically polluted industrial effluents. In the case of anaerobic "activation", degradation of wastewater components is performed according to a concept that has already been realized in nature: biogas is the end product of degradation of organic material in the absence of oxygen.

Aerobic methods only partly solve pollution control problems because up to 50% of the organic matter is transformed into sewage sludge. The problems are thereby shifted in part to the excess sludge produced. However, anaerobic wastewater treatment appears to be a general purification method insofar as the major part of organic pollution (more than 95%) is eliminated under energy-saving conditions. The end-product biogas contributes to the process by making it self-sufficient with regard to energy.

The low carbon incorporation (only 3–4%) into cell mass is associated with the energy limitation of growth under anaerobic conditions. On the other hand, in spite of some prejudices, the metabolic activity of anaerobes is undoubtedly comparable with that of anaerobic microorganisms. This also becomes clear if one considers that evolution under energy-limiting conditions encourages the selection of some mutants that perform at a particularly high substrate conversion rate from the low energy potential available.

PROCESS DEVELOPMENT APPROACH

The anaerobic microbial degradation of complex organic molecules takes place via a rather complicated food chain. To put it simply, at least two groups of microorganisms are involved—namely, the acidogenic and the methanogenic species. Biopolymers are first hydrolytically decomposed and fermentatively converted into low molecular weight fatty acids. This is followed by an acetic acid fermentation with the simultaneous formation of hydrogen and carbon dioxide. Biogas production can begin at this point. The basic scheme is shown in FIGURE 1. Some special features of the anaerobic digestion scheme should be noted: the microorganisms involved have complex interactions between the fermentative acid formation and the methanogenesis steps, in addition to having different generation times; secondly, the critical role played by the hydrogen molecule in the overall scheme is dictated by the narrow thermodynamic window, which means that the hydrogen concentration must be so maintained as to promote the formation of methane while at the same time not hindering the propionic acid (FIGURE 2). Therefore, hydrogen-forming and hydrogen-utilizing microbes must be closely associated (interspecies hydrogen transfer).

However, the central problem, also for anaerobic wastewater purification, is to achieve a high space-time-yield. This is possible in practice if the activity or the biomass concentration or both can be increased simultaneously. Besides pH value, temperature, solid content, diffusional limitation, nutrition composition, complexity of microbial population, etc., microbial activity depends very much on the type of selection and adaptation of the population. By this so-called "preadaptation", a powerful microbial population is obtained according to the concept of optimum selection stress by maximization of the space-time-yield (FIGURE 3). The basic idea is that an optimum composition of a mixed culture with maximum specific activity can be obtained in the shortest possible time if one operates the chemostat on a gradually decreased residence time. In order to increase space-time-yield, the residence time should be reduced. If washout occurs—not only for less active parts from the population, but also for more active parts—the residence time has to be slightly

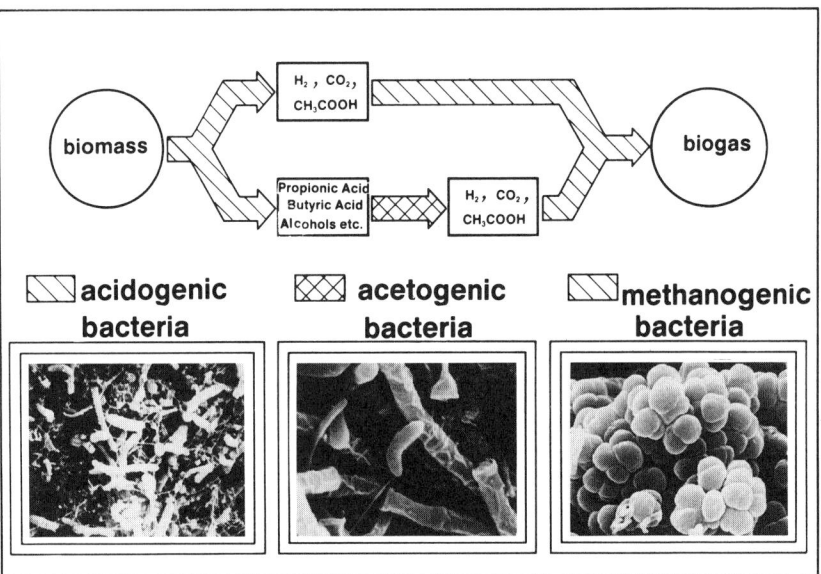

FIGURE 1. General pathway of anaerobic digestion showing methanogenesis by anaerobic degradation of complex organic compounds.

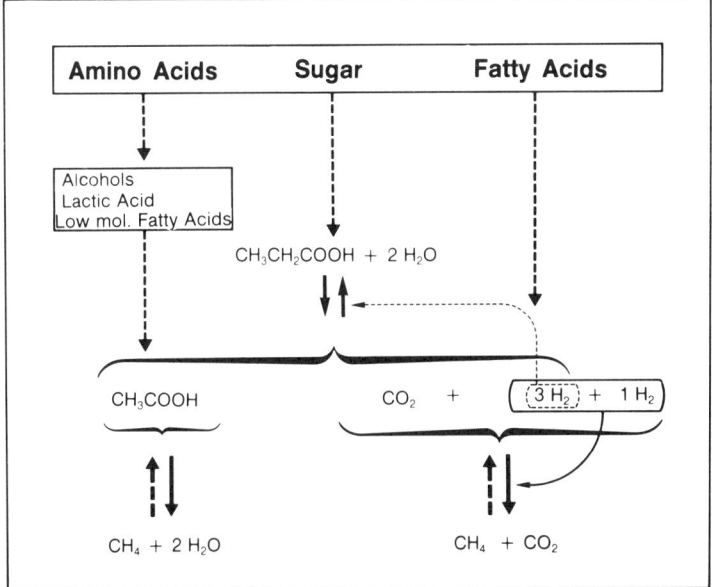

FIGURE 2. Methane formation from acetic acid and H_2/CO_2: influence of hydrogen on propionic acid degradation.

prolonged, but there should be a continual attempt to increase the feed rate again after the population has recovered. In FIGURE 3, it is indicated how there might be a chance of increasing and decreasing residence time in practice (horizontal arrows). Probably the best tool to follow the concept of optimum selection stress is a process computer with a feedforward strategy.

Within six months of operation, the residence time for maximum space-time-yield (5- and 10-liter chemostats) could be brought down to 60 hours. Within three more months, this value could be further reduced to 50 hours in the 150-liter chemostat. With three additional months, the value could be further reduced to 36 hours by use of computer control. In comparison to the first 5-liter chemostat experiments, the doubling time of the population could be decreased to one-fourth of the initial value. In FIGURE 3, the "history" of the increase of biomass activity is given.

However, besides high biomass activity, a high biomass concentration is also required in order to achieve a space-time-yield as high as possible. Due to the energy-limited growth under anaerobic conditions, the biomass concentration is very low. Furthermore, the long generation times of the microorganisms require correspondingly long residence times in continuously operated systems. Because the microorganisms take on a catalytic role, the loss of biomass is to be avoided if possible. This can be performed by effective methods of biomass retention and biomass recycling. In both cases, the increase of particle size as well as the change of particle density in comparison to the liquid phase are performed. FIGURE 4 illustrates different methods of biomass retention.

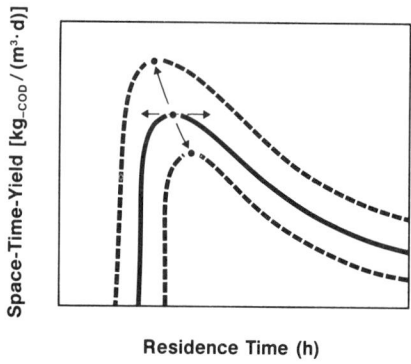

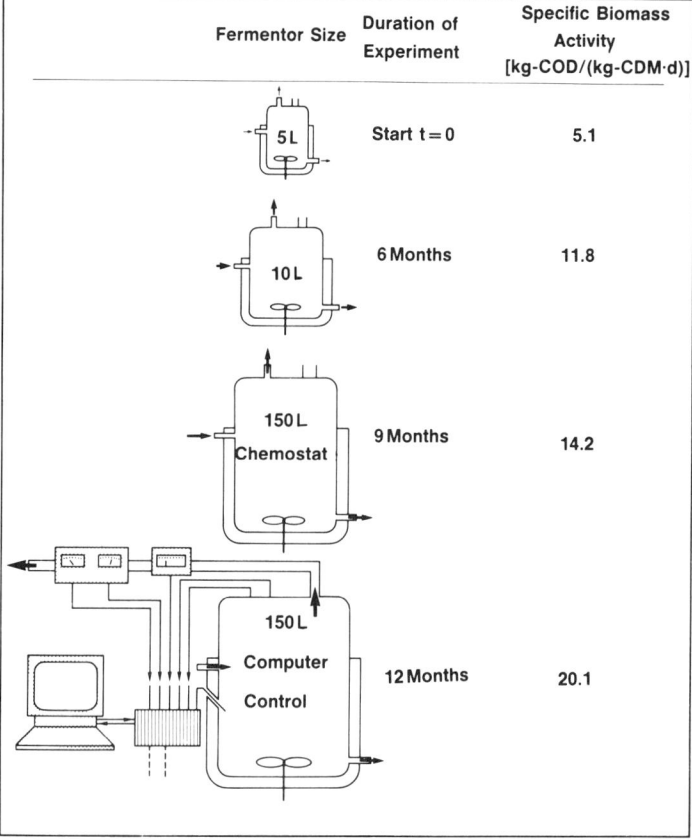

FIGURE 3. Concept of optimum selection stress (top) and development of biomass activity (bottom).

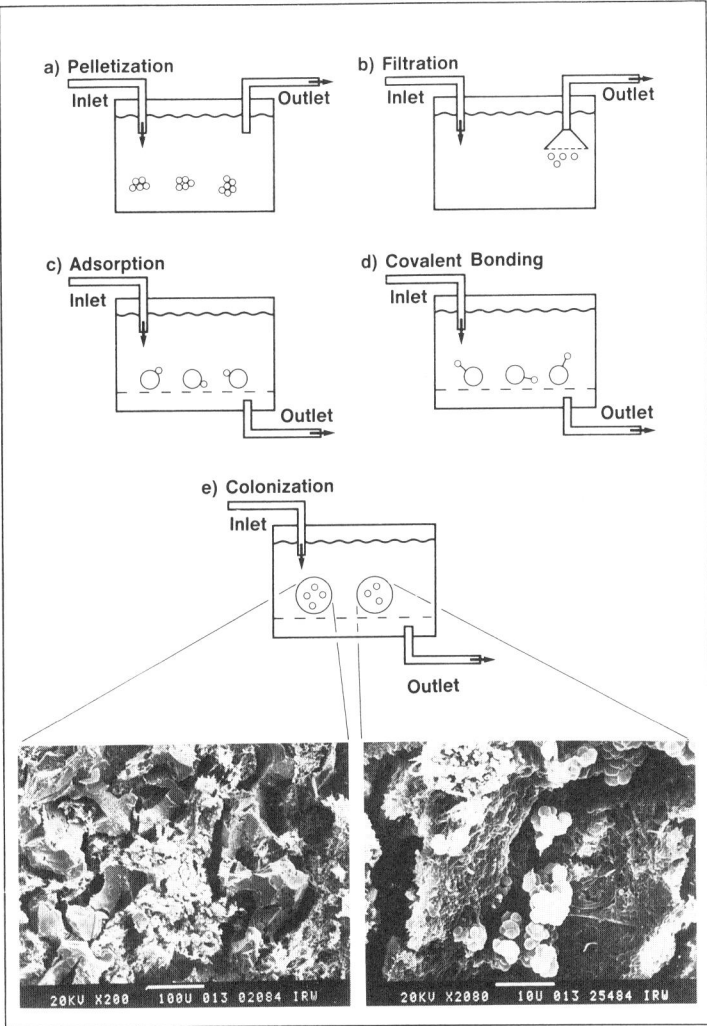

FIGURE 4. Schematic of methods of biomass retention in anaerobic wastewater treatment: (a) flocculation and pelletization, (b) filtration, (c) immobilization by adsorption, (d) immobilization by covalent binding, and (e) immobilization by colonization of macroporous carriers.

Under certain circumstances, usually empirically determined, microorganisms form larger aggregates and pellets with good sedimentation properties. The type of reactor most frequently used for treating wastewater from the food-processing industry, namely, the upflow anaerobic sludge blanket (UASB) reactor, is based on this principle.[1-3] Also, filtration represents a very effective method of microorganism separation in that any cells can be retained depending on the pore size of the filter;

nevertheless, pore clogging can cause considerable difficulties. Furthermore, the energy requirements are so high that only small volumes of wastewater can be treated. By immobilizing microorganisms by adsorption, the effective particle size can be increased easily; however, the biofilm formed can detach from the carrier. This effect can be avoided by covalent binding of microorganisms on a carrier material. Unfortunately, biomass activity decreases and additional costs for chemical fixation limit the potential application of this method to a rather academic significance. Microbial immobilization by colonization of macroporous carrier particles leads to the required increase of particle size as well as to the increase of the density difference compared to the fluid, so a decoupling of residence times for the biocatalyst and the wastewater can be reached without additional chemicals, without additional energy, and without additional equipment.

For the concept of three-dimensional colonization of macroporous carriers, a natural material, volcanic stones, can be employed. Whereas the main pore size distribution for this natural material is very broad (covering a range of 10 to 2000 μm), an alternate man-made material is also available (reticulated sinter "Siran®" glass from Schott Glaswerke, Mainz) with adjustable properties derivable through different manufacturing techniques. This material is obtained by sintering a mixture of glass and sodium chloride powder followed by a washing process to eluate the nonsinterable salt. The resulting reticulated glass sponge has a well-defined pore size distribution depending on the grain size of the salt; porosity is correspondingly a function of the percentage of the filler material. This material can be processed to form tubes, cubes, Raschig rings, and spherical beads. In the left part of FIGURE 5, such a Raschig ring is shown. In the right part (at higher magnification), one can see how anaerobic biomass has colonized the internal volume of this material.

For a practical useful bioreactor, one should achieve high biomass concentrations. This precondition can be fulfilled in a fixed-bed reactor (anaerobic filter), where the interstitial volume is also utilized for microbial retention. Nevertheless, concentration profiles may exist along the reactor length and the biogas release, especially for high efficiency processes, might be limiting. A fluidized-bed reactor avoids these disadvantages and can normally tolerate higher concentrations of suspended solids. However, loss of biomass by biofilm detachment due to the increased shear force in the system provides significant problems in practical reactors. The fixed-bed loop reactor (FIGURE 6) is regarded as a sensible trade-off, combining the advantages of anaerobic filters and fluidized-bed reactors at the same time as canceling out these disadvantages. Scaling up has been performed in our research center in three steps: 1.1-, 11-, and 1000-L working volumes.[4–7]

Due to the low growth rate of the biomass under anaerobic conditions, two aspects become evident:

    (i) time-optimal start-up of a fixed-bed loop reactor and
    (ii) prevention of microbial damage.

Both requirements can be fulfilled by microcomputer-supported on-line data acquisition and process control. For inoculation of a fixed-bed loop reactor containing reticulated sinter glass, a well preadapted mixed population according to the earlier presented concept of optimum selection stress is used. Colonization begins immediately at residence times below the washout value in order to increase selection stress in favor of those mutants with better adherence properties. In this way, the start-up procedure takes place with a high level of space-time-yield. The maximum colonization velocity is then obtained using the concept for maximization of space-time-yield. For this purpose, feedback controls are used, where the set-point values can be changed by a personal computer (set-point control). For process control, the pH, biogas flow, biogas

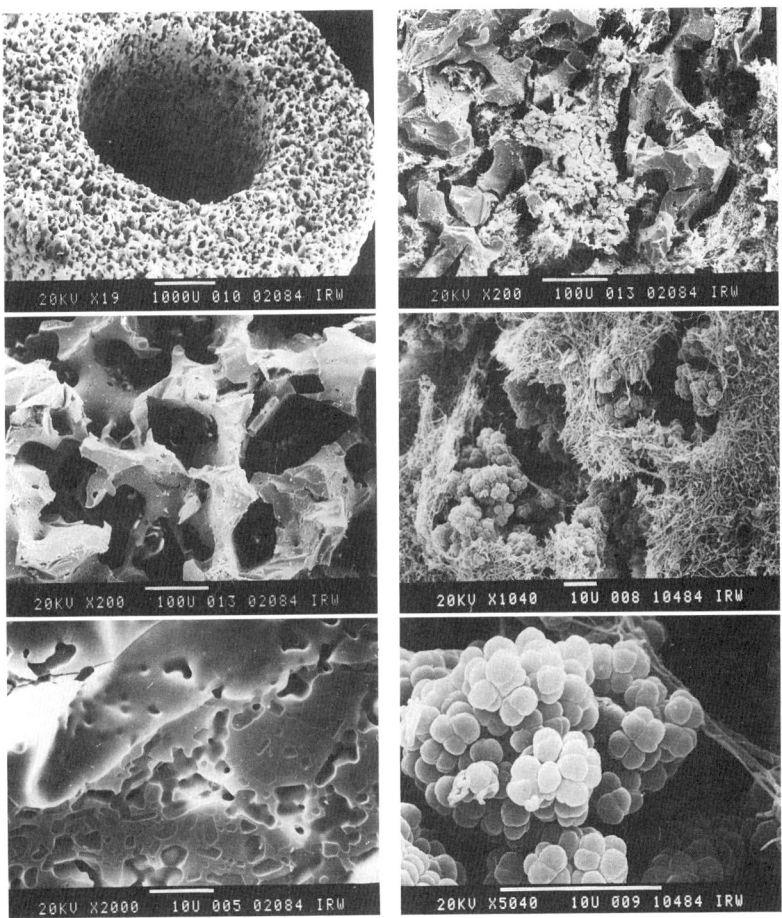

FIGURE 5. Porous glass sponge: scanning electron microscope photographs (SEMs) of a Siran®-Raschig ring (left part) and SEMs of the colonized structure by anaerobic microorganisms (right part) at higher magnification.

composition, as well as their rate change, are used. By a combination of on-line measured variables and the corresponding calculated data, a control matrix can be built up that can then be used in the feedback control loops. An IBM-PC with data input/output interface has been used for this purpose (FIGURE 7). FIGURE 8 shows reactor start-up and residence time reduction (top), as well as space-time-yield and

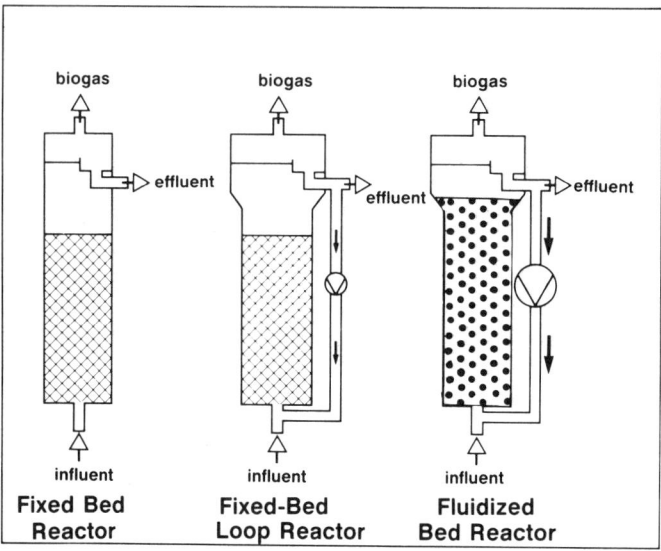

FIGURE 6. Comparison between the fixed-bed loop reactor, the anaerobic filter, and the fluidized-bed reactor.

biogas productivity (bottom) during the time of reactor operation. As can be seen, space-time-yield can be increased in less than one month by a factor of four.

In the course of process development, economical and technical optimization of reticulated sinter glass was carried out. This implemented the replacement of the borosilicate glass by cheaper raw materials as well as the identification of the optimum required porosity, pore size distribution, and particle diameter. The borosilicate glass could be substituted by cheap waste glass fractions. Enlargement of pore size distribution and increase of porosity had a very beneficial influence on colonization behavior. In this way, COD elimination rates of about 160 kg/($m^3 \cdot$ d) at COD loading rates of more than 200 kg/($m^3 \cdot$ d) were possible for treating highly concentrated sulfite evaporator condensates from paper mills. These values correspond to a doubling of the space-time-yield in comparison to the initial manufactured carrier materials at reduced costs. The experimental results correlated very well to the penetration depth of microorganisms in the pore structure as determined by scanning electron microscopy (FIGURE 9). Using the lower porosity and pore size distribution material, only the surface macropores were colonized by the microorganisms (upper part), whereas a total penetration of the biomass using the optimized material can be observed in the lower part of FIGURE 9.

EXPERIMENTAL RESULTS

The described fixed-bed loop reactor as a single-stage process is being used for the treatment of those wastewater components that can directly be transferred to methane carbon dioxide. If more complex wastewater components are presented (for example, carbohydrates, proteins, or higher fatty acids), then a different process configuration should be provided, taking into consideration the stepwise degradation under anaerobic conditions. In particular, a two-stage mode of operation is used in which acidification of the fermentable material is carried out as far as possible during the first stage (continuous stirred-tank reactor, abbreviated CSTR). The discharge then passes into the second stage where methane formation takes place. Because the doubling time of the acidifying biomass in the first stage is far below the corresponding time of methanogenic microorganisms in the second stage, a sedimenter behind the acidification stage for biomass recycling into the CSTR normally provides high enough biomass concentrations.

Having achieved this state of development, we had to move to the field for further scaleup. This was done through a licensing process: Schott Glaswerke (Mainz) and Siemens/Kraftwerk Union (Offenbach) obtained the license for commercializing this

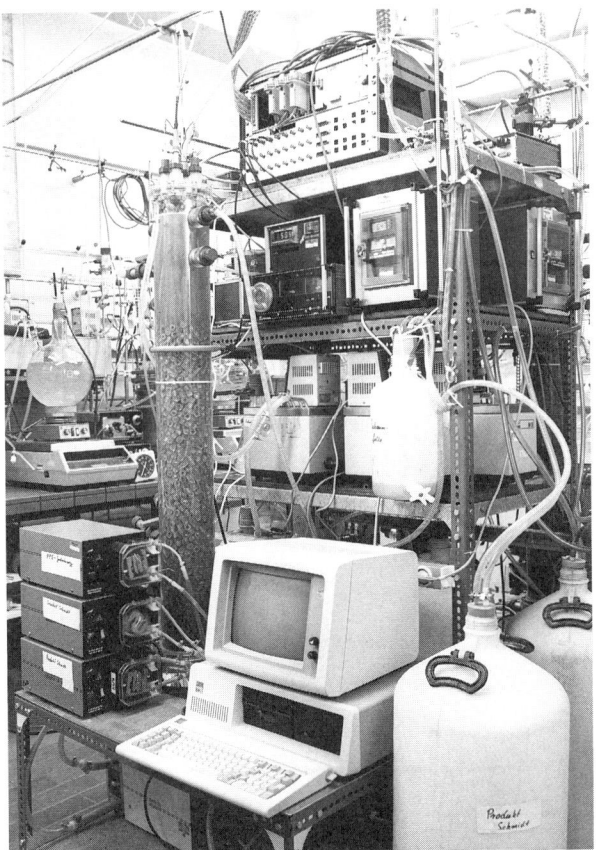

FIGURE 7. Fixed-bed loop reactor with on-line data acquisition and process control by a microcomputer.

high-rate biogas process; furthermore, both companies placed a research and development contract in the Institute for Biotechnology at Jülich. Our licensee Schott employs the Siran® material in anaerobic wastewater reactors up to 20-m^3 fixed-bed size, whereas our licensee Siemens/Kraftwerk Union has decided to use the volcanic stone due to the very favorable price of this material.

Through this cooperation, a further scaleup (3.4-m^3 reactor volume) with Siemens/KWU directly in the paper and pulp industry for the treatment of sulfite evaporator condensates was possible. At a residence time of 2 h and COD loading rates up to 85 kg/($m^3 \cdot$ d), a COD reduction of 80% in the effluent was achieved. In addition,

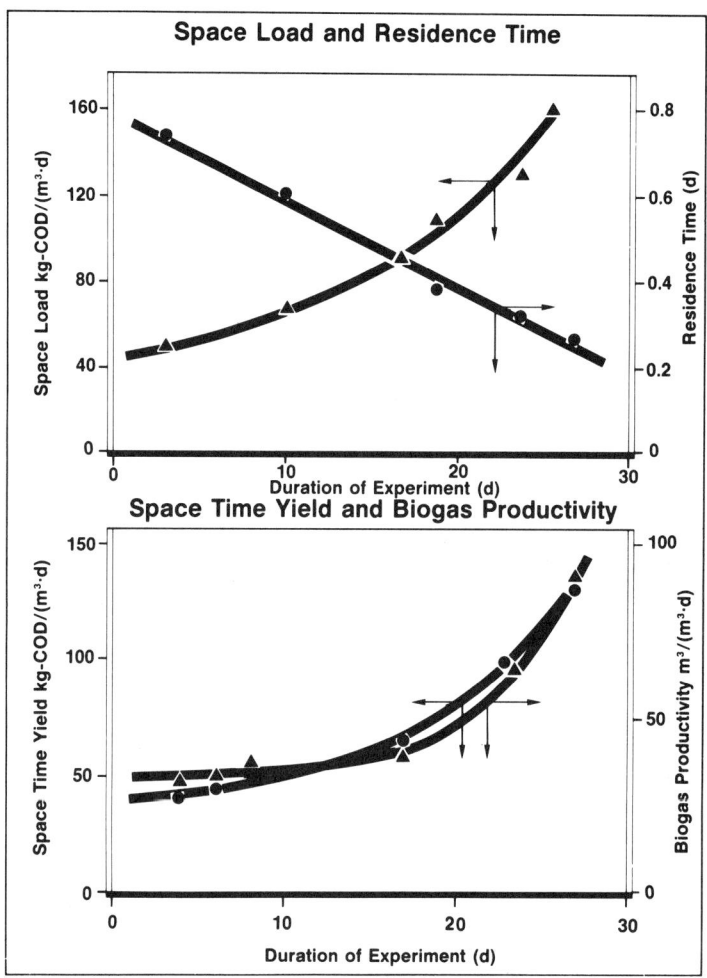

FIGURE 8. Start-up of a fixed-bed loop reactor equipped with on-line data acquisition and process control.

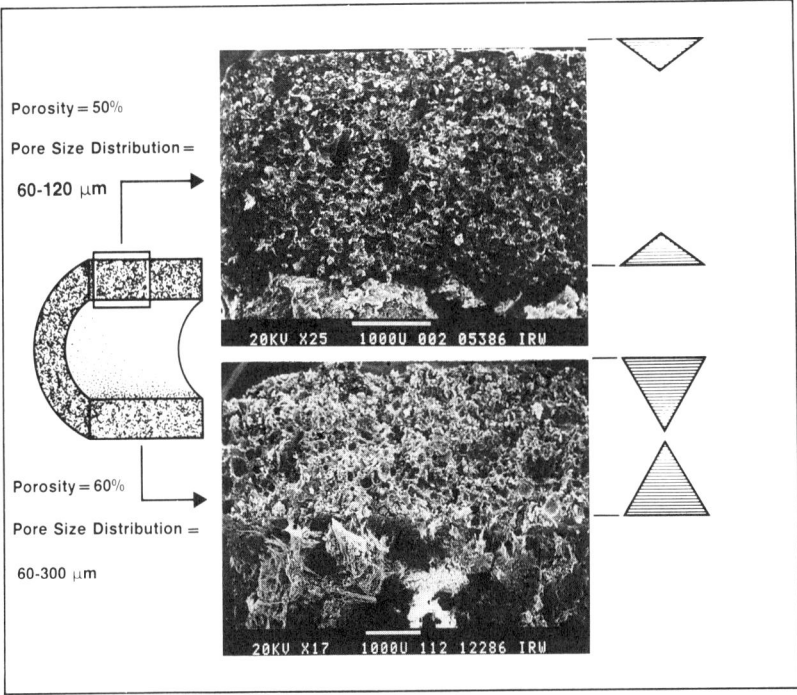

FIGURE 9. Comparison of scanning electron microscope photographs of the microbial penetration characteristics in reticulated sinter glass with different pore structure.

upscaling could also be performed on anaerobic treatment of brewery effluents. In both cases, field results were equal to or better than laboratory-scale results.

From anaerobic treatment of stillage from the ethanol fermentation industry, the limits of single-stage procedure could be clearly demonstrated. The fatty acid spectrum (FIGURE 10, top) was characterized throughout the whole experimental period. Propionic acid dominates during the start-up phase of the experiment. The observed peaks of propionic acid are closely correlated with the hydrogen concentration that underlies the situation described in FIGURE 2. The hydrogen concentration peaks occur a little bit earlier than the maximum of the propionic acid concentration, thus enabling the operator of a biogas plant to take appropriate corrective action (lowering the throughput rate). Immediately after degradation of the propionic acid, the acetic acid concentration rises. The average hydrogen concentration was in the region of 400 to 600 ppm and, in the critical situations mentioned, it achieved peak values of more than 1000 ppm.

Due to this unsatisfactory result with single-stage operation, a two-stage process was instituted. Hence, a separate adjustment of pH value in the acidification stage was possible. A strong influence of the operating pH value on the fatty acid distribution in the effluent of the CSTR could be observed. The propionic acid concentration that can

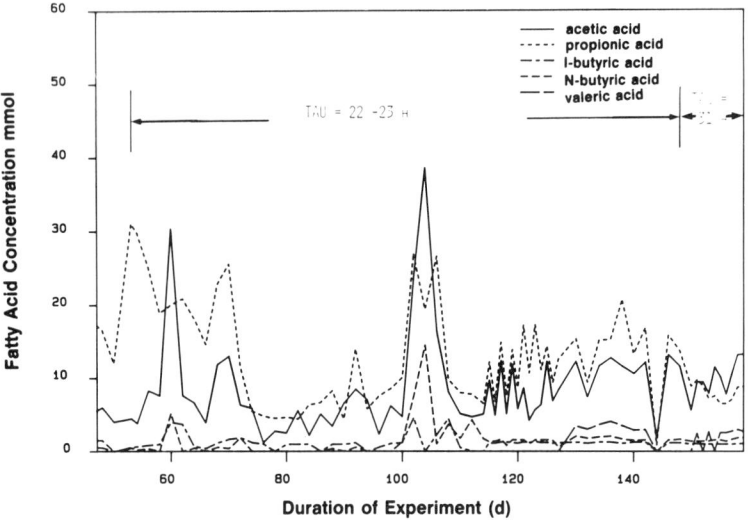

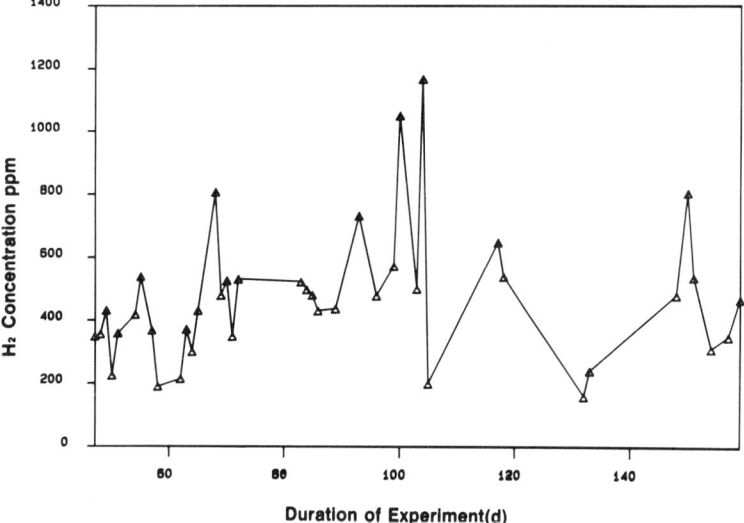

FIGURE 10. Change in the fatty acid and hydrogen concentrations for the single-stage process using fixed-bed loop reactors on anaerobic treatment of stillage.

lead to significant trouble in the following methanation stage was lowest at the lowest pH (4.0) in the acidification stage (FIGURE 11). Over the range of the investigated COD loading rates, the achievable COD conversion for the two-stage system was always higher than that in the corresponding single-stage reactor (FIGURE 12).

Based on these studies, a pilot plant with a 20-m³ fixed-bed methane reactor was

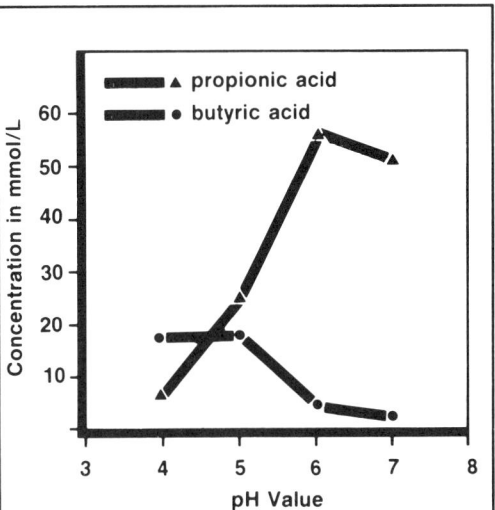

FIGURE 11. Change in the propionic and butyric acid concentrations as a function of the pH value during acidification of the stillage in a continuous stirred-tank reactor.

constructed by our licensee Schott Glaswerke (Mainz) and it went on stream during the summer of 1988. In the autumn of 1987, our partner Siemens/KWU obtained an order from Wendland-Stärke KG, Lüchow, for construction of a full-scale plant for anaerobic treatment of effluents from potato starch production. The following conditions had to be met:

(a) operation during the campaign,
(b) limited time to fulfill the order,
(c) the requirement to start up the waste campaign treatment plant to coincide with the beginning of starch production,
(d) short start-up time required (45 days),
(e) incomplete specification of the wastewater to be treated,

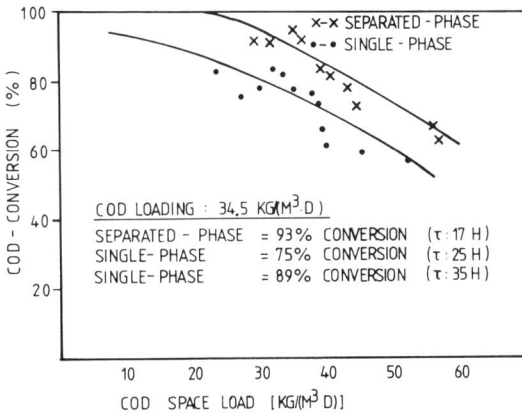

FIGURE 12. Influence of the COD space load on the COD conversion in the single- and two-stage process on anaerobic treatment of stillage.

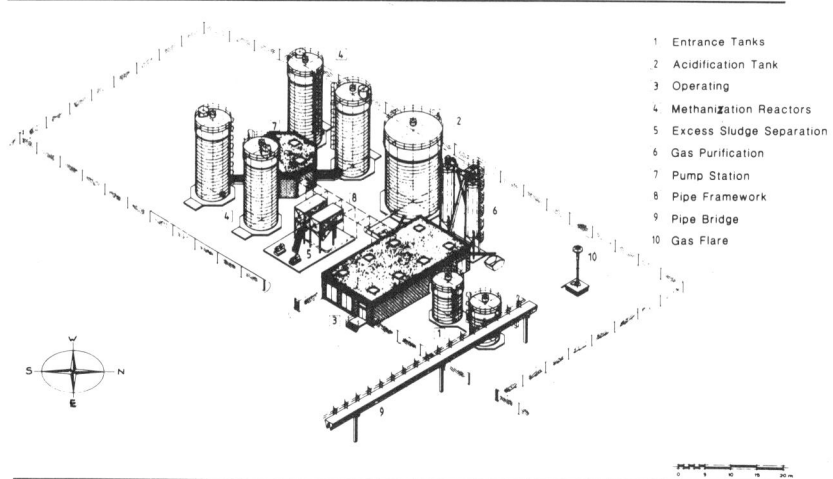

FIGURE 13. CAD flow sheet of the anaerobic full-scale plant at Wendland-Stärke KG for the treatment of wastewater from potato starch production.

(f) wastewater availability only after the start of the campaign,
(g) plant design evaluation without specific experimental data for the starch process.

Due to the fact that the potato starch plant did not exist at the time that the order was issued, literature data about wastewater composition as well as expected values about wastewater flow and COD concentration provided by the potato starch plant designer were used as basic design data for process evaluation. Because a significant part of the COD load consists of proteins and carbohydrates requiring a hydrolysis step prior to methanation, a two-stage process was designed (FIGURE 13). Both main wastewater streams are being mixed in the entrance tanks of the anaerobic plant and are introduced as a total flow of about 86 m^3/h with a 17-kg/m^3 COD load at a temperature of 37 °C in the acidification stage. This reactor is designed as a continuous stirred-tank reactor. The preacidified wastewater then flows after the biomass separation in the methanation stage. This unit has a modular design and consists of four parallel operating fixed-bed loop reactors that contain volcanic stone and that have a common recirculation of the liquid. The COD content of the wastewater should be reduced after anaerobic treatment below 4.6 kg/m^3, corresponding to a conversion degree of 73%. Considering the daily COD load of 34.6 t, this translates to a daily elimination of 25.3 t of COD. At a wastewater flow of 86 m^3/h and a residence time of 12 h in the acidification stage, a tank volume of 1000 m^3 is required. For the methanation step, a residence time of 14 h was chosen, corresponding to a total volume of 1200 m^3 (4 × 300 m^3) and a COD loading rate of 26 kg/(m^3 · d).

Parallel to the construction of the full-scale plant, laboratory experiments using two-stage processes were carried out in our institute as well as in the laboratory facilities of our cooperating partner Siemens/KWU. Pilot-scale trials were also conducted on-site (different location) in the potato starch industry. Reactors from 5-L up to 3500-L volume were used. The obtained data are depicted in FIGURE 14. It can be noted that these data validate the design assumptions for the full-scale plant.

Cell dry mass concentration in the methane stage during steady-state operation is about 20 kg/m³. In order to obtain a comparable biomass concentration within the stipulated start-up period of 45 days, an initial biomass loading requirement of 3000 kg was estimated. This biomass was produced and separated in a 50-m³ pilot plant using a synthetic substrate (acetic acid). After sedimentation of the biomass, the concentrate was used to inoculate the large-scale fixed-bed loop reactors that had, in the meantime, been constructed.

The start-up phase of the acidification stage in the full-scale plant took place within a few days due to the very short doubling time of the biomass present. This could be demonstrated on the basis of the fatty acid composition obtained by gas chromatography. The effluent COD of the methanation stage was far below the required threshold values during the first operational weeks. With increasing operational time and increasing COD loading rate, the excess sludge sedimentation unit behind the acidification tank was unable to handle the high solid concentrations and, consequently, the separation efficiency decreased. These solid overloadings were caused by the simultaneous start-up of the potato processing plant to handle and to remove noncoagulated protein as well as particulate starch. In spite of these initial problems, the start-up phase could be completed within six weeks (FIGURE 15). During the full-scale operation, a hydraulic as well as a COD overloading of the anaerobic treatment plant occurred. The hydraulic overload was up to 20% higher and the COD overload was up

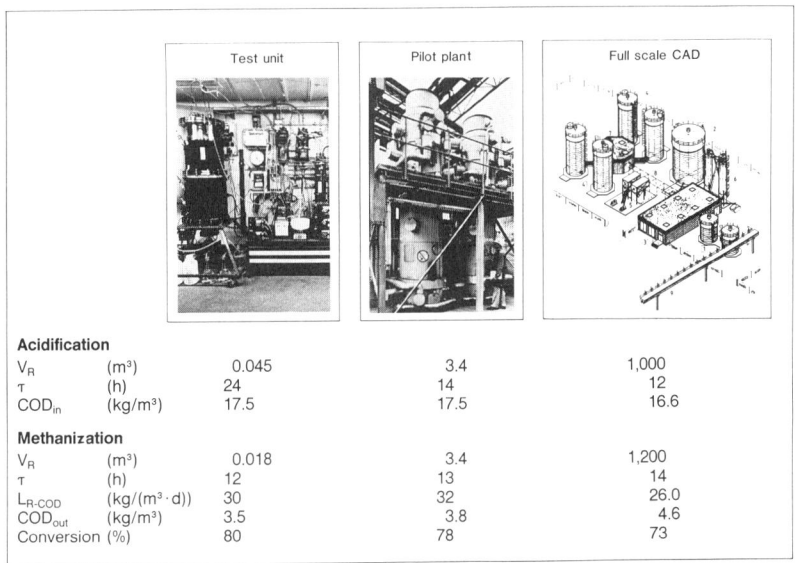

FIGURE 14. Experimental data on anaerobic treatment of potato starch wastewater in laboratory-scale and pilot-scale designs compared to the full-scale design (estimated data).

to 73% higher in comparison to the design. Under these extreme conditions, it was no longer possible to keep the effluent COD values below the required threshold of 4.6 kg/m^3. The COD values of the centrifuged samples were certainly below 4.6 kg/m^3, but the COD of the homogeneous samples was in the range of 5.6 up to 7.5 kg/m^3 only. Nevertheless, the plant could handle this overloading at least with respect to conversion. COD reduction along with space-time-yield were up to 79% higher with respect to the design values. This is mainly due to the steadily increasing biomass concentration in the fixed-bed loop reactor.

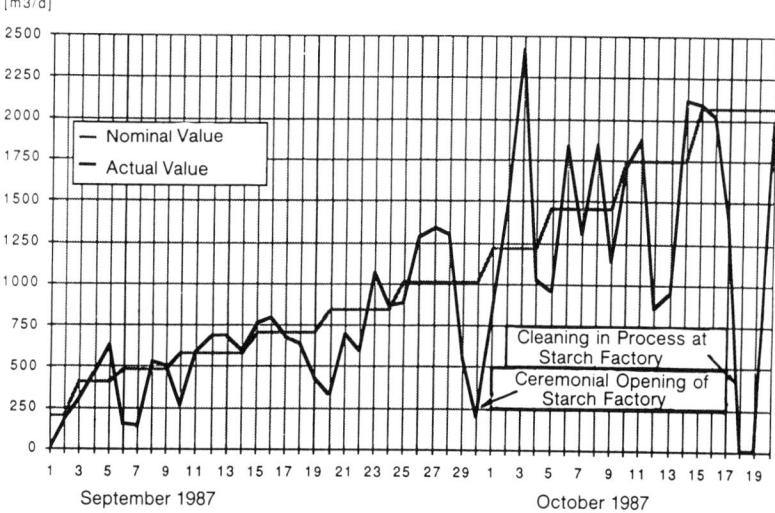

FIGURE 15. Hydraulic start-up procedure of the full-scale plant: comparison of nominal and actual wastewater flow data.

CONCLUSIONS AND OUTLOOK

The case study described in this report demonstrates that it is possible to scale up the anaerobic bioreactors for the treatment of complex organic pollutants. Careful experiments in laboratory- and pilot-scale experiments can provide sufficient data to design a large-scale system and to predict its performance. Crucial to the successful operation of the large scale is the ability to generate adequate amounts of preadapted biomass that can then be used to inoculate and start up the large-scale reactor system. This would facilitate in the reduction of the start-up plant required along with buffering the system against overloading. Wastewater treatment plants should in general be regarded as an integral part of the overall production plant and should be designed and handled in an integrated fashion.

The wastewater treatment plant should not be the catchall system to handle uncontrollable upsets in the production process. Under this imperative, the upsets and disturbances in the production process should be handled in an optimum fashion so that the total plant (including the wastewater treatment plant) can function productively and optimally.

Due to the multidisciplinary character of the bioprocess engineering required for this successful development and operation of the wastewater system, it is essential that the different supplements of microbiology, material sciences, chemical engineering and plant construction, and control engineering should closely collaborate together for careful consideration of all the requirements for a successful operating system.

REFERENCES

1. LETTINGA, G., A. F. M. VAN VELSEN, W. DE ZEEUW & S. W. HOBMA. 1979. Feasibility of the upflow anaerobic sludge blanket process. Reprints from the Proceedings of the National Conference on Environmental Engineering, San Francisco.
2. LETTINGA, G., A. F. M. VAN VELSEN, S. W. HOBMA, W. DE ZEEUW & A. KLAPWIJK. 1980. Biotechnol. Bioeng. **22**: 699.
3. LETTINGA, G., R. ROERSMA, P. GRIN, W. DE ZEEUW, L. H. POL, L. VAN VELSEN, S. HOBMA & G. ZEEMANN. 1981. *In* Anaerobic Digestion. Hughes *et al.*, Eds.: 271. Elsevier. Amsterdam/New York.
4. AIVASIDIS, A. & C. WANDREY. 1984. Ein glasschwamm als bakterienspeicher abwasserreinigung ohne sauerstoff. Jülich-Spez. 1900, Kernforschungsanlage Jülich GmbH.
5. AIVASIDIS, A. & C. WANDREY. 1983. Verfahren und vorrichtung zum kontinuierlichen anaeroben abbau organischer verbindungen. Patentanmeldung Kernforschungsanlage Jülich GmbH no. 3247117.3.
6. AIVASIDIS, A. 1985. Anaerobic treatment of sulfite evaporator condensates in a fixed-bed loop reactor. Water Sci. Technol. **17**(1): 207–221.
7. AIVASIDIS, A. 1986. Anaerobic wastewater treatment in a fixed-bed loop reactor using reticulated sinter glass for microbial storage. Conference papers Aquatech. '86 (Amsterdam), p. 129–143.

Scaleup and Optimization of Oxygen Transfer in Fermentors

Newtonian and Non-Newtonian Systems

V. SINGH,[a] R. FUCHS,[a] W. HENSLER,[a] AND
A. CONSTANTINIDES[b]

[a]*Schering Corporation*
Union, New Jersey 07083

[b]*Department of Chemical and Biochemical Engineering*
Rutgers—The State University of New Jersey
Piscataway, New Jersey 08855

INTRODUCTION

The basic function of scaleup is to determine the operating conditions in equipment of different size and mass transfer characteristics so as to achieve the same process yield. A typical problem is the determination of operating conditions in existing large-scale equipment that will provide similar oxygen transfer as observed in laboratory-scale fermentors.

In most fermentations of industrial importance, the critical substrate is oxygen due to its low solubility in water. For good process productivity, it is important to supply sufficient oxygen to satisfy the needs of the microorganism. However, supplying excessive oxygen is wasteful of power and may even lead to poor performance due to oxygen toxicity in some unusual cases. One way of ensuring adequate oxygen availability is to determine operating conditions that keep the dissolved oxygen concentrations above predetermined critical levels in all parts of the fermentor. In order to minimize power consumption, these operating conditions should be such that the dissolved oxygen levels are not kept higher than necessary.

The work here will be restricted to geometrically similar stirred-tank fermentors. These units are common in the industry because of their versatility and somewhat standardized design. The main operating conditions that must be optimized are the agitator speed and the aeration rate. Suitable methods for the design of equipment for both Newtonian as well as non-Newtonian fermentation broths will be discussed. On-line techniques will be developed for the continuous optimization of oxygen transfer efficiency.

MATERIALS AND METHODS

Oxygen transfer rates (OTR) were measured by offgas analysis using a Perkin Elmer MGA1200A mass spectrometer. The OTR was calculated by the following equation:

$$\text{OTR} = \frac{(O_{2in}F_{in} - O_{2out}F_{out})}{V^m V}, \quad (1)$$

where O_{2in} and O_{2out} are the inlet and outlet oxygen concentrations, respectively; F_{in} and F_{out} are the inlet and outlet gas flow rates; V^m is the gas molar volume at the operating temperature and pressure; and V is the liquid volume.

The exhaust gas flow rate was calculated by a nitrogen balance,

$$F_{out} = \frac{F_{in} N_{2in}}{N_{2out}}, \quad (2)$$

where N_{2in} and N_{2out} are the inlet and outlet gas-phase nitrogen concentrations.

The local mass transfer coefficient, $k_L a$, can be determined from the OTR and the local dissolved oxygen reading, C_L, using equation 3:

$$k_L a = \frac{\text{OTR}}{(C^* - C_L)}. \quad (3)$$

Specific oxygen uptake was calculated by

$$qO_2 = \frac{\text{OTR}}{(\text{cell density})}, \quad (4)$$

where OTR is the overall oxygen uptake rate; note that cell density can be a primary measurement, such as dry cell weight, or an inferential one, such as optical density (Klett).

The experimental setup is shown in FIGURE 1. Details of other procedures are reported by Singh.[1]

OXYGEN TRANSFER—NEWTONIAN SYSTEMS

Oxygen transfer characteristics and dissolved oxygen distributions were extensively studied in a 4.5-m³ fermentor using an *E. coli* strain. It was established from experiments in a well-mixed 10-liter fermentor that, during a batch fermentation, the specific oxygen demand (qO_2) was approximately constant from 6 to 11 hours of fermentation. Experiments were then conducted in a 4.5-m³ fermentor where the aeration and agitation rates were varied over the full operating range of the fermentor during this constant qO_2 period. The OTR and the cell density (Klett units) were determined at each condition, and the specific oxygen uptake rate qO_2 was computed. The data are summarized in TABLE 1. It is clear from these data that the specific oxygen uptake rate qO_2 is not affected much by changes in aeration and agitation rates. This is to be expected for a Newtonian, low-viscosity fluid because the tank is expected to be adequately mixed even at the low end of its normal operating range.

An axial traverse was made using a movable dissolved oxygen probe assembly to determine the dissolved oxygen levels at various depths in the fermentor. A typical traverse profile is shown in FIGURE 2. Although some dissolved oxygen gradients exist,

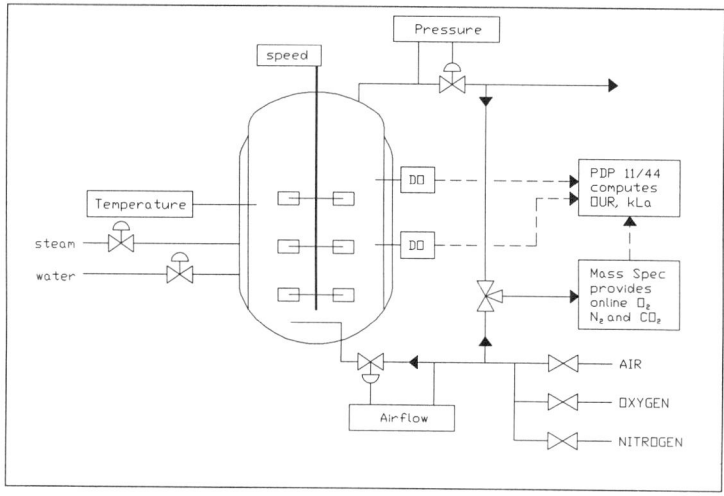

FIGURE 1. Experimental apparatus.

it is apparent that all readings are well above the critical values reported for *E. coli* of 0.005 mMO$_2$/L or about 1% of saturation.[2]

These studies show that mixing effects are not dominant for a typical Newtonian, low-viscosity fermentation and that local mass transfer correlations such as the equations reported by van't Riet[3] (shown below) can be used directly, assuming well-mixed behavior:

$$\text{OTR} = k_L a (C^* - C_L) \tag{5}$$

and

$$k_L a = 0.026 \left(\frac{P_g}{V}\right)^{0.4} (v_s)^{0.5}. \tag{6}$$

The data obtained could explain why most low-viscosity Newtonian fermentations, typical of bacteria and yeast, are in practice easy to scaleup. Caution should be

TABLE 1. Specific Oxygen Uptake Rates in a 4.5-m^3 Bacterial Fermentation

Hours	rpm	vvm	Klett	OTR	$qO_{2\text{expt}}$
6.47	90	0.15	413	10.2	0.024
6.9	108	0.15	547	13.9	0.025
7.2	125	0.15	665	13.1	0.020
7.6	66	0.40	863	24.0	0.028
8.2	76	0.40	1218	29.0	0.024
8.4	85	0.40	1405	33.0	0.023
8.6	95	0.40	1610	42.0	0.026
9.0	104	0.40	2186	45.7	0.021
9.8	113	0.40	3490	77.0	0.022
10.1	125	0.40	4382	64.0	0.014

observed, though, if oxygen demand or critical dissolved oxygen concentration is very high.

OXYGEN TRANSFER—NON-NEWTONIAN SYSTEMS

Two sets of experiments were conducted to study the effect of non-Newtonian fluid rheology on the variation of the overall oxygen transfer rate (OTR) with changes in aeration and agitation conditions. The first set of experiments studied the effect of changing broth rheology on the overall transfer rate in a non-Newtonian batch

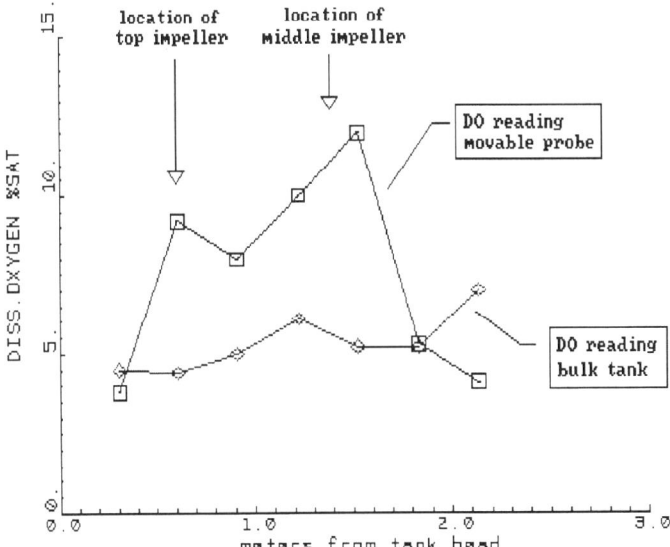

FIGURE 2. Axial dissolved oxygen profile in a 4.5-m^3 bacterial fermentation—9 hours into batch fermentation.

fermentation. The second set provided insight on differences in oxygen transfer characteristics of the non-Newtonian broth with increasing size of the fermentation equipment.

Effect of Rheology on Oxygen Transfer—Non-Newtonian Fermentation

Ten-liter samples were withdrawn from a large-scale mycelial fermentation every 24 hours over the course of a 210-hour fermentation. Because there is very little cell growth during the course of the oxygen transfer experiment, it is not necessary to calculate the specific oxygen uptake rates; the overall oxygen uptake rate is adequate. For each sample, the critical agitation rate was determined by varying the agitation

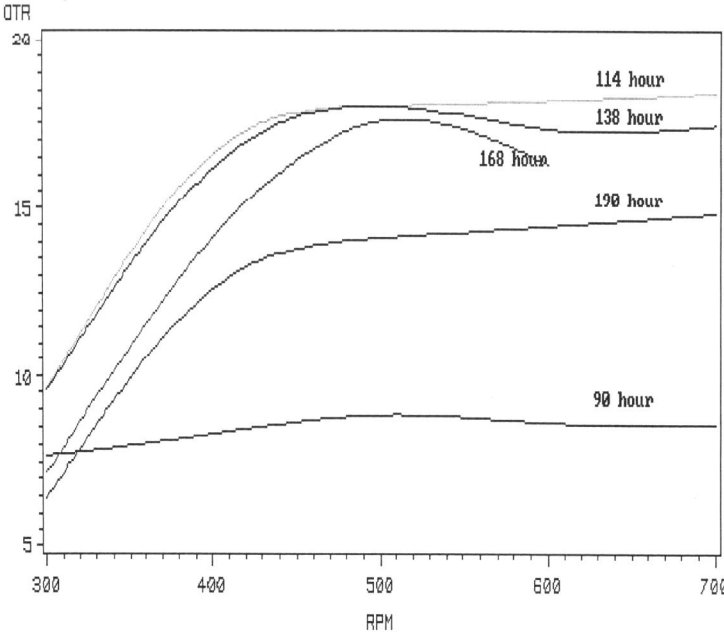

FIGURE 3. OTR as a function of impeller speed for *M. purpurea*: 90, 114, 138, 162, and 190 hours.

speed while aerating with air at a constant 0.6 vvm. The OTR was found to vary considerably as agitation rate was changed, as shown in FIGURE 3.

It can be seen from this figure that there is no effect of agitation speed above 500 to 600 rpm. The speed of 600 rpm seemed to be above the critical value for most samples and measurements of intrinsic oxygen transfer were conducted at this speed. This intrinsic oxygen uptake rate (OUR) is therefore a characteristic of the microorganism and is not a function of fluid mixing. There was no hysteresis in the oxygen measurements, and agitation speed could be raised and lowered over the course of the experiment and the same value of oxygen uptake would result at a given speed. Only in the case of late fermentation samples did too high of an agitation speed (>700 rpm) appear to cause a drop in oxygen uptake rate.

By keeping the unit above its critical speed and at a fixed inlet gas flow rate and by changing the inlet gas composition, it was possible to determine the mixing-independent variation of oxygen demand with dissolved oxygen concentration. The data indicate that the maximal oxygen uptake rate, OUR_{max}, and the critical oxygen concentration, C_{crit}, change as the fermentation progresses. FIGURE 4 shows the intrinsic oxygen uptake profiles for *M. purpurea* taken 90, 138, and 190 hours into the fermentation.

For the 90-hour sample, the OUR appears to be around 10 mMO_2/L-h and is constant over that range of dissolved oxygen investigated. This indicates that the critical dissolved oxygen level at this point in the fermentation is very low and that fluid mixing would have little influence on the overall oxygen transfer (OTR). As the fermentation progresses, the OUR_{max} rises to about 18 mMO_2/L-h and the dissolved oxygen level at fixed aeration and agitation starts to have an influence on the intrinsic OUR. From FIGURE 4, the critical oxygen concentration, C_{crit}, appears to be around 10% of saturation once the non-Newtonian behavior is developed. Below this dissolved oxygen concentration, there is a drop in the intrinsic OUR. The change in OUR characteristics is explained by examining the changing rheology of the broth. The rheological data can be fitted to the classic power law,

$$\tau = \tau_0 + K(\gamma)^m. \tag{7}$$

FIGURE 5 shows the change in power-law coefficients as the batch progresses, showing a transition from Newtonian ($m = 1$) to pseudoplastic behavior ($m = 0.4$).

It is also interesting to note in non-Newtonian fermentation broths, such as the mycelial fermentation studied here, that liquid mixing plays an important role even in small-scale fermentors. From FIGURE 3, it is apparent that there is severe reduction in oxygen transfer at low agitation speeds, indicating partial oxygen starvation.

Effect of Scale on Oxygen Transfer in Mycelial Broths

In order to evaluate the effect of scaleup on the overall oxygen transfer rate, an identical *Aspergillus* fermentation was conducted in fermentors of three different volumes. The fermentation itself was known to be viscous and exhibited non-

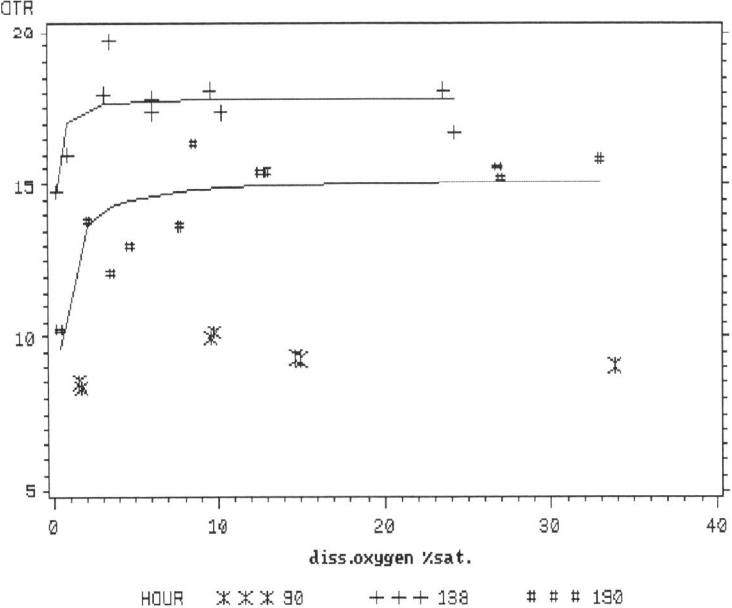

FIGURE 4. Intrinsic oxygen uptake as a function of DO for *M. purpurea*: 90, 138, and 190 hours.

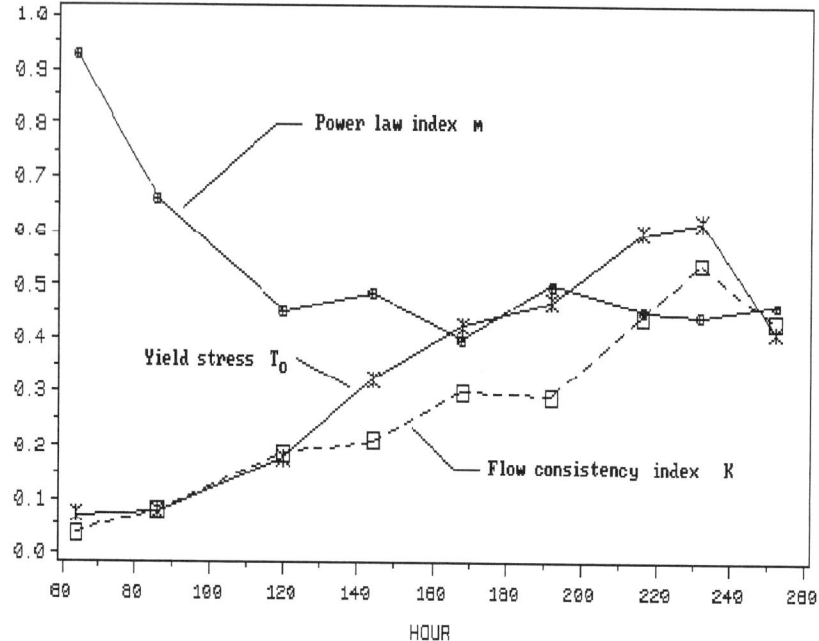

FIGURE 5. Power-law parameters during the course of batch fermentation of *M. purpurea*.

Newtonian rheology similar to the *M. purpurea* data presented earlier. The fermentors used had working volumes of 70, 3700, and 36,000 liters, respectively. The fermentor dimensions are shown in TABLE 2.

The oxygen transfer rates were varied by changing the aeration and agitation conditions at approximately the same cell density at each scale of operation. The data from each tank were fitted individually to a structured model described in earlier papers.[4,5] This model describes inhomogeneity in the fermentor using a tanks-in-series model with recirculation.

TABLE 2. Characteristic Dimensions of the Fermentors Used in Oxygen Transfer Scaleup Studies

	Tank		
Parameter	A	B	C
total volume (m^3)	0.1	4.5	40.0
working volume (m^3)	0.07	3.7	36.0
vessel diameter (m)	0.35	1.4	2.9
impeller diameter (m)	0.13	0.5	1.1
D/T ratio	0.37	0.36	0.39
W/D ratio	0.2	0.2	0.2
no. of impellers	3	3	3

The model consists of a number of equal volume compartments that are each well mixed. The major model parameter is the volume of the highly aerated and agitated compartment immediately surrounding the impeller. The agitator power is assumed to be completely dispersed in this impeller zone, as shown in FIGURE 6.

The other model compartments are assumed to be stagnant with oxygen transfer taking place only from already formed bubbles. The smaller the volume of the impeller compartment, the greater the oxygen gradients in the tank. The value of this parameter is determined by fitting the measured overall oxygen transfer rate to the model equations. The method involves nonlinear regression to a set of oxygen balance equations and is described elsewhere.[5]

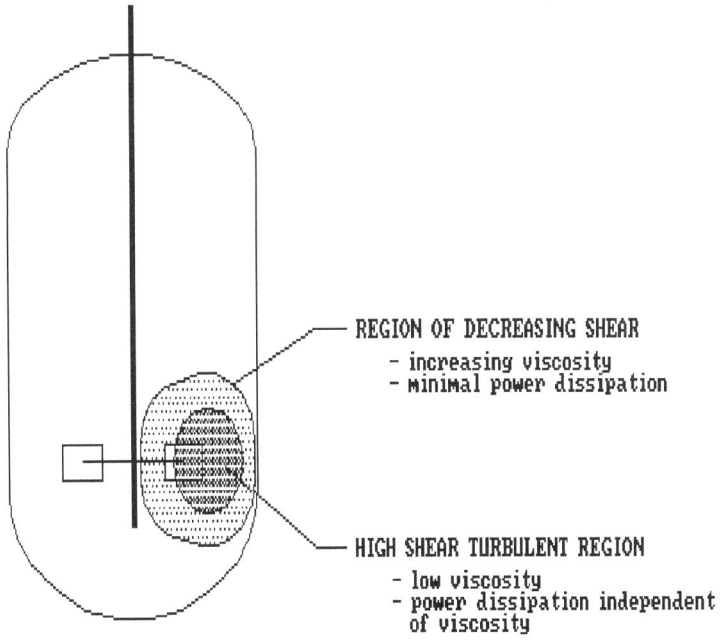

FIGURE 6. Power dissipation in the impeller zone.

FIGURES 7, 8, and 9 show the variation on OTR with changes in aeration (v_s) and agitation (N). The solid lines in each graph are the predicted values based on the fitted model. The fitted model provides estimates of the volume impeller influence zone (V_i) at each aeration and agitation condition. A mixedness index, I_m, which is simply the fraction of the tank that is occupied by the impeller zones, can be computed as follows:

$$I_m = \frac{n_i V_i}{V}. \tag{8}$$

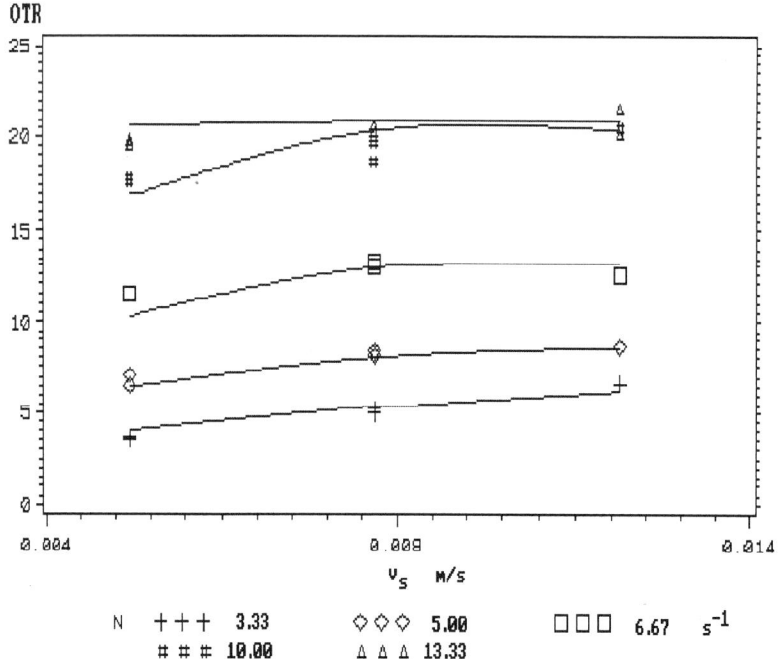

FIGURE 7. Oxygen transfer and fit of the BR model for a 0.1-m³ tank.

TABLE 3 shows that the mixedness index, I_m, appears to increase quite smoothly from poorly mixed ($I_m = 0.1$–0.2) to fairly well mixed ($I_m > 0.5$) as the agitation speed is increased. The mixedness index correlates well ($R^2 = 0.96$) with the average OTR observed at each agitation speed, indicating that it is a useful parameter for the estimation of mixing on the overall oxygen transfer rate.

The impeller zone volume parameters obtained above by fitting the structured model were regressed against various operating parameters. The following correlation was found to predict the impeller zone volume at any given impeller speed for any of the three tanks:

$$\frac{V_i}{\pi ND} = 0.072\, T^{2.32}, \qquad (9)$$

where tip speed $V_{tip} = \pi ND$. This equation fits the data very well as shown in FIGURE 10. The regression R^2 was >0.99.

Substituting for I_m in equation 10 yields

$$I_m = n_i V_i / V = \varphi V_{tip}\left(\frac{T}{H}\right) T^{-2/3}, \qquad (10)$$

where φ = constant and H/T = tank height/diameter ratio. If the mixedness index, I_m, is to be kept constant on scaleup, then the tip speed must be increased considerably as determined by equation 10. This large increase in tip speed may not be permissible due to shear damage considerations. A survey of large-scale fermentors conducted by Einsele[6] indicated that tip speeds are in the range of 5–6 m/s. With this constraint, it is likely that large fermentors will remain more poorly mixed than smaller units and will suffer a consequent loss in oxygen transfer performance.

Although equation 10 provides a powerful and useful correlation, it should be used with caution. It has only been shown to be applicable to geometrically similar tanks and for a particular organism. It is expected that the coefficients of the equation will vary with rheology and that this effect might be difficult to predict from conventional viscosity measurements.

SCALEUP STRATEGY

A two-part approach to scaleup is recommended. First, for design purposes, aeration and agitation conditions should be determined using relatively simple correlations. These estimates should be used to determine the geometry and the required operating range of the equipment. For determining optimal operating conditions, it is recommended that the aeration and agitation be set to low values and that the on-line optimization routines (presented later in this report) be used to continuously match power input to process demands.

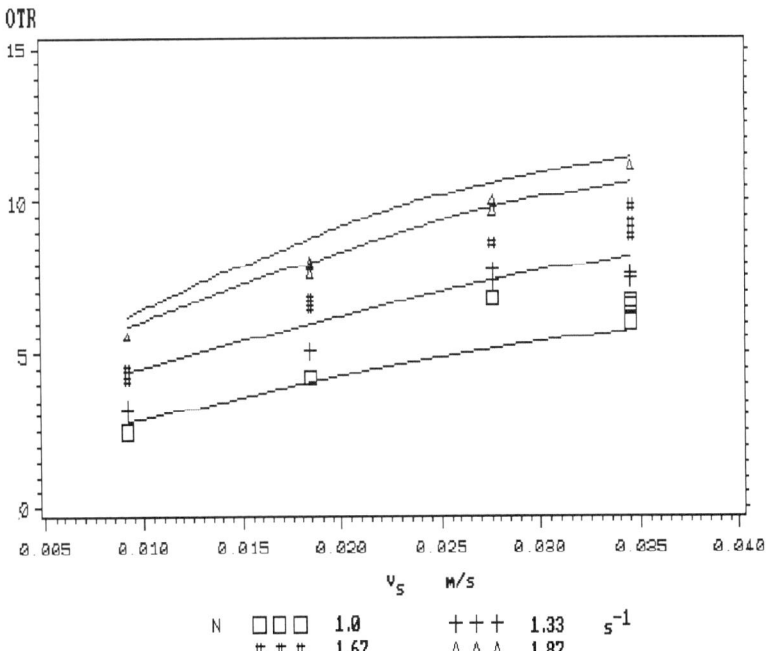

FIGURE 8. Oxygen transfer and fit of the BR model for a 4.5-m^3 tank.

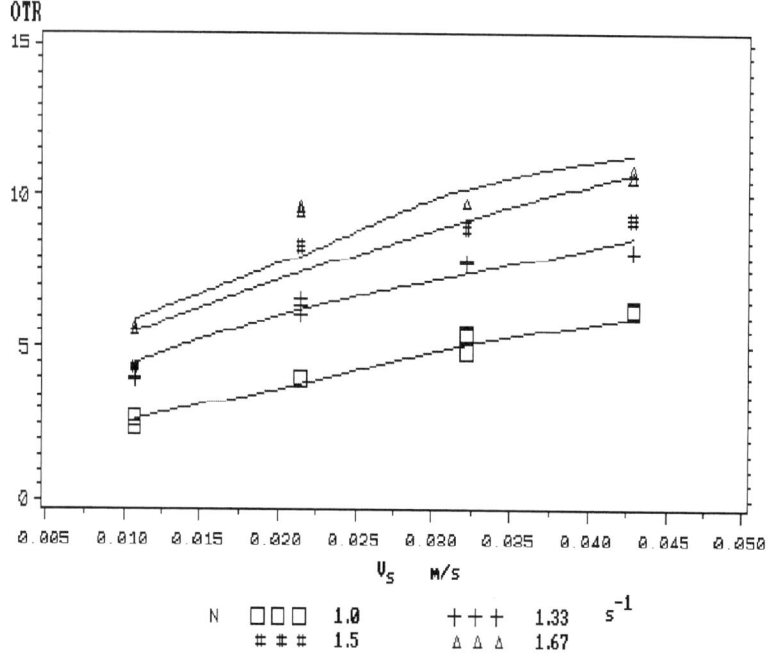

FIGURE 9. Oxygen transfer and fit of the BR model for a 40-m³ tank.

TABLE 3. Values of the Mixedness Index Obtained by Fitting Oxygen Transfer Data in a Non-Newtonian Fermentation (*Aspergillus*)

Tank Volume (liquid m³)	Speed (s^{-1})	Fitted V_i (m³)	I_m	Average OTR
$V = 0.07$	3.33	0.0053	0.23	6
	5.00	0.0083	0.36	8
	6.67	0.0138	0.59	13
	10.00	0.0218	0.93	20
	13.33	0.0223	0.96	22
$V = 3.70$	1.00	0.14	0.11	6
	1.30	0.32	0.26	7
	1.67	0.36	0.29	9
	1.87	0.41	0.33	11
$V = 36.0$	1.00	1.38	0.12	6
	1.30	3.00	0.25	8
	1.50	3.52	0.29	9
	1.67	3.83	0.32	11

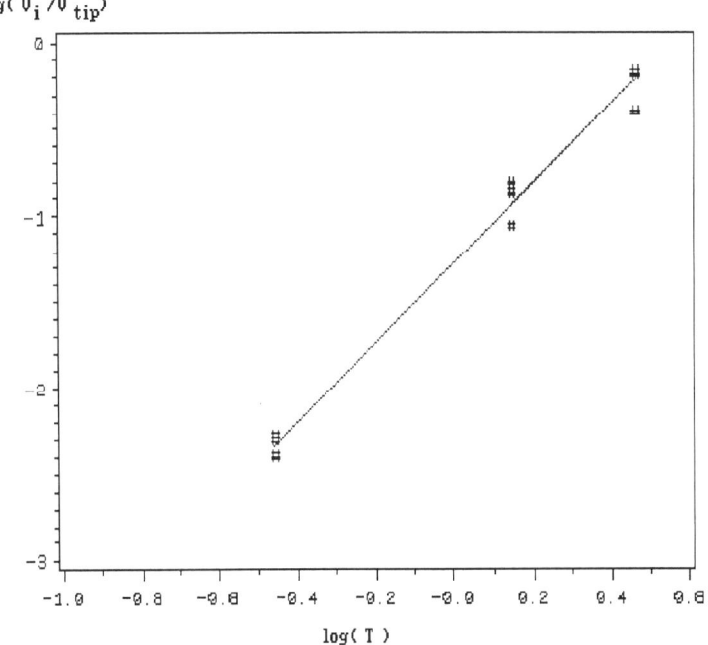

FIGURE 10. Correlation of impeller zone volume to tip speed and tank diameter.

Newtonian Systems

From the oxygen transfer data presented, it is apparent for the purposes of oxygen transfer that Newtonian, low-viscosity fermentations can be considered to be well mixed. The following equations reported in the literature can therefore be used to determine the appropriate aeration and agitation rates required to maintain a specified dissolved oxygen level, C_L, under a given oxygen demand (OUR):

$$\text{OUR} = k_L a (C^* - C_L),$$

$$k_L a = 0.026 \left(\frac{P_g}{V} \right)^{0.4} (v_s)^{0.5},$$

$$P_0 = N_p N^3 D^5,$$

and

$$P_g = C_{MM} \left[\frac{P_0^2 N D^3}{F^{0.56}} \right]^{A_{MM}}.$$

The overall power input into the fermentor can be optimized by proper selection of (i) aeration rate F, (ii) agitation rate N, and (iii) impeller diameter D.

Power input due to agitation is taken to be equal to the gassed power P_g. The power input due to sparged air is taken to be that for an isothermal expansion from a supply pressure Π_0 to atmosphere. This power is not completely dissipated in the fermentation broth due to the usual practice of maintaining some headspace pressure. Power and oxygen transfer efficiency can be calculated using the following equations:

$$P_{sp} = \frac{F^0 R T \rho_g}{M} \ln\left(\frac{\Pi_0}{\Pi_{atm}}\right), \qquad (11)$$

$$P_t = P_g + P_{sp}, \qquad (12)$$

and

$$\text{OTE} = \frac{\text{OTR}}{P_t} V, \qquad (13)$$

where

F^0 is the airflow F at STP,
Π_0 is the supply air pressure (atm),
OTR is the overall oxygen transfer rate = OUR when well-mixed,
OTE is the oxygen transfer efficiency (mMO$_2$/W-h),
M is the molar weight of air (kg/mole),
T is the absolute temperature (K), and
ρ_g is the density of air at STP (kg/m^3).

The optimal aeration and agitation rates can be selected by a search procedure that maximizes the oxygen transfer efficiency (OTE).

Non-Newtonian Systems

Similar correlations can be used to establish operating conditions in non-Newtonian systems. Due to additional intraparticle diffusional resistances present in mycelial systems, it is likely that the coefficients in the mass transfer correlations will depend on the particular organism used and will need to be determined experimentally. Design conditions should be based on the batch conditions where oxygen demand or broth viscosity (or both) is the highest. Tip speed should be used as a primary criterion for selecting agitation rate and impeller diameter either by using equation 10 or by shear rate constraints.

ON-LINE OPTIMIZATION OF FERMENTOR OPERATION

The scaleup methods discussed earlier are designed to estimate the "best" fermentor operating conditions so that the required oxygen transfer is supplied with minimum power input. However, due to variations in media, microorganism performance, and other unforeseen influences, it is usually necessary to make on-line adjustments to the aeration and agitation rates so that the required oxygen is delivered with maximum efficiency during the entire course of the fermentation.

The techniques for determining these optimal conditions are quite different depending on the rheology of the fermentation broth. First, Newtonian, low-viscosity (typical of bacterial and yeast) fermentations will be considered and then a technique for non-Newtonian pseudoplastic mycelial broths will be discussed.

On-line Optimization of Oxygen Transfer in Newtonian Systems

Dissolved oxygen is the most useful measurement for the regulation of oxygen transfer in Newtonian systems because its value is a measure of the balance between supply and demand. Controlling the dissolved oxygen level within a narrow band will ensure sufficient, but not excessive oxygen transfer. The acceptable range of dissolved oxygen for satisfactory operation is obviously process-dependent and must be established by small-scale well-mixed experiments.

Because both aeration and agitation rates influence the dissolved oxygen level in a fermentor, conventional dissolved oxygen control algorithms usually require the manual selection of either aeration or agitation rate as the primary control variable. No attempt is made to achieve optimal efficiency in terms of oxygen transferred per watt of

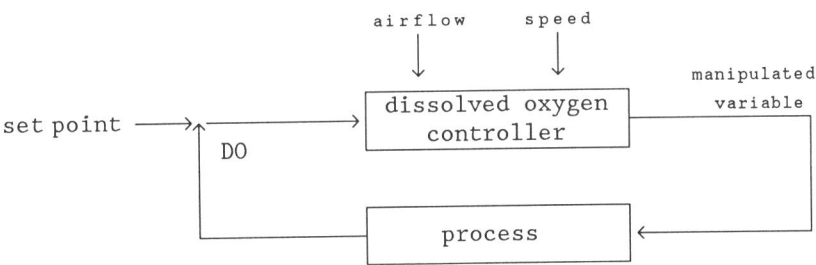

FIGURE 11. Multivariable control of dissolved oxygen.

total power input. However, there are obviously several combinations of aeration and agitation conditions that can provide the same oxygen transfer, but only one combination can do this with minimum total power input. Total power input is defined as the sum of the agitator power input and the power input due to sparged gas expansion:

$$P_t = P_g + P_{sp}, \tag{14}$$

where P_t is the total power input (W), P_g is the power dissipated by the agitator, and P_{sp} is the power dissipated by gas expansion.

The problem of oxygen control is a multivariable one as illustrated in FIGURE 11. The two input variables are airflow rate and agitation speed. Either one or both can be manipulated in order to control the dissolved oxygen level. In fact, it is actually quite simple to incorporate a power-optimizing algorithm into a basic dissolved oxygen control scheme. FIGURE 12 presents a flowchart for an optimizing dissolved oxygen controller (ODOC). The ODOC algorithm operates as follows: The dissolved oxygen measurements (DO) are continuously monitored by a computer or process controller. If the measured DO level goes outside the deadband (db) bounded by an upper limit

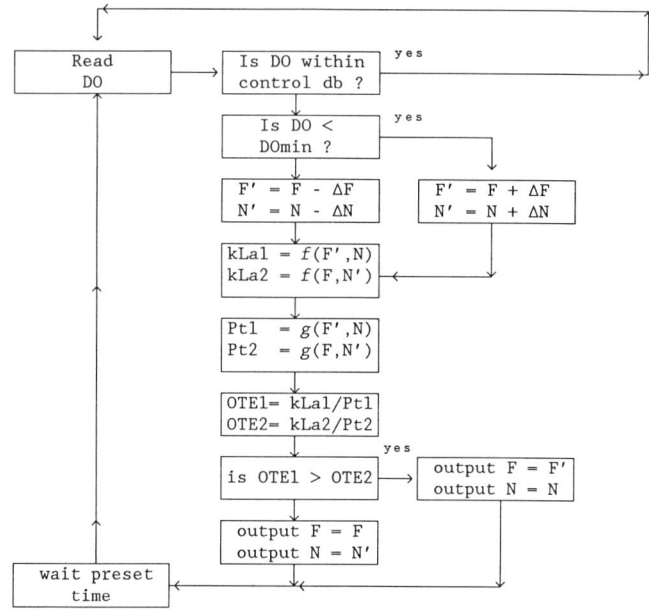

FIGURE 12. Flowchart for optimizing dissolved oxygen control (ODOC).

DO_{max} and a lower limit DO_{min}, then control action is taken. If the DO falls below the preestablished lower limit (DO_{min}), then the controller must increase the aeration or agitation rate in order to raise the DO level. The increase will be accomplished by making a step change (ΔF or ΔN) in either aeration or agitation. The magnitude of the step is determined by tuning constants that reflect the operating and transient response characteristics of the tank. The computer calculates the increase in $k_L a$ along with the overall power consumption (P_t) (agitator power plus sparger power) for the two possible actions—increase agitation rate to N' or increase aeration rate to F'. The power draw and mass transfer rates are calculated using correlations valid for well-mixed systems (TABLE 4). These correlations have been demonstrated to provide

TABLE 4. Equations Used for the Prediction of Power[a] and $k_L a$

$$P_g = C_{MM} \left(\frac{P_0 N D^3}{F^{0.56}}\right)^{A_{MM}} \qquad P_{sp} = \frac{F^0 R T \rho_g}{M} \ln\left(\frac{\Pi_0}{\Pi_{atm}}\right)$$

$$P_0 = \rho N_p N^3 D^5 \qquad P_t = P_g + P_{sp}$$

$$k_L a = 0.024 \left(\frac{P_g}{V}\right)^{0.4} (v_s)^{0.5}$$

$$OTE = k_L a / P_t$$

[a] F^0 is F at STP and Π_0 is the supply air pressure (atm).

reasonably accurate estimates. The efficiency ratio ($k_L a/P_t$) is computed for each combination and compared. The course of action that yields the highest efficiency is the optimal one and the controller raises either the agitation or aeration rate depending on the respective oxygen transfer efficiency (OTE) ratio.

The controller then waits for a preset time to allow the fermentor to reach equilibrium at the new condition. The dissolved oxygen is then rechecked and further adjustments are made if necessary. The cycle repeats until the fermentation terminates.

Similarly, if the dissolved oxygen level were too high, the controller would evaluate whether to decrease the aeration or the agitation rate. If the dissolved oxygen is within the acceptable range, then the controller makes no changes. In this manner, the controller ensures that DO levels are kept above a preset critical level with the minimum power input.

It should be recognized that operating conditions that minimize power input are not necessarily the lowest cost conditions. Minimum cost, though, can be attained by utilizing additional relationships for the estimation of motor and compressor efficiencies and the cost of electricity. These efficiencies tend to be very equipment-specific and also dependent on motor operating speeds. These characteristic efficiencies must usually be obtained from the manufacturer or by measurement. The minimum cost conditions cannot, in general, be estimated from the minimum power conditions; instead, they require a change in the basic criteria for the optimal selection of aeration or agitation rates. The new criteria must be programmed into the ODOC controller. For minimum cost, this criterion is

$$\max\left[\frac{k_L a}{\$_{total}}\right], \quad \text{where} \quad \$_{total} = \$_{air} + \$_{agit}.$$

This type of cost minimization approach also allows the on-line incorporation of other oxygen transfer improvement strategies such as pure oxygen supplementation. The dissolved oxygen controller can compare the cost of each alternative on-line and can select the most efficient option at each level of oxygen demand during the course of the fermentation.

The proposed optimizing dissolved oxygen control (ODOC) algorithm was evaluated by simulation on an IBM AT microcomputer. A PASCAL program was written that simulates both the control action—as described in FIGURE 12—and the process response. The overall power consumption and oxygen transfer efficiency are computed for comparison with other operating strategies.

The process model was derived from the well-mixed *E. coli* studies that were described earlier. The cell growth was modeled by the exponential sigmoid equation. This equation, along with best-fit parameters, is as follows:

$$X = \frac{8800}{1 + 306\exp(-0.96t)}, \tag{15}$$

where X is the cell concentration (Klett units) and t is the fermentation time (hours). The use of this equation allows the prediction of cell concentration during the course of a batch fermentation.

TABLE 5. Comparison of Total Power Expended for DO Control Using Different Operating Strategies

Operating Strategy	Power (W-h)	% of Maximum
(A) maximal N and F	7386	100
(B) N only, then F	4961	67
(C) ODOC algorithm	3251	44

The oxygen uptake rate at any point in the batch cycle is calculated from

$$k_L a(C^* - C_L) = \text{OTR} = \frac{qO_{2\max}XC_L}{(k_{O_2} + C_L)} = \text{OUR (because well mixed)}. \quad (16)$$

Three operating strategies were investigated:

(A) Keeping the aeration and agitation rates at maximal levels throughout the batch fermentation. This corresponds to an uncontrolled case.
(B) Starting at minimum aeration and agitation and increasing agitation only to maintain DO. Aeration is increased under DO control only after agitation is at a maximum. This strategy is commonly used in current industrial practice.
(C) Starting at minimum aeration and agitation and increasing either aeration or agitation as determined by the ODOC algorithm.

In cases B and C, the DO was controlled to between 50% and 60% of saturation. In all cases, the simulation was started 3 hours into the batch cycle. At this point, the OUR is low enough so that DO levels are above 50% even at minimum aeration and agitation rates. The simulation was terminated at 6.33 hours. At this point, in all cases, the OUR is so high that aeration and agitation are both at the maximum allowed values.

The total power (agitator plus sparger) expended in a 0.1-m^3 fermentor between 3 and 6.3 hours of fermentation was computed for each case using the correlations described in TABLE 4. The results are shown in TABLE 5 and, although the conventional strategy (B) decreases the overall power requirements by 33% from the uncontrolled case (A), they clearly demonstrate that the ODOC algorithm is far superior with a reduction in power of 56%. The controlled dissolved oxygen levels were similar using either control strategy.

FIGURE 13a shows how the ODOC algorithm changes both aeration and agitation to maintain as high an oxygen transfer efficiency (OTE) as possible. FIGURE 13b shows the drop in OTE using strategy B. FIGURE 14 compares the total power consumption using either strategy B or the ODOC algorithm (strategy C).

It can be concluded from this study that the ODOC algorithm is superior to conventional dissolved oxygen control techniques in terms of minimizing the overall power requirements for mass transfer in Newtonian systems. It is as simple to implement as conventional single-variable DO control and similarly robust in operation. The ODOC technique is recommended, without reservation, for use in well-mixed systems.

On-line Optimization in Non-Newtonian Systems

Dissolved oxygen measurements are less useful in highly non-Newtonian viscous fermentation because of the presence of stagnant zones in the fermentor. Depending on probe location, the measured dissolved oxygen reading may or may not reflect the average conditions in the tank. This is especially true in large-diameter vessels where the probes are far from the impellers. Hence, it is possible to get poor oxygen supply in certain parts of the tank even if measured DO readings appear to be above critical.

Regulation of mass transfer may still be conducted on the basis of dissolved oxygen control, but the controlled DO level must be set high enough so that even the poorest mixed sections of the tank have oxygen concentrations above the true critical value. This approach typically leads to overaeration and poor oxygen transfer efficiency.

Also, unlike Newtonian systems, the overall oxygen transfer rate in non-Newtonian broths is significantly influenced by the agitation and aeration conditions because these parameters determine fluid circulation rates and patterns in the tank. For these fluids, the correct strategy for mass transfer optimization would be to determine the aeration and agitation conditions that produce the highest overall oxygen transfer rate, thus minimizing detrimental stagnant zones. The secondary objective would again be one of achieving this with minimal total power input.

For purposes of on-line control and optimization, the specific oxygen uptake rate itself can be used as the objective function. Because the oxygen uptake rate is determined from offgas analysis, it is not a local measurement, but a composite picture

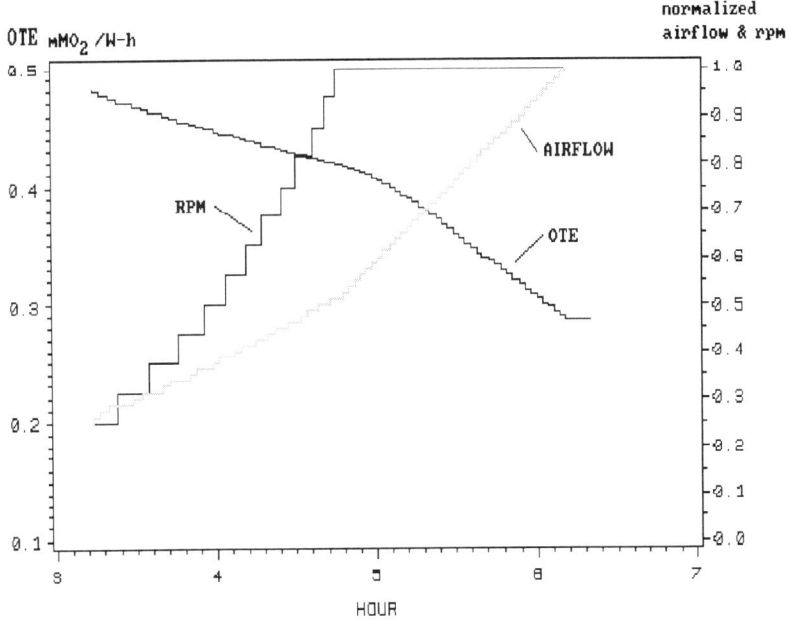

FIGURE 13a. Variation of rpm and airflow—ODOC algorithm.

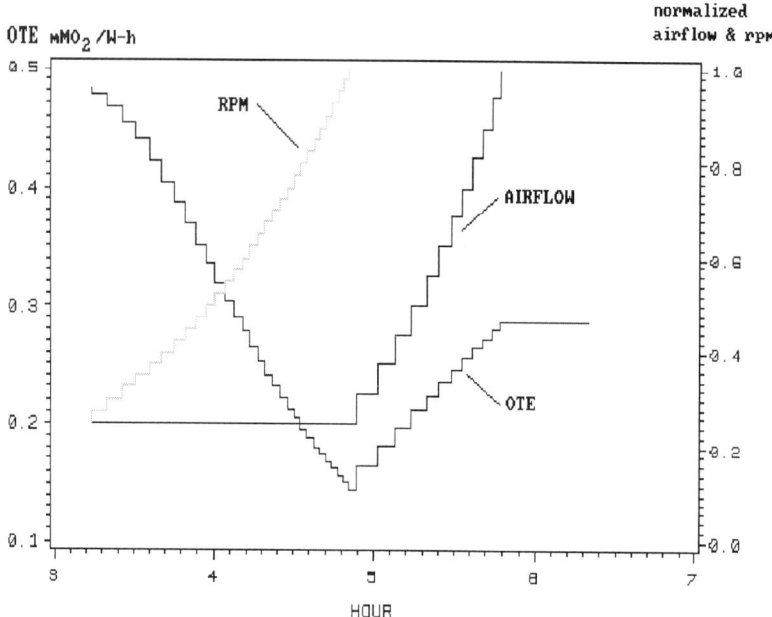

FIGURE 13b. Variation of rpm and airflow—strategy B.

of the performance of the entire tank. For example, the agitation rate can be raised and, if an increase in specific oxygen uptake rate is observed, then further increase in agitation speed is warranted. If, however, no change occurs, then agitation can be lowered until some negative effect on the specific oxygen uptake rate is observed. The aeration rate can also be manipulated simultaneously in a similar manner.

It is clear that optimal aeration and agitation conditions should be chosen such that the overall oxygen uptake rate is the same as that attainable in a well-mixed system. Additional penalty functions and constraints can be applied so that the highest OTR is achieved at the minimal total power input. This type of strategy is not really a scaleup method, but a control technique. It is not expected that the application of this optimization method will overcome limitations due to basic deficiencies in equipment design, but it should be able to elicit the best performance from a given fermentor without the need for extensive trial and error experimentation. The method should also compensate for changes occurring during the course of the fermentation and for batch-to-batch variations.

For the determination of the optimal aeration and agitation rates, the sequential simplex approach[7] is used as follows:

(1) Construct a regular simplex S_0 whose vertices correspond to three different aeration and agitation conditions. The values of the objective function (OTR) are evaluated on-line by setting the operating conditions in the fermentor to each of the three conditions and measuring the OTR.

(2) A new simplex S_p is formed by reflecting the simplex S_0 across a common face. This next vertex Δ^* is computed by

$$\Delta^* = (\Delta_j + \Delta_k) - \Delta_i,$$

where Δ_i is the vertex in S_0 to be discarded and Δ_j and Δ_k are the other two vertices. There are three possible simplexes that can thus be generated depending on the selection of the vertex Δ_i that is to be discarded. This vertex is selected by applying the following rules:

Rule 1: Determine the lowest OTR among the three vertices. Exclude the lowest response vertex Δ_i by reflecting across from it and replacing it by the new vertex Δ^*.

Because there is measurement error associated with the determination of OTR, there is the possibility that the simplex can get stuck around some spuriously high value that is treated as a genuine optimum. To avoid this, apply rule 2:

Rule 2: If a vertex is retained in four iterations, then do not move in the direction dictated by rule 1. Instead, repeat the measurement at this vertex and replace the old value. If the previous measurement was a real optimum, then the repeat observation will also be high. If, however, the result was previously high due to a poor measurement, then the new value will be lower and will be eliminated in due course by rule 1.

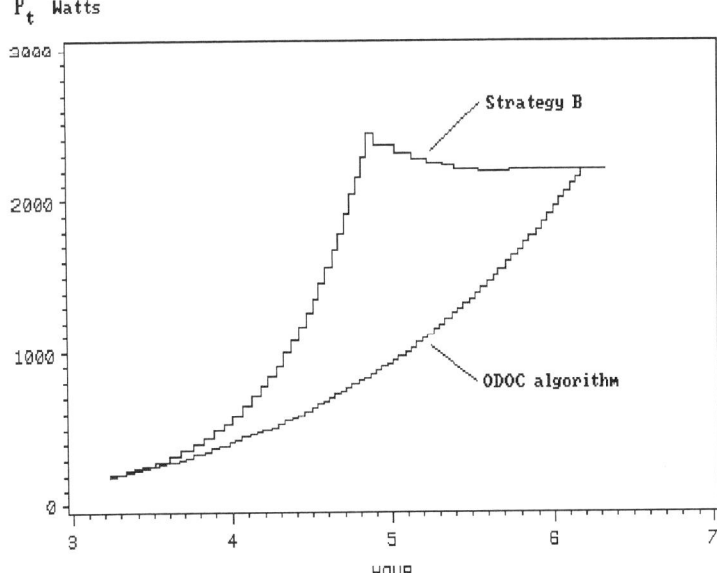

FIGURE 14. Comparison of total power consumption—ODOC versus strategy B.

The optimum can move with time and so the simplex procedure should continually search for improvements. This can be assured by rule 3:

Rule 3: If Δ_i is the worst vertex in the original simplex S_0 and if the new vertex Δ^* in S_p has a response lower than at Δ_i, then do not apply rule 1, which will merely return back to S_0. Instead, reflect across from the next worst value in S_p and replace its value. This will prevent the simplex from oscillating back and forth. Instead, the simplex will indefinitely circle around the optimum. In this manner, if the optimum moves off, the simplex will follow it.

The three rules can be summarized briefly as follows: move by rejecting the lowest observation unless (i) another observation is too old—in which case, renew the latter—or (ii) if such a move would return the system back to the previous simplex—in which case, the next most favorable direction is explored. Given a fixed optimum, the simplexes will approach this optimum with a closeness determined by the step size and will circle continuously around it (straying caused by observational errors are corrected by later observations). If the optimum should move with time, the continuous circle will ensure that information is generated to enable the moving optimum to be followed.

There are several advantages to using this simplex strategy. First, the simplex uses only the last few OTR measurements on which to base its next move. Older, and thus less reliable, readings are automatically discarded. Second, by the use of a regular simplex, the step changes in aeration and agitation can be kept fixed at some value high enough to truly influence the OTR. The method can automatically deal with measurement error and shifting optima. Finally, the calculations involved are very simple.

Certain constraints are required to ensure that the simplex does not try to set aeration or agitation rates that are beyond the capabilities of the fermentor. This is done by applying a penalty function whenever Δ^* is outside operating limits. The penalty is chosen so that Δ^* appears much poorer than the other vertices and the simplex quickly rejects it and returns back to the valid region. If a constraint has been violated, then it should be noted that it is not necessary to evaluate the OTR at all at Δ^*. It is only necessary to assign a response to the illegal vertex that is lower than the other (legal) vertices.

Experimental Results—Non-Newtonian Systems

The sequential simplex method was used on-line to determine how best to operate a non-Newtonian, viscous *Aspergillus* fermentation. The experiments were conducted in the 0.1-m^3 tank and both aeration and agitation were varied automatically by the optimizing program. The objective was to maximize the overall oxygen transfer rate (OTR) during the course of the batch fermentation. Several runs were made. The ranges of operation and program constants were as follows:

(i) airflow range: 20–80 liters per minute with a 5% change per simplex step,
(ii) agitation range: 200–700 rpm with an 8% change per simplex step,
(iii) 10- or 15-minute wait between simplex iterations to allow exhaust gas readings to stabilize.

The following results are typical: During the early stages of the batch fermentation, the growth is quite rapid and the OTR increases substantially as the fermentation progresses. FIGURE 15 shows how the optimizing program continually adjusts the aeration and agitation rates to keep the OTR as high as possible at any stage in the fermentation. On close examination of FIGURE 15, it can be seen in several places that a decrease in aeration or agitation caused a drop in OTR and that the program automatically raised the conditions for the next iteration.

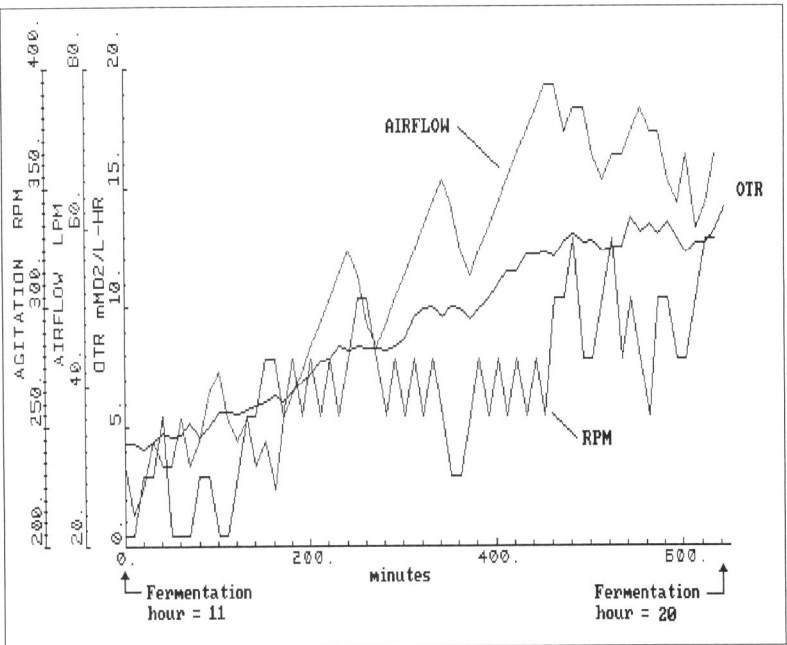

FIGURE 15. Path of the on-line sequential simplex routine—early phase of batch fermentation (*Aspergillus*).

Because of the increase in cell density, it is difficult to verify that the aeration and agitation conditions determined by the program are optimal, that is, in the sense that an increase would not cause a significant increase in the OTR, but a decrease would be detrimental. The efficacy of the optimizing routine was proved by stopping the program once it was circling an optimum and observing the changes in OTR when the agitation rate was raised or lowered above and below this optimum manually.

FIGURE 16 shows the OTR response to changing agitation speed after the optimizing run shown previously in FIGURE 15 (20 hours into the fermentation). The OTR readings on the figure were taken once the exhaust gas concentration was stable (10–15 minutes after the change in the agitation rate). The agitation rate when the optimizer was stopped was 328 rpm. From FIGURE 16, it can be seen that raising the speed to 450 rpm and then to 550 rpm had no significant effect on the OTR. However, lowering the speed down to 250 rpm led to a significant decline. Returning the speed back to near the starting value gave approximately the original OTR.

This study demonstrates that the optimizing routine had successfully found conditions above which no real improvement in OTR took place and that lowering the agitation below the optimum caused a serious drop in OTR. Also important to note is that there is no hysteresis in the OTR measurements because returning to the original

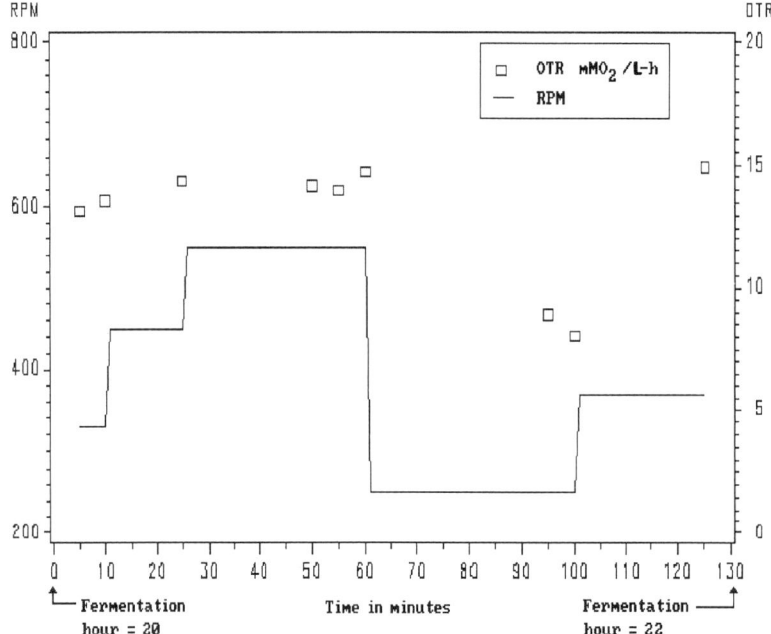

FIGURE 16. Variation in OTR due to manual manipulation of the agitation speed—20–22 hours.

speed after an extended period at other speeds resulted in about the same uptake rate. At the conclusion of this test, the optimizer was restarted and it continued to track the optimum as the batch cycle progresssed. The test was then repeated several times during the course of the batch fermentation with similar results.

In conclusion, the sequential simplex appears to perform extremely well on-line. It is continually able to match aeration and agitation conditions to changing oxygen uptake rates. The technique is very robust and requires no manual intervention or attention. It is recommended for use in fermentations where aeration and agitation conditions affect the overall oxygen uptake rates. This appears to be the case in most non-Newtonian broths. Implementation of the technique is quite simple,[1] but it does require a microcomputer or a programmable controller and an accurate exhaust gas oxygen analyzer.

CONCLUSIONS

Newtonian, Low-Viscosity Fermentations

Mixing effects are not usually dominant in Newtonian, low-viscosity fermentations and, at moderate oxygen uptake rates, the fermentor may be considered well mixed. In that event, the equations of van't Riet may be used directly for scaleup purposes on the basis of providing equal OUR at different scales.

A novel optimizing algorithm for the on-line optimization of oxygen transfer in Newtonian fluids has been proposed and, in simulated studies, has appeared to reduce power requirements significantly. The algorithm controls the measured dissolved oxygen within a set deadband by changing airflow and agitation rates based on a power-minimizing criterion. The method is very robust and is recommended for use in any fermentation than can be considered well mixed.

Non-Newtonian Fermentations

In non-Newtonian fermentations, aeration and agitation rates significantly influence the overall oxygen transfer rate. Scaleup may be effectively conducted by combining the van't Riet equations for well-mixed systems with a tip speed criterion. This tip speed criterion is derived from the mixedness index concept that ensures that a equal fraction of the tank is well agitated on scaleup.

An on-line technique using the sequential simplex algorithm was developed. This technique significantly simplifies the scaleup problem in non-Newtonian systems because it automatically changes aeration and agitation conditions to achieve the highest possible oxygen transfer rate at any time point in the fermentation. The method therefore takes care of changing rheological behavior and oxygen demand.

REFERENCES

1. SINGH, V. 1988. Scale-up and optimization of fermentor mass transfer. Ph.D. thesis. Rutgers—The State University of New Jersey, Piscataway, New Jersey.
2. ATKINSON, B. & F. MAVITUNA. 1983. *In* Biochemical Engineering and Biotechnology Handbook, p. 265. Nature Press. New York.
3. VAN'T RIET, K. 1979. Review of measuring methods and results in nonviscous gas-liquid mass transfer. Ind. Eng. Chem. Process Des. Dev. **18:** 357.
4. SINGH, V., R. FUCHS & A. CONSTANTINIDES. 1987. A new method for fermentor scale-up incorporating both mixing and mass transfer effects—I. Theoretical basis. AIChE Symposium on Biotechnology Processes—Scale-up and Mixing, **200**.

5. SINGH, V., R. FUCHS & A. CONSTANTINIDES. 1988. Use of mass transfer and mixing correlations for the modeling of oxygen transfer in stirred tank fermentors. Second International Conference on Bioreactor Fluid Dynamics, BHRA, Cambridge, paper no. C1.
6. EINSELE, A. 1978. Scaling up bioreactors. Process Biochem., July 13–14.
7. SPENDLEY, W., G. R. HEXT & F. R. HIMSWORTH. 1962. Sequential application of simplex designs in optimization and evolutionary operation. Technometrics 4(4): 441–461.

APPENDIX

Nomenclature

a	interfacial area (m^2/m^3)
A_{MM}	Michel-Miller equation exponent
C^*	saturated dissolved oxygen concentration (mmole O_2/L)
C_{crit}	critical dissolved oxygen concentration (mmole O_2/L)
C_L	dissolved oxygen concentration (mmole O_2/L)
C_{MM}	Michel-Miller equation constant (see equation units)
D	impeller diameter (m)
DO	relative dissolved oxygen level (% saturation)
F	airflow rate (m^3/s)
H	height of tank (m)
I_m	mixedness index
k_{O_2}	Michaelis-Menten constant for O_2 (mmole O_2/L)
K	fluid consistency index (pseudopoise)
k_L	liquid-side mass transfer coefficient (m/s)
m	fluid power-law index
n_i	number of impellers submerged
N	impeller speed (s^{-1})
N_p	impeller power number
OTR	overall oxygen transfer rate (mmole O_2/L-h)
OTE	oxygen transfer efficiency = OTR $\cdot$ V/P_t (mmole O_2/W-h)
OUR	oxygen uptake rate (intrinsic) (mmole O_2/L-h)
P_g	gassed power (W)
P_0	ungassed power (W)
P_{sp}	power input due to sparged gas (W)
P_t	total power input (W)
qO_2	specific oxygen uptake rate (mmole O_2/cell-h)
qO_{2max}	maximum specific oxygen uptake rate (mmole O_2/cell-h)
R	universal gas constant (see equation units)
T	tank diameter or temperature (m or degrees)
v_s	superficial gas velocity (m/s)
v_{tip}	impeller tip speed (m/s)
V	tank or compartment volume (m^3)
V^m	molar gas volume (m^3/mole)
V_i	impeller zone volume (m^3)
X	cell concentration (varies)

Greek Letters

τ	shear stress (N/m^2)
τ_0	yield stress (N/m^2)
γ	shear rate (s^{-1})
ρ	density (kg/m^3)
Π	pressure (atm)
Π_0	pressure at sparger (atm)

Design and Scaleup of an Anchorage-dependent Mammalian Cell Bioreactor

EDWARD L. PAUL, SR.

Merck Sharp & Dohme Research Laboratories
Rahway, New Jersey 07065

INTRODUCTION

In the early 1970s, work was initiated at the Merck Sharp & Dohme Research Laboratories (MSDRL) on the design and scaleup of a bioreactor to produce viral vaccines using anchorage-dependent mammalian cells. The objective was to develop a bioreactor configuration meeting the requirements for growing such cells in tissue culture that could be successfully scaled for production operation in anticipation of increased developmental and manufacturing requirements that would greatly exceed the limited capacity of roller bottles. This work resulted in development of a successful system based on the use of Koch-Sulzer static mixing elements as cell growth surfaces. This paper describes the development of this system and includes data on its performance in cell growth and virus infection on two scales of operation.

DESIGN CRITERIA

The various criteria that have to be satisfied by a successful anchored-cell bioreactor are outlined in TABLE 1. Alternative designs under various stages of development that were evaluated as part of this program were multidisk propagators and packed-bed configurations.[1] Both of these designs were found to be deficient in satisfying one or more of the criteria defined in TABLE 1. The most serious deficiency in both cases was nonuniform distribution of media over all cell surfaces. Two other configurations, suspended beads[2] and hollow fibers, were under active development, but were not sufficiently advanced to meet the restriction of monolayer cell growth; in addition, suspended beads had not received regulatory approval.

KOCH-SULZER STATIC MIXING ELEMENTS

Evaluation of the criteria listed in TABLE 1 led to the conclusion that the principles inherent in the design and operation of the Koch-Sulzer static mixing elements could be beneficially applied to anchored-cell growth and viral infection. These elements, as pictured in FIGURE 1, are made up of a series of baffles arranged to generate uniform blending as well as equal radial and axial flow distribution over a wide range of flow conditions from viscous to turbulent Reynold's numbers. These flow characteristics satisfy the uniform irrigation criteria because viscous low-shear flow can be realized without sacrifice of uniformity and distribution. The inherent design satisfies the surface/volume criteria by providing a ratio of ~11 cm^2/cm^3. These elements are available in titanium, which is a cell-compatible surface. Their use for cell growth and viral infection is described in a United States patent.[3]

Recognition of these design features led to initiation of an experimental program to verify the above suppositions as well as to evaluate scaleup potential and the other design requirements in cell growth and virus replication. TABLES 2 and 3 summarize comparative data on the two sizes of mixing elements studied—5-cm and 20-cm diameter—as well as comparative data with roller bottles. The length to diameter (L/D) ratio of a single unit is independent of diameter.

TABLE 1. Design Criteria for the Anchored-Cell Bioreactor

Uniform irrigation of cell surfaces for:
  nutrient supply
  oxygen supply
  pH regulation

Low shear
High surface to volume ratio   surface compatibility with mammalian cells
Clean in place (CIP)
Sterilize in place (SIP)
Capability for scaleup
Cell growth limited to monolayer[a]
Cell harvest without enzymatic treatment[a]
FDA approval

[a]Additional criteria for the specific viral product candidate.

OPERATIONAL CONFIGURATION

The static mixing elements are incorporated in a recycle configuration as shown schematically in FIGURE 2. The mixers are operated upflow to minimize gas entrapment. Nutrient supply, temperature control, aeration by sparging, and pH control with CO_2 are accomplished in the recycle reservoir. The number of static mixing elements stacked in the vertical reactor can be varied to accommodate residence time and nutrient/O_2/pH depletion criteria. The number of elements used in these experiments was five or six.

FIGURE 1. Koch-Sulzer static mixing elements (5 × 5 cm).

EXPERIMENTAL RESULTS

Flow Distribution

The 5-cm and 20-cm diameter units were compared by residence-time-distribution (RTD) studies to determine their ability to provide uniform mixing. Standard RTD measuring techniques were used to generate the mixing data shown in FIGURE 3. The important conclusion from this study is that the 5-cm unit (pilot scale) mixing capability is exactly matched by the 20-cm unit (production scale) as determined by

TABLE 2. Physical Dimensions of Mixing Elements

	Prototype	Production
diameter × length	5 × 5 cm	20 × 20 cm
total volume	103 mL	7260 mL
void volume	74 mL or 73%	5400 mL or 74%
surface available	12 cm^2/cm^3	11.5 cm^2/cm^3

TABLE 3. Comparison with Roller Bottles

Parameter	Roller Bottle (no. 7000)	Mixing Element (20 cm)
surface	670 cm^2	318,000 cm^2
medium volume to plant	120 mL	29,000 mL
cm^2/cm^3	5.6	11.5
embryo input	1/2	300
cm^2/embryo	1400	1030
plant cell concentration	0.41×10^6	1.02×10^6

the superimposition of the RTD curves. Equal irrigation as well as equal axial and radial distribution are therefore indicated.

Cell Growth

Cell growth studies without viral infection were carried out in both the 5-cm and 20-cm diameter units using primary chick embryo cells, which were the cells of interest throughout this development program. Results for the 5-cm unit are shown in FIGURE 4, where viable cell counts (trypan blue hemocytometer determination) are shown as a function of cultivation time for both the anchored cells and the cells in circulation. Anchored cells were recovered by trypsinization by counting at the indicated intervals. The expected overall doubling of cells from planting to final harvest was achieved. Planting and growth conditions are summarized in TABLE 4. Cell growth results for the

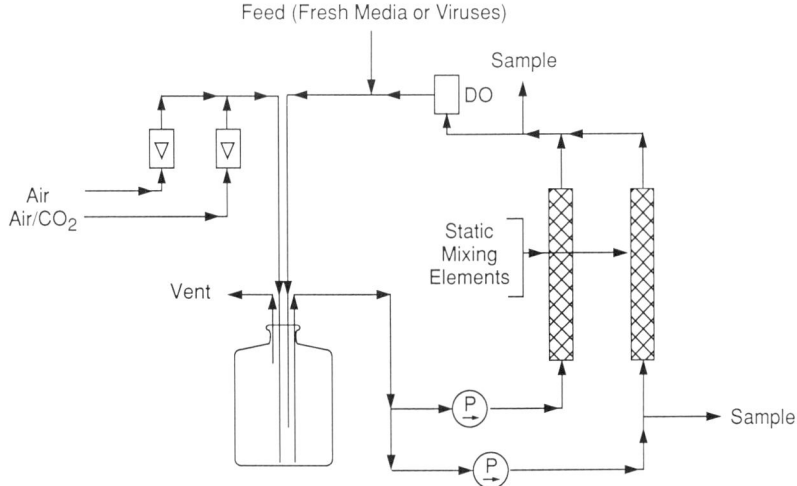

FIGURE 2. The total system incorporating the static mixing elements.

roller bottles and the 5-cm and 20-cm diameter Koch-Sulzer units are summarized in TABLE 5.

Virus Infection and Replication

Virus infection and replication in the chick embryo system were studied using several different viral strains including measles and herpes simplex I.

Results of the measles viral infections are shown in FIGURE 5 for roller bottles compared to the 5-cm unit and in FIGURE 6 for the 20-cm unit. The measles virus is expressed by the cells after infection and the virus-containing fluid is harvested periodically as shown. Viral concentration was determined by a tissue culture infectivity assay. The expected reliability of scaleup was achieved as indicated by the results shown in FIGURES 5 and 6. Equal degrees of successful virus replication were achieved in other viral systems.

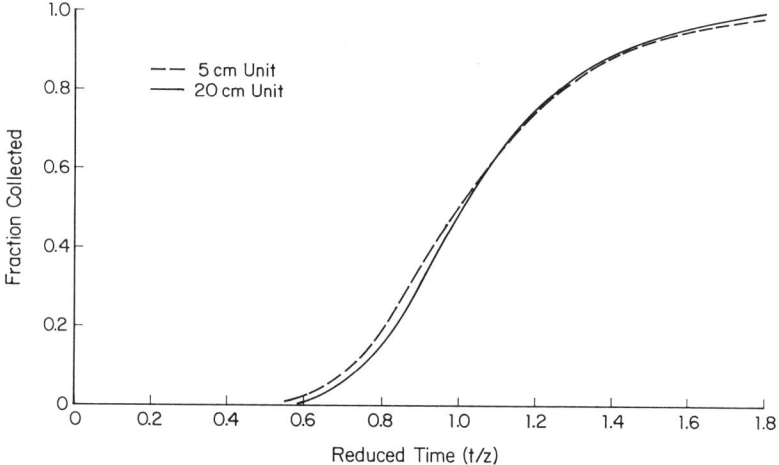

FIGURE 3. Cell attachment and growth in the 5-cm mixing element unit.

Cell Harvest

Two methods of cell harvest were investigated, both of which have potential for successful utilization. In one method, approximately one-half of the liquid is drained from the reactor followed by rotation of the reactor around its horizontal axis to achieve infected cell harvest by the induced shear of the liquid/gas alternating flow. In the second method, the recycle pumping rate is increased severalfold to achieve turbulent flow to remove the infected cells by fluid shear. These methods both require further development for any particular application because it is apparent that different

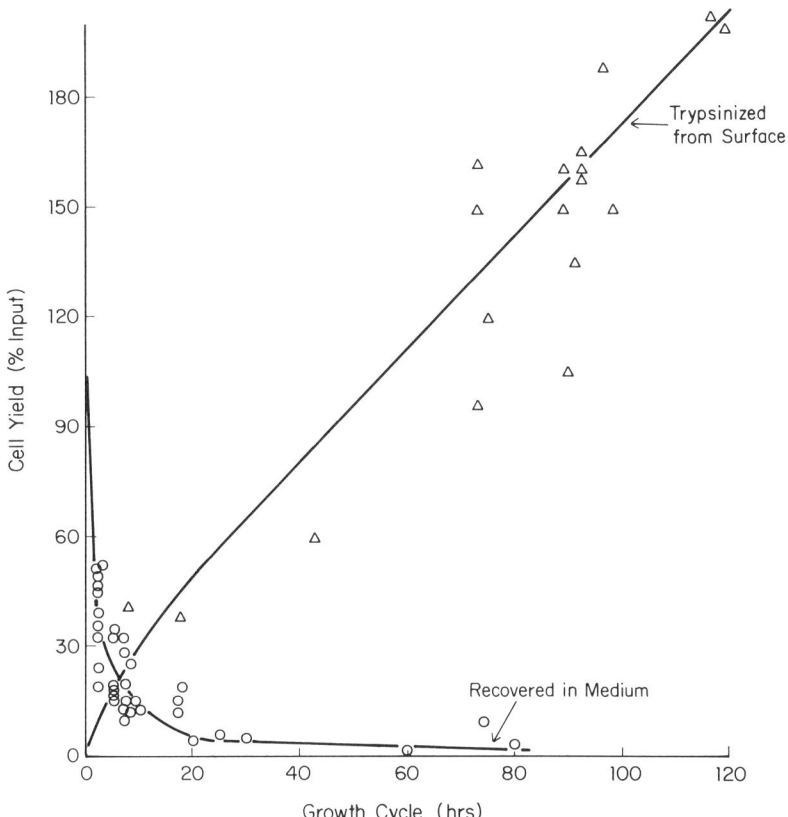

FIGURE 4. Tracer studies of Koch-Sulzer static mixers (5-cm unit).

TABLE 4. Planting and Cell Growth Conditions

Cells:	Primary chick embryo
Cell Preparation:	Washed, minced, trypsinized 2 hours at 30 °C, centrifuged, resuspended in medium "O" with fetal cell serum at 1.8×10^6 cells/mL
Planting Conditions:	Cell suspension charged to the unit in horizontal position, rotated at 6 rpm for 6–17 hours, returned to vertical position and drained (15%–33% lost by draining)
Growth Conditions:	Medium "O" with 2% fetal calf serum circulated with a nominal residence time of 7 min in mixing elements; reservoir aerated with air/CO_2 to maintain both pH (7.0–7.3) and dissolved O_2 (5–6 ppm) and growth continued for 72 hours

TABLE 5. Scaleup of Cell Growth

Size	Cell Density (cells/cm^2)	Glucose Consumption (mg/embryo)
RB[a]	150,000	350
5 cm	200,000	—
20 cm	190,000	350

[a]RB = roller bottles.

cell lines and infecting viruses will cause different degrees of cell attachment and thereby require different levels of shear for harvesting.

CONCLUSIONS

The design and operational criteria established in TABLE 1 have been shown to be achievable by use of the Koch-Sulzer static mixing elements as cell attachment and growth surfaces as well as for virus infection and replication in these cells. The primary reason for the realization of successful scaleup is the inherent nature of these devices to achieve uniform flow and distribution patterns over the cell surfaces while operating at low fluid shear. The chemical resistance and mechanical strength of the titanium elements make them compatible with all required cleaning and flow conditions.

The overall reactor and recycle configuration is ideally suited for clean-in-place and sterilize-in-place procedures without sterile compromise. The sterility record of the system in the experimental program was excellent.

Quantitative comparison of this device with the anchorage-dependent cell methods that are now available is difficult because little has been published on comparable cell

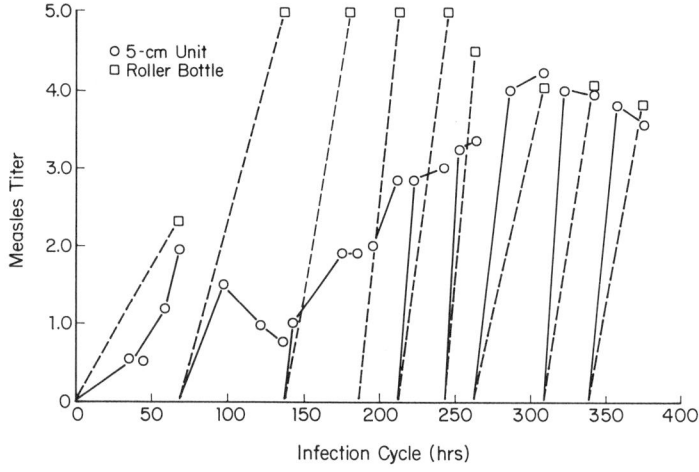

FIGURE 5. Measles infection in mixing elements (5-cm unit and roller bottles).

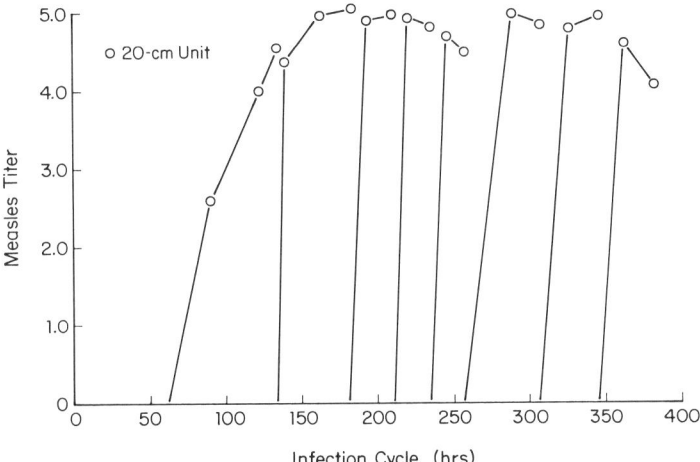

FIGURE 6. Measles infection in mixing elements (20-cm unit).

systems. However, the inherent scaleup potential of this device for other systems is expected to be realized with suitable adaptation to their particular operating requirements.

ACKNOWLEDGMENTS

I wish to acknowledge Ken Posch of MSDRL and Roy Grabner, now with Monsanto, for their developmental and experimental contributions on which this paper is based.

REFERENCES

1. McCoy, X., *et al.* 1962. Glass helix perfusion chamber for massive growth of cells *in vitro*. Proc. Soc. Biol. Med. **109**: 235–237.
2. van Wezel, A. L. & C. A. M. van der Velden-de-Sroat. 1978. Large scale cultivation of animal cells in microcarrier culture. Process Biochem., March 6–8.
3. Grabner, R. & E. L. Paul. 1981. United States patent no. 4,296,204.

Fluctuating Environmental Conditions in Scaled-up Bioreactors

Heating and Cooling Effects

GEOFFREY HAMER AND ARMIN HEITZER

Institute of Aquatic Sciences
Swiss Federal Institute of Technology (Zürich)
CH-8600 Dübendorf, Switzerland

INTRODUCTION

In biotechnology, as in most other process technologies, the design of and operating procedures for commercial- and technical-scale plants are based on the scaleup of process research and process development data generated in laboratory-scale and pilot plant–scale equipment, respectively. The traditional approach to scaleup involves the establishment of relationships between systems of different sizes according to the principles of similarity. Spatial and temporal configurations in systems of different sizes are determined by ratios of specific magnitudes within the systems and are independent of system size.

The process environment in all bioreactors comprises multiple phases that are subject to energy and mass transfer and complex, simultaneous reactions. When the concept of similarity is applied in the scaleup of bioreactors, it involves not only geometrical proportions, but, additionally, features such as fluid-flow patterns, temperature gradients, time-concentration profiles, etc. Similarity can be defined either by intrinsic proportions or by scale ratios, but, as far as bioreactors are concerned, similarity with respect to important variables such as velocity, force, or temperature is defined by an intrinsic ratio for the variable in question, that is, the appropriate dimensionless group.

The three types of similarity that are important in bioreactor scaleup are

(1) dynamic similarity,
(2) thermal similarity, and
(3) chemical similarity.

Each of these states of similarity depends on the condition that the previous state is satisfied.

Dynamically similar systems are defined as geometrically similar moving systems where the ratios of all corresponding gravitational and centrifugal forces that either accelerate or retard moving masses are equal. Thermal similarity is concerned with process systems where heat transfer occurs and hence introduces the dimension of temperature to those of length, force, and time in the concept of similarity. Heat transfer occurs in bioreactors as a result of temperature differences and is frequently enhanced by bulk movement of the bioreactor content. Thermal similarity is of obvious importance both with respect to heat damage of labile biological matter and to questions of bioreactor temperature control. Chemical similarity concerns similarity within reacting process systems in which composition varies either from point to point or with respect to time. The intrinsic criteria that define chemical similarity are the rates of reaction relative to bulk flow patterns and the rates of molecular diffusion. Because of its dependence on both dynamic and thermal similarity, it is often necessary to accept a significant element of approximation with respect to complete chemical similarity, even though it frequently represents the ultimate objective of process scaleup.

For the establishment of novel biological processes, the first step that was normally undertaken was to define process performance criteria in relatively small-scale bioreactors prior to economic evaluation. Although process performance limits were frequently defined, the apparently optimized process operating conditions that resulted were largely irrelevant as far as a scaled-up commercial or technical process, with identical objectives, was concerned.

Two distinctly different situations exist as far as bioprocessing developments are concerned. The first involves situations where redundant bioreactor capacity exists; the second involves situations where new bioreactor capacity, possibly of a novel type and configuration, is to be installed. Whereas the former is the more restricted as far as bioprocess engineering innovation is concerned, the latter clearly implies a greater degree of technological and financial risk, even though purpose-built bioreactors will result. Furthermore, it is also important to differentiate between large-scale technical processes for waste treatment and large-scale commercial processes for product production.

BIOREACTOR CONFIGURATIONS

Probably the most common approach used for the classification of bioreactors is that based on the method of energy input. Energy can be transmitted to the contents of bioreactors by

(1) mechanical agitation,
(2) pumping, and
(3) gas expansion.

However, such a classification evades several very important issues, including that of bioreactor size.

In a recent evaluation of gas-liquid mass transfer in bioreactors, Andrew[1] distinguished two extreme types of bioreactor: free bubble rise systems, on the one hand, and high turbulence systems, on the other hand. Unlike many previous evaluations, bubble columns, tower bioreactors, gas-lift bioreactors, and virtually all large agitated tank bioreactors were grouped as free bubble rise systems, whereas high turbulence bioreactors were restricted to small laboratory-scale agitated tanks. The inclusion of commercial-scale turbine impeller agitated tanks in the former category rather than in the latter must, at first sight, seem surprising, but this was a consequence of the definition of categories in terms of mass transfer phenomena.

Bulk flow patterns in bioreactors are markedly affected by the apparent viscosity of the bioreactor charge. Some process microbes have relatively little effect on the liquid-phase viscosity, but actinomycetes and molds that form complex mycelial matrices frequently impart non-Newtonian characteristics to the bioreactor charge as a result of their physical structure. Moreover, an added complication that can occur is the formation of viscous, water-soluble polysaccharides—by microbes present in the discretely dispersed state—that also impart non-Newtonian characteristics to the bioreactor charge.

Undoubtedly, the biggest problem by far in predicting large-scale bioreactor performance is the paucity of fundamental process engineering data obtained in large-scale bioreactors. However, in the case of large-scale impeller agitated tank-type bioreactors, this has, to some extent, been remedied by the recent work of Oosterhuis[2] in which a 25-m^3 operating volume bioreactor was subjected to extensive investigation. As a result of regime analyses, Oosterhuis and Kossen[3] concluded, in contrast to the conclusions of Andrew,[1] that a two-compartment model could be used to represent the oxygen transfer capacity of the large-scale bioreactor. Simulation of the proposed two-compartment model was possible with two coupled laboratory-scale bioreactors. In these, both low and varying dissolved oxygen concentrations and appropriate residence time distributions for bacterial broths exhibiting low-viscosity Newtonian behavior[4] could be obtained. The scale-down concept employing two-compartment models has been adopted by other researchers, both for broths exhibiting similar low-viscosity Newtonian properties[5] and for non-Newtonian broths containing mycelium.[6] In the case of low-viscosity non-Newtonian broths, it has been predicted by Sweere et al.[7] that process culture heat production and broth cooling will be entirely compatible and will not present major problems in large-scale bioreactors of the impeller agitated tank type. However, what seems to have been overlooked is that an apparently good fit obtained with two-compartment models is not necessarily unique and that the hypothetical two-compartment models may not realistically represent the underlying physical processes involved.

For virtually all medium volume and bulk microbial products, including renovated wastewater, economic criteria dictate that bioreactor design is a matter of technical compromise that inevitably results in nonideal bioreactors.

In a generalized approach to regime analysis for nonideal chemical reactors, van de Vusse[8] proposed that they can be characterized in terms of residence times in parallel zones of plug flow, complete backmixing, bypassing, and dead space (stagnant zone) for each fluid phase present in the reactor, as shown in FIGURE 1.

With the exception of low-aspect-ratio, rectangular, sparged tanks of the type widely used for municipal sewage treatment, bioreactors nearly always contain various internals such as wall baffles, draft tubes, perforated plates, heat exchanger surfaces, etc. Frequently, the presence of internals in bioreactors results in the establishment of stagnant zones, particularly when the bioreactor charge exhibits highly viscous non-Newtonian properties.

Another design feature that has become increasingly common has been the external loop as a means of augmenting bioreactor cooling capacity. Such loops function either by the action of a gas-liquid two-phase pump in the loop or by natural circulation resulting from fluid-phase density differences. As far as biopolymer production processes are concerned, external cooling loops probably represent the only feasible means of controlling the bioreactor operating temperature.

Residence time distribution functions can be generated by methods based on theories that assume a gradient type of liquid diffusion, which can be represented by an effective diffusion coefficient. However, an eddy diffusion model is only valid if the scale of turbulence is small with respect to the space occupied by the turbulent flow.[9] Therefore, difficulties in data interpretation occur when mixing is controlled by the

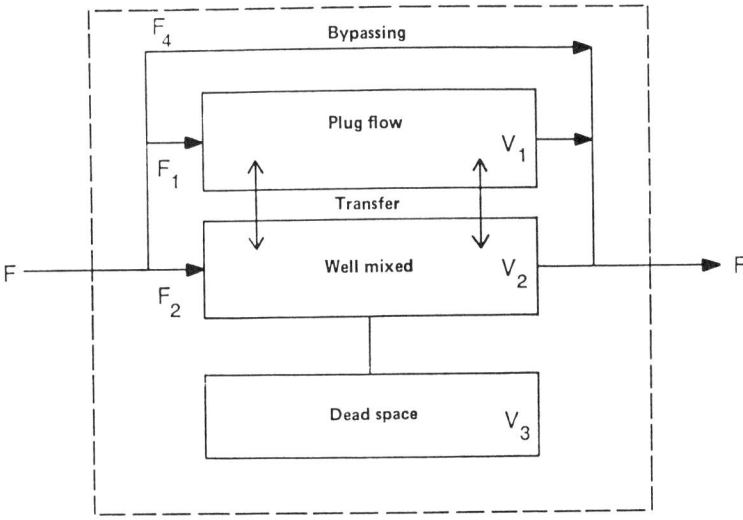

FIGURE 1. Representation of compartments for each phase in a reactor.[8]

development of large eddies and, under such circumstances, a circulation pattern with superimposed eddy diffusion would probably represent the most realistic model of the physical processes in question.

HEAT TRANSFER IN BIOREACTORS

One of the most neglected aspects in process biotechnology is the question of the various heat transfer processes involved in controlling the temperature of the individual microbes in a bioreactor at a level coincident with the temperature optimum for their anabolic and catabolic processes. The optimum temperature range for the growth or product formation by most microbes is narrow—in some cases, only a few degrees.

At temperatures only a few degrees in excess of the optimum for growth, a precipitous decline in the growth rate to zero frequently occurs. The vast majority of microbes used in biotechnology processes are mesophilic and their optimum temperature for growth is somewhere between 20 and 40 °C.

If one considers the production of any bulk microbiological product by an aerobic route, process cooling is a very significant technological problem and a major operating cost when the temperature optimum for maximum yield and productivity is below 40 °C. Currently, it is general practice to use cooling water rather than a refrigerant as the cooling medium for industrial-scale biotechnological processes. The maximum temperature of the cooling water is of critical importance for sizing the cooling surfaces needed for any particular process and the operating costs for cooling systems are governed by the average cooling water temperature at each particular location.

Cooling water temperature is dependent on a number of factors that include the origin of the cooling water, the climatic conditions and variations at the plant location, and the type of cooling water system employed, which can be either a once-through or a recycle employing a cooling tower. The probable future trend for process cooling will be an increased use of recycle systems in order to eliminate both heat and possible chemical pollution as a result of cooling water discharges from once-through systems. Even in temperate regions, the use of mesophilic microbes for biotechnological processes will mean that only relatively small temperature differences will exist between the optimum process temperature and the temperature of the cooling water in large, integrated manufacturing locations.

Most bioreactors used for traditional processes are provided with a cooling jacket or internal cooling coils (or both), but recently, for higher intensity processes, external cooling loops have been introduced irrespective of bioreactor configuration. In bioreactors of equivalent configurations, the provision of internal cooling surfaces would so disrupt essential hydrodynamic operating characteristics so as to make their inclusion impractical. Also, with increasing bioreactor scale, the provision of sufficient internal cooling surfaces becomes increasingly difficult. The main factors that affect the effective cooling of aerobic bioreactors are as follows:

(1) the transfer of heat from the individual microbes to the bulk liquid medium;
(2) the transfer of heat from the bulk liquid, containing dispersed air bubbles and microbes, to the cooling surfaces;
(3) the generation of heat by the mechanical energy dissipated during agitation of the bioreactor charge;
(4) evaporative cooling, temperature equilibration, and expansion work resulting from the continuous passage of air through the bioreactor charge;
(5) the effect, on overall heat transfer coefficients, of fouling of the cooling surfaces by accumulation and growth of microbes in these surfaces or by accumulation of polymeric products at or close to these surfaces.

For each process system, it is important to examine the relative magnitudes and the potential significance of all these factors. In aerobic systems where the design productivity can potentially be achieved, but where the design cooling capacity fails to be achieved, process operation will be nonoptimal and productivity will be markedly reduced.

TEMPERATURE EFFECTS ON MICROBES

When either growing or resting microbial cells are subjected to a temperature change in the form of either a short duration step variation (shock) or a longer duration, more gradual variation (transient), it can be envisaged that the external environment surrounding the cells will be altered and that certain of the macromolecular constituents of the cells might be affected, depending on the rate, magnitude, and duration of the change. The overall effects in growing microbial cultures can be manifested in both changed reaction rates and changed biomass and/or product yield coefficients.

In any analysis of temperature changes on microbes, it is essential to differentiate between changes within the growth-permissive temperature range and those that either extend into or are totally in the nonpermissive temperature range for growth. As far as bioreactor operation is concerned, one should expect interest to be confined to the growth (or product formation) permissive range. However, in most manufacturing locations, economic criteria concerned with process cooling dictate that bioreactor operation, with the exception of wastewater biotreatment processes, should be undertaken at the highest possible temperature; thus, it is probable that the transition region between permissive and nonpermissive temperature ranges and even the lower end of the nonpermissive temperature range cannot be ignored. Furthermore, when postgrowth pasteurization or sterilization is a consideration prior to product harvesting, it is the nonpermissive temperature range that becomes of primary interest with respect to questions of damage, recovery, and death of microbes and/or damage to heat-labile intracellular products.

In the permissive temperature range, the main effect of temperature changes will concern the regulation and activity of genes and enzymes, but, additionally, lipid phase changes in membranes should be expected. Temperature changes from the permissive to the transient region result in the formation of heat shock proteins. Within the nonpermissive range, temperature changes can result in either reversible or irreversible damage to essential functional macromolecules. Approximate temperature ranges for more important site-specific damage are shown for *Escherichia coli*, a typical mesophilic bacterium, in FIGURE 2. This scheme is not claimed to be complete. For example, macromolecules such as mRNA and tRNA that exhibit important structural-functional relationships should not be neglected in consideration of temperature changes.

For effective process design, it is essential to establish quantitative relationships between the effects of temperature on growth and product formation rates. To obtain kinetic expressions that describe temperature-rate relationships, one must differentiate between constant and dynamic (transient) conditions.

KINETIC EFFECTS OF TEMPERATURE

The effect of temperature on the growth rate constant of microbes between their minimum and lowest optimum temperature for growth is usually described by an

Arrhenius equation, that is,

$$\mu = Ae^{-E/RT}. \qquad (1)$$

However, if one wishes to apply an equation that accounts for the effects of temperature on the whole temperature range in which growth occurs, it is obviously necessary to modify equation 1 in order to take account of growth both at optimum temperatures and at superoptimum temperatures.

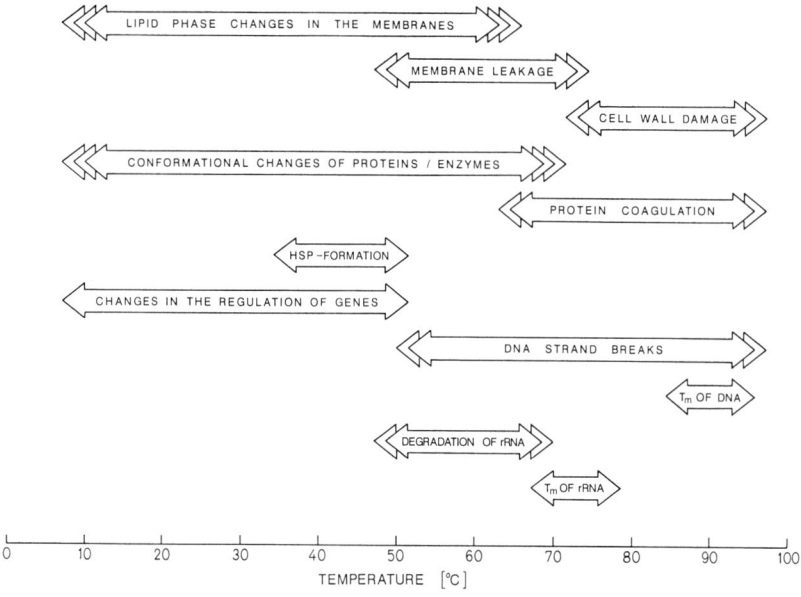

FIGURE 2. Approximate temperature ranges for more important site-specific damage and metabolic changes for *E. coli*.

One such modification is an equation proposed by Fiolitakis et al.[10] for the effects of temperature on the growth of a photoautotrophic, green microalga under a diurnal illumination cycle, that is,

$$\text{GI} = VAe^{-E/RT}. \qquad (2)$$

The growth intensity, GI, was introduced to account for diurnal cycling, whereas V is a dimensionless variable that accounts for temperature-induced changes in the growth activity state. In an alternative approach, Ratkowsky et al.[11] sought to describe the effects of temperature on bacterial growth, over the entire temperature range for growth, by an empirical nonlinear regression model, that is,

$$\sqrt{\mu} = b(T - T_{\min})\{1 - e^{c(T - T_{\max})}\}. \qquad (3)$$

All three equations so far presented apply to constant temperature growth, only describing the effect of temperature on growth at different temperatures. None is suitable for describing the dynamic response of the growth rate constant of cultures subjected to changes in temperature during growth. However, such situations were investigated by Topiwala and Sinclair,[12] who, on the basis of their experiments, introduced a delay function in the Arrhenius relationship to account for the resultant lags in the response of cultures subjected to step changes in temperature, that is,

$$\mu = Ae^{-E/RT'}, \tag{4}$$

where

$$T' = T \pm \Delta T(1 - e^{t/\tau}). \tag{4a}$$

With the exception of the well-known equation for death kinetics, that is,

$$N_t/N_0 = e^{-k_d t}, \tag{5}$$

which applies strictly in the nonpermissive temperature range, no equations have been proposed that describe temperature effects on microbial growth when temperature changes occur in the superoptimum temperature range. However, any such expression that describes the effects of temperature changes in the region between the highest optimum temperature for growth and the initiation of cellular death will clearly be a function of temperature, exposure time, time, and nutrient availability, that is,

$$\mu = f(T, t_E, t, a_n). \tag{6}$$

An overall summary of temperature effects on microbial growth kinetics is shown schematically in FIGURE 3.

TEMPERATURE SHOCKS AND RECOVERY

Whenever a microbe is subjected to a sudden temperature change, irrespective of whether either heating or cooling is involved, the intracellular network of coordinated metabolic reactions will be disturbed. The magnitude of the disturbance depends on the intensity of the shock, that is, the temperature change and the exposure time. In large-scale bioreactors, temperature shocks involving both heating and cooling can be expected; the former in stagnant zones when highly viscous non-Newtonian bioreactor charges are encountered and the latter in external cooling loops, irrespective of charge viscosity. In the specific case of industrial aerobic biotreatment processes, large parcels of high-temperature polluted wastewater are frequently dumped in the biotreaters used for such processes, thereby also resulting in potential heating shocks. The period of time during which process microbes are subjected to temperature shocks in large-scale bioreactors is in the range of minutes rather than in terms of either seconds or hours.

In order to investigate the impact of short-time temperature shocks on the overall performance characteristics of a growing bacterial culture, a test bacterium, *Klebsiella pneumoniae*, was grown in a 2.5-L impeller agitated, continuous flow bioreactor fitted with an external, pumped, heat exchanger loop through which the growing culture

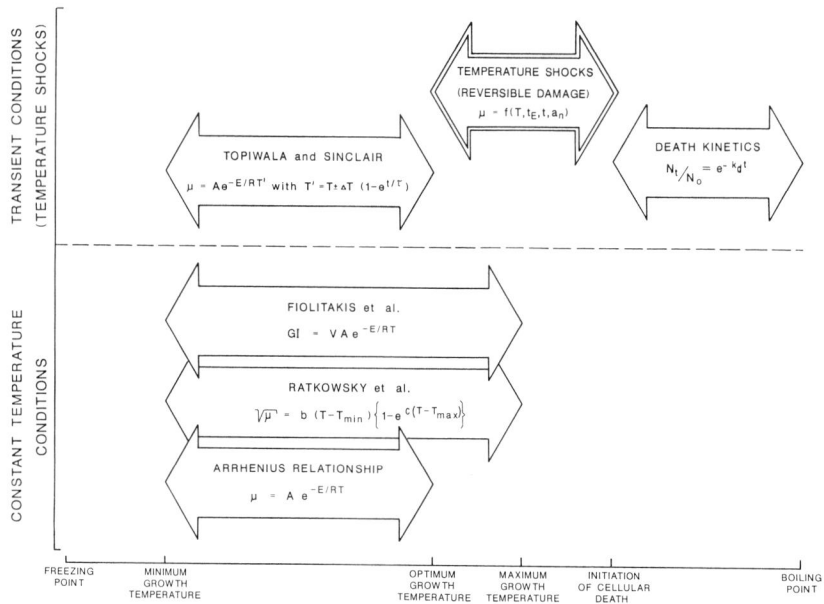

FIGURE 3. Kinetic expressions for temperature effects on microbial growth.

could be recirculated to the bioreactor (shown in FIGURE 4). The growth medium used was a defined mineral salts medium with glucose as the carbon energy substrate. The operating temperature of the bioreactor was maintained constant at 35 °C, the pH was controlled at a value of 6.8, the dilution rate was 0.68 h^{-1}, and the flow rate through the heat exchanger loop was 1.2 L/h, giving an average residence time of 1.15 min. The inlet and outlet temperatures in the heat exchanger were different and are listed in TABLE 1.

In FIGURE 5, the steady-state biomass concentrations in the bioreactor for the different heat exchanger outlet temperatures tested are shown. It can be seen that the suspended biomass concentration in the bioreactor remained essentially constant over a

TABLE 1. Heat Exchanger Temperatures[a]

Inlet (°C)	Outlet (°C)
18.8	3.2
22.1	9.2
24.1	15.3
34.8	35.1
40.2	44.7
44.3	50.0
46.6	53.8
49.2	63.5

[a] *K. pneumoniae*—growth temperature = 35 °C.

wide range of temperatures used in the heat exchanger. However, at approximately 50 °C, a decline in biomass concentration was observed. This suggested that the time necessary for recovery of the bacterial cells from the heating shock became significant relative to the mean residence time for cells in the bioreactor; therefore, a portion of the shocked cells had to be washed out from the bioreactor. At the highest temperature investigated, 63.5 °C, only a pseudosteady state was achieved because of extensive wall growth and hence segregation of the bacterial population into two distinct portions—

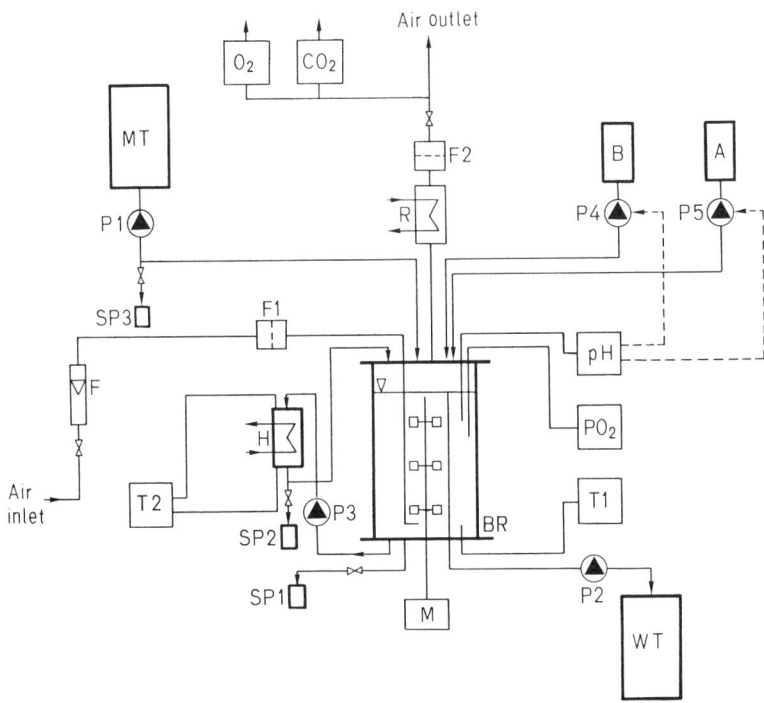

FIGURE 4. Scheme for the bioreactor system: A, acid reservoir; B, alkali reservoir; BR, bioreactor; CO_2, infrared carbon dioxide analyzer; F, airflow meter; F1, air inlet filter; F2, air outlet filter; H, heat exchanger; M, motor; MT, medium reservoir; O_2, paramagnetic oxygen analyzer; P1 to P5, peristaltic pumps; pH, pH control; PO_2, dissolved oxygen measurement; R, condenser; SP1 to SP3, sampling points; T1, temperature controller; T2, temperature measurement; WT, harvest tank.

one that was discretely dispersed and subjected to heating shocks and the other that was attached to the bioreactor walls and thus immune to heating shocks. However, comparison of the results of measurements of INT activity[13] for samples taken directly from the bioreactor and from the heat exchanger loop immediately after temperature shocking for the higher temperatures investigated showed marked differences, indicating the probable loss of metabolic activity during more severe heat shocking (FIGURE 6).

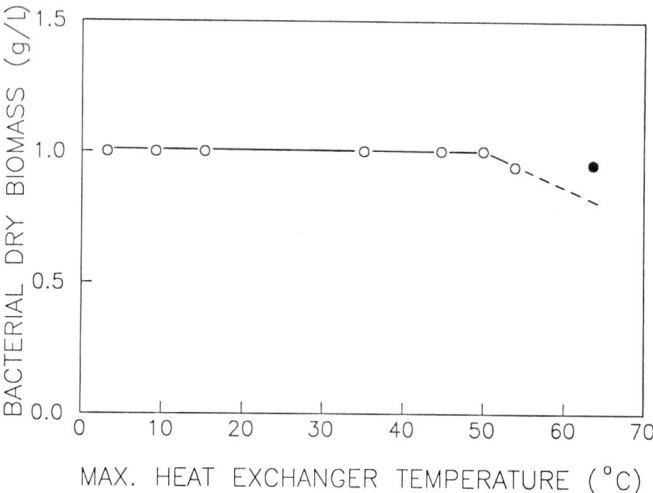

FIGURE 5. Steady-state bacterial biomass concentrations in the bioreactor for various heat exchanger outlet temperatures; ● represents a pseudosteady state where significant wall growth occurred.

In order to study the ability of a culture to recover from particular temperature shocks, samples of the bacterial culture were removed immediately after shocking in the heat exchanger loop and were used as inoculum for batch growth experiments. Samples, subjected to 3.2 °C and 35.1 °C, respectively, in the heat exchanger, initiated growth after an insignificant lag phase, whereas samples subjected to 63.5 °C showed a very marked lag phase, as shown in FIGURE 7.

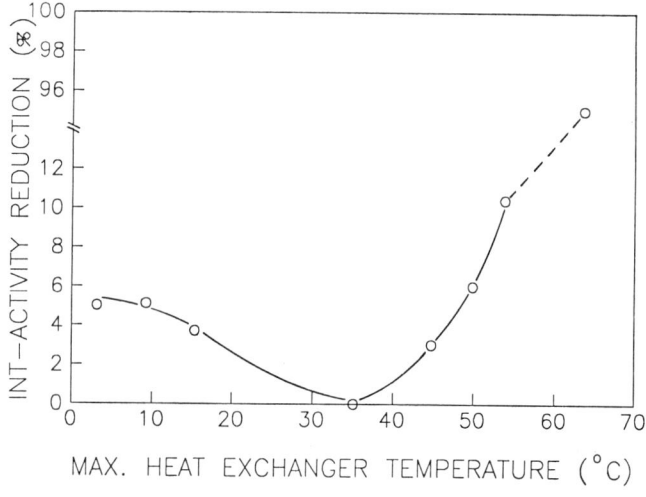

FIGURE 6. INT-activity reduction in samples taken directly after temperature shock relative to the culture in the bioreactor.

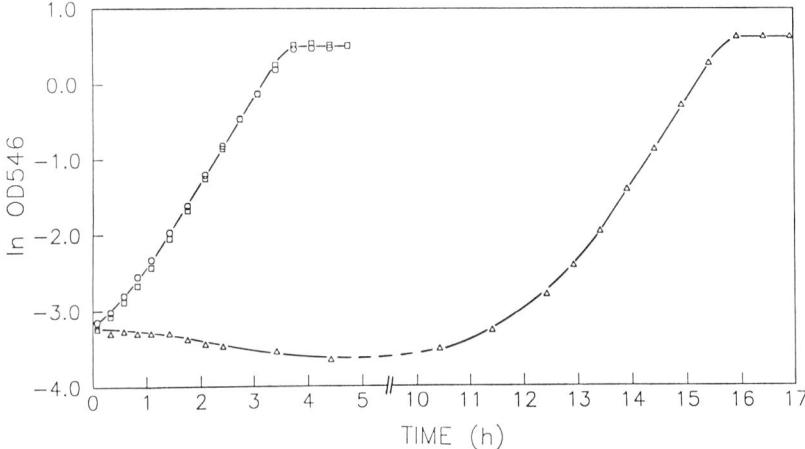

FIGURE 7. Recovery patterns of inocula of *K. pneumoniae* removed directly after temperature shocks from the heat exchanger loop; □ represents 3.2 °C, ○ represents 35.1 °C, and △ represents 63.5 °C outlet heat exchanger temperatures, respectively.

In another experiment, samples from a batch culture of *K. pneumoniae* growing in the exponential phase at 35 °C were inoculated into buffered growth medium at 49.3 °C and maintained at that temperature for 1, 2, and 4 minutes, respectively. After returning the inoculated growth medium to 35 °C, growth was followed as a means of assessing recovery. The responses observed are shown in FIGURE 8 and the delay effect

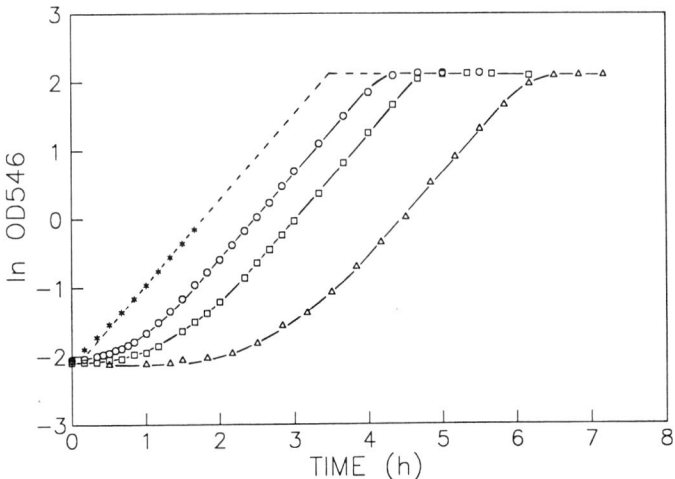

FIGURE 8. Recovery patterns of inocula of *K. pneumoniae* taken from an exponentially growing batch culture at 35 °C after temperature shocking in buffered growth medium at 49.3 °C for 0 (∗), 1 (○), 2 (□), and 4 (△) minutes, respectively.

during recovery with respect to attaining the maximum specific growth rate is shown in FIGURE 9. Two possible hypotheses that explain these responses can be formulated:

(1) the individual bacteria present in the culture require a finite time to reestablish the full range of metabolic functions necessary for rapid growth such that, prior to complete recovery, they grow only comparatively slowly;
(2) segregation of the culture into a population that has lost viability, on the one hand, and a population that has retained full viability, on the other hand, such that the measured specific growth rate for the damaged culture does not represent the typical response of a single bacterium present in the culture.

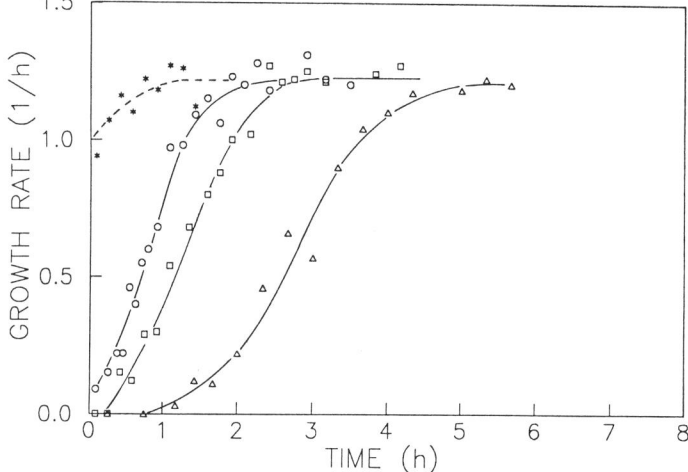

FIGURE 9. Apparent growth rates of the batch cultures inoculated with temperature-shocked inocula treated as in FIGURE 8.

Clearly, the data presented here are incomplete and further experiments on the effects of short-term temperature shocks on whole bacterial populations are necessary before a mathematical model can be established describing the time course of the specific growth rate during recovery with respect to exposure time and temperature.

CONCLUSIONS

Short-term temperature shocks involving cooling below the optimum temperature for growth of portions of a growing culture of *K. pneumoniae* have negligible impact on the overall culture performance. Recovery appears to be virtually instantaneous. In contrast, temperature shocks involving short-term heating of cultures of *K. pneumo-*

niae to temperatures above the superoptimal range for growth result in significant time delays with respect to recovery. Such delays are dependent on the exposure time to the heating shock, suggesting that the damage inflicted as a result of the shock requires energy for its repair and hence occurs at the expense of growth such that the specific growth rate is depressed.

SUMMARY

In large-scale bioreactors that are used either for the manufacture of microbial products or for wastewater treatment, the scale and mode of operation frequently mean that operating conditions vary throughout the bioreactor. Therefore, when considering process scaleup, it is important to establish guidelines that indicate the impact of changing operating conditions on the performance of process cultures. In scaled-up bioreactors, the surface area available for heat transfer reduces with increasing volume and ultimately becomes insufficient to allow easy maintenance of the process temperature within the optimum range if high levels of process intensity are to be maintained. Particularly with viscous non-Newtonian bioreactor charges, portions of the process culture in large-scale bioreactors will be subjected to superoptimal temperatures with respect to their growth such that process performance might be adversely affected.

In many large-scale bioreactors, external cooling loops are now employed for temperature control. When the process culture passes through such loops, it is subjected to a significant temperature gradient that could have both kinetic and physiological consequences.

In this contribution, the effects of subjecting a model process culture to superoptimal and suboptimal growth temperatures have been reported.

REFERENCES

1. ANDREW, S. P. S. 1982. Trans. Inst. Chem. Eng. **60:** 3–13.
2. OOSTERHUIS, N. M. G. 1984. Doctoral thesis (T. U. Delft), p. 151.
3. OOSTERHUIS, N. M. G. & N. W. F. KOSSEN. 1983. Chem. Eng. Res. Dev. **61:** 308–312.
4. KOSSEN, N. W. F. 1984. Proceedings of the 7th International Biotechnology Symposium. T. K. Ghose, Ed.: 365–380. IIT Press. New Delhi, India.
5. PURGSTALLER, A. & A. MOSER. 1987. Chem. Biochem. Eng. Q. **1:** 157–161.
6. LARSSON, G. & S-O. ENFORS. 1988. Bioprocess Eng. **3:** 123–127.
7. SWEERE, A. P. J., K. C. A. M. LUYBEN & N. W. F. KOSSEN. 1987. Enzyme Microb. Technol. **9:** 386–398.
8. VAN DE VUSSE, J. G. 1962. Chem. Eng. Sci. **17:** 507–521.
9. HINZE, J. O. 1955. Am. Inst. Chem. Eng. J. **1:** 289.
10. FIOLITAKIS, E., J. U. GROBBELAAR, E. HEGEWALD & C. J. SOEDER. 1987. Biotechnol. Bioeng. **30:** 541–547.
11. RATKOWSKY, D. A., R. K. LOWRY, T. A. MCMEEKIN, A. N. STOKES & R. E. CHANDLER. 1983. J. Bacteriol. **154:** 1222–1226.
12. TOPIWALA, H. & C. G. SINCLAIR. 1971. Biotechnol. Bioeng. **13:** 795–813.
13. LOPEZ, J. M., B. KOOPMAN & G. BITTON. 1986. Biotechnol. Bioeng. **28:** 1080–1085.

APPENDIX

Nomenclature

A	Arrhenius frequency factor
a_n	availability of nutrients
b	regression coefficient (see reference 11)
c	fitting parameter (see reference 11)
E	activation energy
F	overall volumetric flow rate
F_1, F_2, F_4	partial volumetric flow rates
GI	growth intensity (see reference 10)
k_d	specific death rate
N_0	cell number at time zero
N_t	cell number at time t
R	universal gas constant
T	absolute temperature
T_{min}	minimum absolute temperature at which the growth rate is zero
T_{max}	maximum absolute temperature at which the growth rate is zero
ΔT	temperature difference at a step change
T'	"effective absolute temperature" (see reference 12)
t	time
t_E	exposure time
V	activity state variable (see reference 10)
V_1, V_2, V_3	compartment volumes
τ	first-order time constant
μ	specific growth rate constant

Comparison of Cephalosporin C Production in Stirred-Tank and Airlift Tower Loop Reactors

T. BAYER,[a] T. HEROLD,[b] K. HOLZHAUER,[c] W. ZHOU,[c] AND K. SCHÜGERL[c,d]

[a]*Hoechst AG*
Frankfurt, Federal Republic of Germany

[b]*Sartorius GmbH*
Göttingen, Federal Republic of Germany

[c]*Institut für Technische Chemie*
Universität Hannover
D-3000 Hannover 1, Federal Republic of Germany

INTRODUCTION

Cephalosporins are important antibiotics on the international market because they are more resistant to β-lactamases than the penicillins. For clinical use, cephalosporins are prepared semisynthetically by chemically modifying cephalosporin C, which is produced exclusively by fermentation. In recent years, airlift tower loop reactors have successfully been used for penicillin production,[1-3] indicating possibilities for the use of this reactor type for the production of cephalosporin C. Based on the results of fermentations in a 20-L stirred-tank reactor,[4] a fed-batch process was developed for cephalosporin C production in a 60-L airlift tower reactor with an outer loop[5] and this was compared with the results evaluated from stirred-tank reactors.[4-6]

MATERIALS AND METHODS

Cephalosporin C was produced by *Cephalosporium acremonium* W 53.2.53 from Ciba-Geigy AG (Basel) by fed-batch operation in a stirred-tank reactor (STR) with a total volume of 40 L and in an airlift tower loop reactor (ALR) with a total volume of 80 L using a complex medium containing 30–100 kg/m^3 peanut flour. A glucose/methionine solution with 450 kg/m^3 glucose and 15 kg/m^3 methionine was used as the feed. Methionine serves as the precursor for cysteine and as a stimulator of cephalosporin C biosynthesis.

For on-line analysis, a cell-free sample flow was obtained with a polysulfone membrane (100-kDa cutoff). Glucose concentrations were measured enzymatically and on-line by a flow injection analyzer.[6] On-line HPLC was employed to determine

---

[d]Address all correspondence to K. Schügerl.

the concentrations of β-lactam precursors and products: penicillin N (PENN), deacetoxycephalosporin C (DAOC), deacetylcephalosporin C (DAC), cephalosporin C (CPC), methionine (MET), and its decomposition product of 2-hydroxy-4-methylmercaptobutyric acid (MMBS).[7]

The reactor and the on-line analyzer systems were connected to a computer system, namely, the Computer-Aided System for Fermentation Automation (CASFA).[8] This software package supervised the automatic operation of the analyzers, including cleaning, calibration, evaluation of blank values, and the on-line evaluation of the process parameter (e.g., concentration, $k_L a$, RQ), as well as process control.

The evaluation of the cell mass concentration was particularly difficult because of the presence of peanut flour. Therefore, the cell mass was estimated from the RNA content by the method of Küenzi (1979), which had been adapted to the investigated system.

RESULTS AND DISCUSSION

The operation of the reactor is influenced considerably by the morphology of the mold. Initially, the mold grows as filamentous mycelia, which lead to high broth viscosities and low oxygen transfer rates (OTR). At the end of the growth phase, the morphology changes; the swollen hyphae form arthrospores, resulting in a reduction of the broth viscosity and an improvement in the OTR.

When the glucose fed-batch cycle was optimized and oxygen limitation avoided, 5.5–6.0 g/L cephalosporin C (CPC) was obtained in the STR[6] and 4.5–5.0 g/L CPC was obtained in the ALR with 30 g/L peanut flour. With 100 g/L peanut flour, the final concentration of CPC was 10–13 g/L in the STR (FIGURE 1).

The results obtained from the airlift loop reactor are compared with those from the 20-L[4,6] and 2000-L[9] stirred-tank reactors in TABLE 1. This table indicates that the amount of peanut flour influences the product concentration (CPC) and the productivity (PR) considerably. At a peanut flour concentration of 100 kg/m³, the airlift loop reactor performance would be very poor due to the high viscosity of the broth; thus, no performance data are given. After 150 hours, with 30 kg/m³ peanut flour concentration, CPC concentrations and PR are 20% and 10% less in the ALR than in the STR, respectively; however, the specific productivities (based on RNA and X) are about 30% higher in the airlift reactor than in the stirred tank. The biomass-based yield coefficients, $Y_{P/X}$, are the same, but the substrate-based yield coefficients, $Y_{P/S}$, are higher by a factor of 2 in the ALR. The specific power input (P/V_L) in the airlift reactor is one-third of that in the stirred tank. The OTR is smaller in the ALR, but the efficiency (E_{O_2}) is a factor of 1.6 higher in this reactor than in the STR. Also, the specific product concentrations and productivities based on the power input are higher by a factor of 2.25 in the ALR than in the STR.

The lower volumetric productivity in the airlift reactor is due to the lower cell concentration in this reactor and the higher specific productivity with regard to power input is due to the lower specific power input required in an ALR. The better substrate yield is probably due to the lower maintenance coefficient of the mold in the ALR where the shear stress levels are lower. The higher shear stress levels in the STR caused

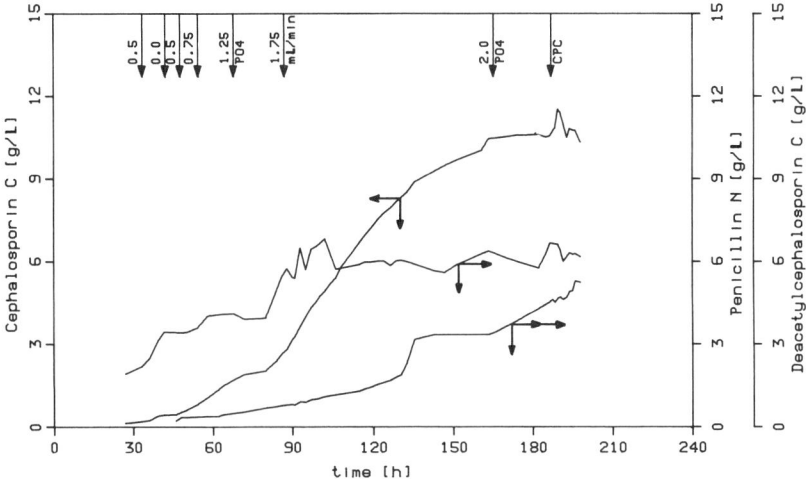

FIGURE 1. Results of fermentation RK 8802 (see text).

a reduction in the branching and in the length of the hyphae and thus a reduction of the broth viscosity and an improvement in the OTR. This makes it possible to attain higher cell concentrations and hence higher volumetric productivities in the STR. This can also be observed in FIGURE 2, in which the volumetric productivities are plotted as functions of the fermentation time for both reactor types. However, these increased volumetric productivities occur at the expense of the substrate yield coefficient, which is lower by a factor of 2 than in the airlift reactor. The energy consumption in the ALR is considerably lower and thus the energy efficiency based on the product formation is higher by a factor of 2 than in the stirred-tank reactor.

TABLE 1. Comparison of the Performances of the 60-L Airlift Tower Reactor (60-L-BS) and the 20-L (20-L-RK) and 2-m³ (2-m³-RK) Stirred-Tank Reactors

	30 kg/m³ ENM		100 kg/m³ ENM	
	60-L-BS	20-L-RK	2-m³-RK	20-L-RK
t (h)	150	150	130	150
CPC (kg/m³)	4.5–5.0	5.5–6.0	6.5–7.0	10–11
PR (g/m³·h)	31–34	35–38	50–54	67–73
SPR_{max} (g/kg·h)	0.08	0.06	—	0.08
RNA (kg/m³)	0.8	1.0	—	1.35
$Y_{P/X}$	6	6	—	7.5
$Y_{P/Sgluc}$	0.026	0.012	0.013	0.018
q_a (min^{-1})	0.7–1.0	1.0	0.8	1.0
P/V_L (kW/m³)	1.0	3.0	3.0	5.5
OTR (kg/m³·h)	0.64	1.2	—	2.2
E_{O_2} (kg/kW·h)	0.64	0.4	—	0.4
CPC/(P/V_L) (kg CPC/kW)	4.5	2.0	2.3	2.0

CONCLUSIONS

Cephalosporin C can be produced by *Cephalosporium acremonium* using a complex medium in an airlift loop reactor.

Difficulties arise from the high broth viscosity during the growth phase because the OTR is reduced considerably. Therefore, growth has to be limited in order to avoid oxygen transfer limitation. This determines the upper limit of the volumetric productivity in an airlift loop reactor.

However, the specific productivities based on cell concentration are comparable. The substrate-based yield coefficient of the product formation is larger by a factor of 2

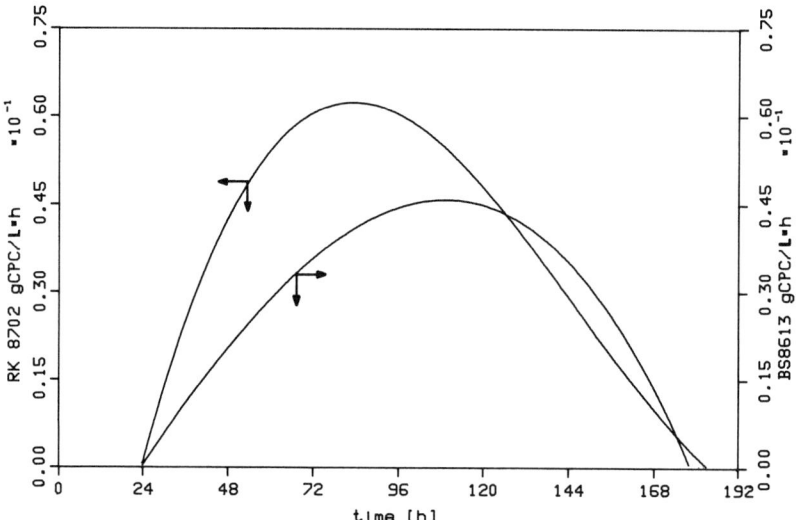

FIGURE 2. Comparison of volumetric productivities (see text).

in the ALR than in the STR. The energy efficiency based on the product formation is also higher by a factor of 2 in the ALR than in the STR.

The optimal strategy is the quick attainment of a high cell concentration and keeping the specific growth rate at a value at which the specific productivity is at its highest point.

REFERENCES

1. KÖNIG, B., K. SCHÜGERL & C. SEEWALD. 1982. Strategies of penicillin fermentation in tower loop reactors. Biotechnol. Lett. **1:** 127–132.
2. WITTLER, R. & K. SCHÜGERL. 1984. Reaction engineering aspects of penicillin production in tower loop reactors. Seventh International Biotechnology Symposium, New Delhi, India (February 19–25). Book of Abstracts—Volume 2, p. 580–581.

3. GBEWONYO, K. & D. I. C. WANG. 1983. Confining mycelial growth to porous microbeads: a novel technique to alter the morphology of the non-Newtonian mycelial cultures. Biotechnol. Bioeng. **25**: 967–983.
4. HEROLD, T., T. BAYER & K. SCHÜGERL. 1988. Cephalosporin C production in a stirred tank reactor. Appl. Microbiol. Biotechnol. In press.
5. BAYER, T. 1987. Reaktionstechnische untersuchungen zur produktion von cephalosporin C im mammutschlaufenreaktor.
6. ZHOU, W. Ongoing dissertation, University of Hannover.
7. HOLZHAUER, K. 1987. Entwicklung und anpassung chromatographischer verfahren zur verbesserung der antibiotikaproduktion. Diploma thesis, University of Hannover.
8. HIDDESSEN, R. 1987. Entwicklung und einsatz eines automatisierungssystem zur on-line überwachung und steuerung pro-zeßrechnergekoppelter bioreaktoren. Dissertation, University of Hannover.
9. AUDEN, J. A. L. Personal communication (Ciba-Geigy AG, Basel).

APPENDIX

Nomenclature

E_{O2}	efficiency of oxygen transfer with regard to the specific power input
$k_L a$	volumetric mass transfer coefficient
OTR	oxygen transfer rate
P	power input
PR	volumetric productivity of CPC
q_a	volumetric aeration rate/broth volume (vvm)
SPR	specific productivity with regard to RNA
V_L	broth volume in reactor

Phase Holdup and Dispersion in a Three-Phase Fluidized-Bed Bioreactor with Low-Density Gel Beads[a]

BRIAN H. DAVISON

Chemical Technology Division
Oak Ridge National Laboratory[b]
Oak Ridge, Tennessee 37831-6226

INTRODUCTION

The most efficient reactor configurations for bioprocessing utilize some form of biocatalyst retention. A particular type of retention is immobilization of the biocatalyst into beads of various natural gelatins. These beads can then be used in a columnar reactor for the bioprocess. Columnar reactors have the potential of high volumetric productivities and simultaneous high conversions. In fact, for ethanol fermentations, columnar reactors have the highest reported productivities.[1]

These columnar bioreactors often have three phases: the solid biocatalyst, the liquid media, and the gas phase, which can be either provided (e.g., oxygen) or produced (e.g., CO_2). At sufficiently high gas or liquid flow rates, the solid fraction of the bed becomes fluidized. This is beneficial because of the increased interphase transport and the improved operational stability (due to a lessened likelihood of channeling or plugging).

The hydrodynamic interactions of these three phases in a columnar reactor are poorly understood and simple plug-flow or continuous-stirred-tank models are often inadequate to evaluate or scale these systems. In order to develop an improved model for the columnar reactor, information is needed on the hydrodynamics of the system, including the phase holdup and the dispersion (or backmixing). More research is needed because the literature correlations for three-phase reactors, which have been developed under different conditions (such as much denser particles), are not in agreement when they are extrapolated using the conditions found in a bioreactor.

This paper builds on an earlier report[2] containing experimental observations in a 2.54-cm-i.d. columnar fluidized-bed reactor (FBR). Phase holdup and axial dispersion measurements were made in a model FBR using small low-density gel beads as the solid phase. Slugging behavior could be observed in this narrow column due to bubble coalescence. In this report, these observations were extended into a larger 7.62-cm-i.d. column. Significantly different behavior (notably, a lack of very large bubbles to cause slugging) was observed in a 2.54-cm-i.d. column and in a 7.62-cm-i.d. column.

---

[a]Research sponsored by the Energy Conversion and Utilization Technologies Program, United States Department of Energy, under Contract No. DE-AC05-84OR21400 with Martin Marietta Energy Systems, Inc.

[b]Operated by Martin Marietta Energy Systems, Inc., for the United States Department of Energy under Contract No. DE-AC05-84OR21400.

BACKGROUND

Previously,[2] it was shown that axial electroconductivity probes could be used for the nonintrusive continuous measurement of the gas fraction within both an actively fermenting and a nonfermenting FBR. In essence, both the broth and the highly porous beads are equally conductive due to the salts present, whereas the gas is not conductive. Therefore, the conductivity decreases with the amount of gas present. A nonfermenting, experimental three-phase system has been used here to characterize and correlate these systems because of the ability to set the gas and liquid flow rates independently. From these measurements and the bed height, the following equations can be used to calculate the phase fractions in the bed:

$$\epsilon_L + \epsilon_G + \epsilon_S = 1, \tag{1}$$

$$\gamma/\gamma_0 = \epsilon_L + \epsilon_S = 1 - \epsilon_G, \tag{2}$$

$$\epsilon_S = \frac{V_S}{V_{bed}}. \tag{3}$$

The first equation is the sum of the phase fractions, ϵ_G, ϵ_L, and ϵ_S, which represent gas, liquid, and solid, respectively. The second equation relates the liquid and solid fractions as the ratio of the overall measured bed conductivity, γ, to the liquid electrolyte conductivity, γ_0. The third equation calculates the average solid fraction across the bed as the ratio of the volume of the solid added, V_S, over the volume of the bed, V_{bed}. This assumes that the solid fraction is uniform throughout the bed, which is generally true for constant flow rate FBRs, but may not be applicable to gas-generating systems.

Multiple axial conductivity probes also were used to sense the conductivity in order to estimate the liquid axial dispersion coefficient from the response curves to a salt pulse.

MATERIALS AND APPARATUS

As reported earlier, the beads were made of 4% k-carrageenan and 3% Fe_2O_3 and had a mean diameter of 0.125 cm. The narrower column used previously had a 2.54-cm i.d. and a 650-mL volume. The larger column was similar to the narrower column, but had a 7.62-cm i.d., a 6-L volume, and conductivity probes placed at ~45-cm intervals (see FIGURE 1). Operating conditions are listed in TABLE 1. The conductivity probes were Pt foil (1.27 × 2.54 cm) mounted on opposite sides of the column. The probes were calibrated with standard solutions and connected to conductivity meters that were interfaced to a computer for data acquisition and subsequent averaging and analysis.

A metered air stream was sparged through a medium glass frit into the bottom of the column filled with a known volume of beads. The liquid (0.05 M KCl) was introduced through four radial inlets directly above the gas sparger. The bed was fluidized with the liquid flow and then conductivity data were collected with and without the gas flow in order to establish the average holdup and a baseline for the dispersion measurements. The bed volume was estimated visually to within 5%. Next, a 10-mL pulse of 3 M KCl was injected into the liquid flow inlet; the tracer response data were collected in the computer as the pulse traveled up the column past the probes. The

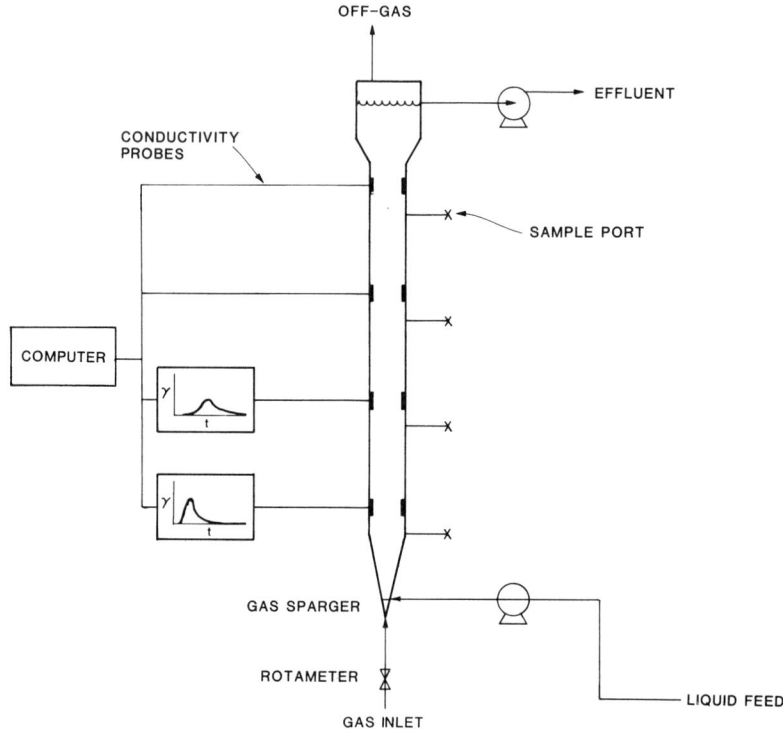

FIGURE 1. Schematic of FBR for holdup and dispersion measurements by electroconductivity.

baseline conductivity data obtained before the injection were averaged and divided by the bulk conductivity (measured without the gas flow) in order to give the average gas holdup. The liquid holdup was calculated from the solid and liquid fractions (see equation 1). The raw conductivity data were smoothed and the first and second moments were calculated at each probe. The moments are time- and concentration-weighted integrals of the response curves that indicate a residence time from the first

TABLE 1. Operating Conditions

Column diameter (cm)	7.62	2.54
Column volume (mL)	6000	650
Particle diameter (cm)	0.11	0.11
Particle density (g/mL)	1.07	1.07
Volume of beads (mL)	650–1500	250
Gas flow rate (L/h)	0–50	0–25
Liquid flow rate (L/h)	14–50	1–5
Liquid	0.05 M KCl	0.05 M KCl & 2% v/v EtOH
Liquid density (g/mL)	1.01	1.01
Temperature (°C)	25	30

moment and a peak width from the second moment. The dispersion coefficient or the Peclet number can be estimated from the change of the second moment between two probes. Further information on this and other experimental details is in the previous report.[2] When indicated, the data were fit to an appropriate equation by a regression routine.

RESULTS

Qualitatively, the larger 7.62-cm-i.d. column operated more smoothly than the narrower 2.54-cm-i.d. FBR at the same superficial gas and liquid velocities. No slugging and only minimal channeling were observed in the larger bed. In the narrower column, bubbles would frequently coalesce to a diameter near that of the column and slugging would be observed. In the larger column, the bubbles reached an apparently stable size of 3- to 5-mm diameter. This indicates operation in a regime where bubble coalescence and bubble breakup due to shear are equal.

Phase Holdup

Data from ~30 runs of various gas and liquid flow rates, each ranging from 0 to 50 L/h, were collected from a 7.62-cm-i.d. column filled with gel beads. Gas holdup measured both by conductivity at each probe and by total volumetric displacement in the entire column was negligible (less than 3%) for all flow rates of gas and liquid considered. With no gas present, there is a steady increase in the liquid holdup as the liquid flow rate increases and a corresponding decrease in the solid holdup (see FIGURE 2). These data can be fit by an equation of the form of Richardson and Zaki:[3]

$$u_L/u_S = \epsilon_L^n, \qquad (4)$$

where u_L is the superficial liquid velocity, u_S is related to the terminal velocity of a falling bead, ϵ_L is the liquid fraction, and n is a function of the Reynolds numbers. Here, u_S is 54.5 cm/min and n is 3.5. The exponential dependence of u_L has been surveyed[3] to be in the range of 2 to 5.

The other data set in FIGURE 2 shows that the liquid fraction rises to near 0.81 for a range of gas velocities between 5 and 18 cm/min for any superficial liquid velocity tested. This phenomenon, an expansion limit, has been observed before in three-phase systems.[4] Expansion limits appear to be due to operation in the transition to the churn turbulent regime, which can be very broad in some systems. This limit was lower for denser systems such as glass, but a limit for ϵ_L of ~0.8 was also observed for the lightest solid studied (Plexiglas with a density of 1.2 g/mL). This limit was reached at relatively low superficial gas and liquid velocities.

FIGURE 3 shows a similar plot for the narrow FBR. Again, the two-phase data can be fit with an equation of the form of Richardson and Zaki[3] with u_S equal to 60 cm/min and n equal to 2.5. However, in this case, the gas flow does not greatly affect the liquid holdup. In addition, because the gas holdup was small (<0.1) for these experiments, the phase holdups are only slightly dependent on the gas flow rate. The two gas/liquid/

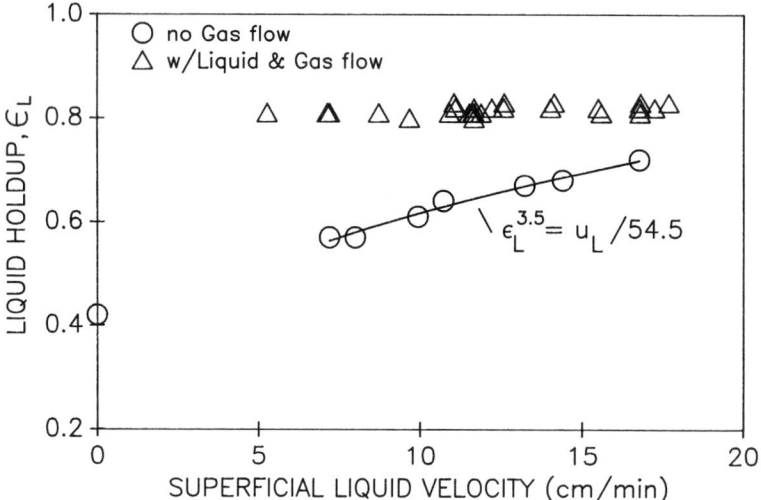

FIGURE 2. Liquid-phase holdup in a three-phase fluidized bed (7.62-cm i.d.). Gas flow at 5 to 18 cm/min. Each data point is from a separate run.

solid data points that lie below the curve are caused by a bed contraction (which can occur at low gas flow rates).[5] This contraction was not observed in the larger column or at higher liquid flow rates in the narrow column.

Furthermore, from the point of view of developing correlations, the large variations in phase holdup that were observed here under similar conditions (due to changing wall

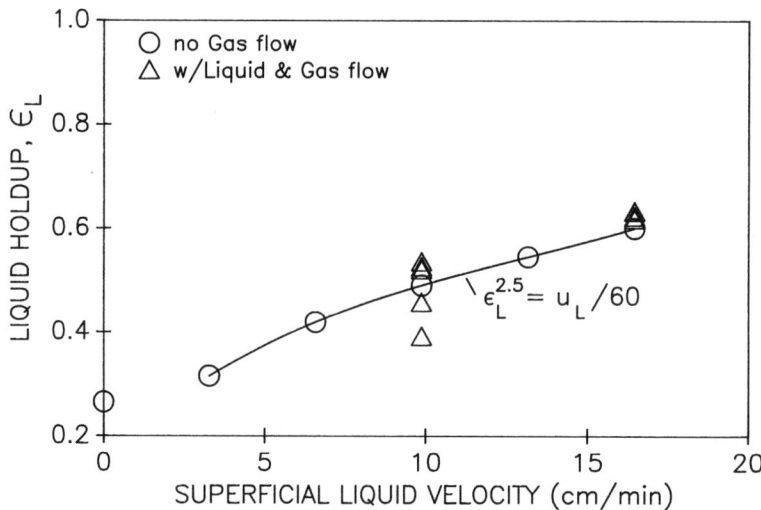

FIGURE 3. Liquid-phase holdup in a narrow three-phase fluidized bed (2.54-cm i.d.). Gas flow at 5 to 75 cm/min.

effects and bubble coalescence) imply that a stochastic nature needs to be considered for the three-phase FBR.

Axial Dispersion

Axial dispersion coefficients were calculated from the response to a salt tracer by the method of moments. Particle Peclet numbers (a ratio of convective to dispersive forces) are defined as

$$Pe = \frac{u_L d_P}{D}, \qquad (5)$$

where d_P is the particle diameter, D is the liquid axial dispersion coefficient, and u_L is the superficial liquid velocity. The Peclet numbers are estimated to be between 10^{-2} and 10^{-1} and are comparable to those measured in the narrow column under similar conditions. FIGURE 4 shows some of these data plotted with the correlations of Kim and Kim,[6] Stiegel and Shah,[7] and Muroyama et al.[8] The error bars indicate the experimental variation between different axial probes and runs at the same conditions. Results are presented for experiments at two liquid flow rates, 11 cm/min (FIGURE 4a) and 17 cm/min (FIGURE 4b). The correlation of Muroyama et al.[8] underestimates the Peclet numbers by an order of magnitude. The data are bounded by the other two correlations in both cases. It should be noted in the use of these correlations for estimation purposes that the equations of Stiegel and Shah[7] and Muroyama et al.[8] are undefined at zero gas velocity. Because the Peclet number without gas flow can be more accurately estimated and because this value is approached as the gas flow is decreased, the form of the correlation that is used should allow a defined value at zero gas flow, such as that of Kim and Kim.[6]

DISCUSSION

The results indicate that larger diameter systems for high productivity bioconversions will have improved operability over the smaller diameter fermentation FBRs that we have successfully tested.[9] Slugging behavior will not be a problem in the larger systems because the bubbles remain small with respect to the 3-inch-diameter column. Likewise, the wall effects, which were seen to be very important in the hydrodynamics and bubble coalescence in the narrower column, are less important in the larger column. Begovich[4] observed little difference between a 7.62-cm-i.d. column and a 15.2-cm-i.d. column, indicating that wall effects should be negligible in the larger column that was used here. If correct, the hydrodynamics of the 7.62-cm column should scale to even larger systems. In addition, if the limit on ϵ_L holds for a broader range of gas and liquid velocities, the gas holdup (which was measured here to be less than 3%) can be neglected in a larger fermenting FBR. This will simplify the modeling and scaleup of the fluidized-bed fermentor system. Nevertheless, the gas flow will have a significant effect on the overall system behavior due, in part, to bed expansion lowering the solids fraction.

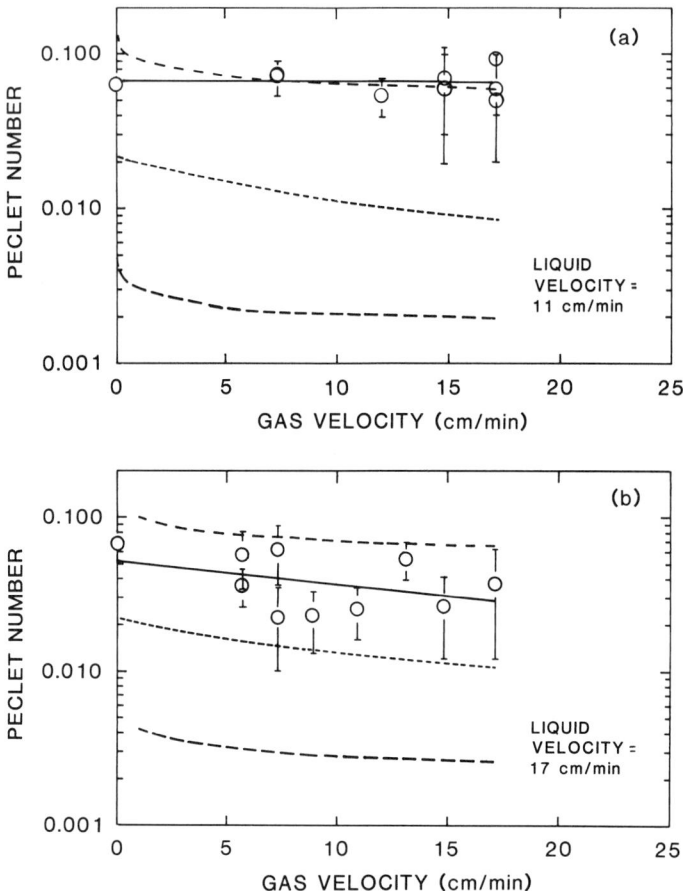

FIGURE 4. Peclet numbers versus (mm) superficial gas velocity in a three-phase fluidized bed (7.62-cm i.d.): (a) $u_L = 11$ cm/min; (b) $u_L = 17$ cm/min. Data were fit to a straight line by regression: (—) data fit, (---) Kim and Kim,[6] (– – –) Stiegel and Shah,[7] and (— —) Muroyama et al.[8]

ACKNOWLEDGMENT

The technical support of Jim E. Thompson is acknowledged in the operation of the experiments.

REFERENCES

1. GODIA, F., C. CASES & C. SOLA. 1987. A survey of continuous ethanol fermentation systems using immobilized cells. Process Biochem. **22:** 43–48.
2. DAVISON, B. H. 1989. Dispersion and holdup in a three-phase fluidized-bed bioreactor. Appl. Biochem. Biotechnol. **20/21:** 449–460.

3. RICHARDSON, J. F. & W. N. ZAKI. 1954. The sedimentation of a suspension of uniform spheres under conditions of viscous flow. Chem. Eng. Sci. **3:** 65.
4. BEGOVICH, J. M. 1978. Hydrodynamics of three-phase fluidized beds. M.Sc. thesis, University of Tennessee, available as no. ORNL/TM-6448.
5. STEWART, P. S. B. & J. F. DAVIDSON. 1964. Three-phase fluidization: water, particles, and air. Chem. Eng. Sci. **19:** 33.
6. KIM, S. D. & C. H. KIM. 1983. Axial dispersion characteristics of three-phase fluidized beds. J. Chem. Eng. Jpn. **16:** 172.
7. STIEGEL, G. J. & Y. T. SHAH. 1977. Backmixing and liquid holdup in a gas-liquid cocurrent upflow packed column. Ind. Eng. Chem. Process Des. Dev. **16:** 37–43.
8. MUROYAMA, K., *et al.* 1978. Axial liquid mixing in three-phase fluidized beds. Kagaku Kogaku Ronbunshu **4:** 662.
9. DAVISON, B. H. & C. D. SCOTT. 1988. Operability and feasibility of ethanol production by immobilized *Zymomonas mobilis* in fluidized-bed bioreactors. Appl. Biochem. Biotechnol. Symp. **18:** 19–34.

APPENDIX

Nomenclature

d_P	particle diameter (cm)
D	liquid axial dispersion coefficient (cm^2/s)
Pe	particle Peclet number ($u_L d_P/D$)
u_G	superficial gas velocity (cm/min)
u_L	superficial liquid velocity (cm/min)
u_S	solids velocity—constant in equation 4 (cm/min)
V_S	volume of the solid added (mL)
V_{bed}	volume of expanded bed (mL)
ϵ_G	gas fraction
ϵ_L	liquid fraction
ϵ_S	solid fraction
γ	overall measured bed conductivity (mS)
γ_0	liquid electrolyte conductivity (mS)

PART X. IMMOBILIZED ENZYME BIOREACTORS

Hydration of Cyanopyridine to Nicotinamide by Immobilized Nitrile Hydratase

JACOB EYAL AND MARVIN CHARLES

Department of Chemical Engineering
Bioprocessing Institute
Lehigh University
Bethlehem, Pennsylvania 18015

INTRODUCTION

Previous studies[1-3] concerning the potential commercial applications of nitrile hydratase have shown that processes based on free or immobilized whole cells, soluble or immobilized purified enzyme, or soluble crude extract are not likely to be competitive except in unique cases. Immobilization of crude extract may provide a viable alternative; therefore, this work was undertaken to:

(1) develop an acceptable immobilization technique,
(2) identify significant problems associated with the immobilization per se and with the use of the immobilized catalyst, and
(3) determine the kinetic and stability characteristics needed for design and economic analyses of proposed processes.

The potential advantages of immobilized enzymes are well known; the reader is referred to several excellent reviews for further information.[4-6] Worth noting here, however, is the point that there is little information available that can guide one directly to a technique suitable for a particular application; one usually must test empirically a variety of methods to find a practical, economical technique that will satisfy particular process constraints (e.g., mechanical properties, product purity).

Several immobilization procedures for nitrilase have been published.[7] These rely mainly on polyacrylamide or alginate entrapment of the enzyme. Preliminary studies[3] associated with the present work indicated that these methods would not be practical; among the reasons are poor mechanical properties of the catalyst and too much loss of enzyme during the immobilization procedure.

An extensive series of empirical evaluations led finally to the choice of entrapment in cellulose triacetate. The method is described in this report; further information concerning entrapment in cellulose triacetate is given in references 8–10.

MATERIALS

A crude extract nitrilase enzyme preparation from *Brevibacterium* R-312 was used throughout this work. The fermentation, recovery, and extraction processes for the enzyme are described in references 1–3.

Cellulose triacetate (Eastman Kodak) was used as the entrapment matrix.

Methylene chloride (Fisher Scientific) was used to dissolve the cellulose triacetate.

Toluene (Fisher Scientific) was used as the coagulation solvent to solidify methylene chloride–dissolved cellulose triacetate containing entrapped enzyme to beads.

METHODS

Catalyst Preparation

The following method was used to entrap the crude extract in cellulose triacetate:

(1) Prepare a crude extract from *Brevibacterium* R-312 and store it at 4 °C until ready for use.[1–3]
(2) Dissolve cellulose triacetate in methylene chloride under gentle stirring at room temperature. Add 0.05 M NaOH to the solution to increase the pH of the solution to 11 and then cool the solution to 1 °C. Perform all remaining steps at 1 °C.
(3) To the cellulose triacetate solution, slowly add crude extract having a protein content of 5–10 mg/mL (determined by the Lowry assay of the crude extract).[2,3,11] Agitate vigorously during the addition until a finely dispersed emulsion forms (about 30 minutes).
(4) Let the emulsion stand for about 10 minutes to eliminate air bubbles.
(5) Add the emulsion dropwise from a syringe into a toluene coagulation bath maintained at 1 °C. Droplets will solidify to beads almost immediately.
(6) Wash the beads formed in step no. 5 (immediately after coagulation) four times with pH 8 phosphate buffer and then store them at 4 °C.

The catalyst beads so formed were very porous, had diameters ranging from 1.5 mm to 3.0 mm, and contained 5–7 mg protein/g beads. No protein was found in the toluene bath after coagulation of the emulsion.

Catalyst Activity Assay

A known weight of beads having a known protein content was added to 96.05 mM (10.0 g/L) β-cyanopyridine in 0.05 M phosphate buffer (pH 8). The assay mixture was agitated by a magnetic stirrer operated at approximately 175 rpm and the temperature was kept at 25 °C. Samples were taken frequently and were analyzed by liquid chromatography[1–3] for β-cyanopyridine, nicotinamide, and nicotinic acid. No enzyme was found to leach from the beads and there was no attrition of the beads during the assay.

The specific activity, v_i, was defined as

$$v_i = \frac{\mu\text{-moles nicotinamide formed}}{\text{mg bead protein–min}}.$$

This assay measures the global rate of reaction and includes the effects of solid-phase diffusional resistance. This resistance becomes negligible in comparison to the intrinsic rate of reaction (in this case) only for particles having diameters of 0.5 mm and less.[3] (See also the discussions below concerning the determination of the diffusion coefficients and the intrinsic rate constants.)

Liquid-film diffusional resistance was negligible for all cases studied in this work.[3]

Protein Measurements

The Lowry assay[11] was used for all protein determinations. A Bausch and Lomb Spectronic 710 system was used to read the absorbances for all colorimetric assays.[3]

Diffusion Coefficients

The β-cyanopyridine and nicotinamide diffusion coefficients in cellulose triacetate beads were determined as follows:

(1) Enzyme-free beads (known, uniform diameter) were added to a 10.0-g/L substrate or product solution (pH 8) at 25 °C and the mixture was agitated vigorously.
(2) The decrease with time of the liquid-phase concentration of the substrate or product was followed by HPLC analyses of the liquid samples.
(3) The Van Heuven method[3,12] was applied to the concentration/time measurements to calculate the distribution coefficients (at equilibrium) and the effective diffusivities.

Intrinsic Kinetic Constants

The intrinsic kinetic constants (actual immobilized-enzyme rate constants unaffected by diffusion) were determined by using the assay technique to determine initial rates for 0.5-mm beads. These beads were prepared by cold-grinding larger ones followed by close screening.[3] Methods for calculating the kinetic constants are given in the RESULTS AND DISCUSSION section.

Storage Stability

Catalyst was stored at 4 °C in pH 8 phosphate buffer and samples were assayed periodically. As can be seen from FIGURE 1, the catalyst had a storage half-life of about 90 days. Preparations used for kinetic studies were stored for no more than 2 weeks so as to insure high, consistent activity levels for obtaining basic kinetic data.

Operational Stability

Substrate solution (in pH 8 phosphate buffer) was passed continuously through two packed-bed reactors in series. Each reactor had a 1″ diameter and was 6″ long; each was packed with 20 grams of catalyst. The reactors were operated at 10 °C, 15 °C, and 25 °C and at an inlet substrate concentration of 10.0 g/L. Effluent samples were

RESULTS AND DISCUSSION

Sodium Hydroxide Treatment

Immobilized crude extract produced at first appreciable amounts of nicotinic acid even at relatively low conversion of 10 g/L of β-cyanopyridine[3] (see FIGURE 2); this occurred despite the fact that very little acid was produced by free, crude enzyme until 95% conversion.[2,3] The reason appeared to be a combination of diffusional resistance (high nicotinamide concentration inside the beads) and enhanced activity of amidase relative to hydratase in the immobilized state.[3]

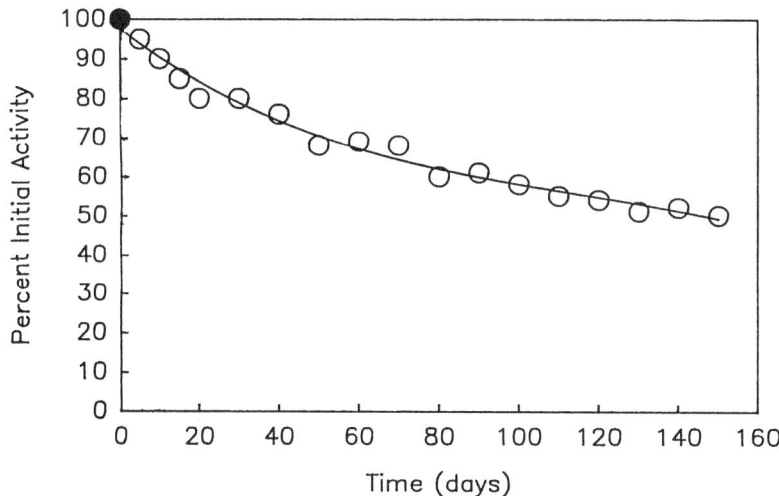

FIGURE 1. Catalyst storage stability at 4 °C.

The most effective method found to overcome this problem was treatment of the cellulose triacetate solution with 0.05 M NaOH before addition of crude extract. This appeared to suppress dramatically the amidase (see FIGURE 2) while leaving adequate hydratase activity.

Temperature and pH Profiles

The temperature and pH profiles for the immobilized catalyst are given in FIGURES 3 and 4, respectively. The maximum temperature is approximately 40 °C (in pH 8 phosphate buffer). This is about 5 °C higher than the maximum temperature found for free, crude extract.[2,3] The temperature profile for the immobilized extract is somewhat

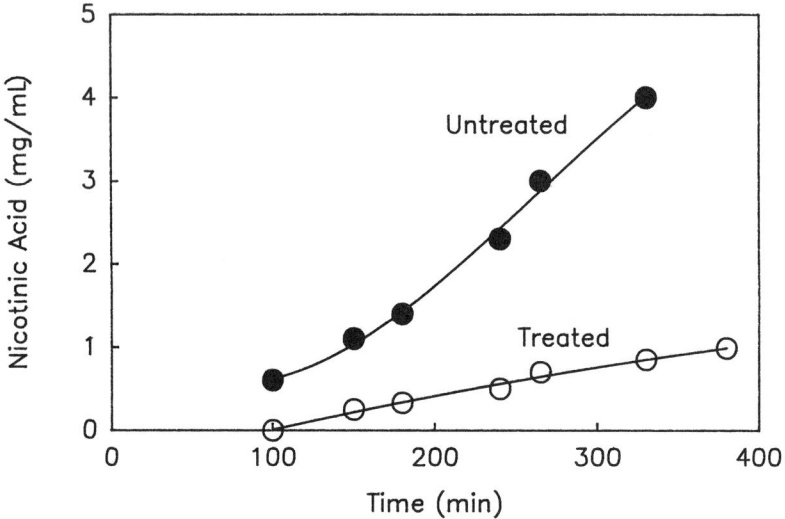

FIGURE 2. Effect on amidase activity of sodium hydroxide treatment.

broader than that for the soluble extract and shows somewhat less sensitivity to temperature changes around the maximum temperature.

The maximum pH is 7.5 (at 15 °C and 25 °C), which is about 0.5 pH units lower than the maximum for free extract.[2,3] The pH curve for the immobilized enzyme is shifted down about 0.5–1.0 units relative to that for the free extract, but there is no significant effect of immobilization on the sensitivity to a change in pH.

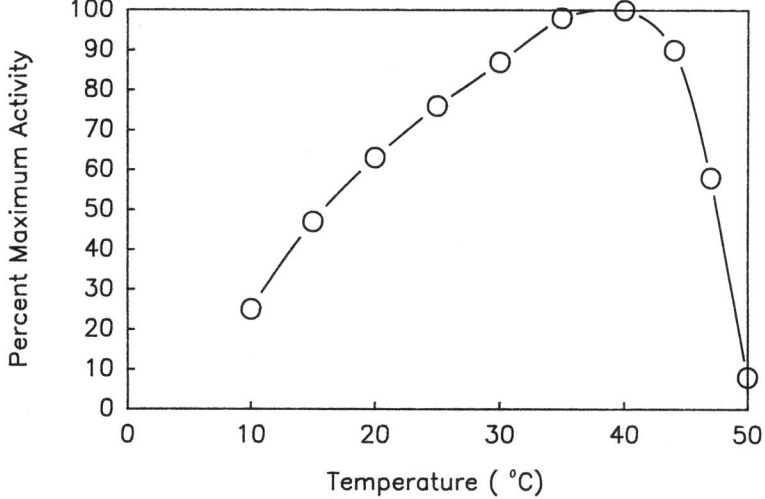

FIGURE 3. Temperature profile of immobilized, crude extract nitrile hydratase (pH = 8).

Operational Stability

The catalyst had an operational half-life of 30 days at 10 °C, of about 10 days at 15 °C, and of 1 day at 25 °C (see FIGURE 5); therefore, most practical processes based on the existing enzyme and immobilization method would have to be run at less than 15 °C.

TABLE 1 gives a comparison of the thermal stabilities of whole cell,[1] free,[2] and immobilized enzyme. From this, it is clear that immobilization improves significantly the thermal stability of the enzyme, but the reason is not clear—it may result from structural stabilization and/or a lessening of possible degradation by *Brevibacterium* proteases;[1-3] clarification is anticipated from studies being planned.

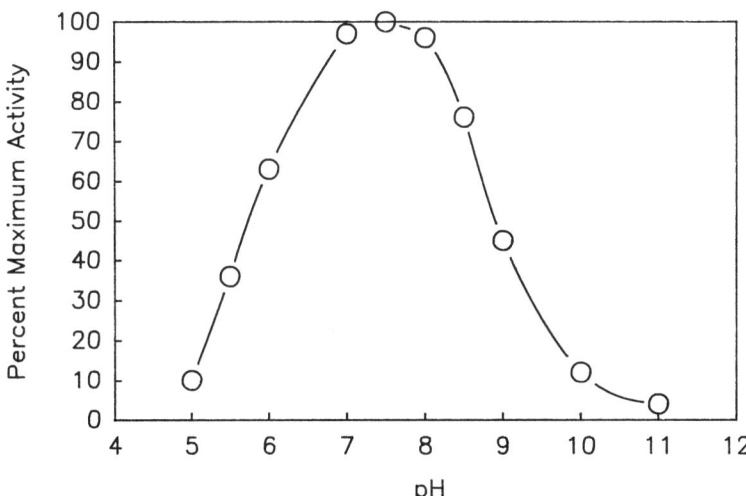

FIGURE 4. The pH profile of immobilized, crude extract nitrile hydratase (15 °C).

Diffusion Coefficients

The curve in FIGURE 6 is typical of those obtained for the decrease of the liquid-phase β-cyanopyridine concentration when beads of cellulose triacetate without enzyme were added to the solution (similar curves were obtained for nicotinamide solutions). Based on these results, the Van Heuven method gave effective diffusion coefficients of 3.1×10^{-7} cm^2/min and 5.2×10^{-7} cm^2/min for β-cyanopyridine and nicotinamide, respectively.

Kinetic Constants

The expression for the specific activity of free, crude extract hydratase hydration of β-cyanopyridine is[2,3]

$$v = \frac{V}{E} = \frac{v_m S}{S + K_m(1 + P/K_P) + S^2/K_S}, \quad (1)$$

where

V = reaction rate (μ-moles/min),
E = weight protein (mg),
v_m = maximum specific activity (μ-moles/min–mg protein),
S = substrate concentration (mM),
P = product concentration (mM),
K_m = Michaelis constant (mM),
K_P = product inhibition constant (mM), and
K_S = substrate inhibition constant (mM).

Note that this model includes both substrate and product inhibitions.[2,3]

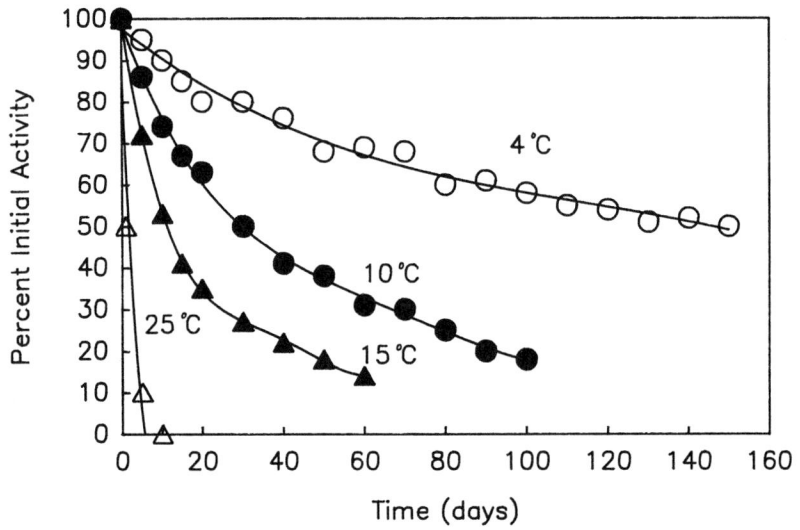

FIGURE 5. Operational stability of immobilized, crude extract nitrile hydratase (pH = 8, inlet substrate concentration = 10.0 g/L).

Equation 1 with different constants also describes the specific activity of the immobilized enzyme when diffusion resistance is negligible (i.e., the intrinsic specific activity):

$$v_i = \frac{V_i}{E} = \frac{v_{mi} S}{S + K_{mi}(1 + P/K_{Pi}) + S^2/K_{Si}}, \quad (2)$$

where the subscript "i" denotes the value of the constant for the immobilized state.

Equation 2 in double reciprocal form is

$$1/v_i = 1/v_{mi} + K_{mi}/v_{mi} \times (1 + P/K_{Pi})/S + S/(K_{Si} v_{mi}). \quad (3)$$

For very low substrate concentration with no product present initially, equation 3 becomes

$$1/v_i = 1/v_{mi} + (K_{mi}/v_{mi})(1/S). \quad (4)$$

TABLE 1. Half-Lives for Whole Cell, Soluble, and Immobilized Nitrile Hydratase[a]

	Half-Life (days)				
	4 °C	10 °C	15 °C	25 °C	35 °C
whole cell	100	12	ND	5	ND
soluble	50	10	ND	<1	0.02
immobilized	150	30	10	1	ND

[a]The 4 °C results were obtained under storage conditions. All others were obtained under operating conditions. ND = not determined.

This is the low concentration asymptote of equation 3 and has the form of the standard Michaelis-Menten equation; hence, v_{mi} and K_{mi} are obtained from the ordinate and abscissa intercepts in the usual way.[3] A double reciprocal plot of the experimental data is given in FIGURE 7, from which K_{mi} = 122 mM and v_{mi} = 1.6 μ-moles/min–mg protein.

K_{Pi} can be obtained by applying the preceding method with various initial product concentrations. For these cases, equation 3 becomes

$$1/v_i = 1/v_{mi} + (K_{mi}/v_{mi})(1 + P/K_{Pi})(1/S). \quad (5)$$

The $1/v_i$ intercepts for all such asymptotes will be $1/v_{mi}$. The slope, ϕ, of each asymptote will be

$$\phi = K_{mi}/v_{mi} + (K_{mi}/v_{mi}K_{Pi})P. \quad (6)$$

A plot of the slopes, ϕ, against the initial product concentration, P, has a slope of $K_{mi}/v_{mi}K_{Pi}$, from which one can obtain K_{Pi}. (Note that one also can obtain K from the

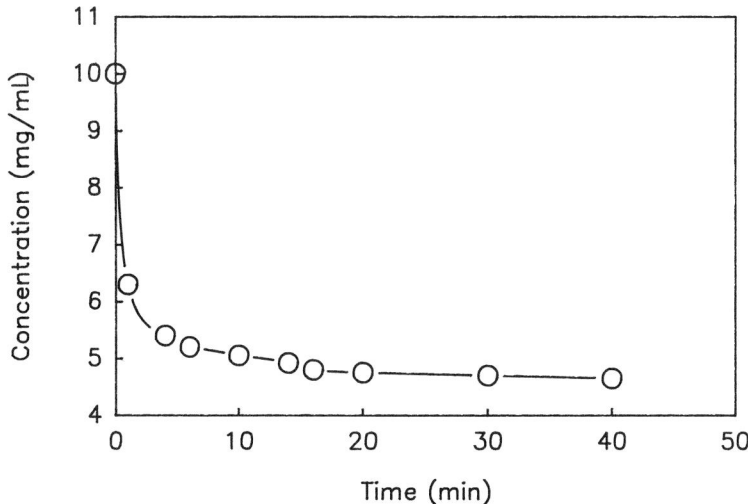

FIGURE 6. Diffusion of β-cyanopyridine into cellulose triacetate beads (25 °C, pH = 8, initial substrate concentration = 10.0 g/L, bead size = 2.0 mm).

P-axis intercept. This is $-K_{P_i}$.) Experimental results are plotted in FIGURE 8, from which $K_{P_i} = 126$ mM.

Finally, by taking the derivative of $1/v_i$ (in equation 3) and setting it equal to zero, one can show that[3]

$$K_{Si} = S_{min}^2/K_{mi}, \qquad (7)$$

where S_{min} is the substrate concentration at the minimum in the double reciprocal plot of equation 3. Then, from FIGURE 7, $K_{Si} = 185$ mM.

The constants for immobilized and free hydratase (crude extract) are compared in TABLE 2. K_{mi} is much greater than the corresponding value for the free enzyme. The result is a lower intrinsic reaction rate in the immobilized state, particularly at low

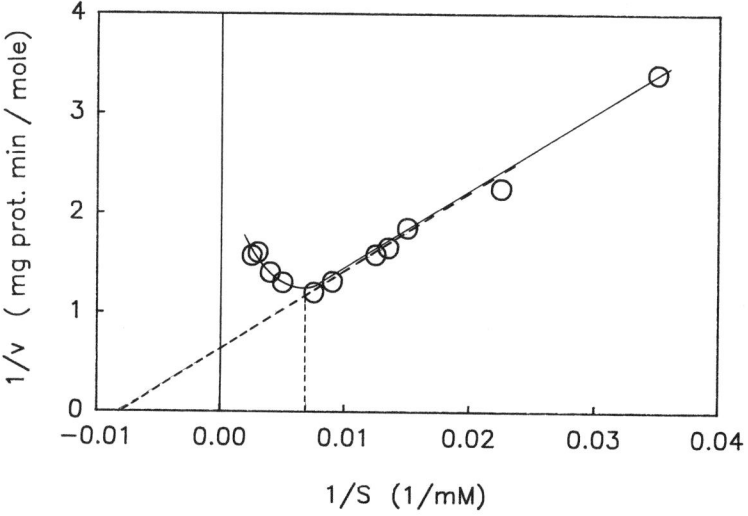

FIGURE 7. Double reciprocal plot for immobilized, crude extract nitrile hydratase (25 °C, pH = 8, bead size = 0.5 mm).

substrate concentration. This negative effect is exacerbated by the fact that the substrate is an inhibitor and must be kept at a relatively low concentration. The negative effect of increased K_{mi} is overcome to some extent by the fact that K_{P_i} and K_{Si} for the immobilized enzyme are considerably greater than the values for the free enzyme. On the basis of maximum specific activities, the immobilization efficiency (v_{mi}/v_m) is approximately 30%. Clearly, it would be desirable to modify the immobilization method so as to decrease the values of K_{mi}, increase further the values of K_{P_i} and K_{Si}, and increase the immobilization efficiency. To do this rationally, one must first determine what causes the changes in the kinetic constants. Such a study is now under way.

Finally, inhibition by nicotinic acid was found to be negligible throughout the work reported here.[3]

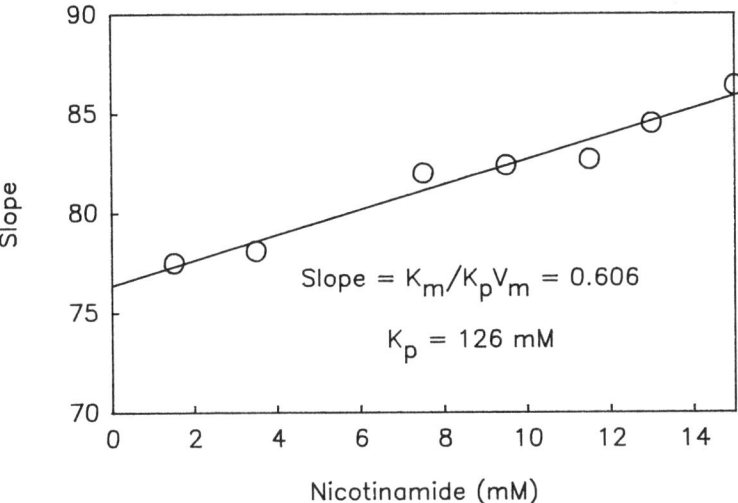

FIGURE 8. Slope of double reciprocal plot versus initial nicotinamide concentration (25 °C, pH = 8, bead size = 0.5 mm).

CONCLUSIONS

Immobilization by entrapment in cellulose triacetate of crude extract hydratase from *Brevibacterium* R-312 provides a catalyst that appears to have some commercial possibilities: it has very good mechanical properties and has an operational half-life of at least 30 days at 10 °C. This temperature is a bit low, but it is feasible in many cases so long as the solubilities of the substrate and product are high enough for commercial applications.

Some treatment (such as with NaOH as described in this report) appears to be necessary for crude extract from the current strain to avoid amidase hydrolysis of the amine. Amidase negative strains of *Brevibacterium* have been reported,[13] but there is some question as to their usefulness and the extent to which they are free of any amidase production.[3]

Further work is necessary to improve the immobilization and, if possible, the strain. Emphasis should be placed on decreasing product inhibition and thermal decay, removing completely and reliably the amidase activity, increasing immobilization

TABLE 2. Kinetic Constants of Immobilized and Free[a] Nitrile Hydratase

Constant	Free	Immobilized
K_m (mM)	28	122
K_S (mM)	155	185
K_P (mM)	36	126
v_m (μ-moles/min–mg protein)	5.8	1.6

[a] In crude extract from *Brevibacterium* R-312.

efficiency, and improving the kinetic constants of the immobilized enzyme. In this regard, it should be noted that there has been a report of a strain of *Rhodococcus*[14] that appears to make a hydratase having greatly diminished product inhibition. Immobilization studies should be done with this enzyme.

REFERENCES

1. EYAL, J. & M. CHARLES. 1988. Hydration of β-cyanopyridine to nicotinamide by whole cell nitrilase. Submitted for publication.
2. EYAL, J. & M. CHARLES. 1988. Hydration of β-cyanopyridine to nicotinamide by free nitrilase. Submitted for publication.
3. EYAL, J. 1987. Bioconversion of β-cyanopyridine to nicotinamide using soluble and immobilized nitrilase. Ph.D. dissertation, Lehigh University, Bethlehem, Pennsylvania.
4. ZABORSKY, O. 1973. Immobilized Enzymes. CRC Press. Boca Raton, Florida.
5. CHIBATA, I. & T. TOSA. 1976. *In* Immobilized Enzyme Technology—Research and Applications. H. Weetall & S. Suzuki, Eds. Plenum. New York.
6. MESSING, R. A. 1975. Immobilized Enzymes for Industrial Reactors. Academic Press. New York.
7. BUI, K., A. ARNAUD & P. GALZY. 1982. New method to prepare amides by bioconversion of corresponding nitriles. Enzyme Microb. Technol. **4:** 195–197.
8. KENNEDY, J. F. 1974. Cellulose derivatives as carriers for enzymatic reactions. Adv. Carbon Chem. **29:** 305–400.
9. PEKKA, L., P. KAISA & W. LARS. 1980. Preparation and kinetic behavior of immobilized whole cell biocatalysts. Biochimie **62:** 387–394.
10. DINELLI, D. S. & V. R. FABLIAN. 1972. Preparation of fibers containing enzyme. United States patent no. 3,715,277.
11. LOWRY, O. H. & N. J. ROSEBROUGH. 1951. Protein measurement with Folin phenol reagents. J. Biol. Chem. **193:** 265–275.
12. VAN HEUVEN, J. W., H. C. H. J. VAN MAANEN & R. LIETERMOET. 1984. Characterization of immobilized enzyme systems. *In* Innovations in Biotechnology. E. H. Houwink & R. R. Van der Meer, Eds. Elsevier. Amsterdam/New York.
13. BERNET, N., A. THIERY, M. MAESTRACCI, A. ARNAUD, G. M. RIOS & P. GALZY. 1987. Continuous immobilized cell reactor for amide hydrolysis. J. Ind. Microbiol. **2:** 129–136.
14. MAUGER, J., T. NAGASAWA & H. YAMADA. 1988. Nitrile hydratase catalyzed production of isonicotinamide, picolinamide, and pyrazinamide from 4-cyanopyridine, 2-cyanopyridine, and cyanopyrazine in *Rhodococcus rhodochrous* J1. J. Biotechnol. **8:** 87–96.

Immobilized Pig Brain NAD Glycohydrolase for the Preparation of NAD Analogues

MARIO PACE, PIER GIORGIO PIETTA,
DARIO AGNELLINI, PIER LUIGI MAURI,
AND SILVIA GHEZZI

Dipartimento di Scienze e Tecnologie Biomediche
Sezione di Chimica Organica
Università di Milano
20133 Milano, Italy

INTRODUCTION

Among the enzymes whose activities are affected by pyridine nucleotides, the dehydrogenases are widely studied under the biochemical and physiological point of view. Their technological importance is based on the development of methods for clinical analyses as well as on the preparation of particular biosensors. In addition to dehydrogenases, other enzymes related to NAD and NADP play an important role in reactions where the pyridine nucleotides work as substrates instead of coenzymes. Some of these reactions are connected with pathological infections, such as bacteria toxins, whereas other enzymes, named NADases or NAD glycohydrolases, catalyze reactions normally occurring in fungi, mammalian seminal plasma, mammalian organs, and other tissues, leading to the breaking of the nicotinamide N-ribosidic bond in NAD(P). The enzymes from bacteria, fungi, and mammalian seminal plasma (EC 3.2.2.5) show a pure hydrolytic activity[1-5] and yield nicotinamide and adenosine diphosphate ribose (ADPR) according to the following reaction:

$$NAD^+ + H_2O \xrightarrow{\text{NADase}} \text{nicotinamide} + ADPR + H^+. \quad (1)$$

On the other hand, NADases from mammalian tissues[6,7] and from snake venom[8] have transglycosidase activity (EC 3.2.2.6) and are able to produce NAD analogues by the exchange of the nicotinamide part in the coenzyme with a pyridine derivative as shown in the following reaction:

$$NAD^+ + X-Py \xrightarrow{\text{NADase}} \text{nicotinamide} + X-PyAD. \quad (2)$$

Most of the NADases from mammalian organs are membrane-bound enzymes and are associated with the microsomal particles of the tissues.[9] Their solubilization is generally accomplished by the use of proteolytic or lipolytic enzymes[6,10] in order to separate the enzyme from the membrane.

NAD analogues, that is, coenzymes modified at the pyridine moiety, have physicochemical characteristics quite different from those of the original coenzyme,[11-14] which leads to a particular behavior toward dehydrogenase isoenzymes. These derivatives are therefore a useful tool for clinical analysis of dehydrogenases when, for instance, damages of liver or heart tissues have to be determined. Kinetic methods using NAD analogues for the measurement of the H-4 and M-4 isoenzymes of lactate dehydrogenase[15,16] as well as a procedure for the determination of alcohol dehydrogenase in liver damage[17] are reported in the literature. Moreover, some pyridine coenzyme analogues are still under study with regard to their inhibitory activity toward enzymes involved in various diseases such as, for instance, cancer and psoriasis. Therefore, the growing technological importance of NAD analogues in clinical, analytical, and preparative chemistry is understandable.

In this discussion, we report the immobilization of pig brain NADase on various supports and the comparative study of these enzyme derivatives for the preparation of a bioreactor suitable for the production of pyridine nucleotide analogues.

MATERIALS AND METHODS

Sepharose 4B was a product of Pharmacia (Uppsala, Sweden) and was activated with CNBr in 1 M Na_3PO_4 for 10 minutes; then, it was reacted with n-propylamine (2.5 mmol/mL gel) for 18 hours at room temperature. Affi-Gel 10 was treated and reacted according to the standard procedure of Bio-Rad (Richmond, California). Porcine pancreas lipase, trypsin, 3-acetylpyridine, 3-aminopyridine, nicotinic acid, 3-acetylpyridine adenine dinucleotide (3-APAD), 3-aminopyridine adenine dinucleotide (3-AmPAD), nicotinic acid adenine dinucleotide (NAAD), and Triton X-100 were all purchased from Sigma (St. Louis, Missouri). Affi-Gel Blue, acrylamide, and N,N'-methylene-bis-acrylamide were obtained from Bio-Rad. All other reagents were of analytical or HPLC grade.

One NADase enzymatic unit (IU) is defined as the amount of enzyme that yields 1 μmole of product(s) per minute at 37 °C and pH 7.3 when assayed titrimetrically according to the reaction in equation 1.

Solubilization of Pig Brain NADase

The method of Walter and Kaplan with lipase[6] was initially used for the separation of the enzyme from the cell membrane, but a better solubilization was then obtained by the original procedure described below. First, 80 g of pig brains, freshly obtained from the slaughterhouse, was homogenized in a Waring blender in the presence of approximately an equal volume (80 mL) of 0.1 M phosphate buffer (pH 7.5) containing 0.5% Triton X-100. After centrifugation at 6000g, the precipitate was suspended in 80 mL of 0.1 M phosphate buffer (pH 7.5) and spun again at 6000g. The supernatants were combined and precipitated with 45% ammonium sulfate. Then, the precipitate was suspended in phosphate buffer and dialyzed against 0.1 M KCl at pH 7.5. The solution was then centrifuged at 18,000g and concentrated in Amicon.

Immobilization of Pig Brain NADase

Hydrophobic Adsorption

A slight modification of the method already described for NAD glycohydrolase from *Neurospora crassa*[18] was used. First, 1.2 IU of NADase was allowed to react for 2 h with 5 mL of *n*-propyl-Sepharose (0.7 mmol/mL of capacity) at 37 °C in the presence of 1 M KCl. Then, the gel was washed with 1 M KCl and water and stored at 4 °C in 0.1 M phosphate buffer at pH 7.3.

One mL of Affi-Gel Blue, suspended in 2 mL of 0.1 M phosphate buffer at pH 7, was added to 0.5 mL of NADase solution at pH 7 (1.3 IU) and was allowed to react for 1 h at room temperature. The gel was then washed with 0.5 M KCl and 0.1 M KCl (10 mL each).

Covalent Binding to Affi-Gel 10

In this case, 3 mL of NADase solution (7 IU) was reacted for 2 h at 4 °C with 2 mL of Affi-Gel 10 suspended in 3 mL of 0.1 M phosphate buffer at pH 7.5. Then, the suspension was treated with a solution of glycine for a further hour and washed successively with water, 1 M KCl, and water. The enzyme derivative was stored at 4 °C.

Entrapment in Polyacrylamide Gel

Fresh pig brain was homogenized with 10 volumes of chilled acetone and, after filtration, the powder was (i) homogenized again with 2 volumes of *n*-butanol, (ii) filtered, and (iii) washed successively with 2 volumes of acetone and ethyl ether in a Waring blender. Acetone powder was filtered, dried under vacuum, and stored at -20 °C until used.

Next, 3 mL of a 40% solution of acrylamide containing 3% N,N'-methylene-bis-acrylamide was added to 0.5 g of acetone powder (1.1 IU) suspended in 5 mL of water; then, 6 μL TEMED and 50 μL of a 1% ammonium persulfate solution were pipetted into the mixture and rapidly stirred. The gel so formed was ground in a Waring blender with water and allowed to settle. Fine particles were removed by suction and the procedure was repeated three times. The resulting derivative was approximately 7 g of wet gel.

Determination of NADase Activity

NAD glycohydrolase activity was routinely measured by the automatic titration of the hydrogen ion produced during the hydrolysis of NAD^+, according to reaction 1, using 5 mM NaOH.

Transglycosidase activity of pig brain NADase was determined by high performance liquid chromatography following the contemporary disappearance of substrate and increase of products in reactions 1 and 2 catalyzed by the enzyme. Chromato-

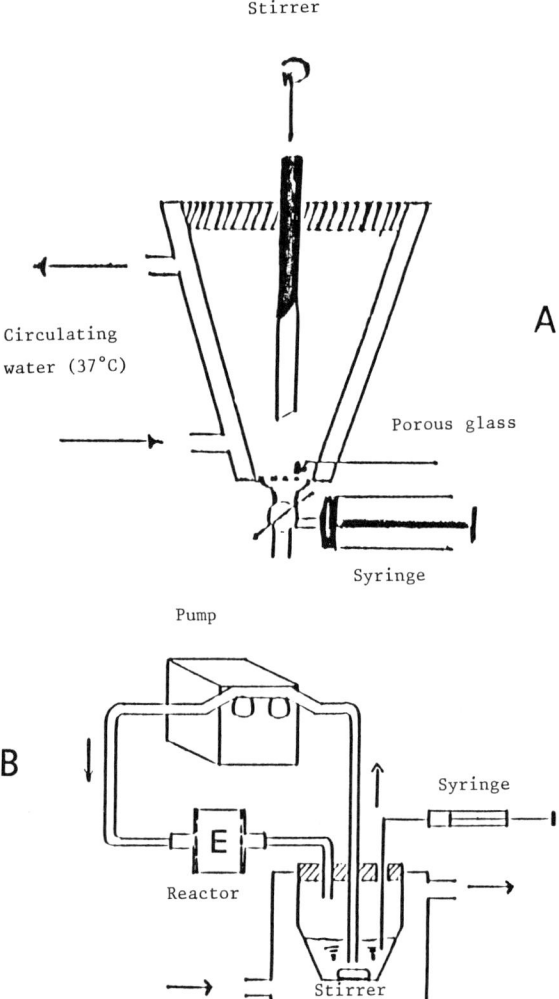

FIGURE 1. Scheme of the bioreactor "in batch" (A) and "in continuous flow" (B). At fixed times, a few microliters of solution were removed by a syringe and, after a proper dilution, analyzed by HPLC.

graphic analyses were performed on a component system consisting of a Model U6K universal injector, a Lambda-Max Model 440 ultraviolet detector, and a Model 730 data module (Waters Associates, Milford, Massachusetts). The column was a C-18 Resolve (300 mm × 4 mm, Waters Associates) with a precolumn of the same type (25 mm × 4 mm). Elutions of NAD^+, pyridine derivatives, and coenzyme analogues were performed with 25 mM phosphate buffer (pH 6.2)–methanol (100:4) as the mobile phase at a flow rate of 1.5 mL/min. Before mixing, the buffer was filtered

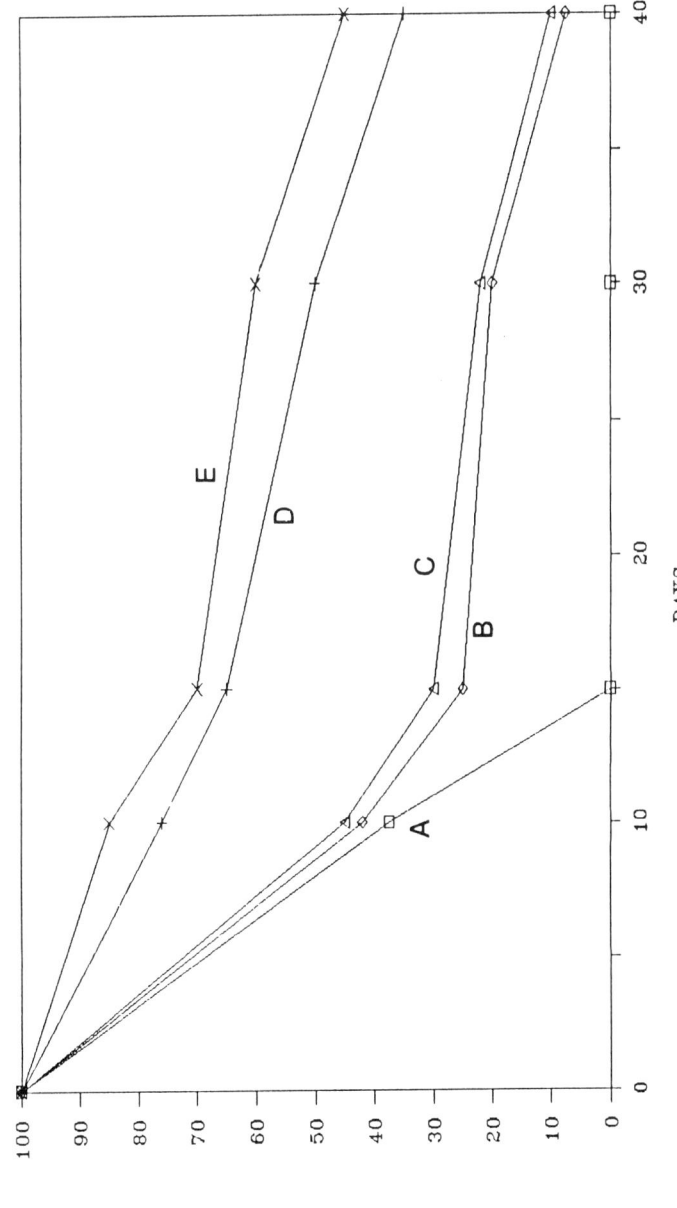

FIGURE 2. Stabilities of free and immobilized pig brain NADase: (A) soluble enzyme; (B) NADase bound to the hydrophobic *n*-propyl-Sepharose; (C) NADase covalently bound to Affi-Gel 10; (D) acetone powder; and (E) NADase entrapped into polyacrylamide gel. Soluble NADase and enzyme derivatives were stored at 4 °C in the presence of 0.1 M phosphate buffer at pH 7.3.

through a 0.45-μm membrane (HA type, Millipore) and the effluent was monitored at 254 nm (0.05 AUFS). The linearity of the components was verified within a range of 1–300 nmol.

The assay was performed either in batch or in continuous flow with the apparatuses shown in FIGURE 1. With the method "in batch", 500 mg of enzyme derivative was suspended under stirring in 2 mL of 0.1 M phosphate buffer (pH 7.3) containing 15 mM NAD^+ and 65 mM pyridine derivative as the substrate at 37 °C. At fixed times, a few microliters of solution were removed by a syringe and, after proper dilution, analyzed by HPLC.

Alternatively, 5 mL of a solution of 15 mM NAD^+ and 65 mM pyridine derivative in 0.1 M phosphate buffer (pH 7.3) was circulated at 37 °C in a column containing 2.5 g of immobilized enzyme and the eluant was assayed by HPLC as already described.

RESULTS AND DISCUSSION

n-Propyl-Sepharose failed to retain NADase solubilized by lipase, but could form stable bonds with the enzyme treated with Triton X-100. In this case, the recovery of the activity in the gel was 0.1 IU/mL. On the other hand, when Affi-Gel Blue was used, no activity was found either in the gel or in the washings. Pig brain NADase covalently bound to Affi-Gel 10 showed an activity of 0.13 IU/mL of gel. Acetone powder entrapped in polyacrylamide gel had an activity of 0.11 IU/g.

The stabilities of the various derivatives were very different depending on the purification of the enzyme. Partially purified pig brain NADase immobilized by hydrophobic or covalent bonds lost about 55% of its activity in 10 days, but entrapped acetone powder was much more stable (the enzyme in solution was completely inactivated in two weeks, whereas acetone powder itself showed an activity comparable with that of the derivative). FIGURE 2 shows the comparative stabilities of the soluble enzyme, the acetone powder, and the NADase derivatives. Organic solvents lowered the activity both of free and immobilized NADase; in particular, the results obtained with the enzyme immobilized hydrophobically were identical to the covalently bound NADase (Affi-Gel 10), indicating that no leakage of protein occurred in the *n*-propyl-Sepharose derivative in the operative conditions.

TABLE 1. Elution of Various Pyridine and NAD^+ Analogues in the HPLC Apparatus[a]

Sample	Elution Time (min)
nicotinamide	5.30
4-methyl-nicotinamide	4.40
nicotinic acid	1.55
3-aminopyridine[b]	13.00
3-acetylpyridine[b]	18.00
ADPR	1.25
NAD^+	2.20
4-MeNAD	2.10
nicotinic acid–AD	1.35
3-AmPAD	3.00
3-APAD	3.00

[a]Chromatographic conditions are reported in the text.
[b]Eluted by applying 80% water/20% methanol after six minutes from the injection.

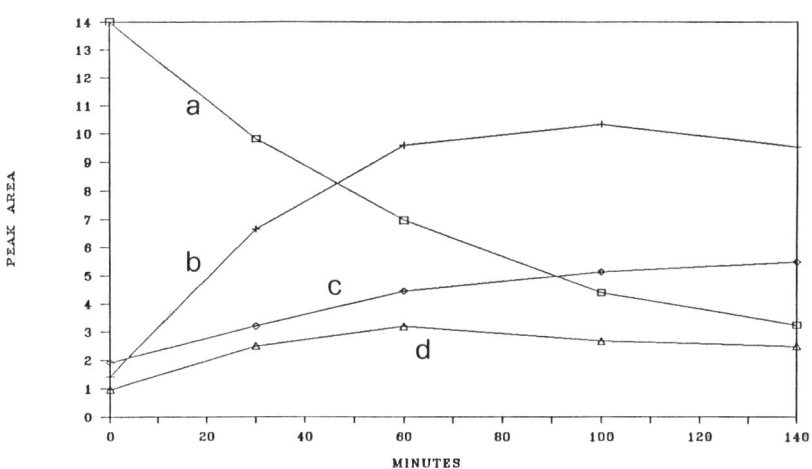

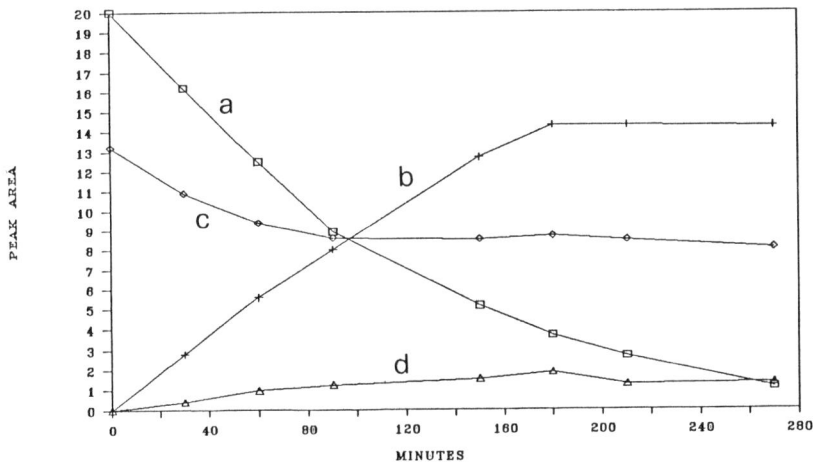

FIGURE 3. Pig brain NADase kinetics assayed by HPLC—panel A: (a) NAD^+, (b) 3-APAD, (c) ADPR, and (d) nicotinamide; panel B: (a) NAD^+, (b) 4-MeNAD, (c) 4-methylnicotinamide, and (d) nicotinamide.

HPLC proved to be a rapid and sensitive technique for the determination of transglycosidase activity. TABLE 1 shows the retention times of various pyridine and coenzyme analogues and FIGURE 3 reports the kinetics of the reactions producing 3-acetylamino adenine dinucleotide (3-APAD) and 4-methyl-nicotinamide adenine dinucleotide (4-MeNAD). The data shown in FIGURE 3 indicate that the transglycosidation stops after a period of time corresponding to an equilibrium between reaction 1 and 2 reported earlier. In fact, whereas the NAD analogue and the nicotinamide reach a maximum of production before decreasing, ADPR increases steadily, meaning that the reverse of reaction 2 takes place in addition to the reaction of hydrolysis of NAD^+ (reaction 1). Therefore, a choice of the reaction time can be useful for the optimization of a bioreactor.

No main difference was found between the two types of bioreactors after a suitable choice of the operative conditions for the continuous-flow apparatus. Experiments carried out with acetone powder entrapped in polyacrylamide gel indicated that a continuative work can be done without significant loss of activity during 24 hours at 37 °C. Polyacrylamide gel entrapped NADase was thus the most suitable derivative with regard to stability and continuative use as a bioreactor.

REFERENCES

1. GRUSHOFF, P. S., S. SHANY & W. BERNHEIMER. 1975. J. Bacteriol. **122:** 599–602.
2. STATHAKOS, D., I. ISAAKIDOU & H. THOMOU. 1973. Biochim. Biophys. Acta **302:** 80–89.
3. EVERSE, J. & N. O. KAPLAN. 1968. J. Biol. Chem. **243:** 6072–6074.
4. LEONE, E. & L. BONADUCE. 1959. Biochim. Biophys. Acta **31:** 292–293.
5. YUAN, J. H. & B. ANDERSON. 1971. J. Biol. Chem. **246:** 2111–2115.
6. WALTER, P. & N. O. KAPLAN. 1963. J. Biol. Chem. **238:** 2823–2830.
7. SCHUBER, F. & P. TRAVO. 1976. Eur. J. Biochem. **65:** 247–255.
8. YOST, D. A. & B. ANDERSON. 1983. J. Biol. Chem. **258:** 3075–3080.
9. MCILWAIN, H. & R. RODNIGHT. 1949. Biochem. J. **44:** 470–477.
10. WINDMUELLER, H. G. & N. O. KAPLAN. 1962. Biochim. Biophys. Acta **56:** 388–391.
11. ANDERSON, B. M., M. M. CIOTTI & N. O. KAPLAN. 1959. J. Biol. Chem. **243:** 1219–1225.
12. YOST, D. A. & B. ANDERSON. 1981. Anal. Biochem. **116:** 374–378.
13. ANDERSON, B. M. & N. O. KAPLAN. 1959. J. Biol. Chem. **234:** 1226–1232.
14. BURTON, K. & T. H. WILSON. 1952. Biochem. J. **54:** 86–94.
15. MINATO, S., H. TAMAOKI & H. MIZUSHIMA. 1976. Clin. Chim. Acta **69:** 243–249.
16. BISHOP, M. J., J. EVERSE & N. O. KAPLAN. 1972. Proc. Natl. Acad. Sci. U.S.A. **69:** 1761–1765.
17. FUJISAWA, K., A. KIMURA, S. MINATO, H. TAMAOKI & H. MIZUSHIMA. 1976. Clin. Chim. Acta **69:** 251–257.
18. PACE, M., D. AGNELLINI, P. G. PIETTA, A. COCILOVO & L. BONIZZI. 1984. Prep. Biochem. **14:** 349–362.

Microbial Decarboxylation of Succinate to Propionate

Kinetic Studies

NISSIM S. SAMUELOV, RATHIN DATTA,
MAHENDRA K. JAIN, AND J. GREGORY ZEIKUS

Michigan Biotechnology Institute
Lansing, Michigan 48909
and
Departments of Biochemistry and Microbiology
Michigan State University
East Lansing, Michigan 48823

INTRODUCTION

Energy transduction through Na^+ transport–coupled decarboxylations is a new concept for biological energy conservation.[1] This mechanism was determined for oxaloacetate,[2] glutamyl-CoA,[3] and methylmalonyl-CoA decarboxylases.[4,5] A recent review summarized their properties.[6] The methylmalonyl-CoA decarboxylase sodium pump occurs in *Propionigenium modestum*[4] and *Veillonella alcalescens*[7] as part of the reaction sequence for succinate decarboxylation to propionate.[8] The small free-energy change of this reaction ($\Delta G° = -20.6$ kJ/mol) does not allow a substrate-linked phosphorylation. In *P. modestum*, the pump is coupled to a Na^+-activated ATPase and this bacterium grows on succinate as a sole source of carbon and energy.[4,9] The ability of *V. alcalescens* (which operates the same sodium pump) to grow on succinate was never assessed reliably,[9] although lactate-grown cells have the potential to decarboxylate succinate to propionate.[10] Fermentation processes for the production of propionate from carbohydrates with pure propionibacteria cultures and in a mixture with lactic acid–producing strains have been extensively studied.[11,12]

The present study represents a new approach in which succinate is decarboxylated to propionate. Succinic acid is the direct precursor of propionate and it is produced by many bacterial strains as an intermediate or an end product (or both) of carbohydrate catabolism.[13] Culture broth of *Anaerobiospirillum succiniciproducens*,[14] which produces succinate with high yields,[15] can serve as a raw material.

In this report, we quantify the relationship between growth rate and the extent and rate of propionate formation from lactate and succinate during *V. alcalescens* fermentation. In addition, the apparent *in vivo* specific activity of succinate decarboxylation is determined from steady-state data in the second stage of a two-stage continuous culture process.

MATERIALS AND METHODS

Bacterium, Growth Media, and Apparatus

Veillonella alcalescens (ATCC 27215) was transferred daily in seed culture medium number ATCC 188 in 158-mL serum vials sealed with black rubber bungs and nitrogen in the headspace. Experimental media were modified by omitting the sodium thioglycolate, altering the amount of lactate, reducing the concentration of glucose (to 5% with respect to lactate), and adding succinate as indicated. Strict anaerobic conditions were established by the addition of 2.5% $Na_2S \cdot 9H_2O$ (10 mL/L) prior to inoculation. The growth temperature was 37 °C and the pH was controlled at 7.3 with 2 N NaOH (in pH-controlled vessels). One-liter batch fermentations were carried out in a 1.4-L MultiGen New Brunswick reactor. Continuous cultures were run in a 0.33-L Bioflow vessel and, for the two-stage continuous process, the 1.4-L reactor served as the second stage.

Analytical Approach

Substrates and products of fermentation were analyzed by high performance liquid chromatography (HPLC) in acidified (1% HCl) samples.[16] Components were eluted with 0.012 N H_2SO_4 from a cation-exchange resin in the hydrogen form. They were detected by a differential refractometer and recorded and quantified by using a Waters 840 integrator. The concentration was computed from the area under the curve and separations were carried out on a 330 mm × 7.8 mm Biorad HPX-87 column. Volatile fatty acids and CO_2 were also determined by gas chromatography and succinic acid was determined by a commercial enzymatic test kit, namely, Boehringer-Mannheim Biochemica Cat. No. 176,281. Protein was determined by the Lowry method.

Enzyme Activity and Production Kinetic Parameters

In vivo activity of succinate decarboxylation to propionate was determined from steady-state process data (see RESULTS section). Rate parameters of propionate formation in continuous cultures were calculated from steady-state process data: $q = PD/X$, where q is the specific rate of production (g propionate/g cells·h), P is the propionate concentration (g/L), D is the dilution rate (h^{-1}), and X is the cell mass concentration (g/L). Rate coefficients were determined using the linearized version[17] of the Luedeking-Piret expression:[18] $q = KD + K'$, where K is the growth-rate-linked propionate formation rate coefficient (g propionate/g cells) and K' is the nongrowth-associated propionate formation rate coefficient (g propionate/g cells·h), which depends on cell density alone.

RESULTS

The continuous culture substrate shift method[19] was applied for determining the ability of *V. alcalescens* to grow on various succinate concentrations at variable growth rates. The cell mass yield on lactate was between 10% to 15% and the maximal specific growth rate was $\mu_{max} = 0.53$ h^{-1} (doubling time of 1.3 h). When shifting steady-state lactate-grown cells to a medium that contained succinate (instead of lactate), a dilution rate–dependent washout occurred (FIGURE 1). These results demonstrate that succinic acid cannot serve as a sole carbon source for growth in *V. alcalescens* independent of its concentration or the specific growth rate imposed upon the culture.

The results shown in FIGURE 2 indicate that lactate-grown cells have the ability to decarboxylate succinate to propionate. The added succinate caused an increase in cell yield (up to 35%) during cometabolism with lactate and had no effect on growth rate. The material balance of *V. alcalescens* fermentations, summarized in TABLE 1, shows that succinate is exclusively decomposed to propionate and carbon dioxide under these

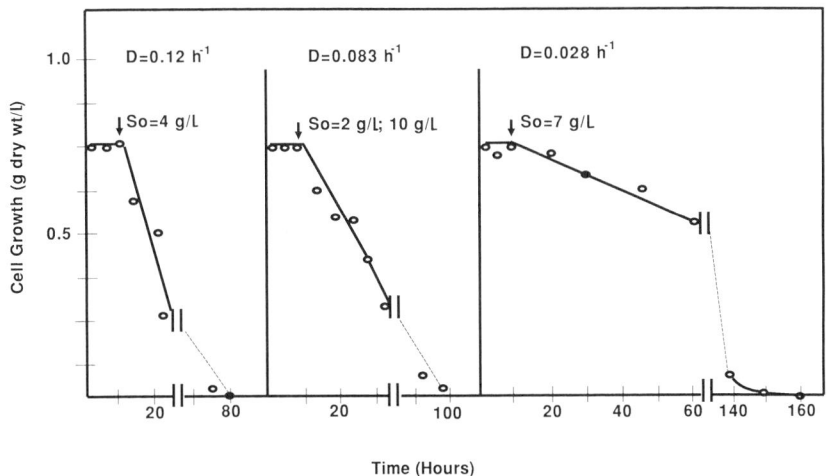

FIGURE 1. Substrate shift experiments at various dilution rates. *V. alcalescens* was grown in a chemostat on lactate (5 g/L). After the establishment of a steady state (in growth), the feed medium was replaced by a fresh medium (see arrows) that contained succinate (So = 4 g/L, 2 g/L, 10 g/L, and 7 g/L) instead of lactate.

mixed substrate batch growth conditions. In continuous culture, only up to 65% of the succinate added to the steady-state lactate-grown culture (at 20% of the maximal specific growth rate) was decomposed (FIGURE 3).

High yield continuous conversion of succinate to propionate was established in a two-stage continuous culture process. In the first stage, optimal growth conditions were maintained for the production of lactate-grown cells that served as the biocatalyst for succinate decarboxylation in the second stage (TABLE 2). The apparent *in vivo* specific activity of succinate decarboxylation to propionate in *V. alcalescens* (0.92 µmol/mg protein·min) was calculated from steady-state process data in the second stage (TABLE 3). This value is similar to that determined by the manometric method (0.108 to 0.8 µmol CO_2/mg protein·min) in whole cells separated from the growth medium (see reference 10).

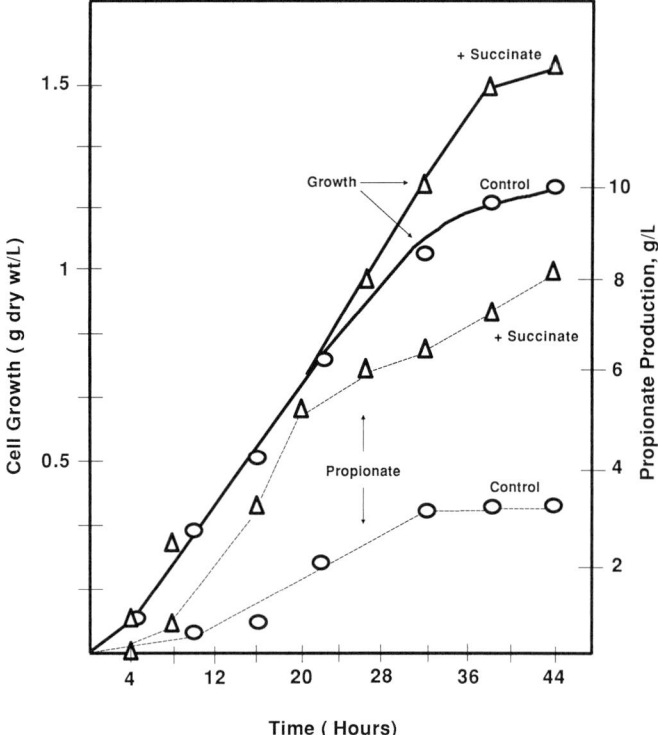

FIGURE 2. Growth (—) and propionate production (---) from lactate (7.4 g/L; ○) and lactate plus succinate (7.4 g/L + 10 g/L, respectively; △) in batch-grown *V. alcalescens*.

Specific rates of formation (q, g propionate/g cells·h) were determined from steady-state process variables (FIGURE 4). The growth-related production coefficient from lactate ($K = 2.35$ g propionate/g cells) was about eight times higher than that calculated for propionate formation from succinate ($K = 0.3$ g propionate/g cells). The nongrowth-linked coefficient of propionate formation (K', g propionate/g cells·h) was five times higher for the decarboxylation reaction (FIGURE 4).

TABLE 1. Material Balance of *V. alcalescens* Fermentations

	Recovery (%)	
Substrates-Products Stoichiometry[a]	Carbon	Hydrogen
1 lactate + 0.06 glucose → 0.48 propionate + 0.4 acetate + 0.13 formate + 0.414 cells + 0.51 CO_2	98	83
1 lactate + 0.05 glucose + 1.7 succinate → 2.2 propionate + 0.5 acetate + 0.11 formate + 0.54 cells + 2.1 CO_2	102	98

[a]Means of triplicate batch experiments. The cell formula was ($C_1H_2O_{0.5}N_{0.2}$).

DISCUSSION

Succinate decarboxylation to propionate ($\Delta G° = -20.6$ kJ/mol) can be carried out with *P. modestum* growing on succinate with a molar growth yield of 2%[9] and with *V. alcalescens* in which succinate cannot serve as a sole carbon source for growth. As succinate decarboxylation in *P. modestum* is growth-associated and the yield of cells is low (due to the small free-energy change), only up to 2.6 g/L propionate has been produced.[4,9] Moreover, this bacterium is unable to ferment lactate as well as a variety

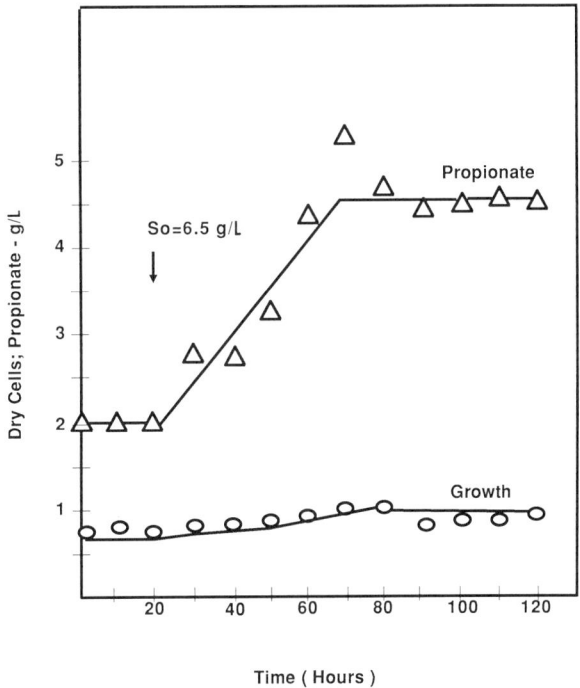

FIGURE 3. Influence of succinate addition on lactate fermentation in continuous culture. *V. alcalescens* was grown on lactate (5 g/L) at $D = 0.1$ h^{-1}. After the establishment of steady-state conditions, succinate (So = 6.5 g/L) was added to the feed medium (see arrow).

of sugars and alcohols.[9] *V. alcalescens* ferments lactate with a molar growth yield of 15% and cometabolizes succinate exclusively to propionate during the exponential and stationary growth phases. The specific activity of succinate decarboxylation by *P. modestum* cell extract, 4.5 nmol/mg protein·min,[9] is much lower than the apparent *in vivo* activity in *V. alcalescens*, 0.92 μmol/mg protein·min. These considerations indicate that the latter strain is the preferable choice when considering a process development for the production of propionate from succinate.

The growth-rate-linked production coefficient (K, g propionate/g cells) refers to energy metabolism diverted for cell mass production and relates the energy gathered

TABLE 2. Two-Stage Continuous Culture Process for Propionate Production from Succinate by *V. alcalescens*

Steady-State Process Variables	First Stage[a]	Second Stage[b]
Dilution rate, D (h^{-1})	0.2	0.02
Dry cell mass, X (g/L)	1.0 ± 0.15	1.13 ± 0.2
Propionate, P (g/L)	3.3 ± 0.2	17.5 ± 0.5
Succinate (g/L)	0	2.25 ± 0.3
Lactate (g/L)	1.7 ± 0.3	0
Productivity, PD (g/L · h)	0.66	0.35
Molar yield (%)[c]	55	99

[a]The first stage was fed with lactate containing medium (9 g/L).
[b]The second stage was fed with the outcoming broth from the first stage to which 25 g/L succinate was added.
[c]Propionate yield calculated from the fermented lactate (first stage) and succinate (second stage).

from product formation to growth rate. The nongrowth-associated production coefficient (K', g propionate/g cells · h) is related to cell mass concentration and refers to the energy diverted for nongrowth purposes.[13] Kinetic analysis of *V. alcalescens* steady-state continuous cultures indicates that the rate of succinate decarboxylations follows a nongrowth-linked pattern, whereas propionate formation from lactate is growth-rate-related. The high ratio between the growth-rate-linked propionate production coefficients from lactate and succinate (K[lactate]/K[succinate] = 7.83) indicates that the energy gathered from lactate fermentation is mainly diverted to growth-related bacterial metabolism. The nongrowth-related coefficient of propionate production from succinate was five times higher than that calculated for propionate production from lactate. This indicates that the decarboxylation energy is mainly used for nongrowth purposes. Succinate is cometabolized in the presence of lactate and enables the diversion of more carbon source for cell mass production, which results in 30–35% higher growth yield. The reduced ability of *V. alcalescens* to decarboxylate succinate at specific growth rates higher than 0.02 h^{-1} (FIGURE 4B) caused the decrease in propionate yield. Two-stage continuous cultures have been employed for studying the formation of extracellular enzymes that are produced in the postexponential growth phase.[17,20] In the present work, two-stage continuous cultures offered the advantage of

TABLE 3. Determination of the Apparent *in Vivo* Specific Activity of Succinate Decarboxylation by *V. alcalescens*

Steady-state process variables:[a]
dilution rate, $D = 0.02$ h^{-1}
dry cell concentration,[b] $X = 1125$ mg/L
propionate concentration,[c] 14.2 g/L
Propionate productivity, PD = 64 μmol/L · min
In vivo specific activity (PD/0.62X) = 0.92 μmol/mg protein · min

[a]Basis: (1) the only products are propionate and CO_2; (2) propionate is not metabolized.
[b]Cell protein content = 0.62 mg/mg cells.
[c]Produced from succinate (see TABLE 2).

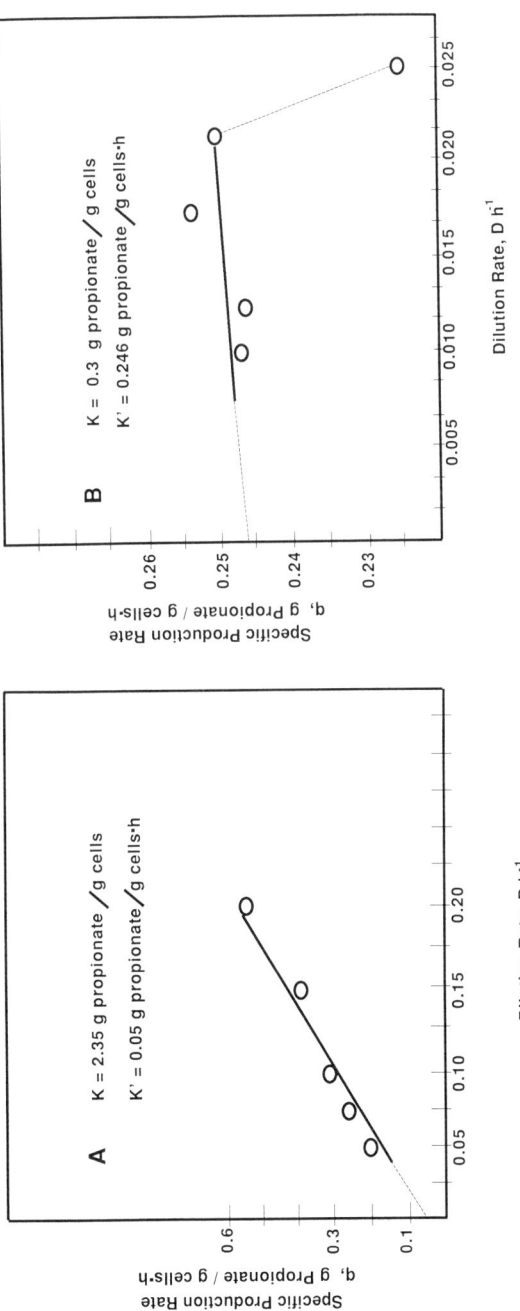

FIGURE 4. Influence of dilution rate on the specific production rate of propionate from lactate (A) and from succinate (B) by *V. alcalescens* in a two-stage continuous culture process.

establishing separate optimal conditions for growth (on lactate in the first stage) and for propionate formation from succinate in the second stage.

ACKNOWLEDGMENT

We wish to thank Keith Strevett for excellent technical assistance.

REFERENCES

1. DIMROTH, P. 1980. FEBS Lett. **122**: 234.
2. DIMROTH, P. & A. THOMER. 1986. Biol. Chem. Hoppe-Seyler **367**: 813.
3. WOLFARTH, G. & W. BUCKEL. 1985. Arch. Microbiol. **142**: 128.
4. HILPERT, W., B. SCHINK & P. DIMROTH. 1984. EMBO J. **3**: 1665.
5. HILPERT, W. & P. DIMROTH. 1983. Eur. J. Biochem. **132**: 579.
6. DIMROTH, P. 1987. Microbiol. Rev. **51**: 320.
7. HILPERT, W. & P. DIMROTH. 1982. Nature (London) **296**: 584.
8. DEVRIES, W., T. R. M. RIEDVELD-STRUIJK & A. H. STOUTHAMER. 1977. Antonie van Leeuwenhoek; J. Microbiol. Serol. **43**: 153.
9. SCHINK, B. & N. PFENNIG. 1982. Arch. Microbiol. **133**: 209.
10. GALIVAN, J. H. & S. H. G. ALLEN. 1968. J. Biol. Chem. **243**: 1253.
11. WILLIAM, P. A., F. A. DALE & E. S. LAWRENCE. 1988. United States patent no. 4,743,453 (May 10, 1988).
12. BODIE, E. A., T. M. ANDERSON, N. GOODMAN & R. D. SCHWARTZ. 1987. Appl. Microbiol. Biotechnol. **25**: 434.
13. GOTTSCHALK, G. 1979. Bacterial Metabolism. Springer-Verlag. New York/Berlin.
14. DAVIS, C. P., D. CLEVEN, J. BROWN & E. BALISH. 1976. Int. J. Syst. Bacteriol. **26**: 498.
15. SAMUELOV, N. S. & R. LAMED. 1988. Abstr. Annu. Meet. Am. Soc. Microbiol. **88**: 215.
16. FISCHER, R. L. & D. C. CHAPITAL. 1987. J. Chromatogr. Sci. **25**: 112.
17. SAMUELOV, N. S. 1988. Biotechnol. Bioeng. **31**: 125.
18. LUEDEKING, R. & E. L. PIRET. 1959. J. Biochem. Microbiol. Technol. Eng. **1**: 393.
19. SAMUELOV, N. S. & I. GOLDBERG. 1982. Biotechnol. Bioeng. **24**: 2605.
20. JENSEN, D. E. 1979. Biotechnol. Bioeng. **14**: 647.

Bacterial Enzymes in Halogenation Processes[a]

WOLFGANG WIESNER, MANFRED KARL OTTO, AND KLAUS DIETER KULBE

Fraunhofer-Institut für Grenzflächen- und Bioverfahrenstechnik
D-7000 Stuttgart 80, Federal Republic of Germany

INTRODUCTION

Halogenated compounds, for example, antibiotics, pesticides, etc., are present as intermediates and products of many biosynthetic pathways. The enzymes responsible for the halogenation of these compounds are largely unknown. In contrast to chemical halogenation, enzymatic halogenation reactions proceed under mild conditions and, due to the specificity of the respective enzyme, yield less by-products. So far, halogenating enzymes from eukaryotes, marine algae, and fungi have been described.[1-3] Difficulties with mass cultivation of these organisms and high peroxidase and catalase side activities of the respective haloperoxidases have hampered their technical utilization. Recently, bacterial enzymes—bromoperoxidases from *Streptomyces* and *Pseudomonas* strains—have been detected.[4-6]

Some of these bromoperoxidases have no peroxidase or catalase activity. These new haloperoxidases have no heme at the catalytic center and have been termed nonheme haloperoxidases. The following scheme shows the main differences between the two groups of halogenating enzymes:

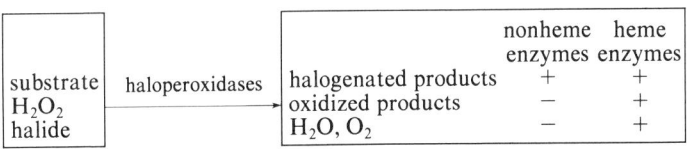

Most of these bacterial enzymes are bromoperoxidases. One of these nonheme enzymes was also found to catalyze chlorination.[7] Systematic screening for such enzymes has been hampered by the lack of a convenient assay. In an attempt to simplify such measurements, we have developed a method to determine halide consumption during enzymatic halogenation with high sensitivity. Some of these enzymes were characterized with respect to substrate specificity and enzyme reactor performance.

---

[a]This work is part of a cooperation with the Institute of Microbiology, Universität Hohenheim, Stuttgart, Federal Republic of Germany, and is supported by the German Ministry for Research and Technology (Contract No. BCT 383).

RESULTS

Characterization of Haloperoxidases

In a screening program for bacterial halogenating enzyme activities in various organisms, 28 haloperoxidases were detected.[8] Of these enzymes, 16 contained heme and 12 were of the nonheme type. The heme enzymes, for example, from *Streptomyces phaeochromogenes*[4] and *Pseudomonas pyrrocinia*,[5] had properties similar to known heme enzymes from eukaryotic sources. None of the nonheme enzymes had catalase or peroxidase activity. TABLE 1 compares these new bacterial nonheme haloperoxidases with heme enzymes.

In the course of a long time incubation at 35 °C of haloperoxidases, the rate of halogenation catalyzed by heme enzymes decreased within minutes mainly due to H_2O_2 consumption in side reactions and irreversible inactivation of the enzyme. However, the halogenation rates of nonheme haloperoxidases were constant over days under these conditions. Also, these enzymes were not inactivated upon storage for several weeks at room temperature. In addition, nonheme enzymes were remarkably stable at elevated temperatures. Chloroperoxidase from *Ps. pyrrocinia*[7] and bromoperoxidase from *Str. aureofaciens*[9] showed constant halogenation activity for two hours at 60 °C or 80 °C, respectively. These properties rendered nonheme haloperoxidases particularly interesting for enzyme-catalyzed halogenation on a technical scale.

Substrate Screening

Little is known about the substrate specificity of these bacterial nonheme haloperoxidases. A screening program was initiated to define their potential application range. Chloroperoxidase from *Ps. pyrrocinia*, bromoperoxidase from *Str. aureofaciens*, and other bacterial bromoperoxidases were used for these experiments.

FIGURE 1 summarizes the enzymatic halogenation reactions catalyzed by these

TABLE 1. Examples of Prokaryotic and Eukaryotic Heme and Nonheme Haloperoxidases

	Fungi	Bacteria		
	Caldariomyces fumago	*Pseudomonas pyrrocinia*	*Pseudomonas pyrrocinia*	*Streptomyces aureofaciens*
Enzyme type	heme	heme	nonheme	nonheme
Heme type	IX	IX	−	−
Mol. wt. × 10^{-3}	42	154	64	90
Subunits	1	2	2	3
Chlorination	+	−	+	−
Bromination	+	+	+	+
Oxidation of *o*-dianisidine	+	+	−	−
Catalase activity	+	+	−	−
Inhibition with NaN_3	+	+	−	−
Inhibition with KCN	+	+	+	+
pH optimum	3.5	5.5	4.5	4.5
Temperature optimum		30 °C	60 °C	80 °C
Cloned (host organism)			(*E. coli*)	(*Str. lividans*)

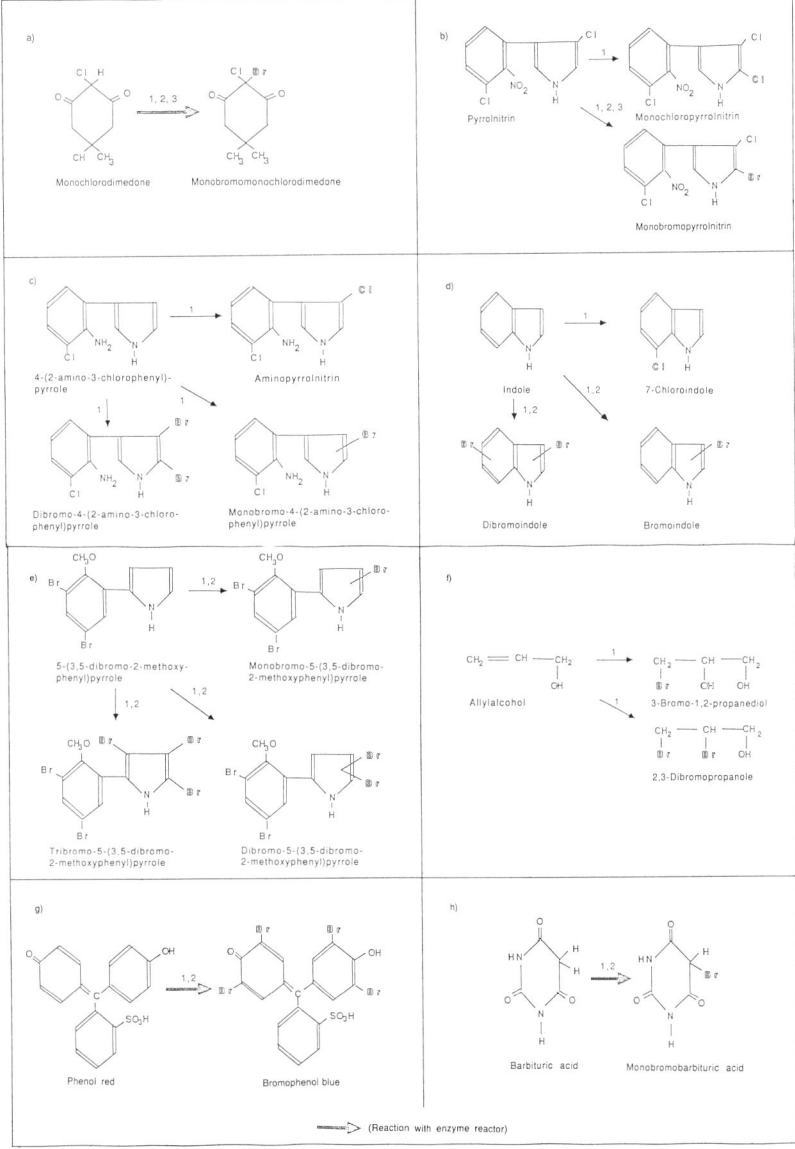

FIGURE 1. Enzymatic halogenations with bacterial nonheme haloperoxidases. The following enzymes were used: chloroperoxidase from *Pseudomonas pyrrocinia*,[1] bromoperoxidase from *Streptomyces aureofaciens*,[2] and other bacterial bromoperoxidases.[3]

bacterial nonheme haloperoxidases. All nonheme enzymes catalyzed the bromination of monochlorodimedone (a) and pyrrolnitrin (b). So far, only one bacterial enzyme, chloroperoxidase from *Ps. pyrrocinia*, chlorinated pyrrolnitrin in position 2 to the monochloro derivative. Moreover, with 4-(2-amino-3-chlorophenyl)pyrrole (c), indole (d), phenol, benzene, and penicillin G as substrates, chlorination by the same enzyme was detected. Bromination by chloroperoxidase from *Ps. pyrrocinia* and bromoperoxidase from *Str. aureofaciens* was detected with the following substrates: indole (d), 4-(2-amino-3-chlorophenyl)pyrrole (c), 5-(3,5-dibromo-2-methoxyphenyl)pyrrole (e), allylalcohol (f), phenol red (g), barbituric acid (h), aniline, phenol, benzene, penicillin G, and ampicillin. From these initial data, it is apparent that bacterial haloperoxidases may allow the development of enzymatic halogenation processes as alternatives to some technical halogenation reactions.

Enzyme Reactor

The immobilization of enzymes can be of advantage for technical processes. Chloroperoxidase from *Ps. pyrrocinia* was immobilized on various polymeric carriers. For example, on Eupergit C (Röhm), up to two units of enzyme (96% of the soluble enzyme) per mL of bed volume was immobilized with approximately 40% of its initial activity. The activity of immobilized chloroperoxidase was fully retained after four months of storage at room temperature or after three weeks under continuous reaction conditions. Bromination of various substances, for example, barbituric acid, uracil, or phenol red, was investigated. Initial experiments with two-phase systems indicate high activity of the immobilized enzyme on Eupergit C in the presence of organic solvent, for example, for bromination of phenol red in 50% *n*-butanol.

Halide Sensor

With a modified flow injection halide sensor, it is now possible to measure halogenation reactions of haloperoxidases by detecting small changes of bromide or chloride (FIGURE 2).[10,11] This setup is convenient for screening of the halogenation of potentially interesting substances and for the reaction control of an enzyme reactor, for example, on-line control of halogenation reactions. The sensor consists of a silver halide electrode combined with a flow injection system. Halides are measured by injection of halide-containing samples (<20 µL) into a halide-free carrier buffer. For protection,

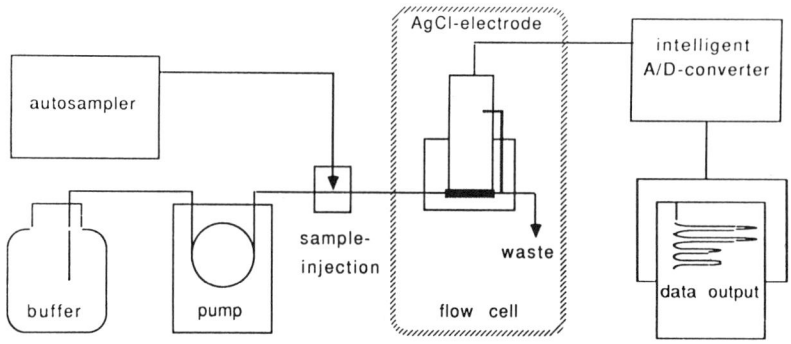

FIGURE 2. Schematic representation of the halide sensor setup.

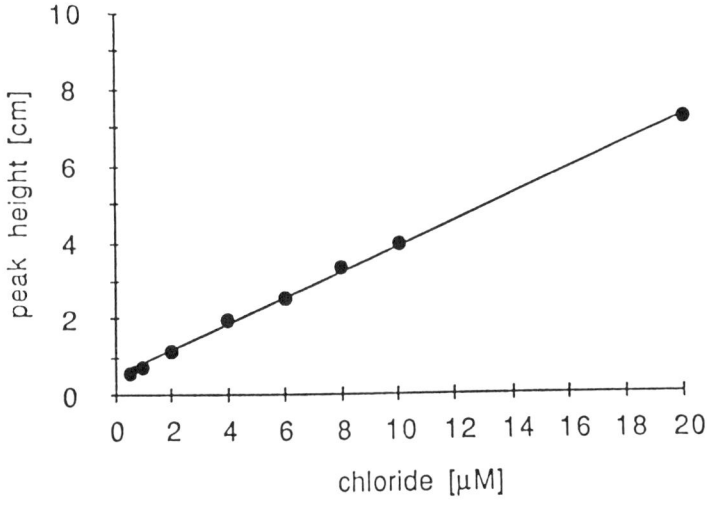

FIGURE 3. Calibration curve.

the sensor surface was covered with a 20-μm dialysis membrane. The response of the sensor is Nernstian at chloride concentrations $> 10^{-4}$ M and linear below this concentration (FIGURE 3). Less than 10 pmoles bromide can be detected.

For measurement of halogenating reactions, a small halide decrease on a sizable halide background must be measured. With the respective background concentration in the carrier buffer, changes of halide concentration of <20 μM chloride (background concentration of 1 mM) or <1 μM bromide (background concentration of 30 μM, FIGURE 4) can be measured.

DISCUSSION

Haloperoxidases are divided into three groups according to their specificity for the halide ions: chloroperoxidases, bromoperoxidases, and iodoperoxidases. Chloroperoxidase, which was isolated from *Caldariomyces fumago* by Hager and co-workers,[3,12] catalyzes the chlorination, bromination, and iodination of various compounds through electrophilic substitution[13] and the addition mechanism.[14] Bromoperoxidases, which are specific for Br^- and I^-, have been detected and purified from several marine algae, such as *Rhodomela larix*[15] and *Penicillus capitatus*,[16] and from bacteria, for example, *Streptomyces phaeochromogenes*[4] and *Pseudomonas pyrrocinia*.[5] These bacterial enzymes as well as the algal enzyme and the fungal chloroperoxidase from *Caldariomyces fumago*[3] are all heme proteins with protoporphyrin IX as the prosthetic group and they also have peroxidase activity with *o*-dianisidine as a substrate. These haloperoxidases show catalase activity as well.

In *Pseudomonas pyrrocinia*,[7] *Streptomyces aureofaciens*,[9] and a few other bacteria,[8] nonheme haloperoxidases have been detected that did not have any catalase or peroxidase activity. These enzymes were not inhibited by sodium azide, which made it very unlikely that they contained heme. Whereas the algal heme-containing bromoperoxidases were inhibited by hydrogen peroxide at low concentrations, that is, above 1 mM for the *Rhipocephalus phoenix*[17] enzyme and in excess of 2 mM for the *Penicillus capitatus*[16] bromoperoxidase, the chloroperoxidase from *Pseudomonas pyrrocinia*[7] was not inhibited by H_2O_2 at concentrations below 50 mM.

In addition, chloroperoxidase from *Pseudomonas pyrrocinia* and bromoperoxidase from *Streptomyces aureofaciens* were thermostable for two hours at 60 °C or at 80 °C, respectively, and could be stored at room temperature in the presence of 5 mM sodium azide without noticeable loss of activity. Their high stability and their broad substrate specificity, which are given in FIGURE 1 with respect to the other haloperoxidases from algae and fungi, make these nonheme haloperoxidases particularly suited for technical application in enzyme reactors.

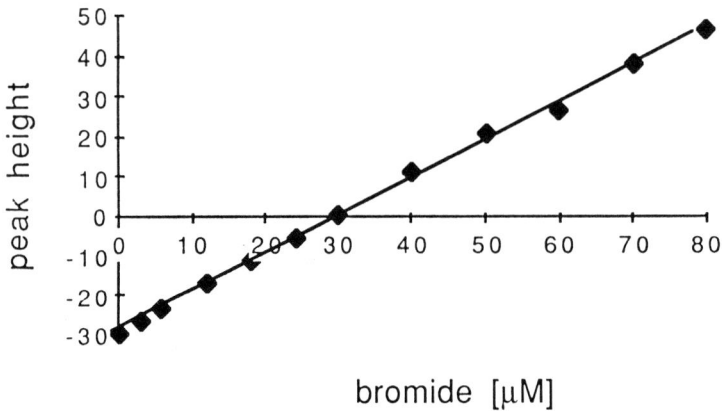

FIGURE 4. Bromide determination on a 30-μM bromide background. Carrier buffer—100 mM sodium acetate, 30 μM KBr, pH 4.5

CONCLUSIONS

The occurrence of many halogenated compounds in nature points towards numerous enzymatic activities that are responsible for the halogenation of these substances. Only a few of these enzymes are known so far. The recent finding of bacterial nonheme haloperoxidases and their favorable properties may allow the development of technical processes based on such haloperoxidases. In an initial screening program, a variety of substances were halogenated by these enzymes. For conversions on a larger scale, conditions need to be optimized.

REFERENCES

1. BAKKENIST, A. R., J. E. G. DE BOER, H. PLAT & R. WEVER. 1980. The halide complexes of myeloperoxidases and the mechanism of the halogenation reactions. Biochim. Biophys. Acta **613:** 337–348.

2. ITOH, N., Y. IZUMI & H. YAMADA. 1985. Purification of bromoperoxidase from *Corallina pilulifera*. Biochem. Biophys. Res. Commun. **131:** 428–435.
3. MORRIS, D. R. & L. P. HAGER. 1966. Chloroperoxidase: I. Isolation and properties of the crystalline glycoprotein. J. Biol. Chem. **241:** 1763–1768.
4. VAN PEE, K-H. & F. LINGENS. 1985. Purification and molecular and catalytic properties of bromoperoxidase from *Streptomyces phaeochromogenes*. J. Gen. Microbiol. **131:** 1911–1916.
5. WIESNER, W., K-H. VAN PEE & F. LINGENS. 1985. Purification and properties of bromoperoxidase from *Pseudomonas pyrrocinia*. Biol. Chem. Hoppe-Seyler **366:** 1085–1091.
6. WIESNER, W., K-H. VAN PEE & F. LINGENS. 1986. Detection of a new chloroperoxidase in *Pseudomonas pyrrocinia*. FEBS Lett. **209:** 321–324.
7. WIESNER, W., K-H. VAN PEE & F. LINGENS. 1988. Purification and characterization of a novel bacterial nonheme chloroperoxidase from *Pseudomonas pyrrocinia*. J. Biol. Chem. **263:** 13725–13732.
8. VAN PEE, K-H., *et al.* Institut für Mikrobiologie, Universität Hohenheim. To be published.
9. VAN PEE, K-H., G. SURY & F. LINGENS. 1987. Purification and properties of a nonheme bromoperoxidase from *Streptomyces aureofaciens*. Biol. Chem. Hoppe-Seyler **368:** 1225–1232.
10. OTTO, M. K., F. MÖRSBERGER, R. MÜLLER, F. LINGENS, H. CHMIEL & K. D. KULBE. 1987. Simplified detection of biological halogenation and dehalogenation with a halide sensor. *In* Biosensors. Volume 10. R. D. Schmidt, Ed.: 279–280. GBF-Monographies.
11. OTTO, M. K., F. MÖRSBERGER, K. D. KULBE, R. MÜLLER, T. SCHENK & J. THIELE. 1987. Simplified detection of biological dehalogenation with a halide sensor. DECHEMA Biotechnol. Conf. **1:** 437–439.
12. HAGER, L. P., D. R. MORRIS, F. S. BROWN & H. EBERWEIN. 1966. Chloroperoxidase: II. Utilization of halogen anions. J. Biol. Chem. **241:** 1769–1777.
13. LIBBY, R. D., J. A. THOMAS, L. W. KAISER & L. P. HAGER. 1982. Chloroperoxidase halogenation reactions. Chemical versus enzymatic halogenation intermediates. J. Biol. Chem. **257:** 5030–5037.
14. GEIGERT, J., S. L. NEIDLEMAN, D. J. DALIETOS & S. DE WITT. 1983. Haloperoxidases: enzymatic synthesis of β-halohydrins from gaseous alkenes. Appl. Environ. Microbiol. **45:** 1575–1581.
15. AHERN, T. J., G. G. ALLAN & D. G. MEDCALF. 1980. New bromoperoxidases of marine origin: partial purification and characterization. Biochim. Biophys. Acta **616:** 329–339.
16. MANTHEY, J. A. & L. P. HAGER. 1981. Purification and properties of bromoperoxidase from *Penicillus capitatus*. J. Biol. Chem. **256:** 11232–11238.
17. BADEN, D. G. & M. D. CORBETT. 1980. Bromoperoxidases from *Penicillus capitatus*, *Penicillus lamourouxii*, and *Rhipocephalus phoenix*. Biochem. J. **187:** 205–211.

Index of Contributors

Agathos, S. N., 372–398
Agnellini, D., 689–696
Aivasidis, A., 599–615
Anders, K-D., 492–507
Andorn, N., 363–371
Antia, F. D., 172–181
Asama, H., 569–579
Ataai, M. M., 82–90

Backman, K., 16–24
Bailey, J. E., 1–15
Balakrishnan, R., 16–24
Baltus, R. E., 245–252
Baneyx, F., 139–147
Barbotin, J-N., 41–53
Bayer, T., 665–669
Bennett, G. N., 67–81
Bentley, W. E., 121–138
Berry, F., 41–53
Betenbaugh, M. J., 111–120
Birnbaum, S., 1–15
Blackburn, J. W., 580–592
Blanch, H. W., 458–475
Blumentals, I. I., 301–314
Bothast, R. J., 25–40
Branstrator, L. E., 25–40
Brierley, R. A., 350–362
Brown, S. H., 301–314
Bryers, J. D., 315–332
Büchmann, A. F., 492–507
Bussineau, C., 350–362
Byun, S. Y., 54–66

Cary, J. W., 67–81
Charles, M., 678–688
Chin, C-K., 54–66
Chmiel, H., 253–260
Cho, T., 399–422
Clark, D. S., 458–475
Clem, T. R., 363–371
Constantinides, A., 616–641
Costantino, H. R., 301–314

Datta, R., 697–704
Davison, B. H., 670–677
Dean, J. F., 593–598
de Gooijer, C. D., 423–430
Dervakos, G. A., 593–598
Dhurjati, P., 111–120

DiBiasio, D., xi
Dien, B. S., 25–40
DiPasquantonio, V., 16–24
Dunnill, P., 157–171
Dutta, B. K., 203–213

Endo, I., 569–579
Erickson, L. E., 431–442
Eyal, J., 678–688

Fass, R., 363–371
Fernandez, E. J., 458–475
Fowler, J. D., 333–349
Fuchs, R., 616–641

Galazzo, J. L., 1–15
Geer, S., 229–244
Georgiou, G., 139–147
Ghezzi, S., 689–696
Glasgow, L. A., 431–442
Goldstein, W. E., xi
Granados, R. R., 399–422
Grandics, P., 148–156

Hamer, G., 650–664
Hammer, D. A., 399–422
Hansma, P. K., 476–491
Hatch, R., 16–24
Hayman, E. G., 443–457
Heitzer, A., 650–664
Hensler, W., 616–641
Herold, T., 665–669
Hirata, A., 569–579
Hirata, M., 569–579
Hoare, M., 157–171
Holzhauer, K., 665–669
Horváth, C., 172–181, 182–191
Howaldt, M. W., 253–260
Hubbell, J. A., 261–270

Jain, M. K., 697–704
Jeong, J., 82–90
Jeong, Y-H., 372–398
Jeong, Y. S., 214–228
Jones, G. T., 431–442
Joung, J. J., 271–282

Kang, W. K., 192-202
Kaufman, J. B., 363-371
Kelly, R. M., 301-314
Kenney, A. C., 529-539
Khosla, C., 1-15
Ko, M., 229-244
Kompala, D. S., 121-138
Kompier, R., 423-430
Kool, M., 399-422
Kosson, R., 350-362
Kulbe, K. D., 253-260, 705-712

Ladisch, M. R., 25-40
Liao, A., 182-191
Lomont, J. M., 25-40
Lu, Z., 245-252

Maegley, K., 229-244
Mancuso, A., 458-475
Maruya, A., 16-24
Massia, S. P., 261-270
Matsuura, T., 214-228
Mauri, P. L., 689-696
McKay, D., 16-24
McMillan, J. D., 283-300
Melle, F., 553-568
Melton, A., 350-362
Moulding, P., 529-539
Müller, W., 492-507
Murphy, M. K., 458-475

Nagamune, T., 569-579
Nasri, M., 41-53

O'Connor, M. J., 16-24
Ogonah, O., 399-422
Okos, M. R., 25-40
Otto, M. K., 705-712

Pace, M., 689-696
Papoutsakis, E. T., 67-81
Parulekar, S. J., 91-110, 508-528
Paul, E. L., Sr., 642-649
Pedersen, H., xi, 54-66
Pellegrino, J. J., 229-244
Petersen, D. J., 67-81
Pietta, P. G., 689-696

Radjai, M., 16-24
Ray, N. G., 443-457

Rivera, R., 229-244
Robertson, C. R., 333-349
Royer, G. P., 271-282
Rudd, E., 16-24
Runstadler, P. W., Jr., 443-457
Ryan, W., 91-110

Samuelov, N. S., 697-704
Sayadi, S., 41-53
Scheper, T., 492-507
Schicho, R. N., 301-314
Schügerl, K., 665-669
Seko, H., 540-552
Senuma, M., 540-552
Shanks, J. V., 1-15
Shiloach, J., 363-371
Shoda, D., 16-24
Shukla, R., 192-202
Shuler, M. L., 399-422
Siegel, R. S., 350-362
Sikdar, S. K., 203-213
Singh, V., 616-641
Sirkar, K. K., 192-202
Skaja, A. K., 301-314
Slininger, P. J., 25-40
Steward, D., 229-244
Szathmary, S., 148-156
Szathmary, Z., 148-156

Takeuchi, S., 540-552
Thomas, D., 41-53
Thompson, P. W., 529-539
Titchener-Hooker, N. J., 157-171
Tosa, T., 540-552
Tramper, J., 423-430
Tranchino, L., 553-568
Tung, A. S., 443-457

Usmany, M., 423-430

Van den End, E. J., 423-430
van Lier, F. L. J., 423-430
Venkat, K., 372-398
Venkatasubramanian, K., 16-24
Vieth, W. R., 214-228
Vlak, J. M., 423-430
Vournakis, J. N., 443-457

Wandrey, C., 599–615
Wang, D. I. C., 283–300
Webb, C., 593–598
Wei, D., 508–528
Weigand, W. A., 508–528
Wickham, T., 399–422
Wiesner, W., 705–712
Wood, H. A., 399–422
Wormald, D., 529–539

Yajima, K., 540–552
Yeh, T., 315–332

Zasadzinski, J. A. N., 476–491
Zeikus, J. G., 697–704
Zhou, W., 665–669